Eighteenth-century Rigs & Rigging

Eighteenth-century Rigs & Rigging

Karl Heinz Marquardt

First published by VEB Hinstorff Verlag under the title
Bemastung und Takelung von Schiffen des 18. Jahrhunderts

First English language edition (revised and expanded)
published 1992 by
Conway Maritime Press Limited,
101 Fleet Street,
London EC4Y 1DE

British Library Cataloguing in Publication Data

Marquardt, Karl Heinz
Eighteenth century rigs and rigging.
I. Title
623.862

ISBN 0-85177-586-1

Designed and typeset by Tony Garrett
Printed and bound in Great Britain by Page Bros (Norwich) Ltd

CONTENTS

III NORTHERN VARIANTS OF SHIP RIG

IV RIGS FOR TWO-MASTED AND SMALLER VESSELS

V DETAILS OF SINGLE-MAST RIG

Foreword

In a lifetime of developing, building and being involved with ship models, I have come to realise that obtaining historically accurate information on masting and rigging often constitutes the most difficult problem in developing model plans. This experience is, I am sure, familiar to many modellers; a consequence is that the majority of ship models deserve higher quality rigging than they actually exhibit. It often seems that the modeller's concern for historical accuracy is abandoned once the finishing touches are put to the hull; once the masts are stepped and the rigging is to be set up, fantasy takes over. Knowledge which is superficial at best seems often to be considered adequate for completion of the rigging.

The reason for this is nearly always the use of inadequate model construction plans. Contemporary draughts of ships, which are sometimes very detailed but which rarely contain any details of rigging, date from the eighteenth century onwards and are extant for a great number of ships. Accurate information on rigging, however, is rare. Only sail plans, giving an outline of the size and sometimes the number of sails, are readily available, supplemented by fragmented details from a variety of sources. These have to be collected, studied and compared before they can be used for reconstructing the rigging plan of a particular ship type.

It is understandable that, in this regard at least, commercial model plans short-change the modeller. The development of historically accurate masting and rigging plans requires much more time and research than is normally possible for the producer of commercial model plans.

A number of ship modelling handbooks, in various languages, have been available for some years now. These provide an abundance of general knowledge, but are usually too broad in scope to provide much specific detail on masting and rigging. It is simply not possible to describe the evolution of the rigging of sailing ships from the early Egyptians to a five-masted full rigged ship in a short book without generalising too much.

Nor, despite the efforts of some authors, can one legitimately seek to describe the rigging of ships by using the terms old and new. What is an old ship? A vessel built in 1970 is considered old today, whereas one built in 1715 was still considered new some decades later. The use of such vague terms for historical periods can only mislead the reader. Any collection of detail drawings, no matter how well executed, loses its value if it is not accompanied by a precise definition of the historical period concerned.

R C Anderson was careful to provide such a definition in 1927, when he wrote *The Rigging of Ships in the Days of the Spritsail Topmast 1620-1720*, and other authors such as H Winter and R Höckel followed his lead. A later edition (1955) of Anderson's work, restricted to British ships, was published under the title *Seventeenth Century Rigging*. Another valuable English work, *The Masting and Rigging of English Ships of War 1625-1860*, written in 1979 by James Lees, Restorer at the National Maritime Museum in Greenwich, covers its subject in great detail and with similar care.

The idea for this book began to take shape in my mind in 1976. During my working life as a restorer, modeller and designer of model plans, I had often felt the need for a handbook which would answer the many questions I had encountered on the subject of eighteenth century rigging. The lack of such a handbook became evident again in the course of development work I had undertaken in 1976 on model plans for an eighteenth century Russian First Rate. My preoccupation with the need for a suitable reference work eventually became so strong that it took precedence over the completion of that project.

The eighteenth century ship, with its carvings and painted decoration, fascinates nearly every lover of ship models. It is not only an object of artistic appreciation in its own right, but also (and perhaps more importantly) provides an insight into the relationship of past sensibilities and technical capabilities to the modern world.

The aim of this book is to give modellers, marine artists and all others who are interested in the marine technology of the eighteenth century the necessary grounding in eighteenth century masting and rigging. All information given here has been compiled from a comprehensive range of contemporary and modern works.

As noted above, masting and rigging was seemingly considered of secondary importance in shipbuilding records until well into the eighteenth century, and it was often the experience of the master attendant, and the particular preferences of the captain, which determined the final rig of a vessel. The first really comprehensive work on rigging was printed shortly before the end of the eighteenth century, in 1794. David Steel, a British Admiralty agent for sea charts and a publisher of nautical books, filled a great gap with his book *The Elements and Practice of Rigging and Seamanship*. If one takes into consideration that this type of publication is necessarily the result of many years of work, it can reasonably be assumed that a considerable amount of Steel's information must already have been valid a decade or so earlier. This may also apply to other, similar early works. When one uses such primary sources of information, it is vital to compare all available contemporary documents: sometimes parts of important older works by other authors were included unrevised, and may provide a distorted picture of the period in question.

For example, Abraham Rees, author of *Naval Architecture 1819-20*, followed Steel closely in many of his comments and thus provided perhaps a clearer picture of the late eighteenth century than the early nineteenth. Similarly, the drawings in E Bobrik's *Handbuch der praktischen Seefahrtskunde* (1848) were mostly direct copies from Darcy Lever's *The Young Sea Officer's Sheet Anchor* (1811-18).

Because of language limitations, I have based my research mainly on English and German literature, and therefore English rigging (increasingly coming into use on the Continent in the course of the eighteenth century) is considered in more detail in this book than the French and other Continental practice; nevertheless, differing French or Continental approaches to particular details of spars or rigging have been highlighted wherever possible. It is interesting to note that construction and rigging differences between the ships of the two main rivals for ultimate sea power in the eighteenth century, Britain and France, are often overlooked by, or not known to, modern ship modellers.

Smaller vessels, of the types likely to be found in northern waters, are detailed in a separate chapter, and further chapters deal with some of the vessels a North European seafarer might have encountered on his journeys over the seven seas. The simple listing of these types alone demonstrates that a study of masting and rigging even for a single century remains an almost impossible task: this book is inevitably limited to certain shipbuilding regions and certain ship types.

I hope that reference to a number of untranslated German works in this book will not prove irksome to English speaking modellers, and that the works listed may in fact provide those who have some knowledge of German with a whole new range of reference works for future rigging projects. As far as possible, subtitles in German and French are provided to make it easier for English speaking modellers to work with French or German model plans.

As the English edition of this book follows closely the revised second German edition, part of the foreword of that revised edition follows; a number of additional contemporary and modern sources were used in the revision.

The often very detailed definitions given in J Röding's *Allgemeines Wörterbuch der Marine* (1793-98) have helped to provide a more precise insight into Continental rigging. Further information on the masts and yards of merchantmen has been taken from Chapman's details of the lengths and diameters of masts and yards, and modern publications by Jean Boudriot on French ships have permitted the verification and amendment of much material originally taken from the works of Admiral Edmond Paris. An anonymous and very early German publication, *Der geöffnete See-Hafen*, printed in 1705, has provided further information on Continental rigging around the turn of the eighteenth century.

John Davis, a boatswain in the Royal Navy, and by his own account also a rigger of ship models, was the author in 1711 of the now very rare *Seaman's Speculum or Compleat School-Master*. This work gives further insights into rigging theory as well as a number of tables for the rigging of English ships at the beginning of the century. These tables are second only to Steel's in volume and give not only length and circumference but also the length proportional to the vessel's beam. *The Ship-builder's Assistant*, published in 1711 by William Sutherland, a shipwright and contemporary of Davis, has provided a great deal of information on early eighteenth century rigging, and some proportional data have also been extracted from J Love's *The Mariner's Jewel* of 1735. T R Blanckley's *A Naval Expositor* of 1750 was checked for additional information, as were a number of modern English and German works.

An additional chapter, 'Northern Variants of Ship Rig', has been included to make even clearer the variety of rigs used in the course of the century.

In conclusion, I would like to thank my wife Sonja and Dave Ferguson of Heathmont, Victoria. Their help in compiling the manuscript made my work a great deal easier. Thanks are also due to my son Andreas, who helped to make the English translation of this book more readable, and to my friend Erhard Schmidt of Rinteln (Germany), whose illustrative material was usefully complementary to my own. My foremost thanks, however, go to all those artists, modellers and authors, past and present, who have recorded ships of the past; without their preparatory work this book could never have been written.

Karl Heinz Marquardt
Montrose, Victoria, Australia
1991

Note: all drawn scales on Figures show English feet, unless marked otherwise.

I

SPARS

Masts

Masten; Mâts

Masts are vertical pole constructions which, in connection with the yards, serve for the setting of sails and thus for the ship's propulsion. A distinction can be made between pole masts and made masts.

Pole masts were made from one piece of timber and their usefulness was limited. They fractured more quickly than made masts and were therefore found only on smaller vessels or as top and topgallant masts on larger vessels.

In contrast, the made mast was assembled from a number of parts. This provided greater elasticity and gave the shipwright the freedom to construct a mast appropriate to the ship's size, irrespective of the size of the available timber.

A mast's total length was usually achieved by interconnecting a number of mast extensions. The first, stepped on the keelson, was the lower mast, followed by the topmast, the topgallant mast and sometimes the royal mast, in that order.

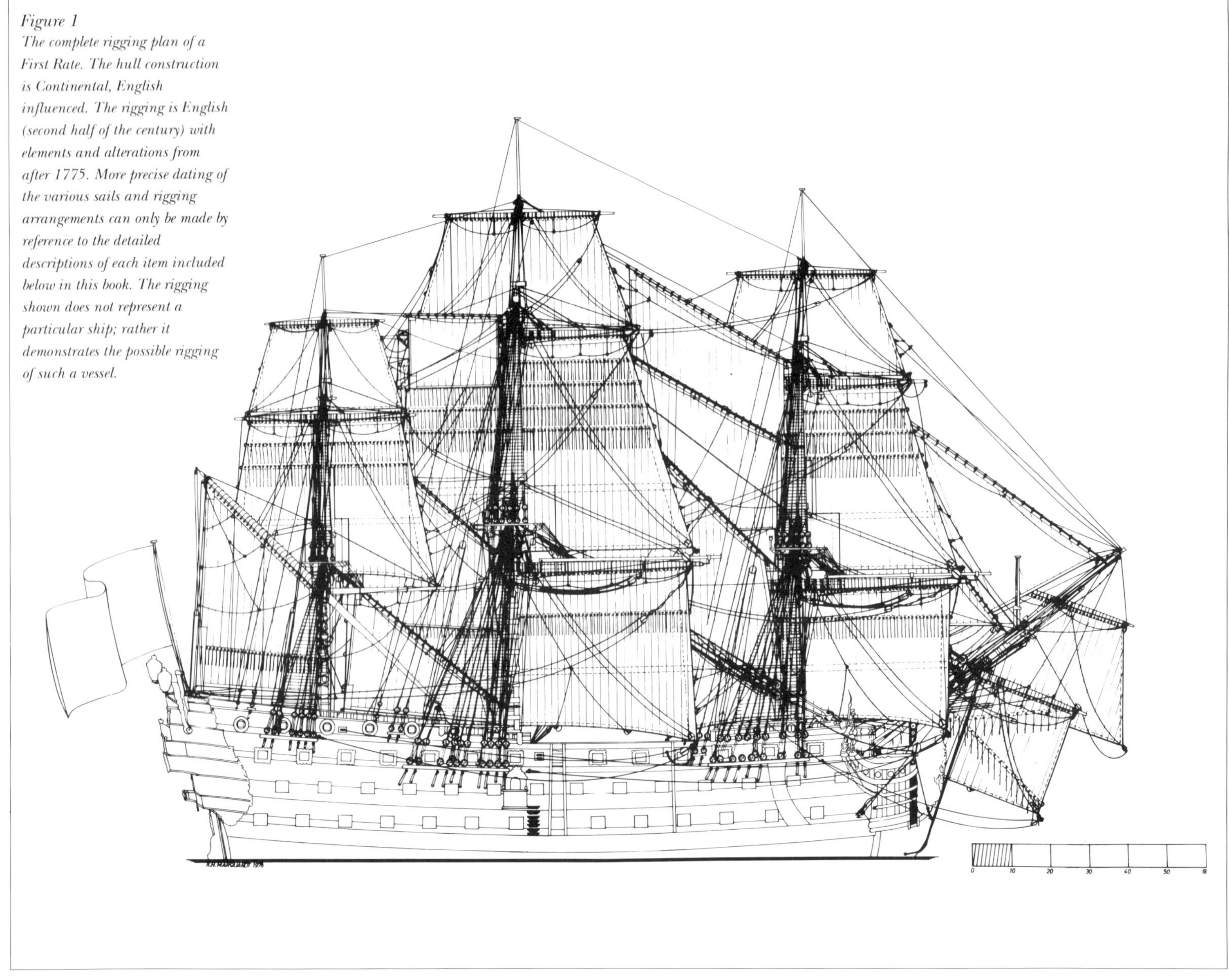

Figure 1
The complete rigging plan of a First Rate. The hull construction is Continental, English influenced. The rigging is English (second half of the century) with elements and alterations from after 1775. More precise dating of the various sails and rigging arrangements can only be made by reference to the detailed descriptions of each item included below in this book. The rigging shown does not represent a particular ship; rather it demonstrates the possible rigging of such a vessel.

Centuries-long experience had led in the seventeenth century to the confirmation of the three-masted design as the optimum for the larger seagoing vessel. These masts, in order from the bow of the ship, were known as the foremast, the mainmast and the mizzen mast (see Figure 2).

Position of masts

Mastzentren; Positions des mâts

The placement of masts was of utmost importance for course stability, the use of sails and speed. It is therefore not surprising that calculations derived from long experience were strongly adhered to. Steel, and Falconer's *Marine Dictionary*, gave the Royal Navy's rules for the calculation of mast centres as follows:

> Measurements were taken from the stem rabbet at the gun deck. The foremast was set at one ninth of the gun deck's length, the mainmast at five ninths, and the mizzen mast seventeen twentieths aft of that point.

H L Duhamel du Monceau noted that for French ships the foremast should step with its fore edge on the lower end of the stemson, or approximately one tenth of the ship's length aft of the stem. The mainmast's centre, he said, should be 7½ to 8 station lines aft of amidships per foot of the ship's length according to some shipwrights, while others calculated 4 station lines aft of amidships per foot of ship length for the mainmast's fore edge. For the mizzen mast, he noted that the fore edge should be placed between the fifth and sixth part of the ship's length, or that the aft edge of the mast should come at two thirds of the ship's breadth forward of the sternpost rabbet.

The merchant service did not adhere to these strict rules. The mainmast was usually close to amidships, and the fore and mizzen masts were stepped in a much more arbitrary way.

Röding's information on main and foremasts was similar to Duhamel's, while for the mizzen mast he specified that the after edge of the mast should occur at a distance equal to the ship's breadth forward of the sternpost rabbet, measured on the lower deck.

Chapman's recommendations were that the centre of the foremast should be a distance equal to 4/31 of the total distance between the rabbets aft of the stem, that the centre of the mainmast should be a distance equal to 2/31 of the total distance between the rabbets aft of amidships, and that the centre of the mizzen mast should be a distance equal to 0.182 of the total distance between the rabbets forward of the sternpost.

The inclination of the lower masts depended on the trimming of the particular ship. Experience showed that some ships

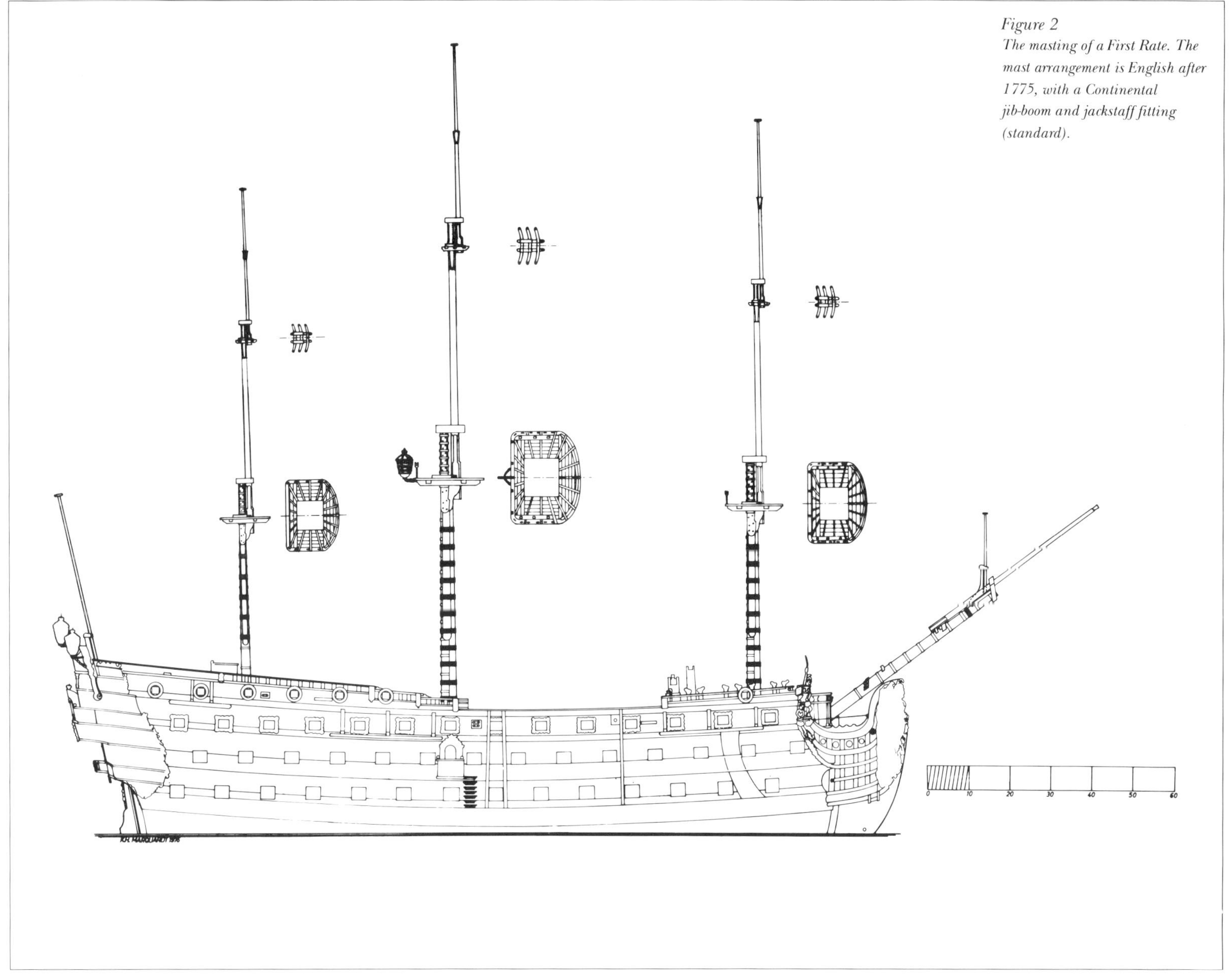

Figure 2
The masting of a First Rate. The mast arrangement is English after 1775, with a Continental jib-boom and jackstaff fitting (standard).

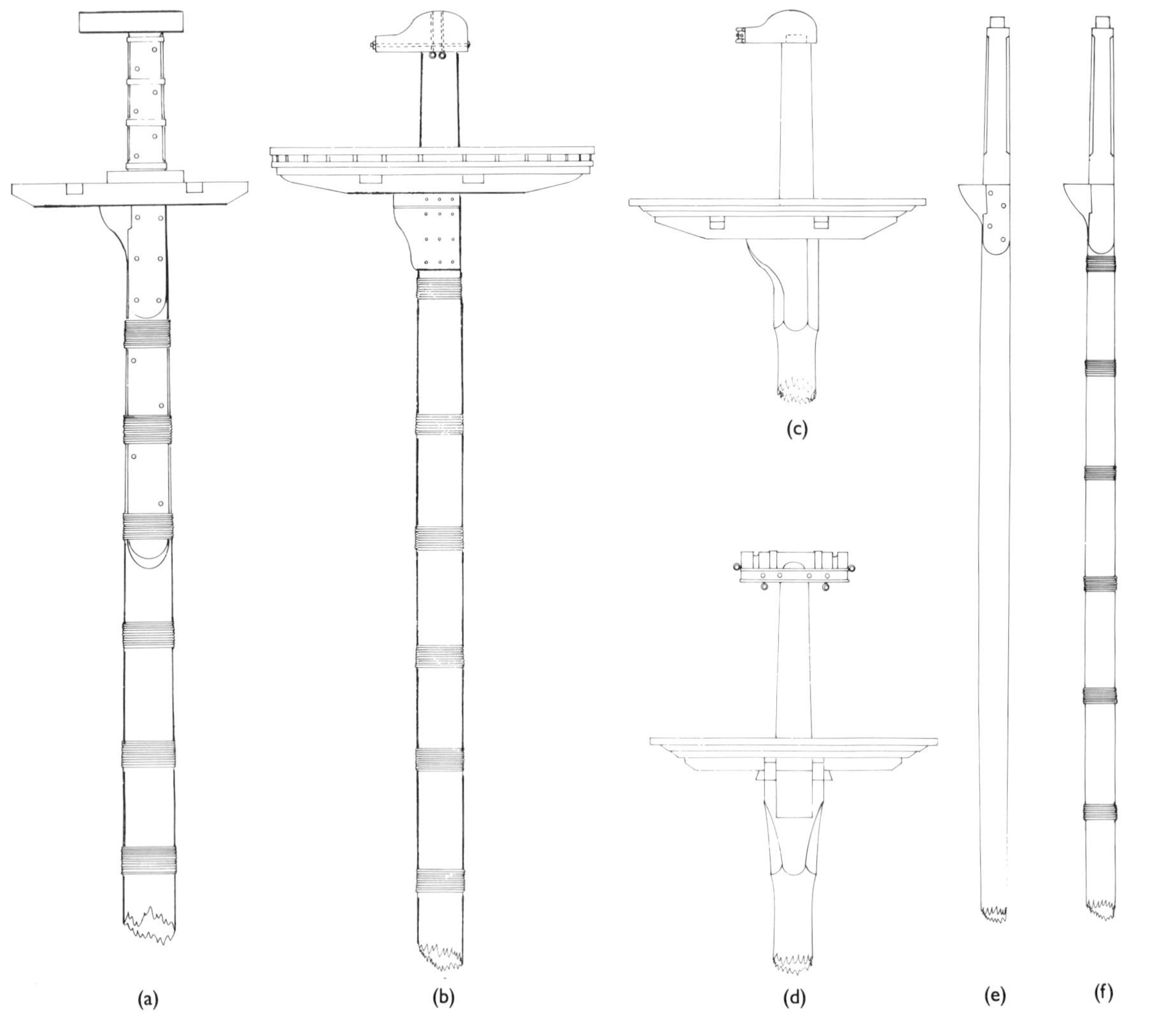

Figure 3
(a) An English mainmast, about 1700. The mast is reinforced by wooldings, and the cheeks and masthead are still relatively short. Bolsters are fitted to the trestle trees and iron hoops reinforce the masthead.

(b) French mainmast, about 1700. Cheeks as on English masts are not used; short bibbs are fitted instead. The masthead is rounded and not only the trestle and cross trees, but also the cap, differ from the English style.

(c) A Dutch masthead and top, about 1700 (profile).

(d) A Dutch masthead and top, about 1700 (forward profile).

(e) An English mizzen mast, before 1730. The mast has neither wooldings nor masthead hoops. Even before 1730, ships of the line sometimes carried a mizzen mast as shown in (f).

(f) An English mizzen mast, after 1730. During the entire century frigates with fewer than 36 guns had a mizzen mast without wooldings.

(a) (b) (c) (d) (e) (f)

sailed better with forward leaning masts while others performed at their best only with masts which were inclined aft – these being in the majority. It was widely believed that for beamier ships the mast standing closest to amidships should have the greatest incline aft. Further, it was believed that long ships managed better with masts perpendicular to the waterline, since it was believed that an inclined mast, already straining the forward part of the partners, might actually break them under wind pressure.

Chapman recorded data on the inclination of masts. He recommended an inclination aft of the mainmast by one thirtieth and of the mizzen mast by one fifteenth of its length (measured at the cap), while the foremast stood vertical.

Brigs and other ships with two masts had the mainmast stepped at a distance of about two thirds of the overall vessel length aft of the stem, and the foremast at about three twentieths of the overall length aft of the stem. The mainmast was inclined aft by about ¾in per yard of its length from heel to cap, and the foremast by about ⅛in per yard of length.

Cutters and other single masted vessels had the mast inclined aft by 1½in per yard of the mast's length, and the bowsprit was nearly horizontal.

Material
Material; Matériel

Pine was customarily used as mast material, as this timber was plentiful in the forests of Eastern Europe and North America.

Abraham Rees declared that before the American War of Independence all large masts for English ships were built from New England pine, since those trees were the tallest and most suitable for mast making. After the loss of that source, the Royal Navy received its mast timber from Riga. There, the largest timber was no more than 24 inches thick, and mostly between 19 and 21 inches; masts had therefore to be constructed from a greater number of pieces.

This fact and the greater specific weight of Eastern European timber led to a weight increase of approximately one quarter over the American masts. Rees also noted that Riga masts were much more durable than those from America. Sutherland had already recorded in 1711 significant differences between Riga, Gothenburg and New England pine; he noted that the strength of 9 inches of Gothenburg pine equalled that of 10 inches of Riga and 12 inches of New England pine.

Much importance was given to timber selection for mast construction, in order to avoid waste and excessive costs. This ultimately resulted in the adoption of procedures for designing masts on the drawing board, before the actual mast construction commenced.

Pole masts
Pfahlmasten; Mâts à pible

Masts produced from a single tree did not differ significantly in overall dimensions from made masts. Pole masts were strengthened by driving on iron hoops, and their construction details (paunches, hounds, etc) were the same as those of made masts.

The lower masts and topmasts of smaller craft such as cutters were often made in one, with the upper third or quarter of a single pole serving as the topmast; such masts had hounds and a squared off masthead. Sometimes, however, the masts of such vessels were built in the normal way, with a separate topmast connected to the lower mast by trestle trees and a cap. Steel commented that this was mainly a question of the preference of the captain or shipbuilder.

Mast making

Mastbau; Fabrication de mâts

The building of a mast began with the spindle, which was made from two pieces coaked and bolted together, with the bolts some 5 feet apart. The side trees of larger masts were also made from two pieces, coaked in the middle and with bolts 10 feet apart.

The heel, made from two shorter pieces, was then connected to the side trees. Once this was done, the length and width of the mast were defined. To produce a round cross-section both the front and back of the mast were covered with long planks, let in and coaked into the surfaces formed by the spindle and side trees. These pieces were called the side fishes.

The next step in assembling a mast was the fitting of the cheeks. These were made from fir, or in earlier years from oak, and their length from 1775 onwards was nine twentieths (or three sevenths for oak) of the mast's length. Up to about 1750 they were only approximately one third of that length and increased to one half between then and 1775. Included in the cheeks were the sides of the squared masthead.

After the mast had been rounded, strong iron hoops were driven on to strengthen the structure. These iron hoops came into use mainly in the second half of the century, at first together with mast wooldings. After 1800 the latter were no longer used.

During the first half of the century iron hoops were used on the masthead only, and the mast itself was generally strengthened by wooldings. Some models of large ships from before 1750, however, provide evidence of some use of iron hoops on the mast as well, either in place of or together with wooldings.

Differences in construction between English and Continental masts were especially obvious in their mastheads. Those of English origin were made square, whereas Continental mastheads were round. Furthermore, the latter had no distinct cheeks, and the paunch finished above the upper deck (see Figures 3a to 4d).

Mastheads

Masttoppe; Tons de mât

In the first twenty years of the eighteenth century the length of the masthead was 4 inches per yard of the mast's length. In the period between 1720 and 1775 this figure increased to 5 inches for the mainmast, $4\frac{3}{4}$in for the foremast and $3\frac{3}{4}$in for the mizzen mast. The last quarter of the century saw a proportional length of 5 inches for the two larger masts and 4 inches for the mizzen mast.

Duhamel du Monceau specified a masthead length of one ninth of the mast's length for French ships, and Chapman recommended a masthead $\frac{5}{36}$ of the mast's length for the mainmast, with the foremast head nine tenths and the mizzen masthead three quarters of that length.

Röding also recommended a masthead length of one ninth

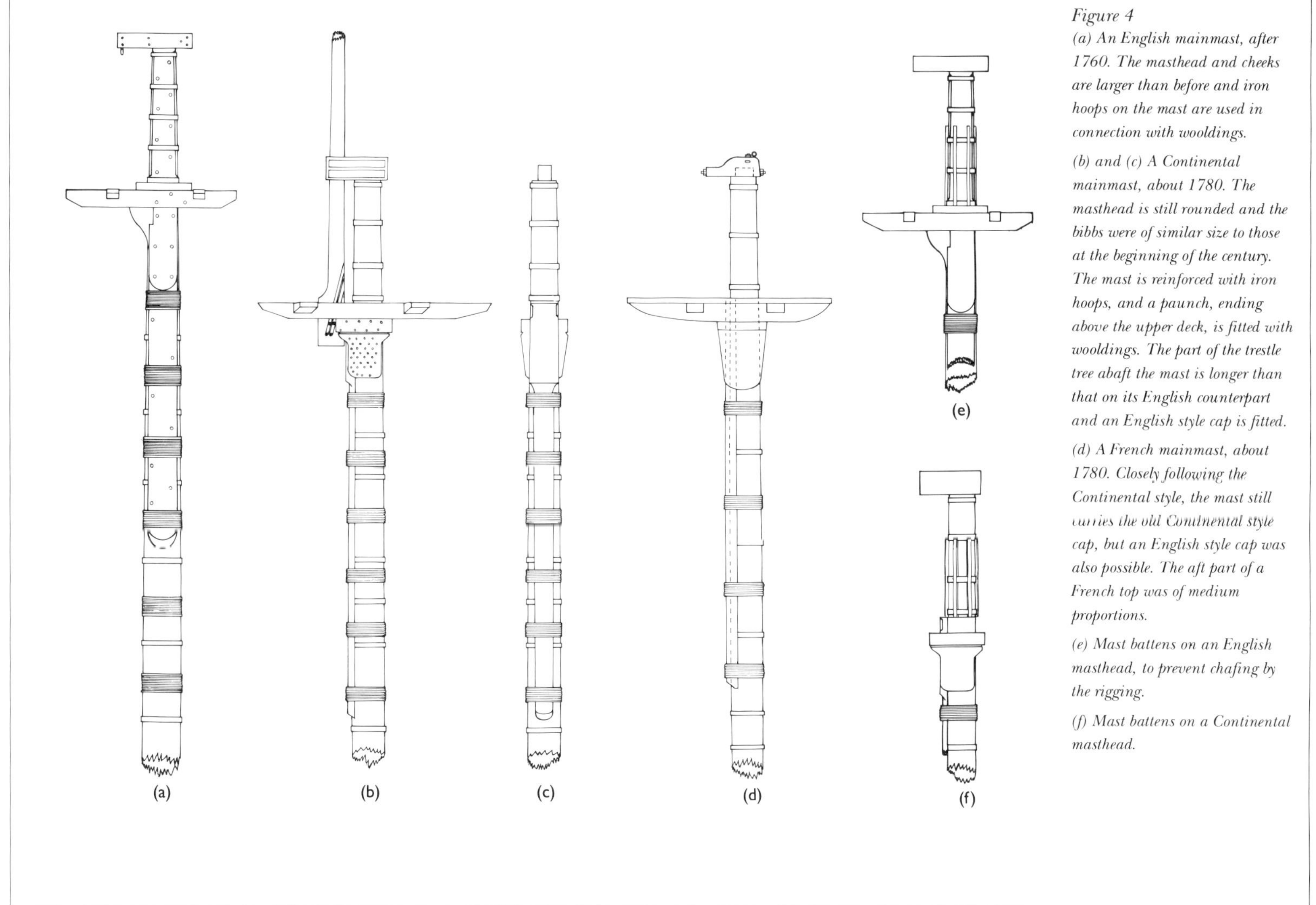

Figure 4

(a) An English mainmast, after 1760. The masthead and cheeks are larger than before and iron hoops on the mast are used in connection with wooldings.

(b) and (c) A Continental mainmast, about 1780. The masthead is still rounded and the bibbs were of similar size to those at the beginning of the century. The mast is reinforced with iron hoops, and a paunch, ending above the upper deck, is fitted with wooldings. The part of the trestle tree abaft the mast is longer than that on its English counterpart and an English style cap is fitted.

(d) A French mainmast, about 1780. Closely following the Continental style, the mast still carries the old Continental style cap, but an English style cap was also possible. The aft part of a French top was of medium proportions.

(e) Mast battens on an English masthead, to prevent chafing by the rigging.

(f) Mast battens on a Continental masthead.

of the mast's length, with the exception of the mizzen mast, where the head should be only one tenth of the mast's length. Similarly, Davis noted a length of one ninth of the mast's length in 1711, at least for some vessels. He later qualified this as applicable only to merchantmen, but his works generally maintain the formula of 4in per yard.

Paunches

Frontfische; Jumelles

Another fish, hollowed to fit to the mast's curvature, was fitted over the iron hoops; this was known as the paunch. Fillings were added to both sides, and the entire fore side of the mast was thus covered. The paunch protected the actual mast from being chafed by the sails and from damage when the yards were hoisted or lowered. It was fastened at its lower end with two iron hoops, and with wooldings elsewhere. In the Royal Navy paunches date from around 1775, but they are evident on some models from the middle of the century.

The Continental paunch did not have the length of its English counterpart, and its lower end was above the upper deck. Röding suggests an overall length of approximately one quarter of the mast's length, which means that the upper end reached up to the upper side of the trestle trees, thus preserving the appropriate distance between the topmast and the lower mast.

Wooldings

Wuhlings; Rousture

Ropes bound tightly around the mast were known as wooldings; they were intended to help to strengthen the lower mast. Their number on a mast differed according to the ship's size. Steel recommended eleven on the mainmast for large ships and nine for frigates. Other sources speak of between six and nine. Steel also specified that each woolding should consist of thirteen tightly wound turns, with each turn nailed to the mast. To prevent them from cutting into the woolding, the nail heads were underlaid with pieces of leather.

Timber hoops 1½in wide were usually fitted above and below the woolding, nailed to the mast and slightly greater in thickness than the woolding rope.

Mizzen masts were fitted with wooldings after 1730, normally two fewer than on the foremast, but the mizzen masts of small ships remained without wooldings or iron hoops until the end of the century. Some large ships may have been fitted with mizzen mast wooldings by 1700 or even earlier.

Lower masts

Untermasten; Bas-mâts

Lengths and diameters

Länge und Durchmesser; Longueurs et diamètres

'The exact height of the masts, in proportion to the form and size of the ship, remains yet a problem to be determined' noted Falconer in 1769. Steel, however, provided a formula which was adhered to in the Royal Navy in the later years of the century: 'the length of the lower deck and the extreme breadth being added together; half of this being the length of the mainmast.' He gave the following example:

> The length of the lower deck of a 74 gun ship around 1790 was 174 feet, extreme breadth 48 feet 8 inches; added together, this makes 224 feet 8 inches. Half of it, 112 feet 4 inches, was the mainmast's length.

This gives a mainmast length of 2.31 times the breadth. Davis suggested 2⅔ times the beam in general and 2½ times the beam plus 1 foot for East Indiamen. Sutherland recorded that in his time some shipbuilders took breadth plus depth times three and divided by five to give the length of the mainmast in yards. He himself advocated the addition of the gun deck length to breadth plus depth (less one sixth of the latter to account for the depth of the mast step), with the sum divided by two to give the length of the mainmast in feet. Love, writing in 1735, gave two thirds of the keel's length plus the beam as the length for the mainmast. Anderson noted that the *Prince George*, a 90 gun ship of 1723, had a mainmast 2¼ times her breadth in length.

Mountaine, in the third edition of his book in 1756, specified a beam equal to seven twentieths of the mainmast's length for merchantmen; this gives a mast length of 2.86 times the breadth. His formula for men-of-war was as follows:

> To find the Length of the Main-mast, take half the Length of the Keel, and the Breadth of the Beam, add them together, and divide them by 3, and that is your Length in Yards.

This gives a mainmast length between 2.4 and 2.5 times the breadth.

Similar figures came from Chapman (1768), who specified 2.43 times the breadth for East Indiamen. For other three-masted merchantmen he advocated a mainmast length of 3.23 times eleven twelfths of the breadth (the latter figure probably standing for beam).

Duhamel du Monceau, in 1752, recommended a mainmast length of 2½ times the beam. In an annotation he recorded that many shipwrights preferred a rule of twice the beam plus the depth in hold for the mainmast's length, and also that for frigates no greater figure should be used then 3½ times the beam, while for three-decked ships the usual figure should be slightly less then 2½ times the breadth. Paris, writing around 1780, noted a mainmast length of 2⅓ to 2⅖ times the beam.

All authors agreed that the length of the foremast should be eight ninths that of the mainmast, with the exception of Sutherland and Mountaine, who both recommended a figure of seven eighths for merchantmen.

Korth, who wrote *Die Schiffbaukunst* in 1826 for Kruenitz's *Encyclopaedia* and based his comments mainly on Duhamel du Monceau and Röding, noted a foremast length of nine tenths that of the mainmast.

The mizzen mast should have, according to Steel, a length of six sevenths of the mainmast. Falconer's *Marine Dictionary* of 1815 dealt with the mizzen mast's length as follows:

> Mizen-Mast, 100 guns, seven-eighths, 90 guns to 50, six sevenths, 40 guns to 20 five sixths of the main-mast. In sloops, the mizen-mast is three-fourths of the main-mast. When the mizen-mast steps on the lower deck, the depth of the hold is to be deducted from the length here given. This method is now generally practised in the Royal Navy.

Six sevenths of the mainmast was also the figure Sutherland gave at the beginning of the century for mizzen masts stepped in the hold. For a mast stepped on the gun deck, two thirds of the mainmast was considered sufficient. In a small ship five sixths of the mainmast was considered the correct length for a mast stepped in the hold.

Mountaine recommended that a mizzen mast measured from the keel should be three quarters the length of the mainmast, or two thirds if measured from the lower deck. The latter measurement concurs with Duhamel's specification (he gave only lengths from the lower deck), and was also Mountaine's figure for merchantmen. Röding measured the mizzen mast only from the lower deck and gave a length of 1¾ times the beam.

On the basis of the other measurements he gave, Paris'} figure for the mizzen mast would have been 1¼ times the beam, making the mast shorter than that of Duhamel du Monceau. Love gave a figure of four fifths of the foremast for a mizzen mast measured from the keel. Davis gave two thirds of the beam as the length for a mast stepped in the hold, but of twice the beam for a mast stepped on the lower deck. For East Indiamen he gave a figure of 2¼ times the beam minus 1 foot.

Korth recommended the beam plus twice the depth of the hold (or the whole length of the mainmast minus the length of the masthead, the depth of the hold, the difference in the ship's draught and the thickness of the step) as the appropriate length for a mizzen mast.

The maximum diameter of a mast was measured on English three-deckers at the partners on the second deck and at the partners on the upper deck on two deckers and smaller ships; on French ships, however, it was measured at the lowest deck.

Mast diameters in proportion to the mast's length were as follows, according to Steel: for ships from 100 to 64 guns, 1/36 (or 1 inch per yard) for fore and mainmasts, with the mizzen mast's diameter being three fifths that of the mainmast; for ships from 50 to 32 guns, 1/40 (9/10in per yard); and for ships with fewer than 28 guns, 1/41 (7/8in per yard). The figure for mizzen masts of 50 gun ships and lesser vessels was two thirds of the mainmast's diameter.

Sutherland gave a ratio of 1 inch per yard for the fore and mainmasts in ships of the largest size, 7/8in or 6/7in per yard for middle sized ships, and 4/5in or 3/4in per yard for small ships. The ratio for mizzen masts was 2/3in per yard.

Mountaine specified a diameter of three quarters of the beam for the mainmast of a merchantman, taking inches for feet, with the other masts being proportional. Duhamel du Monceau gave as a rule for determining the diameter of lower masts that the mast's length should be divided by three, with inches read for feet; this is, of course, equivalent to the figure of 1/36 given by Steel.

Paris also gave 1/36 as the figure for main and foremasts, and gave a figure of 7/288 of the length for the mizzen mast's diameter, and these same proportions were noted by Röding. Davis provided smaller ratios for the first decade of the century: approximately 1/42 for the main and foremasts and approximately 1/56 for the mizzen mast. Korth, who can be quoted for Continental measurements in the final years of the century, specified a diameter of 8 inches for every 10 feet of length (1/45).

Figures for vessels below ship size are given below, at the end of the description of masts.

Taper of masts

Mastverjüngung; Réduction du diamètre des bas-mâts

The sectional diameters shown below (from Steel) were valid, with minor variations, on English ships for the entire century:

First quarter	60/61
Second quarter	14/15
Third quarter	6/7
Lower end of masthead (width)	6/7
(length fore-and-aft)	3/4
Upper end of masthead	2/3
Heel	6/7 of the maximum diameter

Mountaine's comment on the proportions of the lower masts was as follows:

> Every Inch thick at the Partners, requires nine-tenths in the Middle, and two-thirds at the End. Note, the Middle here means the Medium between the Partners, and the very Extremity of the Mast...

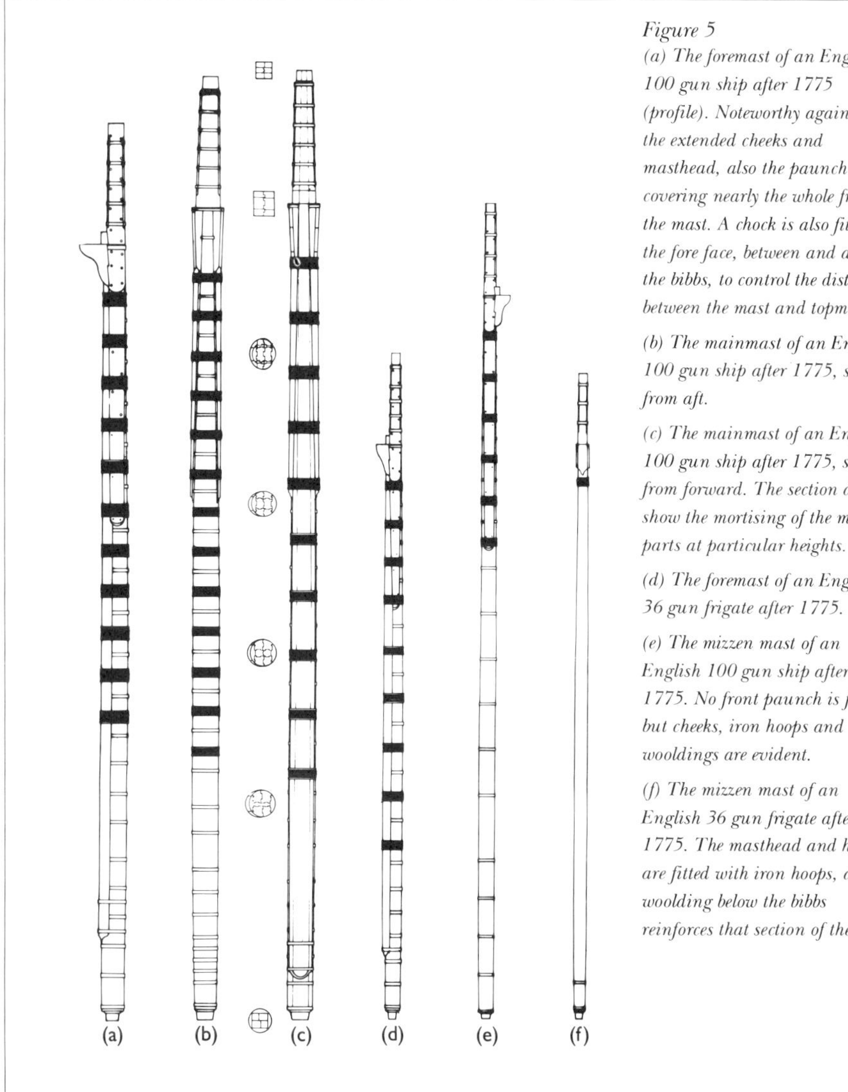

Figure 5

(a) The foremast of an English 100 gun ship after 1775 (profile). Noteworthy again are the extended cheeks and masthead, also the paunch covering nearly the whole front of the mast. A chock is also fitted to the fore face, between and above the bibbs, to control the distance between the mast and topmast.

(b) The mainmast of an English 100 gun ship after 1775, seen from aft.

(c) The mainmast of an English 100 gun ship after 1775, seen from forward. The section details show the mortising of the mast parts at particular heights.

(d) The foremast of an English 36 gun frigate after 1775.

(e) The mizzen mast of an English 100 gun ship after 1775. No front paunch is fitted, but cheeks, iron hoops and wooldings are evident.

(f) The mizzen mast of an English 36 gun frigate after 1775. The masthead and heel are fitted with iron hoops, and a woolding below the bibbs reinforces that section of the mast.

Duhamel du Monceau noted only that the mast's head should be two thirds the maximum diameter of the mast, a view corroborated by Paris and Korth. Röding added that the minimum diameter of a mizzen mast should be seven twelfths of the maximum. As well as specifying 1/36 of the mast's length as its maximum diameter, Love recommended seven eighths of that for the middle and three quarters for the masthead diameter.

Chapman made the following note on mast diameters:

> One has found, that masts below the trestle trees receive their proportional thickness, when they are, at that position, 1/8 thinner than on deck; if the thickness on deck is 128, than it is at the first part 127, at the second 124, at the third 119 and at the fourth or below the trestle-trees 112, directly at the trees 4/5 and at the masthead 5/8 of the thickness on deck.

Hounds and bibbs

Mastbacken; Jottereaux

Immediately below the trestle trees, the cheeks were strengthened by the hounds (see Figures 3a to 5f). Their length was seven fifteenths that of the masthead, and loose extensions known as bibbs were bolted to their forward face to provide further support for the trestle trees. Made from elm and 3 inches to 5 inches thick, they were nine tenths as long as the hounds and two fifths of their length in width. The aft edge was provided with a step to facilitate accurate location, and the fore edge formed an ogee shape (see Figures 3a to 5f). The loose

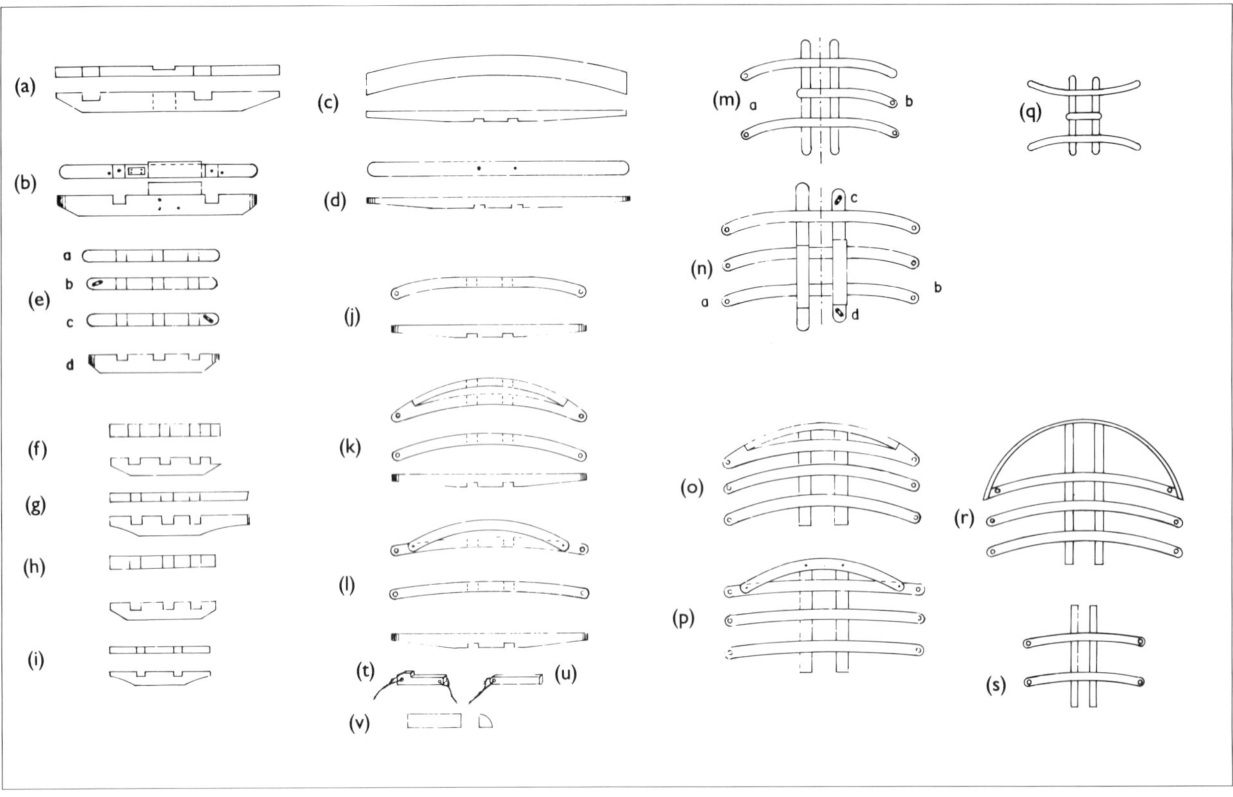

Figure 6
(a) A Continental lower trestle tree, mortised to fit on to the masthead.
(b) An English lower trestle tree, with semi-circular ends, a bolster and an iron plate for the fid.
(c) A Continental lower cross tree, curved aftwards.
(d) An English lower cross tree, straight with semi-circular ends.
(e) An English topmast trestle tree
a. until approximately 1775 (plan view)
b. fore topmast after approximately 1775 (plan view)
c. main topmast after approximately 1775 (plan view)
d. profile.
(f) A French and Continental topmast trestle tree, about 1780.
(g) A French topmast trestle tree, about 1700.
(h) A Continental topmast trestle tree, about 1780.
(i) A French mizzen topmast trestle tree, about 1700.
(j) An English topmast cross tree; the outer thirds are curved aftwards.
(k) A French topmast cross tree, about 1780; it is curved aftwards over its full length. The foremost tree is fitted with chafing protection.
(l) A Continental topmast cross tree.
(m) English trestle and cross trees (plan view)
a. until approximately 1706
b. between 1706 and approximately 1720.
(n) English trestle and cross trees (plan view)
a. between 1720 and approximately 1775
b. after 1775
c. main topmast, after 1775
d. fore topmast, after 1775.
(o) French trestle and cross trees, about 1780 (plan view).
(p) Continental trestle and cross trees, about 1780
(plan view).
(q) Trestle and cross trees for a topgallant mast and a spritsail topmast, late seventeenth century. These were in use until the end of the spritsail topmast era.
(r) French topmast trestle and cross trees, about 1700 (plan view). The chafing protection, here shown as a plank, was later integrated; see (o).
(s) French trestle and cross trees for mizzen topmasts and topgallant masts, about 1700 (plan view).
(t) A wooden fid.
(u) An iron fid.
(v) A bolster for an English trestle tree.

bibbs were connected to the hounds by four bolts of ¾in to 1in diameter. These bolts were secured with rings on the aft side.

French hounds (see Figures 3b and 4d) were either half or two sevenths (minimum) the length of the masthead, and had an upper width of twice the mast diameter below the trestle trees. The lower width was seven fifteenths of the upper, and the thickness of the hounds above was 1⁄144 of the masthead length, with that below being seven sixteenths of the upper.

Mast battens

Mastlatten; Lattes du bas-mât

Battens were frequently nailed over the iron hoops around the masthead to prevent chafing damage to the rigging and the masthead itself. Usually eight battens of three fifths of the masthead length were needed. Their width was one eighth of the masthead diameter, and their thickness half the width (see Figure 4e and f).

Trestle trees and cross trees

Salinge; Élongis et barres traversières

The trestle trees and cross trees rested on top of the hounds and bibbs, and so provided a frame to support the mast top's platform. Within this structure, the fore-and-aft members were known as trestle trees and the athwartships members as cross trees (see Figure 6a to s). The preferred material for the trees was oak.

The most common English measurements for trestle trees were as follows:

Length	¼ of the topmast length
Height	½ of the topmast diameter
Width	⅔ of the height

The sides of each tree were straight and parallel. The ends were rounded in plan view, but chamfered in profile. These chamfers were restricted to the lower half of the ends; the length of the front chamfer was 1½ times its height, and that of the rear was equal to its height. The lower outer edges of the trees

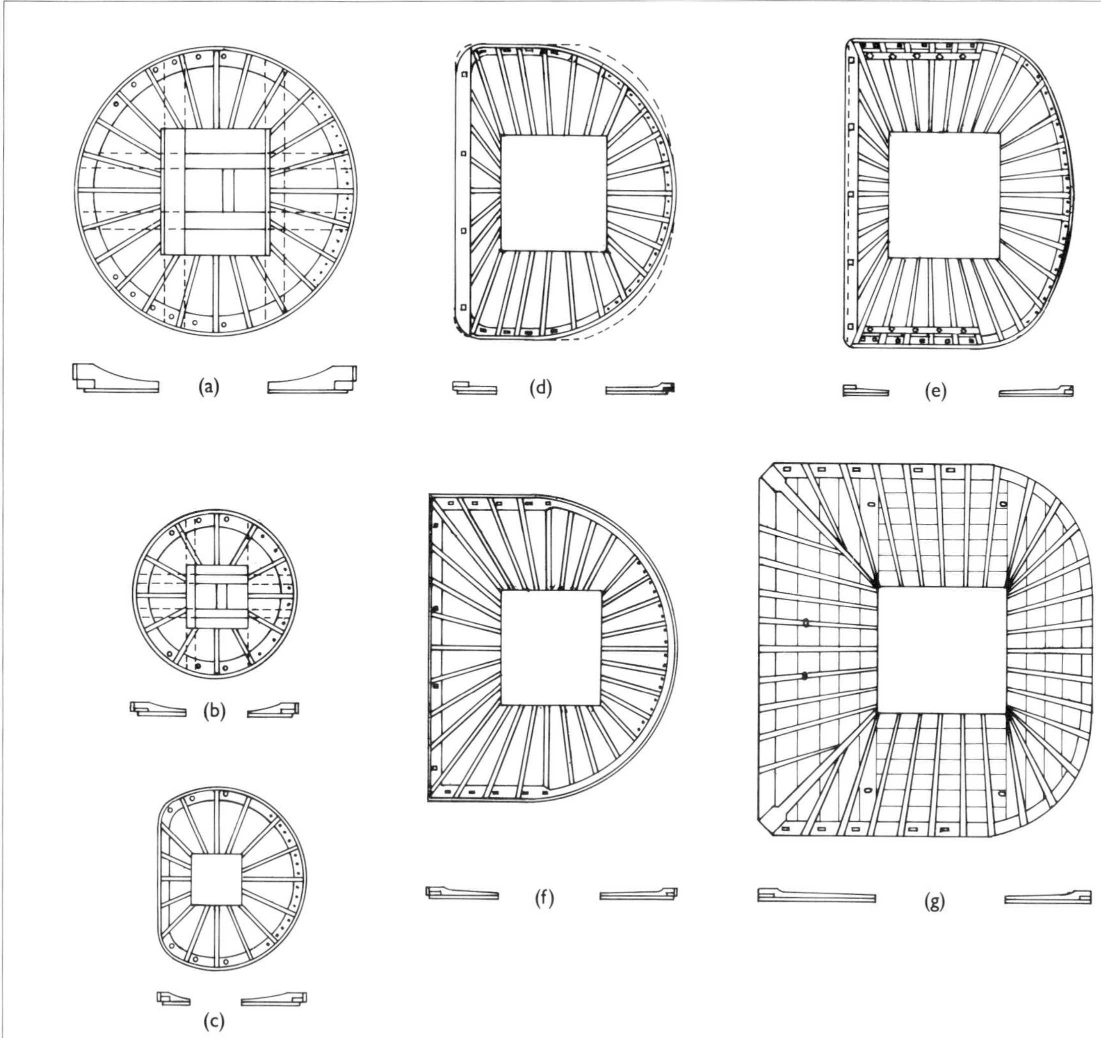

Figure 7
(a)42 A round top for a fore or mainmast until approximately 1700, English and Continental.
(b) A round top for a mizzen mast until approximately 1700, Continental.
(c) A round top for a mizzen mast with a flattened aft side, approximately 1700 (English).
(d) An English top from approximately 1710 to about 1720. The dotted line indicates the shape of a top between 1720 and 1745. Note the tighter radius on the after corners and the more elliptical shape of the semi-circular forward side.
(e) An English top for a ship of the line, about 1750.
A further tightening of the after corners and a flatter ellipse can be seen. In addition, fore and aft timbers for the mounting of swivel guns are bolted to the sides.
(f) A French top, about 1790. This shows a more pronounced semi-circular forward side and a longer aft section than English tops.
(g) A French top for a ship of the line, about 1780. Note the flattened semi-circular forward side and the even longer aft section.

were slightly bevelled over the full length, with the lower inner edges bevelled only between the cross trees (see Figure 6b and e).

On Continental ships the trestle tree chamfers were usually longer and sometimes curved. Chapman noted that a mainmast's trestle tree had a length of one quarter the length of the topmast minus half an inch. The foremast trestle trees were shorter by one fifteenth, and those of the mizzen mast had a length of three fifths of that of the mainmast trees. Topmast trestle trees were three sevenths the size of the lower trees. Their width was either five sevenths or three quarters of the height, though no precise height was given in Chapman's works (see Figure 6a and h).

Cross trees were square timbers with the following proportions:

Length	⅓ that of the topmast minus 6 inches
Width	that of the trestle tree
Height	⅔ of the width

Chapman gave only a hint at the length of the cross trees by supplying a width for the mast top, given as one third of the topmast's length.

On their lower sides, the cross trees tapered upwards towards the ends for approximately one quarter of their length port and starboard, reducing to half the normal thickness. The ends were rounded in the same way as those of the trestle trees, with the lower edges also being bevelled (see Figure 6c and j).

Mortised into each other, the trestle and cross trees together formed a frame. The depth of the mortises on the trestle trees was 1 inch less then the height of the cross trees, and the remaining inch was cut into the lower side of the cross trees. After being joined, the framework was bolted together (see Figure 6m to s).

When the trestle trees were fitted to the mast, it was essential to ensure that the centre of the framework rested on the fore edge of the masthead. The inclination of the mast had also to be considered, since the trestle trees had to sit horizontal. The

Figure 8
(a) An English top for a 36 gun frigate, about 1780. The after corners are now without radii.
(b) An English top for a 74 gun ship, about 1780.
(c) A Continental top, about 1780; this is similar to the French top shown in Figure 6 (g).
(d) An English top for a merchantman, from about the turn of the century; note that the decking consists of an open grating rather than planking.
(e) An English top for a merchantman, from about the turn of the century; note the lubber's holes to each side.

Figure 9
(a) French trestle and cross trees (alternative) for topmasts (1780), with fore and aft iron futtock bands

a. plan view

b. cross tree section

c. profile.

(b) A French main top for a 74-gun ship from 1780; note the flattened elliptical shape of the forward side and the open planking. This top is secured to the trestle and cross trees by toggles.

(c) A French main top of 1780

a. forward profile

b. cross section.

(d) A Continental top for a merchantman, early 1800. This closely planked top has four cross trees and two preventer cross trees.

(e) A Continental top for a merchantman, early 1800

a. The connection of a preventer cross tree to a cross tree is clearly shown here

b. The profile of a trestle tree with four cross trees and two preventer cross trees.

(f) A French mast cap (English style), 1780.

(g) A French topmast cap (English style), 1780.

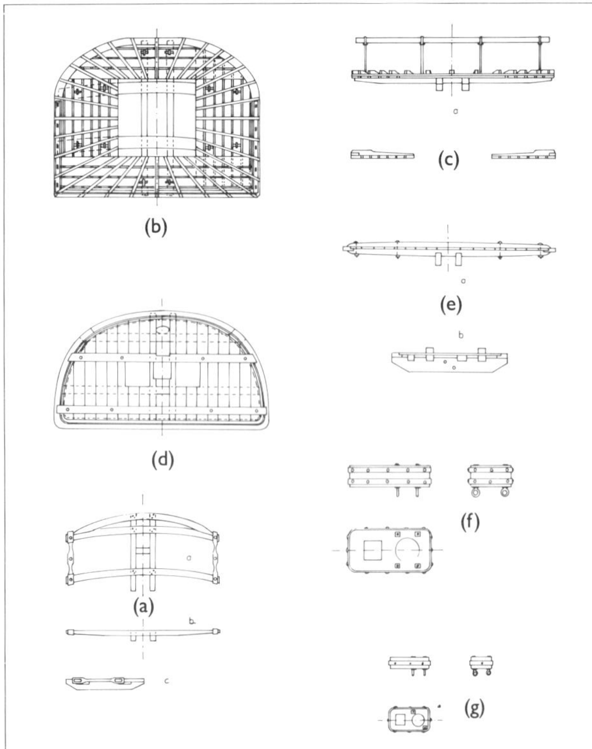

upper edge of the hounds, known as the stop, was bevelled accordingly.

Iron plates some $\frac{3}{4}$in thick were set into the upper faces of the trestle trees aft of the first cross tree mortise. These plates were three quarters the length of the opening for the topmast and two fifths the width of the trestle trees. Their function was to prevent chafing of the topmast fid on the trestle tree surfaces.

Some Continental trestle trees had ends which were rectangular in plan view, rather than rounded as shown in the figures, and the ends of the cross trees were curved on the front face only. The trestle trees, as confirmed by Paris and Röding, were also mortised in the middle for their fitting to the masthead and, because of the top's larger rear overhang on Continental ships, also extended further aft.

Paris provided some French measurements for mainmast trees. The trestle trees, in his view, should have a length of .086 of the ship's length and the cross trees a length of .47 of the ship's breadth. The thickness of the trees should be $\frac{5}{72}$ of their length.

G D Klawitter noted in his *Vorlegeblätter für Schiffbauer 1835* (covering merchantmen of the early nineteenth century) a mast top width of seven twelfths or five ninths of the ship's breadth, and a length half the width. The trestle and cross trees were 1 inch shorter in length at the ends in each case. The thickness of each trestle tree was one twelfth of the mast top's width, and the width of each was three sevenths of the height. He specified $\frac{4}{9}$in per foot of length for the width of each cross tree and a height of nine tenths of the width.

In the context of trestle and cross trees for merchantmen Röding also made mention of preventer cross trees, describing them as 'timbers, being laid over a grating top, above the cross-trees and lashed together with them [serving] to strengthen the cross trees and the top itself.' Such preventer cross trees were in use in the last two decades of the eighteenth century and, because of the two-part construction of larger mast tops, also came into use on English masts. They were then, as Klawitter noted, bolted to the lower cross trees (see Figure 9e).

Bolsters

Kälber; Coussins d'élongis

A piece of fir known as the bolster was nailed to the upper surface of the trestle trees on both sides of the masthead. The bolsters were rounded on their outer sides, so that they obstructed neither the aftmost cross tree aft nor the topmast fid forward (see Figure 6v).

The bolster's function was to lead the shrouds away from the mast and to prevent chafing of the trestle trees due to movement of the mast. For this reason, it was at least $1\frac{1}{2}$ inches wider than the trestle tree. The bolster's height was equal to its width.

Before any rigging was commenced, the bolsters were covered with several layers of old canvas and tarred thoroughly. Continental rigs did not always include bolsters (see Figure 6v).

Tops

Marse; Hunes

The framework of lower trestle and cross trees supported the mast top, a platform which lost its former round shape in the early years of the century (see Figure 7a and b). In the beginning the after quarter only of an English top was cut off and the resulting corners rounded (see Figure 7c); this soon developed into straight sides and rounded corners, with only the fore half rounded. From the 1720s onwards, the rounded fore half of the top became gradually flattened into an ellipse (see Figure 7d and e).

The after corners of the top were still slightly rounded at the middle of the century, and sometimes timbers for the mounting of swivel guns were bolted inside the futtock plates along the top's sides. In the final development of the English top's shape the after corners lost their roundness and became sharp (see Figure 8a and b).

On French and Continental tops, as illustrated by Röding and Paris, the wider platform abaft the mast is especially conspicuous, and the forward part of the top's half-round was often flattened over a length equal to the width of the lubber's hole (see Figures 7g and 8c). Nevertheless, the drawings by Paris of *Le Protecteur*, a French 64-gun ship of 1793-94, shows that even at the end of the century the original half-round shape was also still in use on French tops (see Figures 9b and 7f).

As Lever observed, merchantmen often had semi-circular tops with an open oaken grating, rather than the closely planked tops of men-of-war (see Figure 8d and e, and 9d).

A top's main function was to provide a sufficiently large angle to enable the topmast shrouds to be effective in supporting the topmast. For the futtocks of these shrouds, square mortises were cut into the outer side rims of the top. The top's platform was also of great value when the topsails had to be served or repaired and, in action, it provided an vantage point for sharpshooters. A rail protected the top's after side and provided, in combination with the shrouds, support for a great number of hammocks which formed a parapet for the protection of the marksmen (see Figure 9c). Normally this rail was fitted with netting or canvas.

Dimensions for English tops given by Steel were as follows:

Width	$\frac{1}{3}$ of the topmast length
Length	$\frac{3}{4}$ of the width
Width of the lubber's hole	$\frac{2}{5}$ of the top's width
Length of the lubber's hole	$\frac{13}{14}$ of its width
After edge of the lubber's hole	$\frac{1}{5}$ of the top's length forward of its after edge

Chapman provided very similar dimensions, recommending a top width of one third of the topmast length and a top length of one quarter of the topmast length. In contrast, Röding specified a top width of half the ship's breadth and a top length

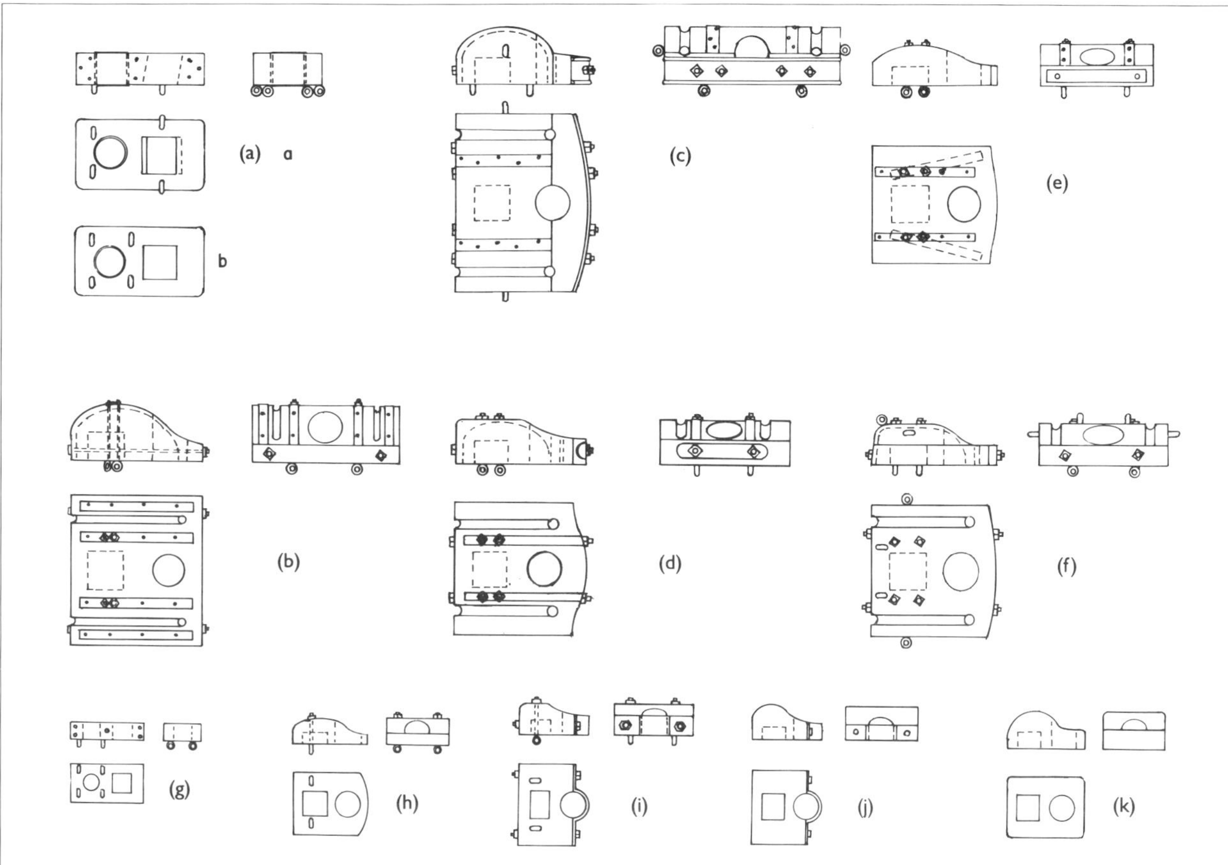

Figure 10
(a) An English mast cap
a. after 1775
b. before 1775.
(b) A French mast cap, about 1700, rectangular in plan view. Two fore and aft bolts on the outer side of the tye channels stabilise the cap.
(c) A Dutch mast cap, about 1700, made from two pieces, with the semi-circular forward piece strengthened by an iron strip. Four bolts on the inner side of the tye channels keep the pieces together.
(d) A French mast cap, about 1780, with two bolts on the inside of the tye channels. The fore side, shaped like a double ogee curve, is also fitted with an iron plate extending between the two bolts.
(e) A Continental mast cap, about 1780, similar to that shown in (b) but without tye channels. The jeers run through blocks at the masthead.
(f) A French mast cap, about 1790, made from two pieces sandwiched together.
(g) An English topmast cap.
(h) A Continental topmast cap; the fore side is curved, similar to French caps from about 1700.
(i) A French topmast cap, about 1780; an iron bracket encloses the fore half of the topmast diameter.
(j) A French topmast cap, about 1760.
(k) A Dutch topmast cap, about 1700.

only fractionally less than its width. His description implied a nearly rectangular top, with the fore corners rounded to prevent damage to the topsail. The opening or lubber's hole, according to his scheme, should have a length and width of seven twentieths of the top's width.

Dimensions given by Paris for a French mainmast top agreed with those of Röding:

Width	½ the beam
Length	width minus 1/20
Lubber's hole	square, with sides 2/5 of the top's length and with the aft edge of the opening 1/3 of the top's length forward of the top's rear edge

Comparison of Continental data even for this one feature of the masting of eighteenth century ships shows significant regional variation in ship design and construction. While Chapman's figures, and therefore Swedish ship-building, tend towards English methods, Röding's in Hamburg were much closer to French practice.

The planking of tops ran fore-and-aft in way of the mast, but athwartships in the forward and aft sections of the top. Elm was used and the planks were 3 inches thick. Where fore-and-aft planks overlapped cross planking, both were reduced to half thickness to maintain the overall thickness of 3 inches. After the planks had been fitted carefully, they were nailed together and the forward edge of the top received its rounded shape. A frame made from elm, the rim, was laid over the outer upper side and nailed to it. Of a thickness of 1⅛ inches and 7 to 8 inches wide, the rim reached extended about 4 inches over the edge of the planking.

In his comments on the planking of Continental tops, Röding noted that a gap of 1½ to 2 inches was left between the planks to enable rainwater to run off. He further noted that closely planked tops were also quite common and that others had a grating rather than planks. The first type of top he saw as a medium between the latter two, and he offered the following explanation for the use of a grating:

> The closely planked top is certainly stronger than a grating top, although this is, however, lighter and does not trap as much wind, which can be considerable, when sailing by the wind.

The battens were fitted next in the construction of a top. These were tapered pieces of timber with a thickness of 4 inches outside and 2 inches inside. In English tops four or five battens each side (depending on the vessel's size) led from the rim to the lubber's hole. Forward of the hole were seven to thirteen battens, with one or two fewer on the aft side, all equally spaced.

French type tops, with a larger platform abaft the mast, had, according to contemporary drawings, one batten more on each side, and the battens either side of the aft corner battens were angled not directly towards the lubber's hole, but rather converged on the corner batten, meeting it slightly inboard of halfway. The corner battens were also thicker and heavier than the others.

Between these battens, filling pieces were nailed to the rim. They were of equal height to the battens and 9 inches wide at the sides of the top, but only 4½ inches wide on the curved fore rim. Two to three holes were drilled into most of the filling pieces in the curved section, giving a total of eighteen to twenty, for the fitting of the stay's crowsfeet. At the aft side of the top a strong plank, the gunwale, was used instead of the filling pieces. The gunwale was 11 to 12 inches wide and 1¼ inches thick. It also had four, or sometimes five, square holes for the rail stanchions.

From approximately the middle of the century onwards, beams 8 inches wide and three times the height of the battens were placed inside the filling pieces on the sides, and bolted to the top. Their purpose was for the mounting of swivel guns.

Röding provided further information about tops:

> Only men-of-war and large merchantmen have tops; smaller merchant ships have double the number of cross trees on their masts instead and the holes for the futtock shrouds are drilled through their ends or through the timber battens which join the cross trees. If the space between the cross trees (now numbering five or six) is still too wide, thin battens are laid across so that, without running the risk of falling through, one can easily walk on them... So that the lower part of the topsail does not chafe, the fore edge of the top is not only rounded and cladded with rope-made chafe mats, but crowsfeet are also fitted from there to the stay.

Caps
Eselshäupter; Chouquets

The head of a lower mast was topped with a cap. This was a rectangular piece of elm with two vertical holes, serving as a locating connection between mast and topmast (see Figure 10a).

The dimensions of the cap of an English lower mast:

Mainmast	length	4 times the topmast's diameter plus 3 inches
	width	twice the topmast's diameter plus 2 inches
	depth	4/9 of the width
Foremast	length	4 times the topmast's diameter plus 2 inches
	width	twice the topmast's diameter
	depth	4/9 of the width
Mizzen mast	length	4 times the topmast's diameter plus 1 inch
	width	twice the topmast's diameter
	depth	4/9 of the width

The depth of a cap, as given by Chapman, was four fifths of the topmast's diameter. For a Continental cap (English style) Röding noted a length of three times the topmast's diameter and a depth of one third of the length. Continental caps were made from two parts joined lengthwise with a swallow tail, bound crosswise with strong iron bands and bolted together.

When an English cap was made from two pieces, these were tenoned and fitted together with six bolts. The hole for the masthead was cut square and that for the topmast round, with the latter having a diameter ¾in larger then the topmast to allow clearance for a leather sleeve which gave the topmast a smoother pass through the cap. The distance between the inner edge of the hole for the masthead and the periphery of the topmast hole was two fifths of the topmast hole's diameter plus half the difference arising from the taper of the masthead.

Four 1¾in diameter eyebolts for the lift blocks and the top rope were driven from the lower side through the cap (see Figure 10a).

English style caps, which were used in the 1770s in increasing numbers on larger French ships and also, as indicated above, by other European nations in a slightly different form, were 3½ to 4 times the topmast's diameter long, half that length wide and, again, half the width high. The vertical corners were rounded to a degree and the cap was secured with two iron hoops around the sides.

A Continental cap's shape varied markedly from that of the English style cap described above. While the latter resembled a brick, French, Dutch and German caps were often rectangular in plan-view, but in profile revealed a rounded hump covering the masthead and a flattened topmast half (see Figure 10b to f). The holes were in a similar position to those on English caps, with the minor difference that the masthead hole, cut from the lower side, was only half the cap's height.

Continental cap measurements were given by Paris. The length and width were three times the masthead's diameter. At the highest point of the rear half of the cap the height was equal to the masthead's diameter, while the flatter forward half of the cap was only half as high.

Smaller holes were drilled into both sides of the topmast hole and grooves were run from these holes over the humpback to the aft edge of the cap. The jeers for the lower yard were led along these channels and through the holes. Topmast caps did not have these holes. Steel, dealing with yachts and similar vessels, described these caps as having sheaves for the jeers of the lower yard on both sides of the masthead hole.

During the last decades of the century, the jeer holes and subsequently the grooves or channels on Continental caps, except the French, disappeared and jeer blocks were fitted to the masthead.

The trend towards English shipbuilding techniques and rigging in the Netherlands increased during the second half of the century and in turn influenced the neighbouring shipbuilding nations. Röding's 1798 *Wörterbuch der Marine* therefore made reference only to English and French style caps.

Cap shores
Eselsstütze; Chandelles de chouquet

On larger ships, especially English vessels, it was common practice to fit a supporting timber post between the fore end of the lower cap and the top, to reduce the topmast's pressure on the cap's overhang. On English ships with more than 50 guns, cap shores can be traced to approximately 1720. These supports were not found on ships with Continental caps (see Figure 11a).

Topmasts and topgallant masts
Stengen (Marsstengen, Bramstengen); Mâts de hune et mâts de perroquet

Top and topgallant masts were loose extensions of the lower mast. In the period covered here, two extensions (that is, both a topmast and a topgallant mast) were common. Both masts fall into the category of pole masts, made from a single piece of timber.

Like the lower masts, top and topgallant masts also underwent a process of development during the century. For example, the lower part of English topmasts above the heel (the footing) was rounded until about the early 1770s, as it was on Continental topmasts throughout the century, but on English masts this section commonly became octagonal after 1770, or as early as 1745 in some individual vessels (see Figure 11b to e).

The heel was the lower, squared part of a top or topgallant mast, sitting between the trestle trees and the forward cross trees. A square hole running athwartships, known as the fid hole, was cut through the heel. The fid, used for stopping the topmast at the trestle trees, was made either from iron or from timber (see Figure 6t and u). An iron fid was 1½ times the topmast's diameter long, one third of the diameter high and two thirds of the height wide. Timber fids varied in height, but about half the diameter of the topmast was common.

The heel of an English topmast had a length of 2½ times its thickness and that of a Continental topmast a length of 3½ times its thickness. Whereas a Continental topmast's heel was its lower end, English topmasts finished with a block underneath the heel. This extension of the topmast footing was made square until the middle of the century (see Figure 11m) and after that octagonal or round (see Figure 11k and l). An iron hoop was pressed on to the lower end of the block, and above that a sheave hole was cut from the port side aft to the starboard side forward. The length of this sheave hole was equal to its

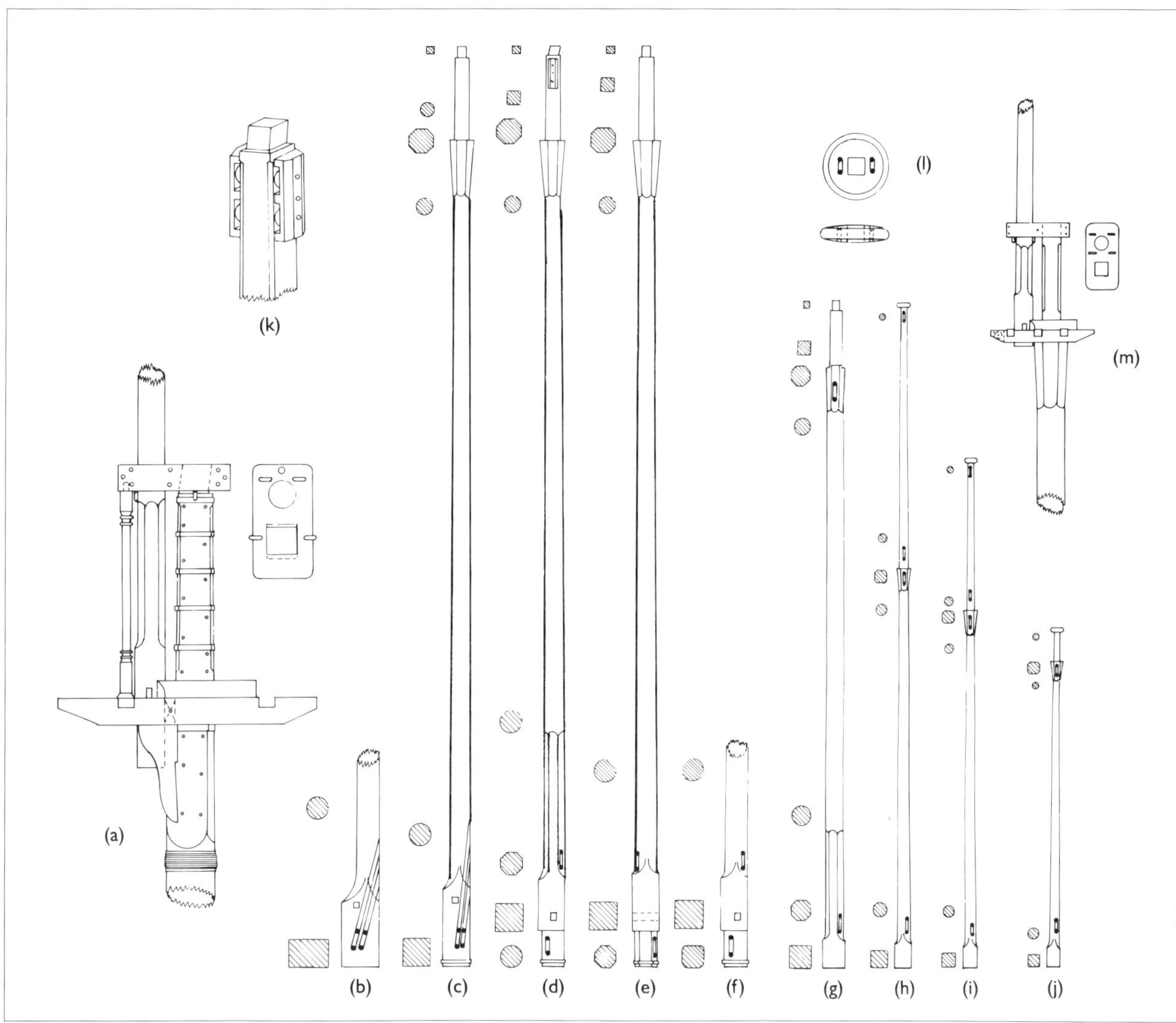

Figure 11

(a) Cap shore on an English lower mast after 1720.

(b) Footing of a Continental topmast, about 1780.

(c) Continental topmast. The hatched cross sections indicate the topmast's shape at that particular position. The masthead was round and an iron hoop was pressed on to the heel.

(d) English topmast, about 1775. The footing above the heel was octagonal and the block beneath rounded. The mast head was, in contrast to a Continental one, squared. After 1775 cheek blocks were fitted.

(e) English topmast after 1745; the footing above the heel was rounded and the block octagonal.

(f) English topmast until 1745; like (e) but with a square block.

(g) English mizzen topmast, without a block but fitted with a sheave in the hounds for the halyard. The footing was shaped according to the period in question.

(h) English topgallant mast with a long pole head.

(i) English topgallant mast with a normal pole head.

(j) English topgallant mast with a stump head.

(k) English topmast head with cheek blocks after 1775.

(l) Topmast or topgallant mast truck.

(m) English main topmast head with topgallant mast footing after 1775, cheek blocks omitted.

depth, and its width was 2 inches per foot of depth (in fact, this rule applied for any sheave opening). The block's length was 1½ times its diameter.

A further sheave opening in the footing was situated 4 inches above the heel. It ran opposite to the block sheave. The topmast's top rope, when fitted, was led over this sheave.

On Continental topmasts both sheaves were set athwartships, side by side into the heel (see Figure 11b and c).

The hounds, used for fitting trestle and cross trees, divided the topmast from its head. These were octagonal and had an approximate length of three fifths of the topmast head.

A sheave for the halyards was set fore-and-aft in the hounds on mizzen topmasts, topgallant masts and on all topmasts of ships with 18 guns and fewer. Contemporary literature warned, however, that this practice might lead to a weakening of the topmast's upper parts and should, if possible, be avoided (see Figure 11g to j).

The topmast head extended above the hound's upper edge, the stop. In English masting practice the topmast head was square and had a length of one ninth of that of the topmast. On Continental topmasts the heads were round, like the lower mastheads, with a trend to octagonal mastheads becoming apparent in the later years of the century. The head length of mizzen topmasts, and topgallant masts if royal masts were carried, was 3½ inches per yard of the relevant mast's length.

Most ships, however, carried so-called long pole heads (mastheads extended into flagpoles) rather than royal masts. Their length was two thirds of the mast's length to the stop. For normal pole heads the length was one fifth rather than two thirds. To calculate the length of a pole head, one therefore subtracted 3½ inches per yard from the mast length, and two thirds or one fifth of the result was the appropriate length for the pole head.

For example, for a 100 gun ship the main topgallant mast length was 35 feet or 11⅔ yards; subtracting 3½ inches per yard gives 31 feet 8 inches as the mast's length to the stop; two thirds of 31 feet 8 inches gives 21 feet 2 inches as the length for a long pole head, and a total topgallant mast length including the pole head of 52 feet 10 inches.

Alternatively, one fifth of 31 feet 8 inches gives 6 feet 4 inches as the height for a normal pole head, and a total topgallant mast length of 38 feet.

Masts with a pole head were rounded above the stop, with a further sheave fitted, 1½ times the sheave hole's size above the stop, for the staysail halyard (see Figures 11h to j and 12c and d).

Korth noted that a mizzen topmast should be one third longer than normal if no topgallant mast was rigged, with this extension forming the long pole head. Mizzen topmasts did not have a block below the heel, and they carried, at least for the first half of the century, a long pole head rather than a topgallant mast.

A mizzen topgallant mast on English ships was first mentioned by Sutherland in 1729. A painting of the Dutch 52-gun ship *Gertruda* of 1720 reveals a mizzen topgallant mast with a yard rigged, and Paris provided dimensions of French mizzen topgallant masts (First to Sixth Rate) from a 1690 manuscript, while the model of the *Royal Louis* from 1690 actually has a rigged mizzen topgallant mast. Therefore it seems that while English rigging was still dominated by the long pole head, Continental ships made extensive use of the mizzen topgallant mast.

Long or normal pole heads were topped with a truck (see Figure 11l). When royal masts were rigged, they were similar to topgallant masts with a normal head and the topgallant masts had squared heads and caps.

Cheek blocks came into use around 1775. They were placed directly below the cap on both sides of a topmast's head (see Figure 11k) and were made from rectangular pieces of elm. With recesses for two sheaves, one above the other, they were bolted to the head, with the bolts serving simultaneously as the axes for the sheaves. Cheek blocks had a length of $2\frac{1}{2}$ times the masthead's diameter plus 6 inches. The thickness was half, and the width two thirds, that of the masthead. Cheek blocks were usually found in English rigging.

Lengths and diameters

Länge und Durchmesser; Longueurs et diamètres

All top and topgallant mast dimensions stood in a particular ratio to each other or to those of the lower masts.

Throughout the century the main topmast length was about three fifths of the mainmast's length, a fore topmast about eight ninths of the main topmast length and a mizzen topmast half until about 1720 and later seven tenths or three quarters of the main topmast length. Sutherland noted in 1711 that 'the Mizon-top-mast is $\frac{1}{2}$ of the Mizon-mast stept in Hold.'

According to Steel, topgallant masts were half the relevant topmast's length, and royal masts seven tenths the length of the topgallant mast.

For main and fore topmasts the diameter was $\frac{1}{36}$ of the length, and a mizzen topmast had a diameter of seven tenths that of the main topmast. Sutherland noted the diameter of the latter as $\frac{3}{4}$in per yard of length.

All topgallant masts had a diameter of $\frac{1}{36}$ of the length, and the diameter of royal masts was two thirds that of the topgallant mast.

Mountaine's figures, at least forty years older than Steel's, differ slightly:

Length (men-of-war)	
main topmast	$\frac{3}{5}$ of the mainmast
main topgallant mast	$\frac{1}{2}$ of the main topmast
fore topmast	$\frac{3}{5}$ of the foremast
fore topgallant mast	$\frac{1}{2}$ of the fore topmast
mizzen topmast	$\frac{3}{5}$ of the mizzen mast
Length (merchantmen)	
main topmast	$\frac{3}{5}$ or $\frac{4}{7}$ of the mainmast
main topgallant mast	$\frac{7}{12}$ of the main topmast
fore topmast	$\frac{7}{8}$ of the main topmast
fore topgallant mast	$\frac{7}{8}$ of the main topgallant mast
mizzen topmast	$\frac{4}{7}$ of the main topmast

The mast diameters were proportional to those of the lower masts.

Even older are Davis' records from 1711. He noted the length of a main topmast as five ninths that of the mainmast or $1\frac{3}{5}$ times the beam minus 1 foot. The diameter was $\frac{1}{42}$ of the length. The fore topmast's length was five ninths that of the foremast or $1\frac{3}{7}$ times the beam, and that of the mizzen topmast either three quarters of the beam plus 2 feet or seven eighths of the beam. The length of topgallant masts was two fifths that of the appropriate topmast and the diameter $\frac{1}{42}$ of the length.

Varying again slightly were Love's measurements from 1735:

Length of main topmast	$\frac{5}{9}$ that of the mainmast
Diameter of main topmast	$\frac{5}{8}$ that of the mainmast
Length of main topgallant mast	$\frac{2}{9}$ that of the mainmast

The dimensions of other masts were proportional to those of their lower masts.

Information given thus far relates to English ships; Korth provided Continental measurements as follows:

Length	
main topmast	$1\frac{1}{2}$ times the beam
main topgallant mast	$\frac{3}{5}$ that of the fore topmast or $\frac{5}{6}$ of the beam
fore topmast	$\frac{1}{10}$ less than that of the main top mast
fore topgallant mast	$\frac{5}{7}$ of the beam
mizzen topmast	equal to the beam
mizzen topgallant mast	$\frac{1}{2}$ of the beam

The lengths of the topmast heads were similar to those of the heads of the lower masts (one ninth of the length), and of the topgallant masts one third of their length, with a further extension of a third when royal yards were rigged.

Röding's specifications are comparable to Korth's. He specified, however, a length for the fore topmast equal to that of the main topmast and for the fore topgallant mast a length of seven ninths of the beam. The (maximum/minimum) diameters he gave were as follows:

Main topmast	$\frac{7}{288}$ of length/ $\frac{7}{12}$ of the maximum
Fore topmast	$\frac{7}{288}$ of length/ $\frac{7}{12}$ of the maximum
Mizzen topmast	$\frac{7}{288}$ of length/ $\frac{5}{12}$ of the maximum
Main topgallant mast	$\frac{1}{48}$ of length/ $\frac{1}{3}$ of the maximum
Fore topgallant mast	$\frac{1}{48}$ of length/ $\frac{1}{3}$ of the maximum
Mizzen topgallant mast	$\frac{1}{48}$ of length/ $\frac{1}{3}$ of the maximum

Pole head lengths were as follows:

Main topgallant mast	$\frac{1}{3}$ of length (or $\frac{2}{3}$ with a royal set)
Fore topgallant mast	$\frac{1}{3}$ of length
Mizzen topgallant mast	$\frac{1}{3}$ of length

On ships without a mizzen topgallant mast, the mizzen topmast was longer accordingly.

Chapman's notes on top and topgallant masts were as follows:

> The length of a main topmast, measured from the upper edge of the trestle trees, amounted to: (naming the mast's length as L) $1\frac{1}{10}$ L divided by 2.73 for frigates, and $1\frac{1}{10}$ L divided by 2.84 for barks. The fore topmast is nine tenths the length of the main topmast. The length of the topgallant masts to the trestle trees is 0.54 that of the topmast... For East Indiamen the main topmast [length is] 0.586 of the mast's [and that of] the mizzen topmast $\frac{3}{4}$ of the fore topmast... All topmasts are $\frac{1}{5}$ smaller below the hounds than at the lower cap; if the thickness at that section is 80, so the thickness at the first division is 79, at the second 76, at the third 71 and at the fourth or below the hounds 64, [and] the thickness directly above the trestle trees is $\frac{2}{3}$ of, and at the head $\frac{5}{9}$ of, the thickness at the lower cap.

Duhamel du Monceau did not give any information on the lengths and diameters of topmasts, but some details of the proportions of French masts were provided for the 1742 First Rate *Royal Louis* by Paris:

Main topmast length	$\frac{3}{5}$ of the mainmast length
Main topgallant mast length	about $\frac{2}{5}$ of the main topmast length
Fore topmast length	$\frac{3}{5}$ of the foremast length
Fore topgallant mast length	about $\frac{2}{5}$ of the fore topmast length
Mizzen topmast length	$\frac{3}{5}$ of the mizzen mast length

Further proportions, also provided by Paris, were as follows:

Main topmast length	1½ times the beam
maximumdiameter	7/288 of the length
minimum diameter	⅔ or 7/12 of the maximum
masthead	1/10 of the length
Fore topmast length	⅔ of a masthead shorter than the main topmast or equal to it
Mizzen topmast length	equal to the beam
maximum diameter	7/288 of the length
minimum diameter	5/12 of the maximum
masthead	1/10 of the length
Main topgallant mast length	⅚ of the beam
maximum diameter	1/48 of the length
minimum diameter	⅔ of the maximum
long pole head length	⅔ of the length
normal masthead	⅕ of the length
Fore topgallant mast length	7/9 of the beam
maximum diameter	1/48 of the length
minimum diameter	⅔ of the maximum
long pole head	⅔ of the length
normal masthead	⅕ of the length
Mizzen topgallant mast length	½ the beam
maximum diameter	1/48 of the length
minimum diameter	⅔ of the maximum
masthead	⅕ of the length

Dimensions given by Paris are not always reliable. In the above extract the diameters shown here as 7/288 were given as 7/188 and the minimum diameter for the main topgallant mast was given as ⅓ of the maximum. Topmast diameters of 1/188 of the length are far too great, and the minimum main topgallant mast diameter of ⅓ does not relate to any of the other minimum diameters provided. No doubt the work has suffered in transcription. More reliable information can be found earlier in his work in connection with the masting and rigging plan of the *Royal Louis*.

Topmast trestle trees and cross trees

Stengesaling; Élongis et barres de perroquet

Topmast trestle and cross trees differed somewhat from those of a lower mast. They consisted of two trestle trees and three cross trees, the latter slightly curved aft in their outer thirds. This shape was not altered during the century.

What did change, however, was the number of cross trees. Until about 1706 only two were carried and, instead of a third, only a short spacer between the masts was fitted. Only two topgallant shroud pairs were rigged until 1720, but after 1706 the spacer was replaced by a complete cross tree and the forward shroud was moved from the first to the second cross tree. After 1720 three shroud pairs were rigged and a bolster was introduced to the topmast trestle trees. Another new development appeared in about 1775: sheaves were fitted into the aft ends of the fore topmast trestle trees for the main topsail bowlines (see Figure 6e-d.). Similar sheaves in the fore ends of the main topmast trestle trees served as lead blocks for the fore topgallant braces (see Figure 6e-c.). The development of topmast trestle and cross trees can be traced in Figures 6e, j, m and n).

Topmast trestle tree dimensions were as follows:

Length	until 1720	approximately 3/7 the length of the lower trestle tree
	1720-1775	⅕ of the topgallant mast's length
	1775-1800	3½in per yard of the topmast length
Height	until 1720	1in per foot of length
	1720-1775	25/26in per foot of length
	1775-1800	1⅛in per foot of length
Width	until 1775	¾ of the height
	1775-1800	⅔ of the height

And topmast cross tree dimensions as follows:

Length	until 1720	½ of masthead diameter plus the length of the trestle tree
	1720-1775	4/15 of the topgallant mast length
	1775-1800	⅓ longer than the trestle tree
Height	until 1775	½ of the trestle tree height
	1775-1800	⅞ of the trestle tree height
Width	until 1775	equal to the width of the trestle tree
	1775-1800	1¼ times the trestle tree width

Only the central two sevenths of a topmast cross tree's length was of consistent height; the wings tapered to half their thickness at the ends. These were semi-circular in plan view, with holes for the shrouds about 4 inches in from the ends. The cross trees were mortised into the trestle trees, in a similar manner to the lower trees. The second cross tree was placed between the topmast head and the topgallant mast's heel. The after cross tree butted directly against the aft side of the topmast head and the forward tree was set with 1 inch clearance to the heel of the topgallant mast.

Continental fore and main topmasts also had three cross trees; the exception was the mizzen topmast which had only two (see Figures 6k, l, o, p, r and s). They were curved somewhat to the rear, and those shown by Paris had a more marked curve than that shown in Röding's illustrations. One of the most substantial differences, by comparison with English cross trees, was the semi-circular chafing board, attached to the fore side of the foremost cross tree. On French ships around 1700 this chafing board was a plank, fitted on edge to the ends of the fore cross tree and forming a semi-circle around the trestle tree ends. The foremost point was approximately one third of the trestle tree's length forward of the first cross tree. Mizzen topmast cross trees did not have this board (see Figure 6s). If topgallant mast trees were fitted, these had the same shape as those of a mizzen topmast (see Figures 6i and s).

During the transition from round to semi-circular mast tops, the shape of topmast cross trees altered as well. The chafing plank became an integral part of the first cross tree and its foremost point was now only one sixth of the trestle tree's length away from the cross tree (see Figure 6k and o).

Röding's illustration differs slightly from the above description, which is based on Paris. In particular, he noted that the chafe protection actually became a fourth, more curved, cross tree, nailed on top of the forward cross tree proper. He agreed that its foremost point was one sixth of the trestle tree's length forward of the forward cross tree (see Figure 6l and p).

Measurements of French topmast trestle trees for a ship with a beam of 46 feet are given by Paris, in a section of his work entitled 'Proportions de la mâture en 1750':

Main topmast	length	2.680m
	height	0.223m
	width	0.110m
Fore topmast	length	2.440m
	height	0.186m
	width	0.096m
Mizzen topmast	length	1.380m
	height	0.112m
	width	0.063m

These measurements were also given as proportions for the topgallant masts:

Length	approximately ¼ that of the topgallant mast
Height	approximately 1/12 of trestle tree length
Width	approximately ½ of the height

Other French sources stipulate one third of the length of

Figure 12
(a) French main topmast (front-view) for a 74-gun ship of 1780, with an English style heel.
(b) French fore topmast (profile) for a 74-gun ship of 1780, with an English style heel.
(c) French mizzen topmast with a long pole head (profile) for a 74-gun ship of 1780, with an English style heel.
(d) French topgallant mast for a 74-gun ship, with an English style heel.
(e) French jib-boom, 1780.
(f) French bowsprit, 74 gun ship 1780
a. Gangboard, connecting forecastle and bowsprit
b. Cross sections with jib boom saddle and bees.
(g) French bowsprit cap, bound with an iron hoop.
(h) Iron bowsprit cap, second half of the century. Used on Russian First Rate man-of-war from 1748 and mentioned by Röding as belonging to small vessels.

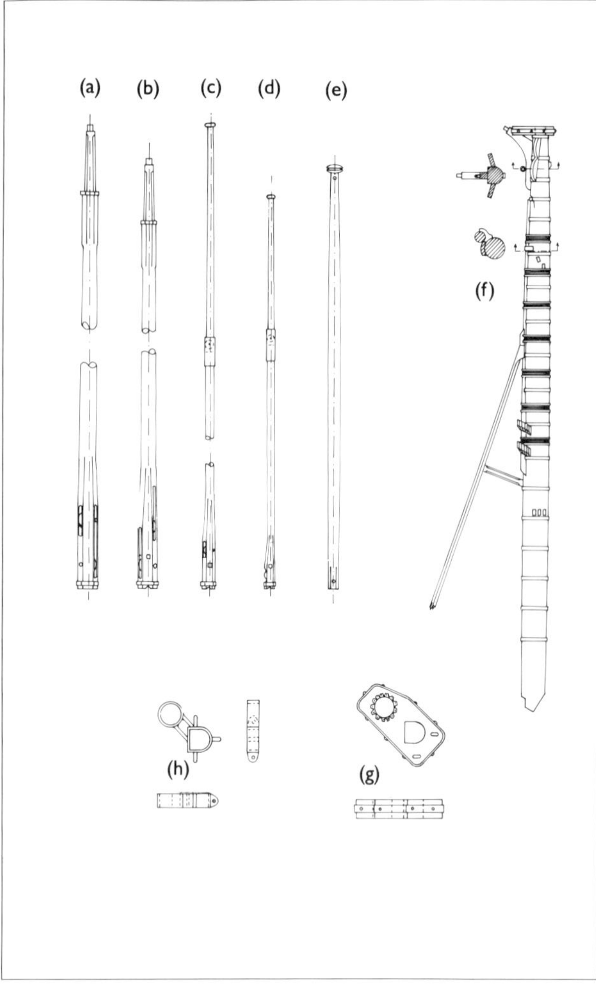

the lower trestle trees and half that of the lower cross trees as measurements for the topmast trees. Height and width were proportional.

Caps
Stengeeselshäupter; Chouquets du mât de hune

English style topmast caps did not differ in shape from the lower mast caps (see Figure 10g). Their dimensions were as follows:

Length	4 times the topgallant mast's diameter plus 1 inch
Width	twice the topgallant mast's diameter
Height (until 1750)	½ the width
Height thereafter	7/15 of the width

Illustrations of Continental style topmast caps indicate an appearance roughly the same as that of the lower mast caps (see Figure 10h). In the early years of the century, however, the topgallant mast hole was completely enclosed, whereas later the hole was cut to half depth only into the forward face of the cap, with the other half of the support consisting of an iron clamp hinged to that face (see Figure 10i and j).

The length of a Continental topmast cap was approximately 3½ times the topgallant mast's diameter. The width in the first decades of the century was 2½ times, and from about 1750 onwards 3 times, the same diameter. The height varied between 1 and 1⅓ times that diameter, with the lower height of the forward part of the cap being half that of the after part (see Figure 10i to k).

A French topmast cap (English style) was half the size of the similarly styled lower cap, with all measurements in proportion. Because of its smaller size it had only one iron hoop around it.

Bowsprit
Bugspriet; Beaupré

Bowsprit construction on larger ships was similar to the making of masts. The bowsprit consisted of at least two pieces coaked together in the middle, first cut eight sided, then chamfered to sixteen sided and finally rounded. Bowsprits were initially reinforced with wooldings, but these were increasingly replaced by iron hoops in the last third of the century. The number of bowsprit wooldings was similar to that of lower masts. A First Rate man-of-war had five iron hoops between the footing and the step, and a further nine outside the stem. Ships of the Second and Third Rates had four and eight hoops respectively.

English bowsprits were angled at nearly 36 degrees to the horizontal. Duhamel du Monceau mentioned an angle of about 35 degrees, Korth noted an angle of 30 to 33 degrees, and Röding 33 to 35 degrees. As a general rule, then, it can be stated that the angle was between 30 and 36 degrees.

Length and diameter
Länge and Durchmesser; Longueur et diamètre

Various bowsprit dimensions have been recorded. Love noted a length of eight ninths of the mainmast length and Davis noted one of 1⅔ times the beam. Davis' contemporary Sutherland specified a bowsprit length of two thirds of the mainmast length, or three quarters of the foremast length for small ships. He gave a figure of nine tenths of the mainmast diameter or more as thickness.

According to Mountaine, a merchantman's bowsprit length was three fifths that of the mainmast, with a diameter 1 inch less than that of the mainmast. The bowsprit length he gave for a man-of-war was eight ninths that of the foremast. Duhamel du Monceau noted a bowsprit length of 1½ times the beam and found the diameter by taking the medium between the fore and mainmast diameters.

The maximum diameter was at the stem and the head tapered to half of this. The steps or partners of the bowsprit were located 1 foot forward of the foremast and were bolted to the beams of the upper and the gun decks.

'The bowsprit is 1⅖ times the beam's length and that part of it, which lies outside the ship, is equal to the length of the beam' noted Röding. He continued:

> The maximum diameter of [a bowsprit] is the medium between that of the main and that of the foremast. The minimum diameter is half of the largest, and the head length is one twelfth that of the whole bowsprit. The bowsprit's declination against the horizon is approximately 35 degrees for large ships, but on smaller ones only 20 to 25 degrees, because their fore staysails and jibs are relatively larger.

Korth also gave a length for the bowsprit outside the stem as equal to the beam, but his total length was 1⅕ times the beam. For the maximum diameter he agreed with Duhamel and Röding, but gave a taper of only one sixth.

Chapman gave a bowsprit length outboard of the stem of 1.115 of the vessel's breadth for the so-called merchant frigate (a large ship-rigged merchantman) and of 1.1 times the vessel's breadth for a bark. The maximum diameter was similar to that specified by Duhamel, Röding and others. For the bowsprit's taper outside the stem (the stem or gammoning being the maximum diameter) Chapman noted 59/60 for the first quarter, 55/60 for the second, 46/60 for the third and half for the end, all proportions of the maximum diameter. Bowsprit dimensions for brigantines and snows were similar to those for a frigate. The bowsprit angle for frigates was 4 feet per 7 feet of length, and for barks 3 feet per 7 feet.

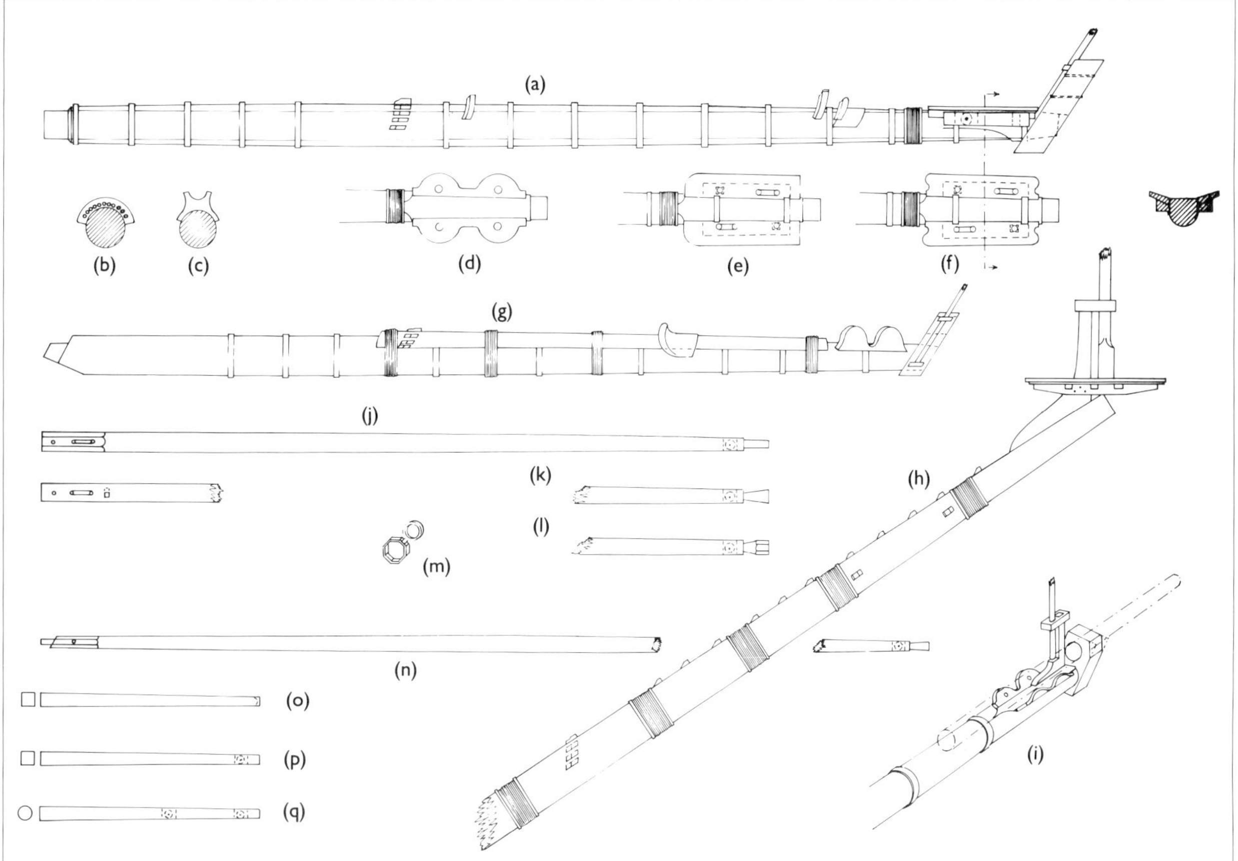

Figure 13
(a) English bowsprit after 1770-75.
(b) Fairlead saddle.
(c) Jib-boom saddle after about 1730.
(d) Bees; this shape was common on English ships until about 1745, and on Continental ships throughout the century. Continental bees were without holes.
(e) English bees after 1745. The sides were straightened and blocks were fitted beneath the bees.
(f) English bees after about 1770, with cross section detail.
(g) A French bowsprit of 1780, differing from the English example (a) in the paunch, the bees, the spritsail sling saddle and the fitting of a jackstaff.
(h) A bowsprit with a spritsail topmast, of approximately 1700.
(i) Continental fitting of a jib-boom with a cocked bowsprit cap, jackstaff and bees.
(j) English jib-boom, up to about 1770.
(k) English jib-boom heel, up to about 1735, and head from approximately 1770 to 1794.
(l) English jib-boom head after 1794.
(m) Flying jib-boom iron, as fitted to the octagonal jib boom head after 1794.
(n) English flying jib-boom after 1794.
(o) Martingale (dolphin striker), with a notch in its lower end; English about 1790.
(p) Martingale with a sheave fitted; English, about 1795.
(q) Martingale with two sheaves fitted; English, about 1800.

According to Paris, the length of a bowsprit was 1.565 times the beam, with a diameter of 1/22 of the length.

Steel gave a bowsprit length of seven elevenths of the length of the mainmast for ships with more than 80 guns, and a length of three fifths that of the mainmast for those with fewer guns. For ships with 64 to 100 guns he specified a diameter 2 inches less than the mainmast's diameter, and for ships with 50 guns and fewer the same diameter as the mainmast.

The taper differed slightly from Chapman's data. For the first quarter Steel gave 60/61, for the second 11/12, for the third 4/5 and for the fore end 5/9 of the maximum diameter. The bowsprit's heel had a diameter of six sevenths of the maximum. See Figure 13a, g and h for illustrations of bowsprits.

Jib-boom

Klüverbaum; Bâton de foc

One of the most important developments in eighteenth century masting was the introduction of the jib-boom, first noted on the English Sixth Rate *Royal Transport* of 1695. Designed by Sir Peregrine Osborne, Marquis of Carmarthen, as a schooner-rigged private yacht for King William III, she was presented as a royal gift to the visiting Tsar Peter I of Russia in 1697.

This first jib-boom was half the bowsprit's length and its introduction necessitated considerable modification of the headgear. During the transition period the sprit topmast and the jib-boom were often carried together, with the jib-boom either lashed or fastened with iron figure-of-eight brackets to the side of the bowsprit and led through an additional opening in the bowsprit top. An early account of the jib-boom is given by Sutherland in 1711:

> There is another Sail call'd a flying Gib, a Sail of good service to draw the Ship forward, but very prejudicial to the Wear of the Ship forward. 'Tis used with a Boom or small Mast extended at the Extremes of the Bowsprit.

This arrangement can still be seen on a number of models from around 1730. After that time, the top disappeared and the spritsail topmast shrank to the size of a jackstaff (see Figure 14a and b).

From then on the jib-boom was mounted directly above the bowsprit in the Royal Navy. Continental ships usually had the jackstaff set up with a standard on the upper face of the bowsprit, and the jib-boom was therefore mounted on the starboard upper face of the bowsprit, with the cap tilted 45 degrees to starboard accordingly (see Figures 13i and 14d).

Initially half the length of the bowsprit, the English jib-boom, early in the century, had a length equal to the beam minus 6 feet. Later the length was seven tenths to five sevenths that of the bowsprit. The maximum diameter was 7/8in per yard of length, constant for the inner third of the length from the heel to the cap. The outer two thirds tapered as follows:

First quarter	40/41 of the maximum diameter
Second quarter	11/12 of the maximum diameter
Third quarter	5/6 of the maximum diameter
Head	2/3 of the maximum diameter

Chapman noted a jib-boom diameter of three quarters the diameter of the main topmast.

For French ships, the jib-boom length of the *Royal Louis* was given by Paris as equal to the beam, with a diameter of 1/48 of the length. Figure 14e shows a ship with a beam of 46 feet, a jib-boom length of 1.2 times the beam and a jib-boom diameter of 1/44 of the length.

Korth and Röding confirm a length equal to the beam and a diameter 1/48 of the length for Continental jib-booms, with Röding giving the smallest diameter as four fifths of the largest. Röding also gave a simple description of the jib-boom fitting:

> For this purpose a cap is located at the end of the bowsprit, with a solid bracket (through which the jib-boom runs) a short distance behind it. Sometimes only a cap or a bracket

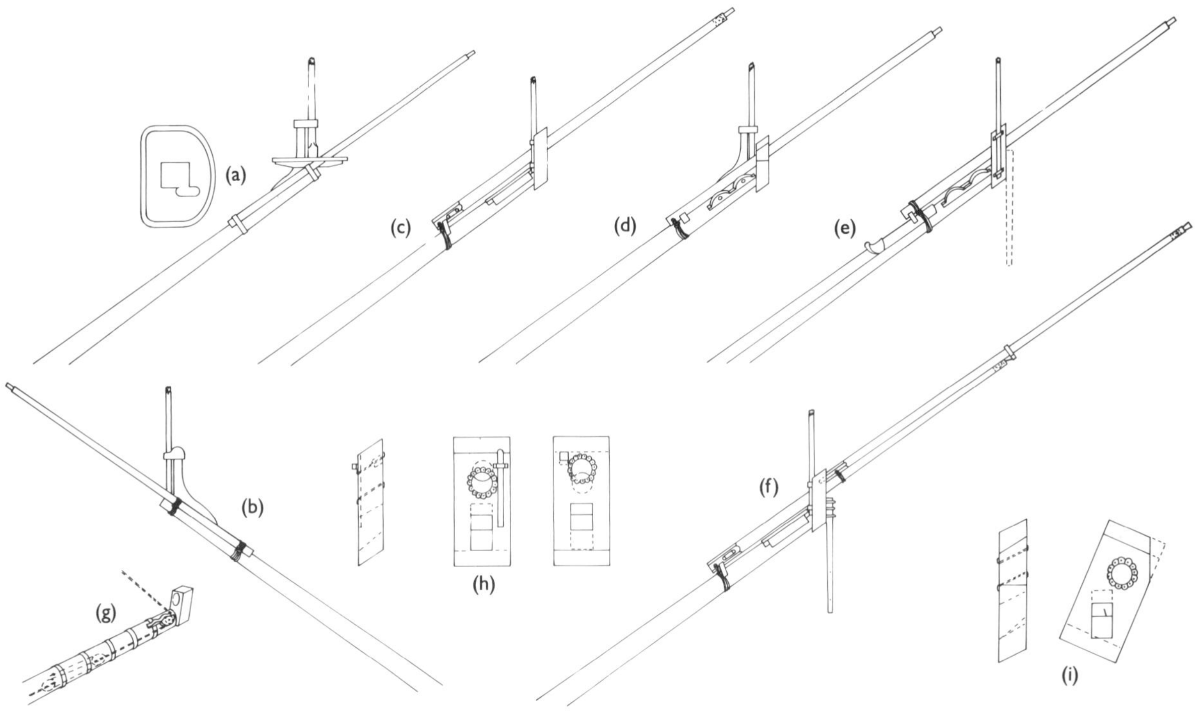

Figure 14
(a) Jib-boom fitting with iron brackets until the end of the sprit topmast period, with a plan view of the sprit top.
(b) Early jib-boom fitting, French, of about 1704.
(c) English jib-boom fitting.
(d) Continental jib-boom fitting.
(e) French jib-boom fitting, 1792.
(f) English jib-boom and martingale fitting.
(g) Leading of a fore topsail stay over a block beneath the bees; Continental, 1790.
(h) An English bowsprit cap; a leather cuff is shown in the jib-boom hole. The jackstaff was fitted to starboard aft, and a square mortise was cut into the starboard face to take the flying jib-boom's heel.
(i) A Continental bowsprit cap, cocked to starboard.

sits on the upper end of the bowsprit, and a woolding is laid around the lower end of the jib-boom, in the middle of the bowsprit... A block is fitted to the head of [the jib-boom], through which the fore topgallant stay leads, and a round or angular truck is fitted to the upper end of the jib-boom.

The English jib-boom was rounded over its entire length until 1735 (see Figure 13k). After that date the heel became octagonal over a length equal to 3½ times the diameter (see Figure 13j). A stop for the fore topgallant stay was notched in 1½ times the diameter from the outer end. Of a constant diameter slightly smaller than the boom until around 1775 (see Figure 13j), this stop was given a reverse taper in later years (see Figure 13k). Continental jib-booms, as Röding noted in the description quoted above, had a head fitted with a truck (see Figure 12e).

Abaft the stop a sheave was inserted vertically for the jib outhauler. Another sheave was set horizontally 1½ times the diameter away from the heel for the jib-boom outhauler. In addition, a horizontal hole was drilled into the aft end for the boom lashing.

The sprit topmast, still a feature of masting during the early decades of the century, had, according to Davis, a length of half to five ninths of the beam, with a diameter of approximately one thirtieth of the length. Sutherland noted its length as half that of the fore topmast, and its diameter as ¾in per yard of length for the smallest spritsail topmast, up to 1 inch per yard for the largest.

Bowsprit cap

Bugspriet Eselshaupt; Chouquet de beaupré

During the early decades of the century, as the jib-boom superseded the sprit topmast but was still rigged with it at times, the boom was fastened with two boom irons (see Figure 14a). Later their place was taken by a cap similar to a mast cap. Such a cap was already in use with the first jib-boom, that of HMS *Royal Transport*, from 1695. It was fitted pointing to starboard and had a hinged iron bracket on its front face, not unlike that of the Continental topmast caps.

An English bowsprit cap in the eighteenth century had a length equal to five times the jib-boom diameter, a width equal to twice the diameter of the jib-boom plus half that of the jackstaff, and a thickness of five ninths of the width. A square hole was cut in it for the bowsprit and a round one for the jib-boom. The hole for the jib-boom was ¾in wider then the boom's diameter, to allow a leather cuff to be fitted (see Figure 14h).

A Continental bowsprit cap was very similar to the English cap. However, to allow for the different placing of the jib-boom, it was set diagonally to the bowsprit's axis, with the starboard edge cut off at jib-boom height (see Figure 14i).

Bees

Violine; Violon de beaupré

Bees were situated on both sides of the bowsprit, directly abaft the cap. They were boards shaped like the letter B, and when mounted as a pair they produced a shape not unlike that of a violin, hence the German and French names. Bees were introduced shortly after the replacement of the sprit topmast by the jib-boom, and retained their characteristic shape at least until the middle of the century (see Figure 13d). Bees on English ships changed in shape around 1745, becoming straight on the outer side (see Figure 13d and e). The curved B shape was maintained much longer on the Continent, in some instances even into the early nineteenth century.

English bees, up to 1740, had a hole in each semi-circle. The strapping of the fore topmast stay and preventer stay led through these holes and around the bowsprit. Paris noted that French bees were without holes throughout the century.

The Royal Navy began to fit blocks beneath the bees, similar to the topmast cheek blocks, in the 1750s. This allowed the sheave on the starboard side to be set into the fore part and that on the port side to be set into the aft part of the particular block. The two other holes in the bees matched square holes in the blocks, without sheaves; these were used only in an emergency.

The bees were not arranged in a single plane, but were each angled slightly upwards, and were bolted to the bowsprit. Their angle upwards was such that the outer edge was higher than the inner by a distance equal to the inner thickness plus 1 inch.

According to a modern authority, the length of the bees was half that of the bowsprit cap; Steel noted a length 2¼ times the bowsprit's diameter. The width, once stated as half the largest diameter of the bowsprit, was two thirds of the bowsprit diameter at the bees according to Steel (which is in fact very similar). The thickness of the elm planks of which the bees were made was one quarter of the given bowsprit diameter at the inner edge, tapering to three quarters or four fifths of that at the outer edge. The holes for the stays were one quarter the length of the bees inside the ends.

As an alternative to Continental bees without stay guiding holes or blocks, Röding illustrated bees with a half-round block mounted beneath the forward semi-circle, with the fore topmast stay running over the forward edge of the bees and the block's sheave. He considered this arrangement characteristic of smaller vessels (see Figure 14g).

Röding also gave a further description of the design and function of Continental bees. They served to keep the fore stay (by which he can only have meant the fore topmast stay) in position because the stay's collar was laid in the indent between the two semi-circles and around the bowsprit. At the same time, the bees provided a safe platform for the sailors working on the bowsprit. The length of the bees was one twelfth that of the bowsprit, the width one third of the length, and the depth one third of the width. A modern French source notes the length as 4½ times the jib-boom diameter, the width as given as one quarter and the depth as one sixteenth of the length. The bees were angled upwards towards the outer edge and ran slightly more horizontal than the axis of the bowsprit so as to provide a more level platform for the sailors to stand on.

Bee blocks

Violinblöcke; Poulies de violon

The length of the bee blocks was seven ninths that of the bees, the width came to 2 inches per foot of length and the depth to seven eighths of the width. The sheave holes were two sevenths of the block's length, were positioned one seventh in from the ends, and their width was one quarter their length. The block holes for emergency stays were square and without sheaves. Bee blocks were similar to cheek blocks and were bolted to the bowsprit, with the bolts providing the axis for the sheaves.

Saddles

Sättel; Taquets

Saddles represented a further item in the furnishing of a bowsprit, and included the jib-boom saddle (see Figure 13c), the spritsail yard sling saddle and the fairlead saddle. The first was one sixth of the bowsprit's diameter high and half the diameter wide. This saddle was introduced at the end of the spritsail topmast period, around 1730, and kept the jib-boom parallel to the bowsprit. The distance of the saddle from the cap was one third of the jib-boom length. Its height was sufficient to keep the jib-boom clear of the second saddle, forward of the first, and to provide the spritsail yard sling with adequate room for movement.

The spritsail yard sling saddle came into existence in about 1775. Its height and width were one eighth of the bowsprit's diameter and it fitted halfway around the bowsprit.

The fairlead saddle (see Figure 13b) bore a certain similarity to the spritsail yard sling saddle and was situated forward of the gammoning. It had, in contrast to the latter, a number of horizontal holes drilled through it to lead the running rigging for the head sails toward the forecastle.

The gammoning cleats also need to be noted. These were short battens nailed to the upper half of the bowsprit to keep the gammoning in position. Their length was half the bowsprit's diameter and their height and width were one quarter of the length.

Flying jib-boom

Aussenklüverbaum; Bâton de clin-foc

The flying jib-boom did not appear before the last decade of the century. It was introduced into the Royal Navy in 1794 and paintings, drawings and models of French ships indicate that no earlier date can be established for these.

Steel mentioned a flying jib-boom in his work of 1794, but in all probability his reference was simply to a boom for the leech of a jib as set in the later part of the century (that is, flying - without a stay). He did not list a flying jib-boom in his mast tables; only a jib-boom is listed, and the measurements provided for this are the same as the proportions he indicated for his flying jib-boom.

Röding elaborated on the *Klüverbaum*, but did not mention the *Aussenklüverbaum* at all in his work, nor in its appendix. This suggests that the flying jib-boom was a novelty in English rigging and had not made much impact on the Continent by the dying years of the century. The flying jib was referred to as a middle jib in the appendix to Röding's English index and it was described as a jib set aft of the actual jib; this would seem to confirm that a flying jib proper was not what was meant.

Dimensions of a flying jib-boom were as follows:

Length	approximately ⅞ of the bowsprit diameter
Diameter	approximately ½in per yard of length

The maximum diameter was at the jib-boom iron and from there the boom tapered towards its head, with the smallest diameter being three quarters of the largest. Shaped like the jib-boom, the flying jib-boom had a vertical sheave in the outer end and a horizontal hole in its heel to enable it to be lashed to the jib-boom. The heel was shaped to butt against the face of the bowsprit cap and was tenoned into it (see Figure 13n).

Doctor W Burney, editor of the 1815 edition of Falconer, described the flying jib-boom as 'a boom extended beyond the preceding [the jib-boom] by means of two boom irons and to the foremost end of which the tack of the flying jib is hauled out.'

The shape of a jib-boom iron resembled that of a studding boom iron and, like a studdingsail boom, the flying jib-boom was mounted at 45 degrees above and to the side of the jib-boom (see Figure 13m).

Ships rigged with a flying jib-boom did not carry a sprit topsail yard.

Martingale or dolphin striker

Stampfstock; Bâton de martingale

Coinciding with the introduction of the flying jib boom was the increasing use of the martingale or dolphin striker. The earliest contemporary illustration known to this author is George Raper's watercolour sketch of the First Fleet anchoring in the harbour of Rio de Janeiro on its way to Australia. This sketch shows fairly new ships rigged with a martingale while older vessels are shown without. Since the building dates of all these ships are known, it can be assumed that first experiments with this boom began in about 1783-4. A slightly later sketch by Raper, dating probably from 1790, shows a French and a Dutch ship with a martingale on the roadstead of Cape Town. Raper, as a young sea officer, has to be seen as a trustworthy reporter of rigging details.

In English literature the martingale received a first mention by Steel in 1794, the year of its official introduction to the Royal Navy. Any earlier dating for the official introduction of the martingale advanced by modern authors cannot be sustained.

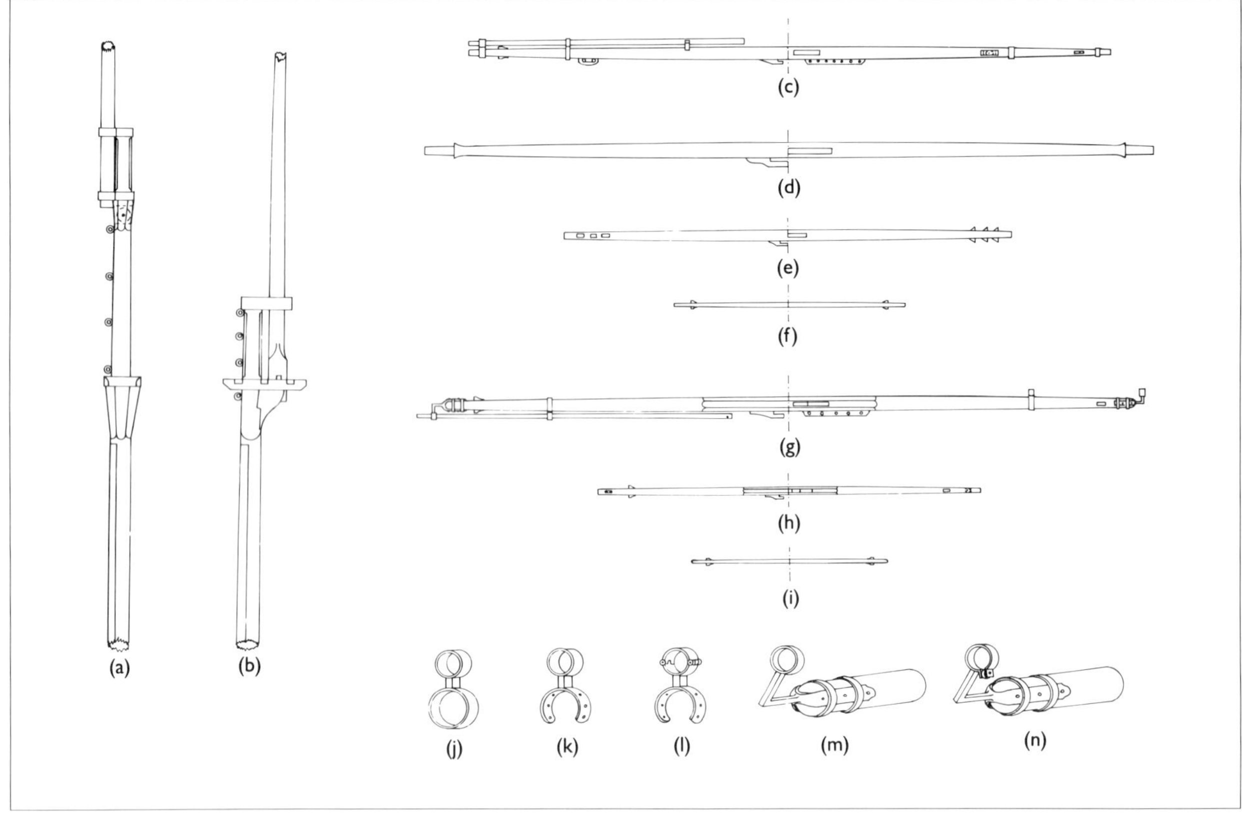

Figure 15
(a) Cutter mast with hounds, bolster and a topgallant mast fitted to the rear. The topgallant mast is set up with iron caps.
(b) A mast for small craft with trestle and cross trees, a timber cap and a topmast.
(c) A Dutch main yard of about 1700 (plan view on the left, forward elevation on the right). Roband strips were nailed beneath the centre. Studdingsail booms were behind the yard.
(d) French main yard, about 1700.
(e) French topsail yard, about 1700.
(f) French topgallant yard, about 1700.
(g) English main yard, 1700. Roband strips disappeared at about this time.
(h) English topsail yard, about 1700.
(i) English topgallant yard, about 1700.
(j) Inner boom iron on an English yard, until 1745.
(k) Inner boom iron on an English yard between 1745 and 1775. The yard hoop had changed to an open band, encircling three quarters of the yard.
(l) Inner boom iron on an English yard after 1775. The boom hoop is hinged.
(m) Outer boom iron on an English yard, until 1775.
(n) Outer boom iron on an English yard after 1775. The boom hoop was fitted with a roller.

Röding noted the martingale in the appendix to his English index, Volume III, published in 1797, coinciding closely with Steel. Paris provided drawings of a model of the *Royal Louis* of 1780 with a sprit topsail and without a martingale, while *Le Protecteur* of 1793-4 was rigged without a sprit topsail but with a martingale. It is possible that the martingale appeared slightly earlier in French rigging but, apart from models, no contemporary evidence is available. Röding's French index (1797) contains the first written reference to the martingale in French practice, and models have to be viewed with caution. All too often the restoration of model rigging has resulted in anachronistic and incorrect additions.

An interesting example is provided by the later development of the double martingale. According to the 1819 appendix to the second edition of the works of Lever, the double martingale in English ships is a novelty of the second decade of the nineteenth century, appearing only after the printing of the first edition of his work in 1811. However, a model of the French ship *Léon* dating from 1780 already exhibits the double martingale. This model, like those mentioned above, was thoroughly restored in about 1870. The rigging's authenticity has clearly suffered in the process. The earliest appearances of the double martingale in pictures found by this author are in watercolours by Roux dating from 1805 and 1807, the first showing an unidentified American frigate and the second the French armed brig *La Tactique*.

The martingale of the early period was a simple round or square ash bar fastened with iron cramps to the face of the bowsprit cap. Steel noted a notch in the lower end to guide the martingale stay, while Röding indicated a hole in the lower end through which the stay led. The latter also made it a point to state that only larger ships were rigged with martingales. Raper, Steel and Röding attested to martingales rigged with a single stay, while Roux and Lever noted the use of two stays. In the last years of the century a sheave was set into the enlarged hole and Lever's illustration of a martingale with two sheaves may have applied only to large vessels. Only after the turn of the century did the martingale undergo a more rapid development (see Figures 13o, p and q, and 14e and f).

Varying dimensions and construction methods for smaller vessels

Abweichende Abmessungen und Bauweisen bei kleineren Schiffen; Dimensions diverses et modes de construction variants pour les vaisseaux des moindres rangs

Small vessels such as cutters were often fitted with a lower mast and topmast made in one piece, with the quarter or third of the mast above the lower hounds representing the topmast. These vessels usually had hounds, bolsters and a squared topmast head. Others had a separate mast and topmast linked by trestle and cross trees and a cap. Steel commented that the choice of mast configuration was merely a matter of preference (see Figure 15a and b).

Cutter masts were often octagonal up to 4 or 5 feet above the deck and rounded from there to the hounds. They were fitted either with bolsters or with trestle and cross trees. When the latter were fitted, the masthead was squared and a cap was fitted for the topmast. Bolsters were only used when the lower and top masts were constructed in one piece. The topmast part above the lower stop was then rounded and the masthead above the topmast hounds was squared.

Above the topmast stop a figure-of-eight iron was fitted, with a second used as a cap on the topmast head. These served to hold the topgallant mast in position, with a bolt through the lower iron acting as a securing fid.

Four eyebolts with shoulders were driven from aft through iron plates set into the mast, through the masthead and clenched on similar plates on the fore side. These bolts were 1¼in or 1½in in diameter. The lower bolt was positioned at the height of the lower hounds, and the other three bolts were equally spaced between the stop and the cap or the topmast hounds.

To prevent chafing of the gaff or boom jaws on the mast's after side, copper sheathing was applied as necessary on the mizzen masts of ships and the mainmasts of brigs and cutters and similar vessels.

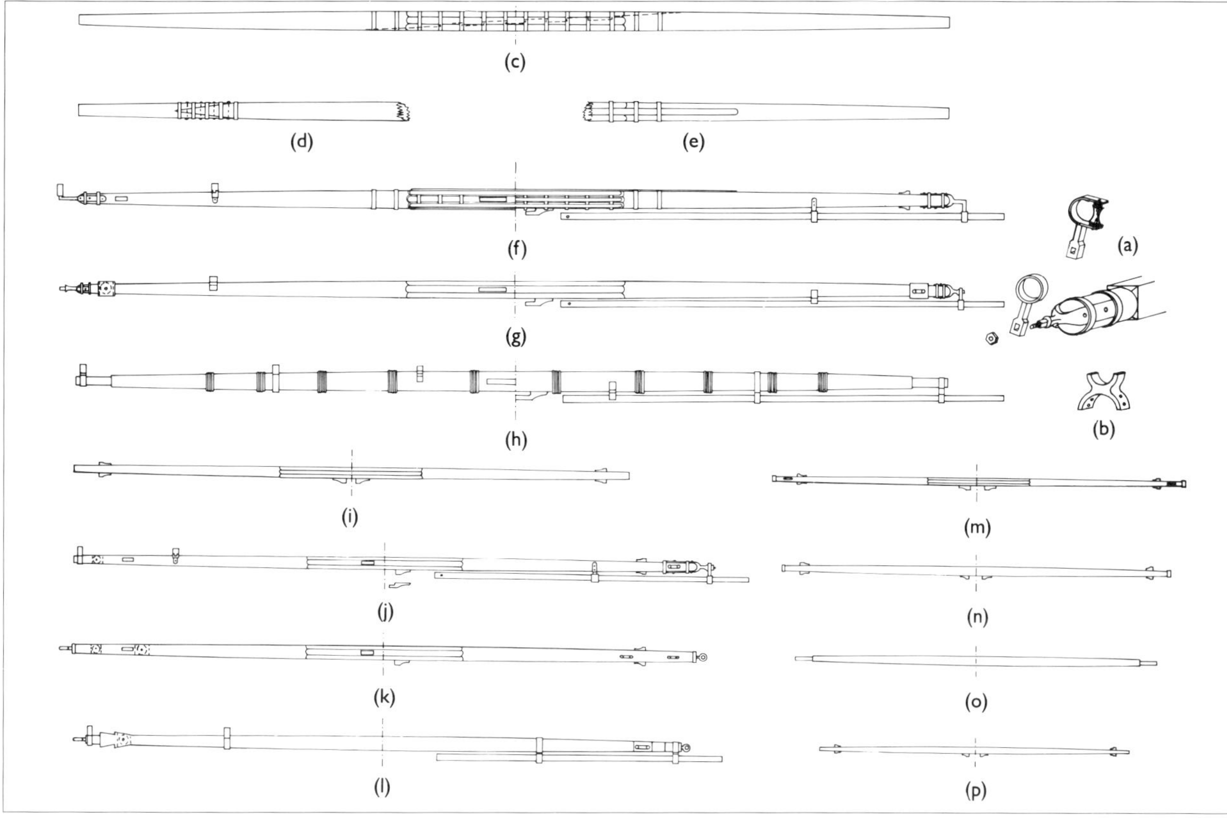

Figure 16

(a) Outer boom iron on yards of English merchantmen and for Continental yards. The horseshoe shaped boom iron with a vertical roller was in use on Continental ships during the last quarter of the century.

(b) Inner boom saddle, in use on English merchantmen and on the Continent.

(c) An English yard after 1773 made from two pieces; the dashed line indicates the scarfing of both parts. A made yard was bolted together and strengthened with a number of iron hoops.

(d) A yard extension at the yardarm. This type of joint was used for temporary repairs.

(e) The flat mast side on an English yard. It was kept longer than the sling octagon.

(f) An English main yard after 1775 (plan view on the right and forward elevation on the left). Timber battens covered the iron hoops on seven of the eight flats of the sling octagon. The front was kept without a batten.

(g) Main yard of an English merchantman, about 1780. Sheaves for the topsail sheets were fitted into the yardarm squares.

(h) A French main yard for a 64-gun ship, about 1790, made and then strengthened with wooldings. This yard contrasts with that shown in Figure 18 (e) but iron hoops were also used on French yards. Outer and inner boom irons were fitted with a full yard hoop (similar to English boom irons until 1745).

(i) English crossjack yard, after 1750.

(j) English topsail yard with studdingsail booms, after 1770. Two methods of fitting the outer boom irons are shown.

(k) English topsail yard without studdingsail booms, after 1780, as used in the Royal Navy on frigates and smaller vessels. The inner sheaves were for the topgallant sail sheets and the outer for the reef pendants.

(l) French topsail yard, about 1790.

(m) English mizzen topsail yard, about 1770.

(n) English topgallant yard, about 1770.

(o) French topgallant yard, about 1790.

(p) English royal yard, about 1790.

Measurements

Abmessungen; Mesures

The length of a brig's mainmast was established by adding the depth of the hold to the length of the vessel and its width and dividing the sum by two. The mast's diameter was 1 inch per yard of length. Foremasts had a diameter of nine tenths that of the mainmast.

Chapman gave the same proportions for a brigantine or snow foremast as for a merchant frigate, but the mainmast of a brigantine or brig was so much higher that the main top was as high as the mast cap of a normal mainmast, and the brigantine's mainmast head was level with the fore topmast's hounds. The mainmast height of a snow was between that of a brigantine (brig is the newer term) and a frigate. For schooners and galeasses Chapman gave a mainmast length excluding the masthead of three times the vessel's breadth, and for howker-yachts a total mainmast length of three times the vessel's breadth.

For the cutter masts the rule for establishing the length was to add the length over deck to the maximum breadth and the depth of hold, and then to take three quarters of the sum as the mast's length. The diameter was 3/4in per yard of length.

In contrast to a ship's mizzen mast, the mizzen mast of a ketch was three quarters the length of the mainmast, and the mizzen topmast was five sevenths the length of the main topmast.

The proportions of schooner masts (according to Paris) were, given a beam equal to 0.266 of the vessel's length:

Mainmast	length	3 times the beam
	diameter	1/46 of the length
	masthead	1/8 of the length
Foremast	length	2.91 times the beam
	diameter	1/45 of the length
	masthead	1/8 of the length
Bowsprit	length	1.5 times the beam
	diameter	1/27 of the length
Main topmast	length	1.6 times the beam
	diameter	1/53 of the length
	masthead	1/4 of the length
Fore topmast	length	1.54 times the beam
	diameter	1/52 of the length
	masthead	1/4 of the length
Jib-boom	length	equal to the beam
	diameter	1/50 of the length

For merchant vessels like sloops, smacks and barges, the proportional lengths were as follows:

Lower mast and topmast in one	3¾ times the beam
Mast to lower hounds	3/4 of the length
Mast to topmast hounds	40/41 of the length
Topgallant mast to stop	1/7 of the topgallant mast length
Bowsprit	5/9 of the mast length

The diameters in fractions of an inch per foot of length were as follows:

Mast	1/4in (1/48 of length)
Topgallant mast	3/8in (1/32 of length)
Bowsprit	3/8in (1/32 of length)

Vessels with a sloop rig followed the above rules, but those rigged with lugsails had mast lengths as follows:

Mainmast	2½ times the vessel's breadth
Foremast	7/8 of the mainmast length
Bowsprit	1/2 the mainmast length

And diameters as follows, again in fractions of an inch per foot of length:

Mainmast	1/4in (1/48 of length)
Foremast	1/4in (1/48 of length)
Bowsprit	5/8in (5/96 of length)

Launches and cutters with a lug rig had the following mast lengths:

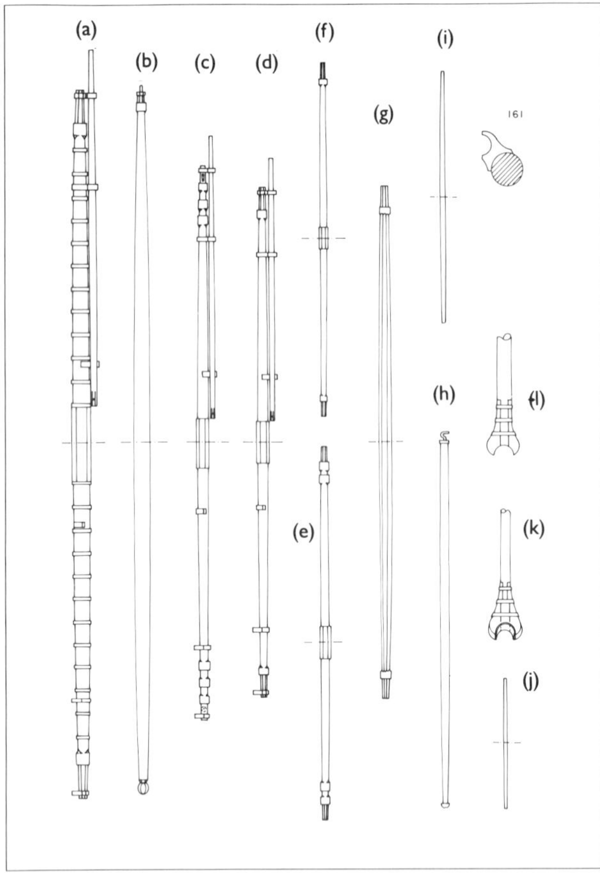

Figure 17
(a) French fore yard for a 74-gun ship, about 1780 (plan view shown on the right). The main yard was similar but larger. Note the short octagonal slings.
(b) French mizzen yard, about 1780.
(c) French main topsail yard for a 74-gun ship, 1780. The number of square stops on the yardarm depended on the number of reef bands in the sail. The fore topsail yard was similar but shorter.
(d) French crossjack yard, about 1780. The spritsail yard was similar but without studdingsail booms.
(e) French mizzen topsail yard for a 74-gun ship, 1780.
(f) French main topgallant yard, 1780. The spritsail topsail yard was similar and the fore and mizzen topgallant yards were slightly reduced in size.
(g) A boom for the lower fore studdingsails (boat boom) on Continental ships.
(h) A boom for the lower main studdingsails (swinging boom) on Continental ships.
(i) Lower studdingsail yard on Continental ships (long alternative).
(j) Lower studdingsail yard on Continental ships.
(k) French method of fitting jaws to a gaff.
(l) French method of fitting jaws to a boom.
(m) Inner boom saddle on French yards.

Mainmast	2¾ times the vessel's breadth
Foremast	8/9 of the mainmast length
Mizzen mast	5/8 of the mainmast length
Sprit	2 feet longer than the mizzen mast
Outrigger	2/3 of the mizzen mast length

And diameters, again in fractions of an inch per foot of length:

Mainmast	¼in (1/48 of length)
Foremast	¼in (1/48 of length)
Mizzen mast	¼in (1/48 of length)
Sprit	1/7in (1/84 of length)
Outrigger	3/7in (1/28 of length)

Launches and cutters with a settee rig had mast lengths as follows:

Mainmast	twice the boat's breadth
Foremast	17/18 of the mainmast length

Diameters for both masts were 3/8in per foot of length (1/32 of the length).

Mast lengths for barges and pinnaces with lateen sails were:

Mast	twice the boat's width plus 8 inches
Topmast	1 1/9 times the mast length
Scarf to the mast	¼ of the mast length

With diameters of 5/16in per foot of length (5/192 of length) for the mast and 1/5in per foot (1/60 of length) for the topmast.

The length of both mast and foremast on barges, pinnaces and yawls with sprit sails was 2¼ times the boat's breadth, and the diameter ¼in per foot (1/48 of length).

For the trimming of sloop, smack, barge and boat masts Steel provided the following rules: the height of the partners should be two thirds of the mast diameter multiplied by twelve; the diameter of the mast halfway between the partners and the hounds should be 30/31, below the hounds seven eighths and at the masthead seven tenths of the largest diameter; the hounds (stop) should be 1½ to 2 inches wider athwartships than the largest diameter.

Topmasts should be two thirds of the largest diameter at halfway between the stops, three fifths at the hounds and three sevenths at the squared masthead 3/7 of the largest diameter. Otherwise they were completed like a cutter's mast.

Boat masts not made in the above fashion had their partners at deck height, and their largest diameter at the partners. They tapered evenly to two thirds of the largest diameter at the head. A sheave for the halyard was fitted slightly below the stop.

Yards

Rahen; Vergues

Yards can be defined as all those spars which were rigged to a mast and served to extend a sail. The middle part of a yard was known as the slings and the extremities as the yardarms, with the distance between divided into first, second and third quarters.

Usually a yard was made from a single piece of timber, but from the early 1770s onwards the main and fore yards of larger vessels were constructed from two pieces. The scarf joining these pieces had to be long enough to extend beyond the inner quarters by 6 inches each side, and was reinforced by a number of iron hoops (see Figure 16c and d).

At first completely rounded, the yards on English ships were made octagonal over their inner quarters from the last years of the seventeenth century onwards, with the mast side of the yard cut flat for another quarter to each side, so that in these sections only seven eighths of the circumference was rounded (see Figures 15g and h, and 16e).

During the last quarter of the century, the flat mast-side of the yard was covered with a 2in thick fish plank and battens were nailed to six of the remaining seven flats. These battens almost entirely covered the flats in length and width, and they were ¾ to 1 inch thick. They were bevelled on the sides and rounded at the ends. All lower and topsail yards, with the exception of the crossjack yard, were furnished with these battens. Topgallant, royal, spritsail and spritsail topsail yards were without battens (see Figure 16f, g, j and k).

The crossjack yard was completely rounded until 1750 and thereafter the inner quarters were sixteen-sided. The royal yards, generally introduced about 1780, were rounded and prepared like the crossjack yard on larger vessels (see Figure 16i and p).

During the greater part of the century the tapering ratio for yards was as follows:

First quarter	30/31 of the diameter at the slings
Second quarter	7/8 of the diameter at the slings
Third quarter	7/10 of the diameter at the slings
Fourth quarter (yardarm)	3/7 of the diameter at the slings

French yards were rounded over their full length throughout the century, as were Continental yards not under English influence (see Figures 15c to f and 16h, l and o).

The exception was the yards of larger ships, especially men-of-war. In the second half of the century these sometimes had octagonal slings five times the maximum diameter of the yard in length (see Figure 17a). French yards also often had squared stops one diameter in length between the yard and the yardarms (the latter were round or sometimes octagonal). These yardarm stops fulfilled the same purpose as the yardarm cleats on English

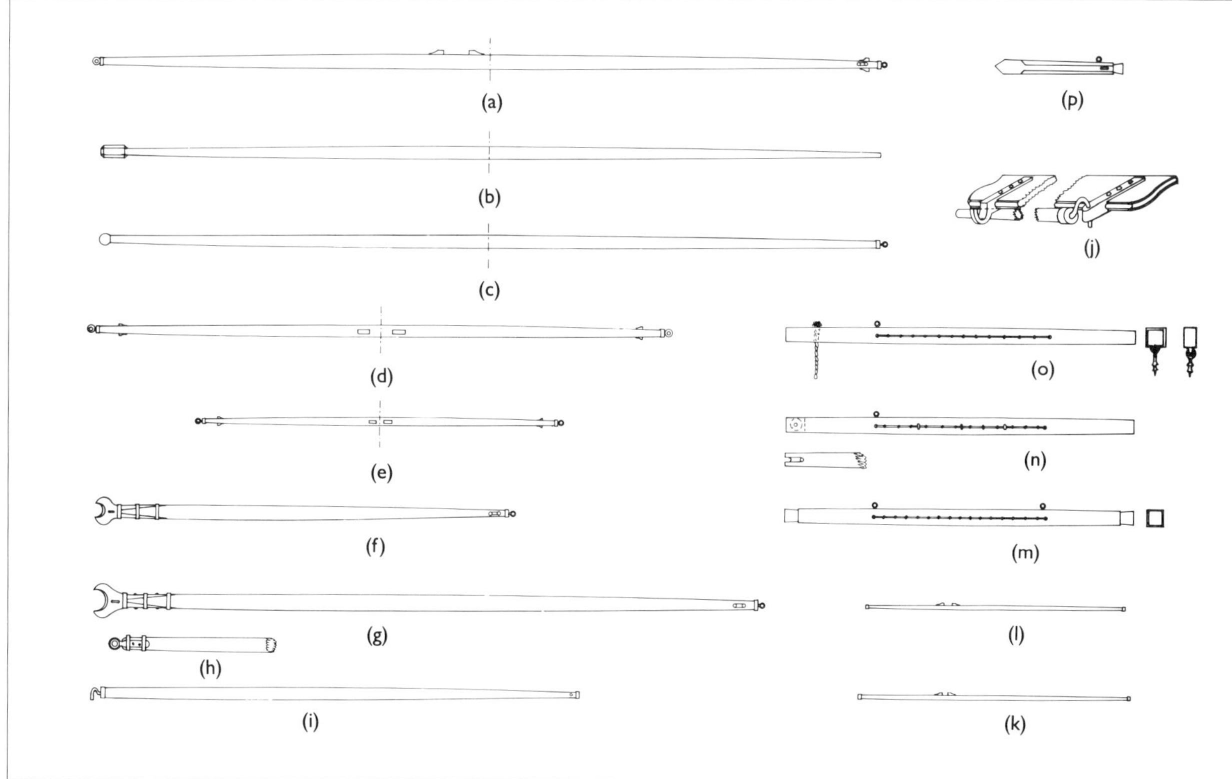

Figure 18
(a) English mizzen yard.
(b) Continental and French mizzen yard at the beginning of the century.
(c) French mizzen yard, about 1790.
(d) English spritsail yard.
(e) English sprit topsail yard.
(f) Gaff.
(g) Boom with jaws.
(h) Inner end of a boom with an eye fitted for a bolt connection.
(i) Lower studdingsail boom with a gooseneck (swinging boom), English version.
(j) A swinging boom fitted to the main channel. French ships and merchantmen from other nations sometimes hoisted their lower studdingsails flying, with only a yard spreading the foot.
(k) Lower studdingsail yard, English.
(l) Topsail studdingsail yard, English.
(m) English long davit, until approximately 1780.
(n) English long davit for smaller vessels after 1780.
(o) Continental long davit. An iron shackle, in which the long davit rested when an anchor had to be fished, is shown beside it.
(p) English short davit, to be placed into steps at the fore channel. These replaced the long davit after 1770.

yards. Whereas on French yards these were carved from the full timber of the yard, Röding mentioned wedge shaped timber pieces being nailed around the yardarms to provide stops. The number of yardarm stops depended on the number of reef bands of a sail (see Figure 17c). A topsail yard might have had up to four stops on each side, placed vertically above each reef cringle. The length of a yardarm was measured from outboard of the inner stop.

Made French yards, prepared from up to four pieces, were reinforced, besides bolts, not only with rope wooldings (see Figure 16h) but also with iron hoops. A plank similar to the fish discussed above on English yards was described by Röding for Continental yards:

> There are also long clamps or shells of 6 to 9 feet nailed to the centre of the rear of a lower yard, to keep it a little away from the mast, so that an even better bracing angle is possible. Cuts are made in this clamp or shell not only for the slings, which are laid around the yard, but also for the block strops placed there.

Before we consider the proportional lengths and diameters of yards, it is worth considering a description of Continental yards at the beginning of the century, as recorded in *Der geöffnete See-Hafen*:

> The main yard is basically a long, rounded and strong piece of timber, slightly tapered to both ends, but a few items on it are noteworthy. First, two cleats can be found right in the middle on the forward face, between which lies the strap for the jeer blocks. Further, a batten approximately 12 to 15 feet long is nailed underneath the middle section, with square holes 1 to 1½ feet apart, to which the main sail is fastened in the middle. This batten is known as a comb.

The anonymous author also explained that boom irons were fitted to the aft face of the yard, with the part of the iron which encircled the yard left 1½ inches open rather than completely closed. One of these irons was positioned on the yardarm and the other about 5 to 6 feet inboard. In addition, iron rings were stapled to the upper face of the yard about 1 fathom apart over the whole length for the furling gaskets. The fore yard was similar to the main yard. The topsail yard was also similar in proportion, with the exception that the boom irons and gasket rings were omitted (the gaskets were nailed direct to the yard). The topgallant yard was also described as similar to the topsail yard, but proportionally smaller.

Proportional lengths and diameters of yards

Proportionale Längen und Durchmesser der Rahen; Longueurs et diamètres proportionnels des vergues

Paris provided the following proportions for French yards:

Main yard	length	0.591 of the ship's length or 2⅙ times the beam
	maximum diameter	1/48 or 1/39 of the yard's length
	minimum diameter	⅓ of the maximum
	yardarm length	1/11 of the yard's length
Fore yard	length	0.543 of the ship's length or twice the beam
	maximum diameter	1/49 or 1/39 of the yard's length
	minimum diameter	⅓ of the maximum
	yardarm length	1/11 of the yard's length
Mizzen yard	length	0.526 of the ship's length or twice the beam
	maximum diameter	1/64 or 1/36 of the yard's length
	minimum diameter	⅓ of the maximum
	length of yard's peak	1/11 of the yard's length
Main topsail yard	length	0.409 of the ship's length or 1¼ times the beam
	maximum diameter	1/53 or 1/39 of the yard's length
	minimum diameter	⅓ of the maximum
	yardarm length	⅙ of the yard's length
Fore topsail yard	length	0.4 times the ship's length or 1⅙ times the beam
	maximum diameter	1/54 of the yard's length or 7/15 of the fore yard's diameter
	yardarm length	⅙ of the yard's length
Crossjack yard	length	0.4 times the ship's length

	maximum diameter	1/60 of the yard's length
	yardarm length	1/8 of the yard's length
Mizzen topsail yard	length	0.297 of the ship's length or 3/4 of the beam
	maximum diameter	1/60 of the yard's length or 1/2 of the mizzen yard's diameter
	yardarm length	1/8 of the yard's length
Main topgallant yard	length	0.256 of the ship's length or 3/4 of the beam
	maximum diameter	1/60 or 1/39 of the yard's length
	minimum diameter	1/3 of the maximum
	yardarm length	1/10 of the yard's length
Fore topgallant yard	length	0.256 of the ship's length or 2/3 of the beam
	maximum diameter	1/60 of the yard's length or in proportion to the fore topsail yard
	yardarm length	1/10 of the yard's length
Mizzen topgallant yard	length	0.185 of the ship's length
	maximum diameter	1/60 of the yard's length
	yardarm length	1/8 of the yard's length
Spritsail yard	length	0.4 of the ship's length or 1 1/4 times the beam
	maximum diameter	1/54 or 1/39 of the yard's length
	minimum diameter	1/3 of the maximum
	yardarm length	1/9 of the yard's length
Sprit topsail yard	length	0.297 of the ship's length or 3/4 of the beam
	maximum diameter	1/60 of the yard's length or 7/16 of the spritsail yard's length
	yardarm length	1/8 of the yard's length

Where alternatives are given, the first is taken from dimensions given by Paris for ships between 1740 and 1780 and the second from a rigging plan for the *Royal Louis* from 1780 headed 'Rules for masting proportions of ships at the end of the last century' [that is, the eighteenth century].

Swedish yard lengths for merchantmen were provided by Chapman in 1768, for merchant frigates as follows:

Main yard	0.52 of the length between stem and sternpost (L)
Main topsail yard	0.79 of the main yard length
Main topgallant yard	0.7 of the main topsail yard length
All yards of the foremast	0.9 of the yards of the mainmast
All yards of the mizzen mast	in proportion to the appropriate mast as for the mainmasts and yards
Crossjack yard	1.22 times the mizzen topsail yard length
Spritsail yard	equal to the fore topsail yard
Sprit topsail yard	equal to the fore topgallant yard

For barks as follows:

Main yard	0.54 of 20/21 times (L)
Main topsail yard	0.8 of the main yard length
All other yards	in proportion as for frigates

And for East Indiamen as follows:

Main yard	0.54 times (L)
Main topsail yard	0.8 of the main yard
All other yards	in proportion as for frigates

On all vessels studdingsail booms were 1 to 2 feet longer than half the length of the yard.

The lengths of yardarms were as follows:

Lower and topgallant yards	1/11 of the yard
Topsail yards	1/7 of the yard

Maximum yard diameters for all vessels were:

Main and fore yards	0.25 [more likely 0.025] of the yard's length
Topsail yards	0.23 [0.023] of the yard's length
Topgallant yards	1/6 [1/60] of the yard's length
Spritsail and crossjack	0.21 [0.021] of the yard's length
Sprit topsail yard	equal to the main topgallant yard
Mizzen gaff	1 inch per 4 feet of length
Lower studdingsail boom	1/72or 1/36 of length

The taper for lower and topsail yards (and for topgallant yards, in brackets) are indicated by the following diameters:

First quarter	26/27 (31/32)
Second quarter	23/27 (7/8)
Third quarter	18/27 (23/32)
Fourth quarter	11/27 (1/2)

Further details of yard proportions on Continental ships were recorded by Röding and Korth (where Korth's figures differ from those of Röding, the former are shown in brackets):

Main yard	length	2 1/4 times the beam
	maximum diameter	1/48 of the yard length (2 1/2in per 10 feet of length)
	minimum diameter	2/5 of the maximum (1/3 less than the maximum)
	yardarm length	1/10 of the yard length
Main topsail yard	length	1 1/2 times the beam
	maximum diameter	3/160 of the yard length
	minimum diameter	2/5 of the maximum
	yardarm length	1/7 of the yard length
Main topgallant yard	length	4/5 of the beam
	maximum diameter	1/60 of the yard length
	minimum diameter	2/5 of the largest
	yardarm length	1/8 of the yard length
Main royal yard	length	8/25 of the beam
	maximum diameter	1/50 of the yard length
	minimum diameter	2/5 of the maximum
	yardarm length	1/12 of the yard length
Fore yard	length	twice the beam
	maximum diameter	1/48 of the yard length
	minimum diameter	2/5 of the maximum
	yardarm length	1/12 of the yard length
Fore topsail yard	length	1 1/3 times the beam
	maximum diameter	3/160 of the yard length
	minimum diameter	2/5 of the maximum
	yardarm length	1/7 of the yard length
Fore topgallant yard	length	7/10 of the beam
	maximum diameter	1/60 of the yard length
	minimum diameter	2/5 of the maximum
	yardarm length	1/8 of the yard length
Fore royal yard	all dimensions	in proportion as for the main royal yard to the main topgallant yard
Mizzen yard	length	twice the beam
	diameter (at 1/3 length)	1/60 of the yard length (equal to the fore yard)
	other diameters	(peak 1/2 and tack 3/4 of the diameter at 1/3 length)
	yardarm length	12 to 15 inches
Crossjack yard	length	1 1/3 times the beam
	maximum diameter	1/50 of the yard length
	minimum diameter	2/5 of the maximum
	yardarm length	19/144 of the yard length
Mizzen topsail yard	length	equal to the beam
	maximum diameter	3/160 of the yard length
	minimum diameter	2/5 of the maximum
	yardarm length	1/8 of the yard length

Mizzen topgallant yard	length	⅔ of the beam
	diameters	similar to the mizzen topsail yard
	yardarm length	similar to the mizzen topsail yard
Spritsail yard	length	equal to the main topsail yard
Sprit topsail yard		equal to the main topgallant yard

Numerous data exist for the lengths and diameters of English yards. Falconer's 1769 *Proportions for the length of yards, according to the different classes of ships in the British navy* gave the following details:

Main yard length	100 guns	0.560 of the gun deck length
	90, 80 guns	0.559 of the gun deck length
	70 guns	0.570 of the gun deck length
	50 guns	0.576 of the gun deck length
	44 guns	0.561 of the gun deck length
Fore yard	100, 90, 80 guns	0.880 of the main yard length
	all others	0.874 of the main yard length
Mizzen yard	100, 90, 80, 60, 44 guns	0.820 of the main yard length
	70 guns	0.847 of the main yard length
	24 guns	0.840 of the main yard length
Main topsail yard	24 guns	0.726 of the main yard length
	all others	0.720 of the main yard length
Fore topsail yard	70 guns	0.719 of the fore yard length
	24 guns	0.726 of the fore yard length
	all others	0.715 of the fore yard length
Main topgallant yard	all rates	0.690 of main topsail yard lengt
Fore topgallant yard	70 guns	0.696 of fore topsail yard length
	all others	0.690 of fore topsail yard length
Mizzen topsail yard	70 guns	0.768 of fore topsail yard length
	all others	0.750 of fore topsail yard length

The crossjack yard and spritsail yard were as for the fore topsail yard, and the sprit topsail yard as for the fore topgallant yard.

Diameters for the yards were as follows:

Main and fore yards	5/7in per yard of length
Topsail, crossjack, spritsail yards	9/14in per yard of length
Topgallant, mizzen topsail sprit topsail yards	8/13in per yard of length
Mizzen yard	5/9in per yard of length
All studdingsail booms	½in per yard of length

Further information on English ships, as given by Davis (1711), Sutherland (1711), Love (1735), Mountaine (1756), Steel (1794) and Falconer's 1815 edition among others, is condensed in the following tables (data applicable to merchantmen are marked M). Lengths of yards were as follows:

Main yard	1711	⅞ or ¾ of the mainmast
	1711 (M)	twice the beam plus 2 feet
	1735	equal to beam plus half the keel length
	1735	5/7 of the mainmast
	1756	3½ times the beam
	1756 (M)	7/10 of the mainmast
	1794	8/9 of the mainmast
Fore yard	1711	6/8 of the foremast
	1711	⅞ of the main yard
	1711 (M)	1⅘ times the beam
	1735	8/9 or 6/7 of the main yard
Fore yards thereafter		⅞ of the main yard
Mizzen yard	1711	equal to the fore yard
	until 1719	⅞ of the main yard
	1711 (M)	1 8/9 of the beam plus 1 foot
	after 1719	6/7 of the main yard
	1735, 1756 (M)	medium between the fore and main yards
Crossjack yard	1711	slightly longer than the main topsail yard
	1711	1¼ times the beam plus 1 foot
	1711 (M)	1⅓ times the beam
	1719	7/10 of the fore yard
	1735	⅗ of the main yard
	1756	equal to the main topsail yard
	1756 (M)	⅔ of the main yard
	1794	⅞ of the main topsail yard
Main topsail yard	1711	8/9 of the topmast
	1711	5/9 of the main yard
	1711 (M)	1¼ times the beam
	1719	7/10 of the main yard
	1735	4/7 of the main yard
	1756	½ of the main yard
	1756 (M)	⅔ of the main yard
	1794	5/7 of the main yard
Fore topsail yard	1711	8/9 of the topmast
	1711 (M)	1 1/7 times the beam
	1719	7/10 of the fore yard
	1735	4/7 of the fore yard
	1756	⅞ of the fore yard
	1794	⅞ of the main topsail
Mizzen topsail yard	1711	⅓ of the mizzen yard
	1711	⅚ of the beam
	1711 (M)	¾ of the beam
	1719	¾ of the crossjack yard
	1735	4/13 or ⅓ of the main yard
	1756	⅗ of the main topsail yard
	1756 (M)	equal to the main topgallant yard
	1794	⅔ of the main topsail yard
Main topgallant yard	1711	½ of the main topmast
	1711	½ of the main topsail yard
	1711 (M)	¾ of the beam
	1735	5/17 of the main yard
	1756	½ of the main topsail yard
	1756 (M)	⅗ of the main topsail yard
Fore topgallant yard	1711	½ of the fore topmast
	1711 (M)	⅔ of the beam
	1735	5/17 of the fore yard
	1756	½ of the fore topsail yard
	1756 (M)	⅞ of the main topgallant yard
All topgallant yards	until 1773	½ of the topsail yard
	after 1773	⅔ of the topsail yard
	1794 (74 guns and over)m	⅔ of the topsail yard
	1794 (fewer than 74 guns)	⅗ of the topsail yard
Royal yard	1779	½ of the topgallant yard
Spritsail yard	1711	5/7 of the fore yard
	1711	1 2/7 times the beam plus 2 feet
	1711 (M)	1⅓ times the beam plus 1 foot
	1719	7/10 of the fore yard
	1735	5/6 of the main yard
	1735 (M)	⅔ of the main yard
	1794	⅞ of the main topsail yard
Sprit topsail yard	1700-11	½ of the spritsail yard
	1711 (M)	¾ of the beam
	1719	⅔ of the spritsail yard
	1735	equal to the mizzen topsail yard
	1794	equal to the fore topgallant yard
Studdingsail yard		4/7 of the studdingsail boom
Driver yard	1794	equal to the fore topgallant yard

The corresponding diameters are as follows:

Main yard	1735	1/36 of the length
	until 1794	¾in per yard of length
	after 1794	7/10in per yard of length
Fore yard	1735	1/36 of the length

	until 1794	¾in per yard of length
	after 1794	7/10in per yard of length
Mizzen yard	1711-1735	½in per yard of length
	after 1735	⅔ of the main yard length
Crossjack yard	1711-1735	½in per yard of length
	until 1794	⅔in per yard of length
	after 1794	⅝in per yard of length
Topsail yard	1711	1/48 of the length
	1735	1/36 of the length
Topsail yard		⅝in per yard of length
Topgallant yard	1711	1/48 of the length
	until 1719	⅝in per yard of length
	after 1719	⅗in per yard of length
Royal yard		½ of the topsail yard diameter
Spritsail yard		⅝in per yard of length
Sprit topsail yard	1700	⅝in per yard of length
	1719	⅗in per yard of length
Studdingsail yard		1in per 5 feet of length
Driver yard		⅗in per yard of length

The diameters specified by Love in 1735, except those for the mizzen yard and the crossjack yard, were all calculated according to the rule of thumb '1 inch per yard of length' and do not always correspond with diameters from other sources, which were generally less. Love provided proportions for the taper of the yards as well: the first quarter was equal to the slings, and the second quarter was ⅔ and the yardarm ⅓ of the maximum diameter.

In contrast to the athwartships yards, those rigged fore-and-aft were of unequal diameters at the ends. On the mizzen yard, which was round in section throughout the period in which it was employed, a distinction was made between the upper and the lower arm. For the first decades of the century, Love provided the following dimensions:

Upper arm	first quarter	⅔ of the maximum diameter
	upper end	¼ of the maximum diameter
Lower arm	first quarter	11/12 of the maximum diameter
	lower end	⅔ of the maximum diameter

For the second half of the century, Steel recorded the following measurements:

Upper arm	first quarter	30/31 of the maximum diameter
	second quarter	⅞ of the maximum diameter
	third quarter	7/10 of the maximum diameter
	fourth quarter	⅖ of the maximum diameter
Lower arm	first quarter	60/61 of the maximum diameter
	second quarter	11/12 of the maximum diameter
	third quarter	⅗ of the maximum diameter
	fourth quarter	⅔ of the maximum diameter

From the middle of the century onwards the mizzen yard slowly became obsolete and was replaced first by the gaff and later by the gaff and boom (see Figure 18f and g). The gradual replacement of the mizzen yard began between 1750 and 1770 on smaller frigates, extended in the 1780s to larger ships and culminated in the last decade of the century with the rerigging of all ships of the line. HMS *Vanguard,* Nelson's flagship at Aboukir in 1798, was the only ship of the line in that action which still carried a mizzen yard.

The gaff originated in the third decade of the seventeenth century, and the first gaff-rigged vessels in the coastal trade were Dutch boeiers. The boom, introduced to full-rigged ships around 1790, was even older than the gaff, and its first use can be found during the development of the Dutch speeljacht around 1600. The gaff and boom rig made its way into the big ships via Dutch coastal craft, yachts, schooners and brigantines.

The gaff's function was to spread and secure the mizzen sail's head. After the mizzen gave way to the spanker, the boom introduced at the same time was used to spread the tack and sheets of that sail.

Only the mizzen or spanker was gaff-rigged on full-rigged ships, but on brigs, schooners, cutters and other small craft gaff-rigged sails included the mainsail, the schooner sail or sometimes the mizzen.

The largest diameter of a gaff was about 4 feet from the inner end, and the outward taper was as follows:

First quarter	40/41 of the maximum diameter
Second quarter	11/12 of the maximum diameter
Third quarter	⅘ of the maximum diameter
Fourth quarter	5/9 of the maximum diameter

A ship's boom had the maximum diameter in the middle and tapered equally towards both ends:

First quarter	40/41 of the maximum diameter
Second quarter	11/12 of the maximum diameter
Third quarter	⅘ of the maximum diameter
Fourth quarter	⅔ of the maximum diameter

The maximum diameter of a boom for cutters, sloops, etc was located close to the stern, at the boom sheets; the inner and outer sections tapered differently from that point:

Inner section	first quarter	40/41 of the maximum diameter
	second quarter	11/12 of the maximum diameter
	third quarter	⅞ of the maximum diameter
	fourth quarter	⅔ of the maximum diameter
Outer section	first half	11/12 of the maximum diameter
	second half	¾ of the maximum diameter

The following dimensions were given by Paris for the gaffs and booms of a schooner:

Length	gaff	0.34 and 0.29 respectively of the vessel's length
	boom	0.66 of the vessel's length
Diameter	gaff	1/52 or 1/50 of the length
	boom	1/57 of the length

Proportional lengths and diameters of gaffs and booms

Proportionale Längen und Durchmesser von Gaffeln und Bäumen; Longueurs et diamètres proportionnels des cornes et guis

Gaff and boom diameters for ships were as follows:

Length	mizzen boom	5/7 of the main yard
	gaff	⅝ of the mizzen boom
	studdingsail boom (lower)	5/9 of the main yard
	studdingsail boom (yard)	½ of the yard length
Diameter	mizzen boom	⅝in per yard of length
	gaff	⅝in per yard of length
	studdingsail boom (lower)	1in per 5 feet of length
	studdingsail boom (yard)	1in per 5 feet of length

For cutters as follows:

Length	main boom	⅔ of the mast
	gaff	⅗ of the boom
Diameter	main boom	9/16in per yard of length
	gaff	¼in per foot of length

For sloops, smacks, barges and boats as follows:

Length	boom	$\frac{2}{3}$ of the mast
	gaff	$\frac{3}{5}$ of the boom
	spread yard	$\frac{5}{8}$ of the mast
	crossjack yard	$\frac{2}{5}$ of the mast
	topsail yard	$\frac{4}{5}$ of the crossjack yard
	topgallant yard	$\frac{5}{6}$ of the topsail yard
Diameter	boom	$\frac{3}{16}$in per foot of length
	gaff	$\frac{1}{4}$in per foot of length
	spread yard	$\frac{1}{7}$in per foot of length
	crossjack yard	$\frac{1}{5}$in per foot of length
	topsail yard	$\frac{1}{5}$in per foot of length
	topgallant yard	$\frac{1}{8}$in per foot of length

For boats with a lug-sail rig as follows:

Length	main and fore yard	$\frac{5}{8}$ of the mast
Diameter	main and fore yard	$\frac{1}{4}$in per foot of length

For launches and cutters with a lug-sail rig:

Length	main and fore yard	$\frac{9}{17}$ of the mast
Diameter	main and fore yard	$\frac{1}{4}$in per foot of length

For launches and cutters with a settee-sail rig as follows:

Length	main yard	$3\frac{1}{2}$ times the boat's breadth
	fore yard	$\frac{9}{10}$ of the main yard
Diameter	main and fore yard	$\frac{1}{4}$in per foot of length

And for barges, pinnaces and yawls with sprit sails as follows:

Length of sprit	$1\frac{1}{8}$ times the mast
Diameter of sprit	$\frac{1}{8}$in per foot of length

Cleats and jaws

Klampen und Klauen; Taquets et mâchoires

Yards were fitted with various cleats of different shapes to suit different purposes. Sling cleats were nailed to the fore side and close to the centre of the yard. They prevented lateral sliding of the slings and jeer blocks. Stop cleats at the yardarms held the braces, horses and lifts in place.

Sling cleats were usually $1\frac{1}{4}$ times the yard's diameter long, one quarter of their length wide and two thirds of their width high. On main and fore yards they had a shoulder of one third of their length. The cleats were generally set one yard diameter outboard from the yard's centre. Exceptions were the crossjack yard and the sprit topsail yard, where the sling cleats were outboard only half that distance. Mizzen yards had the upper cleat placed $1\frac{1}{2}$ diameters below the middle and the lower cleat a further diameter lower than the upper. On spritsail yards and sprit topsail yards the sling cleats were placed on the underside of the yard.

Sling cleats were mainly used on English ships. Paris indicated one on a French lower yard (see Figure 16h) and a model of a Dutch two decker man-of-war from 1660-1670 destroyed in a bombing raid on Berlin had sling cleats on the lower yards; otherwise neither Röding nor Boudriot provided any hint of the use of sling cleats on Continental ships.

Stop cleats were half the yard diameter long, quarter their length wide and two thirds their width high. They were positioned inboard from the yardarm ends by $1\frac{1}{2}$in per yard of length for main, fore, crossjack and spritsail yards, by 3 inches par yard of length for the topsail and topgallant yards, by $2\frac{1}{2}$in per yard of length for the mizzen topsail yard, and on the mizzen yard one diameter away from the peak.

Stop cleats were fitted fore-and-aft on all English yards. On French ships, as noted above, they were worked as a squared part from the yard's timber, and on Continental ships, according to Röding, they were made up from a number of small timber pieces around the yard.

The sling cleats of studdingsail and driver yards were fitted at one third of the length of the yard from its inner end, and one yard diameter apart. The stop cleats were located twice their length inboard from the ends. Sling cleats on boat yards were either in the centre or, in lug, lateen or settee rigs, one third of the yard length from the lower end. They were one diameter apart. The stop cleats were fitted inboard from the ends of the yard by a distance equal to the cleat length.

Gaffs and booms were usually fitted with jaws on the inner ends, made from oak and encircling half the mast's circumference. The thickness of the jaws was about one quarter of the encircled diameter. It should be noted that the inner diameter of the jaws was 1 inch larger than that of the mast, to provide space for a leather lining and for movement. Gaff and boom jaws were similar, with the exception that the gaff jaw sloped by 40 degrees on the inside, to allow the angled setting of the gaff.

Where the jaws joined, the gaff or boom was chamfered to one quarter of its thickness. The flats thus created, and the corresponding tongues on the jaw itself, were approximately 4 feet in length. After being fitted with nails and bolts, the joints were rounded like the rest of the gaff or boom and additionally secured with three or four iron hoops (see Figure 18f and g). The tongues on Continental jaws were often kept shorter and the flats were stepped and recessed. They were only partly rounded (see Figure 17k and l).

Close to the jaw opening, an eyebolt was driven through the tapered boom end from the top and secured underneath with a ring. On gaffs, eyebolts were forced from top and bottom through the tapered gaff end. The upper bolt was for the halyard and the lower for the downhauler and the nock earring.

A vertical sheave for the sail's sheet was set into the boom's outer end, and a ferrule prevented the end from splintering. An eyebolt extended the boom's main axis. A gaff's peak was similarly fitted.

Sometimes, instead of jaws, a boom was fitted with an iron ring with side straps nailed to the inner end and secured with hoops. Two hoops with similar rings were placed around the mast so that the boom ring fitted in between and, secured with a bolt, provided the effect of a hinge (see Figure 18h). Another common connection was the gooseneck (see Figure 18i). Instead of a ring, a gooseneck shaped bolt was driven into the inner boom end and set into rings on the mast. A hole was drilled into the lower end of the bolt in the former arrangement or the gooseneck bolt in the latter, so that it could be forelocked or moused and thus be prevented from dropping out.

When a boom was fitted with jaws rather than ring hoops, a timber saddle was nailed to the mast. This covered the aft half of the mast's circumference, and the boom rested upon it.

One or two holes were drilled horizontally into the jaws to fit a parrel and, where the boom sheets chafed the timber, a leather sleeve was nailed around for protection. A comb cleat was sometimes set over the sheet block's strap as an added precaution to contain the strap.

Studdingsail booms

Leesegelspieren; Bouts-dehors de bonette

The lower studdingsail booms were hooked to eyebolts in the hull planking forward of the fore channels or to a bracket on the main channels. For this reason a gooseneck was fitted to the inner end of the boom (see Figure 18i).

The lower fore studdingsail booms on Continental ships, however, differed from this arrangement. These booms were loose eight-sided boat booms of crossjack yard dimensions, protruding over the side of the forecastle when the studdingsail was set (see Figure 17g; see also Lower studdingsail booms, under Studdingsail booms in Chapter II).

Yard studdingsail booms were placed to the fore side of the yard, on English ships at 45 degrees to the yard's centreline and on French ships sometimes directly forward of the centreline (that is, at 90 degrees). The booms on Dutch ships were fitted at the aft side of the yard until the early years of the century, an arrangement which can also be found on some later models and was a feature of Continental rigging early in the century.

Four boom irons for securing the studdingsail booms were mounted on each yard, one on each yardarm and a second about one third of the boom's length inside the first.

The straps of the inner irons, encircling the yard, were full hoops on English ships until about 1750 (see Figure 15j) and three quarter hoops after that date (see Figure 15k). On Continental ships the full hoop was generally retained throughout the century.

In the Royal Navy these inner boom irons were provided with hinges after 1775, so that the boom could be laid in (see Figure 15l). In merchant service the inner boom irons were sometimes replaced by timber boom saddles (see Figure 16b). Lever noted that 'large ships have an inner Boom Iron, which is fastened round the yard with an iron strap and nailed to it: but in smaller vessels there is a wooden Saddle for the Boom to rest on.' French lower yards carried these saddles in addition to the inner boom irons at approximately two thirds of the boom's length inboard of the yardarm (see Figure 16h).

The fitting of yardarm boom irons differed between English men-of-war and merchantmen. In men-of-war the boom hoop was outboard of the yardarm, welded to an angled square iron rod which was connected to the yardarm by bolted on iron straps and two hoops (see Figure 15m). (French yardarm irons, like the inner irons, were fitted directly to the yardarm itself and were figure-of-eight shaped, except that the yardarm hoop was sometimes octagonal to fit an octagonal yardarm. Lever noted an alternative type of boom iron in English service, fitted directly on to the yardarm as in the French style; he described it as 'driven on the square of the yardarm.')

The iron rod which served as the boom iron's neck was 1 inch longer on each arm than the hoop's diameter, and its thickness was one quarter of its length. The length of the mounting straps was 1⅓in per yard of the yard's length, and their width about half of the yard's diameter. The thickness of the straps at their inner ends was ⅜in, increasing at the outer ends where they joined the neck. The simple outer boom hoop described by Lever was fitted with a roller in its lower half in Royal Navy service around 1775 (see Figure 15n). The roller's size was one third of the diameter of the hoop. The inside diameter of the hoop itself equalled the diameter of the boom, and it was three eighths of that diameter wide; the iron used in the hoop was ⅝in to ¾in thick.

Similar booms were sometimes fitted on topsail yards, following the same fitting principles and with measurements in the same proportions.

Yardarm boom irons were fitted to merchantmen using the method described by Lever or, where the yardarms were rounded, a simple figure-of-eight boom iron, as in the French navy.

Other merchantmen were fitted with a squared bolt extending the main axis of the yard, secured in the same way as the angled neck on Royal Navy yards (see Figure 16a). The boom hoop had a leg with a square hole in the lower end, which was placed on to the squared bolt. A nut or forelock secured the fitting. This method, shown by Steel, is also shown by Röding as a fitting for Continental yards; in his written description, however, Röding dealt only with on the method described by Lever. The only difference in outer boom irons between English and Continental merchantmen was a variation in the boom hoops: Steel noted only a simple hoop without a roller, while Röding showed a horseshoe-shaped hoop with a vertical roller closing the open end (see Figure 16a, upper detail).

Topsail yards on merchantmen, as in the French navy, were usually fitted with studdingsail boom irons using the same method as that used for the lower yards. In the Royal Navy topsail yard studdingsail booms were often not used; when they were, the irons were often driven on to the yardarms as described by Lever. Steel mentioned the angled neck boom iron for topsail yards, but noted that usually only a simple fitting of an eyebolt driven axially into the end of the yardarm was used on ships in the Royal Navy.

The inner ends of all studdingsail booms were provided with a hole to enable the booms to be lashed, when in position, to the yards. Continental booms sometimes also had a heel like that of a topgallant mast, or were octagonal at their inner ends.

Miscellaneous

Sonstiges; Divers

On English ships the main and fore topgallant yards, the mizzen yard, the spritsail and sprit topsail yards were fitted with ferrules driven over the yardarm ends and eyebolts extending the yard axis, as for the main and fore topsail yards. The mizzen topsail and topgallant yards had ferrules but no eyebolts. Driver yards had a sheave set into the peak as well as a ferrule and eyebolt.

In addition to the rigging booms described hitherto, a ship also usually carried a number of fire booms. These were needed to prevent fireships or other burning vessels from coming too close, and were stored in the waist together with the spare spars. Similar in diameter to the lower studdingsail booms, the fire booms were only two thirds the length of these. The inner end of the boom was fitted with a gooseneck to enable it to be hung outboard, and the outer end with a forked iron. Ferrules were driven over both ends to prevent splitting.

Flagstaffs had the following dimensions:

Length	ensign staff	⅓ of the mainmast length above the rough tree rail
	jackstaff	½ the ensign staff length above the rough tree rail
Diameter	ensign staff	½in per yard of length
	jackstaff	¾in per yard of length

A davit was carried by men-of-war, East Indiamen and larger merchantmen for fishing the anchor. Until about 1780 this was a long square beam with a length approximately equal to the forecastle's width, sitting in an iron spanshackle on the forecastle between cleats at the timber heads (see Figure 18m). Röding mentioned the fish davit as being in use mainly on English ships, and Lever noted its use on smaller vessels only. This suggests at least that the fish davit was still extensively used after 1780. The davit became much shorter in the Royal Navy during the 1770s and was used from the fore channels, resting there in a step (see Figure 18p).

Tables

Tables 1 to 16 below are extracted from David Steel's *Elements of Mastmaking, Sailmaking and Rigging* from 1794. Other similar British tables differ only slightly, so the length and diameter specifications printed here provide a near comprehensive view of all spars used on British warships in the late eighteenth century.

James Love produced *An Exact Table of Proportions for the Thickness of Masts according to their Lengths* in 1735 which, from 10 feet to 99 feet of length, gave all diameters at the partners as 1/36 of the length. In the middle the diameters given ranged from nine tenths to seven eighths of the maximum diameter, and at the masthead the diameter given was three quarters of that at the partners. This simplification is of interest here as an early attempt at a comprehensive approach to the description of eighteenth century rigging, but it does somewhat less than justice to the real dimensions used at the time; Love's table is not reproduced.

Tables 17 to 35 are taken from John Davis' *The Seaman's Speculum or Compleat School-Master* of 1711. Tables 17 to 33 are based on his proportional lengths for masts and yards, which he claimed were applicable to all British-built merchant ships, given the keel length and beam. For the remaining two tables he selected specifically an East Indiaman of 84 feet keel length, 28 feet beam and a hold of 12½ feet. These tables provide an overview of the masting of British merchant ships at the beginning of the century.

They are followed by the spar dimensions of the American schooner *Sultana* from 1760, and French dimensions for various classes of ships taken from Admiral Edmond Paris' *Souvenirs de Marine* from 1884.

Finally, Tables 43 to 52 show the dimensions of masts and spars on Continental men-of-war at the close of the eighteenth century. They are based on figures published in 1826 by J W D Korth, in *Die Schiffbaukunst.*

Tables 1 to 16 show the dimensions of masts and yards in the Royal Navy, based on tables published by Steel in 1794. Diameters are measured at the partners, which were at the middle deck in three-decked ships and at the upper deck in all other ships.

Table 1: Ship of 100 guns, 2164 tons

Name of mast/yard	*mast length (ft in)*	*diameter (in)*	*yard length (ft in)*	*diameter (in)*
Main	117 0	39	102 4	24
Main top	70 0	20¾	73 0	15½
Main topgallant	35 0	11⅝	48 9	10
Main royal	0 0	0	36 0	7¾
Fore	103 6	34½	89 1	21
Fore top	62 10	20¾	64 6	13¾
Fore topgallant	31 0	10¾	43 0	8⅝
Fore royal	0 0	0	32 0	6⅞
Mizzen	101 4	23	87 0	16
Mizzen top	52 0	14	49 0	10⅛
Mizzen topgallant	26 0	8⅝	32 9	6½
Mizzen royal	0 0	0	24 0	5
Bowsprit	74 0	37	0 0	0
Spritsail	0 0	0	64 6	13¾
Jib-boom	53 0	15¼	0 0	0
Sprit topsail	0 0	0	43 0	8⅝
Crossjack	0 0	0	64 6	13¾
Ensign staff	45 0	7½	0 0	0
Jackstaff	19 6	4¾	0 0	0
Lower studding boom	56 9	11¼	33 0	6¾
Main top boom	51 2	10⅛	29 6	6
Fore top boom	44 6	9	25 6	5⅛
Main topgallant boom	36 6	7¼	21 0	4¼
Fore topgallant boom	32 3	6½	18 6	3¾
Fire boom	38 0	11¼	0 0	0

Table 2: Ship of 90 guns. 1931 tons

Name of mast/yard	*mast length (ft in)*	*diameter (in)*	*yard length (ft in)*	*diameter (in)*
Main	112 0	37¼	98 0	22¼
Main top	66 0	19¾	70 0	14¾
Main topgallant	33 0	11	46 6	9⅜
Main royal	0 0	0	35 0	7⅜
Fore	100 0	33½	85 9	20
Fore top	59 0	9¾	61 0	12⅞
Fore topgallant	29 0	9⅝	40 6	8¼
Fore royal	0 0	0	30 0	6⅜
Mizzen	96 6	22¼	84 0	15
Mizzen top	49 0	13½	47 0	9¾
Mizzen topgallant	24 0	8	31 0	6¼
Mizzen royal	0 0	0	23 0	4⅞
Bowsprit	71 0	35½	0 0	0
Spritsail	0 0	0	61 0	12⅞
Jib-boom	50 9	14½	0 0	0
Sprit topsail	0 0	0	40 6	8¼
Crossjack	0 0	0	61 0	12⅞
Ensign staff	40 0	6	0 0	0
Jackstaff	18 0	4½	0 0	0
Lower studding boom	54 6	11	31 0	6¼
Main top boom	49 0	10	28 0	3¼
Fore top boom	42 10	8½	24 6	5
Main topgallant boom	35 0	7	20 0	4
Fore topgallant boom	30 6	6⅛	17 6	3½
Fire boom	36 0	11	0 0	0

Table 3: Ship of 80 guns. 1615 tons

Name of mast/yard	*mast length (ft in)*	*diameter (in)*	*yard length (ft in)*	*diameter (in)*
Main	107 0	35⅝	93 0	22
Main top	64 0	19¼	66 6	14⅛
Main topgallant	32 0	10¼	46 0	9¼
Main royal	0 0	0	33 0	7
Fore	95 9	31⅞	81 4	19⅜
Fore top	59 9	19¼	58 0	12¼
Fore topgallant	28 6	9½	40 0	8¼
Fore royal	0 0	0	29 0	6⅛
Mizzen	93 0	21⅛	76 0	14½
Mizzen top	46 0	13¼	43 0	8⅞
Mizzen topgallant	23 0	7⅝	29 0	5¾
Mizzen royal	0 0	0	21 0	4⅜
Bowsprit	68 0	34	0 0	0
Spritsail	0 0	0	58 0	12¼
Jib-boom	48 0	14	0 0	0
Sprit topsail	0 0	0	40 0	8¼
Crossjack	0 0	0	58 0	12¼
Ensign staff	39 0	6½	0 0	0
Jackstaff	17 6	4¼	0 0	0
Lower studding boom	51 6	10¼	29 6	6
Main top boom	46 6	9¼	26 6	5¼
Fore top boom	40 8	8⅛	23 4	4¾
Main topgallant boom	33 3	6¾	19 0	3¾
Fore topgallant boom	29 0	5⅞	16 6	3¼
Fire boom	34 0	10¼	0 0	0

Table 4: Ship of 74 guns, 1199 tons

Name of mast/yard	*mast length (ft in)*	*diameter (in)*	*yard length (ft in)*	*diameter (in)*
Main	111 0	37	97 0	23
Main top	66 0	19¾	70 0	15
Main topgallant	33 0	11¼	46 6	9½
Main royal	0 0	0	35 0	7½
Fore	98 6	32¾	85 0	20
Fore top	58 8	19¾	62 0	13
Fore topgallant	29 4	9¾	40 6	8¼
Fore royal	0 0	0	31 0	6½
Mizzen	95 0	22¼	84 0	15½
Mizzen top	49 0	13½	47 0	9¾
Mizzen topgallant	24 6	8⅛	31 6	6⅜
Mizzen royal	0 0	0	23 0	4⅞
Bowsprit	67 6	35	0 0	0
Spritsail	0 0	0	62 0	13
Jib-boom	50 4	14½	0 0	0
Sprit topsail	0 0	0	40 6	8¼
Crossjack	0 0	0	62 0	13
Ensign staff	40 0	6½	0 0	0
Jackstaff	18 0	4 1/12	0 0	0
Lower studding boom	53 9	10¾	31 0	6¼
Main top boom	48 6	9¾	27 3	5½
Fore top boom	42 6	8½	24 6	5
Main topgallant boom	35 0	7	20 0	4
Fore topgallant boom	31 0	6¼	17 9	3½
Fire boom	35 6	10¾	0 0	0

Table 5: Ship of 64 guns, 1369 tons

Name of mast/yard	*mast length (ft in)*	*diameter (in)*	*yard length (ft in)*	*diameter (in)*
Main	101 0	33½	90 4	21
Main top	58 6	17⅝	65 4	13⅝
Main topgallant	29 3	9¾	39 0	7⅞
Main royal	0 0	0	32 0	6¾
Fore	89 7	29¾	79 6	18½
Fore top	53 0	17⅝	57 6	12⅛
Fore topgallant	26 6	8¾	34 6	7
Fore royal	0 0	0	28 6	6
Mizzen	86 0	19¾	77 0	14
Mizzen top	43 6	12¼	43 3	9
Mizzen topgallant	21 9	7¼	29 0	5¾
Mizzen royal	0 0	0	21 0	4½
Bowsprit	60 4	31½	0 0	0
Spritsail	0 0	0	57 6	12⅛
Jib-boom	43 9	13	0 0	0
Sprit topsail	0 0	0	34 6	7
Crossjack	0 0	0	57 6	12⅛
Ensign staff	38 0	6¼	0 0	0
Jackstaff	16 6	4¼	0 0	0
Lower studding boom	50 0	10	29 0	5¾
Main top boom	45 0	9	25 9	5⅛
Fore top boom	39 9	8	22 9	4½
Main topgallant boom	32 8	6½	18 9	3¾
Fore topgallant boom	28 9	5¼	16 6	3¼
Fire boom	33 3	10	0 0	0

Table 6: Ship of 50 guns, 1044 tons

Name of mast/yard	*mast length (ft in)*	*diameter (in)*	*yard length (ft in)*	*diameter (in)*
Main	92 0	29	82 0	19¼
Main top	53 0	16	60 9	12½
Main topgallant	26 6	8⅞	36 0	7⅜
Main royal	0 0	0	30 0	6¼
Fore	81 6	26⅜	72 0	17
Fore top	48 0	16	52 6	11
Fore topgallant	24 0	8	31 6	6¼
Fore royal	0 0	0	26 0	5½
Mizzen (gaff)	78 9	19⅜	44 0	11½
Mizzen top	40 0	11⅛	39 6	8¼
Mizzen topgallant	20 0	6⅝	26 0	5¼
Mizzen royal	0 0	0	19 0	4⅛
Bowsprit	56 0	29	0 0	0
Spritsail	0 0	0	52 6	11
Jib-boom	39 8	11¾	0 0	0
Sprit topsail	0 0	0	31 6	6¼
Crossjack	0 0	0	52 6	11
Ensign staff	36 0	6	0 0	0
Jackstaff	15 6	3¾	0 0	0
Lower studding boom	45 6	9⅛	26 0	5¼
Main top boom	41 0	8⅛	23 6	4¾
Fore top boom	36 0	7¼	20 6	4⅛
Main topgallant boom	30 5	6	17 6	3½
Fore topgallant boom	26 3	5¼	15 0	3
Driver boom	60 9	11½	31 6	6¼
Fire boom	30 0	9⅛	0 0	0

Table 7: Ship of 44 guns, 879 tons

Name of mast/yard	*mast length (ft in)*	*diameter (in)*	*yard length (ft in)*	*diameter (in)*
Main	88 0	26¼	80 0	18½
Main top	53 0	15¾	57 6	11⅞
Main topgallant	26 6	8¾	36 0	7⅜
Main royal	0 0	0	28 9	5⅞
Fore	78 0	23⅝	70 0	16¼
Fore top	47 0	15¾	52 0	10¾
Fore topgallant	23 6	7¾	32 0	6⅜
Fore royal	0 0	0	26 0	5⅜
Mizzen (gaff)	74 6	17⅞	36 8	11
Mizzen top	40 0	11	39 6	8
Mizzen topgallant	20 0	6⅝	25 9	5⅛
Mizzen royal	0 0	0	19 0	4
Bowsprit	52 6	26¼	0 0	0
Spritsail	0 0	0	52 0	10⅜
Jib-boom	39 0	11½	0 0	0
Sprit topsail	0 0	0	32 0	6⅜
Crossjack	0 0	0	52 0	10¾
Ensign staff	35 0	5⅝	0 0	0
Jack staff	15 6	3¾	0 0	0
Lower studding boom	44 6	9	25 6	5⅛
Main top boom	40 0	8	23 0	4¾
Fore top boom	35 0	7	20 0	4
Main topgallant boom	28 9	5¾	16 6	3¼
Fore topgallant boom	26 0	5¼	15 0	3
Driver boom	57 6	10¾	32 0	6⅜
Fire boom	29 6	9	0 0	0

Table 8: Ship of 38 guns, 951 tons

Name of mast/yard	*mast length (ft in)*	*diameter (in)*	*yard length (ft in)*	*diameter (in)*
Main	90 0	27	81 9	19
Main top	54 0	16⅛	59 0	12¼
Main topgallant	27 0	9	37 0	7½
Main royal	0 0	0	29 0	6⅛
Fore	80 0	23⅝	71 6	16⅜
Fore top	48 0	16⅛	53 0	11½
Fore topgallant	24 0	8	32 9	6½
Fore royal	0 0	0	26 0	5¾
Mizzen (gaff)	75 7	1⅛	38 0	11½
Mizzen top	41 0	11¼	40 0	8¼
Mizzen topgallant	20 6	6¾	27 9	5½
Mizzen royal	0 0	0	20 0	4⅛
Bowsprit	55 0	27	0 0	0
Spritsail	0 0	0	53 0	11½
Jib-boom	38 7	11½	0 0	0
Sprit topsail	0 0	0	32 9	6½
Crossjack	0 0	0	53 0	11½
Ensign staff	35 0	5⅝	0 0	0
Jackstaff	15 6	3¾	0 0	0
Lower studding boom	45 0	9	26 0	5¼
Main top boom	41 0	8⅛	23 6	4¾
Fore top boom	35 9	7⅛	20 6	4⅛
Main topgallant boom	29 6	6	17 0	3½
Fore topgallant boom	26 6	5¼	15 0	3
Driver boom	59 0	11½	32 9	6½
Fire boom	30 0	9	0 0	0

Table 9: Ship of 36 guns, 871 tons

Name of mast/yard	*mast length (ft in)*	*diameter (in)*	*yard length (ft in)*	*diameter (in)*
Main	89 0	26	79 0	18⅝
Main top	53 4	15¾	57 0	12
Main topgallant	25 8	8⅝	34 9	7
Main royal	0 0	0	28 0	6
Fore	79 6	23⅜	69 4	16
Fore top	47 0	15¾	51 9	11
Fore topgallant	22 5	7½	31 6	6⅜
Fore royal	0 0	0	25 0	5½
Mizzen (gaff)	74 8	17¾	36 0	11
Mizzen top	40 0	11⅛	39 4	7⅞
Mizzen topgallant	20 0	6½	27 0	5½
Mizzen royal	0 0	0	19 0	3⅞
Bowsprit	54 0	26	0 0	0
Spritsail	0 0	0	51 9	11
Jib-boom	38 0	11¼	0 0	0
Sprit topsail	0 0	0	31 6	6⅜
Crossjack	0 0	0	51 9	11
Ensign staff	35 0	5⅝	0 0	0
Jackstaff	15 6	3¾	0 0	0
Lower studding boom	44 0	9	25 6	5⅛
Main top boom	39 6	8	22 6	4½
Fore top boom	34 8	7	20 0	4
Main topgallant boom	28 6	5¾	16 6	3¼
Fore topgallant boom	25 10	5⅛	15 0	3
Driver boom	57 0	11	31 6	6⅜
Fire boom	29 3	9	0	0 0

Table 10: Ship of 32 guns, 677 tons

Name of mast/yard	*mast length (ft in)*	*diameter (in)*	*yard length (ft in)*	*diameter (in)*
Main	85 0	24¾	74 4	17⅛
Main top	51 0	15⅛	55 0	11⅜
Main topgallant	25 6	8½	33 6	6⅝
Main royal	0 0	0	27 0	5⅝
Fore	75 0	22	65 0	15
Fore top	45 0	15⅛	48 0	10
Fore topgallant	22 6	7½	29 6	5⅞
Fore royal	0 0	0	24 0	5
Mizzen (gaff)	72 0	17	35 0	10
Mizzen top	38 0	10⅝	36 9	7
Mizzen topgallant	19 0	6¼	24 0	5
Mizzen royal	0 0	0	18 0	3½
Bowsprit	52 0	25	0 0	0
Spritsail	0 0	0	48 0	10
Jib-boom	36 10	10¾	0 0	0
Sprit topsail	0 0	0	29 6	5⅞
Crossjack	0 0	0	48 0	10
Ensign staff	34 0	5½	0 0	0
Jackstaff	15 0	3¾	0 0	0
Lower studding boom	41 0	8¾	23 6	4¾
Main top boom	37 2	7½	21 3	4¼
Fore top boom	32 6	6½	18 6	3¾
Main topgallant boom	27 6	5½	16 0	3¼
Fore topgallant boom	24 0	4⅞	14 0	2⅞
Driver boom	55 0	10	29 6	5⅞
Fire boom	27 3	8¾	0 0	0

Table 11: Ship of 28 guns, 594 tons

Name of mast/yard	*mast length (ft in)*	*diameter (in)*	*yard length (ft in)*	*diameter (in)*
Main	81 4	23⅝	71 3	16½
Main top	48 9	14⅜	52 0	11
Main topgallant	24 4	8	32 6	6½
Main royal	0 0	0	26 0	5½
Fore	72 0	20⅞	62 2	14½
Fore top	43 0	14⅜	46 0	9¾
Fore topgallant	21 6	7	28 6	5⅝
Fore royal	0 0	0	23 0	4⅞
Mizzen (gaff)	69 0	16¾	32 6	9¾
Mizzen top	36 7	10	35 0	7¼
Mizzen topgallant	18 3	6	22 0	4⅜
Mizzen royal	0 0	0	17 6	3⅝
Bowsprit	48 9	23⅝	0 0	0
Spritsail	0 0	0	46 0	9¾
Jib-boom	35 0	10¼	0 0	0
Sprit topsail	0 0	0	28 6	5⅝
Crossjack	0 0	0	46 0	9¾
Ensign staff	30 0	5¼	0 0	0
Jackstaff	14 0	3⅜	0 0	0
Lower studding boom	40 0	8	23 0	4⅝
Main top boom	36 0	7⅛	20 6	4
Fore top boom	31 1	6⅛	17 9	3½
Main topgallant boom	26 0	5¼	15 0	3
Fore topgallant boom	23 0	4⅝	13 3	2⅝
Driver boom	52 0	9¾	28 6	5⅝
Fire boom	26 6	8	0 0	0

Table 12: Ship of 24 guns, 500 tons

Name of mast/yard	*mast length (ft in)*	*diameter (in)*	*yard length (ft in)*	*diameter (in)*
Main	75 0	22¼	65 6	15
Main top	45 0	13¼	47 0	9¾
Main topgallant	22 6	7½	29 6	6
Main royal	0 0	0	23 6	4⅞
Fore	66 6	19½	57 8	13¼
Fore top	40 0	13¼	41 0	8⅝
Fore topgallant	20 0	6½	25 10	5⅜
Fore royal	0 0	0	20 6	4⅜
Mizzen (gaff)	64 0	15	29 4	8⅝
Mizzen top	33 9	9¼	31 5	6½
Mizzen topgallant	16 10	5½	19 6	4
Mizzen royal	0 0	0	15 8	3¼
Bowsprit	45 0	22¼	0 0	0
Spritsail	0 0	0	41 0	8⅝
Jib-boom	32 6	9⅜	0 0	0
Sprit topsail	0 0	0	25 10	5⅜
Crossjack	0 0	0	41 0	8⅝
Ensign staff	28 0	5	0 0	0
Jackstaff	13 0	3¼	0 0	0
Lower studding boom	36 9	7¼	21 0	4¼
Main top boom	33 0	6½	19 0	3⅞
Fore top boom	28 10	5¾	16 6	3¼
Main topgallant boom	23 6	4¾	13½	2¾
Fore topgallant boom	20 6	4⅛	11 9	2⅜
Driver boom	47 0	8⅝	25 10	5⅜
Fire boom	24 0	7¼	0 0	0

Table 13: Ship of 20 guns, 429 tons

Name of mast/yard	*mast length (ft in)*	*diameter (in)*	*yard length (ft in)*	*diameter (in)*
Main	72 0	21⅝	63 0	14½
Main top	43 2	12¾	45 6	9½
Main topgallant	21 7	7¼	28 4	5¾
Main royal	0 0	0	22 9	4¾
Fore	64 0	19	55 0	12⅝
Fore top	38 4	12¾	40 0	8½
Fore topgallant	19 2	6⅜	25 0	5
Fore royal	0 0	0	20 0	4½
Mizzen (gaff)	61 0	14	28 6	8½
Mizzen top	32 5	9	31 0	6¼
Mizzen topgallant	16 0	5¼	19 6	4
Mizzen royal	0 0	0	15 6	3⅛
Bowsprit	43 6	21⅝	0 0	0
Spritsail	0 0	0	40 0	8½
Jib-boom	31 0	9	0 0	0
Sprit topsail	0 0	0	25 0	5
Crossjack	0 0	0	40 0	8½
Ensign staff	28 0	5	0 0	0
Jackstaff	13 0	3¼	0 0	0
Lower studding boom	35 0	7	20 0	4
Main top boom	32 0	6⅜	18 6	3¾
Fore top boom	27 6	5½	15 9	3⅛
Main topgallant boom	22 6	4½	13 0	2⅝
Fore topgallant boom	20 0	4	11 6	2¼
Driver boom	45 6	8½	25 0	5
Fire boom	23 6	7	0 0	0

Table 14: Sloop, 300 tons

Name of mast/yard	*mast length (ft in)*	*diameter (in)*	*yard length (ft in)*	*diameter (in)*
Main	63 0	18⅜	55 0	12¾
Main top	37 6	11¼	39 6	8¼
Main topgallant	18 9	6¼	25 0	5
Main royal	0 0	0	19 9	4⅛
Fore	56 0	16⅜	48 5	11¼
Fore top	33 4	11¼	35 0	7⅜
Fore topgallant	16 8	5⅝	22 0	4⅜
Fore royal	0 0	0	17 6	3¾
Mizzen (gaff)	48 0	12	24 9	7⅜
Mizzen top	26 9	7¼	26 4	5½
Mizzen topgallant	13 0	4¼	16 6	3¼
Mizzen royal	0 0	0	13 2	2¾
Bowsprit	37 6	18⅜	0 0	0
Spritsail	0 0	0	35 0	7⅜
Jib-boom	27 0	8	0 0	0
Sprit topsail	0 0	0	22 0	4⅜
Crossjack	0 0	0	35 0	7⅜
Ensign staff	24 0	4½	0 0	0
Jackstaff	12 0	3	0 0	0
Lower studding boom	30 6	6	17 6	3½
Main top boom	27 0	5⅜	15 6	3⅛
Fore top boom	24 2	5	14 0	2⅞
Main topgallant boom	19 9	4	11 6	2¼
Fore topgallant boom	17 6	3½	10 0	2
Driver boom	39 6	7⅜	22 0	4⅜
Fire boom	20 6	6	0 0	0

Table 15: Brig, 200 tons

Name of mast/yard	*mast length (ft in)*	*diameter (in)*	*yard length (ft in)*	*diameter (in)*
Main	56 0	19	42 0	9¾
Main top	31 0	10	31 6	7
Main topgallant	23 6	6¼	23 6	5
Main royal	0 0	0	15 9	3½
Fore	49 0	17	42 0	9¾
Fore top	31 0	10	31 6	7
Fore topgallant	23 6	6¼	23 6	5
Fore royal	0 0	0	15 0	3½
Bowsprit	34 0	17	0 0	0
Spritsail	0 0	0	31 6	7
Jib-boom	24 0	7	0 0	0
Sprit topsail	0 0	0	23 6	5
Main boom	0 0	0	45 0	10½
Gaff	0 0	0	28 0	7¼
Ensign staff	20 0	3½	0 0	0
Jackstaff	10 0	2½	0 0	0
Lower studding boom	23 6	4¾	13 6	2¾
Main top boom	21 0	4⅛	12 0	2½
Fore top boom	21 0	4⅛	12 0	2½
Main topgallant boom	11 9	2¼	6 9	2
Fore topgallant boom	11 9	2¼	6 9	2

Table 16: Cutter, 200 tons

Name of mast/yard	*mast length (ft in)*	*diameter (in)*	*yard length (ft in)*	*diameter (in)*
Main and top	88 0	22	58 0	9½
			52 0	7½
Topgallant	44 0	9¾	26 0	6
Topgallant (short)	35 0	9½	0 0	0
Royal	0 0	0	13 0	3¾
Bowsprit	64 0	20	0 0	0
Jib-boom	57 0	10	0 0	0
Main boom	0 0	0	66 0	14¾
Driver boom	0 0	0	42 0	8
Gaff	0 0	0	49 6	10¾
Storm gaff	0 0	0	21 0	8¾

Tables 17 to 33 show the dimensions of masts and yards in the British merchant service, based on information published by Steel in 1794.

Table 17: Merchant vessel of 1300 tons

Name of mast/yard	*mast length (ft in)*	*diameter (in)*	*yard length (ft in)*	*diameter (in)*
Main	96 0	31½	86 0	21
Main top	56 0	17½	58 0	14
Main topgallant	27 0	9	38 0	8½
Main royal	20 0	6½	24 0	5
Fore	90 0	30	82 0	20
Fore top	56 0	17½	55 0	13½
Fore topgallant	26 0	9	36 0	8
Fore royal	18 0	6	22 0	5
Mizzen	78 0	21½	72 0	13
Mizzen top	41 0	13	40 0	9½
Mizzen topgallant	21 0	7	26 0	5½
Mizzen royal	12 0	5	16 0	4½
Bowsprit	60 0	31	56 0	11½
Jib-boom	44 0	12½	38 0	8
Driver boom	62 0	12	0 0	0
Crossjack	0 0	0	56 0	11½
Lower studding boom	44 0	9	30 0	7
Main top boom	43 0	8½	24 0	6
Main topgallant boom	29 0	6	18 0	5
Fore top boom	41 0	8½	24 0	6
Fore topgallant boom	27 0	6	17 0	5
Ensign staff	40 0	7	0 0	0
Jackstaff	26 0	5½	0 0	0

Table 18: Merchant vessel of 1200 tons

Name of mast/yard	*mast length (ft in)*	*diameter (in)*	*yard length (ft in)*	*diameter (in)*
Main	91 0	31	84 0	20½
Main top	56 0	17½	58 0	14
Main topgallant	27 0	9	37 6	8¼
Main royal	20 0	6½	24 0	5
Fore	88 0	30	80 0	19½
Fore top	54 0	17½	54 0	13¼
Fore topgallant	26 0	9	36 0	8
Fore royal	18 0	6	22 0	5
Mizzen	78 0	21½	70 0	13
Mizzen top	40 0	12½	39 0	9¼
Mizzen topgallant	21 0	7	26 0	5½
Mizzen royal	12 0	5	16 0	4½
Bowsprit	59 0	30½	55 0	11¼
Jib-boom	42 0	12¼	38 0	8
Driver boom	60 0	11½	0 0	0
Crossjack	0 0	0	55 0	11¼
Lower studding boom	42 0	9	29 0	7
Main top boom	41 0	8½	23 0	6
Main topgallant boom	28 0	6	17 0	5
Fore top boom	39 0	8½	23 0	6
Fore topgallant boom	26 6	5¾	16 0	5
Ensign staff	38 0	7	0 0	0
Jackstaff	24 0	5	0 0	0

Table 19: Merchant vessel of 1100 tons

Name of mast/yard	*mast length (ft in)*	*diameter (in)*	*yard length (ft in)*	*diameter (in)*
Main	90 0	30	80 0	20
Main top	56 0	17¼	57 6	13½
Main topgallant	27 0	9	36 0	8
Main royal	20 0	6½	23 0	5
Fore	84 0	29½	76 0	18½
Fore top	52 0	17¼	52 0	13
Fore topgallant	26 0	9	36 0	7¾
Fore royal	18 0	6	21 0	5
Mizzen	78 0	21	68 0	13
Mizzen top	39 0	12	38 0	9
Mizzen topgallant	20 0	6½	26 0	5
Mizzen royal	12 0	5	16 0	4½
Bowsprit	58 0	30	54 0	11¼
Jib-boom	40 0	12	38 0	8
Driver boom	58 0	11	0 0	0
Crossjack	0 0	0	54 0	11¼
Lower studding boom	41 0	9	28 0	7
Main top boom	40 0	8½	22 0	6
Main topgallant boom	27 0	6	16 0	5
Fore top boom	37 0	8½	22 0	6
Fore topgallant boom	26 0	5½	14 0	5
Ensign staff	36 0	6½	0 0	0
Jackstaff	24 0	5	0 0	0

Table 20: Merchant vessel of 1000 tons

Name of mast/yard	*mast length (ft in)*	*diameter (in)*	*yard length (ft in)*	*diameter (in)*
Main	87 0	29	78 0	19
Main top	52 0	16½	54 0	13
Main topgallant	26 0	9	36 0	8
Main royal	19 0	6½	22 0	5
Fore	81 0	29	72 0	18
Fore top	50 0	16½	52 0	12½
Fore topgallant	24 0	9	34 0	7½
Fore royal	17 0	6	20 0	4½
Mizzen	76 0	20	66 0	12½
Mizzen top	38 0	12	38 0	9
Mizzen topgallant	19 0	6½	26 0	5½
Mizzen royal	10 0	4½	14 0	4
Bowsprit	54 0	29	51 0	11¼
Jib-boom	40 0	12	36 0	8
Driver boom	57 0	11	0 0	0
Crossjack	0 0	0	54 0	11¼
Lower studding boom	40 0	8¾	28 0	7
Main top boom	39 0	8¼	22 0	5¾
Main topgallant boom	27 0	6	15 0	5
Fore top boom	36 0	8½	22 0	5¾
Fore topgallant boom	26 0	5½	14 0	5
Ensign staff	36 0	6½	0 0	0
Jackstaff	22 0	5	0 0	0

Table 21: Merchant vessel of 900 tons

Name of mast/yard	*mast length (ft in)*	*diameter (in)*	*yard length (ft in)*	*diameter (in)*
Main	86 0	28	76 0	18½
Main top	50 0	16	52 0	12½
Main topgallant	26 0	9	34 0	8
Main royal	19 0	5½	21 0	4½
Fore	79 0	28	70 0	17
Fore top	48 0	16	50 0	12
Fore topgallant	23 0	9	32 0	7½
Fore royal	17 0	5½	19 0	4½
Mizzen	76 0	20	64 0	12
Mizzen top	37 0	12	35 0	8½
Mizzen topgallant	19 0	6	23 0	5
Bowsprit	54 0	28	50 0	10½
Jib-boom	40 0	12	36 0	8
Driver boom	50 0	10½	0 0	0
Crossjack	0 0	0	52 0	10½
Lower studding boom	37 0	8½	25 0	6½
Main top boom	35 0	8	20 0	5½
Main topgallant boom	25 0	6	14 0	5
Fore top boom	34 0	8½	19 0	6
Fore topgallant boom	25 0	5½	14 0	5
Ensign staff	34 0	6½	0 0	0
Jackstaff	22 0	4½	0 0	0

Table 22: Merchant vessel of 800 tons

Name of mast/yard	*mast length (ft in)*	*diameter (in)*	*yard length (ft in)*	*diameter (in)*
Main	84 0	27	74 0	18
Main top	49 0	16	50 0	12
Main topgallant	25 9	9	33 6	8
Main royal	18 0	5	20 0	4½
Fore	77 0	27	68 0	16
Fore top	47 0	16	48 0	12
Fore topgallant	22 0	8½	30 0	7½
Fore royal	16 0	4½	18 0	4½
Mizzen	73 0	19	62 0	10½
Mizzen top	36 0	12	34 0	8½
Mizzen topgallant	19 0	6	23 0	5
Bowsprit	53 0	27	50 0	10½
Jib-boom	38 0	12	34 0	7½
Driver boom	49 0	10½	0 0	0
Crossjack	0 0	0	50 0	10½
Lower studding boom	36 0	8	23 0	6
Main top boom	34 0	8	18 0	5½
Main topgallant boom	25 0	5½	14 0	4½
Fore top boom	33 0	8	17 0	5½
Fore topgallant boom	24 0	5½	14 0	4½
Ensign staff	34 0	6½	0 0	0
Jackstaff	22 0	4½	0 0	0

Table 23: Merchant vessel of 700 tons

Name of mast/yard	*mast length (ft in)*	*diameter (in)*	*yard length (ft in)*	*diameter (in)*
Main	82 0	26	70 0	16½
Main top	48 0	15½	50 0	12
Main topgallant	24 0	8½	32 0	7½
Fore	75 0	26	64 0	15
Fore top	46 6	15½	48 0	12
Fore topgallant	22 0	8½	30 0	7½
Mizzen	70 0	18	62 0	10½
Mizzen top	34 0	11	34 0	8½
Mizzen topgallant	18 0	5½	23 0	5
Bowsprit	50 0	25	48 0	9½
Jib-boom	36 0	10½	32 0	7
Driver boom	50 0	10½	0 0	0
Crossjack	0 0	0	49 0	10
Lower studding boom	35 0	8	22 0	5½
Main top boom	33 0	7½	17 0	5
Main topgallant boom	24 0	5½	13 0	4
Fore top boom	31 0	7½	16 0	5
Fore topgallant boom	23 0	5½	13 0	4
Ensign staff	32 0	6½	0 0	0
Jackstaff	21 6	4½	0 0	0

Table 24: Merchant vessel of 600 tons

Name of mast/yard	*mast length (ft in)*	*diameter (in)*	*yard length (ft in)*	*diameter (in)*
Main	78 0	24	62 0	15½
Main top	48 0	15	48 0	12
Main topgallant	23 0	8	32 0	7½
Fore	73 0	23½	58 0	14½
Fore top	46 0	15	46 0	11½
Fore topgallant	22 0	8	30 0	7½
Mizzen	68 0	16	60 0	10½
Mizzen top	30 0	10½	32 0	8
Mizzen topgallant	18 0	5½	22 0	4½
Bowsprit	50 0	24½	48 0	9½
Jib-boom	36 0	10½	32 0	7
Driver boom	50 0	10½	0 0	0
Crossjack	0 0	0	48 0	9½
Lower studding boom	34 0	8	20 0	5
Main top boom	32 0	7½	16 0	5
Main topgallant boom	24 0	5½	12 0	4
Fore top boom	30 0	7½	15 0	5
Fore topgallant boom	22 0	5½	12 0	4
Ensign staff	30 0	6½	0 0	0
Jackstaff	21 0	4½	0 0	0

Table 25: Merchant vessel of 500 tons

Name of mast/yard	*mast length (ft in)*	*diameter (in)*	*yard length (ft in)*	*diameter (in)*
Main	72 0	22	52 0	13
Main top	43 0	14	41 0	10½
Main topgallant	32 0	8	31 0	7
Main royal	15 0	5	20 0	5
Fore	67 0	22	52 0	12½
Fore top	41 0	14	41 0	10½
Fore topgallant	22 0	8	28 0	6½
Fore royal	14 0	5	19 0	5
Mizzen (gaff)	63 0	15	32 0	8
Mizzen top	30 0	10	32 0	8
Mizzen topgallant	17 0	5½	21 0	5
Bowsprit	48 0	23	39 0	8½
Jib-boom	35 0	10	0 0	0
Driver boom	46 0	9	0 0	0
Crossjack	0 0	0	44 0	9
Lower studding boom	30 0	7½	19 0	5
Main top boom	28 0	7	15 0	4½
Main topgallant boom	22 0	5	12 0	4
Fore top boom	28 0	7	15 0	4½
Fore topgallant boom	20 0	5	12 0	4
Ensign staff	30 0	6	0 0	0
Jackstaff	20 0	5	0 0	0

Table 26: Merchant vessel of 400 tons

Name of mast/yard	*mast* *length* *(ft in)*	*diameter* *(in)*	*yard* *length* *(ft in)*	*diameter* *(in)*
Main	70 0	21½	50 0	13
Main top	41 0	13	39 0	10
Main topgallant	23 0	7½	29 0	6½
Main royal	15 0	5	20 0	5
Fore	65 0	21	49 0	12½
Fore top	40 0	13	39 0	10
Fore topgallant	21 0	7½	27 0	6
Fore royal	14 0	5	19 0	4½
Mizzen (gaff)	62 0	15	30 0	7½
Mizzen top	30 0	9½	30 0	7½
Mizzen topgallant	16 0	5	20 0	5
Bowsprit	47 0	21	36 0	7½
Jib-boom	34 0	10	0 0	0
Driver boom	40 0	8½	0 0	0
Crossjack	0 0	0	39 0	8
Lower studding boom	28 0	7	18 0	5
Main top boom	26 0	6½	15 0	4½
Main topgallant boom	20 0	5	12 0	4
Fore top boom	25 0	6½	15 0	4½
Fore topgallant boom	19 0	5	12 0	4
Ensign staff	28 0	6	0 0	0
Jackstaff	20 0	5	0 0	0

Table 27: Merchant vessel of 350 to 360 tons

Name of mast/yard	*mast* *length* *(ft in)*	*diameter* *(in)*	*yard* *length* *(ft in)*	*diameter* *(in)*
Main	68 0	20½	48 0	12
Main top	39 0	12½	36 0	9
Main topgallant	20 0	7	28 0	6½
Main royal	14 0	5	18 0	5
Fore	63 0	20½	47 0	12
Fore top	38 0	12½	47 0	12
Fore topgallant	19 0	7	27 0	6½
Fore royal	13 0	5	17 0	4½
Mizzen (gaff)	60 0	14	27 7	7
Mizzen top	30 0	9½	30 0	7½
Mizzen topgallant	16 0	5	18 0	5
Bowsprit	44 0	20½	34 0	7½
Jib-boom	34 0	9½	0 0	0
Driver boom	38 0	8	22 0	5½
Crossjack	0 0	0	36 0	7½
Lower studding boom	26 0	6½	16 0	4½
Main top boom	25 0	6	14 0	4
Main topgallant boom	19 0	4½	10 0	3½
Fore top boom	24 0	6	14 0	4
Fore topgallant boom	18 0	4½	10 0	3½
Ensign staff	28 0	6	0 0	0
Jackstaff	14 0	4	0 0	0

Table 28: Merchant vessel of 300 to 330 tons

Name of mast/yard	*mast* *length* *(ft in)*	*diameter* *(in)*	*yard* *length* *(ft in)*	*diameter* *(in)*
Main	66 0	19	47 0	12
Main top	38 0	12	35 0	9
Main topgallant	19 0	6½	26 0	6
Main royal	13 0	4½	16 0	4
Fore	61 0	19	45 0	11½
Fore top	37 0	12	32 0	8½
Fore topgallant	18 0	6½	25 0	5¾
Fore royal	12 0	4½	15 0	4
Mizzen (gaff)	58 0	13	26 0	6½
Mizzen top	28 0	9	28 0	7
Mizzen topgallant	15 0	4½	16 0	4
Bowsprit	42 0	19	30 0	7
Jib boom	33 0	9	0 0	0
Driver boom	38 0	7½	20 0	5
Crossjack	0 0	0	35 0	7
Lower studding boom	25 0	6	15 0	4½
Main top boom	24 0	5¾	12 0	4
Main topgallant boom	18 0	4½	10 0	3½
Fore top boom	23 0	5¾	12 0	4
Fore topgallant boom	19 0	4¼	10 0	3½
Ensign staff	26 0	5½	0 0	0
Jackstaff	13 0	4	0 0	0

Table 29: Merchant vessel of 250 to 280 tons

Name of mast/yard	*mast* *length* *(ft in)*	*diameter* *(in)*	*yard* *length* *(ft in)*	*diameter* *(in)*
Main	61 0	18	44 0	11
Main top	36 0	11	32 0	8
Main topgallant	18 0	6	23 0	5½
Main royal	12 0	4½	15 0	4
Fore	58 0	18	42 0	10½
Fore top	35 0	11	32 0	8
Fore topgallant	17 0	6	23 0	5½
Fore royal	11 6	4½	15 0	4
Mizzen (gaff)	54 0	12	25 0	6
Mizzen top	26 0	8	27 0	6
Mizzen topgallant	14 0	5	0 0	0
Bowsprit	38 0	18	32 0	7
Jib-boom	32 0	8	0 0	0
Driver boom	36 0	7½	18 0	5
Crossjack	0 0	0	32 0	7
Main top boom	23 0	5	12 0	4
Fore top boom	22 0	5	12 0	4
Ensign staff	25 0	5½	0 0	0
Jackstaff	10 0	3	0 0	0

Table 30: Merchant brig of 150 tons

Name of mast/yard	*mast* *length* *(ft in)*	*diameter* *(in)*	*yard* *length* *(ft in)*	*diameter* *(in)*
Main	60 0	16½	38 0	9½
Main top	30 0	10	30 0	7½
Main topgallant	16 6	5½	21 0	4½
Fore	54 0	16½	36 0	9
Fore top	32 0	10	30 0	7½
Fore topgallant	16 0	5½	21 0	4½
Bowsprit	32 0	16	30 6	6
Jib-boom	26 7	7	0 0	0
Main boom	34 0	8½	0 0	0
Main gaff	0 0	0	18 6	6
Main top boom	20 0	4½	10 0	3½
Fore top boom	19 0	4½	10 0	3½

Table 31: Merchant ketch of 170 tons

Name of mast/yard	*mast* *length* *(ft in)*	*diameter* *(in)*	*yard* *length* *(ft in)*	*diameter* *(in)*
Main	72 0	19	51 0	12
Main top	36 0	10½	36 0	7½
Main topgallant	20 0	4½	24 0	4
Mizzen (gaff)	49 0	10	18 0	5½
Mizzen top	30 0	5½	21 0	4
Bowsprit	44 0	14½	31 0	6
Jib-boom	28 0	7¾	0 0	0
Crossjack	0 0	0	29 0	5½
Trysail gaff or wing sail sprit	0 0	0	16 0	5
Ensign staff	28 0	6	0 0	0
Jackstaff	8 0	3	0 0	0

Table 32: Merchant schooner of 110 tons

Name of mast/yard	*mast length (ft in)*	*diameter (in)*	*yard length (ft in)*	*diameter (in)*
Main	75 0	15	0 0	0
Main topgallant	32 0	7½	27 0	5½
Main topsail	0 0	0	38 0	6
Main spread	0 0	0	47 0	7
Main boom	47 0	10½	0 0	0
Main gaff	0 0	0	28 0	7
Square sail	0 0	0	38 0	7½
Fore	66 0	15	0 0	0
Fore topgallant	26 0	6½	25 0	5
Fore topsail	0 0	0	30 0	6½
Fore spread	0 0	0	40 0	7
Fore gaff	0 0	0	22 0	6½
Bowsprit	50 0	14½	0 0	0

Table 33: Merchant sloop of 70 tons

Name of mast/yard	*mast length (ft in)*	*diameter (in)*	*yard length (ft in)*	*diameter (in)*
Main	64 0	16	0 0	0
Topgallant	30 0	6½	0 0	0
Square sail or crossjack	0 0	0	40 0	8
Topsail	0 0	0	24 0	6
Bowsprit	40 0	14	0 0	0
Jib-boom	32 0	6½	0 0	0
Main boom	48 0	10	0 0	0
Gaff	0 0	0	28 0	7

Tables 34 and 35, based on information published by Davis, show the dimensions of the masts and yards of a British East Indiaman of 1711 with a keel length of 84 feet and a beam of 28 feet.

Table 34: East Indiaman with short masts and long topmasts

Name of mast/yard	*mast length (ft)*	*diameter (in)*	*yard length (ft)*
Main	71	20	58
Main top	4	17½	36
	[probably 44]		
Main topgallant	18	5	21
Fore	63	18	51
Fore top	40	11½	32
Fore topgallant	16	4½	18⅔
Mizzen	62	13	54
Mizzen top	24½	6	21
Bowsprit	47	19½	38
Spritsail top	14	5½	21
Crossjack	0	0	36

Table 35: East Indiaman with long masts and short topmasts

Name of mast/yard	*mast length (ft)*	*diameter (in)*	*yard length (ft)*
Main	74	21½	55
Main top	40	10½	35½
Main topgallant	17½	5	20½
Fore	65	19	49½
Fore top	36⅔	9¼	32
Fore topgallant	16	4¼	18½
Mizzen	65	14	49½
Mizzen top	22½	5⅔	23
Bowsprit	47	18	38
Spritsail top	14	4½	21
Crossjack	0	0	36

Table 36 shows the dimensions of the masts and yards of the American schooner *Sultana* of 1760.

Table 36: American schooner *Sultana*

Name of mast/yard	*mast length (ft in)*	*diameter (in)*	*yard length (ft in)*	*diameter (in)*
Main	54 7	13¾	0 0	0
Mainmast head	6 0	6 above	0 0	0
		8½ below		
Main boom	0 0	0	37 3	7¾
Main gaff	0 0	0	17 5½	4¾
Fore	53 4	13⅞	0 0	0
Foremast head	5 7	5½ above	0 0	0
		8 below		
Fore gaff	0 0	0	15 0	4½
Squaresail yard	0 0	0	23 3	5
Bowsprit	27 8	11¾	0 0	0
Jib-boom	21 2	4¾		

Tables 37 to 40 give dimensions of the masts and yards of French men-of-war in the eighteenth century, based on figures published by Paris in his *Souvenirs de Marine* of 1884.

Table 37: *Royal Louis* (1740), 128 guns

Name of mast/yard	*mast length (m)*	*diameter (m)*	*head (m)*	*yard length (m)*	*sail length (m)*
Main	38.50	1.10	5.20	38.00	35.20
Main top	23.00	0.63	2.60	27.30	22.00
Main topgallant	9.80	long pole	6.00	18.50	16.80
Fore	35.70	1.00	4.70	32.20	29.20
Fore top	21.80	0.58	2.30	23.00	19.00
Fore topgallant	8.90	long pole	5.50	17.00	15.20
Mizzen	26.70		3.15	35.70	17.00
Mizzen top	16.80		1.58	17.20	13.70
Mizzen topgallant	7.55	long pole	3.45	11.70	10.20
Bowsprit	22.69	1.05			
Jib-boom	14.70	0.44			
Crossjack	0	0	0	0	22.30

Tables 38 to 40 show the dimensions of the masts, yards and trestle trees of a French 80-gun man-of-war of 1750, with a beam of 14.94 metres. Table 41 lists the factors by which these dimensions should be multiplied to obtain the correct dimensions for the masts and yards of French men-of-war from 128 to 26 guns, sloops of 16 and 12 guns and 750-ton flutes.

Table 38: Mast dimensions of a French 80-gun ship (1750), beam 14.94m

Name of mast	*mast length (m)*	*diameter (m)*	*masthead length (m)*	*diameter (m)*
Main	35.356	1.046	3.735	0.676
Main top	22.420	0.523	1.950	0.298
Main topgallant	10.400	0.253	0.974	0.148
Main ensign staff	11.400	0.135	0	0
Fore	33.760	0.996	3.410	0.649
Fore top	20.790	0.473	1.780	0.262
Fore topgallant	9.100	0.225	0.974	0.126
Fore ensign staff	9.740	0.128	0	0
Mizzen	28.910	0.721	2.330	0.338
Mizzen top	14.940	0.271	1.624	0.271
Mizzen ensign staff	9.100	0.121	0	0
Bowsprit	22.420	1.019 maximum		
		0.500 minimum		
Jib-boom	16.240	0.198	0	0
Spritsail top	9.740	0.126	0	0
Ensign staff	14.940	0.225	0	0
Jackstaff	7.470	0.148	0	0

Table 39: Yard dimensions of a French 80-gun ship (1750), beam 14.94m

Name of yard	*yard*		*yardarm*	
	length	*diameter*	*length*	*diameter*
	(m)	*(m)*	*(m)*	*(m)*
Main	32.809	0.671	2.920	0.223
Main topsail	22.420	0.438	3.410	0.162
Main topgallant	9.750	0.148	1.140	0.184
Fore	29.880	0.622	2.600	0.198
Fore topsail	20.790	0.374	3.085	0.148
Fore topgallant	9.100	0.175	1.140	0.175
Mizzen	29.880	0.387	lower 0.249 peak 0.126	
Mizzen topsail	12.990	0.198		
Spritsail	22.420	0.438	2.920	0.162
Sprit topsail	9.096	0.175	0.970	0.175
Crossjack	22.420	0.438	2.600	0.162
Main studding boom	16.240	0.225	0	0
Fore studding boom	14.620	0.198	0	0
Main swinging boom	16.240	0 253	0	0
Fore boat boom	14.620	0.225	0	0

Table 40: Trestle tree dimensions of a French 80-gun ship (1750), beam 14.94m

Name of trestle tree			
	length	*height*	*thickness*
	(m)	*(m)*	*(m)*
Main	6.170	0.496	0.253
Main top	2.680	0.223	0.110
Fore	5.850	0.446	0.223
Fore top	2.440	0.186	0.094
Mizzen	1.085	0.244	0.121
Mizzen top	1.380	0.112	0.083

Table 41: Multiplication factors for mast, yard and trestle tree dimensions of French men-of-war

Vessel	*factor*
128-gun ship	1.109
110-gun ship	1.087
80-gun ship	1.006
74-gun ship (1760)	0.935
74-gun ship (1782)	0.966
64-gun ship	0.893
30-gun ship	0.754
26-gun ship	0.703
16-gun sloop of war	0.448
12-gun sloop of war	0.412
Flute of 750 tons	0.665

Table 42: French schooners of 15.7m (and 21m) in length

Name of mast/yard	*mast*		*yard*	
	length	*diameter*	*length*	*diameter*
	(m)	*(m)*	*(m)*	*(m)*
Main	15.180 (16.710)	0.215 (0.350)	0	0
Mainmast head	1.220 (3.090)		0	0
Main top	(8.950)	(0.170)	0	0
Main topmast head	(2.240)		0	0
Fore	13.700 (16.260)	0.200 (0.360)	0	0
Foremast head	1.850 (2.020)		0	0
Fore top	(8.910)	(0.160)	0	0
Fore topmast head	(2.230)		0	0
Boom (bowsprit)	5.100 (8.380)	(0.310)	0	0
Jib-boom	(5.580)	(0.110)	0	0
Main boom	0	0	9.630 (13.260)	(0.230)
Main gaff	0	0	3.700 (7.140)	(0.130)
Fore boom	0	0	5.180	0
Fore gaff	2.960 (6.090)		(0.120)	
Crossjack	0	0	(9.740)	(0.180)
Topsail	0	0	(7.390)	(0.150)

Tables 43 to 52 show the dimensions of masts and yards on Continental men-of-war at the end of the eighteenth century, based on figures published by Korth in 1826.

Table 43: Ship of 100 guns

Name of mast/yard	*mast*		*yard*	
	length	*diameter*	*length*	*diameter*
	(ft)	*(in)*	*(ft)*	*(in)*
Main	123	38	126	28
Main top	74	20	88	15
Main topgallant	30	11	48	9½
Fore	110	34	102	21
Fore top	66	19	68	13
Fore topgallant	28	10	38	7½
Mizzen	92	24	102	22
Mizzen top	40	13	54	9½
Bowsprit	74	36	0	0
Spritsail yard	0	0	68	13
Sprit topsail yard	0	0	48	9½
Crossjack yard	0	0	68	14½

Table 44: Ship of 90 guns

Name of mast/yard	*mast*		*yard*	
	length	*diameter*	*length*	*diameter*
	(ft)	*(in)*	*(ft)*	*(in)*
Main	110	36	112	25
Main top	72	19	74	14
Main topgallant	26	10½	43	8½
Fore	101	32	98	20½
Fore top	63	18½	65	12
Fore topgallant	22	9½	38	7
Mizzen	85	22	96	20
Mizzen top	36	12½	49	9
Bowsprit	70	35	0	0
Spritsail yard	0	0	65	12½
Sprit topsail yard	0	0	43	8½
Crossjack yard	0	0	65	14

Table 45: Ship of 80 guns

Name of mast/yard	*mast*		*yard*	
	length	*diameter*	*length*	*diameter*
	(ft)	*(in)*	*(ft)*	*(in)*
Main	100	34	108	23
Main top	68	18	72	13
Main topgallant	25	10	41	8
Fore	94	31	94	20
Fore top	60	18	63	12
Fore topgallant	22	9	36	6
Mizzen	80	21	92	19
Mizzen top	35	12	47	8
Bowsprit	64	34	0	0
Spritsail yard	0	0	63	12
Sprit topsail yard	0	0	41	8
Crossjack yard	0	0	62	13½

Table 46: Ship of 70 guns

Name of mast/yard	*mast* *length* *(ft)*	*diameter* *(in)*	*yard* *length* *(ft)*	*diameter* *(in)*
Main	96	32	104	22
Main top	65	17	70	13
Main topgallant	24	9½	39	7
Fore	88	30	91	19½
Fore top	58	17	61	11
Fore topgallant	21	8½	34	6
Mizzen	76	19	88	19
Mizzen top	34	11½	40	8
Bowsprit	60	29	0	0
Spritsail yard	0	0	61	11
Sprit topsail yard	0	0	39	7½
Crossjack yard	0	0	60	13

Table 47: Ship of 60 guns

Name of mast/yard	*mast* *length* *(ft)*	*diameter* *(in)*	*yard* *length* *(ft)*	*diameter* *(in)*
Main	92	30	100	21
Main top	60	16	67	12½
Main topgallant	23	9	37	7
Fore	85	28	88	18½
Fore top	52	16	59	10½
Fore topgallant	20	8	32	6
Mizzen	70	18½	86	18
Mizzen top	32	11	44	7½
Bowsprit	54	28	0	0
Spritsail yard	0	0	59 10½	
Sprit topsail yard	0	0	37	7
Crossjack yard	0	0	59	12

Table 48: Ship of 50 guns

Name of mast/yard	*mast* *length* *(ft)*	*diameter* *(in)*	*yard* *length* *(ft)*	*diameter* *(in)*
Main	90	27	90	20
Main top	58	15	61	12
Main topgallant	22	8½	35	6
Fore	83	24	79	18
Fore top	50	15	54	10
Fore topgallant	19	7½	30	5½
Mizzen	68	18	78	17
Mizzen top	30	10	40	7
Bowsprit	51	26	0	0
Spritsail yard	0	0	54	10
Sprit topsail yard	0	0	35	6⅓
Crossjack yard	0	0	54	11½

Table 49: Ship of 44 guns

Name of mast/yard	*mast* *length* *(ft)*	*diameter* *(in)*	*yard* *length* *(ft)*	*diameter* *(in)*
Main	86	26	85	19
Main top	55	14	54	11
Main topgallant	21	8	33	6
Fore	80	23	75	17
Fore top	48	14	47	9
Fore topgallant	18	7	29	5
Mizzen	64	16½	73	16
Mizzen top	26	9½	38	6
Bowsprit	38	24	0	0
Spritsail yard	0	0	47	9½
Sprit topsail yard	0	0	33	6
Crossjack yard	0	0	47	11

Table 50: Ship of 36 guns

Name of mast/yard	*mast* *length* *(ft)*	*diameter* *(in)*	*yard* *length* *(ft)*	*diameter* *(in)*
Main	80	24	79	18
Main top	50	13	52	11
Main topgallant	20	7½	30	5½
Fore	74	21	71	16
Fore top	44	13	45	9
Fore topgallant	17	6½	26	4½
Mizzen	60	15½	70	15
Mizzen top	23	9	34	6
Bowsprit	45	22½	0	0
Spritsail yard	0	0	45	9
Sprit topsail yard	0	0	30	5½
Crossjack yard	0	0	45	10

Table 51: Ship of 28 guns

Name of mast/yard	*mast* *length* *(ft)*	*diameter* *(in)*	*yard* *length* *(ft)*	*diameter* *(in)*
Main	78	22	72	16½
Main top	49	12	48	10½
Main topgallant	20	7	28	5½
Fore	69	20	63	15
Fore top	40	12½	42	8½
Fore topgallant	16	6½	24	4½
Mizzen	58	15	62	14
Mizzen top	22	8½	32	6
Bowsprit	43	22	0	0
Spritsail yard	0	0	42	8½
Sprit topsail yard	0	0	28	5½
Crossjack yard	0	0	42	10

Table 52: Ship of 20 guns

Name of mast/yard	*mast* *length* *(ft)*	*diameter* *(in)*	*yard* *length* *(ft)*	*diameter* *(in)*
Main	76	20	70	15
Main top	46	11	45	9
Main topgallant	18	6	26	5
Fore	66	18	60	12
Fore top	36	11	40	7
Fore topgallant	15	6	22	4
Mizzen	54	13	56	12
Mizzen top	21	7	30	5
Bowsprit	40	20	0	0
Spritsail yard	0	0	40	7
Sprit topsail yard	0	0	26	5
Crossjack yard	0	0	40	9

II

Rigging of spars (ship rig)

The word 'rigging' encompasses all the cordage on a ship, including that used to secure the masts as well as that used in the operation of yards and sails. These two types are differentiated as standing rigging and running rigging respectively.

Standing rigging is all the cordage that is permanently set with both ends fixed. In contrast, running rigging has only one end (the standing part) made fast. The loose end (known as the hauling part) usually runs over the sheaves of one or more blocks to form a tackle, or through trucks, thimbles or deadeyes, before being belayed.

In the normal process of rigging a ship, according to contemporary literature, the bowsprit was dressed first, followed by the lower masts and topmasts. After this, the lower and topsail yards were hoisted up and rigged. Only after work had been completed on these parts were the topgallant masts stepped and rigged with stays and shrouds. The topgallant and royal yards came last and completed the rigging. This traditional sequence will be adhered to in the discussion of rigging which follows.

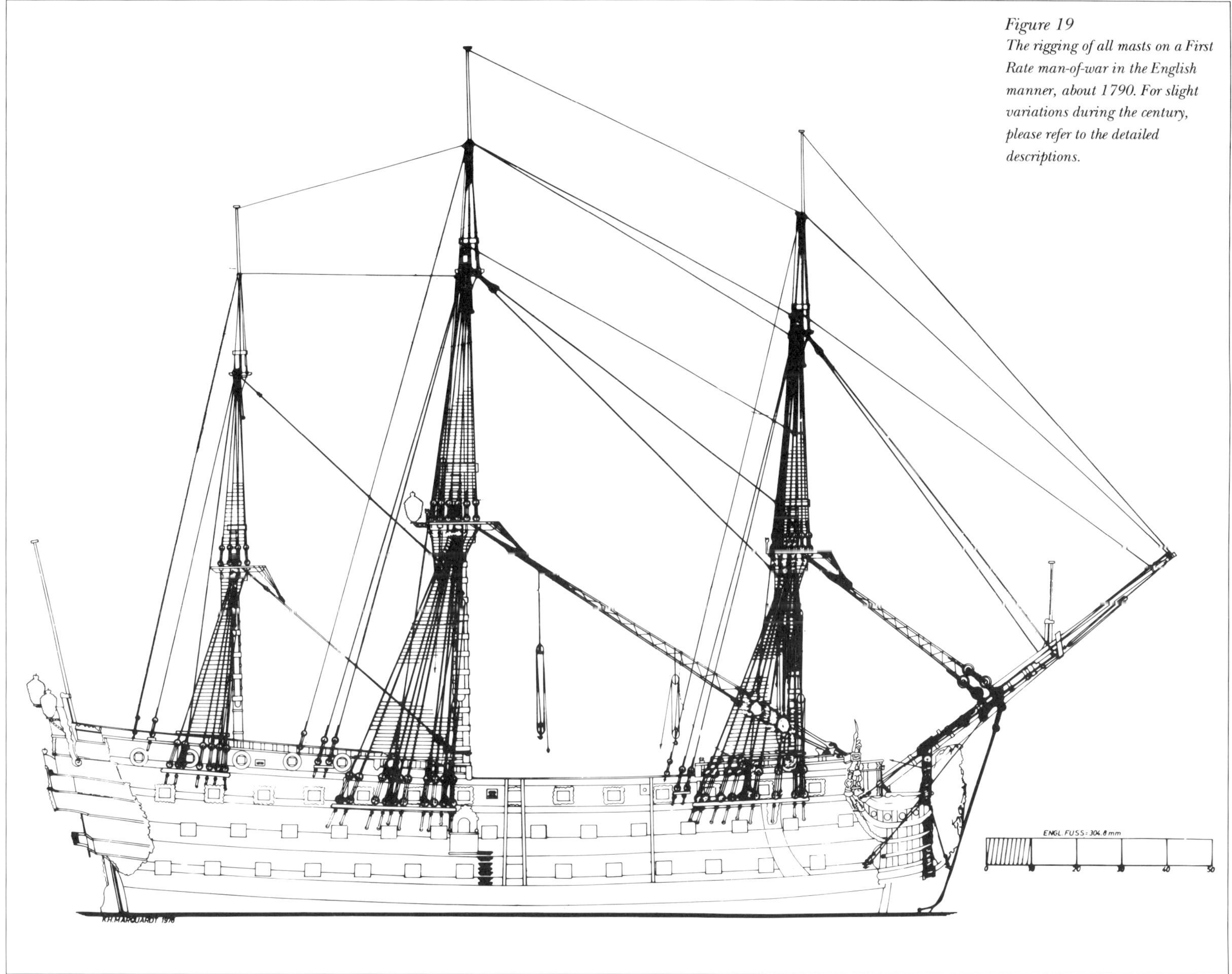

Figure 19
The rigging of all masts on a First Rate man-of-war in the English manner, about 1790. For slight variations during the century, please refer to the detailed descriptions.

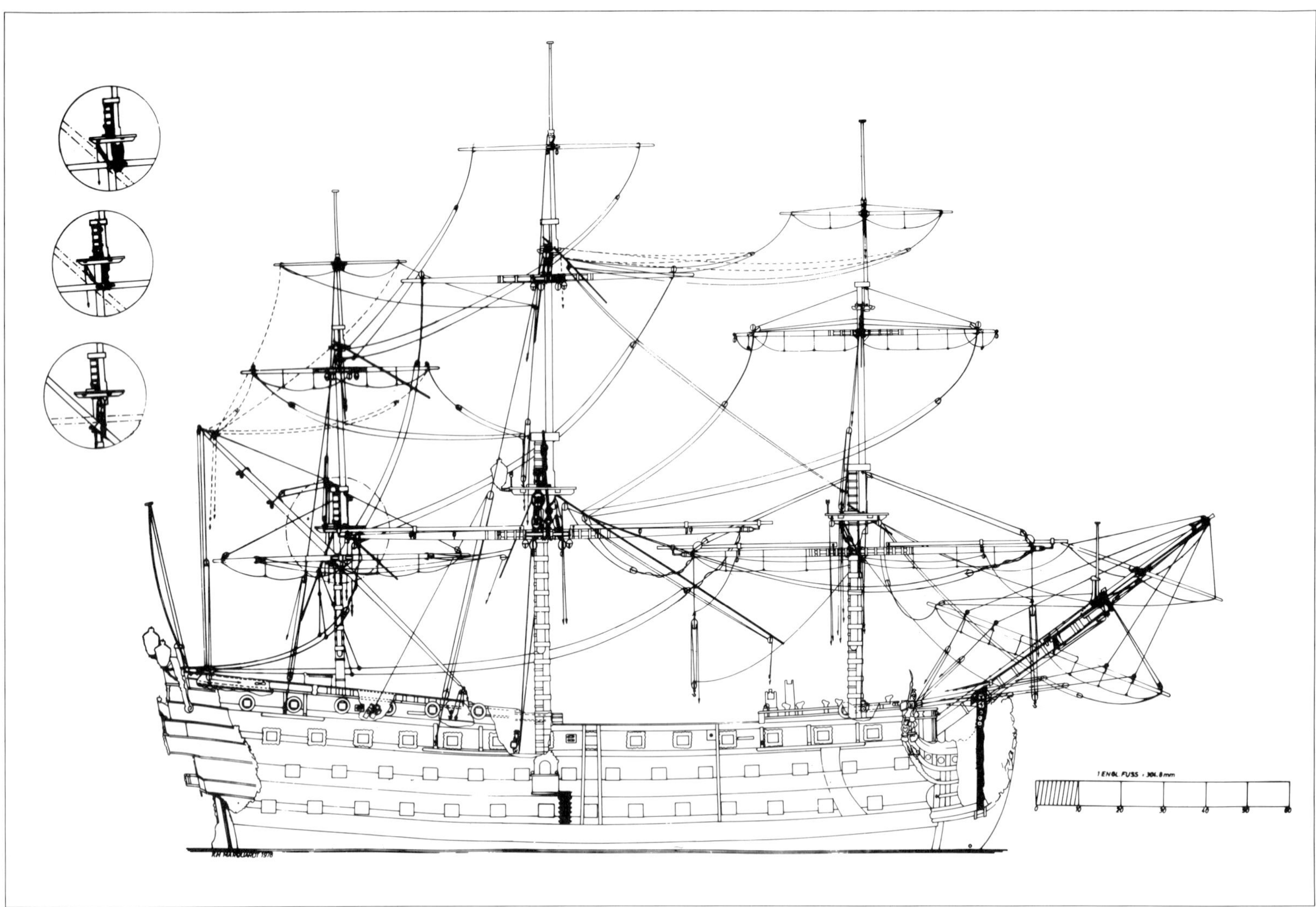

Figure 20
The standing and running rigging of a First Rate man-of-war. The rigging is English and dates from the second half of the century. Introduction dates for the various rigging elements can be found in the text. The dotted lines of the fore topgallant brace show an alternative leading of the braces over sheaves in the main topmast trestle trees after 1775. The dotted lines of the mizzen topsail and topgallant braces indicate their position on a slung mizzen yard or gaff. The details (top to bottom) show a crossjack yard sling before 1775, after 1775, and the mizzen yard jeers.

Bowsprit

Bugspriet; Beaupré

Gammoning

Wuhling; Liure de beaupré

A bowsprit was lashed down to the cutwater, or stem, with single or double gammoning. According to the ship's size, the gammoning was formed with a rope 4½ to 8 inches in thickness, with a length in fathoms 1½ to 2¼ times the bowsprit's length in feet (1½ for single and 2¼ for double gammoning). Davis in 1711 gave the gammoning thickness as one ninth of the bowsprit's diameter.

The gammoning rope was hitched around the bowsprit then led through the gammoning hole in the cutwater's top timber or through a ringbolt in the stem's face. This process was repeated nine to eleven times, with the rope crossing over to the opposite side each time to give the appearance of a twist. Each single turn was hauled tight and nippered. This formed a very strong connection between bowsprit and bow.

When all turns had been passed vertically through the gammoning hole, an equal number of turns were frapped around the vertical turns to tighten them further, and the end of the rope was whipped and seized to an adjacent turn (see Figure 21a-d).

Horses

Fusspferde; Marche-pieds

The bowsprit horses supported crewmen working the bowsprit. The outer end of each was seized to a ringbolt on the upper end of the bowsprit cap, and the other to a similar ringbolt in the knighthead (see Figure 22a-c). On Continental ships the latter fastening point was on the breastwork of the forecastle (see Figure 22d). According to Davis, a bowsprit horse consisted of a single rope running from the sprit topsail standard to the beakhead of the forecastle on each side, which suggests that English practice had originally been to use the inboard fastening used on Continental vessels throughout the century. He also gave the length of each horse as 5½ fathoms, or two thirds of the bowsprit's length.

Bobstay collar

Wasserstagkragen; Collier de sous-barbe

The bobstay collar, a strop with eyes spliced into both ends and a deadeye or heart seized into the bight, was laid round the bowsprit and the eyes were then lashed together above the spritsail sling saddle, with the deadeye hanging downward (see Figure 23a). Until around 1720 British men-of-war carried only one bobstay collar; thereafter two were fitted, and in the early nineteenth century three. Röding reported that larger Continental men-of-war had up to three bobstays as early as 1793, and this is confirmed by evidence from Holland and elsewhere. In France, bobstay collars were the same, or (earlier in the century) were fitted with a double sheaved block instead of the

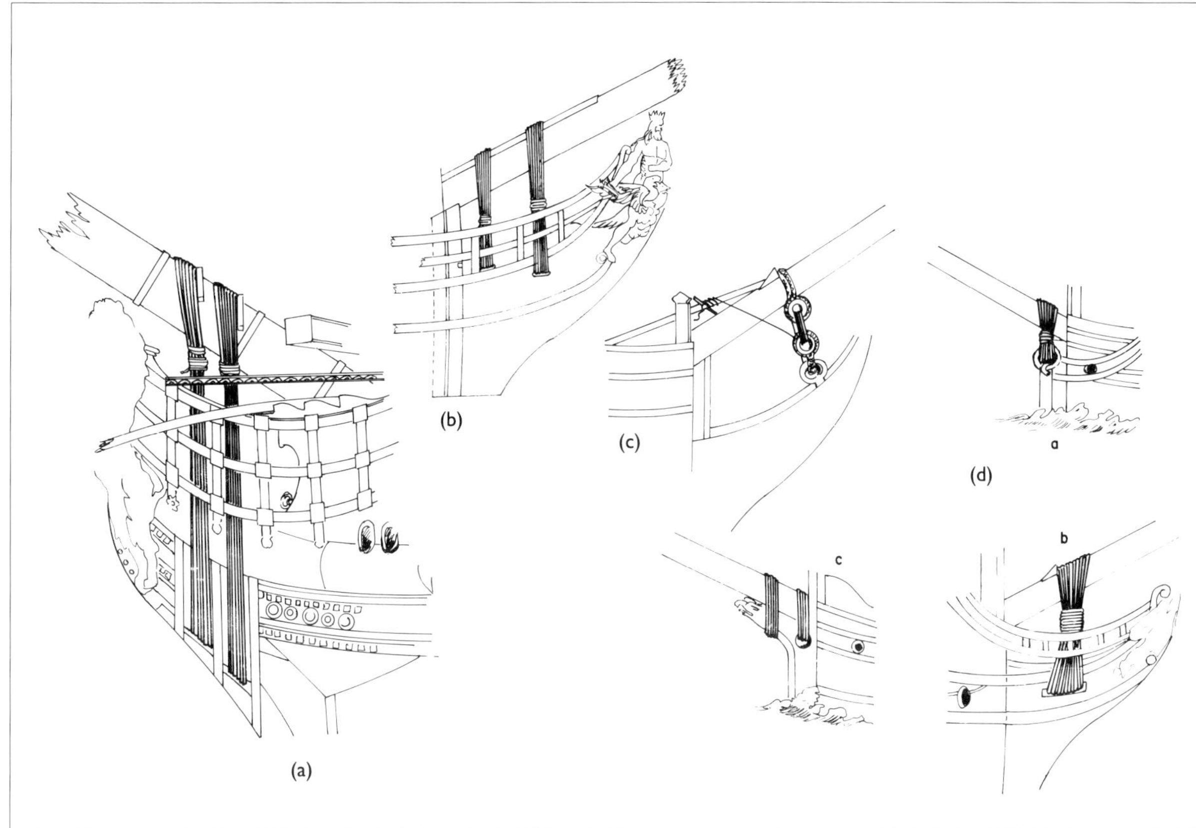

Figure 21
(a) Double gammoning on a Continental ship after 1750.

(b) Double gammoning on an English ship, about 1800, with a gammoning fish instead of gammoning cleats.

(c) Hauling a smaller ship's gammoning tight. In this case hearts are strapped to the bowsprit and head, with a gammoning lanyard run through (from an English illustration of about 1800).

(d) Three different methods of setting up gammoning

a. A large gammoning ring fitted to the stem of a vessel without a head

b. Standard eighteenth century single gammoning

c. Double gammoning on a ship with a short head.

deadeye or heart; vessels with the latter fitting usually had only one bobstay. Merchantmen also had one collar only.

In the early years of the century all bowsprit collars consisted of a closed strop, with the splice on the lower side beneath the deadeye. All collars were usually wormed, parcelled and served, or clad in leather.

Röding provided the following information on the placing of the collars where more than one was fitted:

> The innermost is set up with a lanyard to a deadeye, hanging between the fore stay and the preventer fore stay, the second in a similar manner forward of the preventer fore stay, and the third likewise forward of the fore topmast stay collar.

According to the author of *Der geöffnete See-Hafen* the collar was not placed round the bowsprit as above, but a stropped deadeye was fitted to a hole in the cutwater instead.

Bowsprit shroud collar

Backstagkragen; Collier de hauban de beaupré

A bowsprit shroud collar was made the same way as the bobstay collar, except that two deadeyes or hearts were seized to it, facing outboard on either side when the collar was lashed (see Figure 23b). The shroud collar was placed close to the bobstay collar. Continental vessels were not usually fitted with a bowsprit shroud collar (see Röding's remarks on bowsprit shrouds, below).

Fore stay collar

Vorstagkragen; Collier d'étai de mât de misaine

The fore stay collar was placed above the bobstay collar, with a preventer fore stay collar either afore or abaft it (see Fore preventer stay under Lower masts, below). The collars were laid around the bowsprit and lashed underneath. Until about 1730 a normal heart was seized to a closed single strop collar as for the bobstay collar. From 1730 to 1775 the strop was laid double and lashed underneath the bowsprit, and in about 1775 horseshoe-shaped hearts were introduced, with a groove on the outside and two recesses on the inside to allow them to be seized to the collar (see Figure 23c).

Continental ships retained the normal heart even after the introduction of the horseshoe-shaped version in England, and Dutch and French ships were frequently fitted with double or treble sheave blocks instead of hearts.

The fore stay collar, according to Davis, should be as many feet in length as the bowsprit was inches in diameter. The collar rope's circumference was one third of the bowsprit's diameter. The main stay collar on smaller ships was as long in fathoms as one quarter of the bowsprit's diameter in inches. Its circumference was two fifths of the bowsprit's diameter, with half an inch added for smaller craft.

Bobstay

Wasserstag; Sous-barbe

The bobstay counteracted the upward force exerted by the fore stays upon the bowsprit. Its circumference was 9 inches on large ships. The bobstay was led through a hole in the cutwater and its ends were then spliced together to create a large ring, in effect making the stay double. A deadeye or heart (depending on the type of collar and decade in the century) was seized into the upper bight, thus forming a unit with the lanyard from the collar.

On Continental ships a block was often used in place of the heart or deadeye. In French rigging this block was short-stropped to the stem, with the bobstay then running like a tackle through the stem and collar blocks; in this arrangement, the bobstay was set up to a tackle near the stem, on the starboard side beside the bowsprit (see Figure 22i). As noted above, according to the author of *Der geöffnete See-Hafen* the bobstay was placed forward of the fore stay collar, laid with its bight round the bowsprit and seized underneath. The ends of the bobstay were spliced together to form a ring, with a deadeye seized into the lower bight; a lanyard connected the bobstay to the deadeye collar in the cutwater hole (see Figure 24a).

In contrast to illustrations by Lever and Paris, where the

Figure 22
(a) English bowsprit horses up to about 1740. The horse was long stropped to the fore stay.
(b) English bowsprit horses between 1740 and 1765. Instead of a tackle, thimbles were used for setting up. An additional vertical strop between the horse and the bowsprit maintained the distance between the two.
(c) English bowsprit horses after 1765. The horse was then stropped neither to the fore stay nor to the bowsprit.
(d) Continental bowsprit horses. The horses were stropped to the bowsprit a number of times and secured with a tackle to the forecastle beakhead.
(e) A deadeye seized to a closed collar. The closed collar was used until the 1730s, and the deadeye was sometimes replaced with a heart.
(f) A single open collar with a heart, used until approximately 1730. Between 1730 and 1775 a double collar was used.
(g) Heart lashing.
(h) A double bobstay collar with a block seized to it, frequently used on the Continent.
(i) A bobstay as set up on Continental ships, and especially in France.
(j) A bowsprit shroud.
(k) A jib-boom horse.
(l) An outer jib-boom horse, after 1795;a horse for a flying jib-boom was similar.

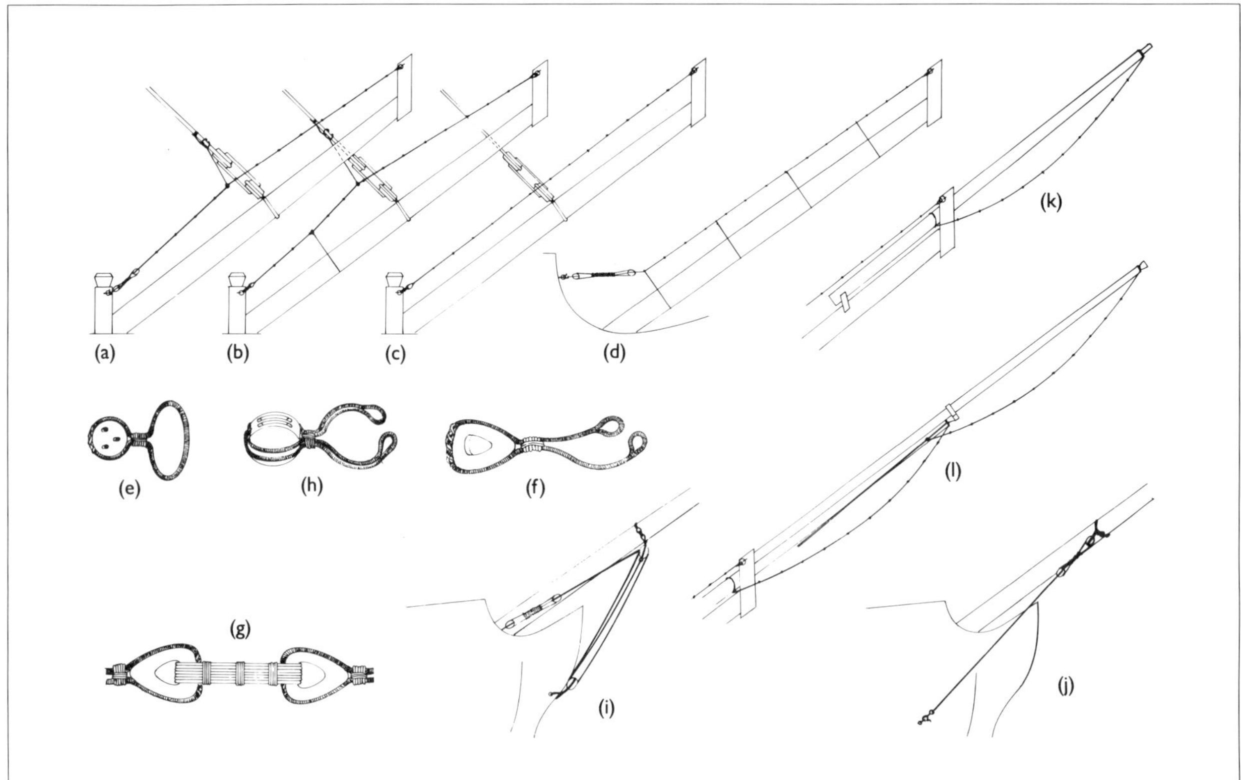

collar block is shown as a normal double block, Röding depicted a fiddle block (see Figure 24b). He noted that large men-of-war were fitted with three bobstays and therefore had three holes in the cutwater, through which these were led. He also described bobstays formed like a winding tackle with a runner and tackle; these were called running bobstays.

It should be noted that Continental rigging sometimes incorporated elements of the English bobstay set-up as well, so the description of Continental bobstays given above should not be taken as prescriptive.

Bowsprit shrouds

Backstage, Bugstage; Haubans de beaupré

Bowsprit shrouds prevented lateral movement of the bowsprit. They were fitted with hooks at the inboard ends, which hooked into rings in the sides of the ship; a deadeye or heart, depending on the type of collar, was spliced into the fore end of each shroud and connected by a lanyard to the collar deadeyes or hearts (see Figure 22j).

Bowsprit shrouds came into general use in about 1706 and can be seen in their earliest form on the St Petersburg model of HMS *Royal Transport* of 1695. Deadeyes were increasingly superseded by hearts after 1770, and in France blocks were generally used instead of either.

A Continental bowsprit shroud was twice the length of an English one. Its bight was laid round the bowsprit head and seized underneath. The ends, of equal length, were fitted with tackles which hooked to the wales, in contrast to the English set-up. Röding remarked that in Continental shipping the bowsprit shrouds had generally disappeared by the later years of the century, because they interfered with the fishing of anchors and when running out cables (see Figure 24c).

Jib-boom

Klüverbaum; Bâton de foc

Traveller

Klüverring, Bügel des Klüvers; Racambeau du grand foc

A traveller was the first item fitted to the jib-boom. It was loose enough to enable the jib stay, connected to it, to move easily. Contemporary literature describes two travellers, differing in their modes of action. While Steel and others referred only to a traveller with a hook, thimble and shackle, Lever noted also a second traveller with a roller in its shackle.

The lower end of the jib stay was spliced round the thimble of the first type of traveller, with the other end of the stay leading, until 1745, over a block beneath the fore topmast cap to starboard (during the later decades of the century the upper sheave of a cheek block took the block's place). This traveller was fitted with an outhauler. Röding noted that travellers of this type were also used on the backstays to bring the topgallant yards down upon deck in bad weather.

The second type of traveller did not require an outhauler. The jib stay was laid round the topmast head, led under the traveller's roller to a sheave in the jib-boom head and then connected to a tackle hooked to the face of the bowsprit cap. The hauling part of the jib stay was belayed at the forecastle. This set-up was introduced at the end of the eighteenth century and prevailed in the early years of the next.

Both types of traveller were fitted with eyelets on either side for the guy pendants (see Figure 23h and i).

Horses or footropes

Fusspferde; Marche-pieds

The outer ends of the jib-boom horses were finished with an eye which fitted over the jib-boom stop, and the inner ends were hitched to the lower part of the jib-boom abaft the bowsprit cap, with the ends stopped. The horses hung about 3 feet below the jib-boom, so that sailors working the jib-boom were supported at the correct working height; diamond or figure-eight knots

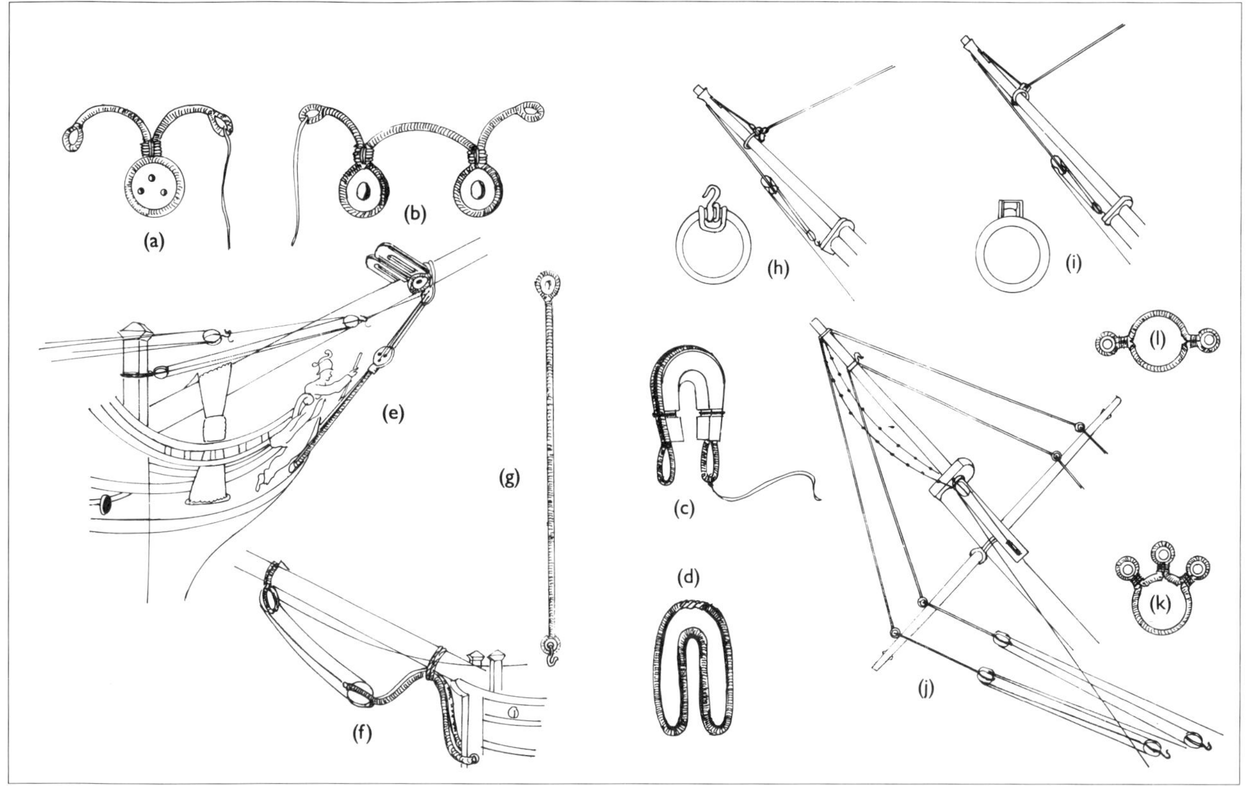

Figure 23
(a) An open bobstay collar with a deadeye.
(b) An open bowsprit shroud collar with hearts, after 1770.
(c) A horseshoe-shaped fore stay heart, English, after 1775.
(d) A double laid collar strop.
(e) An English bobstay being hauled taut.
(f) Hauling out a bobstay on a Continental vessel.
(g) An English bowsprit shroud. The main differences from a Continental shroud were in the use of deadeyes (or hearts after 1770) instead of blocks, and the double length of a Continental shroud.
(h) A traveller with a shackle and hook, and the setting up of a jib stay and outhauler.
(i) A traveller with a reel, about 1800. The jib stay was led through, and no outhauler was needed.
(j) Jib-boom and traveller guys in English rigging after 1775.
(k) A fore topgallant stay collar.
(l) A spritsail topsail yard lift collar.

were worked in every two feet or so to provide a more secure foothold (see Figure 22k and l).

Guy pendants
Geitaue; Cordes retenues

Jib-boom guys were made from one length of rope, with the central bight laid with a hitch around the jib-boom stop. Until approximately 1775 the length of the resulting pendants was half the distance between the spritsail yard and the jib-boom stop (see Figure 25a, left). In later years the pendants became longer and ran through thimbles on top of the spritsail yard. Tackles on the inner ends hooked on to the cathead's fore side, or to rings in the bow (see figure 23j). On the shorter pendants a whip was fitted with an eye over the spritsail yardarm and, after reeving through the pendant block, was led through a thimble on the yard to the forecastle. The shorter pendant was still favoured on the Continent at the end of the century.

Guys, or as they were sometimes called, jib-boom shrouds, first appeared, together with those for the bowsprit, in 1695 on HMS *Royal Transport*, an experimental schooner-rigged yacht of the Royal Navy.

Röding referred to guys as long ropes leading over the yardarms towards the forecastle, to belay there (see Figure 25a, right). When the jib-boom luff needed greater stability, the lee side of the spritsail yard had to be braced slightly more.

Traveller guys
Läufergeitaue; Cordes retenues du racambeau du grand foc

Spliced into the traveller eyelets, the traveller guys ran the same way as the jib-boom guys. Their purpose was to provide extra stability for the traveller itself, since it represented the point at which the jib exerted maximum pressure on the jib-boom (see Figures 23j and 25a).

Topgallant stay collar
Bramstagkragen; Collier d'étai de perroquet

A collar with three thimbles seized into it was fitted over the jib-boom stop to take the fore topgallant stay and bowlines (see Figure 23k).

Jib-boom outhauler
Klüverbaumausholer; Tire-bout

With its standing part secured to a ringbolt at the after side of the bowsprit cap, the jib-boom outhauler rove over a sheave in the jib-boom heel to a block hooked to the opposite side of the cap, then via the fairlead saddle to the forecastle.

Lashing
Laschung; Amarre

A hole in the jib-boom heel enabled the boom to be lashed to the bowsprit (see Figure 25a, b and c).

Flying jib-boom
Aussenklüverbaum, Bâton de clinfoc

Guys
Geie; Étais de bigue

The flying jib-boom came into use in the closing years of the century. Guys and a martingale stay were used to secure and stabilise the boom. The jib-boom guys were fitted over the notched flying jib-boom head, either separately with eyes, or as a single piece of rope with a cut splice. On larger ships the guys were set up to single block tackles and on smaller vessels to thimbles, leading to the catheads or to the bow near the cat-heads.

Horses or footropes
Fusspferde; Marche-pieds

A horse was also fitted to the flying jib-boom. Hanging approximately 3 feet below the boom, it was fitted with an eye over the

Figure 24
(a) Continental bobstay, 1705.
(b) Continental bobstay, 1794.
(c) Continental bowsprit shrouds.
(d) Continental spritsail yard horses.
(e) English spritsail yard horses.
(f) Continental spritsail yard braces, 1705.
(g) Spritsail yard braces as used in English ships from 1711 and French ships in the second half of the century.

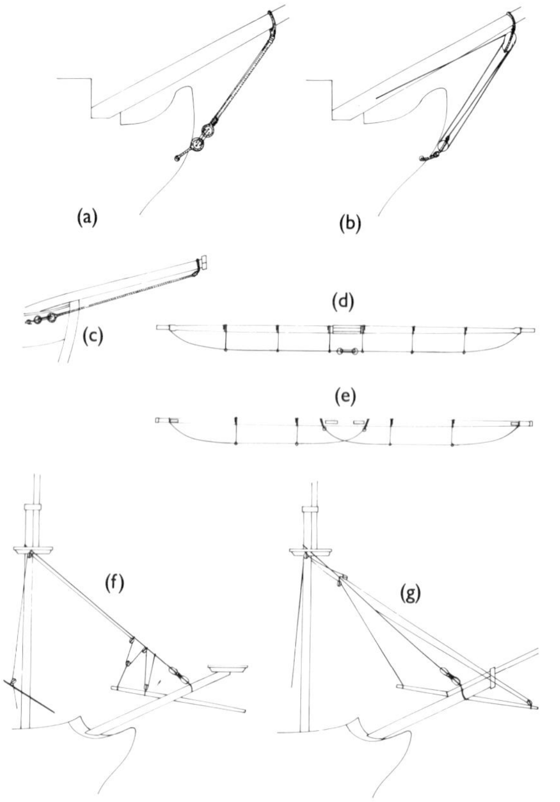

flying jib-boom stop, with the inner end fastened near the jib-boom head to the jib-boom guy. This footrope was without stirrups, but every 2 feet a diamond or figure-eight knot was hitched in to provide footholds for the men working the boom (see Figure 22l, right).

Martingale or dolphin striker

Stampfstock; Arc-bouton de martingale

In the final years of the century a short spar known as a martingale or dolphin striker was fitted perpendicularly to the face of the bowsprit cap to provide the necessary angle for stays designed to counteract the upward force on the jib-boom when the jibs were set. Early experimental use goes back to approximately 1784 (see Figure 25b and c).

Martingale stay

Stampfstag; Martingale de foc

The earliest martingale stay fitting was described by Steel. He noted a stay fitted with one eye over the outer end of the jib-boom, then led over a score in the martingale's lower end, and with a double block seized or spliced into the inner end. A tackle formed with a single block was hooked to an eyebolt in the bowsprit head (see Figure 25b).

When two martingale stays were set, according to Lever, the outer was put over the jib-boom stop, rove through the lower sheave hole in the martingale and then through a block stropped close to the fore stay collar on the starboard side of the bowsprit, then to the forecastle, where a block was spliced into the end to form either a luff tackle or a whip.

The inner stay, spliced to the traveller, rove through the upper sheave hole in the martingale and followed on the port side the same route followed by the outer stay on the starboard side (see Figure 25c).

Spritsail yard

Blinderah; Civadière

Horses or footropes

Fusspferde; Marche-pieds

The rigging of a spritsail yard commenced with the horses. Both ends of the horse had eyes spliced in; the outer eyes fitted over the yardarms, while the inner eyes were lashed to the yard approximately 3 feet from the centre on the opposite side (see Figure 24e). French horses, and those of some other Continental vessels, had deadeyes seized to the inner ends, which were lashed together underneath the spritsail yard sling (see Figure 24d). The horses hung some 3 feet below the upper side of the yard, a little more than half a man's height. A constant distance was ensured by two or three supporting ropes, known as stirrups. A thimble was spliced into the lower end of each stirrup, and somewhat more than the upper half of the total length was plaited; this plaited part was given 1½ or 2 turns round the yard and nailed to it. The horses passed through the thimbles.

Brace pendants

Brass-Schenkeln; Pantoires de bras

Next came the brace pendants. These had a single block spliced into one end and an eye into the other, which was fitted over the yardarm (see Figure 26a-c and f).

Braces

Brasse; Bras

The standing part of the spritsail yard brace went through a number of changes in position during the course of the century. English ships of the first two decades had spritsail yard braces fastened to the outer quarters of the yard itself and reeving through blocks halfway up the fore stay before running through the pendant blocks. The braces then rove through lead blocks on the upper part of the stay to single blocks forward and aft beneath the fore top, and from there to the belaying point as detailed below (see Figure 26a). Towards the 1730s the standing part of the brace was secured to the upper part of the stay rather than the yard, and the brace led down to the pendant block before following the same lead as above (see Figure 26b).

From about 1735 until 1760 the standing part of the brace was secured to the middle of the stay (shown as a dashed line in figure 26b), and after 1760 to the very top of the stay beneath the collar. The braces then passed through the pendant blocks and the inner sheaves of double blocks arranged fore-and-aft beneath the fore top to be belayed at deck level (see Figure 26c). Up to 1735 the belaying points were cleats or belaying pins on the forecastle breast rail. Later the braces led to the fore jeer bitts.

The rigging of spritsail yard braces on Continental vessels varied slightly from that on English ships. During the sprit topsail period the standing parts were made fast to the fore stay just below half height. After passing through the pendant blocks, the leads rove through blocks on the fore stay about one third of the stay's length above the bowsprit, then vertically down to blocks on the bowsprit and aft from there to belaying points on the beakhead rail (see Figure 26f). The author of *Der geöffnete See-Hafen* (1705) noted in detail a somewhat different rig: the braces were made fast to the fore stay at a height of 2 to 3 fathoms above the bowsprit and led from there, via blocks on the spritsail yard a quarter of the distance towards the yardarm, to blocks on the fore stay slightly above the standing part. They then led down to reeve through the brace pendant blocks and up once more to blocks on the fore stay, 1 feet above the first blocks. The braces then led upwards along the stay to blocks beneath the mast top

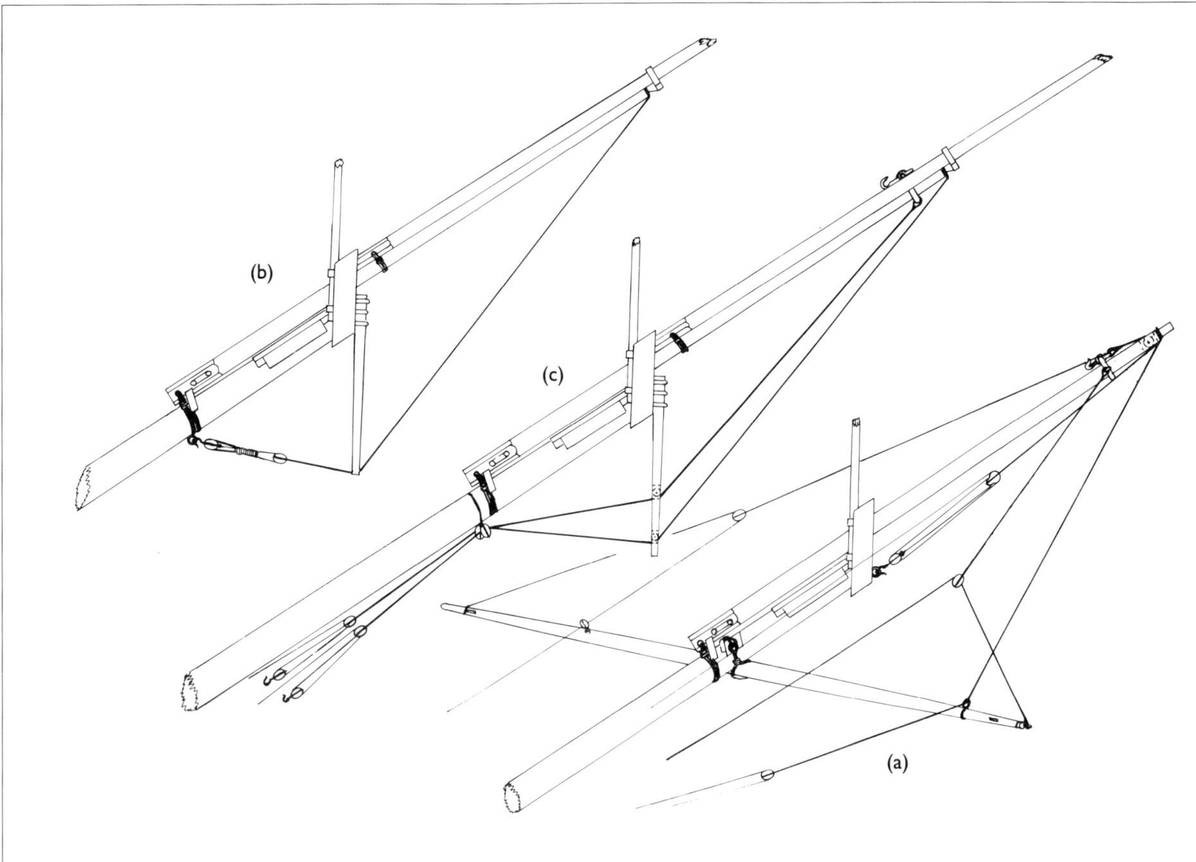

Figure 25
(a) Jib-boom and travelling guys (port side, English jib-boom guy until approximately 1775; starboard side, Continental jib-boom and travelling guys until the end of the century).
(b) English martingale stay, about 1790.
(c) English martingale stay, about 1800.

and, after reeving through another set of blocks on the main stay just a few feet below the fore brace blocks, they belayed to the sides of the forecastle (see Figure 24f).

The method of rigging braces on Continental vessels given first above was also used in the Royal Navy before 1700. The only variation was that the blocks at the bowsprit were replaced by blocks on the head rails.

Davis did not mention the rigging of spritsail yard braces, but he gave brace lengths, from which it can be inferred that braces around 1710 were commonly rigged both with and without pendants; the latter method therefore required a block stropped to the yardarms. He noted:

> Two Braces double 40 Fathom, or 6 times and a half the Length of the Yard, or 5 times the Length of the Boulsprit. Two Braces triple 46 Fathom, or 7 times and a half the Length of the Yard, or 6 times the Length of the Boulsprit.

The latter type of brace was longer by the length of the yard or bowsprit, which is approximately the length of two pendants.

The use of brace pendants became uncommon in the later years of the eighteenth century and the brace block was increasingly stropped directly to the yardarm. Röding and Paris agreed on this, though they disagreed on the point at which the standing part was made fast. On the *Royal Louis* of 1780 (as described by Paris) the fastening point was at half height on the fore stay, and the hauling part led through a block beneath the fore end of the fore top (see Figure 26d). Röding placed the standing part at the lower third of the stay, and his hauling part, after passing through the yardarm block, rove through a block on the stay about a foot above the point at which the standing part was secured, led from there to a second block hitched to the same stay at approximately two thirds height, and then led downward forward of the mast to the topsail sheet bitts (see Figure 26e).

Etchings by Ozanne suggest that the braces, as in Paris' description, were secured at half height to the fore stay and led from the yardarm blocks to blocks at three quarter height on the stay (see Figure 24g), or that they were fitted like English braces of about 1800.

The various leads described above do not exhaust the possibilities for the rigging of spritsail braces on eighteenth century ships: other minor variations to the methods described can be found in contemporary sketches and paintings.

During the eighteenth century and later the German term for a spritsail yard brace was *Trisse*, but the author of *Der geöffnete See-Hafen* provided a detailed description of the Trisse as a type of preventer brace:

> The Trisse is used to reinforce the braces when the yard is to be braced and topped. A block is placed on each side of the spritsail yard at the first quarter, and a violin block with a large sheave above and a smaller below is placed on the fore stay below the point at which the spritsail yard braces mentioned above are secured; further blocks are placed on both sides of the bowsprit, close to the fore stay. The end of each Trisse is made fast to the block on the yard and runs up to the smaller sheave on the fore stay block, down again to the block to which the end is made fast and over its sheave, then up once more to the fore stay block and over the larger sheave. It is then led through the block on the bowsprit and straight to the forecastle, from where the Trisse can be controlled simultaneously with the braces according to the wind [see Figure 28a).

Lift blocks

Toppnantenblöcke; Poulies de balancine

The stropped lift block was fitted outboard of the brace pendant on each side of the spritsail yard, and was secured to the yard by an eye over the yardarm (see Figure 27a, left).

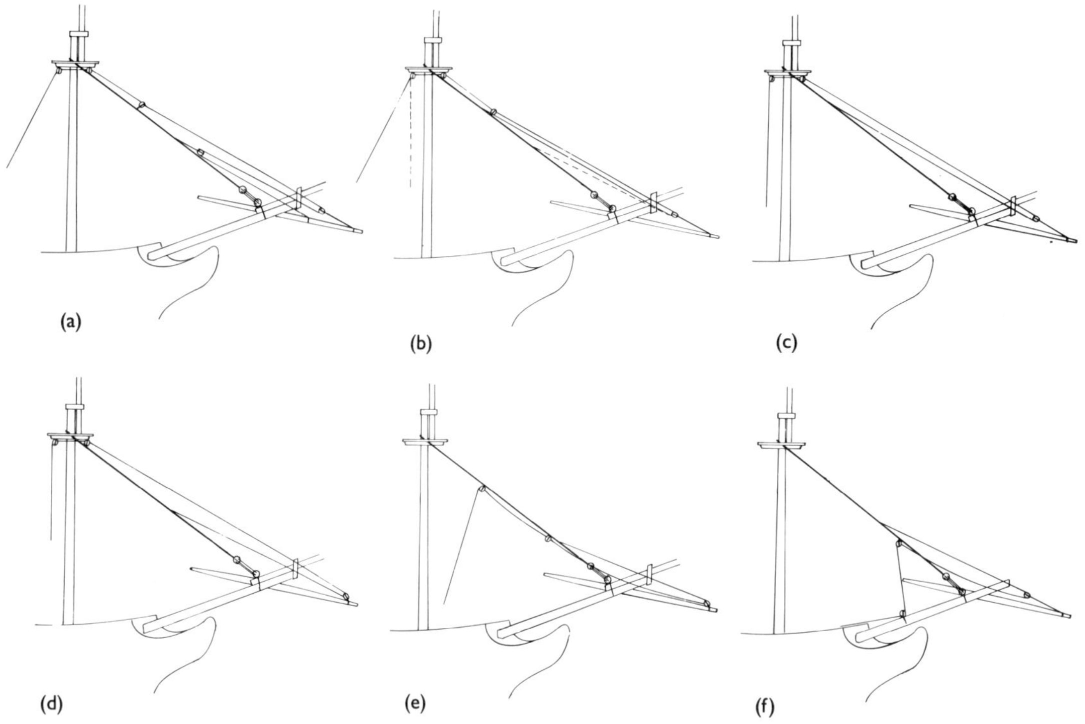

Figure 26
(a) Spritsail yard brace on an English ship between 1700 and 1720.
(b) Spritsail yard brace on an English ship between 1720 and 1735. The dotted line shows the rig in the period 1735 to 1760.
(c) Spritsail yard brace on an English ship after 1760.
(d) Spritsail yard brace on a French ship, about 1780.
(e) Spritsail yard brace on a Continental ship, about 1790.
(f) Spritsail yard brace on a Continental ship up to the end of the sprit topmast period.

(a) (b) (c) (d) (e) (f)

Lifts

Toppnanten; Balancines

Spritsail yard lifts were rigged single or double, depending on the ship's size. A single lift was fitted with an eye over the yardarm and led through a block on the bowsprit cap to the beakhead rail. The lead block on the cap dated from approximately 1760. Hooked or spliced to an eyebolt in the bowsprit near the sprit topsail standard for the first three decades of the century, the lead block was seized to the preventer fore stay collar between 1730 and 1760 (see Figure 27a, right).

Double lifts were hooked to ringbolts on the bowsprit cap and led through the yardarm blocks then back through blocks on the cap or seized to a strop behind the top, and belayed in the same manner as single lifts. If a sprit topsail was rigged, then the double lifts often acted as sprit topsail sheets as well (in such cases, the standing part of the lift was hooked into the sprit topsail clew rather than secured to the bowsprit cap). In Continental rigging toggles were frequently substituted for the hooks (see Figure 27a, left). The author of *Der geöffnete See-Hafen* noted that the lift was fastened to the clew of the spritsail topsail, then led through a block on the yardarm and another block on the bowsprit beneath the top. It then rove through an iron thimble on the fore stay collar towards the forecastle. When the spritsail topsail was not set, the ends of the lifts were made fast to a strop on the outer end of the bowsprit.

French rigging plans of the last decades of the eighteenth century show a lift with the standing part spliced onto a stropped cap block, and thus unable to be used as a sheet. They also show standing lifts, fitted one fifth of the yard's length outboard of the sling and rigged to the yard with deadeyes. Short-stropped deadeyes were used at the yard, and deadeyes with longer strops at the bowsprit behind the cap; the strops of the latter were sometimes long enough to necessitate only a short lanyard. The author of *Der geöffnete See-Hafen* noted a similar standing lift each side consisting of an 8 to 9 fathom long pendant secured to the upper part of the bowsprit and ending in the upper of a pair of deadeyes, with the lower stropped to the spritsail yard near the Trisse blocks (see Figure 28b, right).

Establishment rigging plans for English men-of-war from 1719 and 1745 also indicate the use of standing lifts. These lifts were spliced round the yard halfway between the sling and the yardarm, with long pendants to the lower deadeyes and a short strop from the upper to the bowsprit (see Figure 28c). The bowsprit deadeye strop in 1719 was secured a distance abaft the cap approximately equal to one tenth of the bowsprit length outboard of the ship; in 1745 this distance had increased to one sixth of that length. Davis (1711) noted 'two standing Lifts with Eyes over the Yardarms 6 Fathom, or the Length of the Spritsail Yard'.

Steel's description of the spritsail yard still made mention not only of lifts, but also of standing lifts. These were fitted round the yard with an eye approximately one quarter outboard of the sling. The other ends had thimbles spliced in and were set up to stropped thimbles within the bees. Steel's reference shows that standing lifts, if not always fitted, were at least still known in English shipping at the end of the century.

That this relic of the sprit topsail era also survived also on Continental ships is evident not only from the French rigging plans noted above, but also from Röding, who called them 'Spanish lifts' and described them thus:

> Spanish lifts were also rigged to the spritsail yard in bygone times, consisting of a short rope laid round the bowsprit fore end with a timber hitch, with a deadeye seized to each end. One end reached to port and the other to the starboard side of the spritsail yard. At one eighth of the total length of the yard to both sides from its centre a similar deadeye was secured, and a lanyard rove through this and the former deadeye. The whole was known as a Spanish lift, and one can find them still on large ships to hold the yard in place if the sling should break, but they are very inconvenient in action.

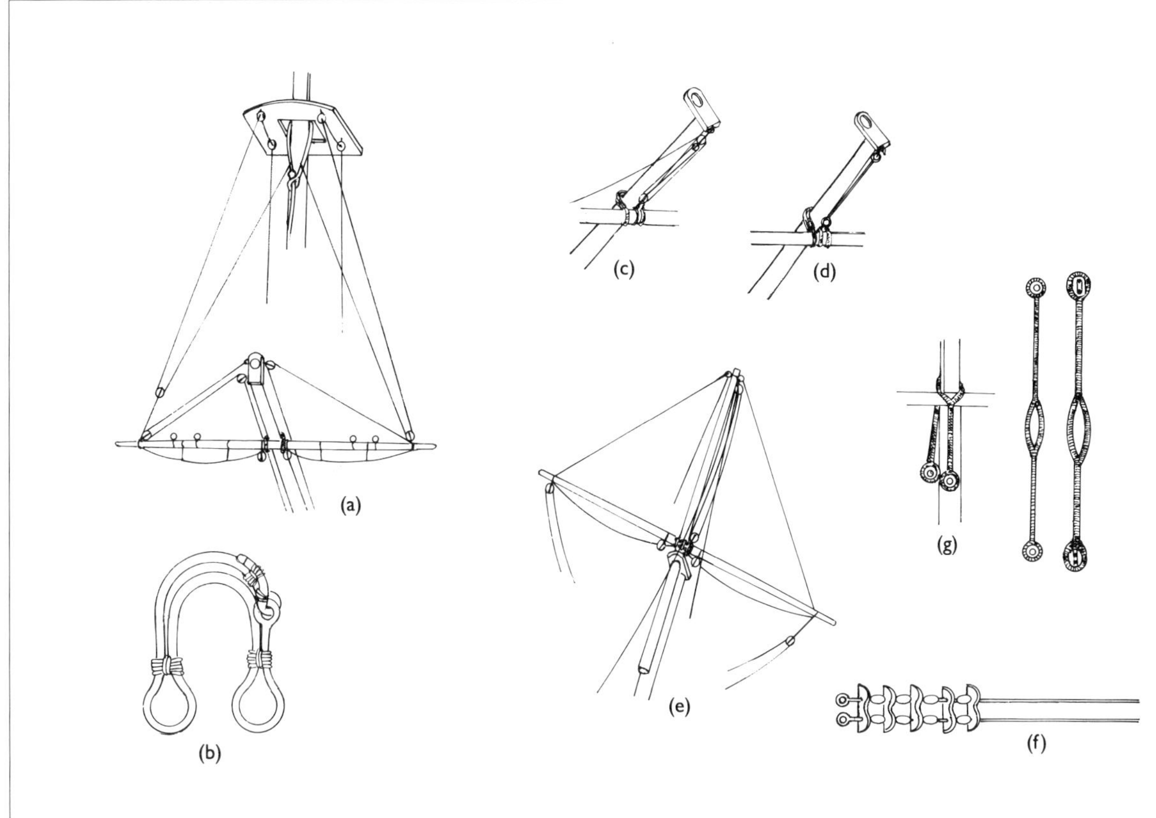

Figure 27
(a) Spritsail yard braces and lifts on an English ship, about 1800. At about this time the brace pendant shown on the left was replaced with a block (right). The left side of the drawing shows a double lift, which might also be used as a sprit topsail sheet, and on the right the single lift of a smaller ship is shown.

(b) A spritsail yard sling.

(c) A spritsail yard halyard or outhauler.

(d) Rigging of a spritsail yard standing halyard or preventer sling after 1760, used only in harbour by men-of-war but generally in the merchant service.

(e) A sprit topsail yard with halyard and lifts, about 1780. The braces are fitted as in (a).

(f) A sprit topsail yard parrel.

(g) Mast tackle pendants, used until 1780 with seized-in blocks (detail right) and thereafter with spliced-in thimbles (detail left), to take the tackle hooks.

Clewline blocks

Geitaublocken; Poulies de carguepoint

The clewline blocks were stropped and seized to the spritsail yard approximately 3 feet outboard of the sling.

Sling or parrel

Rack; Suspente

A sling, sometimes rather confusingly referred to as a parrel, was used to secure the spritsail yard to the bowsprit. It was laid around the yard and seized to it in such a way that the eye in one end was located somewhat above the yard. The other end passed over the bowsprit saddle and around the yard on the other side, where it was seized again. After passing a second time over the saddle, the running end rove through the eye of the short end, then turned back on itself and was seized thoroughly with a throat and round seizing (see Figure 27b).

The spritsail yard was not always secured with a sling. The author of *Der geöffnete See-Hafen* noted the use of a normal parrel, distinguished from those used on the masts only by the fact that bowsprit's spritsail yard section was covered with copper and greased to allow easier movement and to prevent damage.

Halyard

Vorholer; Drisse

At the centre of the fore side of the spritsail yard, a single block for the halyard was hooked to a strop with a thimble, and the hook was moused. The opposite block, a fiddle block, was hooked and moused to the lower aft side of the bowsprit cap. As an alternative to the fiddle block a double block was occasionally used.

The standing part of the halyard was hitched to the strop around the yard block and, passing through both blocks, formed a tackle belayed on the beakhead rail, or led via the rack over the bowsprit to the forecastle (see Figure 27c). Röding made the following comment on the belaying point of the halyard:

> With this tackle, the spritsail yard can be hauled out to the end or head of the bowsprit, and the hauling part can even be secured there, because the spritsail yard itself is not raised or lowered while the vessel is at sea.

The author of *Der geöffnete See-Hafen* noted that the outhauler was made fast to the block beneath the bowsprit top, then led through a block on the yard and, after returning to the first block, led in above the yard, to be belayed on the gammoning.

'These Halyards are now generally left off, and are never used in the Merchant Service,' noted Lever, and Röding's description continued with the note that 'for precisely this reason, the spritsail yard is now usually hung on a solid strop; these days a halyard is seldom used.'

From about 1760 onward the halyard was replaced by a standing halyard when a vessel was in harbour. This halyard was also called a preventer sling, and it soon became widely used, as Lever's and Röding's comments show. Steel noted that preventer slings were used when the halyards were taken in. The outer end of this sling was hooked to the lower side of the cap, and the inner end rove through the thimble of the yard strop noted above, and was hitched with two half hitches or spliced (see Figure 27d).

Thimbles

Leitkauschen; Cosses

Two thimbles were stropped to each side of the upper face of the spritsail yard after 1775, in order to provide a lead for the jib-boom and traveller guys (see Figure 23j above).

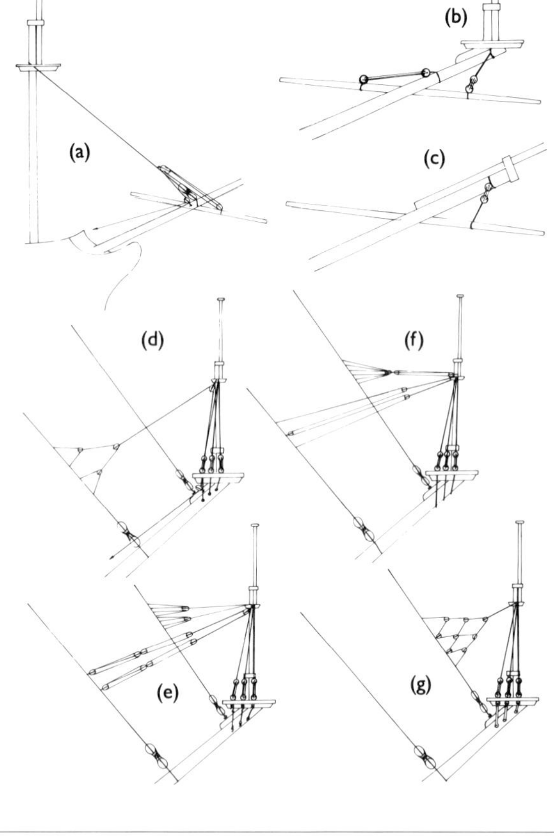

Figure 28
(a) Spritsail yard trisse, German 1705.
(b) Continental spritsail yard standing lifts from the beginning of the century (port side Dutch and starboard side German).
(c) English spritsail yard standing lift.
(d) French sprit topmast shrouds and backstay, 1700.
(e) Dutch sprit topmast shrouds and backstay, 1700.
(f) German sprit topmast shrouds and backstay, 1705.
(g) English sprit topmast shrouds and backstay, 1711.

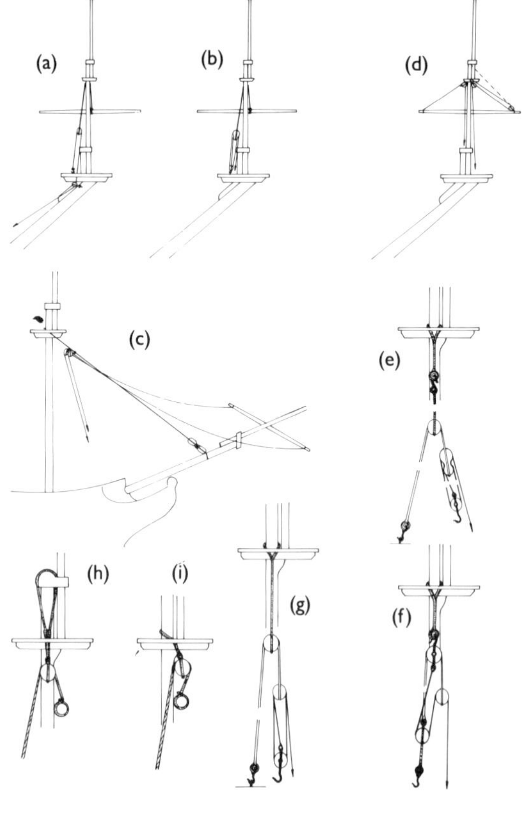

Figure 29
(a) Continental sprit topsail yard tye and fall, 1705.
(b) English sprit topsail yard tye and fall, 1680 to 1719.
(c) French sprit topsail yard braces on larger ships after the end of the sprit topmast period.
(d) Sprit topsail yard lifts, single and double; the dashed line shows Dutch practice.
(e) Mast tackle of the type known as a 'Spanish tackle', 1705.
(f) Mast tackle: the so-called 'French tackle', 1705.
(g) English mast tackle, 1711.
(h) and (i) Continental jeer block pendants, 1705.

Sprit topmast

Sprietmast; Mât de perroquet de beaupré

Shrouds

Hoofdtaue; Haubans

Two or three pairs of shrouds provided lateral stability for the sprit topmast. With deadeyes turned into their ends, these shrouds were lashed to futtock deadeyes on the bowsprit top, with the futtock irons bolted or nailed to the bowsprit (see Figure 28d, e and g) or frequently led round its lower side (see Figure 28f).

Backstay

Pardunen; Galhauban du mât de perroquet de beaupré

The sprit topmast backstay counteracted forward pressure on the sprit topmast; it was set up with crowsfeet or bridles at its inboard end. At the beginning of the century these were seized to the fore topmast stay on English ships, and to the fore stay on Continental ships (see Figure 28g and d respectively). The latter method was used on English ships only up to about 1670.

The author of *Der geöffnete See-Hafen* noted that the backstay was set up with two pendants of 1 fathoms length, each with a block turned into its end. Another block was placed on the fore stay slightly below its mid point. A lanyard was fastened above this position, led through the first pendant block, back to the fore stay block, then through the second pendant block to be hauled taught and set up to the fore stay at the same distance below the block as the other end was above (see Figure 28f).

Additional support was also provided as follows:

> A crowsfoot is an oblong piece of timber, pierced with holes, which ...is secured to the fore topmast stay by several ropes about 3 feet long. A small block is secured to the opposite face of this piece of timber, and a further block is set up above the sprit topmast trestle trees. A rope leads from the crowsfoot block to the block on the head, returns to the crowsfoot block, runs over its sheave and back to the head, where the end is made fast; the crowsfoot is thus kept in position. It is used more for decoration than for any practical support it provides for the sprit topmast.

Sprit topsail yard

Bovenblinderah, Schiebeblinderah; Contre civadière

Tye and halyard

Drehreep und Fall; Itague et drisse

Sprit topsail yards set on a sprit topmast had a tye which was 1 foot shorter than the length of the sprit topmast measured from the top, according to Davis. For the halyard, he noted a single pair (or whip) twice the length of the mast, or halyards with two blocks which were either 7 fathoms in length or 2⅔ times the length of the sprit topsail (see Figure 29b). The author of *Der geöffnete See-Hafen* noted that the halyard's hauling part led through the bowsprit top and, passing a block underneath, led along the bowsprit towards the forecastle, where it was belayed (see Figure 29a).

Horses or foot ropes

Fusspferde; Marche-pieds

Horses, without stirrups, were rigged from the yardarm to about 3 feet beyond the centre of the yard on the opposite side, and were set up as described above, under Spritsail yard. Continental rigging was similar.

Braces

Brassen; Bras

The braces of a sprit topsail yard as set on a jib-boom were double only on larger ships; otherwise they were single. In English rigging, the standing parts of double braces were made fast to the fore stay collar, and after passing the brace pendant blocks the braces ran through the second sheaves of the double blocks beneath the fore top described under Spritsail yard braces, above. These blocks were, however, fitted as singles until about 1773, and each brace had its own pair of blocks. The braces were belayed to the fore jeer bitts (see Figure 30b).

The braces of a sprit topsail yard as set on a sprit topmast were set up just beneath the backstay block at the mid point of the fore stay, and ran through the pendant blocks and blocks somewhat below the standing part at the fore stay, vertically downward to a double block on the bowsprit and from there to the beakhead rail (see Figure 30a). They disappeared on smaller ships soon after the turn of the century, on ships up to 70 guns around 1720, and on all other ships around 1745 at the latest. Davis noted the following lengths for these braces in the first decade of the century: 'Two single Braces 24 Fathom, or 4 times the Length of the Topmast and Yard added together, add 4 Foot. Two Braces 28 Fathom, or 4 times and a half the Topmast and Yard added together.'

After 1775 brace pendants were no longer used on the yard.

Sprit topsail yard braces on Continental ships differed from those on English vessels. In Continental practice, braces were usually single and were set up with an eye directly over the yardarm, or with a toggle to a strop (see Figure 30c, detail). They passed through a set of single blocks at the lower quarter of the fore topmast stay, and another at the upper quarter of the fore stay, before being belayed on the beakhead rail or the fore topsail sheet bitts (see Figure 30c). On larger French men-of-war they were led directly to blocks on the fore stay collar and belayed to the beakhead rail (see Figure 29c).

Lifts

Toppnanten; Balancines

Sprit topsail lifts were single and passed through thimbles or single blocks at the jib-boom stop, then via the fairlead saddle to the forecastle, to be belayed on the rail.

When a sprit topmast was stepped, the standing parts of these lifts were fastened to the topmast head, then ran via the yardarm blocks through blocks at the topmast head and were belayed at the bowsprit top (see Figure 29d). Davis noted both single and double lifts for the sprit topmast period.

Clewline blocks

Geitaublöcke; Poulies de carguepoint

Clewline blocks were stropped to the lower side of the yard, a distance of 2 feet from the centre.

Parrel

Rack; Racage

The sprit topsail yard parrel was made from ribs and two rows of trucks (see Figure 27f), as was a topsail yard parrel. A Continental alternative to this parrel during the last few decades of the century was a simple sling (see Figure 27b).

Halyard

Ausholer; Drisse

The halyard of a jib-boom rigged sprit topsail yard was a single block tackle, set up to the yard sling and the jib-boom stop, with the hauling part belayed at the forecastle.

Lower masts

Untermasten; Bas-mâts

Bolsters

Kälber; Coussins d'élongis

Before the actual rigging of the lower mast began, the bolsters (see Bolster in Chapter I) were covered with several layers of old canvas and tarred over. At the beginning of the century a grommet was noted in contemporary literature as an antichafing measure: 'a thick rope is first laid round the head, with the ends together, and then nailed thoroughly to the mast' (*Der geöffnete See-Hafen*).

Mast tackles

Masttakeln, Seitentakeln; Caliornes de mât

Once this task was completed, the mast tackle pendants were the first items to be fitted over the masthead. Their lengths at the beginning of the century, according to Davis, were:

Foremast	one pair	18 feet, or ¾ of the mast length between deck and cross tree.
Mainmast	one pair	20 feet, or ¼ of the mainmast length plus 1 foot
Mizzen mast	one pair	4 fathoms, or ⅓ of the length of the first shroud pair.

The runners which ran through the pendant blocks on the fore and mainmast tackles had a length of 2⅔ times the distance from the deck to the cross tree, and the falls were eight times that length. For the mizzen mast Davis noted Burton tackles with single blocks and a fall length six times the beam plus 1 foot. For long tackle blocks, the length was seven times the beam plus 8 feet.

Continental ships, according to the author of *Der geöffnete See-Hafen* had a tackle pendant 2 to 2½ fathoms long on each side of the mast, with an iron thimble spliced into each of the lower ends.

> A strop or strong rope, fitted at its upper end with a hook, is now hooked into this thimble; at the lower end of this strop is a block through which a rope known as a runner passes. This runner leads downwards on one side to the channel, where it is hooked through an iron chainplate, fitted especially for the purpose. The other end of the runner is fitted with a double block, with a large sheave above and a smaller below...

The hooked block for the tackle purchase was spliced into the lower end of a fall which ran over the smaller sheave of the double block, back through the purchase block, and over the larger sheave of the double block to a further block stropped to the channel; the whole arrangement was known as a 'Spanish tackle' (see Figure 29e). Such tackles were used for taking in and out boats, guns, and other such loads, and were found either on both sides of the fore and mainmasts, or only on the starboard side, while on the port side a so called 'French tackle' was used. This was a single block hooked to the mast pendant, and a strong

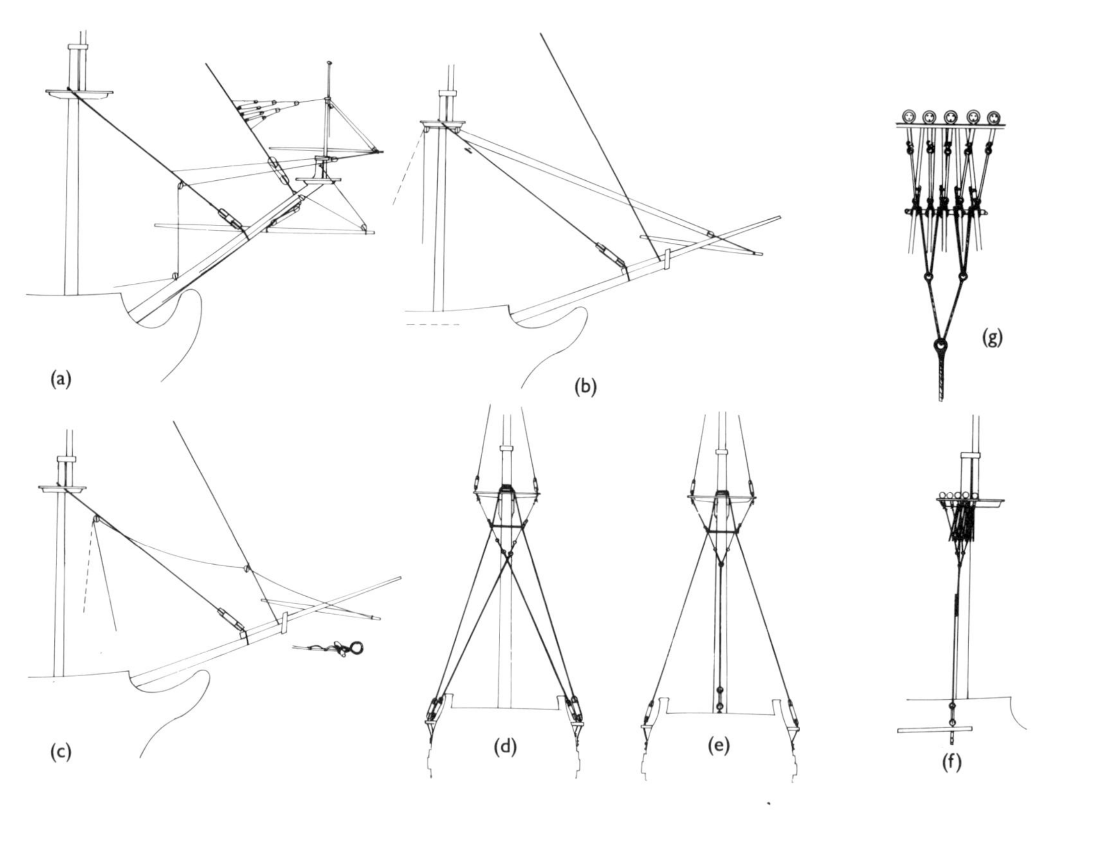

Figure 30
(a) Sprit topsail yard lifts and braces yard and spritsail yard lifts up to the end of the sprit topmast period.
(b) Sprit topsail yard braces on English ships. Until 1735 these were belayed to the forecastle waist rail or belfry (as shown by the dotted line); thereafter they were belayed to the fore jeer bitts (solid line).
(c) Sprit topsail yard braces on Continental ships after the end of the sprit topmast period.
(d) Bentinck shrouds, seen from aft.
(e) A standing single Bentinck shroud, seen from aft.
(f) Double Bentinck shrouds, side view; arrangement as in (d).
(g) The rigging of a Bentinck shroud to the futtock stave and shrouds.

runner with blocks at both ends, one of which was long-stropped with a hook. The fall ran through the upper of the two blocks upon the runner, then the through the lower, and was hitched to the pendant block (see Figure 29f). These, then, were the tackle arrangements for English and Continental ships around 1700.

Steel listed the following lengths for English lower mast tackles in the last decades of the eighteenth century:

Foremast	one pair	1/10 of the mast length
	two pairs	1/5 of the mast length
Mainmast	one pair	1/12 of the mast length
	two pairs	1/6 of the mast length
Mizzen mast	8 fathoms for large and 6 fathoms for small ships.	

Information about the diameter of these and other ropes on various sizes of ships is provided in the rigging tables in the Appendix below.

Where one pair only was fitted, mast tackle pendants were usually made from one piece of rope, cut in the middle and spliced with a cut splice to form an opening big enough to go over the masthead. Thimbles were spliced into the lower ends and the whole pendant length was served. Blocks of considerable size were hooked to these thimbles, or, in English rigging up to 1780, spliced directly into the pendants. A pair of tackle pendants was normally rigged on Continental ships and English merchantmen as part of the setting up of the foremost shroud pair. For further information see Swifter below.

Where two pairs were fitted, the pendants were laid round the masthead in shroud fashion. They were of equal length in English rigging, but on the Continent the pendant closer to the waist was longer than the other (one quarter of the shroud's length).

The circumference of such pendants was similar to that of the adjacent shrouds.

The runner, a rope about two thirds the thickness of the pendant, ran through the pendant block, and its standing part, which had a thimble and hook spliced in, was hooked to a ringbolt in the channel. The other end of the runner was fitted with a double block slightly smaller than the pendant block. Together with a large, long stropped, single block fitted with a hook, this runner formed the mast tackle, hooking into another set of ringbolts in the channels (see Figure 29g). The connecting tackle falls were approximately half the runner's thickness. These tackles were similar to Spanish tackles at the beginning of the century.

Mast tackles (Röding also called them side tackles) on larger vessels, which were fitted with two pairs per mast, had varying roles. The after tackle at the foremast and the forward tackle at the mainmast were set up as winding tackles, with treble and double blocks, for moving heavy goods and boats. The other tackle pair in each case was the usual long tackle block and single block combination, used for normal loads, the setting of shrouds and the fishing of anchors.

Jeer block pendants

Kardeelblock-Hänger; Pantoires de la drisse de basse-vergue

An unusual jeer block fitting is mentioned in a very short note in *Der geöffnete See-Hafen*, and since the jeers followed the mast tackles in rigging routine, it is worth taking note of here (see Figure 29h and i). The anonymous author noted: 'a further rope leads forward over [the mast tackles], with the large jeer block at its end.'

Shrouds

Hoofdtaue; Haubans

Shrouds were large ropes supporting the masts laterally, and followed the tackle pendants over the masthead in the normal sequence of rigging. Their number depended on the diameter of the mast and the size of the sail to be carried; with the

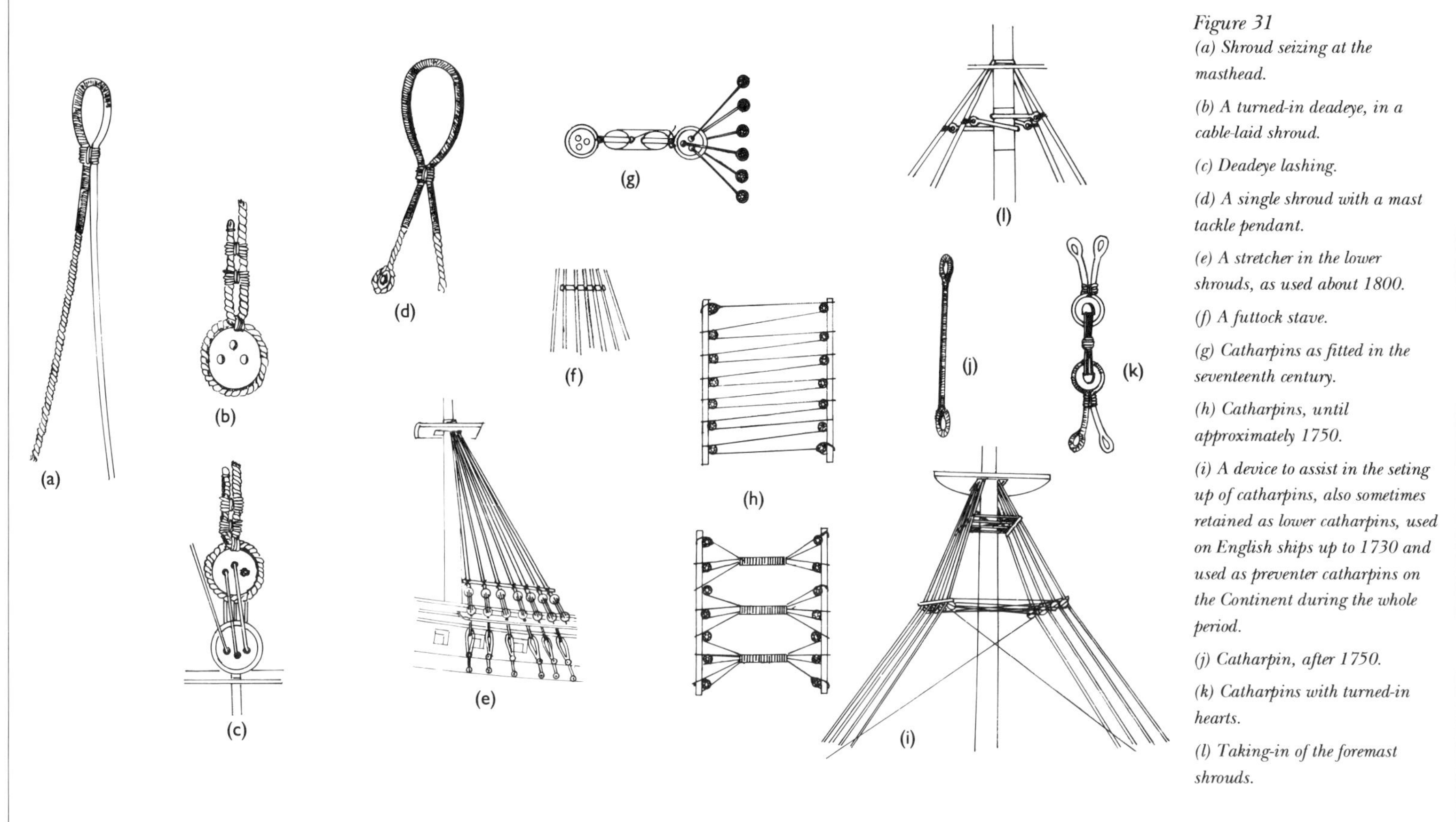

Figure 31
(a) Shroud seizing at the masthead.
(b) A turned-in deadeye, in a cable-laid shroud.
(c) Deadeye lashing.
(d) A single shroud with a mast tackle pendant.
(e) A stretcher in the lower shrouds, as used about 1800.
(f) A futtock stave.
(g) Catharpins as fitted in the seventeenth century.
(h) Catharpins, until approximately 1750.
(i) A device to assist in the seting up of catharpins, also sometimes retained as lower catharpins, used on English ships up to 1730 and used as preventer catharpins on the Continent during the whole period.
(j) Catharpin, after 1750.
(k) Catharpins with turned-in hearts.
(l) Taking-in of the foremast shrouds.

increasing size of ships during the eighteenth century the number of shrouds per mast grew as well. For example, around 1700 a First Rate man-of-war had nine shroud pairs on the mainmast and eight on the foremast. Around 1750 these figures were already ten and nine, and towards the end of the century this had grown to ten on both masts.

The shrouds were always set up in pairs (known as a span), with the first going to starboard forward, the second to port forward, the third to starboard again and so on. A single shroud was always last, except in the merchant service, where the odd shroud (called a swifter) was combined with the mast tackle pendant and was generally the foremost (see below). The foremost shroud was usually completely served to protect it from chafing by the sail; on masts with gaff sails this serving was applied to the aftmost shroud as well. On all other shroud pairs, the bight and the upper quarter of the shroud length was served.

The seized bight of the foremost shroud pair lay immediately on top of the bight of the mast tackle pendants, on top of the trestle tree bolster. The second shroud pair lay on top of the first, the third on top of the second, and so on. Each shroud pair thus hung clear of the preceding pair, and a further advantage of this order of fitting the shrouds was identified by Steel:

> By this method, the yards are braced to a greater degree of obliquity, when the sails are close hauled, which could not be, were the foremost shrouds fitted last on the mast-head.

Deadeyes were turned into the lower ends of the shrouds with a throat seizing, and two or three round seizings a little further up the shroud further secured the doubled end. The deadeyes were then set up with lanyards to their counterparts at the channels. These channel or chain deadeyes were bound with iron and linked to the chain plates, which reached from the channels to the wales to which they were bolted.

The doubled end of the shroud at the deadeye on cable-laid shrouds (used on English men-of-war, large merchantmen and sometimes on French men-of-war) passed around the deadeye from right to left, crossed behind the standing part of the shroud, and was lashed to the lefthand side of the standing part, as seen from outboard (see Figure 31b). Smaller merchantmen very often had hawser-laid shrouds, which were seized to the right. Continental ships normally had hawser-laid shrouds. Röding noted: 'Usually the shrouds are hawser-laid, but the English also use cable-laid ropes'. In the same article he also gave a rule of thumb for the diameter of main shrouds, suggesting one seventy-second of the beam.

Swifters

Borgwanttaue; Faux haubans

A single shroud was usually named a swifter, and on English warships it was set up after the shroud pairs, with an eye splice over the masthead. On French men-of-war, and on English merchantmen, it was the foremost shroud, combined with the smaller mast tackle pendant (see Figure 31d). Falconer referred to swifters as providing additional support to the masts, and suggested that they were not confined by catharpins; otherwise they were set up exactly like shrouds. Röding saw the swifter as a preventer shroud (see below for further details).

Deadeye lashing

Juffern-Zurring; Amarre de cap de mouton

The shroud lanyard was passed through the foremost hole in the upper deadeye from the back, with a Matthew Walker knot as a stopper, according to Lever. Steel, however, noted that 'the end of the lanyard is thrust through the after-hole of the upper dead-eye and stopped with a walnut-knot, to prevent its slipping'. Röding noted only one hole in the upper deadeye. The lanyard then passed through the corresponding hole in the lower deadeye from the face, and the centre hole in the upper deadeye again from the back, and so on until all holes were filled (see

Figure 31c). It was then hitched to the mast tackle and hauled taut. After that, the hauling part was nippered to the nearest standing section, then passed between the throat seizing and deadeye, hitched, and taken round the shroud until the whole was extended; the end was then lashed to the shroud.

The lanyards were smeared with tallow or grease to enable them to slide more easily through the holes in the deadeyes, and thus to distribute the strain more evenly.

It was customary for a newly rigged ship to have the shrouds first set up with temporary lanyards of worn rope and spun-yarn seizings, which were replaced when the ship set out for sea.

Stretchers and futtock staves

Spreizlatten und Würste; Bâtons de trélingage

In the later years of the century stretchers, or squaring staffs, were sometimes seized to the lower shrouds, just above the deadeyes, to keep them from twisting (see Figure 31e). Such a stretcher can be seen on a Russian ship model from as early as 1748, where it served also as a pin rail. The French rigging plan of the *Royal Louis* from 1780 indicates such stretchers in the topmast shrouds, but not in the lower.

Another stretcher was seized to the upper part of the shrouds. The futtock stave, as this was called, should be positioned, according to Steel, as far below the mast top as the mast cap was above (see Figure 31f). Its purpose was to secure the futtock shrouds and to set up the catharpins, which pulled the upper part of the shrouds inwards, thus counteracting the pull of the topmast shrouds as well as maximising the possible bracing angle of the lower yards.

The material for these futtock staves was a strong piece of rope, wormed, served and tarred, although they were sometimes also made from timber (according to Lever). Röding described the futtock stave as extending from the second shroud from forward to the aftmost, and seized to either the inside or the outside of the shrouds, or sometimes even to both sides. The first two shrouds were often left free on French ships to make it easier to slacken these when the yard had to be braced sharply. On English ships the stave was lashed to all the shrouds, or, as an illustration by Lever shows, all but the foremost and the aftermost. Lever noted that the stave '...only seized to those Shrouds, which are to be catharpined in'. In English rigging, futtock staves were seized to the inside of the shrouds.

Catharpins

Schwichtungen; Trélingage

During the seventeenth century catharpining was carried out by means of a tackle, the blocks of which were seized to deadeyes, with the catharpin legs leading through the deadeye holes and fastened to the shrouds (see Figure 31g); a simpler method was introduced in about 1700.

A rope was seized to the first shroud at the height of the futtock staves and, after being laid round the opposite shroud (and the stave), was laced back and forth between the shrouds and made fast to the last. The lacings were then frapped together into three equal bundles (see Figure 31h). Noted by the author of *Der geöffnete See-Hafen* in 1705, this catharpining method was still used on French men-of-war in the second half of the century, and a very similar description given by Röding suggests that it was common practice on the Continent even at the end of the century.

As an aid to setting up the catharpins, temporary stretchers were set up parallel to, and approximately 6 feet below, the futtock staves, with a block hitched to each shroud. The shrouds were hauled in by means of a rope reeving through all the blocks, with both ends on deck (see Figure 31i). After the catharpins had been fitted, this makeshift appliance was removed. This method of tensioning the shrouds was, however, retained on many ships in the early decades of the eighteenth century as lower catharpins, set up at half height on the shrouds.

Such lower catharpins, fitted in England only until about 1730, were a normal part of Continental rigging throughout the century, and were known as preventer catharpins.

During the third decade of the eighteenth century a new type of catharpin appeared in English rigging. Instead of a single rope which had to be hauled taut and frapped in position, prefabricated catharpins were used. One such catharpin was initially used to link each shroud to its opposite, but they were reduced to four in total in 1750. These had an eye spliced into each end and were served all over (see Figure 31j). Steel noted that these catharpin-legs were of differing lengths: the length of the foremost varied between 4 and 8 feet, according to the vessel's size, but the foremost was always the shortest, and each successive span increased by 1 inch in length to account for the layering of the shroud bights over the masthead. With the shrouds temporarily hauled in, these catharpins could simply be lashed into place.

Lever also mentioned cross catharpins:

> The foremost Shroud formerly, was never catharpined in, on account of its being so much abreast of the Mast, that the leg would chafe it; but it is now customary, in the Merchant Service, to have both the foremost, and aftermost Shrouds catharpined. This is done by an additional leg on each side, one Eye is seized to the aftermost Shroud on one side, and the other to the foremost one opposite, above the other legs. These are called cross catharpins and are of great use in keeping the lee-rigging well in, when the ship rolls.

Another method, coming into use at the end of the century, involved a heart seized into the bight of a catharpin, thus creating two legs, the eyes at the end of which were then lashed to two shrouds at one side (see Figure 31k). The hearts from opposing sides were then pulled together with a lanyard. For the foremost (single) shrouds catharpins without hearts were used. Parcelled in leather and lashed to the foremost shrouds, these catharpins were led round the mast and secured at the other end to the aftermost shroud on the same side; the mast was coppered where these catharpins might chafe (see Figure 31l).

Bentinck shrouds

Bentinck-Wanten; Haubans de Bentinck

Bentinck shrouds (named after their inventor, Captain William Bentinck) were only rigged at sea and provided additional support to the lower masts when the ship rolled heavily. They were not an alternative setting of futtock shrouds, as has been suggested by some modern authors, nor merely an extension of the futtock shrouds; they were, however, set up around the futtock stave and the shrouds in a similar fashion to the futtock shrouds.

A thimble was spliced into the upper end of a Bentinck shroud with a span leading through. Both span ends were then also fitted with thimbles, and another two spans leading through these were made fast round the futtock stave and the closest shroud (see Figure 30g). The lower end had a deadeye turned in and was set up to a spare deadeye at the opposite channel (that is, the starboard Bentinck shroud led to the port channel). Modern authors sometimes show the Bentinck shrouds set up to the ship's waterways but Steel and Lever made it clear that they were fitted to extra deadeyes in the channels (see Figure 30d).

Where Bentinck shrouds were rigged permanently (as was sometimes done) the bridles led abaft the mast, like an extension of the futtock shrouds, where a single shroud was set up vertically (see Figure 30e).

Bentinck shrouds came into being after 1780 and were only

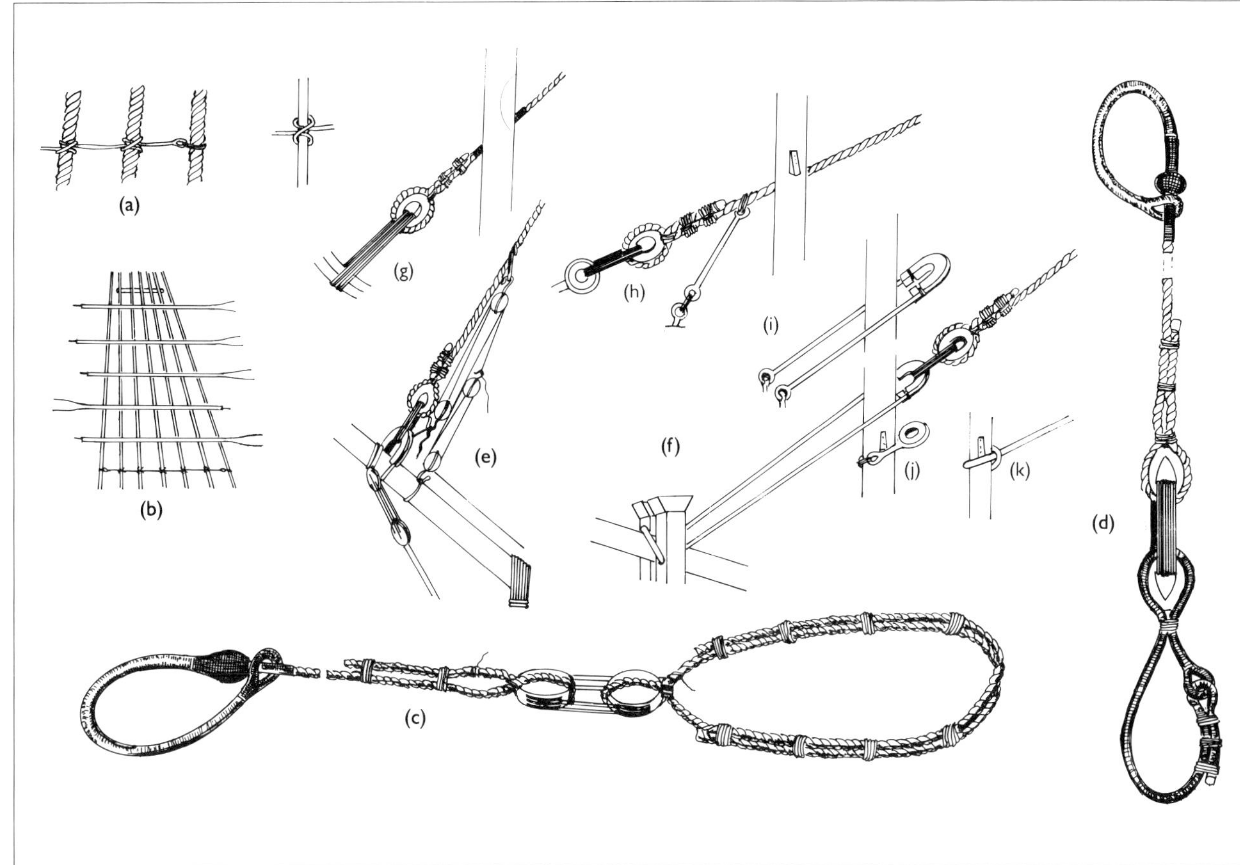

Figure 32
(a) A ratline, clove-hitched.
(b) Oars temporarily seized to the shrouds for the rattling of ratlines.
(c) A Continental main stay; note the use of blocks and the seizing of the main stay collar.
(d) An English main stay. The main difference from the Continental example is in the use of hearts and a different type of stay collar.
(e) The setting-up of an English fore stay after 1775.
(f) The main stay of an English merchantman, about 1800.
(g) A main stay set up to the breast hook without a second heart and collar; English, about 1800.
(h) A main stay set up to an iron-bound heart on deck; English, about 1800. To prevent the stay from chafing the fore mast, a pendant, called a jumper, was seized to the stay and to an eyebolt a short distance from the mast to starboard, keeping the stay from working too much.
(i) The main stay collar of an English merchantman, about 1800, with eyebolts fastened to the breast hook.
(j) The collar of a main preventer stay on English ships.
(k) The setting-up of a main preventer stay on smaller English vessels, about 1800. An eye was spliced to the lower end and the stay laid about the fore mast. The upper end, divided into two parts with eyes spliced into each and passed around the mainmast head, was lashed together with a lanyard.

used for a few decades. Lever noted in 1819 that a new system replacing the catharpins with an iron hoop round the mast, to which the futtock shrouds were set up, rendered Bentinck shrouds obsolete.

Preventer shrouds

Borgwanten; Haubans de fortune

Preventers were additional ropes rigged at times of abnormal stress to provide extra support to the rigging. Preventer shrouds were laid over the masthead rigging between the masthead and the topmast heel and seized like other shrouds. A deadeye was turned into each end and set up with a lanyard to a spare deadeye in the channel. Röding noted that the preventer shrouds were set up with a winding tackle inboard.

Steel noted that there was little difference between a preventer and a Bentinck shroud, as follows:

> Bentinck shrouds are additional shrouds to support the masts in heavy gales. Preventer shrouds are similar to Bentinck shrouds, and are used in bad weather to ease the lower rigging.

One was therefore as good as the other, and the choice between them depended for the most part on the preference of the individual captain.

Preventer shrouds represented the older of the two methods and were generally used on Continental ships, while Bentinck shrouds were mainly English.

Ratlines

Webleinen; Quaranteniers, Enfléchures

Ratlines, running across the shrouds like the rungs on a ladder, began 13 inches below the futtock staves and were set a distance of 13 inches apart. They were fastened to each shroud with a clove hitch, except at the ends, where an eye was spliced in and seized to the shroud (see Figure 32a).

The measurements quoted for the distance between ratlines varied greatly from one author to another. Steel noted 13 inches, Lever 12, Anderson 15 to 16, and Boudriot 13 to 14 inches. In view of the extra effort required from a sailor in running up ratlines which were widely spaced, a distance greater than Steel's 13 inches seems unlikely.

As a temporary measure to facilitate the rattling down of the shrouds during the rigging of a ship, oars or spars were seized across the shrouds on the outboard side, about 5 to 6 feet apart (see Figure 32b). These provided a firm support for the riggers.

Ratlines did not always run across all shrouds. Illustrations of French ships show the foremost and aftmost shrouds omitted, or only every sixth ratline (that is, ratlines 2 meters apart) running across all shrouds. Boudriot noted that the foremost shroud (and sometimes also the second) was not rattled, but his drawings indicated that ratlines normally ran to the aftmost shroud.

On Continental and early English ships it can be seen that the foremost shrouds were normally rattled, but the aftmost shrouds were reached only by every sixth ratline. After 1730 the lowest six ratlines on English vessels tended to omit the foremost shroud, and in the last quarter of the eighteenth century neither the lower nor the upper six ratlines extended to the foremost or the aftmost shroud. All other parts of the shrouds were fully rattled.

Falconer noted in 1769 that all shrouds were rattled down without exception, and Lever confirmed this.

Stays

Stage; Étais

This description of lower mast rigging has hitherto generally been applicable to all masts. With the stays, however, a distinction must be made between the fore, main and mizzen stays, which all varied in their leading.

Some comments nevertheless are generally applicable. An

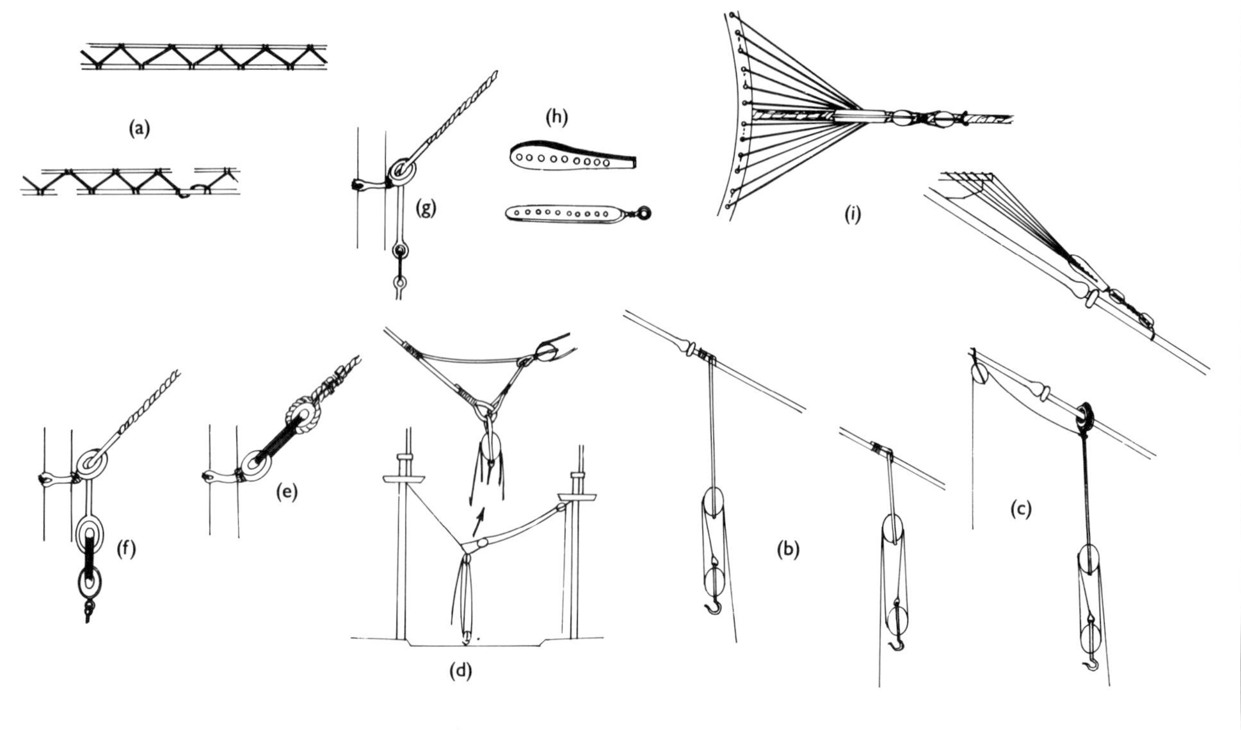

Figure 33
(a) Stay snaking and snaking on a broken stay.
(b) English stay tackle, used as main and fore hatch tackle after about 1740.
(c) Running stay tackle, English.
(d) Garnet tackle, used in English shipping until approximately 1730 and on the Continent during the whole of the century.
(e) Mizzen stay lashing.
(f) English mizzen stay lashing, about 1800.
(g) English mizzen stay lashing with thimbles, for smaller vessels.
(h) Euphroe.
(i) Crowsfeet with a euphroe and tackle on a stay.

eye was spliced into the upper end of all stays, just large enough for the stay itself to lead through to create a large loop. Before the stay was set up, it was stretched a few times with the windlass until the middle strand or heart was broken in several places. A pear shaped mouse was then raised on the stay to prevent the loop or collar from closing, about one third of its length down from the eye (or, according to other sources, twice the width of the masthead plus twice the length of the trestle trees down from the eye). From the shoulder to the beginning of its tail the length of the mouse equalled one third the circumference of the stay, and its diameter was three times that of the stay.

After the mouse had been shaped with spun-yarn, it was parcelled with old canvas, tarred and pointed over with 1 inch or ¾ inch rope, with the pointing often extended, especially on the Continent, in both directions beyond the mouse itself. Usually, however, the eye, the collar (the part of the stay between eye and mouse) and up to 1 fathom below the mouse were wormed, parcelled and served. The remaining length of the stay was also wormed, to make it more round and give it more strength (see Figure 32c and d).

A heart was turned into the lower end of the stay and seized in the same manner as the shroud deadeyes (see Figure 32d). On French and some Continental ships a double or treble block was often used instead of a heart (see Figure 32d).

Stays stabilised all masts in the forward direction, but as the position and size of each mast varied, so did the size and method of setting up each stay. Common to all lower stays was the way in which they were set up to the masthead: the stay was passed through the eye in its own upper end, until the mouse formed a stop, thus creating a large collar. This collar was then laid over the masthead above the shroud bights.

Fore stay

Vorstag, Fockstag; Étai de misaine

The fore stay on English ships was a strong, four- stranded cable-laid rope; on the Continent a three-stranded cable-laid stay was more common. After its upper end had been laid around the masthead, the stay passed down through the fore top lubber's hole and the heart turned into its lower end was set up with a lanyard to the fore stay collar (see Figure 32e; for details of this collar see Bowsprit in Chapter II).

Fore preventer stay

Vorborgstag; Faux étai de misaine

Fore and main stays were usually rigged double, with the second stay known as a preventer or spring stay. This functioned as a back-up stay in case of damage, and was about two thirds the diameter of the stay it supported.

Lever discussed the setting up of the preventer stay, though his comments apply mainly to the end of the century, suggesting that sometimes the preventer stay was laid over the masthead before the stay, but more frequently after. The preventer stay then led through the stay collar so as to run under the stay, to facilitate the bending of a staysail to it.

The fore preventer stay was therefore set up to the stay collar furthest inboard on the bowsprit. Steel amplified Lever's comments by noting that on ships of 20 guns or fewer the fore preventer stay was sometimes placed below the stay so that a staysail could be bent to it, but that as a rule the preventer stay was set above the stay. He was quite specific that the fore preventer stay was only set below the fore stay on smaller vessels, and only in the last decades of the century.

Röding noted that the fore preventer stay was rigged above the stay, but qualified his remark thus:

> In England, however, [the fore preventer stay] is rigged below the actual stay, and thus serves at the same time as a staysail stay. This usage is now very widespread, because it saves setting an extra staysail stay under the stay.
>
> Large ships have only four loose stays (preventer stays), namely the main loose stay, the loose stay of the foremast, the loose stay of the main topmast and also the loose stay of the fore topmast, of which the last can only be found on very large ships.

His remarks, made at the same time as Steel's, suggest a very wide acceptance of the preventer stay as staysail stay in the 1790s and probably even earlier. The emergence of preventer stays has been dated by Anderson to the last decades of the seventeenth century, but neither *Der geöffnete See-Hafen* of 1705, Davis' *The Seaman's Speculum* of 1711, nor Sutherland's *The Ship-builder's Assistant* of 1711 made mention of preventer stays.

Main stay

Gross-Stag; Étai de grand mât

The main stay was slightly thicker than the fore stay and led down from the mainmast head to the aft side of the foremast, or past its starboard side, ending with a large turned-in heart or block set up with a lanyard to a stay collar.

The fitting of this collar was not standardised. On the contrary, Steel, Lever, Röding, Boudriot, Lees and others each provided a different description of the setting-up of a main stay and its collar. Generally they agreed, however, that men-of-war had a stay collar which was either hooked to the gammoning knee, led through a hole in the cutwater or in some other way fastened around or beneath the bowsprit.

Since Röding offered a clear description of the differences between Continental and English methods of fitting a main stay collar, his comments are relevant here:

> In French practice the collar or span is led around the reversed knee in the cutwater and the foremast, in such a way that the block lies abaft the foremast [see Figure 34a, b and c]. In English practice, by contrast, no such heavy span or collar is needed: the stay leads on the starboard side past the foremast and is set up with hearts, one on the stay and the other on the collar, to a much shorter collar in the head which goes through a hole in the ship's cutwater [see Figure 34d).

Steel gave a more detailed description of this second type of collar:

> The Collar reeves from the starboard-side through a large hole in the standard in the head (or a large triangular eyebolt is driven through the stem in some merchant ships), then reeves through the eye in the other end, and is brought down to its standing part, and securely seized and crossed in two or three places, and the end capped; the heart is then seized in the bight.

English main stay collars extended abaft the foremast only in the seventeenth century. The most convincing explanation of Lever's illustrations which appear to contradict this is that setting up the main stay to a collar heart abaft the foremast was revived in merchantmen in the last years of the eighteenth and in the early nineteenth century (see Figure 32f).

Main preventer stay

Grossborgstag; Faux étai de grand mât

The main preventer stay differed from the fore preventer stay only in its diameter, and was set up in the same way to a collar above or below the main stay.

On English ships this collar was short, with a heart seized into the bight and eye splices in each end. It was laid round the foremast and lashed at the fore side of the mast. Cleats nailed to the mast prevented it from slipping upwards (see Figure 32j). Steel noted an alternative to the main preventer stay collar in merchantmen: '...or through a large eyebolt in the head, the same as the main-stay-collar'. The preventer stay passed the foremast on the port side.

A method of setting up a main preventer stay without hearts and a collar on small ships was given by Lever:

> The Stay is first reeved through Hanks, for bending the Sail to; and then an Eye is spliced in the lower end, taken round the Foremast under a Cleat [see Figure 32k), and the other end reeved through it. At the upper end of the Stay, there is another Eye: and a Pendant of the same size as the Stay, having an Eye in one end, is spliced into the Stay at where the Mouse would be: these are set up with a Lashing, or Lanyard, reeving alternately through the Eyes, abaft the Mainmast Head above the Rigging, by a Spanish Windlass.

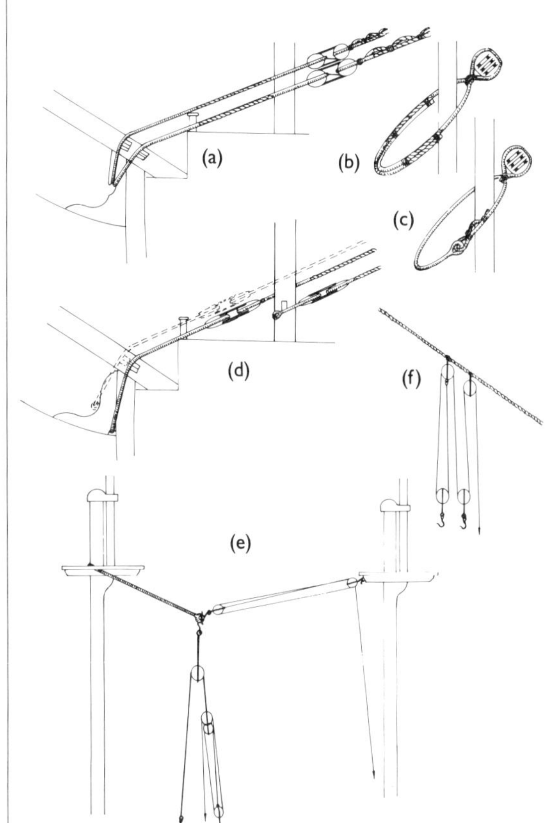

Figure 34
(a) Fitting of a French main stay and main preventer stay.
(b) A Continental main stay collar.
(c) An alternative type of Continental main stay collar.
(d) Fitting of an English main stay and main preventer stay; the dashed lines show a main preventer stay according to Steel.
(e) The use of a Spanish tackle as an alternative to a stay tackle; Continental, 1705.
(f) Stay tackle on an English merchantman, 1705.

On the Continent the main preventer stay collar was set up in the same way as the main stay collar. While the main stay collar was led directly inboard between the beakhead bulkhead and the beakhead rail, the preventer stay collar was led past cleats on the bowsprit to the forecastle above the rail.

By another method, both main and preventer stay collars were prepared in the English fashion but with a long collar and the blocks abaft the foremast. Each collar therefore had a long and a short leg, with an eye spliced into the long leg through which the short leg was led, reversed and seized two or three times to the incoming part.

The Continental method of setting up a main preventer stay collar in a similar manner to the main stay collar was used in some English ships in the last decade of the eighteenth century only.

Snaking

Verschlingung; Serpentage des étais

Frequently during times of war the main and fore stays were snaked together with their respective preventer stays. Röding added '...this is usually done before action, so that if either the stay itself or the loose stay [that is, the preventer stay] as well as the snake line is shot through by a cannon ball, it will remain in place, held by the other'. The stays were snaked by reeving a thinner line between them from the mouse down to the heart (see Figure 33a). Each zigzag connection was seized to the respective stay. Snaked stays obviously could not carry staysails, since the snaking prevented the staysail hanks from moving up and down; staysails were therefore bent to staysail stays rigged additionally for the purpose.

Stay tackles

Stagtakeln; Palans d'étai

The main stay frequently carried one or two tackles, which hung down over the hatches to facilitate the loading and unloading of goods and the hoisting heavy objects such as boats. These tackles consisted of a double (long tackle) block and a single block, with the double block spliced to a pendant which in turn was seized to the stay (see Figure 33b). The lower, single block was iron-bound and provided with a hook.

If two tackles were set up on the stay, then one was above the main and the other above the fore hatch. The fore hatch tackle came into use around 1740 and was initially rigged to a pendant like the main hatch tackle. This pendant disappeared around 1775 and the upper block was then lashed directly to the stay. Steel noted that a main stay tackle pendant had an eye spliced in one end and a double block in the other, with both splices served. The eye was seized to the stay.

It is interesting to note that the author of *Der geöffnete See-Hafen* made mention as early as 1705 of two large single blocks, lashed direct to the main stay above the main hatchway, with two other single blocks, stropped with thimble hooks, forming the lower part of the tackle. The fall was secured to the lower end of one of the stay blocks, passed through one lower block, the stay block, downwards again to the second lower block, back to the second stay block and finally upon deck (see Figure 34f). The same description included the comment that this tackle served on English merchant ships to lift goods in and out of the main hatchway, but that German crews preferred to use the Spanish tackle of the mainmast for this purpose. A tackle was hooked into the pendant's thimble and to the aft side of the foremast top to haul the Spanish tackle into position (see Figure 34e).

Tackle pendants were not always rigged in this manner. Towards the end of the century a running pendant was also in use, according to the rigging plan of a 20-gun ship provided by Steel. In this plan the pendants were not seized to the stay but were fitted with tricing lines, which points to the movable nature of these pendants. Such pendants were spliced around thimbles on the stay. The tricing line led through a block at the stay collar, under the main top (or, as Steel indicated, on the mainmast head) and then upon deck (see Figure 33c). In this way the position of the tackle could be controlled from the main deck. The Royal Navy abandoned this idea after a time, because of chafing damage to the stay from the thimbles.

Until about 1730 English ships were rigged with garnet tackles, and these were preferred on Continental ships for most of the century. Falconer described the garnet as 'a sort of tackle fixed to the main-stay of a merchant ship, and used to hoist in and out the goods of which the cargo is composed'. A garnet pendant had eyes spliced into both ends. The pendant led through the upper eye in the manner of a stay, and the loop so formed was put around the mainmast head. This loop had a length of approximately half that of the main stay collar. Seized into the lower eye was a double or long tackle block, and a rope some 2 fathoms long was seized to the block. Another thimble stropped single block went over this rope, and the end was then hitched and seized to the pendant, thus forming a triangle. This rope was mostly dispensed with in English rigging. On English vessels the garnet guy was slipped with an eye on to the lower eye of the pendant and led through a block on the foremast head down to the fore jeer bitts. On Continental ships the standing part of the garnet guy was fastened to the foremast head, and only after passing through the single block in the triangle did the hauling part follow the English lead to the bitts (see Figure 33d).

Boudriot noted that stay tackles were fitted temporarily to French ships only when the need arose. This is much in line with the description given in *Der geöffnete See-Hafen.*

Mizzen stay

Besanstag; Étai d'artimon

For most of the eighteenth century the mizzen stay was set single. Where it was deemed necessary, a preventer stay was rigged in the Royal Navy only after 1793. Continental rigging made no use of mizzen preventer stays during the eighteenth century.

The mizzen stay was laid round the mizzen masthead and led down abaft the mainmast. Like the other stays, the mizzen stay had a block, heart or deadeye turned into the lower end and was set up to a stay collar lashed to the mainmast (see Figure 33e).

This rig, dating from the seventeenth century, remained largely unchanged in the eighteenth; a modification of the mizzen stay lashing appeared only in the last decades, and only then in English vessels. In this modification, the stay collar heart was replaced by a smaller heart or large thimble, and the stay, leading through the thimble, was set up on deck abaft the mainmast to small hearts or thimbles (see Figure 33f). Large ships used deadeyes. Röding, describing this mizzen stay setting, noted that the stay was set up to the second deck.

The height on the mainmast of the mizzen stay collar above deck was given as 6 or 12 feet. It is probable that 12 feet was correct when the quarterdeck began abaft the mainmast, and 6 feet when the mainmast passed through the quarterdeck.

Crowsfeet

Hahnepooten. Stagspinne; Araignée

The lower stays were connected to the mast tops with a number of lines to prevent the foot of the topsail from chafing on the rim of the top. Some eighteen to twenty holes were drilled into the rim of the top to take these lines, and a euphroe tackle was set up to the stay below the mouse. A single line led from the holes in the top rim to the euphroe (see Figure 33h)and back until extended, thus forming the crowsfeet; the euphroe tackle was then hauled in to take up the slack (see Figure 33i). This tackle can be seen on contemporary English models throughout the century, but it was first mentioned in Blanckley's *Naval Expositor* of 1750. According to Röding, in Continental rigging the euphroe was seized direct to the stay and a tackle was not used.

Topmasts

Marsstenge; Mâts de hune

The topmast was the first extension of a lower mast, and it rested on a fid at the trestle trees. Topmasts were rigged as follows.

Burton tackle

Hanger, Burtontakel; Palanquins des mâts de hune

First the Burton pendants, cut spliced together, were placed over the topmast head and hung down at the sides of the mast forward of the crosstrees (see Figure 35a). Davis noted a length of 3 fathoms, or half that of the topmast, for these pendants, but later the length was reduced to one ninth of the topmast. Up to 1780 a block was spliced into the lower end of the pendant, but this was later replaced by a thimble to which the tackle was hooked. When single blocks were used, the length of the falls was four times that of the topmast, according to Davis, 'but if you have long Tackle Blocks aloft, the Falls must be 5 times the length of the Topmast'. One of the main duties of a Burton tackle was to set up the topsail shrouds (see Figure 35b). The tackle itself was

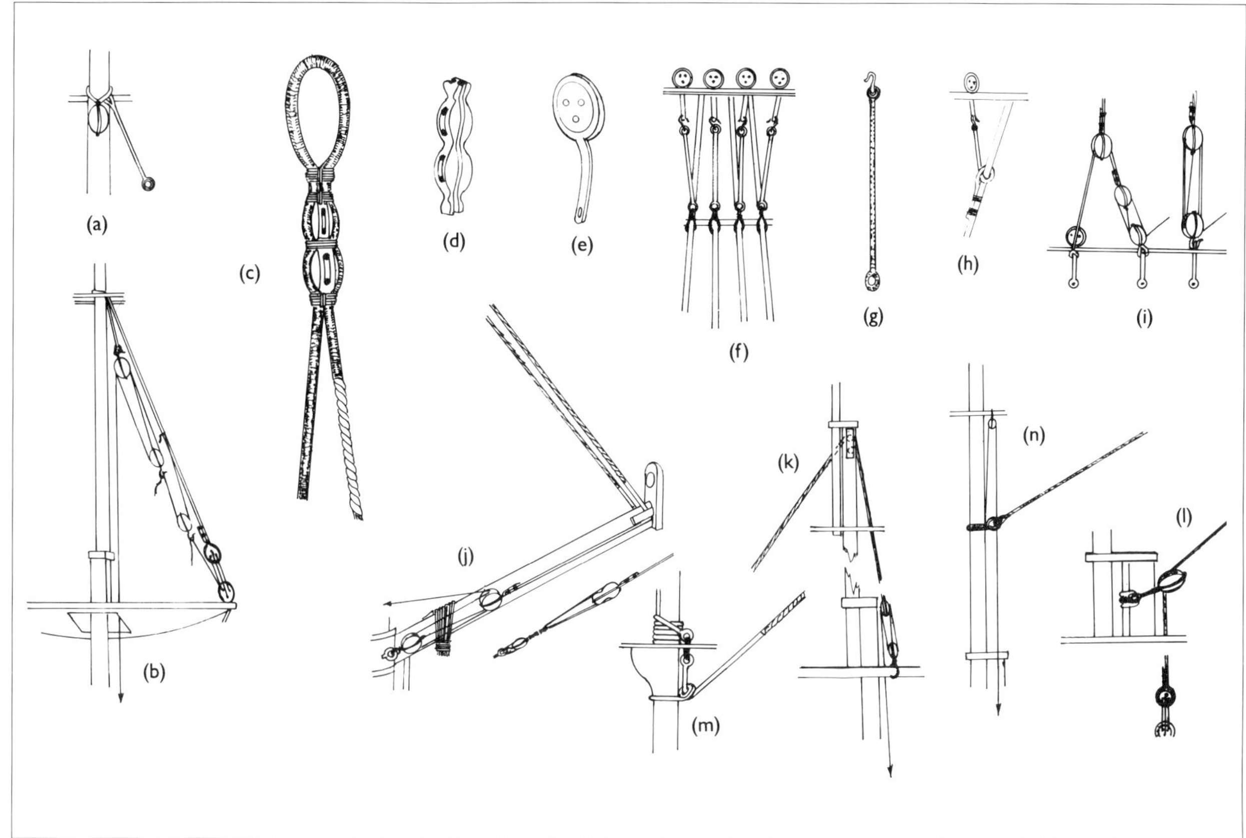

Figure 35
(a) Burton tackle pendant.
(b) A Burton tackle used in setting up a topmast shroud.
(c) A pair of topmast shrouds with a seized-in sister block. The block came into use as a lead block for the topsail yard lift and the reef tackle after 1790.
(d) A sister block.
(e) An iron-bound futtock deadeye.
(f) Setting up of futtock deadeyes and futtock shrouds to the lower shrouds. The futtock shrouds had thimbles at the lower ends which were seized to the shrouds below the futtock stave.
(g) A futtock shroud with a hook and eye.
(h) The securing of a futtock shroud without an eye or lower thimble. The futtock shroud was laid round the futtock stave and shroud and lashed to the latter.
(i) Two methods of setting up a breast backstay.
(j) The rig of a fore topmast and a fore topmast preventer stay on English ships after 1745.
(k) The setting up of a jib stay, leading over a cheek block, on English ships after 1750.
(l) The setting up of a main topmast stay abaft the foremast on English ships.
(m) The setting up of a main topmast preventer stay on English ships, about 1800.
(n) The fitting of a middle staysail stay to a grommet, on English ships after 1775.

a luff tackle, consisting of a double and a single block. When studdingsails were set, the Burton tackle also acted as an additional lift.

When not in use, the Burton tackles were either hooked to the mast top inside the first topmast shroud or removed altogether, in which case the pendants were lashed to the first shroud.

Top rope and top tackle

Stengewindreep; Guinderesse et palan de guinderesse

Röding described the top rope as 'a strong rope, which served for hoisting and lowering the topmast'. Rigged to all topmasts up to about 1800, the top rope on English ships normally consisted in fact of a top rope and a top tackle.

Davis, whose book drew on his experience as a boatswain on English men-of-war, gave the length of the top rope as one and a half times that of the topmast. For a large ship's top tackle, with a treble and a double block, the fall was five times the length of the mast between top and deck or six times the length of the topmast. On tackles with double blocks the fall was shorter by one topmast length.

The top rope was seized to an eyebolt on the starboard side of the lower mast cap and rove through the topmast above the heel (through a sheave if the weight of the topmast necessitated one). It then ran over a block on the port side of the lower mast cap and downward; the top tackle was hooked into a thimble at the end of the top rope and an eyebolt on deck, and the fall ran through a lead block on deck (see Figure 36a).

'The French fit a top rope to heavy topmasts which goes twice through the topmast footing, over two sheaves' noted Röding, and he explained further that these double top ropes passed through lead blocks on deck towards the capstan (see Figure 36b-b.). Top tackles were not used on French ships. The top rope was also noted by the author of *Der geöffnete See-Hafen*, who implied that it was removed after the topmast had been hoisted into position and secured by the fid.

Double top ropes were used on English ships from about 1675 on (see Figure 36b-a.). Davis' rigging lists from 1711 prove the existence of double top ropes on ships of the Third Rate and above at that point. Love underlined this in 1735, by listing single top ropes only for a ship of the Fourth Rate.

A single top rope and tackle was used for mizzen topmasts for all Rates, and it should also be noted here that Davis listed a single top rope and tackle for all topmasts for HMS *Royal Sovereign*, a First Rate in the Royal Navy.

Shrouds

Hoofdtaue; Haubans

Next in the rigging sequence for topmasts came the shrouds, as for the lower masts. The number of shroud pairs varied according to the size of ship, ranging from four to six, and smaller vessels than ships had fewer. The precise number for vessels of each size can be found in the rigging tables in the Appendix below.

The foremost of these shrouds was again fully served to prevent damage through chafing from the topsail.

The size of the topmast shroud deadeyes depended upon the diameter of the shrouds, and are given in the rigging tables in the Appendix.

During the last decade of the eighteenth century (and mainly on English ships) sister blocks were seized to the foremost shroud pairs, just below the shroud pair seizing (see Figure 35c and d), to act as lead blocks for later rigging.

Futtock deadeyes

Püttingjuffern; Caps de mouton de revers

To connect the topmast shrouds with those of the lower mast, a number of deadeyes were set into the strengthened side rims of the lower mast top. These so-called futtock deadeyes were bound with a flat-section iron strop which extended through the rim of the top and was welded together on the lower side, with a hole through which the futtock shrouds were hooked (see Figure 35e). These irons were known as futtock plates. Earlier, the

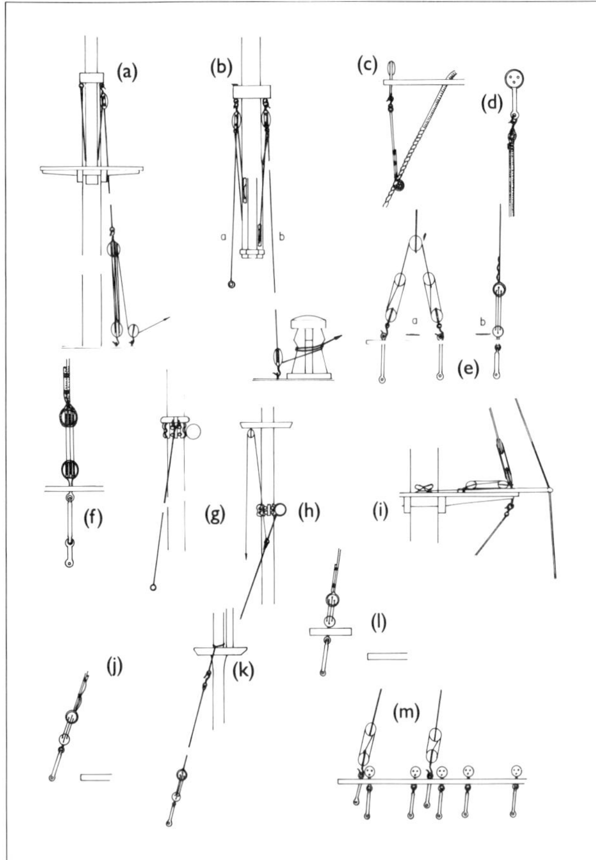

Figure 36
(a) Toprope and toptackle of an English ship of less than Third Rate.
(b) Double toprope
a. English method
b. French method
(c) Futtock shroud fitting, according to Röding.
(d) French futtock shroud, according to Boudriot.
(e) English breast backstay, 1711
a. According to Sutherland
b. According to Davis
(f) French breast backstay, 1780.
(g) Travelling backstay, on a traveller.
(h) Travelling backstay, on a span.
(i) Breast backstay outrigger.
(j) Continental backstay on smaller ships, 1705.
(k) Continental backstay on larger ships, 1705.
(l) Backstay on a stool.
(m) English backstays, 1711, according to Sutherland.

futtock plates had been made from round-section iron and forged flat at the end, and this feature was retained in Continental rigging throughout the century. Where English influence can be observed, as for example on the model of the Russian First Rate *Zacharii i Elisavet* of 1748, flat section plates were used.

Futtock shrouds

Püttingwanten; Haubans de revers

Each futtock shroud consisted of a short rope with a hook and thimble spliced into the upper end and sometimes a thimble or eye into the lower. Those without the lower thimble were set up with the lower end turned once around the futtock stave and seized to the nearest shroud (see Figure 35h). When a thimble was spliced in, a lanyard was fastened to it and lashed alternately round the futtock stave and the lower shroud, with the end frapped round the lashing (see Figure 35f). Lever noted that the latter method had the greater merit, since it did not strain the lower shrouds as much and was not likely to chafe.

The shrouds were wormed and occasionally served. The hooks were moused to prevent them slipping out.

Alternative methods of fitting futtock shrouds were given by Röding and Boudriot. According to Röding, the lower end of the shroud was hitched to the futtock stave and seized to itself (see Figure 36c). Boudriot described the futtock shroud as a middled rope with the thimble and hook seized in the bight, but otherwise set up as described above (see Figure 36d). The foremost shroud was served all over for protection against chafing, but the others were served only around the thimble. This method was also used on the *Zacharii i Elisavet*. This small point suggests again that Continental ships were not necessarily rigged strictly according to one method or another, but often combined the best features from several schools of thought.

Ratlines

Webleinen; Quaranteniers, enfléchures

Ratlines on topmast and futtock shrouds were set up to the same distances as on the lower shrouds, but with all shrouds rattled.

Breast backstays

Brustbackstage; Galhaubans volants, galhaubans à croc

Next the breast backstays, the backstays and the stay were put over the topmast head.

Breast backstays, or running backstays, led down to the middle of the channels and were not always in use. They served to provide additional lateral stability to the topmasts when the ship sailed upon the wind. They were, on models, frequently set up to deadeyes, but Steel and Lever noted the use of blocks. Since both authors wrote about the English rigging, one might easily assume that deadeyes were more common in Continental rigging. In many instances, however, they were also used in English ships in the early years of the century, as Davis noted in his rigging lists: 'Measure out ½ the Pair of Breast Stayers and Half the Pair of Back Stairs with deadeyes in the Chains' (see Figure 36e-b.). Sutherland, writing in the same year, gave a contradictory illustration of a breast backstay, set up to a span but with two block tackles (see Figure 36e-a.).

Röding, though he did not precisely identify breast backstays, noted that all backstays were set up to deadeyes; his only detailed reference to breast backstays was a note that backstays were also used 'to secure them [topmasts and topgallant masts] laterally to the ship, since the shrouds fastened to the mast tops or cross trees are not sufficient for this purpose.'

Blocks and deadeyes can be found on French ships. The rigging plan of the *Royal Louis* shows deadeyes for the breast backstays, while Boudriot's roughly contemporary *74 Gun Ship* has treble blocks (see Figure 36f). It is possible to interpret the latter fitting as running backstays, while the deadeye set up indicates a standing breast backstay.

Though he did not identify breast backstays as such, Röding did take specific note of shifting backstays; they are also mentioned in Falconer's 1815 edition: 'and the third kind [of backstay] takes that name from being shifted or changed from one side to the other, as occasion requires'. Shifting backstay was therefore not synonymous with running backstay. Running backstays came into use on English ships in the fourth decade of the eighteenth century, and shifting backstays are only found later, though the term 'shifting' was used by Blanckley in 1750: 'backstays... are sort of Shrouds, which go up to the Topmast-head, hath Lanyards reev'd through deadeyes, and Backstay Plates at the Ship's Side, are called standing or shifting, and are for succouring the Topmasts'. The term 'shifting' may, however, have been used here to describe the action of pushing the luff breast backstay away from the masts top with an outrigger so as to increase the backstay's angle (see Outrigger below). Only a few years later Falconer makes no mention of a shifting backstay, but does note the use of an outrigger; the 1815 edition of his Dictionary described the fitting in its improved form, calling it a travelling backstay.

In his *Art of Rigging* of 1818 Steel described shifting backstays thus:

> Shifting-Backstays are fitted as other backstays, and have a thimble spliced in the lower end, to which is hooked a luff-tackle, the lower block of which is hooked to an eyebolt without-board, and frequently shifted from place to place.

In Röding's use of the term, a shifting backstay was an additional backstay, set up in conditions of strong wind or heavy rolling on the luff side and moved as necessary when the vessel tacked.

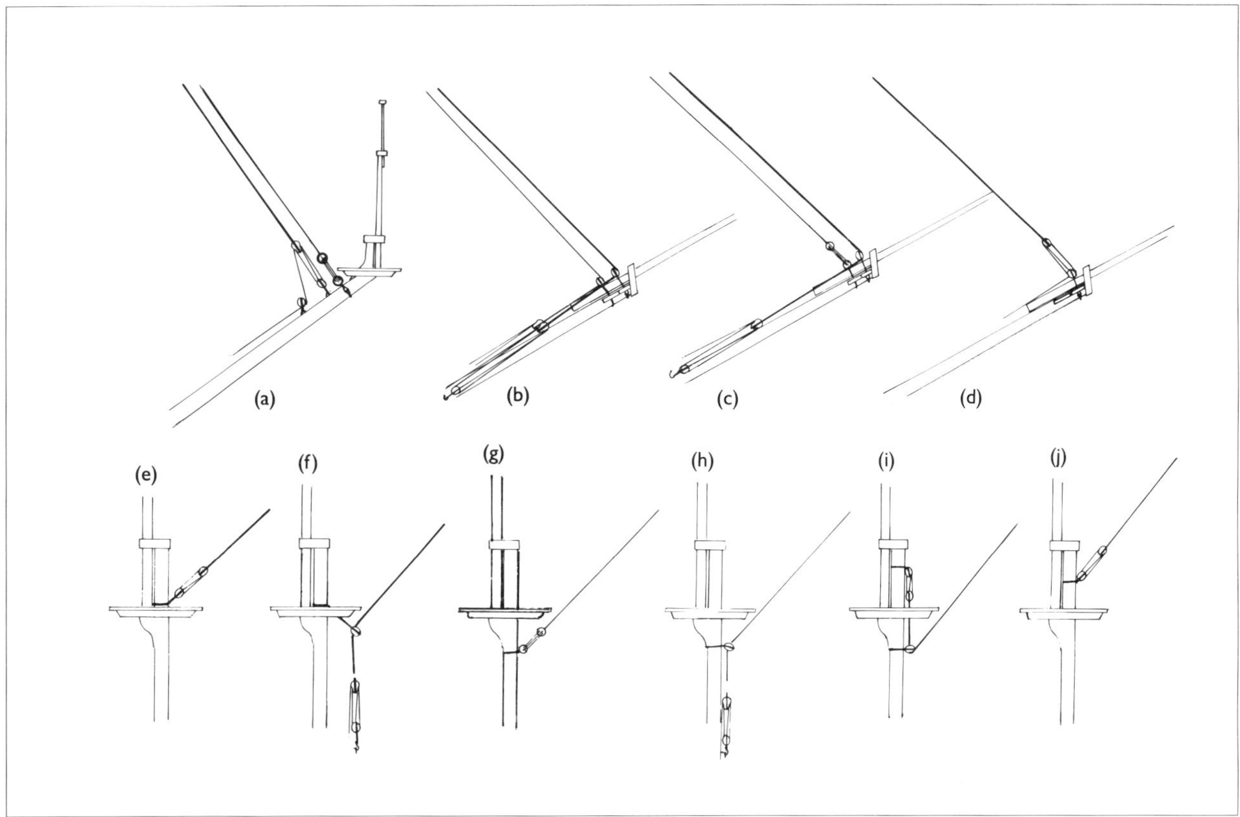

Figure 37
(a) Fore topmast stay and preventer stay until the end of the spritsail topmast period.
(b) English arrangement of fore topmast stay and preventer stay 1700 to 1745 (leading over blocks and stropped to the bees).
(c) Alternative English arrangement of the same period, with the preventer stay set up to deadeyes.
(d) The Continental method of setting up a fore topmast stay, up to 1800.
(e) The lower mounting of a Continental main topmast stay after 1725.
(f) The same stay on Continental ships before 1725.
(g) English main topmast preventer stay, before 1720.
(h) English main topmast preventer stay, after 1720.
(i) Alternative rigging of an English main topmast preventer stay fitting after 1745.
(j) Continental main topmast preventer stay, after 1725.

The 1815 edition of Falconer's *Dictionary* defines travelling backstays as follows:

> Travelling Backstays are so denominated from their having a traveller upon the topmast, which slides up or down according to the reefs in the topsail, thereby confining the principal support of the backstay to that part of the mast immediately above the top-sail-yard [see Figure 36g].

In 1818 Steel described the same stays spliced to a span beneath the parrel and set up to the chains with a luff tackle. Tricing lines were spliced into thimbles on each side of the span, and these passed through blocks on the trestle trees and led down to the top (see Figure 36h).

Outriggers

Ausleger, Jütte; Arcs-boutants de galhauban volant

The outrigger, used in connection with the breast backstays, was described clearly in Falconer's *Dictionary* of 1769:

> [The] Out-Rigger is also a small boom, occasionally used in the tops to thrust out the breast back stays to windward, in order to increase their tension, and thereby give additional security to the topmast.
> This boom is usually furnished with a tackle at its inner-end, communicating with one of the top-mast-shrouds and has a notch on the outer end to contain the back-stay, and keep it steady therein. As soon as the back-stay is drawn tight, by means of its tackle in the chains, the out-rigger is applied aloft, which forces it out to windward, beyond the circle of the top, so as to increase the angle which the mast makes with the back-stay, and accordingly enable the latter the better to support the former [see Figure 36i].
> This machine is sometimes applied without any tackle; it is then thrust out to its usual distance beyond the top-rim, where it is securely fastened; after which the back-stay is placed in the notch, and extended below.

Backstays

Pardunen; Galhaubans

Backstays counteracted the forward pull of the stays, thus further stabilising the topmast. They were set up in similar fashion to the shrouds.

At the end of the seventeenth century the chainplates for the backstay deadeyes were either bolted to the wales or set up to channels. Davis noted for the three lowest Rates that 'their deadeyes are commonly higher than the Chains', and the higher Rates 'with deadeyes in the Chains'. The author of *Der geöffnete See-Hafen* offered a slightly different description: he noted that only backstays on smaller ships were set up like shrouds (see Figure 36j), and that those on larger ships were set up to a pair of short pendants with an iron thimble spliced into the lower ends, which reached downwards just below the trestle trees on both sides. The upper end of the backstay was fitted with a thimble and hook which hooked into the thimble on the pendant, while the lower end was set up to deadeyes as noted above (see Figure 36k). Two backstays were fitted to each side.

At about the same time, new extra channels for the backstays were introduced in English ships, usually placed abaft and somewhat above the channels. Whether Davis' note (quoted above) on deadeyes for smaller men-of-war had any connection with this innovation cannot be established with certainty. These small channels, aptly known as backstay stools, disappeared around 1770 on English men-of-war and backstays were then set up to an extended channel. Other sources indicate the survival of backstay stools until about 1800. Röding listed backstay stools only as a part of rigging for merchantmen: 'on some merchantmen a small, special channel is fitted so that the ropes [backstays] can be led a little further aft'. Steel still listed them, and even the 1815 edition of Falconer included a specific entry on backstay stools; Lever associated the stool only with the rigging of small vessels (see Figure 36l).

As in many other cases, this survival of the term in contemporary literature suggests that the stool was still in use long after its disappearance from Royal Navy rigging. Long established methods were not immediately replaced by new ones.

The number of standing backstays varied according to the size of the vessel and between shipbuilding nations. While French ship models, paintings and drawings often reveal only one pair, two were specified generally in Röding's work, and in fact he noted three to four for the main topmast of large ships and one or two for the topgallant mast, though of course this includes breast and shifting backstays. Information about French ships of the line of the same period corresponds with Röding.

English sources note three pairs on ships of the largest order, and Steel specifically listed three pairs for ships above 74 guns, two pairs for ships of 74 to 20 guns, and one pair for ships of 18 guns or fewer. Until about 1720 two pairs were set up to the stools and the third to the channels; later only one pair was set up to the stool and the others to the channel (see Figure 36m).

While the standing backstays on English ships were usually rigged abaft the shrouds, and the topgallant backstays abaft those of the topmast, French rigging differed. On French ships the topmast backstay was aftmost, with a second backstay set up between the shrouds about two or three shrouds forward, and the topgallant backstays were set up between and forward of those of the topmast.

Stays

Stage; Étais

Last to be fitted over the topmast head were the stays, first the topmast stay and then the topmast preventer stay, with the latter sometimes led through the eye of the stay to run below it. Whereas the arrangement of the topmast stays in their upper sections was very similar to that of the lower stays, the set up of the lower ends differed greatly.

Fore topmast stay

Vormarsstengestag; Étai du petit mât de hune

Concurrent with the change in head gear necessitated by the substitution of a jib-boom for the sprit topmast came alterations to the rigging of a fore topmast stay.

Up to about 1720 the fore topmast stay led down to the sprit topmast standard or abaft it, where the stay was set up to a long tackle and a single block (sometimes hooked to a bolt in the bowsprit close to the bowsprit top), with the hauling part of the tackle fall also passing through a lead block on the bowsprit (see Figure 37a). With the introduction of the jib-boom and its associated cap and bees (after 1700 on smaller and about 1720 on medium sized ships), the fore topmast stay passed through a lead block stropped through the foremost holes in the bees and around the bowsprit, and positioned above the jib-boom. A long tackle block was spliced to the lower end of the stay, and a single block with a hook and thimble was hooked to an eyebolt near the stem (see Figure 37b).

Around the middle of the century the stay was led through the forward starboard bee hole and over the newly-established bee block towards the bow, were the aforementioned long tackle was hooked either to an eyebolt near the stem, or to a strop round the bowsprit abaft the gammoning (see Figure 35j).

Continental ships did not follow English rigging of fore topmast stays in every case. Chapman, for example, showed the fore topmast stay of a frigate set up to deadeyes, and the model of the Russian First Rate *Zacharii i Elisavet* of 1748 also has the stay set up to deadeyes; the rigging plan of the *Royal Louis* of 1780, as seen in Paris' *Souvenirs de Marine*, shows the stay set up in the bees, and Röding's illustrations also clearly show the stay tackle set up in the bees (see Figure 37d). His description was as follows:

> [The stay] is set up to a collar at the fore end of the bowsprit. Merchantmen and small vessels usually have no collar, and the stay reeves on the starboard side over a cheek block nailed to the bowsprit, and is set up with a tackle to the stem [see Figure 38b).

Common to all three authors is the absence of a preventer stay, and the provision of a staysail stay for the fore top staysail. The model of the *Zacharii i Elisavet*, however, differs.

Boudriot's drawings in *Le Vaisseau de 74 Canons* of these stays need some explanation. The rigging plans included as a frontispiece in all four volumes show all stays rigged in English fashion, while the detailed stay drawings in Volume 3 follow Continental practice and agree in many points with Röding's description.

In English rigging the fore topmast stay led over a block stropped abaft the bees to the bowsprit, and was set up to a tackle abaft the gammoning which consisted of a long tackle and single block. Boudriot described this as an alternative to the Continental rig, in which the stay collar was stropped around the middle of the bees. A single block was turned into the stay and a long tackle block was stropped to the collar. The end of the fall, spliced to the single block, was frapped round and stopped to the stay (see Figure 38a).

Fore topmast preventer stay

Vormarsstenge-Borgsstag; Faux étai du petit mât de hune

On ships with sprit topmasts the fore topmast preventer stay was set up to deadeyes forward of the fore topmast stay, with the lower deadeye seized to a collar (see Figure 37a). On ships with jib-booms the preventer stay ran in a similar way to the stay, except that the lead block strops were led through the aftmost holes in the bees (see Figure 37b). In some ships the lead block was replaced by a deadeye, to which the preventer stay was set up (see Figure 37c).

Continental ships were rarely rigged with a fore topmast preventer stay. Röding noted that only the largest ships carried one. On English ships the preventer stay was common from the beginning of the century onwards, according to Lees, but the works of Sutherland and Davis from 1711 made no reference to it; nor was a fore topmast preventer stay included in any of their rigging lists.

After the introduction of the bee blocks, in the late 1740s, the preventer stay led through the aftmost hole of the port side bee. The foremost hole on the port side and the aftmost one on the starboard, both without a bee block sheave underneath, were kept free for emergency stays.

On Continental ships the fore topmast preventer stay, when rigged in English fashion, ran through a lead block stropped to the bowsprit abaft the fore topmast stay and was set up in the same manner. Normally preventer stays were set up to deadeyes, as in the alternative English rig described above.

Jib stay

Klüverstag; Étai du grand foc

The setting up of a jib stay depended on the type of traveller used, and details are given Chapter II under Traveller. The jib stay came into use around 1720, about 15 years after the official introduction of the jib; before this the jib was set flying. In the last quarter of the century jibs were often rigged flying again.

In the years before the introduction of the topmast cheek blocks around the middle of the century, the upper end of the stay led over a single block fastened to the starboard side of the topmast cap; after the introduction of the cheek blocks the stay led through the starboard one of these. On smaller ships the upper end of the stay was hitched directly to the aftmost cross tree, while on larger ships it was set up with a tackle composed of a double and a single block, with the single block at the cross tree as before and the fall leading upon deck to belay on the fore jeer bitts (see Figure 35k).

In this set up an outhauler was clinched to the shackle of the traveller. The outhauler ran through the sheave hole in the outer jib-boom end, and a double or long tackle block was turned into its end; the fall of this tackle rove through a single block hooked to an eyebolt in the face of the bowsprit cap, and the hauling part led to the forecastle.

The inhauler was secured to an eyebolt in the side of the bowsprit cap and ran through a block lashed to the traveller, with its hauling part leading again to the forecastle.

At the end of the century the jib stay was also sometimes laid with an eye over the topmast head. The lower end then led over a reel on top of the traveller, through the sheave hole in the outer jib-boom and was finally set up like the outhauler described above.

Travelling guys were connected to both types of traveller, as described in Chapter II under Jib-boom.

Continental ships were sometimes rigged with the former type of traveller for the outer jib and a travelling grommet for the inner jib. Both had an outhauler and inhauler, with the outhaulers reeving through sheaves in the jib-boom truck and the inhaulers being clinched to the travellers and leading in to the forecastle. Boudriot also mentioned trucks on the lower section of a traveller ring to allow easier movement.

Röding did not elaborate on jib stays, writing of jibs both with and without a staysail stay. A Continental staysail always required a stay; it was fastened near the stay's mouse and its lower end was set up to a tackle. The English type of jib stay, as described by Steel, received a short mention in Röding's description of a staysail stay. According to Röding, it was fastened to the traveller and ran through the jib's double sheaved halyard block upon the forecastle, where it was set up with a tackle. Cheek blocks as used in the English rigging did not feature on Continental masts and Röding consigned them to the appendix of his English index. His description of the traveller noted only that the jib tack was hitched to it, which implies a flying outer jib. Chapman's rigging plan of a frigate also shows a flying outer jib, as does the Paris sail plan of the *Royal Louis*. In general, however, the outer jib (also called the large or standing jib on Continental ships) was hanked to a staysail stay, and the inner jib was also referred to as the second or flying jib.

Boudriot's description of the standing jib of a 74-gun ship in the French fleet of around 1780 included English-style cheek blocks, but such blocks were not shown in his detailed drawings, nor indeed anywhere else in his work. An extension of the staysail stay and the jib halyard towards the topmast cap, below which the cheek blocks were placed, cannot be proven from other sources either, even from Ozanne's excellent illustrations or the rigging plans included in Paris' *Souvenirs de Marine*. In all these illustrations the staysail stay and jib halyard run without exception to the topmast trestle trees. Boudriot's description therefore is unsupported by illustrative sources (even his own) for French ships.

Main topmast stay

Grossmarsstengestag; Étai du grand mât de hune

The main topmast stay on English ships ran to the forecastle through a block stropped above the rigging at the aft side of the foremast head (see Figure 35l). At the forecastle it was set up with a tackle or deadeyes to an eyebolt abaft the foremast.

On French and other Continental ships in the early years of the century the stay led over a long-stropped block below the aft side of the fore top and was set up with a tackle comprising a long tackle and a single block to the forecastle deck (see Figure 37f). Later it was set up to a tackle fastened to the masthead rigging of the foremast (see Figure 37e). Röding's note was: 'the eye of the [main topmast stay] is laid round the main topmast head and the collar round the fore masthead'. Evidence that both set-ups described were in use in some ships as early as the beginning of the century is provided by *Der geöffnete See-Hafen*.

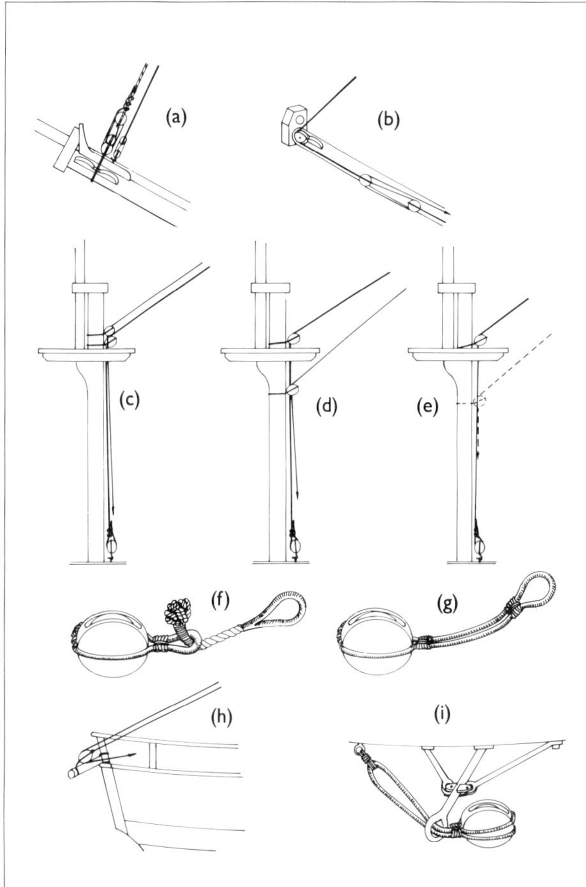

Figure 38

(a) French fore topmast stay and preventer stay, 1780.

(b) Continental fore topmast stay for merchantmen and small vessels, according to Röding (1794).

(c) French main topmast stay and preventer stay according to Boudriot, 1780.

(d) French main topmast stay and preventer stay according to Boudriot, 1759.

(e) French mizzen topmast stay according to Boudriot; the solid line shows the arrangement in 1780, the dashed line the arangement in 1759 and the alternative rig in 1780.

(f) A Dutch brace pendant from the beginning of the century.

(g) A French brace pendant.

(h) An English early brace spreader on an brig, 1790.

(i) Brace spreader (spider) according to Nares, 1862.

The English method of rigging a main topmast stay and main topmast preventer stay was used as an alternative in the French fleet; this can clearly be seen in Ozanne's many illustrations of eighteenth century French ships.

Main topmast preventer stay

Grossmarsstenge-Borgstag; Faux étai du grand mât de hune

In the early eighteenth century the main topmast preventer stay was set up to deadeyes below the fore top (see Figure 37g). After 1720 a block instead of a deadeye was stropped to the foremast below the hounds and the preventer stay was led like the stay upon the fore deck, set up in a similar way (see Figure 37h). An alternative rig appeared near the middle of the century: the block stropped to the foremast was replaced by a thimble, and another thimble was turned into the end of the stay, which now led upwards; this latter thimble was lashed to another thimble stropped to the foremast head (see Figure 35m). Instead of thimbles, blocks were also sometimes used.

When a main topmast preventer stay was rigged to a Continental ship, it usually ran above the main topmast stay and was set up to the foremast head in the same manner (see Figure 37j). Boudriot, however, noted an alternative method for French ships (see Figure 37c). He also noted that instead of a preventer stay, a staysail stay was often rigged below the main topmast stay, and this was set up to the foremast below the hounds (see Figure 38d).

Middle staysail stay

Mittelstagsegelstag; Draille de contre-voile d'étai

The middle staysail stay came into use during the 1770s. A grommet with a thimble turned in was laid around the fore topmast at half the mast's height. The stay, seized to the thimble, passed over the upper sheave of the port side cheek block on the main topmast, and was set up like a jib stay. The grommet was fitted with a tricing line which led over a block at the aft side of the fore topmast trestle trees and upon deck (see Figure 35n).

During the 1790s a jackstay (or horse) rigged between the fore topmast trestle trees and the mast top increasingly replaced the grommet. It led downwards through an eyebolt in the aft face of the foremast cap, and a thimble was turned into its lower end. This thimble was set up with a lanyard to another thimble seized to the lower cross tree or to the masthead. A tricing line was fitted to the lower end of the stay and, running on the thimble turned into its end, the stay could be moved up and down the jackstay (see Figure 39a).

Röding described the middle staysail stay as leading over a block seized to the main topmast cap. This again confirms that cheek blocks were not used in Continental rigging.

Mizzen topmast stay

Besanmarsstengestag; Étai de mât de perroquet de fougue

At the beginning of the eighteenth century the mizzen topmast stay was generally set up to the mainmast in the same way as the main topmast preventer stay was set up to the foremast. The only difference was in the longer survival of the lashing to the deadeyes, which could still be found as late as 1760 on mizzen topmast stays (see Figure 39b). On mizzen topmasts with long pole heads the stay was set up to the lower aft side of the mainmast head. The stay deadeyes of large ships were replaced by blocks in the 1770s.

This set-up can be seen in the rigging plans of ships rigged according to the British establishments of 1719 and 1745, which had a flagstaff fitted to the topmast instead of a long pole topmast. Rigging plans of First to Sixth Rate ships in Deane's *Doctrine of Naval Architecture* from 1670 also show mizzen topmast stays rigged in this fashion, though on other ships at that time the stay was still set up with crowsfeet to the main shrouds. A plate showing standing and running rigging in Sutherland's *Ship-builder's Assistant* of 1711 shows a similar type of mizzen topmast, but the stay is set up to the mainmast below the top. A certain amount of variation in the rigging of this stay therefore seems characteristic of the first half of the eighteenth century.

Various new methods of rigging the stay developed in English rigging from about 1760 onwards. The stay collar deadeye below the hounds became a thimble or heart and the stay ran through it either upward (to be set up to deadeyes or thimbles on the mainmast head, a fitting which Lever illustrates in 1811; see Figure 39c), or downward, as advocated by Steel. In his description the stay was set up to another collar with a thimble just below the catharpins (see Figure 39d). When the topmast had a long pole head, the stay ran through a block at the mainmast head, and, with a thimble turned into the end, was set up to a span made fast to the main trestle trees.

Around 1773 further developments occurred, in both English and Continental rigging. An eyebolt was driven through the main top and aftmost cross tree and a thimble turned into the lower end of the stay, and the stay was then hauled taut with a lanyard (see Figure 39g). Otherwise mizzen topmast stays on Continental ships were set up to the lower part of the mainmast head with deadeyes or blocks (see Figure 39f), as Röding confirmed by his mention of a collar for this stay round the mainmast head. Other Continental sources also support this, including *Der geöffnete See-Hafen* as early as 1705 (the author notes blocks in the end of the stay, and the stay set up to the mainmast just above the main top). Boudriot described two methods of setting up a mizzen topmast stay for his 74-gun ship and a third for an East Indiaman of 1759. In the first the stay was set up like the main topmast stay and passed through a block at the mainmast head; in the second the lead block was lashed to the aft side of the mainmast, at about two thirds height between deck and top. In both cases the stay was set up to a tackle on deck (see Figure 38e). The third method was identical to the English rig noted first in this section.

Staysail stays, when set, usually ran below the stay either down to the main top or to the mainmast at the height of the catharpins.

Fore yard and main yard

Fockrah und Grossrah; Vergue de misaine et grand vergue

The main yard and the fore yard were used to set squaresails. In fitting out both yards were identical; they differed only in size.

Horses

Fusspferde; Marche-pieds

An eye in the outer end of the horse was fitted over the yardarm (see Figure 39k). A deadeye was turned into the inner end of the horse, and early in the century this was set up to another deadeye on the yard just inboard of the near sling cleat (see Figure 39h). Later, until about 1760 in the Royal Navy, the deadeyes from both horses were seized together to create a continuous foot rope (see Figure 39i); this practice was already in use on the Continent in 1705, Röding still noted it as a fitting for Continental ships after 1790 and French horses towards the end of the century were rigged similarly but with thimbles spliced in instead of deadeyes (see Figure 39k, right). In English rigging after 1760 thimbles were also used, but the horses crossed in the middle and were lashed to eyebolts underneath the yard or to the yard itself outside the opposite sling cleat (see Figure 39j).

For details of life-lines see Main and fore topsail yard, Horses (below).

Stirrups

Springpferde; Étriers de marche-pied

To maintain an equal distance of about 3 feet from the upper side of the yard, (Falconer mentioned 2 feet beneath the yard) each horse ran through two or three stirrups, depending on the yard's length; larger French ships had four. At the beginning of the century one fewer stirrup was used for all vessels.

When stirrups first came into use cannot firmly be established. They were mentioned neither in *Der geöffnete See-Hafen* in 1705 nor by Sutherland or Davis in 1711 and illustrations from as late as 1720 do not show them. It can be assumed that they did not generally feature in rigging in the first quarter of the eighteenth century.

The stirrup itself was a strop with a thimble spliced into the lower end. Above a 3ft length of normal, served rope, a longer end of approximately three times the yard's circumference was plaited; this was then laid three times around the yard and nailed to it (see Figure 39l).

Yard tackles

Rahtakel; Palans de bout de vergue

Next over the yardarm was the yard tackle pendant. It also had an eye spliced into one end for this purpose, and at the other a long tackle block (fiddle block) was turned in. Together with a hook and thimble-fitted single block this formed a tackle which was used to lower or hoist boats and for many other types of work. Davis noted the pendant length as 8 feet, and that of the tackle fall as three times the mast length from deck to the trestle trees (see Figure 40a). Röding's description of a yard tackle was: 'this is either a whip hanging on a pendant, or else a tackle, and on very large ships a winding tackle.' At the end of the century

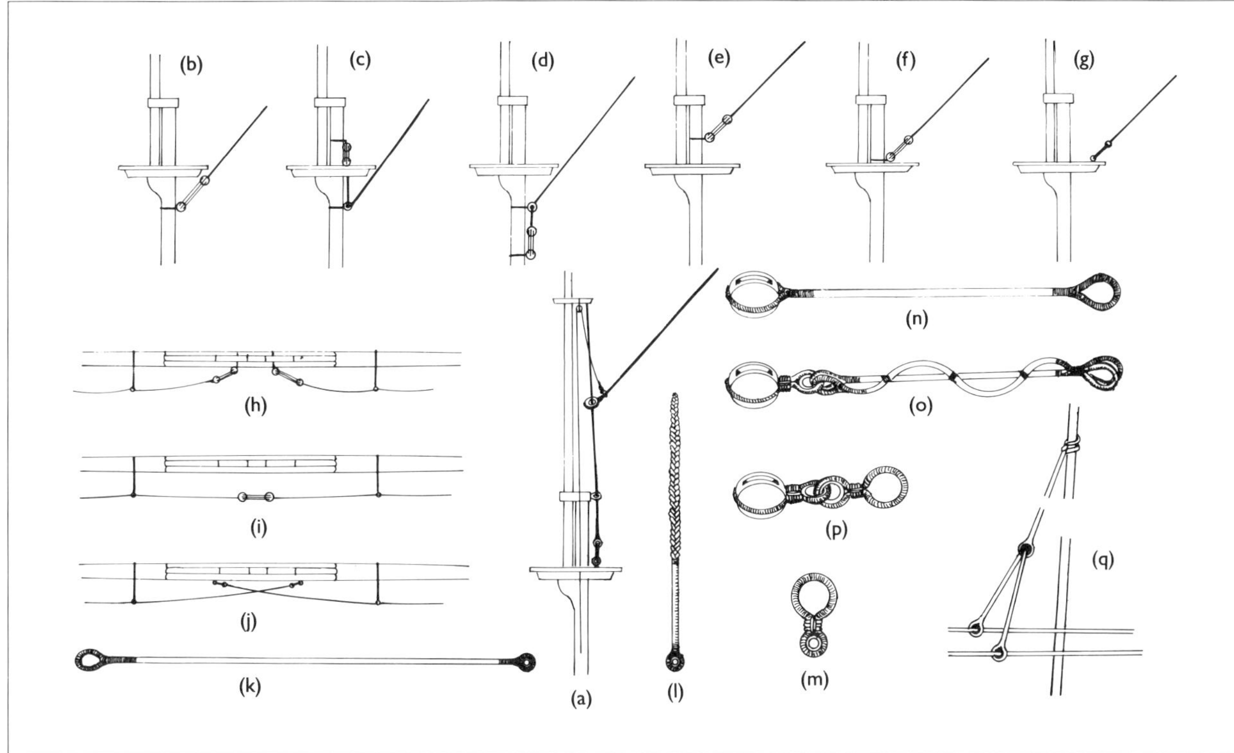

Figure 39
(a) An English middle staysail stay on a jackstay, after 1785-90.
(b) An English mizzen topmast stay, up to approximately 1760.
(c) An English mizzen topmast stay, after 1760.
(d) An alternative method of setting up an English mizzen topmast stay after 1760.
(e) An English mizzen topmast stay for topmasts with poleheads, after 1775.
(f) A Continental mizzen topmast stay.
(g) An alternative set up on English and Continental ships after approximately 1775. The mizzen topmast stay was seized to an eyebolt abaft the main top.
(h) The set up of horses to the sling of an English main yard, about 1700.
(i) The set up of horses on an English yard until approximately 1760, and on Continental yards throughout the century.
(j) The set up of horses on an English yard after 1760.
(k) An English horse after 1760. Before this date, and later also on Continental ships, a deadeye was seized into the inner end.
(l) A stirrup with a plaited tail.
(m) A round strop with a thimble for a yard tackle.
(n) A brace pendant.
(o) A brace pendant with a preventer, as rigged to English ships between 1720 and 1775.
(p) A brace block with a thimble seized round the strap (the so-called 'dog and bitch' connection.
(q) A span hitched to a mizzen shroud to lead the main braces.

a thimble was often stropped to the yardarm instead of the tackle pendant (see Figure 39m).

Tricing lines

Einholer; Halebreus, lève-nez

When the yard tackle was not in use, it was secured with an inner and outer tricing line to the yard. The outer tricing line was spliced to the strop of the upper block of the yard tackle and ran through a lead block fastened to the yard one pendant length (one sixth of the yard length) inside the yardarm, then through another at catharpin height lashed to the shrouds, to belay finally in the lower shrouds. The tricing line was also sometimes found running from the first lead block up to the top, and making fast to a cleat.

A thimble was spliced into the end of the inner tricing line, and this was slipped over the hook on the lower tackle block. The hauling part passed through a lead block beside that for the outer tricing line in the shrouds, and belayed like the outer tricing line to a cleat in the lower shrouds (see Figure 40a).

Brace pendants

Brass-Schenkel; Pantoires de bras

The brace pendants came next in rigging a yard. These also were placed over the yardarms with an eye splice. The most common fitting consisted of a single block spliced or turned into the lower end of a pendant (see Figure 39n), but this was neither universal nor used consistently throughout the century.

In Dutch ships of the late seventeenth and early eighteenth centuries the brace block was seized to a round strop and the pendant, passing through the eye of this strop, had a stopper knot or a crown in its end (see Figure 38f). The same kind of block stropping can be seen on the model of HMS *St George* of 1701 noted by Anderson. The pendant here was spliced around the eye in the block's strop. Anderson expressed some doubts about this fitting, but in the light of the Dutch use of this kind of brace block strop, and its use about twenty years later in the Royal Navy for the fitting of preventer pendants, it probably has some validity.

It was noted in contemporary literature that these blocks were sometimes stropped to the yardarms, directly or without pendants, on naval as well as on merchant ships, but this probably did not occur much earlier than the last decade of the century. Steel mentioned it, Röding also noted the use of blocks without pendants, and Lever wrote as follows:

> Brace Pendants are now seldom used; in their stead a stout strap, with a Thimble, is seized in: another Thimble is placed round this, and the strap of the Brace-block lies over it, so that it keeps snug to the Yardarm.

This so-called dog and bitch connection became official in the Royal Navy in 1815 (see Figure 39p).

In paintings, models and drawings from the later years of the century a number of different fittings can be found. Paris, for example, showed a short block to yard connection in his drawing of the *Royal Louis* of 1780, while his yard drawing for *Le Protecteur* of 1790 showed short double pendants. These were actually long stropped blocks with the strops seized to the yardarms and above the blocks, so that the middle parts formed pendants of close to half a normal pendant's length (see Figure 38g). This was very much a French arrangement.

Between 1720 and 1775 brace pendants on English ships were often reinforced by preventer pendants to provide additional security during fighting or other calamities. A preventer pendant was longer than a normal pendant and also had an eye at its upper end which was put over the yardarm. The lower ends of both pendant and preventer pendant were spliced around thimbles to which the stropped brace block was fitted. The loose middle part of the preventer pendant was then seized about three times, in serpentine fashion, to the brace pendant (see Figure 39o). Steel wrote of preventer brace pendants, but probably considered them part of preventer braces, notwithstanding his contradictory statement that they had a block stropped to the yardarm when rigged.

Fore braces

Fockbrassen; Bras de vergue de misaine

The standing parts of the fore braces were hitched to the main stay at approximately half height at the beginning of the

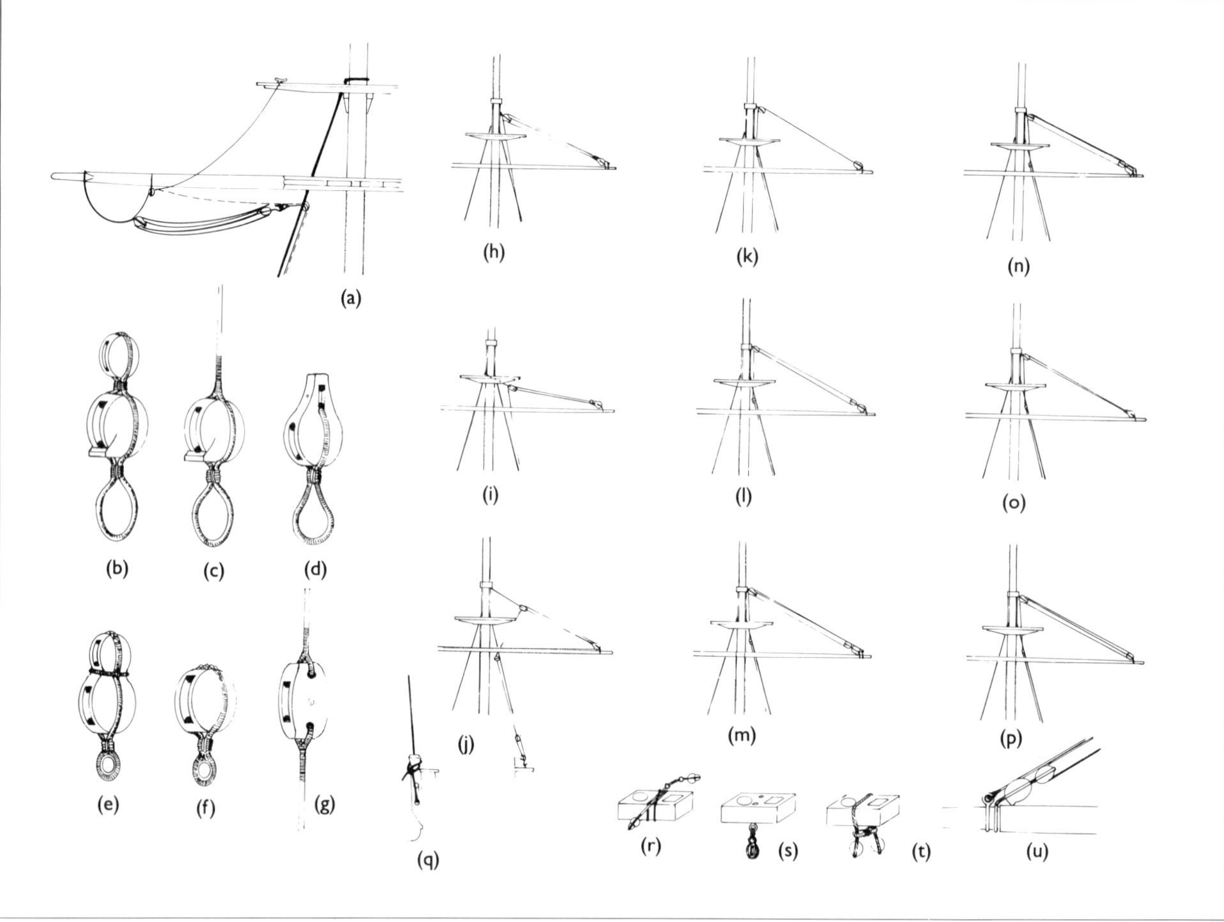

Figure 40
(a) Yard tackle with an inner and outer tricing line. The dotted line shows an alternative rig for the outer tricing line.
(b) An English topsail sheet and lift block.
(c) An English topsail sheet block with a single lift.
(d) A Continental topsail sheet and lift block (pear shaped).
(e) A lift block (fiddle block).
(f) An English lift block.
(g) A Continental lift block.
(h) The rig of a lift on a French lower yard, about 1700.
(i) The rig of a lift on a Dutch lower yard, about 1700.
(j) The rig of a lift on a Dutch lower yard, after 1720.
(k) A single lift on small English ships.
(l) The rig of an English lift between 1700 and 1720.
(m) The rig of an English lift between 1720 and 1760. A fiddle block was set up to the cap and the standing part of the lift was secured around the yardarm.
(n) An English lift with a single block at the cap, after 1760.
(o) The rig of a Continental lift on a lower yard, about 1790.
(p) The rig of a lift on a French lower yard, about 1780. A fiddle block was used, but the standing part of the lift was spliced into the strap of the block.
(q) A fore yard lift belayed inboard to a kevel (knight).
(r) An English cap with a span and lift blocks, up to 1760.
(s) An English cap with a lift block hooked on, after 1760.
(t) An English cap with a span and blocks, about 1800.
(u) A detail of the set up of a lift on an English lower yardarm from 1720 to about 1760.

century, and later at about two thirds height. From the 1770s onwards, the hitching point was either just below or just above the stay's mouse. From this point the brace ran through the blocks on the yardarm, with or without pendants (see above), and then through lead blocks on the stay, which were usually set below or above the standing part (after 1760 mostly above) (see Figure 55a below).

The author of *Der geöffnete See-Hafen* commented as follows on this point:

> One end of the brace is made fast at to the main stay at half height and the brace is led to the pendant block, from there to a block at the main stay some 6 or 7 feet lower than the standing end, and then upon deck, where it is led through a block and belays at the side of the ship [see Kevels below].

Steel explained that in English rigging the standing parts were hitched to each side of the main stay collar and their ends were seized. After passing through the yard blocks, the leading parts ran through single blocks above the hitched part, close to the masthead rigging, down to sheaves in the bitts on the fore part of the quarterdeck, and were belayed there.

Chapman's sail plan of a frigate confirmed Röding's description of the way in which the fore yard braces were fastened:

> They are fixed to the main stay, somewhat below its collar, and reeve through the single block at the rear of the fore yard, from there again through a double block either above on the main stay or fastened to the cheeks of the main mast, or also beneath the main top, and lead from there down to the ship. Over the other sheave of the afore mentioned double block goes the fore topsail yard brace.

If the braces were belayed otherwise than as Steel suggested, then a foot block was needed to lead them to the belay point (before 1760). Chapman's only variation was in showing brace pendants. Boudriot's description of French methods also differed slightly from Röding's. He noted an alternative fastening point at the bibbs (not cheeks), returning the fore braces via single blocks on the main stay (sometimes omitted), and double blocks on the bibbs, to the ninepin bitts or cleats nearby.

During the sprit topmast period the fore yard brace was belayed to a knight (kevel) in the waist, situated either vertically down from the stay lead block or further forward, directly abaft the forecastle.

Main braces

Grossbrassen; Bras de grand vergue

The standing part of a main brace was hitched to an outboard ringbolt at the aft side of the quarterdeck with a fisherman's bend. After reeving through the pendant block, the brace passed through a block at the planksheer above the ringbolt, and this block, a snatch block, led to an inboard cleat on the quarterdeck. Alternatively, the standing part might be fastened to a ringbolt at the planksheer or taffrail and passed through a sheave hole in the side (see Figure 55a below).

Modern English authorities state that a main brace spreader was introduced to the Royal Navy around 1805 (see Figure 38i). The practice of using a spreader in fact goes back at least to 1790, or even earlier (see Figure 38h). Raper shows a brace spreader in his watercolour *His Majesty's Brig Supply* of about 1790 and Röding commented as follows:

> The English frequently use brace pendants; they also keep the aft blocks away from the ship with a spreader, so that these braces have a less acute angle, and therefore can act with increased force on the yards.

Brace spans

Brassenspanne; Pattes d'oie de bras

In connection with the main braces, Steel mentioned a span, intended to reduce the swinging motion of these braces in bad weather.

> A Span for main-braces has two legs, with a thimble spliced in the end of each leg, which reeves the standing and leading part of the brace, and the span makes fast with a half-hitch, and the end seized up round the mizen-shrouds [see Figure 39q].

Falconer, adding to Steel's description, noted that such spans were sometimes also used on other braces, bowlines, etc, and that they were usually half-hitched round the corresponding stay, with the ends pointing left and right.

Preventer braces

Hilfsbrassen; Contre-bras, faux-bras

On the subject of braces, Steel included some interesting comments on preventer braces:

> Preventer-braces, in war, are reeved through a block lashed round the yardarm, and reeve through a block in a span, round the bowsprit-cap; they then lead in upon the forecastle, and the standing parts make fast round the cap. The main-brace reeves through the block on the yardarm, then through a block lashed to the fore-shrouds, close below the catharpins; they then lead down upon the forecastle, and the standing-part makes fast to the shrouds above the block with a hitch, and the end seized. Brigs reeve the same.

Röding described the preventer brace as a single rope, laid with a strop around the yardarm when strong winds made it necessary, running alongside the normal braces to increase their effect on the lower yards. He noted, however, that some seamen used the same term for forward-leading braces (as in Steel's description above), which he named counter braces (see Figure 56a, dashed line lower left). Falconer (1769) regarded these temporary braces very much as did Röding and added, in his 1815 edition, that they took the place of the usual braces in the event that these were shot away in action.

Topsail sheet and lift blocks

Marssegelschot- und Toppnantblöcke; Poulies d'écoute de hunier et de balancine

Next on the yardarm came the topsail sheet block, seized together with and above the lift block. Both were single sheave blocks, and the sheet block had a shoulder to prevent the sheet from jamming.

These two blocks were stropped by forming a round strop, serving it and seizing in first the lift block, followed by the sheet block, and then placing the remaining eye over the yardarm (see Figure 40b). On smaller ships the lift was rigged single and a yardarm lift block was not necessary (see Figure 40k).

In contrast to English rigging, topsail sheet and lift blocks were used in Continental ships for the greater part of the century. These blocks had the shape of a pear. The full round part of the block housed the sheave for the sheet, and the lift sheave was set at a right angle to the former into the smaller part (see Figure 40d). 'Now the lifts are usually single and fast to the yardarm; they reeve only through a block below the cap' wrote Röding, after having explained the pear shaped block. A typical exception to the rule can be seen on the *Royal Louis* rigging plan of 1780 included in *Souvenirs de Marine*. Instead of the aforementioned block, two single blocks are drawn, each separately stropped to the yardarm.

Lifts

Toppnanten; Balancines

The lifts on larger ships were usually double rigged. Blocks for these were fastened to the foremost eyebolts in the cap at the end of the century (see Figure 40s), but until 1760 these blocks were fixed to a span round the cap. They were either spliced into the span legs, or thimbles were turned into the span legs, and the upper lift blocks were lashed to these (see Figure 40r). Large English and Continental ships were also fitted with fiddle blocks for this purpose (see Figure 40e).

During the early years of the century, and according to Röding still at a later date too, special upper lift blocks were used on Continental ships. These were flat, single blocks with slightly extended blunt ends and holes in both ends to take the standing part of the lift and a span (see Figure 40g). At the beginning of the century the eye of the span was laid around the masthead and the block hung below the mast top (see Figure 40i); by 1720 it was fastened to the cap (see Figure 40j), first with a longer span and later, when the special lift block was replaced with a single block, as described above. The author of *Der geöffnete See-Hafen* noted as early as 1705 that on larger ships the upper lift blocks hung from the lower caps; the reason for this was that the longer yards could then be topped right up to the cap to prevent accidental damage in an anchorage.

Large Continental ships, mainly French, were rigged with a fiddle block instead of an upper lift block from the beginning of the century. This block was attached first to the masthead below the cap and later to an eyebolt in the cap (see Figures 40h and 40p). Röding did not refer to fiddle blocks, but mentioned a double or triple lift for large ships, with a two-sheave block hanging below the cap. Larger English ships were rigged with fiddle blocks between 1720 and 1760 (see Figure 40m).

The standing part of a double lift was generally secured either to the cap block or the upper lift block (if it was a single block, which was usually the case on English ships up to 1720). For a fiddle or double block the lift was spliced into an eye strop on, or with an eye around, the yardarm (see Figure 40m).

After reeving through the blocks the lift ended either in a tackle which was hooked to an eyebolt in a channel and belayed to a fife rail (see Figure 40j), or ran down to a knight (kevel) within board, passed over the sheave and fastened to its head (see Figure 40q). The author of *Der geöffnete See-Hafen* noted only that the lifts were led down beside the foremost shrouds. Single lifts were either stropped to the topsail sheet block, or, in the Continental manner, as Röding noted, 'fastened to the yardarm, and rove once through a block below the cap', and belayed as above (see Figure 40k).

A third method of rigging a lift was frequently practised on merchant ships. The single lift did not reeve through the block at the cap, but went over the cap, without a block, and was set up to a double and single block tackle at the opposite channel. Since the hauling part of this tackle came from above, a further lead block, stropped to a bolt inboard in the side of the ship, was needed before it could be belayed. The part of the lift above the cap had to be parcelled in leather (see Figure 42a).

Quarter blocks

Marsschotleitblöcke; Conduite-poulies d'écoute de hunier

The topsail sheet lead blocks or quarter blocks were lashed to the yard inside the sling cleats, or, if no cleats were fitted, outside the jeer blocks, hanging downward. Lever noted the use of large single blocks and Steel double blocks with a wide and a narrow sheave (the wide sheave was used for the sheet and the narrower for the clew line; Steel considered that the latter was without much value and that the rigging of double blocks was not very successful as the narrow sheave obstructed the clew line). Those blocks were only used at the end of the century and only on a

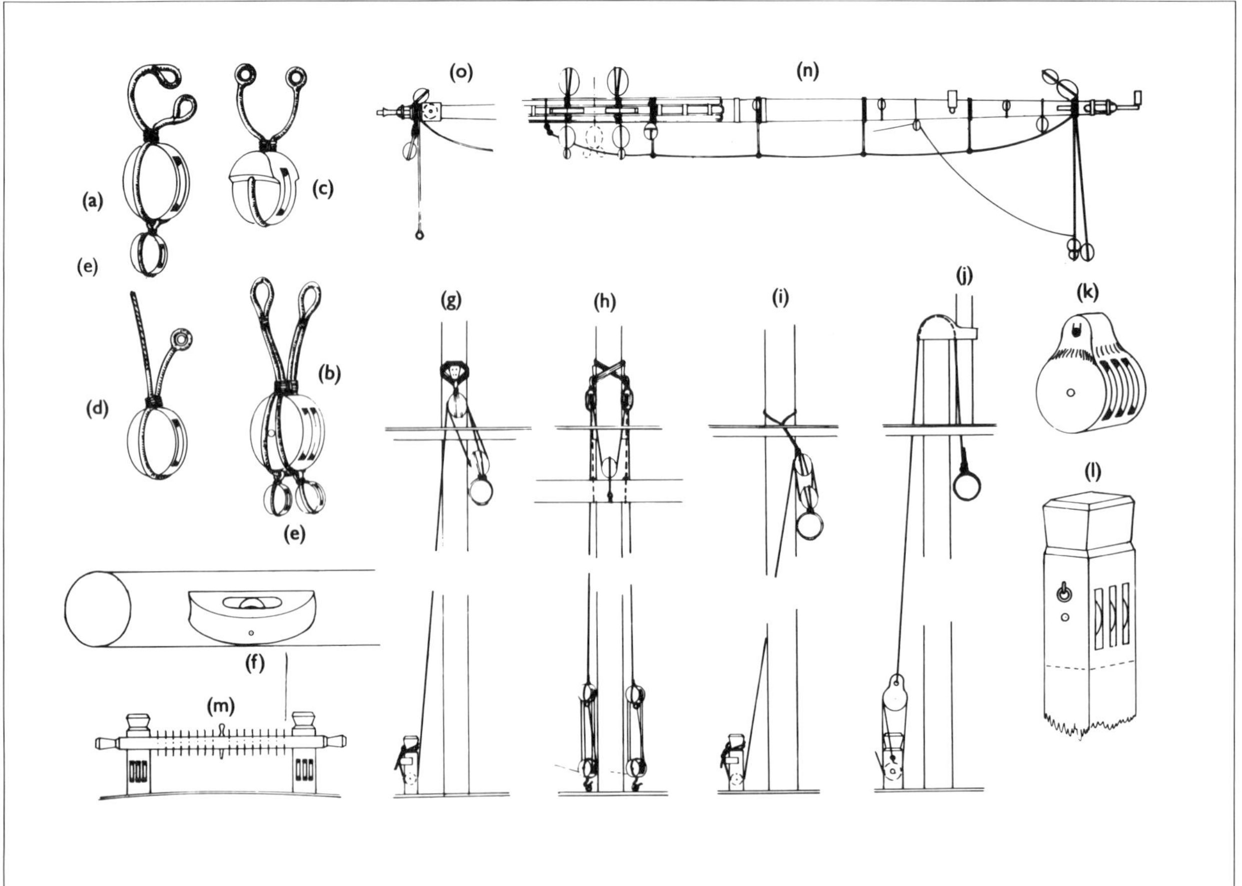

Figure 41
(a) A topsail sheet lead block with a slab line block.
(b) A double topsail sheet lead block with two slab line blocks lashed to it.
(c) An English clewgarnet block up to 1775; a similar block with a different strop was used as a buntline block in French rigging.
(d) A common buntline block, also used as a clew block after 1775 in the Royal Navy. The type of strop depended on the method of fitting for different blocks.
(e) Slab line blocks; in English rigging these were always lashed to the quarter block.
(f) A cheek block for leading the leechline on the fore side of a Dutch yard.
(g) Jeers on an English mast. The mast jeer block is somewhat too high in this illustration.
(h) A jeer set up with tackles and a single yard block on a small English ship after 1775.
(i) Jeers on a Continental mast after the middle of the century.
(j) The tye and halyard of a Continental ship, with the tye leading over the cap. This arrangement was typically French, in the second half of the century.
(k) A ram's head on Continental ships.
(l) A knighthead on Continental ships.
(m) Mast bitts.
(n) An English lower yard after 1775, showing a brace pendant with a block, yard tackle with a fiddle block, a single or double (dotted) quarter block and a topsail sheet block with a stropped lift block.
(o) The lower yard of an English merchant ship, about 1790. The topsail sheet block is integrated in the yardarm as a sheave, and a single lift block and a yard tackle pendant with a thimble for the tackle are rigged. The brace block is fitted in the dog and bitch method.

small number of ships.

Small merchant ships more frequently had a normal double block stropped to the middle of the yard instead of two single blocks (see Figure 41b).

Clewgarnet blocks

Geitaublöcke; Poulies de cargue-point

Until about 1775 English clewgarnet blocks had a shoulder, increasing the width of the block by nearly half above the pin, through which the strop was passed and seized at the top of the block (see Figure 41c). Rees (1819-20) and Lever (1811) still noted the use of this type of block. After 1775, common single blocks were used in the Royal Navy, as noted by Steel. Such blocks were also used on Continental yards. The clewgarnet blocks hung below the yard directly outside the sling cleats, or, as Steel specified, 4 feet out from the centre.

On French ships at the beginning of the century clewgarnet blocks were placed one eighth of the yard length out from the centre, while French ships around 1780 show clewgarnet blocks positioned either as noted above, or placed one fifth to one third of the yard's length inside the yardarms. Röding noted for Continental ships that clewgarnet blocks were fitted to the fourth or third part of the yard, which is similar to the French position. One quarter of the yard's length out from the centre was the position given in *Der geöffnete See-Hafen*.

Slab line blocks

Schlappleinenblöcke, Kerkedortjenblöcke; Poulies de fausse-cargue

A slab line block was a small single block lashed below a topsail sheet lead block or quarter block (see Figure 41e). Röding noted only that the block was stropped directly to the yard, but gave no details.

Buntline and leechline blocks

Bukgording und Nockgording blöcke; Poulies de cargue-fond et de cargue-bouline

Buntline and leechline blocks were stropped to the fore side of the yard. According to Steel, leechline blocks were placed 10 feet inside the yardarm cleats and the buntline blocks an equal distance between the former and the slings. On Continental ships early in the century the leechline block was placed about one quarter of the yard's length inside the yardarm, with buntline blocks in the middle of the yard, according to the author of *Der geöffnete See-Hafen*; later the position of the leechline blocks remained the same but the buntline blocks were placed a few feet inside the leechline block and outside the jeer block (according to Röding), or the leechline block about one third of the yard length inside the yardarm and the buntline blocks one quarter of the yard's length in, with another at the slings. This last set up was French, and Boudriot noted an alternative: the outer buntline block was one eighth of the yard length inside the yardarm and the inner at the end of the yard sling.

That this does not exhaust the possible positions can be seen from Paris' spar drawings of *Le Protecteur* and the rigging plan of the *Royal Louis*. In the former, leechline blocks are placed one sixth of the yard's length inside the yardarm, the outer buntline block at one quarter, and the inner at one tenth, or one eighth of the yard's length out from the sling. In the latter, two leechline blocks are shown on the fore yard, one quarter and one sixth of the yard's length inside the yardarm, and one on the main yard, about one fifth of the yard's length inside the yardarm. Buntline blocks are only one on each side, close to the slings.

In the first decades of the century on Continental and especially Dutch ships, a kind of cheek block was frequently nailed to the yard's fore side instead of leechline blocks (see Figure 41f). Leechline and buntline blocks in English rigging were normal single blocks, but the French used blocks similar to the English clewline blocks. During the last decades differently shaped blocks came into use on merchant ships. 'Monkey Blocks are sometimes used on the lower yards of small merchant ships, to lead (into the mast, or down upon deck) the running rigging belonging to the sails' stated Falconer (1815), and Lever specified a block 'for the Bunt-lines to reeve through... Sometimes it has a Swivel above the Saddle, to permit the Block to turn, when used for a Leech-line.' These blocks looked, according to Lever, like a single block with a built-in wooden saddle, and Falconer as well as Röding described them as small single blocks, attached by a strop and swivel to an iron strop encircling and nailed to the yard. Others were made completely from wood, shaped nearly octagonal, with a roller working in the middle and a saddle underneath (see Figure 184a, b and c below).

Jeers

Kardeele, Rahtakel; Drisses de basse-vergue

Jeers on the lower yards of large ships consisted of two heavy tackles with triple and double blocks. On Second and Third Rate ships they consisted of double blocks on the mast and the yard, and on Fourth and Fifth Rates double blocks on the mast and single on the yard. Sixth Rate ships and smaller vessels had two single blocks fixed to the mast and one double block to the yard (see Figure 41h).

Blocks (usually the larger, except on Sixth Rate ships) were lashed to both sides of the masthead above the masthead rigging, with the blocks positioned close together and very close below the top (see Figures 42b and 42c). Pieces of wood, the jeer cleats, were bolted to the masthead to support the lashing (see Figure 42d). This was a practice particularly characteristic of English rigging.

Methods used by other nations usually involved stropping the blocks around the masthead, with the strops above the masthead rigging and the blocks hanging lower below the mast top (see Figure 41i). Röding gave the following description:

> A pendant is laid over the masthead rigging on each side of the mast, with a threefold block seized in it which hangs down below the bibbs. Similar blocks are stropped to the middle of the yard, at both sides of the mast. These four blocks are called the jeer blocks, and two falls are reeved through them. Each is fastened to the pendant and reeves over the three sheaves of both blocks, then downward beside the mast and through a foot block or over the sheave of a knight. These falls are actually called the jeers. They are wound on to a capstan when the yard has to be hoisted.

According to Boudriot, the jeer blocks were not seized into pendants, but had a thimble fitted to their ends and were lashed to the masthead.

The yard blocks in English rigging were normally double stropped smaller blocks. The standing ends of the jeers were hitched or seized to the yard, and the jeers rove over the first sheave in the mast blocks, then back and forth between the blocks and down to lead over a sheave in the bitts abaft the masts, and finally belayed to the bitt head (see Figure 42c). On smaller ships during the last quarter of the century the double block on the yard was replaced by a single block, and both ends of a single jeer fall were set up to double and treble block tackles, with the lower treble blocks hooking into eyebolts on the deck (see Figure 41h). Lever quoted variations to this for small ships in the Royal Navy and East Indiamen. According to him, no yard block was used and two jeer tyes were made fast directly to the yard, either by being laid round the yard with Flemish eyes, or hitched with a sheet bend to round strops placed inside the sling cleats.

He also noted that some smaller vessels were fitted with a single hanging block, hung with a short and a long leg over the masthead like the slings. Alternatively, sheaves were fitted in stout chocks strongly secured to the trestle trees in place of hanging blocks; Steel noted that 'the caps of yachts and similar vessels have a sheave-hole on each side for the jeers of the lower yards'. It can thus be seen that a wide range of jeer fittings were used, even simply within English rigging.

When masts were fitted with Continental caps, which in some countries and for particular ship types was the case for much of the century, then a tye and halyard combination was used (see Figure 41j), with a ramshead in the bight of the tye (see Figure 41k). Both ends of the tye led over the grooves in the cap and were made fast to the yard. As a counterpart to the ramshead a knighthead was placed abaft the mast, and the halyard rove through both (see Figure 41l). Around 1700 large single blocks were also often long-stropped and hung below the tops, with the strops going over the aforementioned grooves in the cap. The tye was then led through the blocks instead of over the cap. Another variation, also referred to in *Der geöffnete See-Hafen*, has the two long-stropped blocks hanging down from the cap and a double block lashed to the centre of the yard. With one end fastened to the masthead, the tye rove through one of the double block sheaves on the yard, then through one of the mast blocks, back through the other yard sheave, and, after passing the second mast block, upon deck over a sheave in the knighthead. 'This type is harder to work then the aforementioned, and is more often used by the English than in German practice.'

Slings

Hanger; Suspentes

To relieve the jeers and to provide additional support to the yard, slings were introduced to the Royal Navy after 1770. Slings were also first mentioned for Continental ships about the same time.

Röding (1794) noted that some men-of-war and nearly all merchantmen had entirely dispensed with jeers, and that their yards were now hung in slings only, though preventer slings were often rigged as well, as Röding and Lever pointed out. As in so many respects, English and Continental methods of rigging these slings differed. Röding described preventer slings reeving beside the fitted jeers through a thimble in the middle of the yard and laid around the mast, with the ends made fast to the mast. These preventers were replaced by chains in times of action.

Lever's explanation was: 'Ships which carry no Jears, have frequently two Pairs of Slings: one of which is called the preventer Slings' (see Figure 42g).

The slings were usually of two parts, the sling and the strop. The former was a served rope, not very long and with an eye in one end. At approximately one third of its length a thimble was seized to it, dividing the sling into a long and a short leg. With the thimble hanging in the middle beneath the top, the long leg went over a cleat at the upper aft side of the masthead, or sometimes over the cap, and passed through the eye of the short leg: it was then turned back on itself and securely seized in several places (see Figure 42e, top). Outside England, the slings led neither over the cap nor over a cleat, but were laid around the masthead on top of the masthead rigging.

The strop for the yard was a rope ring, served over and with a thimble seized to it. It was laid double around the centre of the yard with the thimble passing through the bight of the strop, so that the thimble was on the upper side. A lanyard was spliced to the sling thimble and reeved alternately through both yard and sling thimbles until fully extended, then frapped around the turns and fastened with two half hitches (see Figure 42e, bottom). The jeers for the yard were then loosened, so that the yard was largely supported by the slings.

A number of ships had slings made in one part rather than two. Like the yard strop, a single-part sling was laid around the

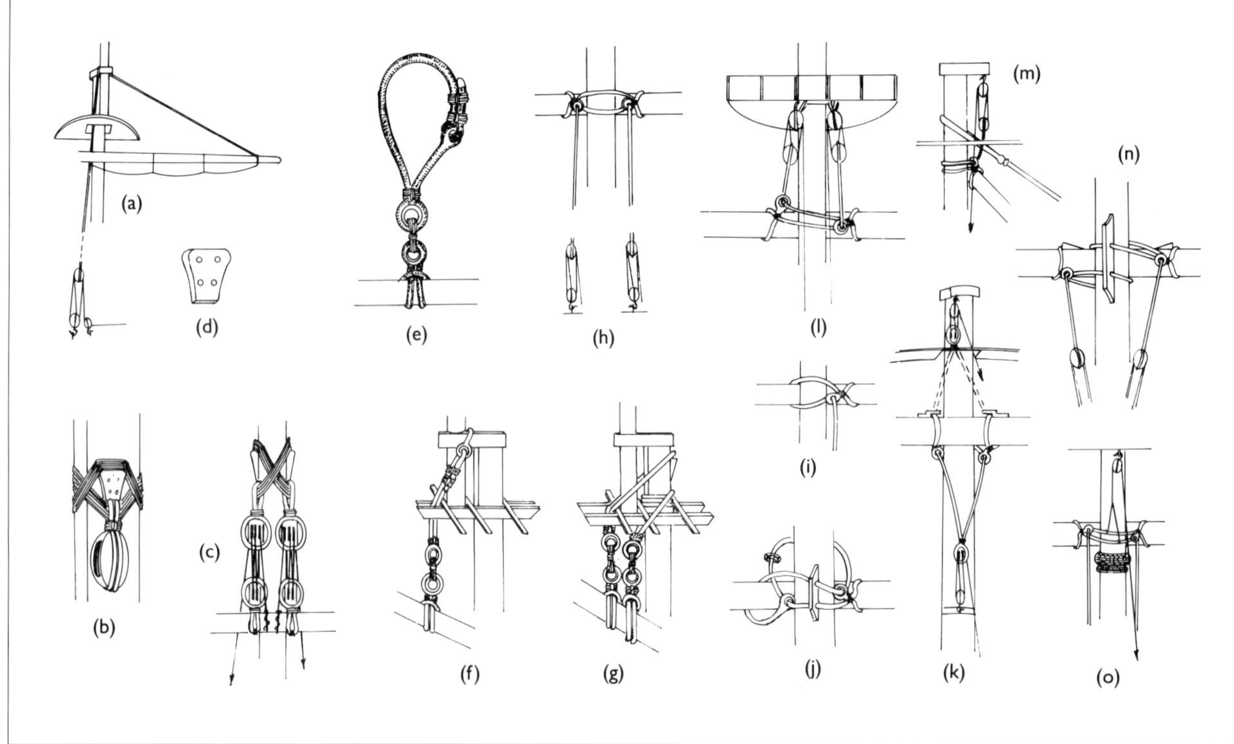

Figure 42
(a) The rig of an English lift without a block at the mast cap, on merchant ships about 1800.
(b) English jeer block stropping on main and fore masts.
(c) English jeer block lashing on mast and yard, seen from forward.
(d) A jeer cleat for the hanging of a jeer block in the English method.
(e) Slings, about 1770.
(f) The English method of rigging slings.
(g) Two sets of slings, rigged when no jeers were carried.
(h) English trusses, as in the first set-up described in the text.
(i) English trusses as in the second set-up described in the text.
(j) English trusses as in the third set-up described in the text.
(k) A further alternative method of rigging English trusses, not described in the text; instead of a running thimble as in (k), a double block is seized into the bight of the truss and set up to a tackle either stropped to the mast below the yard (solid line) or hooked to the cap (dotted line). The trusses are further held in place by yard cleats.
(l) English trusses as in the fourth set-up described in the text.
(m) English trusses as in the fifth set-up described in the text.
(n) English trusses as in the sixth set-up described in the text.
(o) A nave line connected to the trusses; the puddening and dolphin are shown beneath the yard.

yard and the two legs of equal length, with eyes spliced in their ends, were lashed together behind the masthead (see Figure 43a). If a yard was without jeers then two slings were rigged. The first led upwards before the topmast and between the trestle trees and was made fast as above, while the second sling passed outside the trestle trees and forward of the cross trees close to the mast, and rested on the masthead rigging.

Lever identified another method of securing the strop to the yard. An eye was spliced into each end of the strop, and both eyes were seized to the upper side of the yard; each leg of the strop was then additionally seized to the yard close to the seized thimble. This strop, applied to the upper side of the yard only, was supposed to have provided more play for bracing (see Figure 43b).

The chains mentioned by Röding and Falconer (1815) – 'to Sling the Yards for Action is to secure them close up by means of iron chains, which are not so liable to be cut through by the enemy's shot as rope' – were illustrated and also described by Lever:

> Sometimes the lower Yards are slung with Chains; for which purpose, a Bolt is driven into the lower mast between the Bibbs, or hooped to it: a Link of the Chain is fastened to the Bolt, and reeved through an iron Strap on the Yard: the other End is hooked to the Bolt.

Parrels and trusses

Perlenracke und Trossenracke; Racages de vergue et drosses de vergue

The lower parrel underwent significant change in the course of the eighteenth century. The parrel, consisting of ribs and trucks, had been the major connection between yard and mast some centuries, and also predominated for the greater part of the eighteenth century. The open parrel, with a rope and tackle leading upon deck, was the more common type for larger ships on the Continent at the beginning of the century (see Figure 43c). In England, by contrast, the parrel was rigged standing from the early seventeenth century (a more detailed description of both types can be found below under Main and fore topsail yards, Parrels; see also Figure 43f). That the latter type of parrel connection was not confined to English rigging is shown by detailed sketches both in a Swedish work from 1691 and in *Der geöffnete See-Hafen*, though in these cases the parrels shown are for smaller ships.

Both Röding and Paris described and illustrated parrels for Continental ships of the late eighteenth century; these are identical to the standing English parrel, except that they have trucks instead of ribs at the outside, and the mast face of the ribs is not entirely flat but slightly curved. Another type of parrel, with three rows of trucks but without ribs, can also be seen in both works; these were generally used on lateen-rigged vessels.

Röding's comments with regard to the use of parrels on the lower yards of Continental ships at the end of the eighteenth century were as follows:

> Such a parrel, made from three rows of trucks, supports the main and fore yards only on men-of-war; it is now very seldom used alone; instead, since both yards are rarely lowered, they are connected to the mast with slings and trusses.

While Röding (1794) associated the lower yard parrel only with men-of-war in his time, Falconer (1769) found them in use not on warships but 'frequently for the lower yards, on merchant-ships'. He noted that in the Royal Navy in about 1760 the lower yard parrels were replaced by trusses: 'these are peculiar to the lower-yards, whereon they are extremely convenient'. Just as so many other English methods influenced rigging on the Continent, so this innovation also soon became part of the rigging practice of other nations.

Trusses became the subject of many experiments, and ships used various versions. Six common types are detailed below.

(1) To keep the yard close to the mast, a truss was laid around the yard on each side of the mast, and seized in such a way that a thimbles spliced into the end hung freely but close to the aft face of the yard. Both trusses were fastened inside the sling cleats. The other end of each truss was led across the aft side of the mast and passed through the opposite thimble and then down upon deck, where a block turned or spliced into the end formed the upper part of a double and single block tackle hooked into eyebolts near the mast. The trusses were served,

with the parts round the yard also leathered and greased (see Figure 42h).

(2) A single truss was laid around the yard as above. The long leg went around the mast and yard at the opposite side, back across the mast and through the thimble, and then upon deck as described above (see Figure 42i).

(3) The truss was seized to the yard as above, and a cleat with two holes, one above the other, was nailed to the aft side of the mast. The truss passed through the lower hole, two thimbles were then put over, and after leading above the yard around the mast and through the seized thimble, it went through the upper hole in the cleat and around the yard from above. The end was then spliced around the thimble nearer the cleat. A single block was seized to the second, free thimble and set up as part of a tackle which was either hooked to the cap at the fore side of the mast or stropped around the mast itself (in the former case the double tackle block was the cap block, and in the latter the truss block). In both cases the tackle fall led upon deck and was belayed to the bitts (see Figure 42j).

(4) The trusses were rigged as in (1) but the long legs were much shorter; they led upwards rather than down upon deck, and were set up as tackles to double blocks fastened to the aft part of the trestle trees, and the tackle falls led upon deck. This truss arrangement also fulfilled the function of preventer slings should the slings give way, holding the yard in position until it could be secured (see Figure 42l).

(5) The trusses were rigged as in (4), but the tackles were placed differently: here they were hooked to the cap. Lever noted that both the stay collar and the trusses needed to be well leathered to prevent chafing, which eventually would have led to damage (see Figure 42m).

(6) A cleat with two holes was nailed to the mast as in (3), and two further cleats were nailed to the top of the yard, outside the sling cleats. The trusses were rigged to the yard as in (1), and the long legs passed around the forward side of the mast and back through the cleat holes, one above and one below the yard, to return through their own thimbles upon deck, where they were set up to tackles (see Figure 42n).

Trusses rigged as in (3) and (6) are characteristic of the closing years of the eighteenth and the first decade of the nineteenth centuries.

Nave lines

Naveleinen, Aufholer des Racks; Halebreus de racage

Both parrels and trusses often carried a nave line (or tricing line). With regard to parrels, Blanckley (1750) offered the following description:

> [A] Naveline, Is a Rope reeved through a Block made fast to the middle rib and another Block made fast at the Mast-head; the Line goes through them, which makes a Tackle to hoist the Parrel.

And on trusses Lever commented as follows:

> In order to over haul these Truss-pendants, a rope called a Nave Line with a Span, is reeved through a Block under the after part of the Top and the ends are spliced to Thimbles on the Pendants [see Figure 42o].

Anderson noted that during the seventeenth century the line was named a knave line, a spelling which was still used in Falconer (1815) together with navel line and naval line, but all authors from Davis (1711) to Steel (1794) used the term nave line. It would seem that the alternative spellings reflect nothing more significant than careless pronunciation.

The nave line block was at the masthead according to Blanckley, was hooked below the aft part of the top according to Röding, or was fastened to a staple 3 feet below the top according to the author of *Der geöffnete See-Hafen*). The fall was belayed to the lower shrouds, or simply 'on deck'. The purpose of the nave line was to keep the parrel horizontal abaft the yard, since a parrel jammed either upward or downward greatly inhibited the bracing of the yard.

Pudding or puddening and dolphin

Maus (Leguan, Mitis) und Delphin; Bourrelet de mât et baderne des mâts majeurs

Where yards were rigged with trusses or parrels leading upon deck, additional support was provided by a pudding and dolphin. Their function is occasionally described very fancifully in modern books: a pudding was neither a part of a parrel or truss, nor an alternative English type of parrel.

As noted above, trusses which led upwards also acted as preventer slings and could keep the yard in place until damage to a broken sling was repaired. Since a downward-leading truss could not fulfil this function, a pudding was lashed around the mast beneath the yard to prevent the trusses or parrels sliding downward. Falconer offered the following description of the function of a pudding:

> [A pudding was] fastened about the main-mast and fore-mast of a ship, to prevent their yards from falling down, when the ropes by which they are usually suspended are shot away in battle.

Röding's explanation was similar, and he added that a pudding was lashed to the fore side of the mast, so that its main bulk faced aft. He also noted that the use of puddings on masts was a thing of the past by his time, except on smaller vessels where they provided a stop at the lower section on which a gaff sails could rest. These puddings were also clad in leather.

The smaller dolphin, fitted underneath the pudding, provided additional support.

The pudding was a piece of rope of a length slightly less than the circumference of the mast, with an eye spliced into each end. It was served over with spun-yarn, with the turns increasing towards the middle to give it a tapering effect. The whole was then pointed over (see Figure 44a). The pudding was fastened to the mast with a rose lashing (see Figure 44c).

A dolphin (see Figure 44b) was strictly speaking a pudding without a raised middle part. Its preparation and fastening were identical to those of a pudding.

Besides their use on masts, Steel noted that puddings were also used on yards, 'to prevent the sheets from chafing the rope-bands etc' (see Figure 43h).

Crossjack yard

Bagienrah; Vergue sèche, vergue barrée

The crossjack yard on square rigged masts differed from all others yards in that it was not used to set a sail; its primary function was to take the clews of a squaresail set on the topsail yard above it. On fore-and-aft rigged vessels however a flying squaresail was sometimes set on a crossjack yard at the foremast.

The crossjack yard horses, braces, lifts and topsail sheet blocks were fitted to the yardarms in the same manner as those on other yards, with the slight difference that the brace blocks

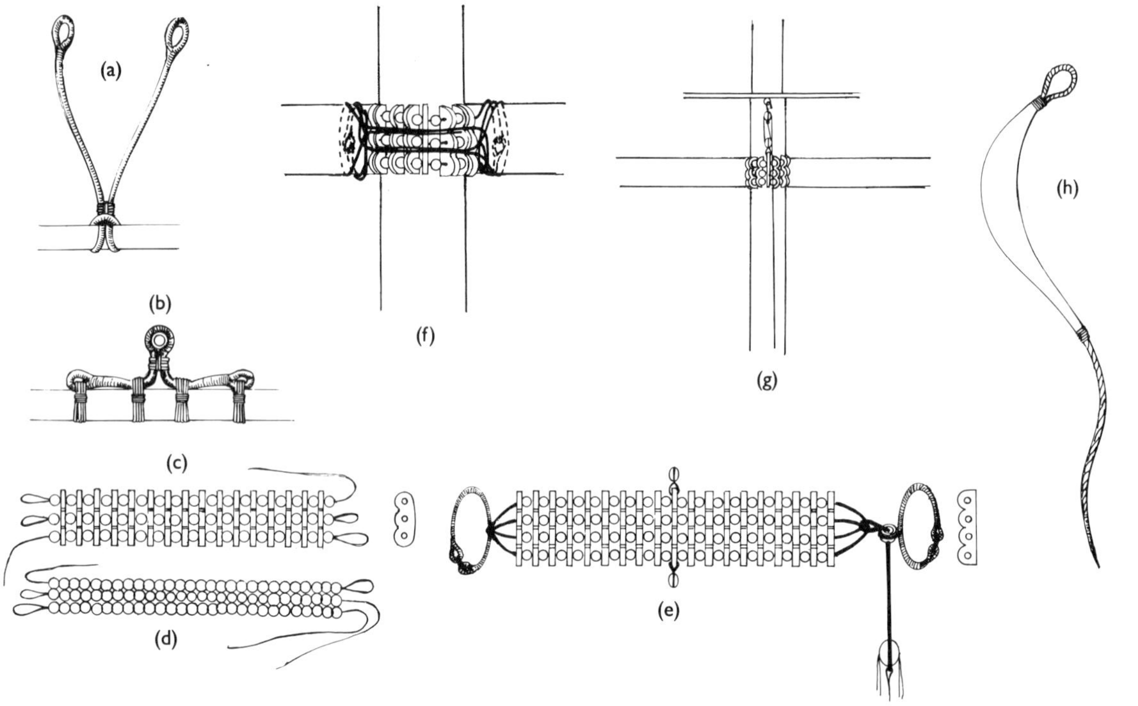

Figure 43
(a) An alternative set-up for slings.
(b) An alternative method of seizing the sling strap to the yard.
(c) A French parrel.
(d) A French parrel without ribs, mostly used in the Mediterranean Sea on vessels such as galleys.
(e) A Continental parrel for a large yard, with a downhauler and nave line blocks.
(f) An English parrel.
(g) A nave line fitted to a parrel.
(h) A pudding for a yard.

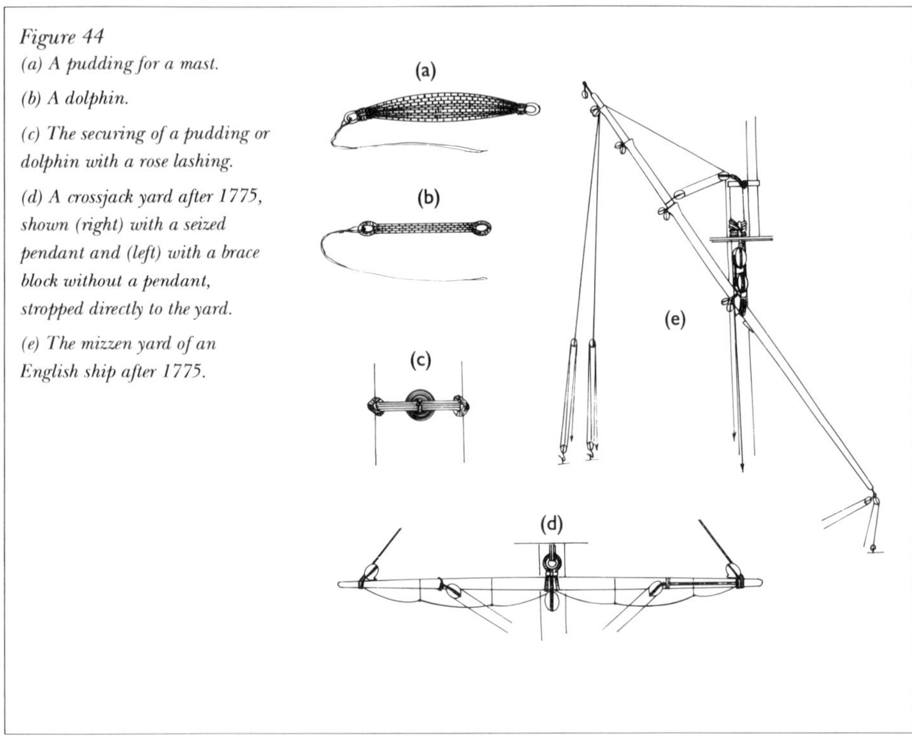

Figure 44
(a) A pudding for a mast.
(b) A dolphin.
(c) The securing of a pudding or dolphin with a rose lashing.
(d) A crossjack yard after 1775, shown (right) with a seized pendant and (left) with a brace block without a pendant, stropped directly to the yard.
(e) The mizzen yard of an English ship after 1775.

were arranged so that the braces could be led forward (see Figure 44d). Towards the end of the century, topsail sheet blocks were frequently replaced by sheave holes in the yardarms.

Quarter blocks

Marssegelschotleitblöcke; Conduite-poulies d'écoute de hunier

A quarter block was a double-stropped, double-sheaved lead block, lashed to the lower side of the yard. At the beginning of the century two single blocks were used instead of a double block.

Slings

Hanger; Suspentes

Crossjack yard slings were rigged throughout the century; jeers were not used on the this yard. Until the 1770s the sling passed through a block stropped around the centre of the yard (see Figure 45b). In the following period slings were set up in similar fashion to those of the other lower yards.

Brace pendants

Brass-Schenkeln; Pantoires de bras

The crossjack yard brace pendants were seized about 4 to 6 feet inside the yardarms to the fore side of the yard. At the end of the century, pendants were frequently not rigged and brace blocks were stropped directly to the yard in the same position. Instead of a rope strop, iron strops were occasionally used. Crossjack yard braces were crossed; that is, the brace ran from the starboard brace block to the port side main shrouds and vice versa, and it was from this peculiarity that the yard received its name.

The standing part of the starboard brace was hitched and seized to the rear port side main shroud at about two thirds height, near the catharpins. After passing the yard block, the brace passed through a single block and then through a lead truck, both seized to the same shroud beneath the standing part, and belayed to a pinrail. The port side brace ran the opposite way.

According to Boudriot and to Ozanne's illustrations this was also the way French crossjack yard braces were rigged, but this is not confirmed by photographs of French models in *Souvenirs de Marine.* There, the rigging followed the method described in *Der geöffnete See-Hafen* (1705) and also by Röding at the end of the century: in neither case were the braces crossed. The only difference between these two authorities was that single braces were advocated at the beginning and double braces without pendants at the end of the century.

Lifts

Toppnanten; Balancines

On English ships at the end of the eighteenth century crossjack yard lifts were usually single, and were fitted with an eye over the yardarms. Before this, the lifts of smaller ships only were single; those for larger vessels were made fast at the mast cap, ran through lift blocks at the yardarms, through blocks on a span hitched around the cap, and led down upon deck (see Figure 45f). Single lifts ran from the yardarms to the cap and down upon deck.

A single lift might also be spliced around the strop of the topsail sheet block and hitched, with the end seized back on itself, to an eyebolt at the cap (see Figure 45e). Such single lifts mentioned by Steel, however, were standing lifts, of which he noted: 'Standing lifts are made fast, and belong to yards that never require to be topped.'

The blocks on the span around the cap usually hung about 2 feet away from the cap, and blocks on the yardarm were either seized to the topsail sheet block strop or, when the yard had sheaves instead of sheet blocks, were fitted over the yardarm with a round strop.

The lifts on Continental crossjack yards were rigged in similar fashion to the fore and main yard lifts, with certain exceptions, according to the author of *Der geöffnete See-Hafen*. The yardarm lift block was not the smaller sheave of a pear-shaped topsail sheet block, but rather a smaller block stropped to the topsail sheet block; the lift blocks, on fore and mainmasts below the top, were stropped to the cap (see Figure 45g).

Crossjack yard lifts did not lead, like those of other yards, to the channels or knights (kevels) at the sides of the vessel, but directly (without tackles) to cleats on the lower mast or to a pin rail.

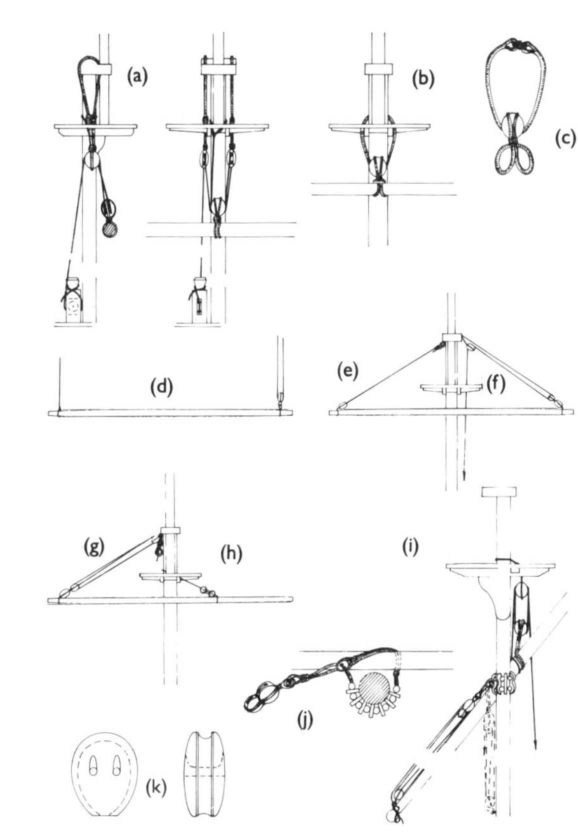

Figure 45

(a) English jeers according to Der geöffnete See-Hafen, 1705.

(b) Early hanging of a Continental crossjack yard sling.

(c) A Continental crossjack yard sling with a yard block.

(d) Some Continental crossjack yard braces did not cross over.

(e) A crossjack yard standing lift, according to Steel.

(f) An English crossjack yard lift; French ships used a pear-shaped block on the yardarm.

(g) A Continental crossjack yard lift, from 1705.

(h) A French crossjack yard standing lift.

(i) Continental mizzen yard jeers and parrel after 1730; the dashed line shows the rig of the jeers before 1730.

(j) Plan view of the mizzen yard parrel.

(k) Mizzen yard parrel deadeye.

> A different account is given in *Der geöffnete See-Hafen*:
> The ends of the lifts are now fastened to the outer lift blocks; they lead to blocks on the cap, over their sheaves and then back to the lift blocks, and finally again to the cap, where the lift is hauled on in such a way that a small number of men can get up to the top, bring the yard into position and make the lifts fast behind the blocks to the cap.

Standing lifts, in some respects similar to those Steel described, were used on English ships until about 1720. These lifts were placed around the yard about one eighth of the yard's length inside the arms; deadeyes were turned into the other ends and into a cap span, and these were set up with lanyards (see Figure 46b). Davis (1711) listed '2 Standing Lifts to the Yardarms' for all ships, except the First Rate, and noted for the First Rate HMS *Royal Sovereign*: '2 Lifts Treble, with 1 end of each of them fast to the Quarter of the Yard, and the other made fast in the Mizen Shrouds.' If the standing lift for the largest type of ship was rigged threefold, then blocks must have been fitted to the yard and to the shrouds. The First Rate drawing in Deane's *Doctrine of Naval Architecture* (1670) indicates double standing lifts from the yardarms to the shrouds.

From the different fittings on these two First Rates it is clear that no hard-and-fast rules for the rigging of standing crossjack yard lifts existed. Davis' contemporary Sutherland, for example, described them as fastened at one quarter of the yard length, leading to the mizzen shrouds, while a 1719 establishment rigging plan has the lifts leading from the same position to the cap. On a model of a 1719 establishment 80-gun ship, the standing lifts were rigged to the middle of the masthead, with additional double running lifts from the yardarms to the cap. The standing lift is not rigged on models of English ships after about 1730, but the 1745 establishment rigging plan of a 60-gun ship confirms the survival of the rig on the 1719 model noted above.

On French ships these standing lifts, together with running lifts, were still rigged as late as 1780, and Röding (1798), as well as Falconer (1815), included *fausses balancines* (preventer lifts) in their French indexes. These lifts were placed above the masthead rigging around the masthead and were set up to deadeyes on the yard just outboard of the futtock shrouds (see Figure 45h). The author of *Der geöffnete See-Hafen* made no mention of standing lifts.

Truss or parrel

Rack; Racage

A truss, though not mentioned in contemporary literature, was as necessary on the crossjack yard as on any other. The only contemporary mention of this item is in Steel's rigging tables, which cannot be surpassed in their comprehensiveness. He provided data for ships from First to Sixth Rate and for merchantmen down to 330 tons. All had truss pendants and falls with tackles. Models indicate the introduction of this type of truss around 1770; before this date a parrel was used, similar to that on spritsail yards. French yards were generally rigged with a truck parrel of two rows, but this was not the case with crossjack yards. Boudriot noted that the crossjack yard of a 74-gun ship of 1780 was rigged with neither a parrel nor a truss, but that trusses were sometimes fitted.

Halyard

Fall; Drisse

Halyards were not used on crossjack yards.

Figure 46
(a) A crossjack yard sling, up to about 1775.
(b) An English crossjack yard lift, up to about 1720.
(c) Mizzen yard rigging up to about 1710.
(d) Alternative rig of a mizzen yard derrick with crowsfeet in place of bridles, up to about 1710.
(e) An English mizzen yard derrick, up to about 1720.
(f) An English mizzen yard derrick, up to about 1775.
(g) Continental mizzen yard derricks
(a) According to Röding
(b) According to Paris

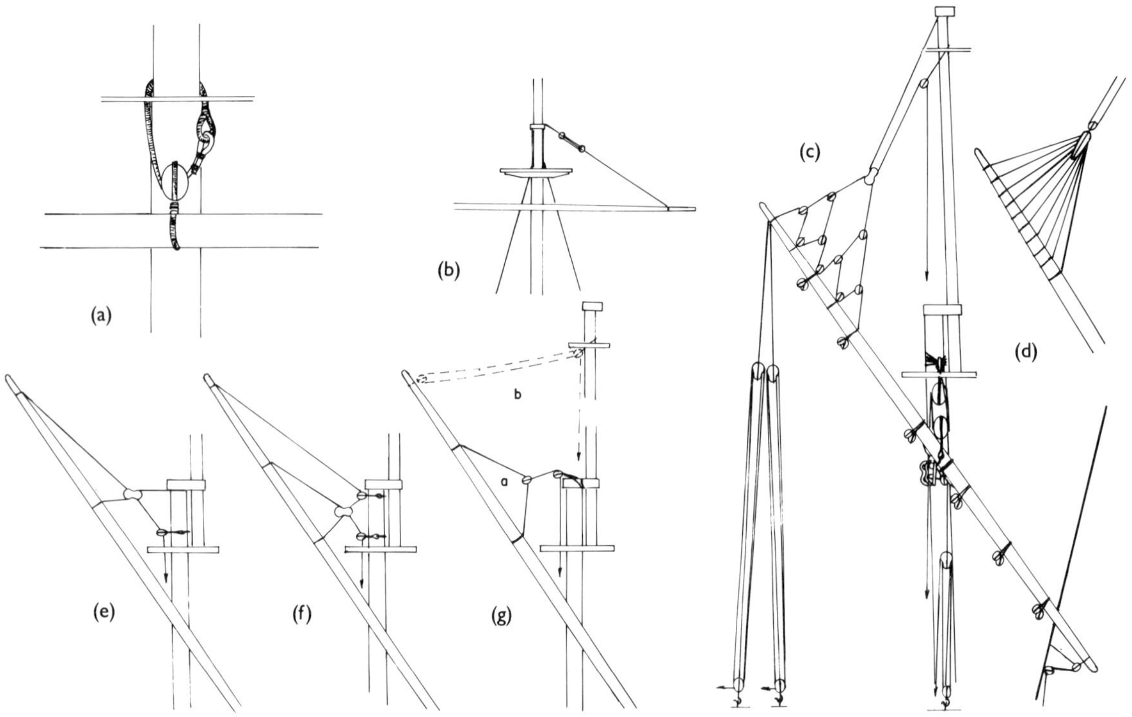

Mizzen yard

Besanrute; Vergue d'artimon

The mizzen yard was used to set a fore-and-aft sail. During the eighteenth century a significant change occurred in the yard's rig. While the lateen sail, or in England after about 1680 the settee sail (that is, a lateen sail with an integrated bonnet), characterised the first few decades of the eighteenth century, by about 1740 the part of the sail forward of the mast had begun to disappear, turning the triangular sail into a trapezoid shape, with its new mast leech laced to the mast.

In the 1760s the gaff became a regular part of the rigging of three-masted ships, but smaller ships, up to frigate size, were fitted with gaffs before this. The sail plan of a frigate in Chapman's *Architectura Navalis Mercatoria* from 1768 shows the gaff clearly, but not a boom. The latter began to appear in ship rig only some two decades later.

The longevity of the mizzen yard is vouched for by Steel's and Röding's remarks in 1794: 'it is not often used, except in ships above 50 guns, and in East India Ships' (Steel) and 'since a gaff sail is far more convenient, the mizzen yard has been nearly completely done away with' (Röding). Even twenty years later the yard was not entirely a thing of the past, judging by Lever's remark: 'these yards were formerly used by all Line of Battle Ships, and East Indiaman, but they are now entirely laid aside.'

Jeers

Kardeel; Drisse

The jeer block, a double-stropped double block, was seized to the upper side of the yard between the jeer cleats. A similar or treble block hung at the starboard side on the masthead. It was fitted like other lower yard jeer blocks (taking account of the differences between Continental and English rigging). The description of mizzen yard jeers in *Der geöffnete See-Hafen* included a double block at the aft part of the masthead and a single block at the middle of the yard.

The jeer fall was either spliced around the strop of the upper block or hitched around the masthead above the masthead rigging. On Continental ships, and occasionally also on English ships carrying a treble upper block, the jeer fall was spliced around the yard on the down side of the jeer block, or made fast to the yard block according to the author of *Der geöffnete See-Hafen.* It passed through the blocks and came down to the starboard mizzen channel, where it was hitched to a ringbolt or around one of the deadeyes. The remaining slack was coiled up and tied to the shroud lanyards. On Continental vessels the hauling part frequently ran through a lead block on the deck or side, and belayed there to a cleat or to a pin rail. Or, according again to *Der geöffnete See-Hafen,* 'abaft the mast to a knighthead with a sheave on the side of the ship, over which the fall reeves and is hauled forward.'

Derrick, peak halyard

Rutendirk, Rutenpiekfall; Martinet d'artimon

The tackle which kept the mizzen yard peak at the right angle was known as a derrick, though a number of other names were also used. During the greater part of the century it was referred to as the 'mizzen peak halyards' or a pair of 'pendant halyards to the mizzen peak', but late eighteenth century authors like Steel, Lever and Röding preferred the term 'derrick' and modern authors generally use the term 'mizzen lift'.

Until about 1710 the halyard was a single whip, or a whip upon whip (with the standing part made fast around the topmast head below the cap), set up to the yard with either bridles or crowsfeet (see Figure 46c and d). Soon after 1710, a shoe block carrying a span was substituted for both alternatives. The halyard, fixed to the mast cap or slightly below, passed over the second sheave of the shoe block, then over a further block at the lower masthead and upon deck (see Figure 46e).

After 1720 the upper end of this span was fastened at one third the length of the upper part of the yard down from the peak, with the lower end two thirds down. The standing part of the halyard was made fast around the peak of the yard, then passed through a block below the mast cap, through the second sheave of the shoe block and then as above (see Figure 46f). During the last three decades of the eighteenth century the derrick was modified again. A block was stropped to the upper side of the yard halfway between the peak and the jeers and a double block was either hooked to an eyebolt in the aft side of the cap or stropped around the cap in such a fashion that the block itself lay on top of the cap.

The derrick fall, fitted with an eye over the peak of the yard, passed through the cap block, the yard block, then the cap block again, and ran at the port side of the mast down to the mizzen channel, where it was belayed similarly to the jeer fall on the starboard side.

Röding noted two variations on this last set up, both of which were common on the Continent during the second half of the century. In the first, the span was as described above for the period after 1720 but fitted with a single block around which the lanyard was spliced; the lanyard rove through the cap block and upon deck (see Figure 46g (i)). In the second variation the yard block was made fast to the peak, and the mast block was long-stropped to the topmast head so that the block hung below the trestle trees. The derrick fall was spliced around the strop of the mast block, passed through the yard block and then ran via the mast block upon deck (see Figure 46g (ii)).

At the beginning of the century, according to the author of *Der geöffnete See-Hafen*, a Continental mizzen yard derrick had two sets of crowsfeet, linked by a span running through a block. The lanyard ran from this block through a block at the mizzen topmast trees and upon deck. A French manuscript of as late as 1728 indicates a derrick with crowsfeet, of the type that disappeared from the Royal Navy in about 1710.

In the second half of the century, according to Paris, larger French ships such as the *Sans Pareil* of 1760 or *Le Protecteur* of 1790, had derricks rigged similarly to English practice after 1770, but with the standing part spliced around the yard, closer to the yard block, and an additional single topping lift going to the topmast head. The *Royal Louis* of 1780 had a span with a single block fastened to the third quarter of the upper half of the yard and a long-stropped block hanging from the topmast head. The derrick, spliced around the block, passed through the long-stropped masthead block and upon deck.

Ozanne (1737-1813) made a great number of sketches, some of which show ships still rigged with a bridled peak halyard and others rigged as above, with the slight difference that these sketches all show a double topping lift instead of a single as shown on the models named above. Ozanne's sketches also reveal that smaller frigates and merchantmen did not carry a lower derrick, only a peak topping lift as in Röding's second variation. Ozanne's work suggests that ships with bridled halyards must still have existed in the late 1750s or early 1760s.

The setting up of peak halyards on merchantmen (described above as topping lifts, to distinguish between lower derricks and peak halyards on Continental, especially French ships), was identical in the works of Ozanne, Röding and Boudriot. The last named differed only in the positioning of the same in his description of a 74-gun ship: he placed the peak halyard not on the topmast, but on the cap of the mizzen mast, with no derrick like those of the French models noted above.

Brail blocks

Dempgordinge; Poulies de cargue

Three pairs of brail blocks were seized to the lower side of the mizzen yard on English ships, the lower pair just outside the upper jeer cleat, the second beneath the derrick block, and the uppermost half way between the second and the peak of the yard. Röding noted three to four brails, and French ships had up to five, spaced evenly between the upper quarter of the yard and the jeers. If the yard carried a lateen or settee sail, then a number of block pairs on the yard's lower half were equally spaced between the tack and the lower jeer cleat. Large ships had four, medium-sized three and smaller vessels two pairs of foot brail blocks.

Brace blocks

Brassenblöcke; Poulies de bras

On a mizzen yard with jeers (that is, a hoistable yard) there were no brace blocks because the braces of the mizzen topsail yard and topgallant yard were led forward. Where the yard hung in slings and was without jeers, however, these braces ran aft and two double or four single blocks were lashed to the peak. The latter was mainly characteristic of English rigging.

Vangs

Rutenger; Palans de garde

Vang pendants were doubled and fitted over the peak end of the yard with a cut-splice or an overhand knot. A double block, often a fiddle block, was spliced into the lower end of the vang. A single block was hooked or stropped into an eyebolt at the quarter pieces on each side of the ship, and the fall, spliced into the back of the single block, ran between the two blocks and was belayed to a cleat at the taffrail or around the single block.

Röding described vangs as having a single block spliced to the pendants and another single block stropped to the rails. The latter might also be hooked to a ringbolt, so as to be available like a yard tackle for lowering boats, etc. When vangs were used for this purpose the pendant block was a double block.

Vangs on French ships can be seen on the models listed under Derrick above and in Ozanne's pictures, generally on smaller ships such as corvettes and on merchantmen. The rigging plan of the *Royal Louis* does not include vangs, and Boudriot made no mention of vangs for the 74-gun ship.

Flag halyard block

Flaggleinenblock; Poulie de drisse de pavillon

After 1770 an eyebolt was frequently fitted to the peak of the mizzen yard to take the Ensign halyard block. A signal flag halyard block stropped to the peak was, however, mentioned in *Der geöffnete See-Hafen.*

Mizzen bowline

Halstalje, Pispotten; Bouline de verque d'artimon

A similar eyebolt was driven into the lower end of the mizzen yard and two blocks were stropped to it. Before 1770 these blocks were stropped directly to the foot of the yard, a practice to which Continental ships adhered somewhat longer. On Continental ships at the beginning of the century an iron thimble was seized into the tack of the mizzen, and another was seized to a strop fastened inboard on the quarterdeck beside the main shrouds. A tackle (double and single block) was then hooked to both thimbles according to the wind. Up to 1790, the standing part of the bowline was usually made fast to the aftermost main shrouds, passed through the foot blocks of the yard, then through lead blocks a few feet below the standing part on the shrouds, and to the side of the ship and where it was belayed. During the later years of the century the standing part was spliced around a thimble on an eyebolt near the sides, passed through the yard blocks and was belayed to cleats at the side. Steel also mentioned lead blocks, either hooked to an eyebolt abreast the foot of the lower yard at each side or lashed to the mizzen shrouds.

Parrel or truss

Rack; Racage

Lever and Steel made no mention of a mizzen yard parrel or truss, but ships certainly had parrels until the beginning of the last quarter of the century. Davis and Love included parrel lines and truss-tackle falls in their rigging lists, the author of *Der*

Figure 47
(a) Slung gaff.
(b) Hoisting gaff.
(c) Hoisting gaff, about 1800.
(d) Hoisting gaff with a tye, for vessels such as brigs.
(e) The jaws of a gaff with slings.
(f) The jaws of a gaff with a throat halyard block.
(g) A boom sheet block and guy pendant.
(h) A boom sheet block on an iron horse.
(i) A boom topping lift with a crane line, foot ropes, sheet and a guy pendant.
(j) A boom topping lift with a span.
(k) A single boom topping lift.
(l) A double boom topping lift.

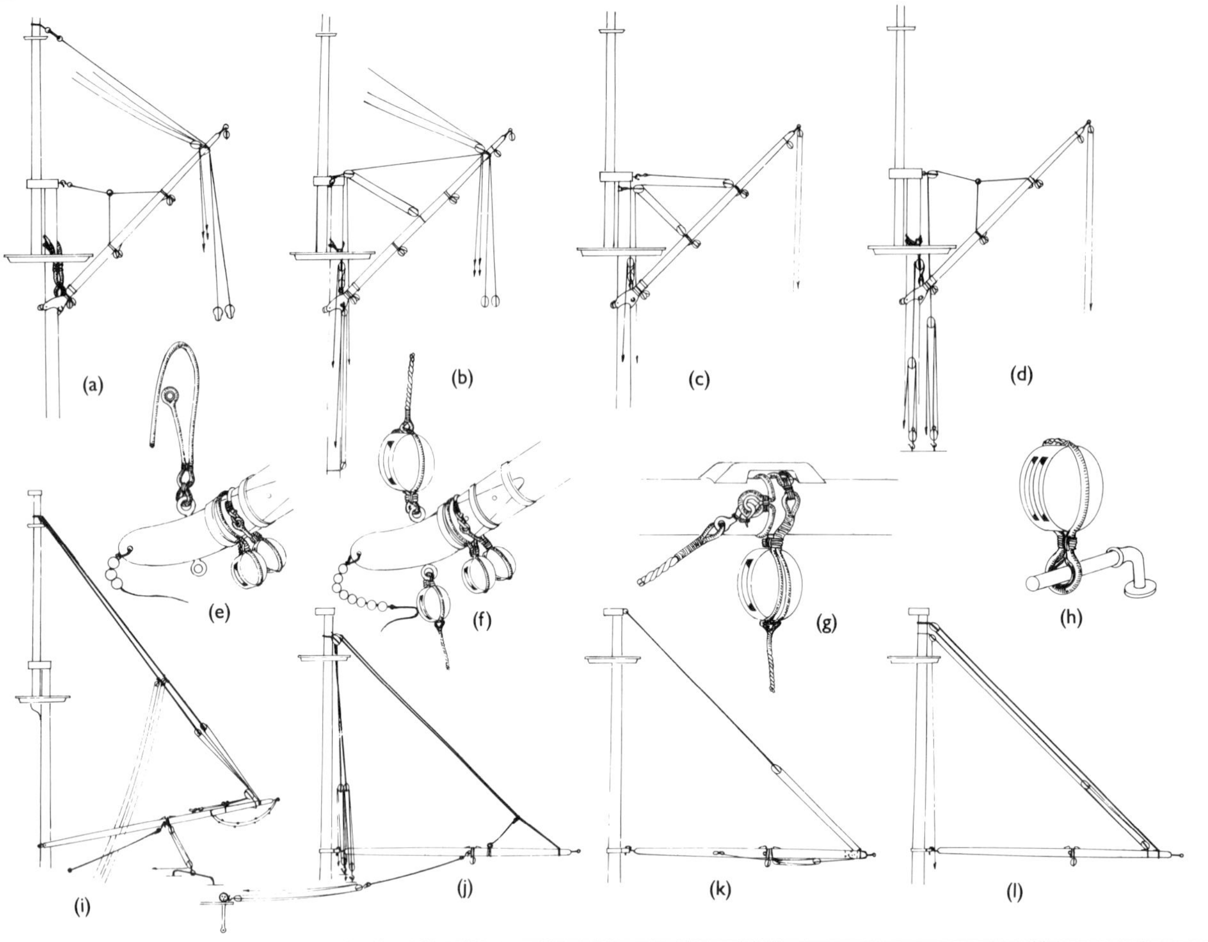

geöffnete See-Hafen included a detailed description, and Blanckley offered a general note about parrels and the following comment on trusses:

> [A truss is] a Tackle fastened to the Parrel at the Yard, which binds it fast when the Ship rowls, lying either a-hull or at Anchor, and the Fore, Main and Mizon Yards have them.

Parrels survived even into the last few decades of the eighteenth century on the Continent, whereas English ships of the later period had simple trusses.

Mizzen yard parrels were usually made of two rows of trucks. A two-hole deadeye (see Figure 45k), or in England usually a thimble, was seized to the jeer block strop, and the parrel ropes, made fast to the same point, passed through the two holes and were spliced together about 3 feet outside the deadeye. French parrels were not fastened to the jeer block strop, as Paris and Boudriot indicated, but were laid around both mast and yard, with the deadeye (or block at first, if the model of the *Royal Louis* of 1692 is a guide) seized into a bight in the parrel rope. The opposite ends were either spliced together directly behind the parrel (where a block was used), or passed through the deadeye's holes as above.

A thimble was spliced into the joined ends. A truss tackle, with a fiddle block above and a single block below, was hooked to the thimble and lashed or hooked to the foot of the mast. A similar parrel was described in 1705.

Shortly after 1730 the lower truss tackle block shifted from the foot of the mast to the foot of the yard, with the tackle placed at the upper side of the yard (see Figure 45i). French ships were rigged thus right up to the end of the mizzen yard's existence, while on English vessels the yard was hung permanently in slings after the mid-1780s, and a truss was therefore no longer required.

Gaff

Gaffel; Corne

The gaff, first introduced to three-masted ships in about the middle of the century and completely replacing the mizzen yard in the following four decades, was used, like the mizzen yard, to set a fore-and-aft sail. Its rigging varied somewhat. An important distinction can be made between a slung or standing gaff and one which could be raised or lowered (see Figure 47a and b).

Parrel

Rack; Racage

It was common to both types of gaff that a parrel formed of a single row of trucks kept the gaff jaws close to the mast (see Figure 47f).

Slings

Hanger; Suspentes

The slings for a standing gaff were either similar to those of a lower yard, or taken around a thimble at a ringbolt in the upper part of the jaws (see Figures 47a and e). Above the thimble a throat seizing was put on, with a round seizing a little higher up. The sling legs were usually of equal length, with an eye splice in their ends, and were seized together above the masthead rigging, around the masthead. The slings were wormed, parcelled and served all over.

Span

Spann; Patte d'oie

A span, served, leathered and with eyes on both ends, was pushed over the centre of a slung gaff and held in place with stop cleats at about a third of the gaff's length in from both ends. A pendant was spliced around a thimble which ran freely on the span, and was hooked to an eyebolt at the aft side of the mizzen cap (see Figure 47a).

Tye

Drehreep; Itague

The eye of another pendant, the peak tye, was laid around the peak end of a standing gaff. A thimble was spliced into its other end, and another was seized to the topmast head; these were then seized together with a lanyard (see Figure 47a).

Brail blocks

Dempgordingblöcke; Poulies de cargue

Brail blocks were fitted to a slung gaff in a similar manner to those of a mizzen yard. Traversing gaffs did not have brail blocks.

Brace blocks

Brassenblöcke; Poulies de bras

Two double or four single blocks were fitted to the peak of a standing gaff for the mizzen topsail and mizzen topgallant braces; in the case of a gaff which could be raised and lowered, however, the braces led forward, according to Lever, though not all authors agreed on this simple rule. In Steel's illustration of the rigging of a 20-gun ship the braces lead toward the peak of the gaff, even though a throat halyard is mentioned. His accompanying text made no mention of braces leading forward. Falconer in 1769 and Röding in 1794 noted that the braces led, forward and in the 1815 edition of Falconer's *Dictionary* the mizzen topgallant and royal braces led aft but the heavier topsail braces led forward. Ozanne's and Roux's many pictures of French ships support the implication of Paris and Röding that Continental ships were rigged with the braces leading forward. For further details see Mizzen topsail yard, Braces (below).

Vangs

Geere; Palans de garde

Lever noted that 'when the Gaff is rigged to hoist, there are no Vangs'; in other words, only a standing gaff was rigged with vangs (the fitting of these is described under Mizzen yard above). In contrast to Lever, other sources like Steel, Falconer and Rees made no distinction between a slung or hoisting gaff, and all noted vangs as part of gaff rigging. However, a close study of contemporary pictorial material makes it clear that Lever's description is simply more detailed, especially when one takes into account his note that 'ships which have their Gaffs traverse, carry Mizzen booms'. Röding confirmed Lever's description by noting that 'on gaff sails, the gaff usually remains in position; ... on boom sails it [the gaff] can be lowered'. These two elements, a hoisting gaff and boom, were usually found together, and vangs, acting upon the gaff like braces, were not needed with a mizzen sail or spanker controlled by a boom. Furthermore, the lee vang restricted the boom's movement and both vangs had to be disconnected when the gaff was lowered. These points confirm the unsuitability of vangs rigged to a hoisting gaff.

Throat tye, throat halyard

Klau-Drehreep, Klaufall; Itague de corne, drisse de corne

For the halyard of a hoisting gaff, a single or double block with hook and thimble (depending on the gaff's size) was hooked to a throat eyebolt (see Figure 47f), and the strops of another double-stropped double block were taken up between the trestle trees and laid around the mizzen masthead, where they were lashed together (see Figure 47b). Sometimes these blocks also had a thimble and hook, and were hooked to an iron ring strop or bolt at the masthead. The connecting fall was usually belayed to a mast cleat.

Larger ships, as well as brigs and other vessels with a fore-and-aft mainsail, were frequently rigged with a tye hooked or spliced into the throat eyebolt, instead of the arrangement described above. It passed through a large single block at the masthead, hung below the trestle trees, and had a double block turned into its leading end. A single block, with the halyard spliced to it, was hooked to a deck eyebolt near the mast at the starboard side. The halyard was belayed as above (see Figure 47d).

Peak tye, peak halyard

Piek-Drehreep, Piekfall; Martinet de corne, drisse de pic

Various methods of rigging the peak of a gaff also existed. One was similar to the rigging of a mizzen yard derrick after 1775 (see above). A second consisted of two single blocks stropped to the upper side of the gaff, where the brail blocks were fitted, and a double block seized to the rear upper part of the masthead. The standing part of the halyard was either hooked to the aft part of the mast cap or hitched around the masthead beneath the cap, reeving through the upper gaff block, the double block, the lower gaff block and the double block again, to be belayed like the throat halyard.

The third method displayed a certain similarity to the span of a slung gaff, except that instead of a pendant a tye was spliced round the thimble and a block hooked to the aft part of the mast cap. The tye passed through the mast block and was set up on the port side of the mast to a tackle, duplicating the throat tye on the starboard side (see Figure 47d).

Röding provided much the same explanation of Continental rigging of gaff peaks, with the difference that tackles at the leading ends of throat and peak tyes were not mentioned and the span was set up over the upper third of the gaff. Chapman's sail plan of a frigate gives the same impression as Röding's description. The rigging of French gaffs followed very much that of French mizzen yards, with a peak topping lift set up to the mizzen topmast head, and sometimes a peak halyard to the mast cap as well. Other illustrations reveal topping lifts leading toward the mast cap.

Peak downhauler

Gaffelniederholer; Hale-bas de corne

Hoisting gaffs were rigged with a peak downhauler, fitted like a flag halyard to the mizzen yard (see Figure 47c and d).

Throat downhauler

Klauniederholer; Hale-bas de mâchoire de corne, hale-bas de croissant

These gaffs also had a throat downhauler, the upper block of which was stropped to an eyebolt at the lower side of the gaff's throat. The lower block of this tackle was hooked to an eyebolt in the base of the mast, or to an eyebolt driven into the boom throat. Both blocks were single sheaved (see Figure 47b, d and f).

Mizzen boom

Besanbaum; Gui de voile d'artimon

The mizzen boom was used in conjunction with the gaff for setting a fore-and-aft sail. It was connected to the mast in a number of ways, all of which are described in Chapter I under Yards, Cleats and jaws. The use of a boom depended on the type of gaff rigged. Lever's comment that ships with a hoisting gaff had a mizzen boom and that those gaffs had neither vangs nor brace blocks but only a peak downhauler implies that for a normal mizzen sail a slung gaff sufficed, while for the rigging of a spanker sail a hoisting gaff was needed. One of the reasons for this can be found in reefing. A reef in the mizzen shortened the sail at the foot without any need to lower it; a reef in the much larger spanker, however, with a boom extending the sheet over the taffrail, required the gaff to be lowered in order to be effective. At the time of its introduction, in the late 1780s, the spanker was known as a driver. It was a fine weather sail, not always set, and replaced the mizzen in the Royal Navy from about 1805.

It is obvious that the lowering of a gaff with such a large sail would have been hampered by braces led aft and vangs led upon deck. Lever's note on this point differed from Steel's largely in that the former's was more detailed.

In contemporary English documents the mizzen boom was known as a driver- (and later spanker-) boom. Its forerunner on ships was a boom or pole, which projected over the lee quarters, to take the sheets of the original driver, a trapezoid mizzen studdingsail, hoisted on a yard to the peak of the mizzen yard or gaff. This type of driver was still common in merchant shipping early in the nineteenth century. The permanent rigging of a boom, with jaws or a gooseneck, on a full rigged ship, dated from the early years of the nineteenth century, while the temporary setting of a boom in connection with a large driver can be seen in English paintings as early as 1786.

Roux provided some information on the fitting of permanent booms on French ships. His watercolour of the *Commerce de Paris*, painted in 1809, clearly shows a boom rigged spanker, while his watercolour from 1811 of *Le Wagram* still has the trapezoid driver noted above hoisted to the gaff's peak. This indicates a transition period for large ships during the first decades of the nineteenth century. The history of the gaff and boom rig goes back, however, to the early 1600s, when the boom was first used in Holland to spread the foot of a shoulder of mutton sail.

Röding mentioned the boom mainly in connection with gaff-rigged smaller vessels. Only the boom is described in his work under 'driver boom', but he noted also under 'boom sail' that this, when used as a mizzen, was called a 'running mizzen' in contrast to the normal mizzen, of which he noted that 'nearly all mizzens are gaff sails today, with their lower part to be reefed'. Korth (1826) described the boom as still a temporary measure: '... the driver yard or the driver boom, since these are rarely used and only in calm weather... are placed with the spare spars'. Since Korth's sources were mainly late eighteenth century works, his comments can be taken as applying to the situation before the turn of the century.

All in all, it can be concluded that the Continental use of a fore-and-aft driver with a boom began in the 1790s, slightly later than in English rigging.

Parrel

Rack; Racage

A single row of trucks, like those of the gaff, held a boom fitted with jaws against the mast.

Topping lift

Dirk, Baumtoppnant; Balancine de gui

Several methods of rigging mizzen boom topping lifts were in use and no hard and fast rules determined which was applicable for any particular vessel. The methods described below were used not only on ships but also on lesser vessels such as brigs, sloops, cutters and schooners.

(1) The topping lifts were doubled and hitched around the mizzen topmast head above the masthead rigging, then seized together. They were served up to about 3 to 4 feet below the trestle trees. A single block was spliced or turned into each end and a double block was stropped to the boom end, resting against a stop cleat or shoulder. The fall was middled and also laid around the boom end, aft of the double block. Both ends of the fall passed through the single blocks, then over the sheaves of the double block and through lead thimbles at the boom, to be belayed to cleats just inside the inner seizing of the boom horses.

Each of these topping lifts had two Turk's heads worked in about halfway up, and a small single block stropped to the topping lifts between these knots. With a double wall knot at one end, lines known as crane lines were stapled to the fife rails on each side, passed through the blocks and belayed to cleats on deck. They were used for hauling the lee topping lift over to clear the sail (see Figure 47i).

(2) It was more common to hitch and seize the bight of the doubled topping lifts around the boom end. A larger single block was lashed to each side of the mizzen masthead, with the topping lifts reeving through. Double blocks were spliced into the lower ends and connected by falls to single blocks hooked to eyebolts on deck. The hooks were finally moused with spunyarn to prevent them from slipping out (according to Lever). Steel stated that these single blocks were hooked to eyebolts in the mizzen channels and the leading parts of the falls were belayed to cleats at both sides of the mizzen mast.

Occasionally a span was fitted to the topping lifts. This had a thimble spliced in each end and, under slight tension, was lashed to both the topping lift and the boom, several feet inside the boom end. The reason for fitting this span was to provide a margin of safety for the topping lifts in situations of extreme stress: before the lifts broke, the somewhat thinner span snapped and warned of the danger (see Figure 47j).

(3) A pendant, hooked to an eyebolt in the aft part of the mizzen cap, led downward and had a single block spliced into the other end. A runner was fitted with an eye over the boom end, and then passed through the single block on the pendant and a sheave hole in the end of the boom, ending with a double block. A single block was stropped around the middle of the boom, resting against a stop, with the standing part of the tackle fall bent to a becket behind it and the hauling part leading through to a cleat further in on the boom (see Figure 47k).

(4) The topping lift was put over the boom end with an eye and led through a single block at the aft side of the mizzen masthead. It ran back towards the boom end, and a single block was turned in or spliced to the running end. Inside the standing part, a double block was stropped around the boom end. The fall, bent as before to the single block, passed through both blocks and up again to the masthead, and from there through a lead block upon deck, where it was belayed to a cleat at the mizzen mast (see Figure 47l).

(5) On ships in the North and Baltic Sea trade with fore-and-aft mainsails, and on schooner sails on vessels such as North American schooners, the topping lift of the respective boom ran not to the masthead, but to the fore side of the next mast aft. A pendant was spliced to an eyebolt in the end of the boom, and had a thimble spliced into its other end. Below the bibbs of the next mast aft (the mizzen mast generally, but the mainmast for a schooner) an eyebolt was hooped to the front. A tackle was hooked into both the pendant thimble and the mast eyebolt, and its fall was belayed to a cleat at the mast (see Figure 48a).

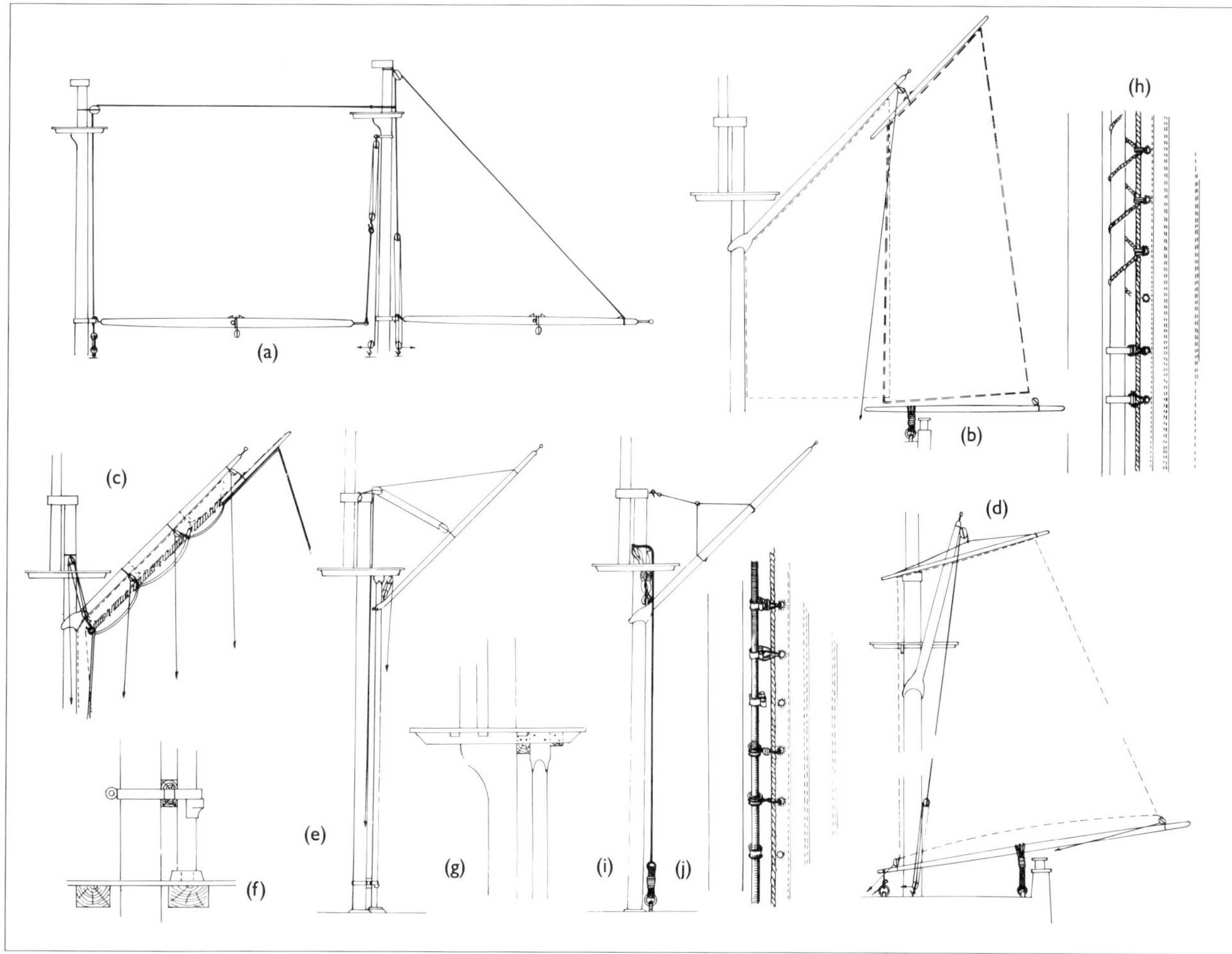

Figure 48
(a) Topping lifts and a stay on ships with fore-and-aft rigging; this rig was used on vessels in the North and Baltic Sea trade.
(b) A driver for a merchantman, about 1800. The mizzen is without a boom, but the driver is provided with a short, loose driver boom.
(c) A large driver for a man-of-war, about 1800; this was set temporarily instead of a mizzen, and the mizzen sail was brailed up (dotted line).
(d) A driver up to about 1770; hoisted to the gaff peak, the sail was sheeted to a driver boom reaching over the lee quarter. This driver was not an extension of the mizzen, but a square sail abaft it.
(e) A snow mast.
(f) Detail of a snow mast's footing.
(g) The upper end of a snow mast secured to the mast's trestle trees.
(h) Two different methods of bending the sail to a snow mast:
laced (above) and with wooden hoops (below).
(i) A horse, rigged to some men-of-war instead of a snow mast.
(j) Two different methods of bending a trysail to a horse: the upper section is shown with hanks, and the lower with thimbles.

Further information on topping lifts, boom sheets, etc, can be found in Chapter V below.

Boom sheet

Baumschot; Écoute de gui

The boom sheet block, a double-stropped single block, was hung on the mizzen boom, inside the taffrail. Two methods existed for securing it to the boom:

(1) A round strop was doubled, laid around the block and seized. The ends which formed eyes went from both sides around the boom and were lashed together, keeping the block firmly below the boom (see Figure 47g).

(2) The block was turned into two strops which were spliced around the boom and seized tightly together between the block and the boom.

The sheet was bent to a becket at the boom block's strop and passed through a double block which slid with a thimble on an iron horse (see Figure 47l), then through the boom block and double block again; the hauling end was belayed to a cleat or a pin through the lower block.

A comb cleat or simple stop cleats were nailed over the sheet block strops to the upper side of the boom, to retain these in place (see Figure 47g).

It is not clear whether the early type of temporary driver boom had a boom sheet. These booms were probably held in place only by boom guys.

Boom guy

Baumgei, Bullentau; Retenue de gui

Guys were used on booms to prevent the sail from accidental gybing (Falconer defined gybing thus: 'to Gybe, to shift any boom-sail from one side of the mast to the other'). A very early illustration of such guys on a ship can be seen in Raper's watercolour of HMS *Sirius* of 1787. The pendants of the boom guys are hooked to the wing-transom and the tackles to the boom ends, with the leading part coming in over the taffrail. A number of lithographs by Serres from 1805 indicate the reverse arrangement, with the pendants fastened to the boom end and the tackles to the sides above the quarter galleries; this latter arrangement was the more common on English ships. The falls in this arrangement came in over lead blocks on the rails.

Steel's description of the driver or spanker boom included the following note:

> Guy-pendants have a hook and thimble spliced in one end that hooks to the thimbles seized in the strap. They are spliced around the boom perpendicular to the lower block fixed around the horse within the taffarel and there stopped by cleats nailed on the foreside. A thimble is spliced in the inner ends of the pendants with a luff-tackle hooked in them on each side, and are used where most wanted.

The running rigging plan for a 20-gun ship which accompanies his description shows the lower tackle block hooked to the quarterdeck, level with the aftmost main chain. He did not describe boom sheets and they are not shown in the plan, but his rigging tables include boom sheets for all sizes of warships, though not for merchantmen. Lever gave a similar description, except that the lower tackle block hooked to the main chains.

Röding considered the boom guy a part of the rigging of smaller vessels only, and also described it as fastened to the middle of the boom, with the tackles hooking to the main chains. This suggests that mizzen booms were introduced later to Continental ships.

Boom horse

Baumpferd; Marche-pied de gui

Horses, to enable crewmen to work outside the taffrail, were spliced around thimbles in the eyebolt at the end of the boom. Where no eyebolt was used, the horses were doubled, laid around the end of the boom and seized underneath. The inner ends of the horses were spliced around thimbles also, lashed near the taffrail to the boom. These horses had diamond or overhand knots worked in to provide footholds for the sailors (see Figure 47i).

Driver boom

Treiberbaum; Bout-dehors de tape-cul

The boom of a driver, when this sail was set in addition to the mizzen sail, was not connected to the mast. Drivers were used in the Royal Navy from about 1730 until the late 1780s, but longer in the merchant service and on Continental ships (see Mizzen boom above). Like the studdingsails, a driver was set only during calm weather (for further details see under Early driver and New driver in Chapter XII below).

Falconer's description of the boom's use in 1769 and earlier ('the lower corners of [the driver] are extended by a boom or pole, which is thrust out across the ship, and projects over the lee-quarter') are echoed in Röding's text and illustrations at the end of the century (see Figure 48d). Lever noted a differently shaped fore-and-aft driver in addition to the mizzen and described the 'pole run out over the Tafferel'. The same type of driver boom and rigging can be found in Hutchinson's *Treatise on Naval Architecture* of 1794.

Though a block for the driver sheet was clearly lashed to the outer end of the boom, details of the fastening of the inner end of the boom cannot be found either in text or illustration. Since the sail was hoisted at one third of the yard's length to the gaff peak, it must be assumed that the inner end of the boom was lashed to the taffrail or to an eyebolt inside it, to prevent the peak of the sail from dropping off (see Figure 48b).

Driver yard

Treiberrah; Vergue de tape-cul

A driver yard was the studdingsail yard for the gaff or mizzen yard and shared many features with the other studdingsail yards. Where a driver was hoisted in addition to a gaff, the yard was about two thirds the length of the gaff; for a driver hoisted instead of the mizzen, the yard was only three eighths the length of the gaff (see Figure 48b and d). When the latter type of driver was set, the mizzen had to be furled. Lever noted that it was more common on merchant ships for the driver yard to be long enough to extend over the whole length of the top leech of the driver sail, rather than being an extension of the gaff (see Figure 48c, shaded line); with this type of yard the sail stood better, since the pull on the middle, inner and throat halyards all acted upon the yard, and the top leech was not liable to bag. The halyard for a driver hoisted in addition to a gaff, originally connected to the centre of the driver yard, subsequently moved to the inner third (see Figure 48b).

Snow mast, trysail mast

Schnaumast; Mât de senau, baguette de senau

In the eighteenth century, a snow mast was found only on a snow; its introduction to three-masted ships did not occur before the second decade of the nineteenth century. Falconer noted that a snow mast was a small mast abaft the mainmast, with its foot resting in a block of wood on the quarterdeck and its head attached to the after part of the main top (see Figure 48e, f and g).

Sometimes, however, a horse was used instead of a small mast. He added that 'When the sloops of war are rigged as snows, they are furnished with a horse, which answers the purpose of the try-sail-mast' and noted, in another connection, 'those sloops of war which occasionally assume the form of snows, in order to deceive the enemy'. This horse was laid around the masthead with an eye and set up with deadeyes to an eyebolt on deck abaft the mainmast (see Figure 48i). The gaff, which could be raised and lowered on a snow mast, was slung to the mainmast when a horse was used.

Not much information is available about the diameter of a snow mast or its distance from the mainmast. Generally it could not have been more than half the diameter of a mast, and it had no taper.

Main and fore topsail yards

Gross- und Vormarsrahen; Vergues de grand hunier et de petit hunier

The main and fore topsail yards were rigged identically, except for the braces (see below).

Horses, foot ropes

Fusspferde; Marche-pieds

The horses were fitted with an eye splice over the yardarms, to rest against the stop cleats. The inner ends had thimbles spliced in, which were seized to the parrel rope, though some sources mentioned the quarter block or tye block strops instead of the parrel. Horses which were seized to the quarter block strops had a span spliced into both, to prevent them from slipping outwards when the topgallant sheets were not hauled taught. The inner eye splices might also be seized to the yard just outboard of the sling cleats. The first three methods were all noted by Lever, while Steel mentioned the last, which was introduced around 1770. Before this the first method was most common in English rigging.

French ships, in the early days, had horses set up to small deadeyes close to the yard sling, as shown by the *Royal Louis* of 1695; late in the eighteenth century, as the spar plan of *Le Protecteur* (1793) shows, the horses were set up to eyebolts in the same position. Boudriot noted a fitting similar to that on the lower yards for a 74-gun ship of about 1780, but with thimbles turned into the inner ends and lashed together to create one large foot rope. Röding noted only this type of fitting, but with deadeyes instead of thimbles. This fitting was already in use on Continental lower yards in 1705; on all other yards, however, a lashing to the parrel was advocated.

In addition to normal horses, English ships, from the middle of the century on, had short yardarm horses (called Flemish horses) rigged to the topsail yards. These had an eye spliced into each end; one was put over the yardarm eyebolt when studdingsail booms were carried over the outer boom iron, and the other was seized to the yard within the arm cleats. Sometimes the inner end was spliced to the outer part of the foot rope instead.

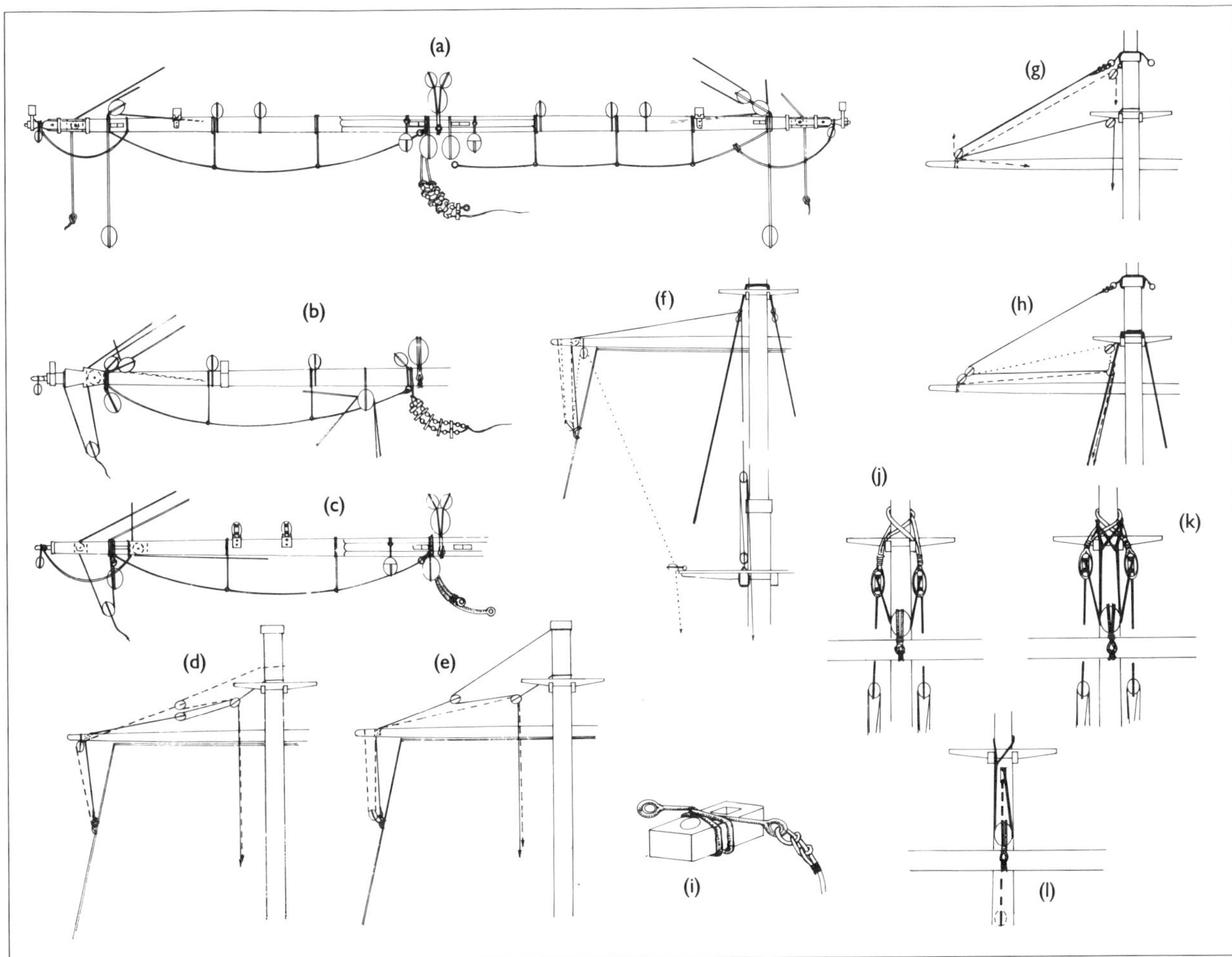

Figure 49

(a) An English topsail yard of about 1775. The left side shows the topgallant sheet as a lift (until 1790) and the Flemish horse seized to the yard with a free hanging jewel block. In contrast the right side has a double lift in addition to the sheet, the Flemish horse runs with a thimble on the foot rope and the jewel block is spliced into the outer end of the horse itself.

(b) A French topsail yard of about 1790. Differences from the English yard include: double reef tackles, no Flemish horse, different placing of the sheet, lift and clewline blocks, no brace pendants, dog and bitch lift blocks.

(c) An English topsail yard for a merchantman, about 1790. It is rigged with a double reef tackle, and the topgallant sheet reeves through a yardarm sheave hole. The buntline blocks are nailed to the yard. The parrel has been replaced by slings (rope parrel).

(d) English reef tackle, up to 1750; the dotted line shows the Dutch arrangement of the same tackle.

(e) English reef tackle, 1750-1790; the dotted line shows the French arrangement of the same tackle.

(f) English reef tackle after 1790; the dashed line shows a double tackle and the dotted line an alternative arrangement.

(g) Continental lifts, up to about 1790; the dotted line shows the lift of a French ship of the line of about 1700.

(h) English topsail yard lifts; the solid and dotted line shows the lift from 1715 to 1735, the solid line a double lift after 1790, and the dashed line a single lift after 1790. Before 1715 and between 1735 and 1790 the lift was also used as a topgallant sheet, and was rigged like the solid and dotted line.

(i) A lift span around an English cap.

(j) A topsail yard tye with a single sheaved tye block at the yard.

(k) A topsail yard tye with a double sheaved tye block at the yard.

(l) A mizzen topsail yard tye, also used for other topsail yards on smaller ships.

It has generally been said that French ships did not have Flemish horses, but a number of drawings by Ozanne do show them; Röding, not apparently considering these horses a specifically English feature, commented as follows:

> Sometimes a special horse is hung from the outer end of the yardarms to the inside of the arm cleats; this is called a yardarm horse. It provides a support for sailors working on the yardarm, for example when they bend the reef tackle pendants and or fit the lifts, brace pendants and yard tackles over the yardarm.

Stirrups came into use on topsail yards not before the second quarter of the century. The number needed on each side for larger ships was three, for medium sized ships two and for smaller ships one.

A rope very rarely mentioned in connection with foot ropes is the lifeline. Röding again provides a description:

> For this reason, especially in bad weather, back horses were also rigged, against which the yardmen could lean so as not to be thrown backwards by heavy pitching and rolling; these led usually from the lift to the yard tye. Now they are seldom found.

Although regarded as old-fashioned by Röding, these life lines can still be seen twenty years later in an appendix to Lever's work, in connection with a jackstay fitting: 'a Life line is rigged from the Lift to the Strap of the Tye Block'. Steel mentioned them in his article 'Lines' in a similar manner. They were a feature of rigging for most of the century, and beyond.

Jewel blocks

Leesegelfallblöcke; Poulies de bout de vergue

The block lashed to the yardarm eyebolt, or the outer boom iron outside the Flemish horse, was known as a jewel block. Sometimes this block was seized into the outer eye of the Flemish horse itself. On Continental ships these blocks were stropped to the yardarms and were in existence as early as 1705. Jewel blocks functioned as studdingsail halyard blocks.

Reef tackles

Refftakel; Palanquins de ris

A reef tackle was employed to pull the skirts of a reef nearer to the topsail yardarms, to lighten the sail when reefing commenced.

The reef tackle pendant on English ships passed through a sheave or hole in the yardarm and was made fast to a cringle 3 feet below the lowest reef band. Until about 1790 the part of the pendant above the yard was much shorter than that below and had a single block spliced to it. A further single block, stropped to a pendant around the topmast head, hung below the trestle trees. The standing part of the tackle fall was at first bent to a becket at the back of the masthead pendant block; in the second half of the century it was made fast to the topmast cap (see Figure 49d and e, solid lines). After reeving through the blocks the fall led downward upon deck.

Röding's description was similar. The standing part was made fast to the upper part of the masthead, with the topmast block stropped to the masthead slightly below, and the single block was not spliced but hooked to the pendant. The description given by Falconer in 1769 of this masthead block agrees with Röding's. Dutch ships in the first half of the eighteenth century had blocks rather than sheave holes at the yardarms, and the standing part of the tackle was made fast above the topmast head rigging; otherwise it was rigged like English tackle (see Figure 49d, dotted line).

Several methods of rigging a reef tackle can be found on French ships. The standing part was sometimes spliced around the yardarm and the fall passed through a block fastened to the lower reef cringle, then through the yard sheave hole and another block below the trestle trees or beside the tye block, and upon deck. A single fall, bent to the lower reef cringle, led through the yard sheave hole, through a block near the parrel, and upon deck through a nine-pin block, to be belayed at the bitts.

Boudriot offered another description, broadly similar to those of Röding and Falconer, but with the addition that the running part came down through the lubber's hole and was belayed to the bitts, after passing over one of the topsail sheet bitt sheaves. The running part might also pass through holes in the after part of the top, then through trucks at the catharpins, and be belayed to mast cleats. Reef tackles for the *Royal Louis* of 1780 consisted of pendants with single blocks as in English practice, a long span with two blocks laid around the middle of the topmast head (double blocks for the fore and main topsail yards and singles for the mizzen topsail yard). The fall at the larger yards was made fast to the pendant block, and at the mizzen topsail yard to the span block, and passed through these and lead blocks or trucks at the catharpins upon deck.

After 1790 these topmast head blocks were replaced by sister blocks, seized just below the shroud seizings to the foremost pair of topmast shrouds, with much longer pendants leading over the upper sheaves of these blocks. The tackle consisted of a double block, turned in or seized to the hauling end of the pendant, and another (or sometimes a single block) seized to the aft part of the top (or trestle tree). The running part of the fall led upon deck (see Figure 49f).

Merchant ships frequently had double reef tackle pendants, rigged according to the first French method described above. On merchant ships with only a few deck hands, the pendants were stropped around the yardarm, passed through blocks at the cringles, other blocks at the yardarms, down to lead blocks at the sides of the top and from there upon deck. With this rig, a taut weather tackle acted simultaneously as a rolling tackle (see Figure 49f, dotted line).

Brace pendants

Brassen-Schenkel; Pantoires de bras

Brace pendants were similar to those of the lower yards. In merchant shipping in general, and on French men-of-war during the last few decades of the century, a long-stropped or thimble-stropped brace block without a pendant predominated.

Fore topsail braces

Vormarsbrassen; Bras de vergue de petit hunier

Until the 1730s fore topsail braces were hitched to the middle of the main topmast stay (two thirds of the way up according to *Der geöffnete See-Hafen*), and passed through the pendant blocks and a set of blocks on a short span about 5 feet below the standing part. They then passed through a set of lead blocks vertically below on the main stay, through foot blocks on deck, and were belayed to cleats. (The author of *Der geöffnete See-Hafen* noted a lead from the main stay to the fore deck, to be belayed near the fore shrouds).

After the 1730s and until about mid-century, the standing parts of the braces were made fast to the main stay about one fifth of its length down from the stay collar, and the braces passed through the pendant blocks, the short spanned blocks a few feet below the standing part and through lead blocks at the lower main stay directly above the belfry; they led upon the forecastle and were belayed to cleats beside the belfry. On some Continental ships the lead blocks were seized to the foremast catharpins, and the braces came down, reeving through the inner sheaves of the fore jeer bitts.

After about 1745 the standing parts were hitched to the main stay collar, the lead blocks were slightly below this and the braces led as before to the belfry. In the later years of the century, according to Steel, the standing parts were hitched below the lead blocks to the collar, and, after passing the pendant and collar blocks, the braces came down to other lead blocks at the stay, abreast of the fore hatchway. They passed through a fourth set of lead blocks, were stropped with thimbles to eyebolts in the aft part of the forecastle and finally belayed to iron pins at the boat skid.

A trend in English rigging around 1800 towards moving the standing part of the braces to the main topmast stay collar can also be seen on some French ships of this period. The two parts of the braces were thus further separated, and the braces could therefore act more evenly on the yard, rather than pulling it down to lee when it was hoisted to its highest position. The running parts led through further lead blocks high on the main stay collar, down to the main topsail sheet bitts.

A few decades earlier, the standing parts of French fore topsail braces had already been made fast either to the main or main topmast stay collar, or alternatively to the main topmast shrouds above the catharpins. After passing blocks just below the main stay mouse, the braces passed through the second sheaves of the double blocks mentioned under Fore yard and main yard, Fore braces (above), and over bitt pin sheaves of the bitts on the quarterdeck. Röding, though considering that the braces acted upon the main stay like English braces, also described the leading parts running over the second sheaves of these same double blocks and belayed to the quarterdeck bitts.

Main topsail braces

Grossmarsbrassen; Bras de vergue de grand hunier

Up to 1775 the standing parts of the main topsail braces were made fast to the inner section of a span around the mizzen mast beneath the crossjack yard, according to Lees. This is in some respects corroborated by a Continental description dating from the first decade, in which the standing part was described as fastened to the back of a lead block on the foremost mizzen shroud just below the crossjack yard. A closer examination of English contemporary rigging plans (Sutherland 1711, the 1719 and 1745 Establishments, and Steel 1794) suggests that the span was above the crossjack yard. Steel also noted that 'the leading-part reeves through a block in the span around the mizen-mast head below the hounds.' The outer ends of this span were spliced around the brace lead blocks. From 1775 to the end of the century the standing parts, in English rigging, were made fast to the mizzen stay collar.

The span was still used on the Continent, with slight variations, well after 1775. Röding placed the span above the parrel of the mizzen yard, which is much the same position as given in the English description. Chapman's rigging plan of a frigate also shows a span above the crossjack yard, though the position at which the standing parts of the braces are made fast are not clear. The span shown on the *Royal Louis* (1780) is slung around the masthead and hung beneath the mizzen top. Boudriot described the span as fitted above the mizzen top at the same date, though he noted that the English set up of braces after 1775 and 1805 provided alternative fitting points for the standing part. (The set up he gave for 1805 differed from the preceding English practice in that the standing part was not hitched and seized to the mizzen topmast stay collar, but to the mizzen topmast head. This fastening position was probably in use in France not much earlier than in England.) The length of one span leg was approximately 6 feet.

The leading parts of the braces passed through yard blocks (with or without pendants), the blocks in the ends of the span or in the upper part of the foremost mizzen shrouds, and upon deck, where they were belayed either to cleats at the mizzen mast or to the rails abreast the mizzen mast. Steel noted that the braces led through sheave holes in the mizzen topsail sheet bitts, abaft the mizzen mast, and belayed there.

Röding noted that instead of leading through the span blocks, the main topsail brace 'commonly passed on English ships through a block or a sheave at the outer ends of the mizzen mast cross tree'.

This statement is not confirmed by contemporary English authors, but he must have based it on observations of English merchant ships visiting Hamburg. If their main topsail braces were rigged conventionally, he would not have made this particular note.

That such deviations from the rule were possible at the end of the century is confirmed by Lever's description of the rigging of these braces. He described the leading part of the brace reeving through a block at the mizzen stay collar, or a block stropped to an eyebolt hooped to the mizzen masthead, or, more commonly, through lead blocks spliced to each side of a pendant laid around the mizzen masthead and seized to the mizzen stay collar.

> This leading of the Main Topsail Braces to the Mizen masthead, has the effect of canting the Yard when up at the masthead... In the Merchant Service they often have, on this account, the Mizen Topmast made stouter than usual, and lead the Main Topsail Braces to the Mizen Topmast Head, which causes it to traverse in a more horizontal direction; an additional Backstay is also frequently used. At all events it would perhaps be better for the standing part of the Brace to be taken to the Mizen Topmast Head, like that of the Fore Topsail Brace to the Main Topmast Head, mentioned in the former page.

In the context of the run of the leading part of the braces from the lead block positions as described, it is worth noting that later Continental ships had another lead block fitted to the foremost mizzen shroud. While Röding merely noted the existence of these blocks, Paris described them as fastened to about the middle of the shroud, and Boudriot to the lower third.

Lifts

Toppnanten; Balancines

Until 1790, topsail yard lifts usually also functioned as topgallant sheets. The only notable exceptions to this rule were French ships of around 1700, on which single lifts for the topsail yards existed beside topgallant sheets. These lifts passed through single blocks at the topmast trestle trees, according to Paris (see Figure 49g). On other Continental ships, as confirmed in *Der geöffnete See-Hafen*, these lifts had the standard double function. Many English ships between 1715 and 1735, however, also carried lifts and sheets separately. The lift block was then stropped to the sheet block at the yardarm and the sheet ran through an additional sheet lead block near the yard slings (see Figure 49h).

The most common method of rigging lifts throughout the century was to bend or toggle the standing part to a span around the cap (English) or around the topmast head slightly below the cap (Continental).

Boudriot described the standing parts of French lifts as set up to eyebolts at the cap, using toggles when the lifts were used as topgallant sheets. Since a toggle fits better into the eye of a rope span than into an eyebolt, Röding's description of the lifts toggled to a span (donkey's ears) around the topmast head may have more merit.

Falconer offered much the same description:

> There are also toggles of another kind, employed to fasten the top-gallant sheets to the span, which is knotted around the cap at the top-mast head. For as the lifts of the topsail-yard are out of use when the topsail is hoisted, they are always converted into top-gallant-sheets, to render the rigging at the mastheads as light and simple as possible. Before the topsail-yards can be lowered so as to be sustained by their lifts, it therefore becomes necessary to transfer that part of the lift to the top-mast head, so that the whole weight of the yard may be sustained by its masthead, and no part thereof by the top-gallant-yard, which would otherwise be the case. This is performed by fixing the double part, or bight of the lift, within the eye of the span above mentioned, and inserting the toggle through the former, so as to confine it to the latter, which operation is amongst sailors called putting the sheets in the beckets.

From the masthead, the lifts passed through the yardarm lift blocks and the blocks on the span over the topmast head beneath the trestle trees, then ran through the lubber's hole and through trucks made fast to the lower shrouds, to belay to timberheads. On French ships they were belayed to a pin in the pin rail, and Paris and the author of *Der geöffnete See-Hafen* noted that they came down on the third shroud.

The last decade of the eighteenth century brought a number of changes. Steel and Lever no longer described the lifts as toggled, but noted instead that the standing parts were hooked to beckets around the cap (Steel), or to eyebolts, additional to the span, in the cap (Lever). Both authors omitted the blocks beneath the trestle trees and noted that the lifts passed over the lower sheaves of sister blocks seized to the upper part of the foremost topmast shroud span. Both also mentioned separate topgallant sheets.

Single lifts were not seized to the topgallant sheet blocks when the yard had sheave holes for the sheets, but went over the yardarm with an eye splice like single lower lifts. They passed over the lower sheave of the sister blocks and the ends were made fast around one of the lower shrouds. Lever noted also that several large ships, by the end of the century, had blocks turned into the lower ends of the lifts, and these were set up with a single whip hitched or clinched to the chain plates or lower shrouds.

Röding (1794) noted in one place that lifts were sometimes still used as sheets for the topgallant sails, with the topmast lead blocks below the donkey's ears rather than beneath the trestle trees, and in another that topsail yard lifts were mostly rigged single by his time of writing, and therefore extra topgallant sheets were needed. He also mentioned the use of sister blocks, and noted that earlier sister blocks had three sheaves, with the third used for the studdingsail halyard.

It is clear from this that the last decade of the eighteenth century was a time of great changes in Continental rigging as well as in English.

Clewline blocks

Geitaublöcke; Poulies de cargue-point

Clewline blocks were placed outside the sling cleats and hung downward. Their placement throughout the century in both English and Continental rigging is discussed in detail under Main topsail in Chapter XII below. English clewline blocks, like the clewgarnet blocks of the lower yards, had a shoulder up to about 1775, but were normal blocks after this date. Continental clewline blocks were usually normal blocks, but the yard drawings for *Le Protecteur* of 1793 indicate double sheaved clewline blocks which also function as sheet lead blocks (quarter blocks), similar to the thick and thin sheaved quarter block mentioned by Steel; see Quarter blocks under Fore and main yard above.

Quarter blocks

Bramschotführungsblöcke; Conduite-poulies d'écoute de perroquet

Where lifts and sheets were rigged separately, lead blocks for the topgallant sheets hung inside the sling cleats or beside the slings, facing downward, on each side of the yard (see Figure 49a).

Tyes, halyards

Drehreepe, Fallen; Itagues, drisses

Röding noted that in nearly all languages the terms for 'tye' and 'halyard' were frequently confused.

The hoisting of a topsail yard with a tye depended on the size of the yard, and the rig of the tye underwent some development during the eighteenth century. At the beginning of the century ships on the Continent were still rigged with single or double tyes bent to the topsail yards. These passed through one or two sheaves in the hounds, and the bight of a double tye was led around the upper sheave of a leg and fall block (see Figure 50b and f). The standing part of the runner in the lower sheave was hooked to an eyebolt at the aft end of the port side fore channel for the fore topsail yard, or to a similar position on the starboard main channel for the main topsail yard. The hauling part, set up to a long tackle, was hooked into the opposite side (French).

Instead of the leg and fall block, Dutch topsail yard tyes featured a shoe block with two sheaves in the lower part (see Figure 50e) and a double block hooked into a strop around the aftmost cross tree, with the fall coming down abaft the mast.

Single tyes were set up, according to the size of the yard, either to a long tackle or to a gun tackle (see Figure 50c), rigged as above. Occasionally the tackle fall was taken through a lead block on deck before belaying to the bitts (see Figure 50c, dot and dash line).

The practice of fitting tye sheaves in the topmast hounds was discontinued on larger ships of the Royal Navy before the turn of the eighteenth century, and only on smaller ships are such sheaves found right through into the nineteenth century. From about 1680 onward, blocks were used on the main and fore topmasts and the sheave hole only on mizzen topmasts of larger ships. Several methods of rigging tyes existed:

(1) A double tye block was double-stropped to the centre of the yard. Two single blocks of equal size were lashed to the sides of the topmast head, hanging just underneath the topmast stay collar. The standing parts of the tyes were clinched around the topmast head and passed through the double block and the single blocks, crossed so that the tye coming from the starboard side passed over the foremost sheave and the port block, and that from the port side went over the other sheave and the starboard single block. Double-sheaved fly blocks for the topsail halyards were turned into the ends of each tye. The lower single halyard blocks were long-stropped to clear the gunwales when hooked to a swivel eyebolt at the aft end of the channels (see Figure 49k). The leading part, for the fore topsail halyards only, passed through an additional foot block at the aft side of the forecastle and was belayed to a pin in the rails. The main topsail halyards led over the sheave of a kevel within board and was belayed to its head (see Figure 50c). Steel mentioned a foot block on the quarterdeck instead.

(2) When a single tye block was stropped to the yard, the tye passed through it and the topmast head blocks, with both ends set up as before. During the first decade the fly block was a fiddle block; later a normal double block was used (see Figure 49j).

(3) Tyes of smaller vessels had a single tye block stropped to the yard. With one end clinched around the topmast head, the tye passed through the block and the topmast sheave hole and led down as before (see Figure 49j).

(4) The author of *Der geöffnete See-Hafen* described a main topsail yard tye consisting of a double block stropped to the yard and two blocks to the topmast head, hanging below the trestle trees. One of these was a single and the other a double block. The upward tye, with a block in its lower end, rove through the single block, over one sheave of the yard block, the first sheave of the masthead double block, back over the second sheave of the yard block and then over the other sheave of the masthead double block and downward. A block was spliced into the lower end of the tye. A runner was hooked to an iron staple abaft the main shrouds on both sides of the ship, with a further block spliced into it after it had passed through the block at the lower end of the tye. A fall was made fast to the back of the tye block, led through a block hooked into the staple, then through the block on the runner and upon the quarterdeck. It passed through a hole to the main deck and finally through a foot block to be belayed.

(5) The same author also described a second set up, in which one end of the tye was belayed to a knighthead and the tye was passed through the blocks noted above, after which the other end was made fast to the yard beside the yard block.

The two methods listed here as (4) and (5) of leading a topsail yard tye applied only to main topsail yards; fore topsail yards had a simple tye leading through a sheave in the topmast hounds as described above. However, the German author noted that large English ships used the same method for the fore topsail yard as well.

(6) No block was used at the yard at all, but a strop similar to a sling strop was laid around the yard and the tye bent to it with a sheet bend, the end also being seized (see Figure 50a). The tye led through the sheave hole, and a single block was turned into the end, taking a runner with a fly block in each end. The halyards led as before, via single blocks to the channels. Some ships had only one pair of halyards, with the standing part of the runner hitched to the trestle tree and the other leading down at the opposite side to a channel, 'in order to avoid the necessity of shifting the Topsail Halyards over to windward, when there is but one pair, they lead thus,' wrote Lever.

> The Fly-block is spliced into the end of the Tye, a little below the Topmast Catharpins: the single Block instead of leading to the Channel, is hooked to a Strap around the aftermost Part of the Trestle-trees. This Tye, which is short, is served the whole Length: and the Fall of the Halyards leads down abaft the Mast, and through a leading Block below.

Röding's description is very much the same as (2) with a double or treble block as the fly block and an iron bound block with a swivel-hook, usually hooked to the sides of the forecastle and quarterdeck or to the channels. French ships had an iron topsail halyard traveller fitted 2 or 3 feet above the fly blocks, between the tye and nearest backstay, which prevented twisting of the halyard.

Buntline and leechline blocks

Gordingblöcke; Poulies de cargue-fond et cargue-bouline

In English rigging two smaller single or one double block were seized to the upper part of the tye block strops. Up to 1745 these blocks served directly as leechline blocks, with the leechlines leading from there upon the deck. After 1745 lead blocks were lashed at about the midpoint of each side of the yard, and the leechlines rove through the blocks above the tye block, over the inner sheave of double blocks beneath the topmast trestle trees, and upon deck.

Up to about 1750 blocks were stropped to the yard for the buntlines and lead blocks, spliced to a span around the topmast head, were hung below the trestle trees. After 1750, the outer sheave of the double blocks noted above took the place of the span blocks. Steel noted single blocks rather than double, lashed

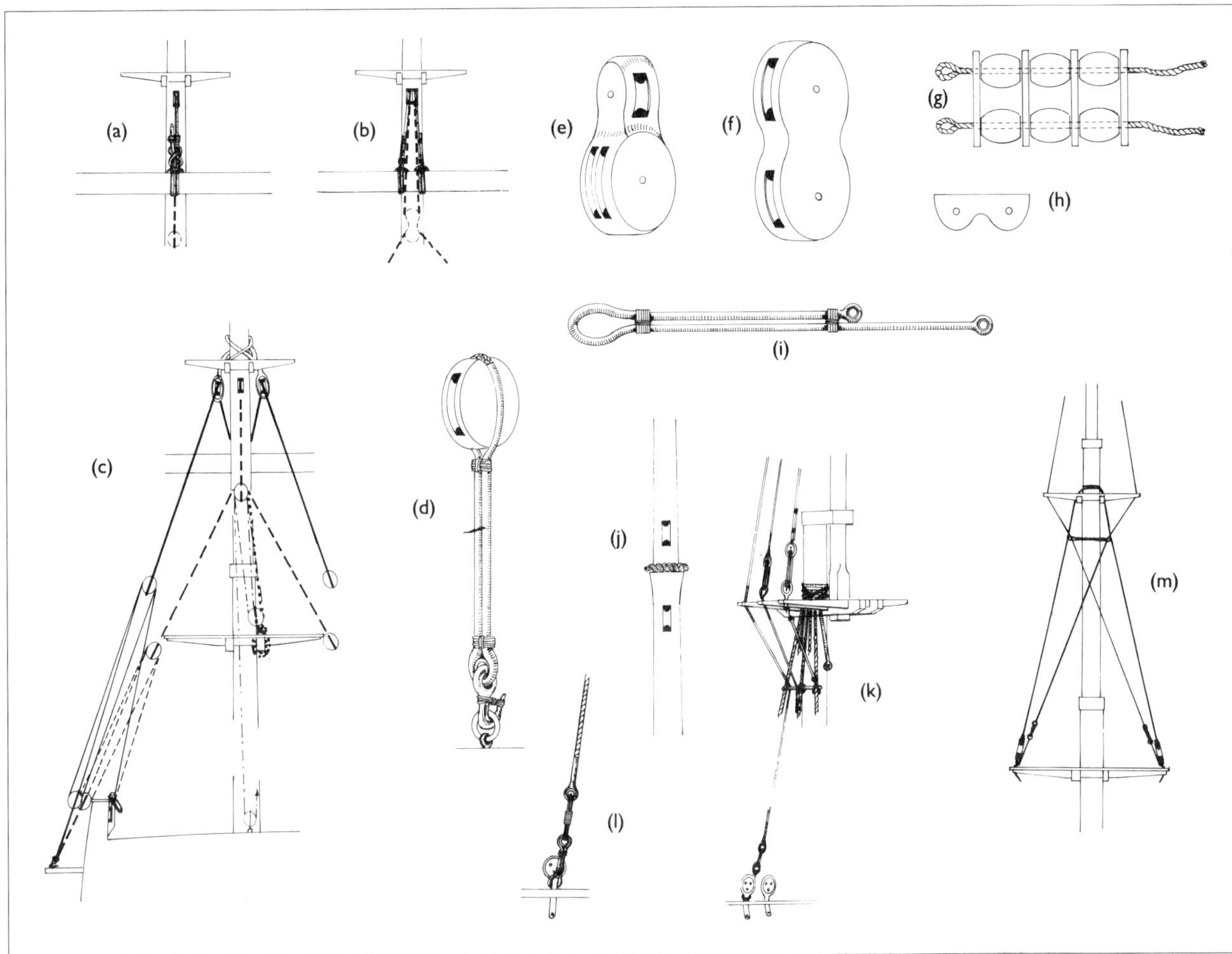

Figure 50
(a) Topsail yard tye on small ships.
(b) A Continental double topsail yard tye from the beginning of the century.
(c) Various methods of leading the topsail yard tye tackle on English and Continental ships, seen from abaft the mast. The solid line shows English practice and the various broken lines Continental rigs and those for small vessels.
(d) A long-stropped block, as used for a tye tackle.
(e) A double sheaved shoe block, used on Dutch tye tackles.
(f) A leg and fall block for French topsail yard tyes.
(g) An English topsail yard parrel.
(h) Detail of a rib of an English topsail yard parrel.
(i) An English topsail yard rope parrel or sling for a merchantman, after 1780.
(j) A grommet positioned on a topmast stop to prevent chafing from the shrouds and stay.
(k) Three different methods of setting up topgallant shrouds. The foremost shroud (right) is set up with deadeyes and the futtock shroud is hooked in (used on larger ships until about 1780); the centre shroud is set up with a thimble, with the futtock shroud spliced around a second thimble and turned around the futtock stave, and the shroud seized to this (used on smaller ships until about 1780); the aftmost shroud (left) passes through a hole in the cross tree, over the futtock stave and down the topmast shroud to be set up with thimbles to the futtock deadeye (used on English ships after 1780).
(l) Detail showing thimble lashing of a topgallant shroud to a topsail futtock deadeye.
(m) An alternative set up of topgallant shrouds on English ships after 1780.

close underneath the topmast cross tree. Leechlines were not mentioned in his text, but did feature in his tables.

Leechlines were not fitted after 1790 and the blocks above the tye block were often used as lead blocks for the buntlines. Large ships at this stage were rigged with two buntlines to each side, and the right number of blocks had to be lashed to the yard, two double blocks to the tye block and a double block to each side of the topmast head. Single buntlines ran directly to the tye lead block, without passing another block at the yard.

At the end of the century merchantmen had monkey blocks nailed to the upper side of the yard or, if the buntline was led through an eyelet hole just below the third reef, a block was stropped to the back of the sail, like a clewline block.

Continental ships did not have these lead blocks above the tye block. The French *Royal Louis* of 1692 had a buntline block placed close to the slings, and a leechline block a quarter of the yard's length inside the stop cleats. Both blocks rested on the yard. Single lead blocks for both lines could be found hanging to each side of the topmast stay collar.

Continental buntlines were described in 1705 as fastened to the footrope (boltrope), one third of the sail's width apart, then led through thimbles up to blocks on the yard, through further blocks below the topmast trestle trees and then upon deck. Where the topsail yard had a tye only, a block was seized to tye above the yard and the buntlines rove through this block to the mast top.

The centre buntline rove through a block above the yard and through another block fastened below the trestle trees, then upon deck.

Leechlines run from both leeches above the bowline bridles toward a block in the middle of the yard, and then into the mast top.

On early Dutch ships the leech- and buntlines, without yard blocks, passed directly through the stay blocks. Later French ships still had leechlines as above, but the buntline stay collar block had moved to the topmast head. French buntline blocks were similar to English clewline blocks. On some ships the buntline led as noted here for Dutch ships, but through a double block, and the topsail yard of *Le Protecteur* from 1793 indicates a leechline block, an outer buntline block and an inner buntline double block close to the tye block. The outer buntline block might have been used for what Röding called *Schmiergordinge* (preventer leechlines) and of which he commented:

> Because of their depth, the topsails also have Preventer leech-lines, which are rigged in similar fashion to the leech-lines but further below to the standing leech. Usually these are only needed at a anchor; at sea they are unreeved and serve as buntlines.

Topgallant sheet blocks

Bramsegelschotblöcke; Poulies d'ecoute de perroquet

When the lifts were not also employed as topgallant sheets an additional single block was put over the yardarm (see Lifts, above).

Parrels

Racke; Racages

A parrel consisted of a number of ribs and trucks, with each rib having two to four holes for the parrel ropes, depending on the yard's size and type. The topsail yards were held close to the topmasts by a two-row parrel. Topsail yard parrels on Continental ships differed considerably from those on British vessels.

On Continental ships the open parrel predominated for topsail yards as for lower yards. The Dutch method of rigging an open parrel for the lower yards at the beginning of the eighteenth century, described by Anderson and illustrated by H

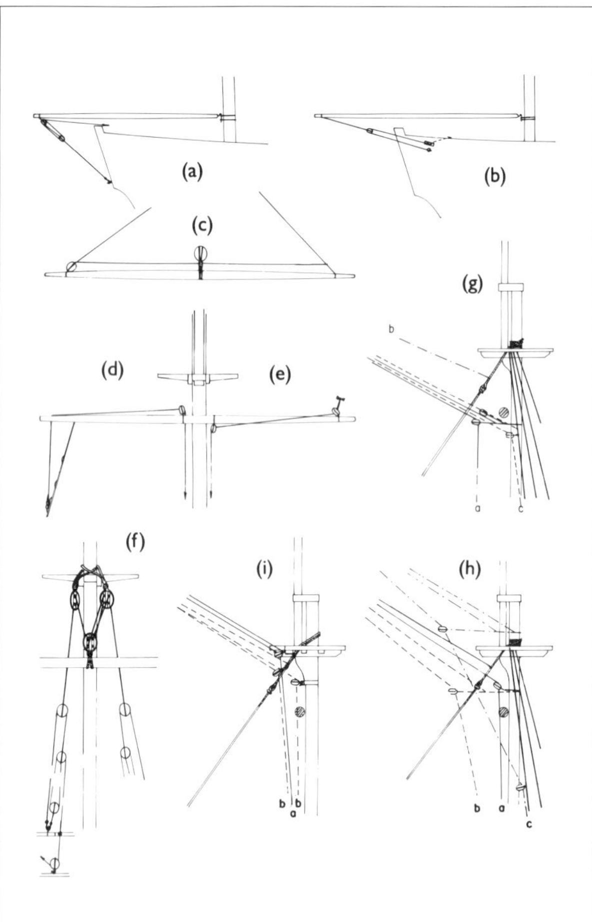

Figure 51
(a) An English boom guy of about 1787.
(b) An English boom guy of about 1800.
(c) A lifeline.
(d) French reef tackle, 1780.
(e) A French topsail yard lift and topgallant sail sheet, 1780.
(f) A Continental topsail yard tye, 1705.
(g) The lead of main topsail yard braces (I)
a. English, up to 1773 (according to Lees)
b. English, after 1773 (according to Lees)
c. Continental, 1705.
(h) The lead of main topsail yard braces (II)
a. English, 1711 and 1719 (according to Sutherland and the 1719 Establishment)
b. Swedish, 1769 (according to Chapman)
c. French, 1780 (according to Boudriot)
(i) The lead of main topsail yard braces (III)
a. Continental, 1794 (according to Röding)
b. English, 1811 (according to Lever)

Ketting, involved two separate parrel ropes. The bight of the lower rope went around the yard and the ends ran behind the mast through two sets of trucks and the two lower sets of holes in the ribs, then passed through a double thimble (or two-hole deadeye) stropped around the yard at the opposite side of the slings; the two parts of the bight were seized together a few feet beyond the thimble to make an endless rope, and a fiddle block for the truss tackle was lashed to the end. The upper parrel rope, equal in length to the lower and similarly doubled, passed through the upper holes in the ribs (and a set of trucks) and along the upper notch; the two parts of the rope were seized together just outside the outer parrel ribs on each side, the doubled rope passed over the yard to the fore side, and a fiddle block was lashed to each end. These two fiddle blocks thus hung over the fore side of the yard, but were passed (presumably via thimbles stropped to the lower side of the yard) to the aft side of the sail.

A Dutch open two-row topsail yard parrel was rigged as follows. An eye was spliced into one end of a parrel rope, and the rope was passed through the upper holes in the ribs and the intervening trucks; on the opposite side of the mast a bight was taken around the yard and closed by a seizing. The parrel rope was then led back through the lower holes in the ribs and the lower trucks, rounding the fore side of the yard from underneath, then passing through the eye splice and back around the fore side of the yard. It then crossed the parrel again via the notches in the ribs, and was turned again around the yard at the opposite side from underneath, before going through a thimble stropped to the yard (see Figure 52i).

Topsail yard parrels were not always rigged open, even on Continental ships. The author of *Der geöffnete See-Hafen* mentioned open parrels only for large ships, noting that other parrels were rigged closed or standing, in what was known as the English way. Unchanged for nearly two hundred years, this method was first described by the unknown author of *A Treatise on Rigging* published in 1625. In 1769 Falconer gave a useful description (see Figure 43f above): two parrel ropes each had an eye spliced into one end, and these eyes were seized together on the fore side of the yard. The parrel ropes were led through the ribs and trucks across the aft side of the topmast (one through the upper holes in the ribs and the other through the lower), passed around the yard on the opposite side, then returned over the notches in the ribs to the fixed side. They were then passed around the yard again and back over the parrel until the ropes were spent. To complete the process, the turns of rope were marled together, confining them to the notches of the ribs. When the parrel was set up, sufficient clearance was allowed to enable the yard to be hoisted and braced.

Röding offered a slightly different description of the rigging of parrels in general, but his comments and those of other authors confirm that, during the second half of the eighteenth century, the trend in Continental rigging practice was towards English standing parrels. Röding described a parrel with three rows of trucks (thus a lower yard parrel), which had a separate short parrel rope led through the middle holes with thimbles spliced close to the outside trucks. The upper and lower ropes were longer, with an eye spliced into one end, set on opposing sides. These ran through the ribs and trucks, then the long ends were turned a few times around the yard (from above for the top rope and below for the lower), passing through a thimble on the other side, then the remaining ends were led several times across the back of the ribs and marled together. The two eyes were seized together in front of the yard.

According to Röding, such large parrels were used only on the fore and main yards of men-of-war; topsail yards, and also topgallant yards, had parrels with two rows of trucks. A similar picture is given by the yard drawings for *Le Protecteur*, which show a three-row parrel for the main yard, and by parrels shown in another plate in Paris' work (taking the diameter of a yard as a yardstick for the length of the ribs, the topsail yard shown has only a two-row parrel). In Boudriot's work, on a slightly larger ship, a three-row parrel is noted for the fore and main topmasts and a two-row for the mizzen topmast and the topgallant masts. Of these three authors, Röding is the only contemporary; Paris was a near contemporary, having served in the sailing navy from 1820, and Boudriot is a modern author.

A two-row truck parrel as illustrated by Röding and Paris (see Figure 49b) had a doubled parrel rope with the bight going around the yard, and with both ends led through a single truck before passing through the two rows of ribs and trucks; at the other side of the mast the two parts of the rope again passed through a single truck, and were then turned around the yard from above, with the ends then passed back over the rib notches until extended. The rope turns were finally marled together.

Ribs were made from ash and had a length equal to the diameter of the yard. Their width was one third their length. The inside face of the rib was flat, while the outside was formed into two semicircles with a notch between. Lower parrels, with more rows of trucks, accordingly had more semicircles in the outer profile of the ribs. A hole for the ropes was drilled through the centre of each semicircle (see Figure 50h).

Trucks were also made from ash or from harder timbers and were equal in diameter to the width of a rib. The length of a truck was about one third greater than the diameter, and the hole ran through the length of the truck.

The length of a parrel rope was given by Davis (1711) as equal to the length of a topmast without the head, and by Steel as one fifth of the yard length, taking fathoms for feet.

By the end of the eighteenth century merchantmen had ceased to rig parrels. Lever noted that 'these Ribs and Trucks are seldom used in the Merchant Service, except in the largest ships'. Instead a standing truss parrel (see Figure 50i) was used, wormed, parcelled and served or leathered. An eye was spliced into each end, then a round seizing put on to form a bight to go over the yard. A quarter seizing was then put on close to the eye on the shorter part (two thirds the length of the longer), the longer end was taken around the yard and the two eyes were seized together. Röding gave a similar description of these parrels, and added that sometimes the longer end had no eye and was thinned out, so that it could pass through the other eye and be seized to itself. Smaller vessels had truss parrels like those used on topgallant yards (see below). Naval vessels were rigged with truss parrels only from the early nineteenth century on.

Mizzen topsail yard

Besanmarsrah; Vergue de perroquet de fougue

The mizzen topsail yard was rigged, with minor differences, like the other topsail yards. These differences affected the lifts, braces and halyards.

Lifts

Toppnanten; Balancines

Before 1790, mizzen topsail lifts and topgallant sheets usually corresponded to those on the other masts, if topgallant sails were set. In the last decade lifts were rigged single as a rule; only on large ships were they occasionally double, in which case blocks identical to those on the other topsail yards were fitted to the yardarms.

Braces

Brassen; Bras

Mizzen topsail braces led either forward or aft, as noted above under Mizzen yard and Gaff. When led forward, 'and commonly... single' (Lever), they passed through a double block stropped to an eyebolt at the aft side of the main cap.

'On smaller ships [the braces] are always single, and made fast to the mizzen topsail yardarm. From there they reeve through a block positioned at the aft side of the main top, down to the ship,' commented Röding. He described double mizzen topsail braces on larger ships, set up with their standing parts made fast to the aftmost main shrouds, near the catharpins. From here they led through brace blocks at the yardarms and blocks fastened to the shrouds below the standing parts, and upon deck. Boudriot added that the leading part also ran through a lead block seized to a bolt in the spirketting and belayed at the quarterdeck pin rails.

The aft-leading mizzen topsail brace was a feature of English rigging, and was rarely found on Continental ships. It was characteristic of the rigging of English men-of-war and East Indiamen for the whole of the eighteenth century, though there is some indication that East Indiamen began to change to forward bracing in the last decade. The only divergences from the great number of rigging plans, from Deane's *Doctrine* of 1670 to the early nineteenth century, underlined by many paintings and models of the period, are Falconer's 1769 sail plan of an English two-decker with forward-leading mizzen topsail braces and the description in the 1815 edition of his work, which also mentioned forward leading topsail braces, together with a drawing by Rees indicating the same. The latter two are perhaps only indicative of a trend which had begun with the ships of the East India Company.

Steel's description of the usual rig of these braces was as follows:

> Mizen topsail braces reeve through the block in the pendent, and the standing-part makes fast around the peek-end, and the leading-part reeves through single blocks at the peek, and comes down to the fore-side of the taffarel

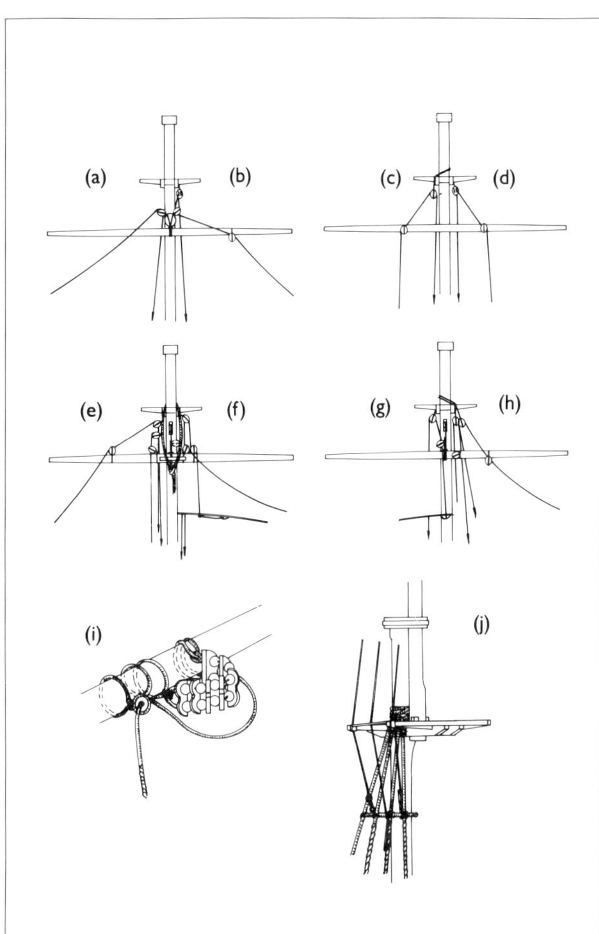

Figure 52
(a) English leechline block before 1745.
(b) English leechline blocks after 1745.
(c) English buntline blocks before 1745.
(d) English buntline blocks after 1745.
(e) French leech and buntline blocks, about 1700.
(f) Continental leech and buntline blocks, about 1705.
(g) Continental centre buntline blocks, about 1705.
(h) French leech and buntline blocks, about 1780.
(i) A Dutch method of fitting a topsail yard parrel.
(j) The French method of setting up topgallant shrouds in the later part of the century.

His description is applicable to English ships throughout the century.

Tye, halyard

Drehreep, Fall; Itague, drisse

Mizzen topmasts usually had a sheave hole in the hounds for the topsail yard tye. The methods of rigging a tye on such topmasts are described in detail above under Topsail yards, Tye.

Topgallant masts

Bramstengen; Mâts de perroquet

As noted at the beginning of this chapter, topgallant masts were only brought on board when the lower rigging was completed.

Top ropes

Stengewindreepe; Palans de guinderesse

For raising the topgallant mast a top rope was employed; with its standing part hitched around the polehead, this rope led down along the topgallant mast and passed through the sheave hole in its foot. It then led upwards again, and was securely stopped at various points to the topgallant mast, and after reeving through the mast hole formed by the topmast trestle and cross trees, went through a top rope block hooked to an eyebolt at the port side of the topmast cap. The hauling part led down upon the deck, passed over a snatch block, and was laid with a few turns around the capstan.

After the topgallant mast (with its pole head) had been hoisted sufficiently high into the cap's hole, the end of the top rope at the pole head was cast off, so that the mast now hung only by its stops, and was hitched to an eyebolt at the starboard side of the cap. All stops were then cut and the mast could be hoisted freely into place.

When the mast had just passed through the cap, a grommet, shrouds, backstays and a stay were lifted on to the masthead with the help of girtlines. The mast was then swayed into position, the fid thrust into the fid hole, and the top rope eased and removed. Top ropes were thus only temporarily rigged, for hoisting and lowering a topgallant mast.

Girtline
Jolltau; Cartahu

A girtline block was the very first item lashed temporarily to the head of any mast in the process of rigging. A rope known as the girtline rove through it, and was used to rig any part of the standing rigging. A sailor also could hoist himself up on the mast by securing one end of the girtline around his waist (or hitching it around a short piece of timber to sit on) and pulling the other part down with his hands. Like top ropes, girtlines were removed when the rigging was completed.

Grommet, gromet
Grummet; Bague

The actual rigging was commenced by driving a grommet down to the topgallant stop. A grommet was made by unlaying one strand of a length of rope and forming it into a circle. The remaining part of the rope was turned around the single strand until a three-stranded rope ring had been created. The ends were spliced into the ring, like those of a long splice. Grommets were used to prevent chafing of shrouds and the eye of the stay on the sharp edges of the stop (see Figure 50j).

Shrouds
Hoofdtaue; Haubans

Next the topgallant shrouds were put over, with the first pair again set up to the starboard side. Only two pairs were rigged until about 1720 and three after this, with the odd shrouds paired with the backstays. All shrouds were served as for the lower shrouds, and also in those sections passing through the holes in the cross trees.

The setting up of topgallant shrouds underwent considerable changes during the period in question. For the seventeenth and early eighteenth centuries, Anderson noted that the shrouds were set up with deadeyes or thimbles to futtock shrouds. This was still the case several decades later; and according to Röding even until the closing years of the century. Korth (1826) also mentioned futtock shrouds in his passage on shroud rigging.

French ships of the last quarter century differed from this set up by omitting futtock shrouds. The topgallant shrouds passed through the cross tree holes and each had a thimble spliced into the end. Thimbles were also made fast to the futtock staves, and the corresponding thimbles were seized together with lanyards. The shrouds were also sometimes set up directly to the staves, without thimbles. Because of this arrangement the topmast shrouds of French ships were often catharpinned, whereas on other vessels this was normally the case only for the lower shrouds. These French methods, described by Boudriot, are corroborated by the nineteenth century author Bobrik (see Figure 52j).

Steel and Lever, describing English topgallant shrouds in the last decades of the eighteenth century, both noted that the shrouds were passed through the holes in the cross trees and between the topmast shrouds over the futtock staves. They ran down on the topmast shrouds and each had a thimble turned into its end; these were set up to thimbles seized to strops around the futtock-plates (see Figure 50k). In some instances the topgallant shrouds were

> ...set up to Thimbles in the opposite Futtock Plates: this is a great ease to the Futtock Stave and Topmast Rigging and the lee Rigging, by these means, is kept taught, as they act like Bentinck Shrouds below [Lever; see Figure 50m].

Exactly when the rigging of topgallant shrouds began to change is not clear, but it was probably during the last two decades of the century. The 1815 edition of Falconer's *Dictionary* noted that the preceding thirty years had seen considerable changes in the way ships were rigged.

In the 1790s a block or thimble was seized between and just below the bight seizing of the first and second shroud, for the topgallant yard lift to reeve through.

Ratlines
Webleinen; Enfléchures, quaranteniers

Up to about 1720 ratlines were usually not rigged on topgallant shrouds; a notable exception is some large French ships. Thereafter, until the 1750s, larger ships in general were rigged with topgallant ratlines, and these disappeared again in the second half of the century. On English ships only about six ratlines were usually rigged above the deadeyes or thimbles. Larger French ships again provide a contrast: their topgallant shrouds were rattled to a sufficient height to enable the yardmen to reach the foot ropes of the topgallant yards, but again these ratlines had disappeared by the end of the century. Paintings of ships such as the *Commerce de Paris*, painted in 1809, and *Le Wagram*, painted in 1811, by Roux do not show ratlines, but his painting of the American Letter of Marque *Grand Turk* of 1815 (a brig with additional royal masts and shrouds) had the topgallant shrouds rattled.

Backstays
Pardunen; Galhaubans

Topgallant backstays were often rigged as the aftermost topgallant shroud pair. Lever noted that men-of-war had two topgallant backstays on each side, but commonly only one standing pair was rigged; only the largest ships in the Royal Navy occasionally had two.

In the first decade of the eighteenth century Davis listed a backstay and a breast backstay at the main topgallant mast for a Sixth Rate, a backstay only at the main topgallant mast for a Fifth Rate, a backstay at the fore and main topgallant masts for a Fourth Rate, a backstay and a breast backstay at the fore and a backstay only at the main topgallant mast for a Third Rate, a breast backstay at the fore and a backstay at the main topgallant mast for a Second Rate, and both stays rigged to both masts for a First Rate. The standing backstay was usually set up, together with the topmast backstay, to a backstay stool (a small channel abaft and above the actual channel) or to a chainplate above and abaft the stool. Standing backstays were set up to deadeyes (see Figure 53a).

The author of *Der geöffnete See-Hafen* made no mention of topgallant backstays and they are not shown in German or Dutch rigging diagrams from the turn of the century, but French ships were often rigged with a standing and a shifting (breast) pair, and Röding considered it one backstay pair normal and sometimes two for larger ships. He described shifting backstays differently (see Topmast, Breast backstays, above).

Shifting backstays as specified in Steel's tables were rigged to both sides for ships down to 24 guns; for ships below that size only one was set up, to the lee side. Each had a thimble turned into the end, and ships down to 24 guns had double and single block tackles hooked in, while smaller vessels had a lesser tackle of two single blocks.

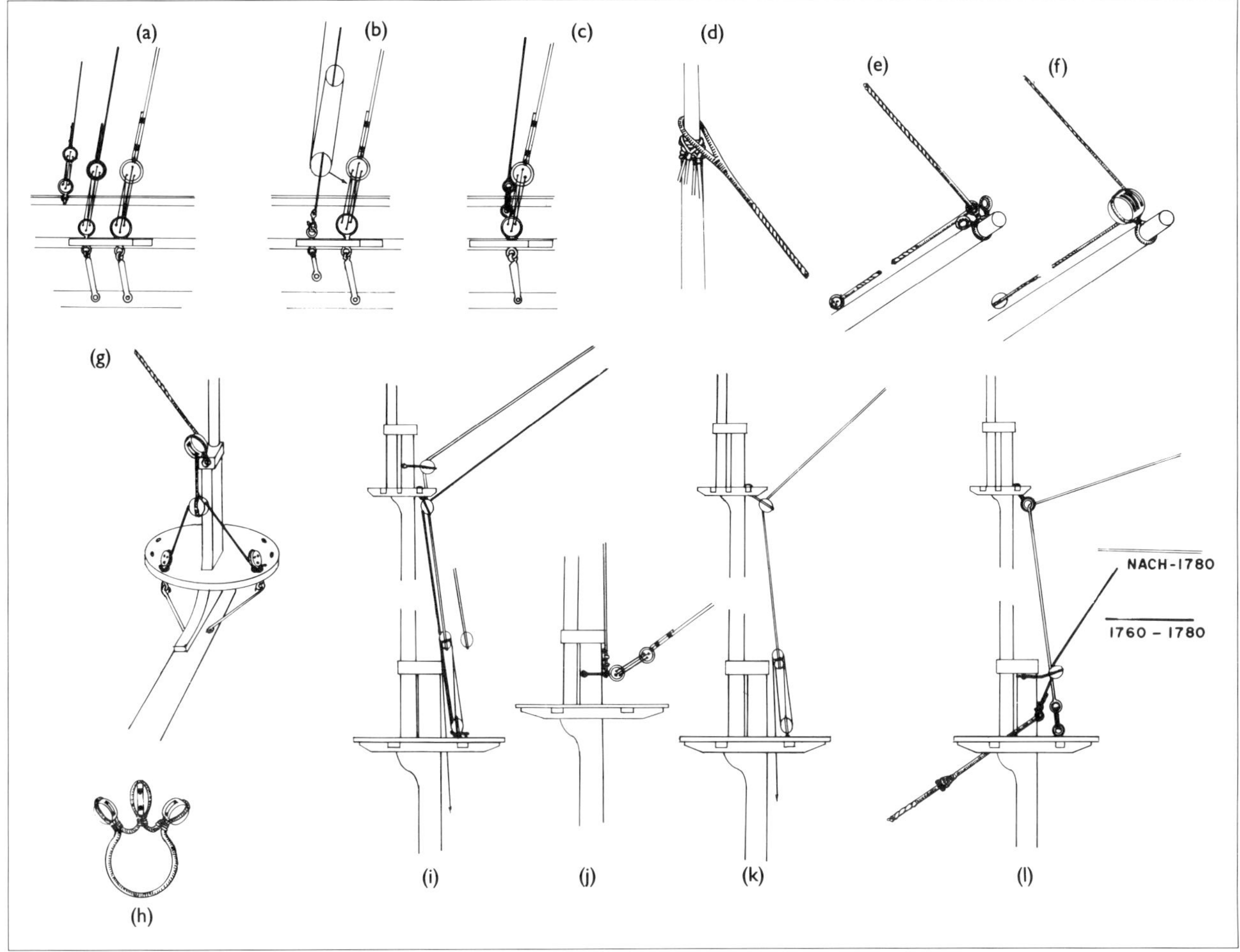

Figure 53
(a) Backstays set up to a stool; from right to left: topmast backstay, topgallant backstay, royal backstay.
(b) Backstays set up to a stool; a flying topgallant and a royal backstay on a merchantmen.
(c) The set up of topgallant backstays on smaller vessels.
(d) The eye of a topgallant stay.
(e) The lead of a fore topgallant stay at the jib boom.
(f) An alternative set up of the fore topgallant stay at the jib-boom, using a treble block.
(g) The lead of a fore topgallant stay during the spritsail topmast period.
(h) A fore topgallant stay collar with three blocks.
(i) A main topgallant stay (light line) set up with a main topgallant staysail stay (dark line).
(j) English set up of a main topgallant stay after 1735.
(k) The set up of a main topgallant stay on Continental ships.
(l) An English mizzen topgallant stay up to 1780 (dark line) and after 1780 (light line).

On smaller ships, thimbles instead of deadeyes were used for setting up standing topgallant backstays, so that a topgallant backstay was set up to a topmast backstay in a similar way to that in which a topgallant shroud was seized to a topmast shroud chainplate (see Figure 53c).

On merchantmen the topgallant backstay was occasionally set flying. A longstropped double block was hooked to an eyebolt in the side or to a chainplate at the stool, and a single block was turned into the backstay's end. The royal backstay passed through the double block, then the single block on the backstay, and again through the double block, to be belayed within board. The royal backstay in this arrangement was a part of the topgallant backstay, with the pressure on both backstays taken up equally (see Figure 53b).

Topgallant stays

Bramstengestage; Étais de perroquet

In contrast to the lower stays, the topgallant stays were fitted neither with an eye nor with a mouse. The collar was spliced directly into the upper end of the stay, and served. It rested on the topgallant mast stop above the shroud masthead rigging (see Figure 53d); this, at least, was the case in English rigging. A different rig is indicated by the author of *Der geöffnete See-Hafen*: he stated that a topgallant stay was a 'rope which, like the aforementioned stays, encircles the topgallant mast; it is doubled for 1 fathom, but thereafter is single to the top of the fore topmast head'.

Röding offered a similar description: '[The stay's] eye is laid around the head of the main topgallant mast'. Whether these two contemporary references are to open or spliced-in eyes is not clear, since pictures usually do not give enough detail and most contemporary models no longer have original rigging. By way of comparison, it is worth noting that Steel recorded that 'the [stay] has an eye spliced in the upper end to the circumference of its masthead', whereas the author of *Der geöffnete See-Hafen* (see above) described a 6ft doubling of the upper end of the stay, a very long eye indeed.

Fore topgallant stay

Vorbramstengestag; Étai de petit perroquet

Leading to the forward end of the jib-boom, the fore topgallant stay passed there through the middle thimble (see Figure 53e), or through the middle block (see Figure 53h), of the fore topgallant stay collar, or over the middle sheave of a treble block stropped to the jib-boom end (see Figure 53f). The stay ended in a tackle, a deadeye seizing or a thimble seizing, fastened to the bowsprit gammoning, fore stay collar or an eyebolt at the head.

On smaller vessels the end of the stay ran without a tackle through the bowsprit fairlead saddle to be belayed on deck. In a few cases at the end of the century the topgallant stays (both fore and main) had no collar and were spliced around a thimble on a traveller. Lever noted that for this reason pole heads had to be made strong enough to take the strain, since the royal yard was hoisted with, and the topgallant staysail bent to, the traveller, and the staysail halyard block was also seized to it.

In the days of the sprit topsail the stay passed through a block at the aft face of the sprit topsail cap, and a second block was turned into the end; this was set up with a whip which had one end secured to an eyebolt at the starboard side of the topmast standard (or around a futtock-plate at the sprit top), passed through the stay block and was made fast similarly to the opposite side (see Figure 53g). The author of *Der geöffnete See-Hafen* noted that the stay went through a block at the aft side of the sprit topmast head below the shrouds and was fastened to the sprit top trestle trees without a whip.

Main topgallant stay

Grossbramstengestag; Étai de grand perroquet

The main topgallant stay passed through a block stropped to the aft side of the fore topmast head and was set up to the fore top. A fiddle block was turned into the end and a single block was seized to a strop around the aftmost cross tree. The connecting fall, running from the single block, was belayed on deck (see Figure 53k). Where a tackle was not used, the stay was made fast directly to the fore top trestle trees. After 1735 the stay was bent without a tackle to the collar strop of the main topmast stay lead block (see Figure 53i, light line). It sometimes also had a thimble spliced into the end and was set up to another thimble stropped to the foremast trestle trees.

Other methods of setting up a main topgallant stay can be found in Continental literature. Röding noted that the stay came down from the lead block at the fore topmast head, through the lubber's hole in the top and was belayed on deck. Boudriot described three methods. In the first the stay led as Röding described, but down only to the catharpins, were it was set up. His second method was like the English rig after 1735, with the slight difference that a block was seized to the aftmost cross tree, and the stay passed through this before its end was seized to itself. Boudriot's third method followed exactly the English method before 1735. He appears to have taken nearly all possibilities into consideration, and his detail drawings provide certain hints, but his frontispiece drawing of a complete rigging plan does not show a main topgallant stay at all, though a royal stay is clearly depicted.

Staysail stays

Stagsegelstage; Drailles de voile d'étai

About 6 feet below the topgallant mast stop the staysail stay was spliced into the main topgallant stay. It passed through a block lashed to the aftmost cross tree of the fore topmast and down to the fore top, where it was hitched around the aftmost cross tree. This stay was not often used, and usually only then on large ships. It was mentioned only in passing in English literature (Steel, Lever) and then exclusively in connection with the main topgallant stay. Continental ships did not carry this stay (see Figure 53i, dark line).

Mizzen topgallant stay

Besanbramstengestag; Étai de perroquet de perruche

Introduced no earlier than 1760, the mizzen topgallant stay was set up at its lower end to a block at the aft face of the mainmast cap and hitched to the collar of the main stay. In Continental rigging it came down upon the deck and was belayed abaft the mainmast.

After 1780 the stay passed through a thimble or block at the main topmast head or trestle trees and, with a thimble in its end, was seized to another thimble in the aftmost cross tree, or made fast to the main topgallant stay. This latter lead was not described in Röding's text, but illustration 91 in his volume of plates shows it clearly rigged (see Figure 53l).

Flagstaff stay

Flaggenstockstag; Étai de bâton de pavillon

Flagstaff stays, also known as royal stays, were fitted with a running eye around the pole head directly below the trucks. To prevent them from sliding down the mast, a cleat was nailed to each side (see Figure 54a). The fore flagstaff stay of the early eighteenth century passed through a block at the sprit topmast head and was set up with deadeyes to the bowsprit top (see Figure 54c). After the introduction of the jib-boom the stay led through a thimble at the jib-boom stop, outboard of the topgallant stay collar and, after leading along the bowsprit, was made fast to the fore stay collar (see Figure 54b).

The main flagstaff stay passed through a block on the fore topgallant mast above the topgallant rigging and was seized with thimbles to the aftmost cross tree of the fore topmast (see Figure 54d). The mizzen flagstaff stay followed the lead of the mainmast, but was not used before the early nineteenth century.

Topgallant yards

Bramrahen; Vergues de perroquet

The rigging of topgallant yards was similar to that of the lower yards, but was very simplified because of the smaller size of the topgallant sails (see Figure 54e).

Lifts

Toppnanten; Balancines

Lifts were single or double, but later in the century almost always single, as on topsail yards. Davis, for example, listed single and double topgallant lifts, while Steel and Lever listed single lifts only. In Continental rigging double lifts are indicated for the first few decades; a Brest manuscript of 1747, mentioning lifts but no lift blocks on the yards, suggests single lifts. Other contemporaries from Ozanne to Röding also show single lifts, though Dutch prints from around 1780 still show double lifts on most Dutch ships.

Until 1790 lead blocks were seized to a short span at the topgallant mast stop, and the lifts passed through these to the mast top (see Figure 54f). Single lifts were fitted with an eye splice around the yardarm, while double lifts were made fast to the topgallant mast stop and passed through a lift block at the yardarm. Early in the eighteenth century they were made fast to the cap, ran through the yardarm blocks and then through further blocks below the topgallant trestle trees, to be belayed at the top.

After 1790 topgallant lifts led through a thimble seized to the upper part of the topgallant shrouds, and came down to the top to belay around a deadeye attached to a futtock-plate (see Figure 54g).

Alternatively, the lift ran downward through the lubber's hole, through a lead truck in the lower shrouds and to a pin in the pinrail or a cleat in the shrouds. This lead was mainly characteristic of Continental ships.

Braces

Brassen; Bras

In accordance with the size of a yard, topgallant braces were either single or double. Single braces were mainly used on smaller ships (and more often on Continental vessels), and double braces were more common on all vessels earlier in the eighteenth century. Steel, however, took no account of ship size in his description of topgallant braces, noting that all were double except mizzen topgallant braces, which were single; this was contradicted by Falconer and partly by Lever.

Topgallant brace blocks were usually spliced to pendants, but in the last few decades of the eighteenth century English merchantmen and Continental ships in general were more often fitted with brace blocks stropped directly to the yardarms. In the French Navy long-stropped blocks were occasionally used, as is shown in the models of the *Sans Pareil* of 1760 and *Le Protecteur* of 1793, or an eye strop into which the single brace could be toggled, as Paris indicated in his drawings of the *Royal Louis* of 1780 (see Figure 54i).

The leading of topgallant braces was again dependent on

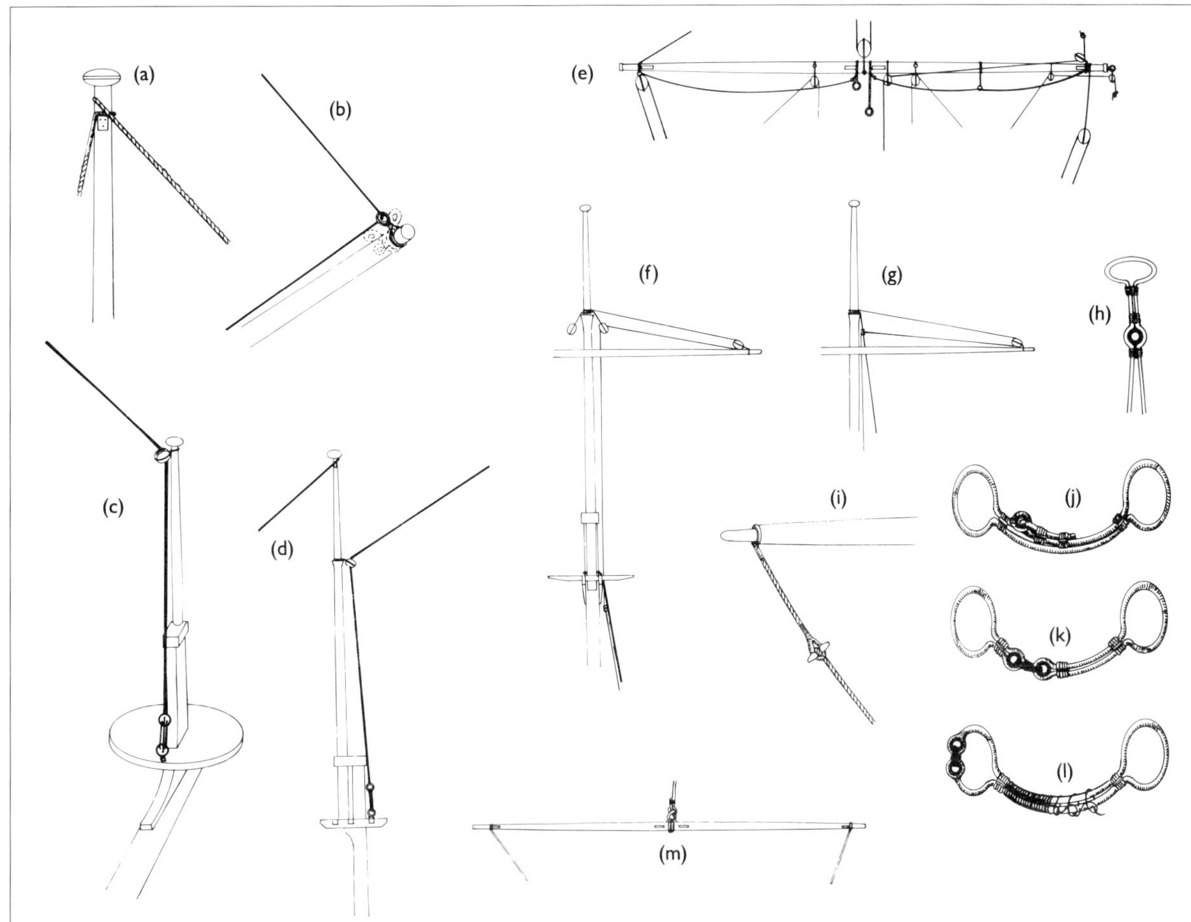

Figure 54
(a) A royal stay and backstay.
(b) A fore royal stay led through a jib-boom thimble.
(c) A fore royal stay set up at a sprit topmast.
(d) A main royal stay set up at the fore topgallant mast.
(e) A topgallant yard, showing the arrangement on an English man-of-war near the end of the century (right), and on a merchantman (left).
(f) A topgallant lift, up to 1790.
(g) A topgallant lift, after 1790.
(h) Topgallant shrouds with a seized-in thimble for leading the topgallant lift, after 1790.
(i) A French topgallant brace.
(j) A spritsail yard parrel or truss, sometimes also used for topgallant yards.
(k) A topgallant yard truss parrel as described by Lever.
(l) A topgallant yard truss (or rope parrel) as described by Davis.
(m) Royal yard.

their position in the ship's rig and to some extent on the provenance of the ship.

Fore topgallant braces

Vorbrambrassen; Bras de petit perroquet

Until about 1730 the standing parts of fore topgallant braces were made fast to the main topgallant stay, between a quarter and a third of the way up. They passed through the pendant blocks at the yardarms, returned through blocks on a short span fastened a few feet (4 feet in 1705) below the standing part on the same stay, then rove through lead blocks fitted to spans on the main topmast stay and the main stay, or the main stay only, and were belayed near the belfry (see Figure 55a, dashed line, left). Rigging plans according to the English 1745 Establishment show that 70-gun ships and larger still had fore topgallant braces rigged this way in 1745, while a similar plan of a 60-gun ship indicates fore topgallant braces rigged as detailed below for ships after 1730. Single braces were put with an eye splice over the yardarms and then followed the same lead as double braces, but usually without the extra lead blocks.

This general description applies to Continental ships as well as English, but several Continental variations can be found. A German illustration dating from 1700 of a Dutch ship shows the braces further up (at two thirds height) on the main topgallant stay, with the leading part going to the belfry without lead blocks. A Dutch ship drawn by Wenzel Hollar and a model of 1700 both show the standing part at two thirds height on the main topmast stay, and lead blocks a few feet below. The braces then passed through lead blocks lashed to the aftmost fore shrouds below the fore top. They might also have led to the main topgallant stay as in the general description above, and down to lead blocks at the topmast stay and to the fore top, as an etching by van Yk of 1697 reveals. Seventeenth century models and early eighteenth century etchings of Dutch ships (by A van der Laan) show another arrangement, with the braces leading from the fore topgallant stay lead blocks (fitted at two thirds height on the stay) up to the aftmost fore topmast shrouds, then belayed to kevels or cleats at the forecastle or in the waist. These variations suggest that no absolute standardisation can be assumed for the period.

The following years brought a change in the positioning and the lead of topgallant braces. Standing parts were hitched, with their ends seized, to both sides of the fore topmast stay collar. They rove through the yard pendant blocks and returned through blocks seized to the collar below the standing parts, then followed the topmast stay through single blocks at the aft side of the fore top, and were belayed to cleats beside the belfry (see Figure 55a, solid line, left). In some cases, instead of the blocks at the fore top, a double block at the aft part of the foremast cap was used.

As noted above, these braces were double on larger merchant ships and on many English men-of-war, but usually single on smaller ships and on most Continental ships after the first few decades. Precise definition of the usage of double and single braces is complicated further by Falconer. An author with great nautical knowledge and personal experience at sea of more than three decades, he noted in the 1760s that 'all the braces of the yards are double, except those of the top-gallant and spritsail-topsail yards.' Serres, marine painter for George III, followed the same rule in many of his pictures.

Single braces were spliced around the yardarm or toggled to short pendants, as noted above. They were led the same way as the double braces until about 1775, when the introduction of fitted sheaves at the fore ends of the main topmast trestle trees on Royal Navy ships meant that the braces could pass through these, instead of the collar lead blocks, upon the deck. Merchant ships still used collar lead blocks, together with a second block lashed to the upper part of the foremost main topmast shroud, to lead the brace downward beside the mainmast. Lever's description was as follows:

> In the Merchant Service, the Fore, Fore Topsail and Fore Top-gallant Braces, are generally led down by the Mainmast and belayed together there but in small vessels they are led through a treble block, seized to the foremost Main Shroud, and belayed to a pin in the block. So that in 'hauling off all' they are let go together.

Röding gave a more conservative description of the lead of fore topgallant braces. He noted that they passed through a lead block just below the main topmast stay collar, through another at the main topmast trestle trees, and came down along the topmast stay to the fore top, reeving there through one or more blocks and upon deck (see Figure 55b, solid line, left). Boudriot made no mention of a block at the main topmast trestle trees, but detailed the lead of the brace upon deck. According to him, after passing the block at the aft part of the fore top, the brace went vertically downward and through a block on the main stay, to be belayed at the forecastle rail. The *Royal Louis* rigging plan of 1780 has the brace leading through the collar block and, by-passing the fore top, through a block at the catharpins and upon deck.

Main topgallant braces

Grossbrambrassen; Bras de grand perroquet

Main topgallant braces were double or single as for those on the foremast. In English rigging during the 1790s lead blocks for these braces were fitted at the mizzen topmast stay collar (according to Steel), and others were seized to the upper part of the foremost mizzen topmast shrouds (according to Lever). The standing part of the brace was hitched to the collar and the leading part belayed to the mizzen shrouds (see Figure 55a, solid line, centre and right).

At the beginning of the century the standing part was made fast to the mizzen topmast head at the stop, and after passing the pendant block, the brace rove through a lead block on a short span (about 2 feet in length) below the mizzen topmast head and upon deck (see Figure 55a, dot and dash line, centre and right). In the 1750s this span block was replaced by a block at the upper part of the foremost mizzen topmast shroud; this is the block mentioned by Lever at the end of the century (see Figure 55a, solid line, centre and right).

Continental ships at the beginning of the century seem to have had the lead blocks stropped to the mizzen topmast, just below the trestle trees (see Figure 56a, solid line, centre and right). This is clearly indicated on Paris' rigging plan of *Royal Louis* of 1697, and can also be assumed from several etchings and paintings in which the leading part of the brace clearly runs below the trestle trees. The author of *Der geöffnete See-Hafen* noted that the lead blocks were made fast to the upper section of the foremost mizzen shroud, as was the case four decades later in English rigging. For the single braces generally used later in the century, Röding noted a block, perhaps double, stropped to the mizzen topmast above the stop, with the brace leading downwards, probably to a cleat in the mizzen shrouds (see Figure 55b, solid line, centre and right). Boudriot gave a similar description, with two single blocks seized to the mizzen topmast masthead rigging and the brace belaying at the quarterdeck pin rails (see Figure 56a, dashed line, right).

Mizzen topgallant braces

Besanbrambrassen; Bras de perroquet de perruche

Mizzen topgallant braces were single, and were fitted with an eye splice over the yardarms. The same two alternatives existed for their rigging as for the mizzen topsail braces.

When the mizzen topgallant braces were led aft, blocks or thimbles were stropped to the mizzen yard or gaff peak and the braces were belayed to cleats at the inside of the taffrail (see Figure 55a, right). Forward leading braces (when the gaff or mizzen yard was hoistable), more common on Continental ships, had lead blocks seized to the upper part of the aftermost main topmast shrouds, the hauling part being belayed to cleats in the lower shrouds or to a pin in a pinrail. In 1705 the lead block was placed very high up on the aftmost main shroud (see Figure 56a, dashed line, right and Figure 55b, solid line, right and centre).

Clewline blocks

Geitaublöcke; Poulies de cargue-point

Clewline blocks, stropped with two lashing eyes, were lashed to the lower side of the topgallant yard, about 3 feet outside the slings (Steel), just inside the sling cleats (Lever), or at about one third of the yard length (Continental) (see Figure 54e).

Tyes, halyards

Drehreepe, Fallen; Itagues, drisses

A single tye block was stropped to the slings and the tye, fastened around the topgallant masthead, passed through the block and then through the topgallant mast sheave hole (from about 1765 onward, as an alternative, the tye was hitched to the yard as discussed above under Topsail yard tye). Turned into the other end of the tye was a double block, which, together with a connecting fall and a single block lashed to the aft of the lower trestle trees, formed a halyard and belayed around the cross piece of the bitts abaft the mast.

This method, dated by Lees after 1765, was actually mentioned by Davis in 1711 for topgallant yards. He noted 'a Tye with one end fast about the Yard' and 'a single Tye', and gave as an alternative: 'a Tye with a Block upon the Yard'. The latter applied only to men-of-war of the First and Second Rates, with the exception that for the main topgallant yard of a Third Rate a double tye was noted.

In regard to halyards Davis noted 'hallyards for Ditto with 1 Block' or '1 Pair of Hallyards with one end fast to the Cross-trussel Trees or Back Stay'. For Second and First Rate ships he listed '1 Pair of Hallyards with two Blocks, that is 1 of them in the Top'. Davis made it clear in his rigging lists that only large ships, early in the century, had a tye block and a halyard tackle down to the top. All others had the halyard set up as a whip, with one end fastened to the top or a backstay.

For the same period on Continental ships a similar set up was followed, with two single blocks according to the author of *Der geöffnete See-Hafen*. The hauling part in this case ran to a knight-head on the centreline on deck.

The tye on small vessels was also sometimes rigged without a halyard, as Röding noted. It passed either through the sheave hole, or through a block at the topgallant mast top if a sheave hole would have weakened the masthead too much, to the deck.

Lever also described a single tye:

> Frequently the Tye and Halyards are in one, the Tye (a) being hitched around the Strap of the upper (single) block and toggled, to reeve alternately through the lower block (b), at the Tressle-tree, and the block (a). When the Tye leads thus, the Toggle being taken out and the Fall unreeved, it answers for a Top-rope to send the Yard up and down by.

Another method, used in France around 1780, was to hook the tye into the strop around the yard slings. The tye then had a fly block turned into, or seized to, the other end. The halyard, a whip, passed through the block and at both sides through lead

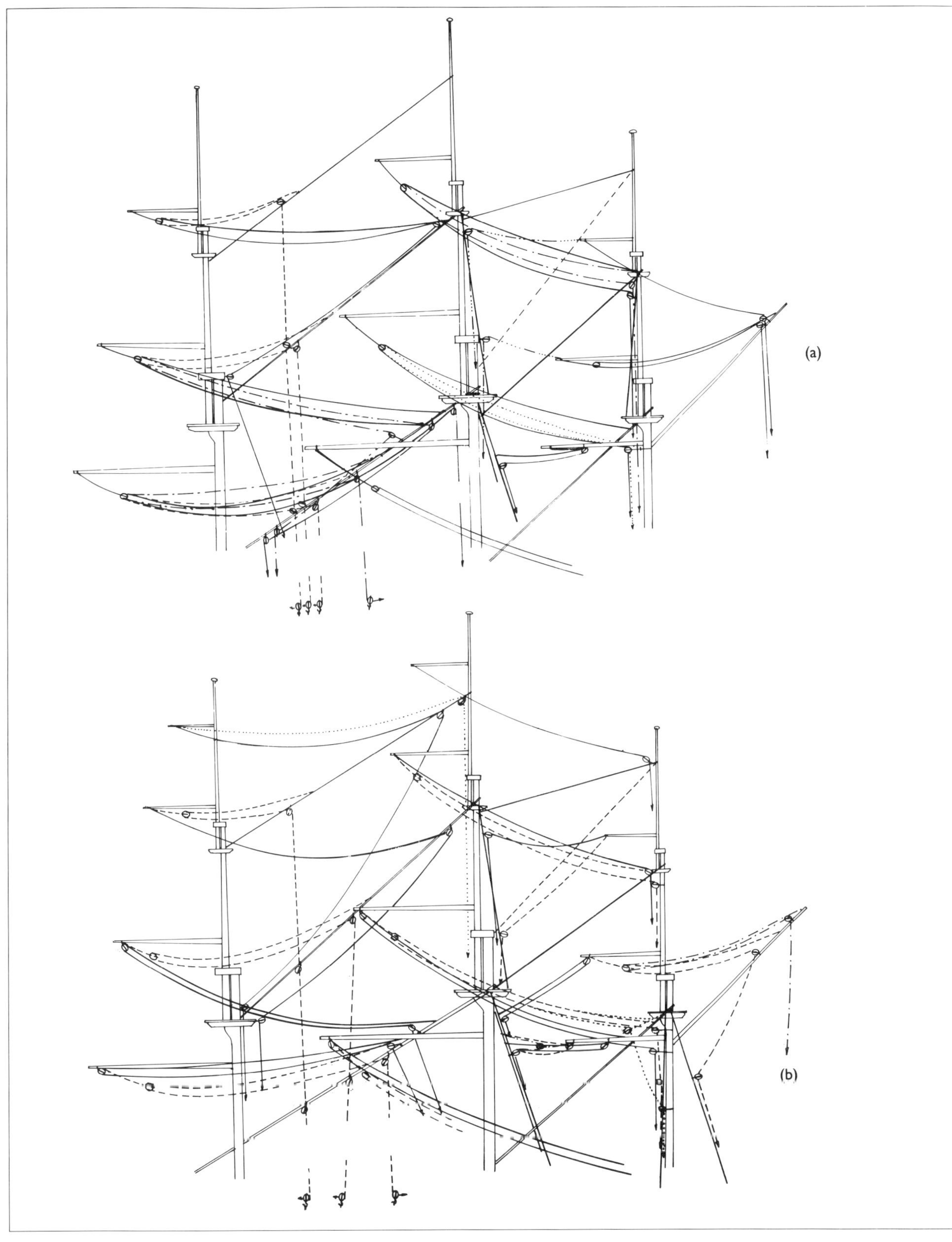

Figure 55
(a) The arrangement of English braces. Variations over time are indicated by different lines: up to about 1730 (dashed line), up to about 1750 (dot and dash line), up to about 1775 (dotted line), up to 1805 (solid line). The line with dashes and multiple dots shows the alternative lead with a raised gaff. Where a brace is indicated by a dashed and a solid line, the dashed line indicates the lead up to 1730 and the solid line the lead from 1730 to 1800.
(b) The arrangement of Continental braces. The dashed line shows French braces at about 1700, and the dot and dash line contemporary Dutch variations on the French rigging. The solid line indicates German rigging at about 1790, and the dotted line French variations on the German rigging.

blocks hooked to eyebolts on deck, to be belayed to pinrails or cleats.

Occasionally one side was turned into a standing part by hooking it to the eyebolt. This method was a slightly more refined variation of that mentioned by Davis for the early years of the century.

Parrels

Racke; Racages

Simple trusses were often used as parrels on topgallant yards, though larger ships retained two-tier parrels of ribs and trucks for a long time. A more precise division of the rigging of topgallant parrels by period and size of vessel cannot be established. Davis (1711) noted for Sixth and Fifth Rates one parrel rope of only 6 to 8 feet in length, which indicates a simple truss parrel, but for the Rates above these he provided two parrel ropes of a length four to five times greater, clearly implying a rib and truck parrel.

Falconer (1769) noted that truss parrels were used for topgallant yards, and he offered a description of a parrel for lesser Rates which accords closely with that of Davis:

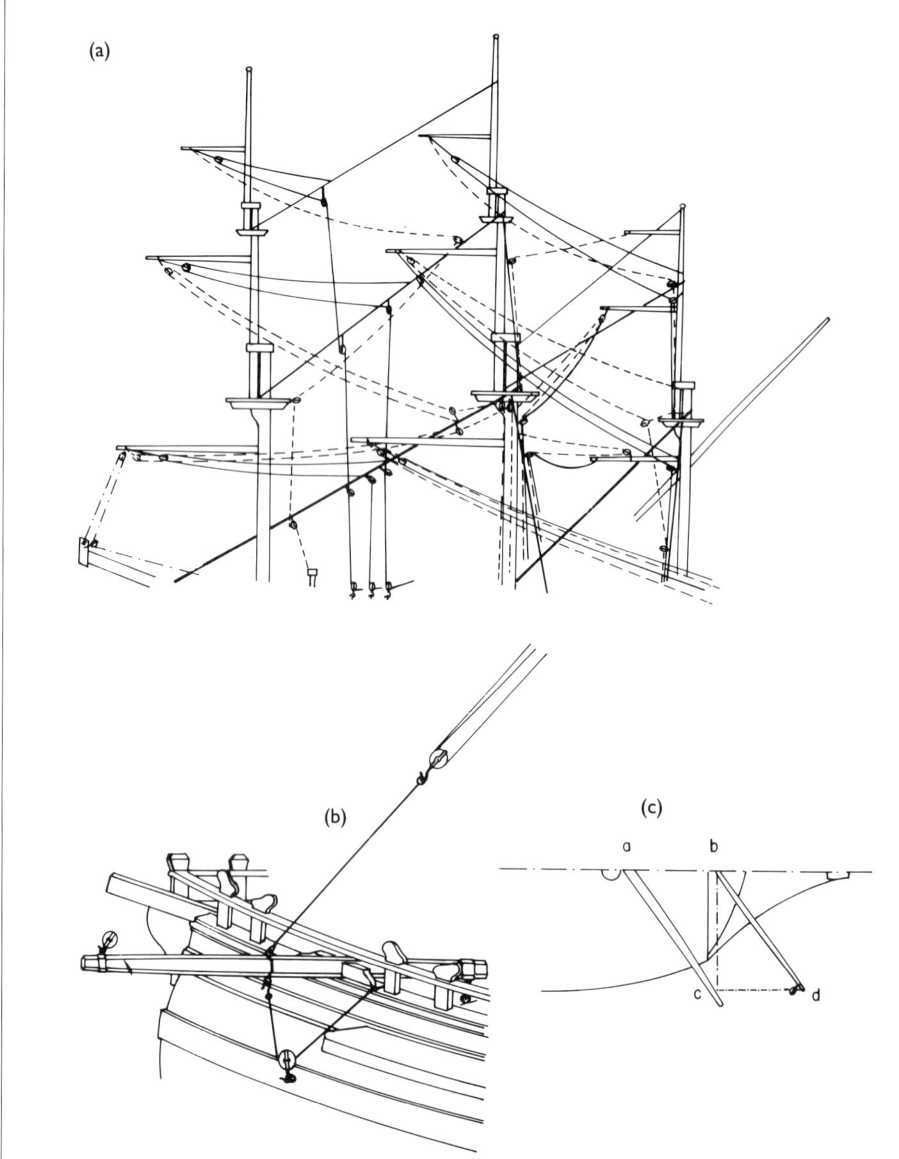

Figure 56
(a) The arrangement of Continental braces. The solid line indicates German rigging from 1705, the dashed line French practice from about 1780, and the dot and dash line indicates the lead of a preventer brace towards the bowsprit cap.
(b) The rigging of a boat boom as a fore studdingsail boom on Continental ships and also in English ships during the early years of the century.
(c) The positioning of a bumkin, according to Bobrik.
***a.** The fore yard at the 36° bracing position.*
***b.** The bumkin, parallel to the fore yard at a point established by drawing a rectangular line between the ship's centreline and the yardarm.*
***c.** The fore course clew position.*
***d.** The length of the bumkin.*

The first of these [parrels], which is also the simplest, is formed of a piece of rope, well covered with leather, or spun-yarn, and furnished with an eye at each end. The middle of it being passed around the middle of the yard, both parts of it are fastened together on the after-side of the yard, and the two ends, which are equally long, are passed around the after-part of the mast and one of them being brought under, and the other over the yard, the two eyes are lashed together with a piece of spun-yarn on the fore-side thereof, while another lashing is employed to bind them together, behind the mast, according to the manner described [see Figure 54l].

As an alternative to this, the truss parrel of a spritsail yard (see Figure 54j) might also be used for topgallant yards. Lever provided yet another alternative: a short and a longer round strop were seized to the yard slings, each side of the tye, with a thimble seized to each. The long strop went around the back of the mast and was lashed to the shorter strop (see Figure 54k).

Steel (1794) noted that a topgallant yard parrel was similar to that of the topsail yard, and he described the latter as a two-tier parrel with ribs and trucks. Röding concurred with this statement. The Brest manuscript of 1747 noted above also specified a truck parrel for topgallant yards, and Boudriot gave both a truck and a truss parrel as possibilities for his 74-gun ship of 1780.

The conclusion to be drawn is that in the Royal Navy larger vessels were rigged with truck parrels for the topgallant yards at least until the middle of the century, and some probably until the end. The truss parrel was more common on merchantmen, and Continental ships mostly had truck parrels, even at the end of the century.

Horses, foot ropes
Fusspferde; Marche-pieds

Topgallant yard horses were fitted in the same manner as those of the topsail yards. Only a few topgallant yard horses had stirrups, and then usually not more than one to each side; as a rule, topgallant yard horses had no stirrups.

Royal masts
Royalstengen; Mâts de cacatois

When a topgallant mast had a normal masthead with a royal mast stepped, instead of a long polehead, then the royal mast was rigged similar to a topgallant mast. Steel noted, however, that East Indiamen often carried a royal mast stepped abaft the topgallant mast.

Royal stays
Royalstage; Étais de mât de cacatois

Royal stays are described above under Flagstaff stays, to which they were identical.

Royal yards
Royalrahen, Oberbramrahen; Vergues de cacatois

Royal yards were not often used. They were used to set fine weather sails only, and were rarely a permanent part of the rigging of a fully rigged ship even at the end of the century. Lever noted that royal yards 'are seldom rigged across. When they are, they have a Royal Mast fidded on the Tressle-trees at the Top Gallant masthead: the Masts and Yards are rigged like the Top-Gallant ones'. Röding did not mention royal masts at all; he noted only that royal yards were hoisted right up to the long polehead trucks. His passing mention of a mizzen royal merely noted that even on the largest ships it was very rarely used.

Royal masts may therefore not have existed earlier then the late eighteenth century; the royal yard, however, was already an established part of seventeenth century rigging. It has often been assumed that the royal yards shown rigged on the *Sovereign of the Seas* of 1637 were unique, but at least two contemporary documents provide evidence to the contrary. The *Treatise on Rigging*, written in about 1620-25 by an unknown author, noted the use of 'top-topgallant' sails for the fore and mainmasts, and Davis' *The Seaman's Speculum or Compleat School-Master* of 1711 includes an intriguing document, dated 'October the 2d, 1676' and entitled 'A Demand for the Boatswain's Sea Stores for the *Degrave* Frigat, Capt. Will. Young Commander, by me John Davis, Boatswain'. In this he asked for '1 Fore Topgallant Royal and 1 Main Topgallant Royal'. Given these three earlier references to royal sails, Falconer was clearly describing not a novelty but an established part of the rigging of ships when he commented as follows in 1769:

Royal, a name given to the highest sail which is extended in

> any ship. It is spread immediately above the top-gallant-sail, to whose yardarms the lower corners of it are attached. This sail is never used but in light and favourable breezes.

His note that the lower corners of the sail were attached to the topgallant yardarms indicates that the royal sail was set flying.

Halyards

Fallen; Drisses

Royal yard halyards were bent to the yard in the same manner as a topgallant yard tye, and led upon deck through a sheave hole in the royal masthead or a block lashed to the long polehead beneath the truck.

Braces

Brassen; Bras

Royal yard braces were rare; 'when these Masts are not stepped, the Royals are set flying; that is, they are not rigged across, having neither Lifts nor Braces (though sometimes the latter)' [Lever]. When braces were rigged, they were single and led through blocks at the next forward topgallant masthead and down upon the deck, according to Steel. As was often the case, Continental rigging practice was slightly different: Röding noted lead blocks for the fore royal braces at the upper part of the main topgallant stay, with additional blocks at the aft side of the foremast head. The braces then led down upon the forecastle. Braces for a main royal yard were seldom needed, according to Röding, but when used they passed through a thimble at the mizzen topmast polehead stop down to the quarterdeck (see Figure 55b).

Parrels, clewlines and lifts

Racke, Geitaue und Toppnanten; Racages, cargue-points et balancines

Parrels, clewlines and lifts were not needed when a royal yard was set flying. On permanently rigged yards, they were all rigged as on the topgallant yards.

Studdingsail booms

Leesegelspieren; Bout-dehors de bonette

A studdingsail boom was necessary to extend the foot of each studdingsail. Note that, like most booms, studdingsail booms therefore take their names from the sail set above, in contrast to yards.

Lower studdingsail booms

Leesegelbäume; Arcs-boutants de bonette de misaine

The first step in rigging studdingsails was to hook the goosenecks at the inner ends of the lower studdingsail booms into strong eyebolts outboard on the ship's side. On English ships these eyebolts were positioned between the fore channel and the cathead for the fore booms, and usually at the fore part of the main channels for the main booms. On French ships the lower studdingsails were set partly flying, without a boom, and this method was also used on most merchantmen. Lees postulated the use of such flying, loose-footed studdingsails on all English men-of-war before 1745, and on smaller ships throughout the eighteenth century, but repeated Anderson's comment that little is known about the rigging of studdingsails before 1790. The foot of a studdingsail rigged flying was spread by a light yard, and a guy, fastened with a span to the inner third of this yard, performed the functions of sheet and tack (see Figure 57a, left).

A clear indication that swinging lower studdingsail booms were used in the Royal Navy long before 1745 was, however, given by Davis in 1711. He listed 'two Fore studding-sail Halyards 32 Fathom' and 'two Tacks 4 times the lengths of the Booms' for the foremast, and 'two Sheets, if the Booms hook without board' as well as 'two Tacks, each of them twice and a half the Boom' and 'two Guyes with a Hitch about the Boom' for the mainmast rigging. He mentioned booms for all Rates, but only for the mainmast in his lists for the three higher Rates. He also made a distinction between booms at the foremast and booms hooking outboard at the mainmast. The author of *Der geöffnete See-Hafen* noted the use of studdingsail booms on the Continent, where studdingsails were triangular like staysails. The foot of these sails was extended by a boom: *unten aber am Segel wird ein Spaar eines Gickes gleich ausgestecket.* The German term *Gick* or *Giek* was used for a studdingsail boom, later described by Röding as swinging for the mainmast and extended over the side for the foremast. *Der geöffnete See-Hafen* also included a description of the fastening of a tack to the boom, for both main and foremasts.

The very first mention of a lower studdingsail boom in English sources was that of Captain John Smith in *A Sea Grammar* of 1627. His description of studdingsails was as follows:

> ...and in a faire gaile your studdingsailes, which are bolts of Canvasse or any cloth that will hold wind, wee extend alongst the side of the maine saile, and boomes it out with a boome or long pole which we use also sometimes to the clew of the maine saile, fore saile and spret saile when you goe before the wind, or quartering, else not.

This application of a boom or long pole suggests the origin of the swinging studdingsail boom in the loose spread boom, used from Viking times onward to steady the shivering leech of a sail.

An interesting titbit of information on studdingsail booms is given in another document included in Davis' *Seaman's Speculum*, entitled 'A Demand of Stores for present Use to compleat the *Degrave* for the Sea now she is Rig'd, Captain Will. Young Commander, by me John Davis, Boatswain, August the 22th 1676'. He listed 'wood for Studding-sail Booms, 4'. It seems that such booms were not necessarily part of the normal rigging of a ship, but might be fabricated on board.

Röding also described swinging studdingsail booms, and his description of the fore boom probably relates to the same type of fitting Davis had hinted early in the century:

> A boom at the forecastle, extended outboard between the timberheads, is used to secure the longboat or pinnace should such a boat lie alongside the ship. The same boom is also used for setting out the fore studdingsail sheet.

This use of the boat boom as fore studdingsail boom can also be seen in many of Ozanne's drawings, while pictures by Roux sometimes show the flying boom described by Lever. Boudriot's description of a fore studdingsail boom for a 74-gun ship closely followed that of Röding (see Figure 56b).

Smaller vessels such as schooners and cutters, which carried a squaresail boom and on which the squaresail yard or crossjack yard was fitted with studdingsail booms, had the same type of booms rigged to the squaresail boom (see Figure 88a below).

Two strops, with two thimbles each, were seized to the middle of a lower studdingsail boom. One pair of thimbles faced up and down, and the other fore and aft, and these had various functions which are described below (see Figure 57a, bottom right).

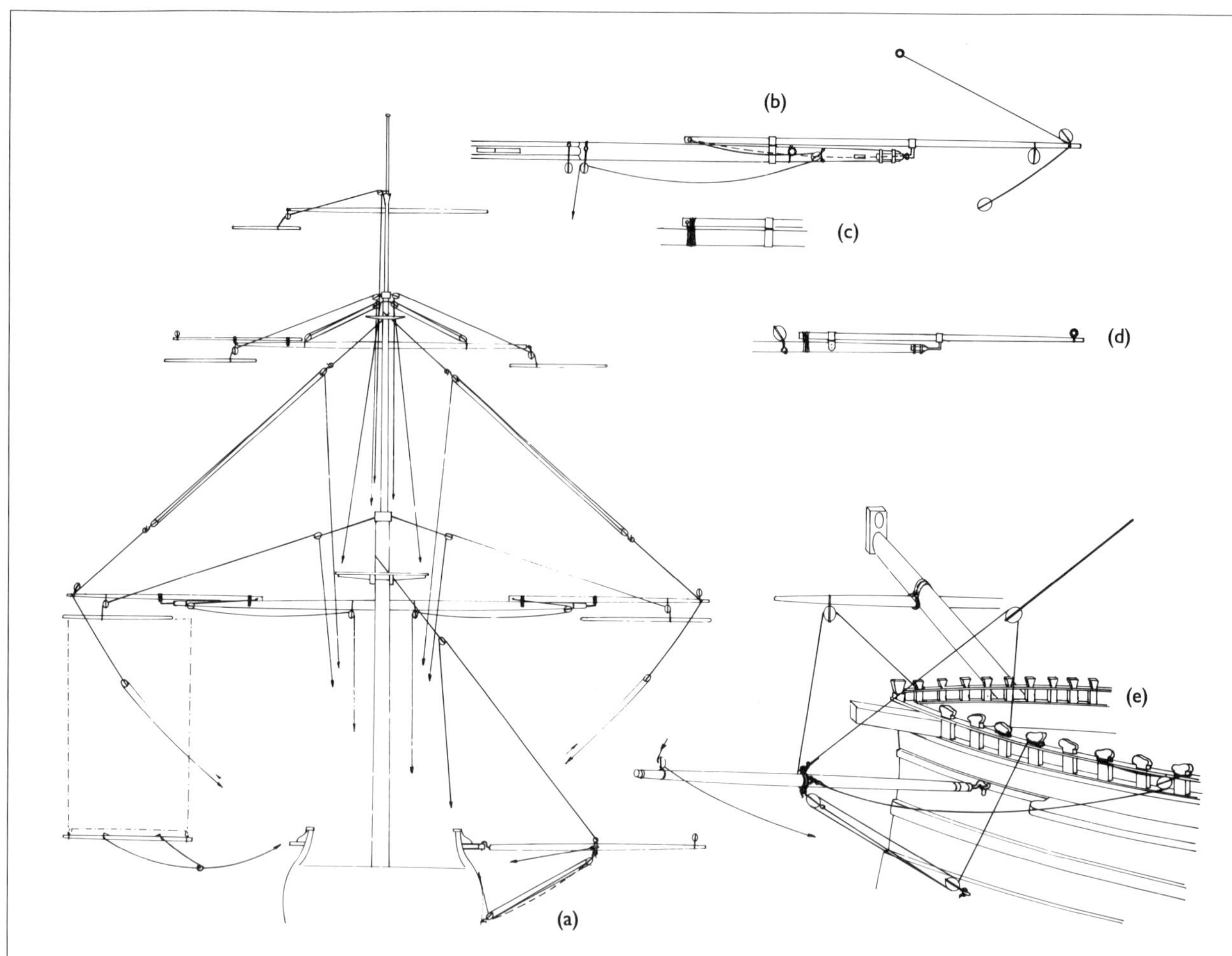

Figure 57
(a) Studdingsail boom rigging. The right side of the drawing shows a rig with a lower (swinging) boom, no topgallant boom and no topgallant studdingsail yard. On the left side is a topgallant studdingsail yard, a topgallant boom and a loose footed lower studdingsail (as preferred on some Continental ships).

(b) A topmast studdingsail boom with its rigging.

(c) Detail of the lashing of a studdingsail boom.

(d) A topgallant studdingsail boom in position, with accessories.

(e) The rigging of a lower (swinging) fore studdingsail boom. The main boom was rigged in similar fashion.

Lifts
Toppnanten; Balancines

The lift for a lower studdingsail boom was hitched to the upper thimble noted above. It passed through a block on a long span hanging from the masthead, and was belayed to a timberhead. No lifts were mentioned by Davis, and it would seem that they were used only later in the century.

Martingale
Wasserstage; Martingale d'arc-boutant de bonette de misaine

The lower thimble was used for the martingale. Leading through blocks hooked to the wales, the martingales for the fore studdingsail booms were belayed to timberheads at the forecastle, and those for the main boom to a cleat on deck. Men-of-war were frequently rigged with a tackle instead of a single martingale (see Figure 57a). There is no mention of these martingales in the early part of the century.

Guys
Geie, Kehrtaue; Retenues

The fore and aft facing thimbles were used for guys which kept the boom at right angles to the ship. Davis described these guys as hitched around the boom about two thirds of its length outboard, and he noted that 'you must carry one end of these Guyes forward, and the other on the Quarter-deck'.

The forward guy for the fore boom passed through a block stropped to the outer quarter of the spritsail yard, and on to the forecastle; the after guy led through a block seized to the aftermost timberhead at the forecastle and was belayed to a timberhead nearby (see Figure 57e). The forward guy for the main boom ran over the second sheave of the fore course sheet channel in the side, while the after guy came in through a gunport.

Studdingsail tack blocks
Leesegelhalsblöcke; Poulies de point d'amure de bonette

A single block for the studdingsail tack was fitted to the outer end of the boom (see Figure 57e, left).

Topmast studdingsail booms
Unterrahspiere; Bouts-dehors de bonette de hunier

After the topmast studdingsail boom was run out, the inner end had to be lashed to the yard. Booms on smaller vessels could be extended by hand, but for larger booms an outhauler was needed. The standing part of the outhauler was made fast to the yardarm, and it passed through the hole in the inner end of the boom (later also used for the lashing), a block at the yardarm and another at the inner quarter of the yard, and upon deck; the boom could therefore be run out from the deck (see Figure 57b).

Tack blocks
Halsblöcke; Poulies d'amure

A single block for the studdingsail tack was stropped to the upper side of the outer end of the boom (see Figure 57b, right).

Halyard blocks
Fallblöcke; Poulies de drisse

A block, facing downward, was stropped to the boom slightly inboard of the tack block, to be used for the halyard of the lower studdingsail yard (see Figure 57a and b).

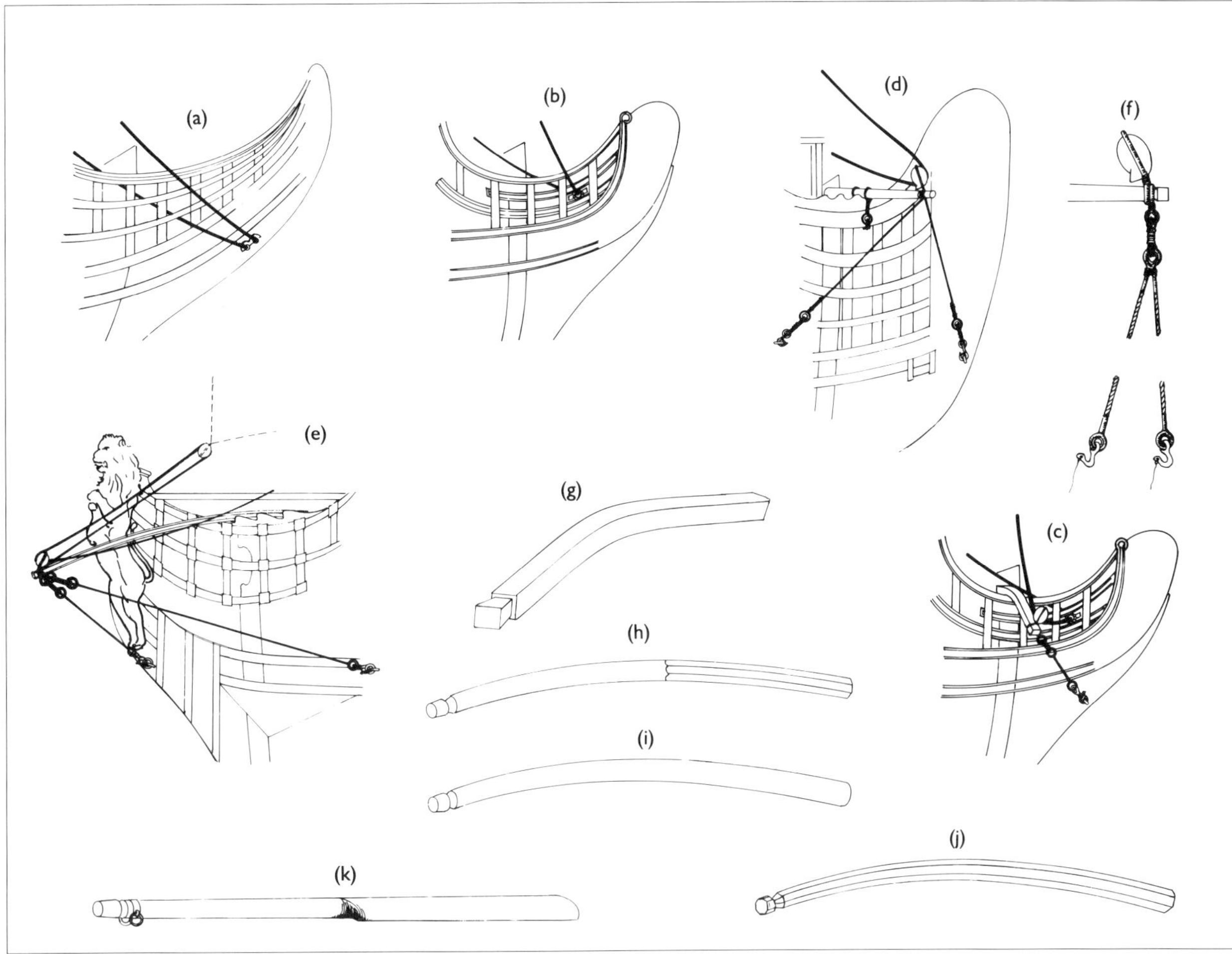

Figure 58
(a) The lead of a fore course tack through holes in the cutwater (seventeenth century).
(b) The fore course tack led through a deadblock between the head rails (English, from 1670 to approximately 1710).
(c) The lead of a fore course tack over a bumkin and through a deadblock, between 1710 and approximately 1720.
(d) The lead of a fore course tack through a bumkin block (English, after 1770).
(e) The lead of a fore course tack (Continental); note the double tack.
(f) The outer end of a bumkin with a tack block and an alternative stay seizing. Note the set up of the stays: the English method was to use thimbles turned into the lower or upper ends for seizing (as shown) while the Continental method was to use deadeyes or blocks at the upper end.
(g) An English bumkin of about 1710, square in section and markedly curved.
(h) An English bumkin after 1730; the outer half is round in section and the inner octagonal, and the curve is slight.
(i) An English bumkin after 1780, rounded in section throughout its length and slightly curved or straight.
(j) A Continental bumkin, octagonal in section with a slight curve.
(k) An English bumkin after 1800; the outer half is round with rings on an iron hoop, and the inner half is wider and half round.

Braces

Brassen; Bras

A brace pendant with a single block for the boom brace also went over the outer boom end (see Figure 57a and b).

Topping lifts

Toppnanten; Balancines

A further pendant, with a thimble spliced into the free end, extended upward from the outer boom end. The lower block of the topmast Burton tackle was hooked into this thimble so that it could act as a topping lift for the studdingsail boom (see Figure 57a and b).

Topgallant studdingsail booms

Marsrahspiere; Bouts-dehors de bonette de vergue de perroquet

Topgallant studdingsail booms were introduced around 1770, together with the topgallant studdingsails.

Tack blocks

Halsblöcke; Poulies d'amure

When the boom was used, it usually served only to spread the topgallant studdingsail tack and therefore was fitted only with a tack block at the outer end. The boom was also secured with a lashing to the inner end (see Figure 57d).

Top Burtons

Topptakel; Palans de fausse balancine

As supports for the topsail yard lifts when studdingsails were set, additional Burton tackles were hooked to a strop or eyebolts at the topmast cap and to strops around the yard, half way in between slings and yardarm (see Figure 57a).

Halyard blocks, jewel blocks

Fallblöcke, Leesegel-Fallblöcke; Poulies de drisse

A halyard block for the topgallant studdingsail yard was fastened to the topgallant yardarm, with an additional lead block seized to the topgallant mast stop (see Figure 57a).

Bumkins or boomkins

Butluve; Boutelofs

One of the most difficult problems in the rigging of the seventeenth and eighteenth centuries was the leading of the fore tack. Guide holes in the cutwater were used in the seventeenth century, despite the changing shape of the head (see Figure 58a). A change began to occur in the last quarter of the century with the introduction of a deadblock (see Figure 58b). This culminated, as Anderson noted, in about 1710 in another improvement: the introduction of a bumkin for the first time in English ships. Its origin can be seen in the single bumkins of late seventeenth century Dutch fly-boats, which protruded forward from the prow and had two lead holes for the tacks. The bumkin was at first used in addition to the deadblock, an ornamental fairlead between the head brackets and the rails, and it took more than a decade until the bumkins replaced the deadblock entirely. Continental ships had no bumkins before 1735-40.

In this transitional decade bumkins were relatively short, reaching only about 6 feet outboard; they were pieces of timber square in section and bent downward, with a width and height of 1 inch for each foot of length. The outer part was notched so that the tack lead block could be stropped to it (see Figure 58c).

During the 1730s the bumkin became longer, and octagonal in section inboard and rounded outboard (see Figure 58h). Curved slightly downward, it tapered towards the fore end by about one quarter, and its length was determined by the degree to which the fore yard could be braced (that is, the outer edge of the bumkin had to be parallel to the braced yardarm. Bobrik explained the position of a bumkin as follows: a line at an angle of 36 degrees forward was drawn from the centre of the foremast on a plan draught, with a length equal to half the fore yard length marked on it. A line drawn perpendicular to the centreline from this point indicated on the latter the point from which the bumkin departed at the same angle as the fore yard. This placed it just above the middle head bracket, with its aft end resting against the knighthead and its outer end as far out as the yardarm intersection point. The outer end also had a notch for the block and the bumkin stays. From around 1780 English bumkins were completely rounded in section, and just slightly bent or completely straight (see Figure 58i).

On Continental ships the bumkin was still octagonal in section almost to the end of the century, and, according to Röding, still somewhat bent downward. The standing part of a double fore course tack was bent around the fore end of the bumkin, and the tack lead blocks were also stropped to it (see Figure 58e).

At the very end of the century the inner half of the bumkin began to change again. It became double in width and only the upper half was rounded, while the lower side remained flat. In addition, the eye or deadeye strops at the fore end were replaced by iron hooped rings (see Figure 58k). Before 1800 the inner end butted against the beakhead, the stem or knighthead, depending on the ship's bow construction, with the latter used more at the end of the century. When the bumkin butted against the beakhead, an iron band secured the end to a deck beam.

Bumkins were usually only mentioned in connection with ship rig, and modern authors tend to discount the use of bumkins on any other vessel, encouraged perhaps by Röding's statement that 'merchantmen do not often carry a bumkin, and in consequence generally use the cathead' and Lever's that 'small vessels in the Merchant Service... carry no Boomkin; the Tack is taken under the aft side of the Cat-head, and belayed to the Timberhead before it', or Bobrik's that 'smaller vessels, which have no bumkin, lead the fore tack to the end of the cathead'. It is, however, more likely that the fitting of bumkins was determined by the placement of the foremast rather than the size or precise rig of the vessel. From the middle of the nineteenth century onwards, for example, nearly all larger ships also used the catheads rather than bumkins, precisely because their foremasts had moved further aft.

Particularly noteworthy among smaller vessels which were fitted with bumkins are Chapman's kray, a small three masted Baltic Sea trader, and his Dutch flute; these two drawings actually contain a third of all the bumkins shown in his work. His depiction of many extended and shaped hawse timbers on bluffheaded ships confirms their use, as do the illustrations in Lescallier's and Röding's works. In these, a cat, a three-masted galiot, a snow and a brig are all fitted with bumkins. Schooners and cutters, which only hoisted a squaresail before the wind, were fitted with a squaresail boom running athwartships forward of the mast instead of bumkins.

It can therefore be said that the apparent distinction in the use of bumkins between men-of-war and merchantmen was not necessarily so clear. Certainly a majority of merchantmen were not fitted with bumkins, but a very large variety of rigs were employed in the sea trade and the square rigged kind was anyway in the minority.

Tack blocks

Halsblöcke; Poulies d'amure

A large single block, usually a shoulder block, was stropped to the outer end of a bumkin to lead the tack toward the forecastle.

Stays

Stage; Étais

The bumkin, secured to the rail, had two stays fitted outside the tack block to counteract the upward force of the tack (see Figure 58c, d and e). These usually consisted of a single long rope doubled, with the bight seized to the bumkin. One leg led down forward to the cutwater and the other down aft to the bow, where they were set up with thimbles or small deadeyes (see Figure 58d). Röding described both leading to the bow, and referred to them as bumkin pendants.

These stays were sometimes hooked to ringbolts at the same locations. In this case, a thimble was seized to the bight and an additional thimble strop was laid around the bumkin end. Both thimbles were then seized together beneath the bumkin (see Figure 58f).

Continental ships had a strop with deadeyes or blocks turned in seized to the lower side of the bumkin end. The stays were hooked to or spliced around a thimble at the bow and cutwater ringbolts and had a matching deadeye or block turned into the top end; these were hauled taught with a lanyard (see Figure 58e).

Until the beginning of the 1770s only one forward leading stay (or pendant) was normally in use.

III

Northern variants of ship rig

Cat

Katt, Katschiff; Chat

'[A cat is] a three-masted vessel, especially employed in Norway, Sweden and Denmark, ... [it is] roughly built like a Fly-boat and only fitted out for trading... The English also use Cats in the coal trade'. This was Röding's description of one of the major merchant ship types of eighteenth century Northern Europe. Falconer provided more details:

> [A cat is] a ship employed in the coal trade, formed from the Norwegian model. It is distinguished by a narrow stern, projecting quarters, a deep waist, and by having no ornamental Figure on the prow. These vessels are generally built remarkably strong, and carry from four to six hundred tons or, in the language of their own mariners, from twenty to thirty keels of coals.

The importance of cats in North and Baltic Sea shipping is underlined by Chapman, who included ten draughts of the type in his *Architectura Navalis Mercatoria* of 1768, six of which he called ship-rigged, one each a snow and a brigantine, and the remaining two sloop-rigged. What Chapman classified as ship-rig was somewhat more closely defined by Steel: '[a cat is] rigged similar to an English ship, having, however, pole-masts and no topgallant-sails... The mizen is with a gaff'. Modern authors have stated that such a rig cannot be confirmed for English cats, but Steel's descriptions of other rigs have been confirmed as detailed and precise, so there seems no reason to question his description of the rigging of these vessels, which he would have observed almost daily on the Thames.

Röding and Lescallier agreed with Steel's observation that cats were polacre-rigged, having masts without topmasts, and therefore very rarely carried topgallant sails. The sails had a shape similar to those of three-masted ships, and, as Korth noted, though cats were able to carry large cargoes, they performed poorly when sailing.

The cats described by Chapman contrasted markedly with Falconer's statement that cats carried between 400 and 600 tons. Their capacity, from as little as 43 to a maximum of 1097 tons, was of a far greater range.

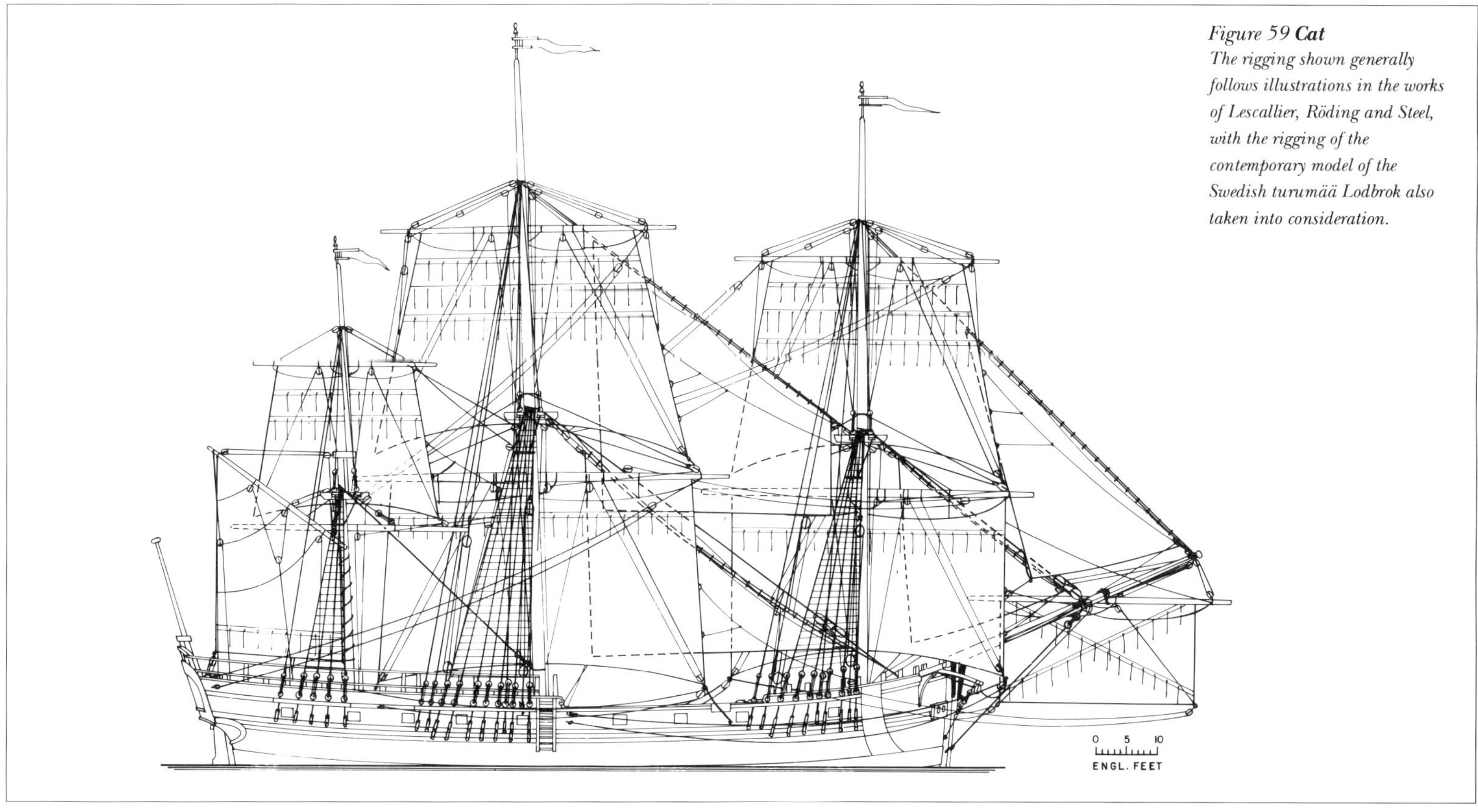

Figure 59 ***Cat***
The rigging shown generally follows illustrations in the works of Lescallier, Röding and Steel, with the rigging of the contemporary model of the Swedish turumää Lodbrok also taken into consideration.

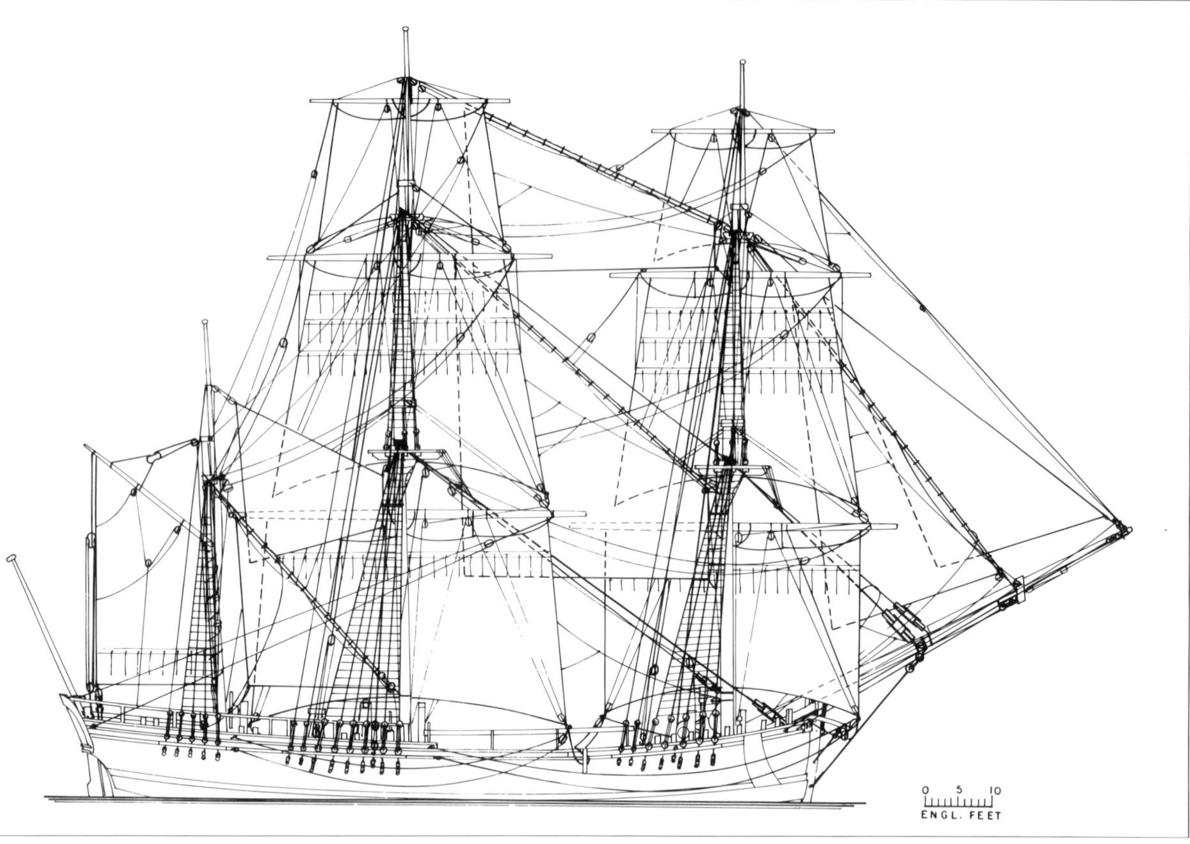

Figure 60 ***Bark***
The basis for the rig shown is Gwin's watercolour A Collier of about 1780, except for a royal yard on the main mast, which is not shown here. Chapman's drawings and spar dimensions for barks are also taken into account. Barks also sometimes rigged a spritsail and sprit topsail, and also studdingsails.

Cats were full-bodied vessels, frequently differing in rig, but always in stern construction, buttock shape and deck arrangement from the generally similar bark. With a very curved wing-transom, the cat's stern was narrower than a bark's and the vessel did not have the latter's prominent lower counter. Moreover, on larger cats the decks were always arranged similarly to those of a ship, with a normal forecastle and quarterdeck. The term 'cat' was therefore used mainly as a hull type specification. A light square rig was common, but as Chapman made clear, a cat could might have other types of rig as well; an etching by Groenewegen, for example, shows a cat with a full ship rig, including with topmasts, etc. The rig most commonly associated with a cat, however, was that of a polacre (see Polacre in Chapter VIII).

Röding's comment on rigs for cats was as follows:

> Pole masts are normally used on cats, and the rig is set up similarly to that of any other large ship. At the required height grommets or a sennit are laid round the pole mast to take the shrouds, stays and side tackle pendants.
>
> Since the masts have neither top nor cap, but are made from one piece (except occasionally the mizzen mast), they have no top rope, no futtock shrouds and no topmast shrouds. In order for the crew to reach the masthead, a storm ladder hangs at both sides down from the head to the capelage. The sails are all square sails and are hoisted like these on other ships. Since nothing obstructs the lowering of the topsails and topgallant sails, both can be struck right down to the lowest yard, which can be of great benefit in a sudden wind gust. Furthermore this rig has the advantage of being very light. The disadvantage of a polacre rig, however, is that when an upper mast part is carried away or breaks, it cannot be repaired other than by replacing the whole mast, in contrast to a mast with topmast and topgallant mast, where the part so damaged easily be substituted by a reserve spar. For this reason pole masts are not very high and are of adequate thickness; they are also cut from carefully chosen timber,
>
> Pole masts are often made from two pieces, fitted together at the capelage with a very long scarf, wooldings and iron hoops.

In addition to Röding's description, it is worth noting that in a polacre rig for a cat the bowsprit was without a jib-boom, and that a spritsail yard was carried. The mizzen mast did often carry a separate topmast, and was therefore rigged like a ship's mizzen (initially with a mizzen yard and from the 1760s on with a gaff, which according to contemporary illustrations was probably slung).

In contrast to Röding's text, the illustration of a cat in his work showed trestle and cross trees on all masts, and therefore also two topmast shroud pairs and the necessary futtock shrouds. The fore and main stays were single, and a staysail stay was fitted with each.

The polacre rig for cats was also used on Swedish turumääs, with the exception that a jib-boom was rigged. This rig can be observed on a model of a turumää of 1771 in Stockholm.

Bark

Bark; Barque

In a similar manner to the term 'cat', the term 'bark' referred to a particular hull shape for the first half of the eighteenth century, rather than to a particular rig; it was only after about 1750 that the most common rig for such vessels became known as a bark rig. Falconer referred in 1769 to the term 'bark' as

> a general name given to small ships; it is however peculiarly appropriated by seamen to those which carry three masts without a mizen top-sail. Our northern mariners, who are trained in the coal-trade, apply this distinction to a broad-sterned ship, which carries no ornamental Figure on the stem or prow.

Figure 61 ***Kray***
Kray rig according to Chapman. The dotted lines indicate the slightly different rigging on Swedish naval udenmää.

Steel's definition was even shorter: 'small English ships, having no mizen-topsail, are called barks'. Both these statements suggest, therefore, that the distinguishing features of eighteenth century barks were small size and the lack of a mizzen topsail.

Chapman, however. made it clear (by providing no fewer than twenty-one draughts of differently sized barks in his *Architectura Navalis Mercatoria*) that ship-rigged barks could be quite large vessels. His draughts include six such three-masted barks, a further two rigged as snows, three as brigs, five with a single-masted sloop rig and three with a 1½-masted galeas rig. In addition one was rigged as a kray (that is, with a three-masted polacre rig) and one as a schooner.

The differentiation made in contemporary English literature between the smaller bark and the larger cat was not maintained in Chapman's work. Just as his cats ranged from very small to large, so his barks ranged from 41 to 1257 tons. They were therefore no smaller in general than any other eighteenth century merchant vessel. Röding similarly made no mention of small size as a characteristic of barks, describing a bark as 'a large three-masted ship, fitted out for trade and, to increase her carrying capacity, therefore built less sharp than a frigate; otherwise a bark has a ship's rig, and commonly a flat deck and a stern but no gallery or head.'

(Röding's use of the terms 'ship' and 'ship rig', both here and in the discussion of cats above, is somewhat idiosyncratic: 'to [the category of] Ships therefore belong in fact only the man-of-war and the frigate, the bark, hagboat, fly-boat, cat, pink, East Indiaman and galleon; all other types are considered as Vessels...')

Though it may initially have been only a general term for small vessels, by the eighteenth century 'bark' had become, as Chapman proves, the term for one of the eight major northern hull types. A bark was, as noted above, very similar in appearance to a cat; the main differences were a larger stern, a nearly straight wing-transom and a very distinct lower counter. Barks were also generally flush-decked, with no real quarterdeck or forecastle (where these decks existed, they formed only low steps in the deck), and this provides a further point of distinction between barks and cats since cats had a deep waist, a distinct forecastle and a quarterdeck of normal height.

A good account of a bark's sailing qualities was provided by Hutchinson in 1794, in his *Treatise on Naval Architecture*:

> The good effects of deep and narrow squaresails, cannot be better recommended as answering this purpose, than by the performance of ships in the coal and timber trades to London, though the designed properties in building and fitting these ships, are burden at a small draft of water, to take and bear the ground well, and to sail with few hands, and little ballast; yet these ships perform so well at sea, that government often makes choice of them for store ships, in the most distant naval expeditions; and in narrow channels among shoals, and in turning to windward, in narrow rivers, there are no ships of equal burden can match them, which I attribute a great deal to their deep narrow squaresails, which may be perceived to trim so flat and fair, upon a wind, that all the canvass stands full, at a proper angle from the direction of the keel, so that the wind goes freely off from the lee leech of these sails, without being much altered in its direction from one sail to the back of the next, which is not the case when a ship's squaresails are so broad as to overlap each other much; one sail then shakes the next to it, and they extend so far to leeward, that the lee sheets make the after part of the canvass, or lee leeches, stand rather as so much back sail to stop the ship's way, or only to press the ship's side down and to leeward, which is the effect of all the canvass in a ship's sail, when it does not stand in a proper angle with the direction of the keel, when a ship is sailing close by the wind.

Illustrations of comparable eighteenth century merchant ships are not many, but those available reveal that bark rigs had their variations too, a point to be kept in mind when considering the rig of a particular vessel. A brief example is provided by Edward Gwyn's watercolour *A Collier* of approximately 1780, which shows a bark-rigged cat, with her main topsail and topgallant yard braces leading aft, a normal gaff-rigged mizzen and no spritsail yard, to mention a few obvious features. A picture of 1803 (artist unknown) of the Danish bark *De Kinds Kinder* at Hallig Hooge shows a number of rigging variations. The same braces lead forward, instead of the mizzen a spanker with a gaff and boom is set, and the bowsprit carries a spritsail. Both vessels are approximately equal in size, considering the number of shrouds shown. A possible reason for the forward lead of the braces on the latter ship is the documented rerigging of this vessel from a brig to a bark, so that certain features of the brig

rig may have been retained.

Further notes on the theme of barks can be found under Bark in Chapter VIII below.

Kray

Kraier, Krayer; Crayer

A kray was a three-masted merchant vessel of the Baltic Sea. Korth described the type as a North German and northern (mainly Swedish and Danish) merchant ship, while Chapman noted a Finnish origin. Contemporary references to the type are, however, unfortunately few. Röding described the kray thus: 'a three-masted ship with a polacre rig, common to the Baltic Sea... it carries squaresails and staysails, similar to those on three-masted ships'. Korth added only that the masts were without topmasts and tops.

Röding and Chapman provided illustrations of a polacre rig for a kray. That included in Röding's work differed very little from the rig of a small cat. Three pole masts are shown fitted with trestle and cross trees, but no topmast shrouds; instead a ladder is set up to starboard between the cross trees and mast-head. Beside the main and fore courses a short mizzen is rigged to a mizzen yard and a simple bowsprit carries a spritsail yard. Topsails are shown at every mast, as is a flying jib, but no topgallant sails. Further staysails are suggested by a main and a main topmast staysail stay.

Chapman's kray has a much simpler rig. Instead of lower and topmast stays for every mast, only three stays are shown, leading down from the upper rigging stops. Topsails are shown only on the main and foremasts, with the topsail on the latter shown without braces. The vessel has no spritsail yard, but is shown with a fore top staysail and a flying jib. In place of a mizzen is a spanker, rigged with gaff and boom.

A similar rig, but slightly heavier, with additional stays and a jib-boom, was used on the oared fighting ships known as udenmääs developed by Chapman for the Swedish inshore fleet.

A close study of the frontispiece of Chapman's *Architectura Navalis Mercatoria* reveals another kray in the medium ground. Shown in three quarter stern view, it displays a flat stern with two windows and three Swedish crowns on the taffrail above. The vessel's rig is similar to that shown in Plate LXII of the work, except that topsail yard lifts are shown.

Chapman's kray rig drawing provides the basis for the drawing included here, with the extra stays of the udenmää marked as dotted lines.

Skerry-boat

Schärenboot; Bateau de skerry

Skerry-boats were generally fore-and-aft schooner rigged, but a Norwegian sail plan, dated 'Nyeholm den 20d Apriil 1768' at the Rigsarkivet, Copenhagen, shows an alternative three-masted square rig. It is on account of this rig that the skerry-boat is considered in this chapter.

A skerry-boat's square rig was very simple, without topsails. The short masts were without topmasts or flag-poles and the fore and main courses were cut to a trapezoid shape in Norwegian fashion.

The mizzen mast had only a small spanker with a gaff and boom. The draught noted above also shows two jibs set to the horn bowsprit, both flying.

A similar rig was proposed by Chapman in 1799 for battery transport vessels; the only difference was that he proposed a leg-of-mutton or sliding gunter mizzen instead of a gaff mizzen.

The Swedish index of Röding's work described the skerry-boat in a few words: '*Skaerbaat*: a small armed vessel, several of which patrol between the skerries of Stockholm to protect the harbour entrance'. The Swedish inshore fleet was not alone in employing skerry-boats for coastal defence, which could not have been effectively maintained without a great number of such small and agile armed vessels; other Scandinavian nations used them as well.

Skerry-boats superseded Mediterranean type galleys in the Baltic fleets, where galley squadrons had formed part of the navies for a whole century (though they had declined to a largely defensive role by the later half of the eighteenth century). Early skerry-boat development dated from the 1760s. They were quite similar to swift Levantine brigantines, with a length of between 54 to 66 feet and a height above the waterline amidships of about 3 feet. Besides being rigged to sail, each vessel carried ten to twelve pairs of sweeps; their armament consisted of eight to ten swivel guns and howitzers. Skerry-boat crews consisted of about fifty officers and men.

Anderson suggested that skerry-boats did not appear before 1786, but the Danish boats to which he referred were already an advanced development of the original type, broader in the beam and equipped with two 18-pounders; this places them in the class of gunboats, a new type of vessel of the late eighteenth century.

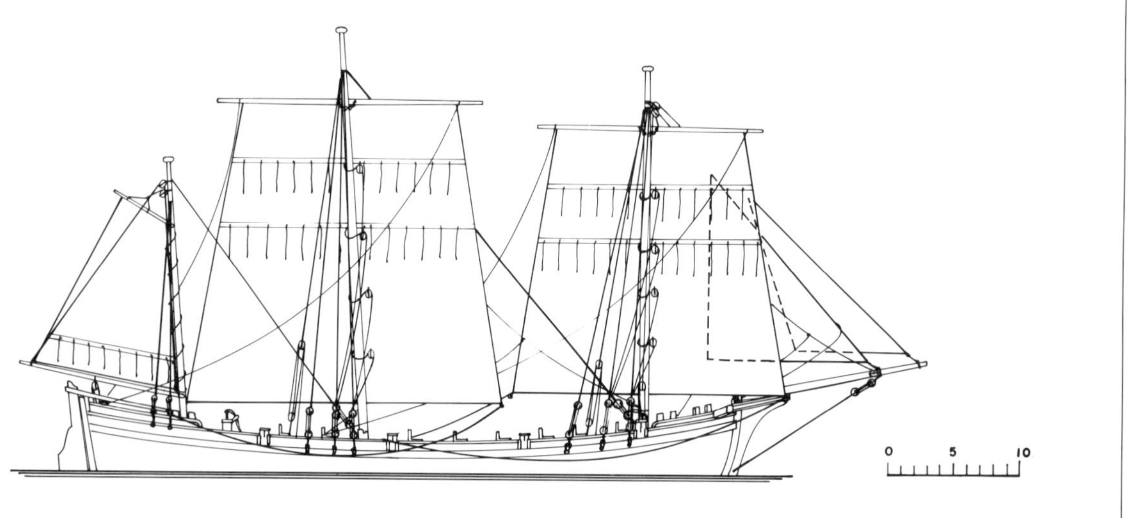

Figure 62 ***Skerry boat***
Three-masted skerry boat rig, considered an alternative to fore-and-aft schooner rig, from a Danish sail plan dated 20 April 1768.

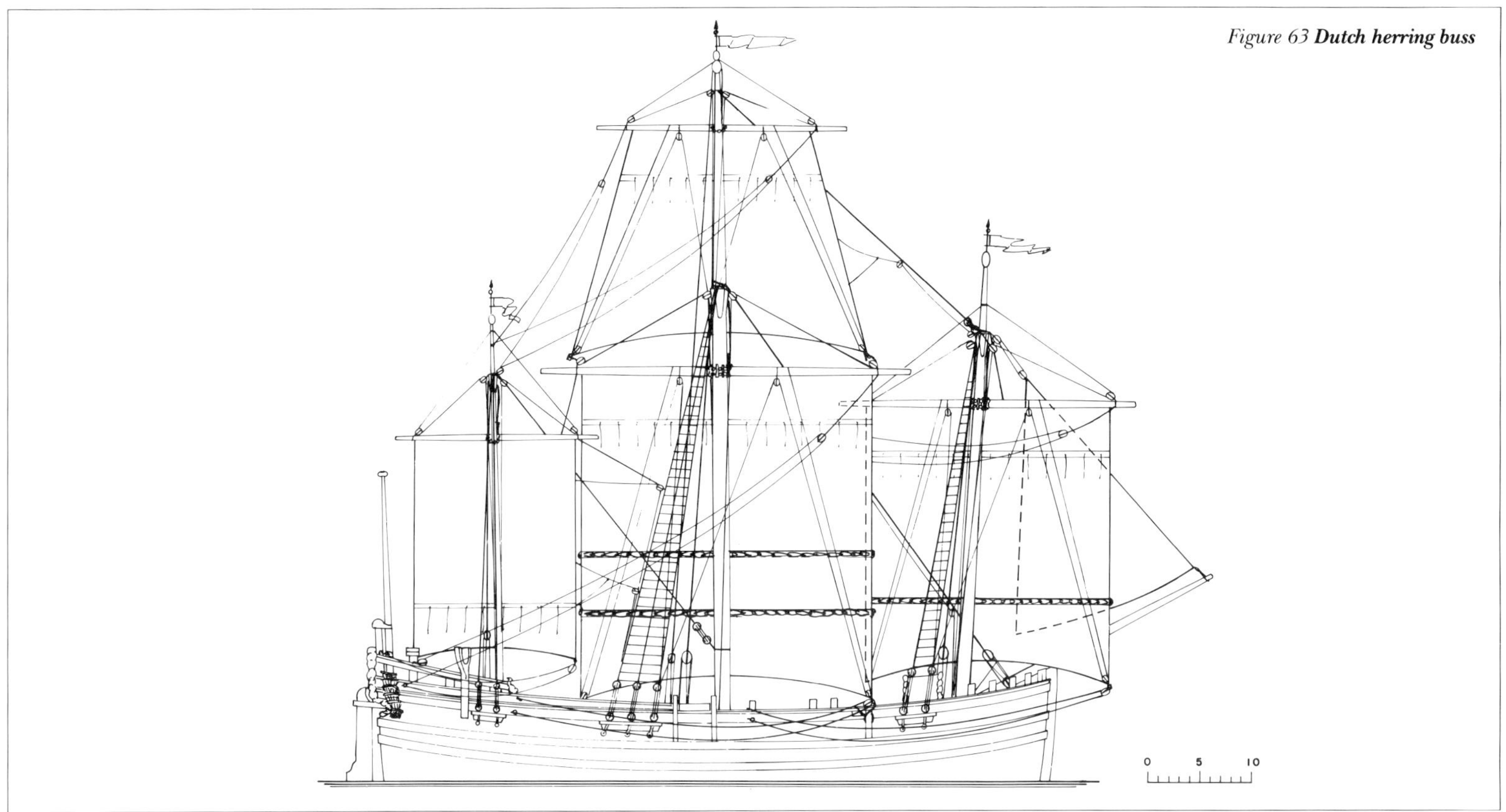

Figure 63 **Dutch herring buss**

Dutch herring-buss

Holländische Heringsbüse; Buche de hareng hollandais

The herring-buss as described by Falconer and Röding was two-masted, but Steel (1794) and etchings by A van der Laan (1690-1742) and C J Visscher (1587-1652), as well as other pictorial evidence, confirm the addition of a foremast; the type can therefore be considered three-masted, at least for the early part of the century. The *Beschrïvende Catalogus der Scheepmodellen and Scheepsbouwkundige Teekeningen 1600-1900* of the Nederlandsch Historisch Scheepvart Museum, Amsterdam, (published in 1943) commented as follows:

> These old vessels were used right up to the second half of the 19th century for herring fishing. They originally had three square-rigged masts, but changed at the beginning of the 18th century to a two-masted dogger rig.

Groenewegen's *Versameling van Vier ent achtig Stucks Hollandsche Scheepen* of 1789 showed several herring busses with a kind of dogger rig, except that the mizzen mast carried a squaresail instead of a fore-and-aft sail. A merchant buss was rigged as a howker.

Like the skerry-boat described above, the buss differed from other three-masters of this period by having only short masts with a simple square rig.

Dutch and English fishing vessels were collectively referred to as 'busses' and Röding's note on the rigging of this type - 'a large sail, a running topsail, a fore staysail and a small half mizzen, sitting on a mizzen yard' - makes it clear that by 1790 the term 'buss' had become a general name for fishing craft, and not only in English: his description is not of a buss *per se*, but a dogger.

A very good impression of these two fishing vessels side by side is given in the painting *De Haringvisscherï* by H Kobell. In a 1781 etching by M de Sallieth both types are shown fishing together and the differentiating features are clear, not only the square versus fore-and-aft rigged mizzen, but also the hull structure (see also Dogger in Chapter IV below) and the fact that the buss had a pink stern above the tiller and a hut-like building on the fore deck. Falconer explained this superstructure in his short description of the buss: '[the buss] is generally from fifty to seventy tons burthen, being furnished with two small sheds or cabins, one at the prow and the other at the stern, the former of which is employed as a kitchen.' Falconer's description has been confirmed by Korth, who added that larger busses were three-masted, of a tonnage between 40 and 100 tons, and from eighteen to twenty-four crew. Sometimes they were armed with a few small guns and side-arms for defence.

A description of the rigging, neatly complementing the Sallieth etching, was given by Steel. He referred to a buss as follows:

> A Dutch fishing-vessel with three short masts, each in one piece. On each is carried a square-sail, and sometimes a topsail above the mainsail. In fine weather they add a sort of studdingsail to the lower sails, and a driver. Occasionally they add a jib forward, upon a small bowsprit or spar. To shoot their nets they lower the main and fore masts, which fold on deck by large hinges, and stow aft upon crotches.

Fly-boat, Dutch flight

Fleute, Fleutschiff; Flûte hollandais

A fly-boat was one of the principal ship types in seventeenth and eighteenth century Holland, instrumental in making that country one of the leading trading nations of the period. A characteristically capacious merchant ship, the fly-boat carried grain from the east, timber from Norway and more exotic goods from the Mediterranean.

Fly-boats were also employed in the service of the Dutch East India Company, where they collected cargo for the large East

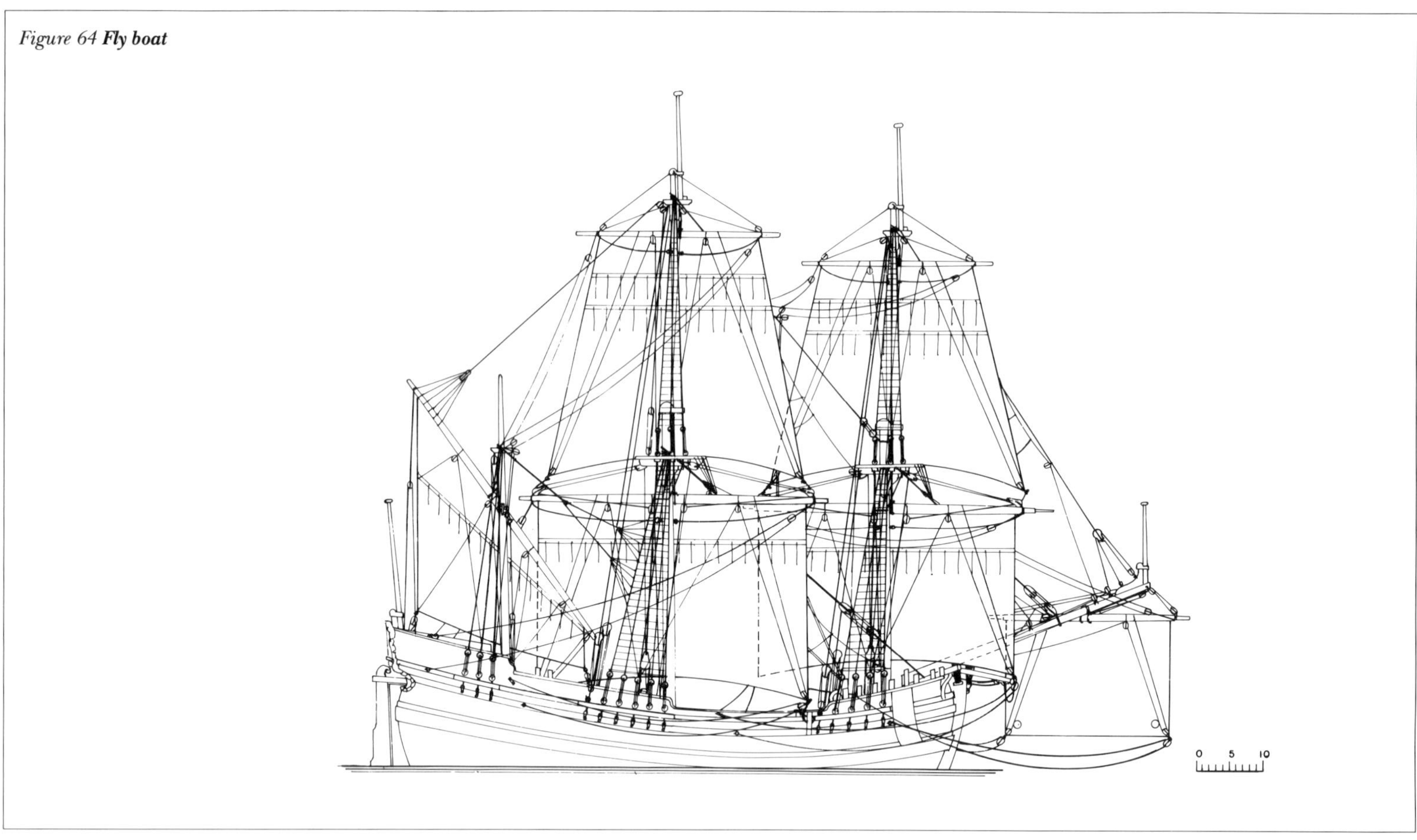

Figure 64 **Fly boat**

Indiamen. One of these, engaged in their search for new sea lanes and the exploration of new trading places, achieved a lasting place in the history of maritime exploration: the fly-boat *Zeehaen* accompanied Abel Tasman's pinnace-ship *Heemskirk* during his voyage of 1642-3, in the course of which he discovered Tasmania and New Zealand.

The eighteenth century fly-boat was, in Röding's words:

> ...a large three-masted vessel, very broad both forward and aft, and with a very flat bottom. In proportion to its large body a flight has somewhat low masts and it is therefore only a very slow sailer. They are still employed by the Dutch and Hamburgers for trading and especially for whaling. Their rigging is similar to that of men-of-war, and they also have a coach at the aft end of the deck; in this they differ from the now much more common merchant barks. Abaft they have very pronounced buttocks and a wide stern. Flights are built from 300 to 900 lasts. Since these vessels are broad and slow-moving, they are only in limited use nowadays [1 last is approximately equal to 2 tons].

Large fly-boats had a full ship rig, but on smaller vessels, especially during the first half of the century, masts without topgallant sails and a mizzen yard rigged with a lateen mizzen were common.

In Groenewegen's *Versameling van Vier ent Achtig Stucks Hollandsche Scheepen* of 1789, a galliot and a Greenland whaler both retain a similar rig, with the exception that a gaff-rigged mizzen has replaced the mizzen yard's lateen sail.

Since the term 'flute' is sometimes applied to a Dutch flight in English sources, it is worth noting that a French flute had nothing but its name in common with the Dutch vessel.

French flutes were fleet transport (supply) ships and were neither flat bottomed nor slow sailers. In French this vessel was known as a *flûte*, while the Dutch flight was referred to as a *flûte hollandais*.

Jackass bark, jigger bark, hermaphrodite bark

Barkentine, Schonerbark, Dreimast-Toppsegelschoner; Barquentin, barque-goélette

Contemporary literature does not reveal a specific name for this rig, so the nineteenth-century name has to be taken. Probably the nearest contemporary term was 'hermaphrodite bark'. It was not considered unusual to have an unnamed new rig (for instance, the term 'schooner' evolved at least two decades after the first appearance of the schooner rig).

The later expression 'barkentyn' was already in use by Groenewegen, but only in connection with a merchant brig's rig (for example 'een hoeker met een barkentyns tuyg'). Other Dutch literature explains that the terms 'brik' and 'barkentyn' in eighteenth century Dutch language were applied to the same type of rig, with brik (brig) generally being applied to those in naval service.

The type of rig later known as jackass bark or barkentine rig was developed during the last quarter of the eighteenth century in England. C F Steinhaus noted in *Die Construction und Bemastung der Segelschiffe* of 1869 that although the rig had gained popularity in more recent times, it was already known in the later years of the eighteenth century. F L Middendorf referred in 1903 to this rig in his book *Bemastung und Takelung der Segelschiffe*, and gave it the now rarely used name 'polka bark', and David MacGregor noted pictorial evidence of the rig from 1785 and 1807. A painting by C Roussel from 1813 entitled *The Corsair Alligator* shows the same rig with an additional fore gaff sail rigged to a horse and a mizzen stay leading upon deck at the starboard side of the mainmast.

Without providing a name for the vessel, Lever also gave an illustration and an excellent description, which is worth repeating in full:

Figure 65 ***Jackass bark***

Ships in the Baltic and Coasting Trade, which carry fore and aft Mainsails, have the Mizen Mast as taunt as the Mainmast, and seldom carry any Mizen Topmast, but a Flag Staff, to hoist the Ensign, &c. This is sometimes made strong enough to hoist a small Topsail with a Gaff, or Yard slung by the third, the Clew hauling out to the Mizen Peak. They carry no Cross Jack Yard, and consequently no square Topsail, but the Mizen is very large in proportion.

The Mizen Stay is taken to the Mainmast Head, and reeved through a Block, strapped to an Eye-bolt just under the Main Cap. The Main Topping Lift, instead of leading to the Main, is taken to the Mizen Mast Head. A double Block is strapped to an Eye-bolt hooped round the Mizen Mast; the single Block is strapped with a Hook and Thimble, and connected by the Topping Lift Fall, with the double one. The Pendent has a Thimble spliced in the upper end, to which the single Block is hooked: the lower end is spliced to a Thimble in the Eye-bolt, in the Boom end. The Peak and Throat Tyes, or Halliards, reeve as described for the Gaff, which traverses. The Sheet hooks to an Eye-bolt amidships; if the Sail be large, there are double Sheets, one on each side.

The Guy Pendent is hooked to a Strap round the Middle of the Boom, the double Block, of a Luff Tackle is hooked to it, and the single one to the Main Chains. Gaffs which traverse have a Throat Down-hauler - the lower Block is hooked to a Strap round the Mast, and the upper one to an Eye-bolt under the Jaws of the Gaff.

Ships rigged in this Manner can sail with a Hand or two less, and answer very well for working through Narrows, there being no after Sail to brace about, but the Main Topsail and Main Top-Gallant Sail, and their Braces lead forwards.

The Main or Cross Jack Brace leads through a Block at the aftermost Fore Shroud, and the Main Topsail Brace, through a Block strapped to an Eye-bolt in the after Part of the Fore Cap, the standing Part being hitched round the Foremast Head. The Top-Gallant Brace leads to the after Part of the Fore Topmast Cross-Tree.

The aftermost Shroud of any Mast rigged with a Boom, and Gaff which traverses, the Sail having Reefs in, should be served the whole Length: and the aftermost Topmast Backstay should go with a Block and Runner, as mentioned for the Breast Backstay. The former, that it may not be chafed when going large, and the latter to make room for the Boom to be guyed forwards.

Jigger mast

Papageienmast, Jagermast; Mât de tape-cul

A jigger mast in the eighteenth century was an additional mast which, without any alteration to the existing mast positions and the rigging, allowed a two-masted vessel to be turned temporarily into a three-masted one. The name of a rig was not affected by the stepping of a jigger mast; one might therefore have a three-masted brig or a four-masted ship. The term 'jigger', used for this supplementary mast in the eighteenth century, became in the later nineteenth the designation for a fourth mast.

Jigger masts were traditionally characteristic of vessels of the French Atlantic coast, fitted in vessels of which the length-rigging ratio necessitated an extra mast for good handling, but where the extra length did not warrant the permanent rigging of such a mast. A jigger mast was set into the step for the ensign staff. The approximate length of such a mast was that of a main topmast, with its thickness governed by the rules for a pole-headed topmast.

However, to hoist an additional sail (a boat sail) temporarily at the stern, thus adding to the mizzen and assisting in certain manoeuvres, an eighteenth century captain often made use of the ensign staff itself. Röding described this practice thus: 'small

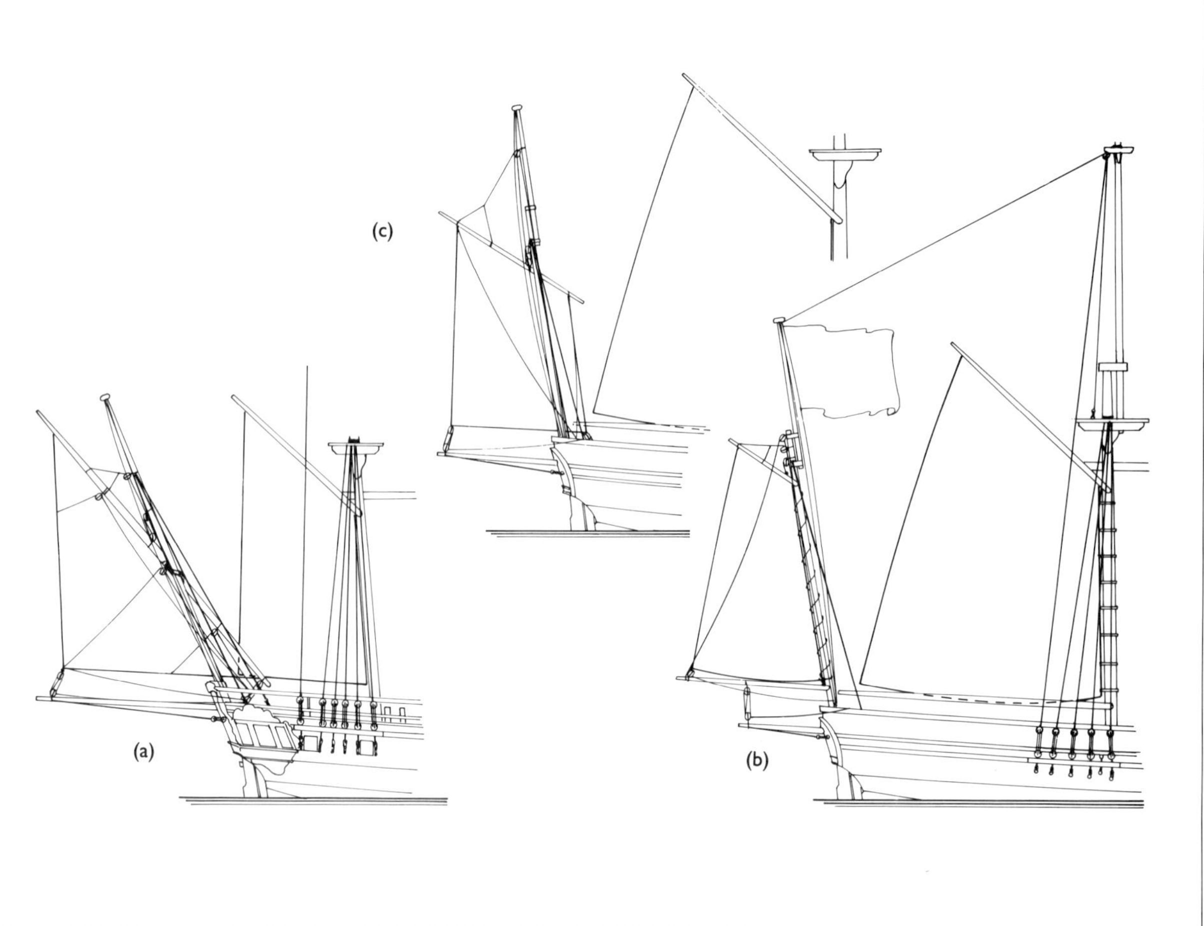

Figure 66 **Jigger mast details**

(a) The lateen-rigged jigger mast of a French corvette as depicted in an etching by Ozanne, approximately 1785. A similarly rigged jigger mast can be seen in a Dutch etching of 1779 by J Punt, after an original by H Kobell.

(b) A gaff and boom-rigged jigger mast as seen on a French brig in an etching by Groenewegen of 1793.

(c) A lug-rigged jigger mast on the French brig Ste. Marguerite (1803), from a watercolour by Roux.

rusails [*Rusegel* is an old German term for a sail hoisted on a mizzen yard] are now sometimes used on the ensign staff', and Falconer noted a 'ring-tail' as

> ...a small triangular sail, extended on a little mast, which is occasionally erected for that purpose on the top of a ship's stern. The lower part of this sail is stretched out by a boom, which projects from the stern horizontally. This sail is only used in light and favourable winds, particularly in the Atlantic Ocean.

These temporary sails were superseded in the 1770s in English ships by the fore-and-aft driver (see also Mizzen boom and Driver in Chapter II).

The use of a jigger mast on French ships has been documented by several artists. Ozanne, for example, depicted a three-masted French corvette with a lateen sail hoisted to a jigger mast, the sail about two thirds of the mizzen's size (see Figure 66a). Groenewegen (1793) showed a French merchant brig with a gaff-rigged jigger mast (see Figure 66b), and Roux (1803) a picture of the brig *Ste. Marguerite* with an additional lugsail (see Figure 66c).

Common to all these sails which extended over the taffrail was the jigger bumkin, a spar projecting aft over the stern to secure the sail's sheet.

IV

Rigs for two-masted and smaller vessels

As in other centuries, a large number of smaller vessels with a diversity of rigs were in existence beside three-masted ships. It is impossible to mention all of them within the framework of this book, so the following chapters will concentrate on the characteristic features of the rigging of the main types. This chapter and Chapters V to VII cover Northern European rigs; those from South Europe and the rest of the world are covered in Chapter VIII.

Snow

Schnau; Senau

Every contemporary source considered the snow to be the largest two-masted vessel (see Figure 67). The sails and rigging on the main and foremasts were arranged similarly to those of a full-rigged ship; the snow differed from a full-rigged ship in the following details.

Braces

Brassen; Bras

The main topsail and topgallant braces led forward to the foremast head and fore topmast trestle trees respectively.

Preventer stays

Borgstage; Faux étais

Preventer stays were not rigged to snows.

Snow mast, trysail mast

Schnaumast; Mât de senau

The snow's main feature was the snow mast, also named a trysail mast. Stepped on deck abaft the mainmast, this thin mast had its head blocked and bolted between the aft parts of the main trestle trees (see Figure 48g above), or fastened with an iron band to the lower part of the masthead. The mizzen or trysail was seized to wooden hoops encircling the snow mast, or was laced directly to it.

Gaff

Gaffel; Corne

The snow mast gaff had a mizzen bent to it, large enough at the foot to extend from the mainmast to the inside of the taffrail.

Boom

Baum; Gui

No trysail boom was rigged on snows during the eighteenth century. This boom appeared after 1800, on vessels with a larger trysail than that described above.

Horse

Jackstag; Filière d'envergure d'un mât

Armed snows in the service of national fleets were commonly known as corvettes or sloops of war. They did not often step a snow mast but instead were fitted at the aft of the mainmast with a horse or jackstay, spliced with an eye round the masthead and set up with deadeyes to an eyebolt on deck (see Figure 48i above). The trysail was bent to thimbles or hanks on this horse (see Figure 48j), with the gaff slung to the mast. For further details see Snow mast in Chapter II.

Brig

Brigg; Brick

It is not easy for a layman to distinguish between a snow and a brig. The main differences between the two types can be found in the mainmast rigging. On a snow the mainmast was rigged with a main course and a trysail set to a trysail mast or horse (which was effectively an integrated mizzen mast); the term 'brig', however, applied to a vessel with a main and foremast, but with the mainsail bent not to a yard, but to a gaff, and with its aft clew extended over the taffrail by a boom. The lower yard on a brig's mainmast was a crossjack yard like that on a ship's mizzen mast. The mainmast itself was larger by comparison with the foremast on three masted ships than on brigs and snows.

Crossjack yard

Bagienrah; Vergue sèche

A crossjack yard instead of a main yard was rigged on brig mainmasts until the last years of the century (see Crossjack yard in Chapter II), and the mainsail was set up to a gaff and a main boom in the same manner as a spanker to the mizzen mast of a ship. In fact the brig sail was the forerunner of a ship's spanker, being of much larger size then an ordinary mizzen. Its mast leech was bent to wooden hoops which fitted loosely around the mainmast, to allow the sail to be hoisted. On Royal Navy brigs the crossjack yard was replaced in about 1800 by a main yard with an additional square mainsail set to it.

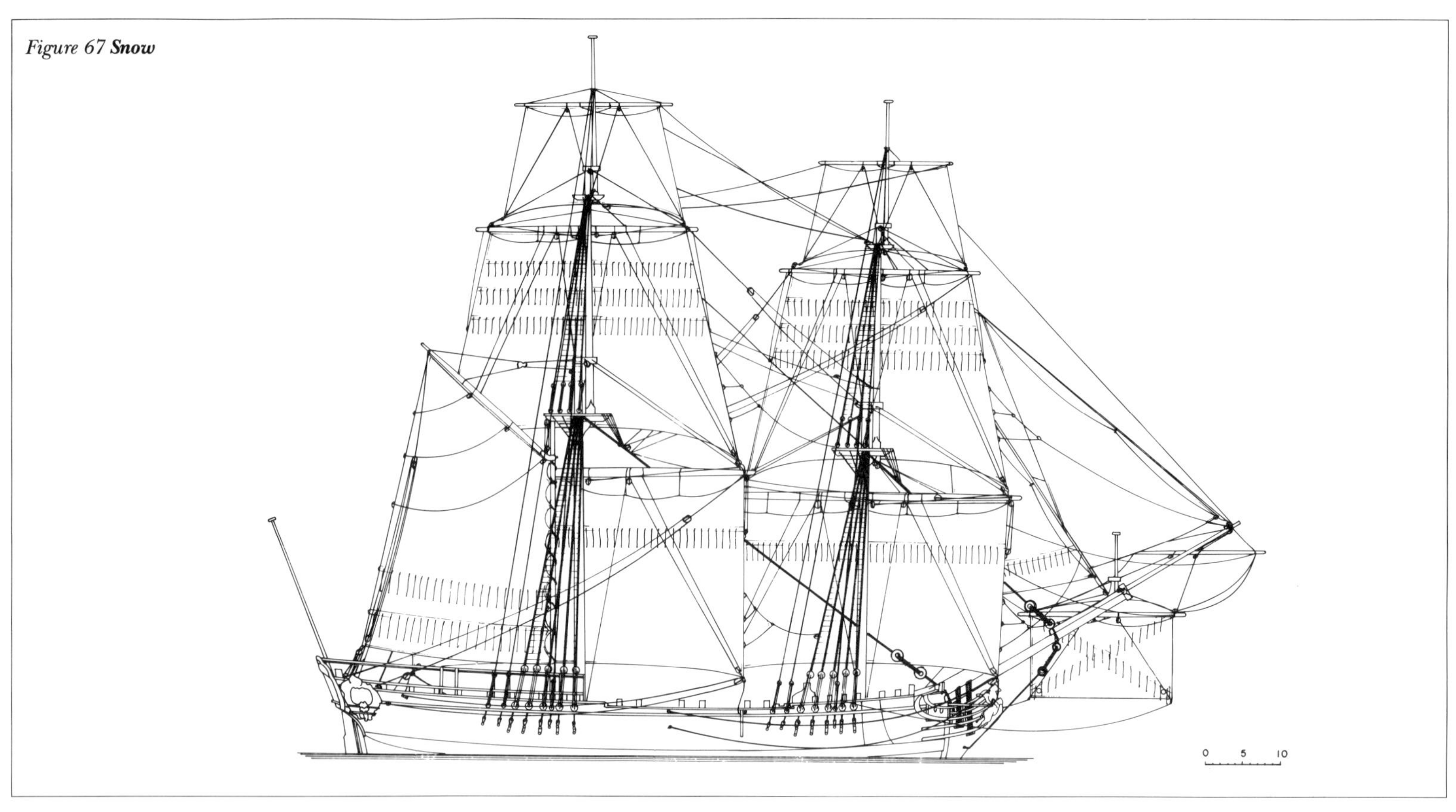

Figure 67 **Snow**

Figure 68 **Brig**

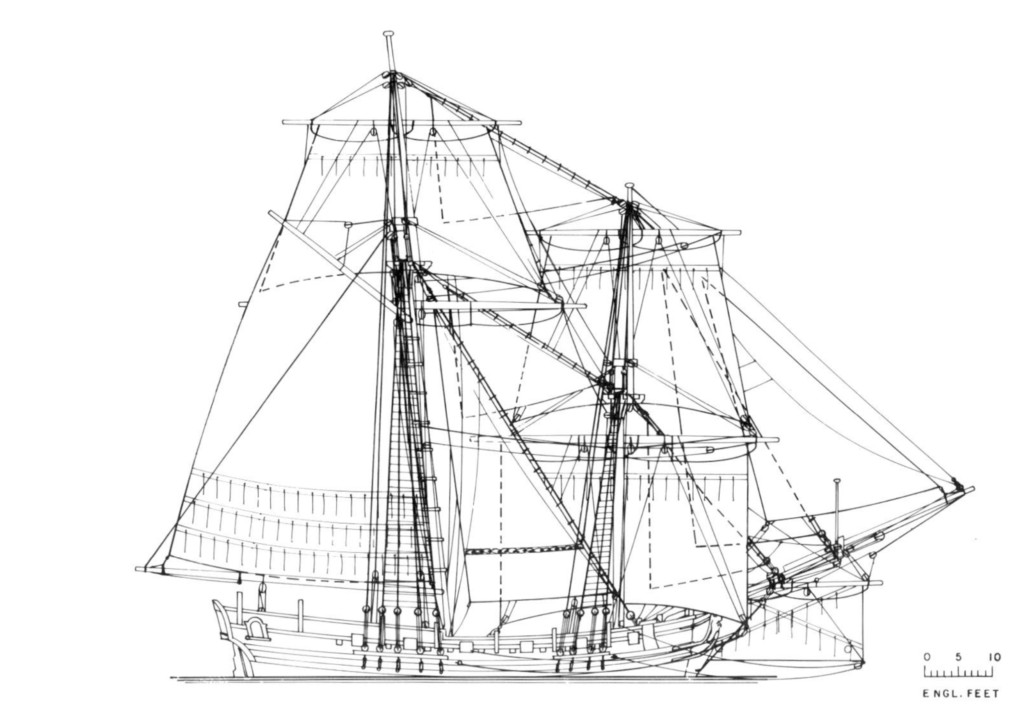

Figure 69 ***Brigantine***
The rigging plan of this early brigantine is taken from the sources mentioned in the text.

Braces
Brassen; Bras

While the crossjack yard braces usually led forward on Continental merchant brigs as well as on naval brigs, those on British merchant brigs led either forward or aft, and on Royal Navy brigs generally aft. All other mainmast braces led forward.

Shrouds, backstays
Hoofdtaue, Pardunen; Haubans, galhaubans

The aftmost main shrouds were wormed, parcelled and served over their full length to counteract chafing from the brig sail. The aftmost backstay was set up to a tackle so that it could be slackened when it interfered with the boom.

Brigantine
Schonerbrigg, Briggschoner; Brigantin

The brigantine (or brig, as the vessel was somewhat confusingly also called in the eighteenth century – brig being an abbreviation of brigantine) was more a parallel development to the bilander and the snow than an offshoot of her earlier namesake. In 1750 Blanckley noted that brigantines were 'not now used, but were built light for rowing or sailing and had two Masts and square Sails'. The small drawing accompanying his text was of a vessel with a bowsprit, a main and a foremast. A spritsail, main, main topsail, fore, and fore topsail were shown and six pairs of oars visible. English galley-brigantines were rigged in this fashion up to the early eighteenth century.

The term 'brigantine' was probably transferred from these defunct two-masted square-rigged vessels to two-masted gaff-rigged sloops, which can be traced back to about 1715-16. One of the earliest groups of gaff-rigged sloops built for the Royal Navy was the *Ferret* class of 1710-11, though HMS *Ferret* was re-rigged in 1715-16 with two masts. Nobody knows what sails these two masts carried but Anderson, in *The Sailing Ship*, makes the reasonable assumption that the rig was similar to the two-masted sloop *Drake* in an English engraving of 1729. This vessel was fitted with a gaff and topsail rigged mainmast and a square-rigged foremast, producing a rig which later became known as brigantine rig.

A similar rig can be seen in the etching *Boston, 1725* by Burgis, with the mainmast rigged with a gaff and boom. Earlier illustrations by G du Pas (1709) and Labat (1720) show a leg-of-mutton sail with a sail spreader in the peak instead of a gaff, and the du Pas etching also shows a square main topsail.

In 1769 Falconer defined a brig or brigantine as a merchant ship with two masts, and added:

> Amongst English seamen, this vessel is distinguished by having her main-sail set nearly in the plane of her keel whereas the main-sails of larger ships are hung athwart, or at right angles with the ship's length, and fastened to a yard which hangs parallel to the deck; but in a brig, the foremost edge of the main-sail is fastened in different places to hoops which encircle the main-mast, and slide up and down it as the sail is hoisted or lowered; it is extended by a gaff above, and by a boom below.

He illustrated his description with a sketch of a brigantine, which was a pure sailing vessel with a topmast and topgallant mast stepped on the foremast, and a mainmast as long as the three foremast parts together with only a flag pole stepped. The mainsail was of similar height to the fore course and fore topsail together, with a topsail rigged above the gaff sail.

Other contemporary illustrations of the type were by Ozanne, who depicted a Breton brig with pole masts and topsails but no topgallant sail, a brig from 1762 with topmasts and sails but no topgallant sail, a fully rigged naval brig (corvette), and a brigantine, which differed by having only a gaff-boom sail rigged to the mainmast. The rig of the last occurred also in a watercolour by Edward Gwyn (c1780) entitled *A Yarmouth Brig.* Chapman's sail plan of a brig from 1768 was already fully developed and showed topgallant sails as well, and another Gwyn watercolour, entitled *A Coal Brig,* had flying royals set in addition.

From these sources it seems likely that despite an earlier development of the brigantine rig (or, as it was later known, the brig rig) the transfer of the defunct galley-brigantine's name to the new rig type must have occurred sometime between about

Figure 70 **Bilander**

1740 and 1770. By about 1770 this common merchant rig had also been used on naval vessels; Röding commented as follows:

> Those [brigantines] fitted out for war carry ten to twenty guns and, as a result of being lean built, are very suitable for cruising. Most brigantines are only employed as merchant ships and have no guns. The English use them more then any other nation.

One of the earliest occurrences of the name 'brigandine' is in *A Sea Grammar* of 1627 by Captain John Smith, and Furttenbach's *Architectura Navalis* of 1629 offers a definition of a 'bergantino', which is described as a swift two-masted vessel with ten pairs of oars, generally employed as a Turkish privateer. The author of *Der geöffnete See-Hafen* denoted a 'brigantin' in the same way. Seventeenth and early eighteenth century brigantines were oared sailing ships smaller than galleys, lateen rigged on two masts and with not much more in common with the later vessel than the name.

Steel noted that some sloops of war were rigged variously as snows or brigs to deceive the enemy. He commented as follows:

> An Hermaphrodite is a vessel so constructed as to be, occasionally, a snow, and sometimes a brig. It has therefore two mainsails: a boom mainsail when a brig, and a square mainsail when a snow, and a main-topsail larger than the fore-topsail. Sometimes the boom mainsail is bent to the main-mast, as a brig or on a trysail-mast, as a snow.

Bilander

Bilander; Bilandre

A bilander was a small merchant vessel with two masts; it was set apart from other two-masted vessels by the shape of its mainsail. The sail was bent to a yard similar to a mizzen yard, and had the shape of an early eighteenth century settee mizzen. The yard hung at an angle of about 45 degrees to the ship's waterline, with its peak directly over the stern and its fore end approximately amidships. The aft and fore clews of the sail were set up with tackles to ringbolts in the taffrail or on the round stern, and in line with the fore leech amidships on the deck respectively.

Blanckley described a bilander somewhat differently, and perhaps was referring to a single-masted rig:

> [A bilander] has Rigging and Sails not unlike a Hoy, but is broader and flatter; the covering of the Deck is raised up half a Foot higher than the Gunwale, between which, and the Deck, there is a Passage left free for the Men to walk... [These vessels] are seldom above twenty-four Tun, and can lie nearer the Wind than a Vessel with cross Sails can do.

A reference to a 'bellande' in *Der geöffnete See-Hafen* was clearly to a similar vessel: '...a Nordic kind of ship, flat below and fitted with a mast and sail similar to a hoy'. A description of the type by Korth indicated a small vessel with lee-boards of approximately 10 feet long and 6 feet wide, with two masts, where the mainsail was a settee sail. The maximum size was about 80 tons, and the vessel was manned by four men.

'It may not be improper to remark, that this name, as well as brigantine, has been variously applied in different parts of Europe to vessels of different sorts,' noted Falconer, and both he and Steel commented that at their time the rig was no longer much in use because of its relative inconvenience in comparison to others. Steel wrote in 1794 that bilanders were then seldom employed, except in Holland. However, Röding noted at the same time that these vessels were mainly in use in England and Sweden. The development of the bilander seems to have run parallel to that of the snow and brigantine rigs, and the bilander did not survive beyond the eighteenth century. Seventeenth century two-masted vessels with a main and foremast configuration were usually square rigged without an additional fore-and-aft sail for the mainmast.

Ketch or howker

Ketsch oder Huker; Quaiche, ketch ou hourque

In contrast to the rigs described above, the two-masted type under consideration here comprised a main and a mizzen mast,

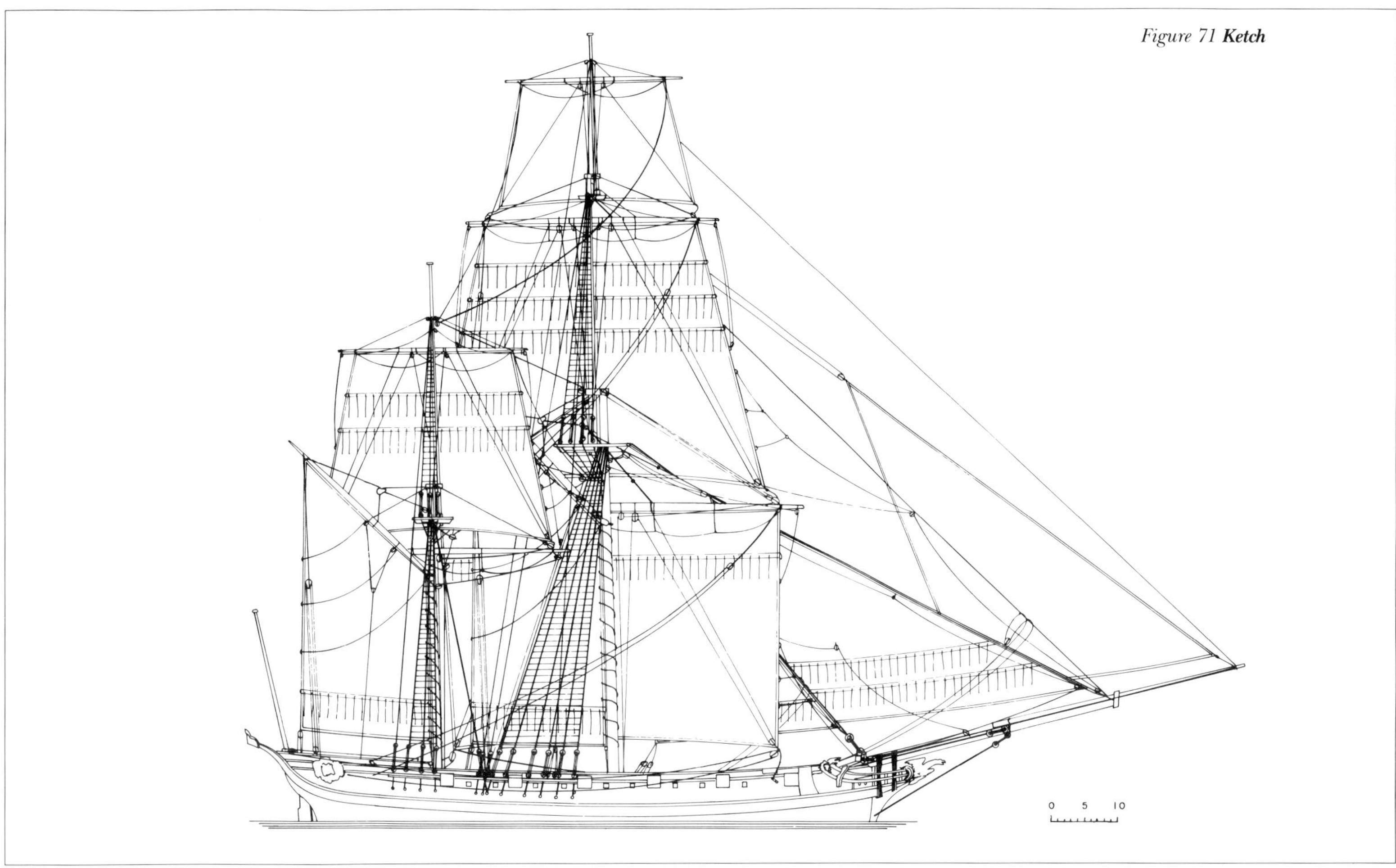

Figure 71 ***Ketch***

thus leaving the fore half of the ship's deck unobstructed. The two terms 'ketch' and 'howker' cover the same type of rig, differentiated only according to the vessel's employment. Ketches in general were naval vessels, while similarly rigged merchant vessels became known as howkers. According to Röding, however, the ketch was a vessel found mainly in England and Sweden, while the howker was found in Holland, Denmark and Sweden. Steel confirmed this view, and he also associated the bomb-ketch with France, noting that English bomb vessels were formerly rigged as ketches, but by his time (1794) were rigged as ships. Blanckley recorded that bomb vessels were rigged either as ketches or ships.

Two types of ketches were commonly found in the navies of Britain and the Continent. They were classified as 'bomb-ketch' and 'ketch', with the latter employed either as a yacht or a letter of marque.

Bomb-ketch

Bombardierketsch; Bombarde, galliote à bombes

> [Bomb ketches] are built remarkably strong, being fitted with a greater number of riders than any other vessel of war, and indeed this reinforcement is absolutely necessary to sustain the violent shock produced by the discharge of their mortars, which would otherwise, in a very short time, shatter them to pieces.

To this short and precise description in the 1780 edition of Falconer, the 1815 edition added: 'a bomb-ketch is generally from sixty to seventy feet long from stem to stern, and draws eight or nine feet of water'. The later edition also included a note that bomb-ketches were said to have been invented by a Monsieur Reyneau and first used in action at the bombardment of Algiers (1682). A modern work claims that Admiral Du Quesne's *galliotes à bombes* were designed by a Monsieur d'Elicagaray. Both works were partially right; the designer was the Chevalier Bernard Renau d'Eligaray. These new additions to the fleet proved an effective weapon in the quest for French naval superiority in the Mediterranean Sea.

The rigging of a bomb-ketch's was described by Steel. The masts were positioned and rigged like a ship's mainmast and mizzen mast, with staysails and a very large jib set between the mainmast and bowsprit. With the mortars placed in the fore half of the ship and the bombs discharged forward, a normal mainstay would quickly have been destroyed by the vessel's own fire; an iron chain mainstay was therefore used instead. The vessel was also fitted with preventer shrouds, the backstays were doubled and the yards rigged with additional support, all to withstand the shock of firing. Röding stated that the mainstay had to be removed before the mortars could be fired.

Ketch-rigged yacht

Jacht; Yacht

Ketches first began to appear in naval lists around 1650. They were small vessels of not much more then 40 feet in length and of between 40 and 60 tons. Ketch-rigged yachts can be dated from 1682, with the building in that year of the English royal yachts *Fubbs* and *Isabella*; at this time these were the largest yachts of Charles II. In the early eighteenth century, when the three-masted *Peregrine Galley* became the Royal yacht *Caroline*, the ketch-rigged yacht was usually employed for the transport of royal princes, ambassadors and other high ranking officials. In the ranking of yachts, the ketch took second place after the ship-rigged yacht; the latter belonged to the sovereign only, and all other officials and wealthy private persons were permitted single-masted craft only.

Letters of marque or privateers (that is, private ships commis-

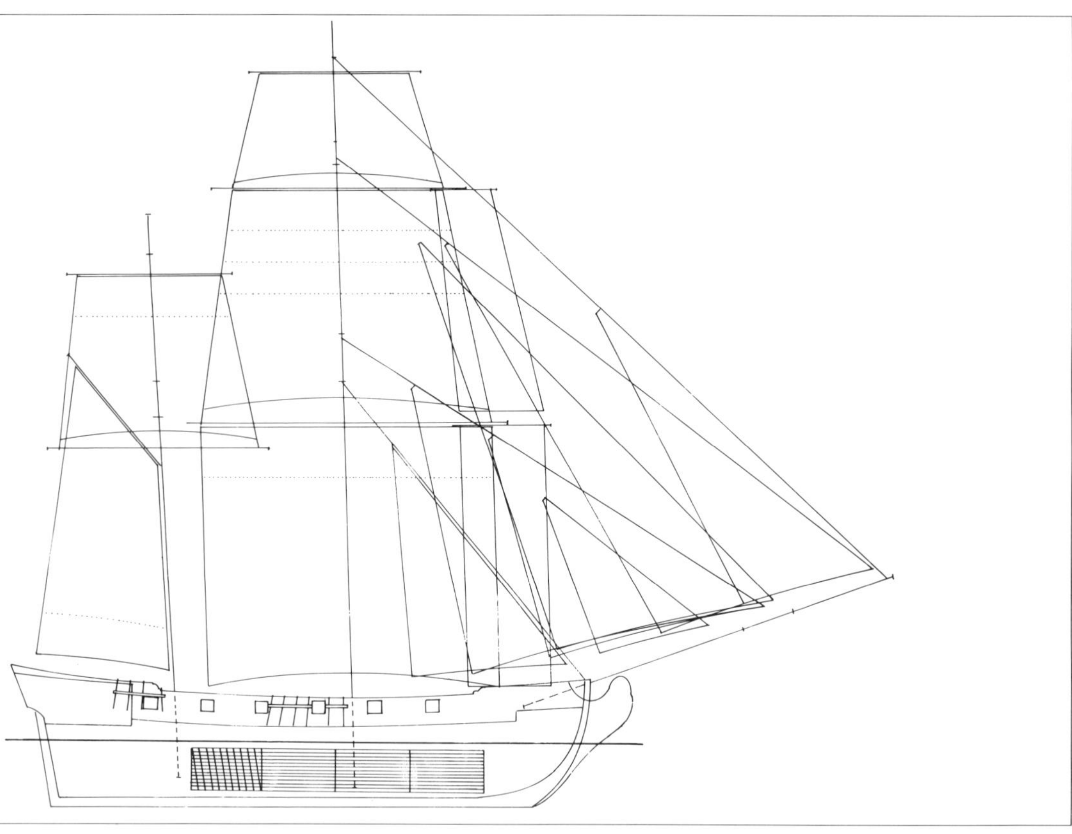

Figure 72 ***Danish howker***
The sail plan shown is taken from a draught dated 'Nyeholm, 16 February 1759' for a 12-gun sloop of war, following the best English sloop of war practice for ketch (hukkert) rigging. The vessel did not rig a wingsail, but note the various sizes of headsails and the studdingsails.

sioned by the Admiralty for independent action against enemy ships) were sometimes also rigged as ketches. Chapman's *Architectura Navalis Mercatoria* contains three draughts of vessels with this rig. Two were privateers with eleven and twelve guns and the third was a howker-rigged pink.
Danes have vessels of war, called howkers, which are a sort of sloop-of-war with three masts'. This observation was, however, a little wide of the mark. The Danish vessel of war with a hukkert rig was a ketch. A draught at the Rigsarkivet, Copenhagen, of a 12-gun howker of 1759, drawn by the shipwright F M Krabbe after the 'beste Engelske Sloops', gives full detail of the sail plan (see Figure 72).

The ketch's mainmast was fully rigged, but, as Chapman indicated, the topmast was often stepped abaft the mainmast. An etching by Sbonsky de Passebon in 1682 entitled *French bomb ketch* and another illustration by Lescallier from 1791 confirm this characteristic arrangement. The mainmast was also often rigged with a standing gaff sail (without a boom), called a wingsail. Such a wingsail was similar in shape to the mizzen course of a ship, and was made of canvas Number 6 or 7; its mast leech was seized to hoops about the mast. A wingsail was not always part of ketch rig, however. Steel noted the possibility of other sails:'... and sometimes, abaft the mainmast is a large gaffsail, called a wingsail'. Wingsails were not part of the rig of a bomb-ketch, and can only safely be assumed to have been a regular part of the rig of other ketches for the last 30 to 40 years of the eighteenth century. The Danish hukkert of 1759 did not have a wingsail; Chapman's 1768 sail plan of a ketch did.

During the eighteenth century the mizzen mast of a ketch usually carried a topmast and a mizzen topsail yard. In some cases, though, only a mizzen bent to a gaff (or to a mizzen yard before the introduction of the gaff) was set on the mizzen mast.

According to Chapman, when a ketch had a wingsail hoisted at the aft of the mainmast, mizzen stays were set up to both sides just abaft the main shrouds, like forward-leading mizzen shrouds. However, it is possible that this arrangement represents Chapman's own solution to the problem of leading the mizzen stay rather than a reflection of general practice; it is not confirmed by other contemporary illustrations, which mostly show the mizzen stay led in the same manner as that of a ship even when a wingsail was rigged.

With a long bowsprit, a ketch's headsails consisted of a stay foresail and two jibs. The foresail and inner jib were often hanked and rigged with a downhauler and reef points. The Danish sail plan of 1759 noted above (see Figure 72) shows seven different combinations of these three sails. Chapman's drawing of a ketch rig showed a bowline bent to the aft leech of the large stay foresail, with two bridles fastened about one third up the leech, leading through a block at the foremost shroud to a pin rail. The inner jib had a lift line fitted to the aft clew to lift it over the main stay when needed. This line rove to a block at half height on the aft leech, through another in the middle of the jib, and down to the bowsprit, where a further lead block directed the line into the fore ship. In addition to the normal sails, the Danish plan also showed studdingsails.

Howker

Huker; Hourque

Röding's description of a howker's rig is very similar to that of a ketch, and thus disagrees with Steel's view of the two types. While the former noted that a howker carried a main top and topgallant mast, the latter described the mainmast as a pole mast, usually carrying three squaresails but sometimes only a mainsail and topsail. Röding also mentioned the wingsail, which was not described by Steel. The head sails were similar to these of a ketch, though the bowsprit was not always fitted with a jib-boom. For further information see Dogger below.

Schooner

Schoner; Goélette

The schooner rig came into existence in the last few years of the seventeenth century. It evolved from the rigging of the two-masted Dutch speeljacht and the single-masted gaff rig, which

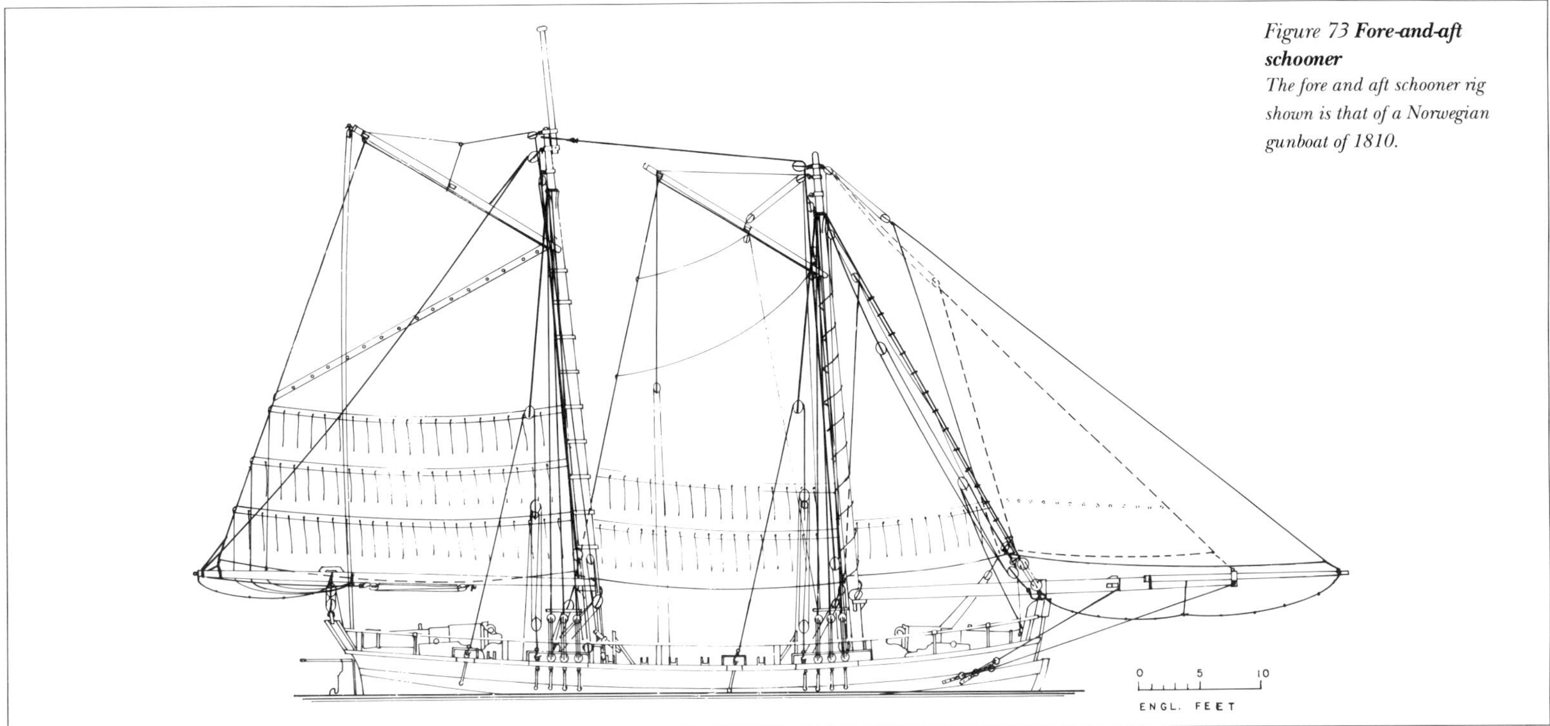

Figure 73 ***Fore-and-aft schooner***
The fore and aft schooner rig shown is that of a Norwegian gunboat of 1810.

also originated in Holland. The development of the schooner occurred in England; the first vessel rigged as a schooner was the private yacht of King William III, HMS *Royal Transport* of 1695. During the visit of Tsar Peter I to England in 1698 the yacht was presented to him as a gift by the King.

The name 'schooner' first appeared about two decades later, in 1717, in the *Boston News-Letter* in America. It heralded an increasing American adaptation of that rig for regional shipping. The rig was seemingly ideal for the conditions in colonial America, and received rapid and widespread acceptance there, becoming America's national rig within a few decades. It was there that between approximately 1720 and 1735 an amalgamation of the original fore-and-aft schooner rig with that of the topsail-carrying sloop took place. The schooner rig allowed a vessel to sail very close hauled and a schooner of equal size could be worked with fewer hands than a sloop. After the topsail schooner had evolved, a distinction was made between this and the fore-and-aft schooner, which carried gaff sails only.

Fore-and-aft schooner

Gaffelschoner; Goélette franche

As noted above, the original schooner rig consisted only of gaff sails plus two or three head staysails or jibs. Like nearly all other gaff rigged vessels, a fore and-aft schooner was also sometimes rigged with a squaresail on the foremast. This was merely a temporary sail, described by Röding as follows:

> This [sail] is referred to as a square or yard sail, and is bent to the crossjack yard of those vessels which carry only a gaff, sprit or boom sail, for example smacks, koffs, sloops or yachts... One should not confuse this [sail] with [that hoisted on] the crossjack yard in three-masted ships; the square sail is used when sailing before the wind, but is lowered as soon as the vessel is sailing by the wind.

To differentiate easily between a schooner's gaff sails, the foremost was called a schooner sail and that of the mainmast a mainsail. The gaff topsail was mainly a development of the last quarter of the eighteenth century, while the boom for the schooner sail was an innovation dating from the mid century. An early indication of this boom can be found in a 1747 sketch by Ashley Bowen of the schooner *Peter & Mary* off Marblehead; the artist himself sailed in this vessel.

Topsail schooner

Toppsegelschoner; Goélette carré

The difference between the topsail schooner and the fore-and-aft schooner lies in the setting of additional square topsails to the masts. While fore-and-aft schooners had pole masts only, often with the extension of a flagpole, the rigging of yard sails necessitated masts similar in proportion to those in ships for topsail schooners; alternatively, cutter masts were stepped. Schooner masts of both types usually had some amount of aft rake, up to an extreme 17 degrees.

The shape of a schooner's topsails depended on the choice of masts. For a ship mast with a stepped topmast the topsail was identical to that of a ship, but where a cutter mast was fitted the topsail required an enlarged roach; the depth of this roach depended on the position of the mast's rigging stop, from which the stay led forward.

Contemporary written sources agree that the topsails on schooner foremasts were permanently rigged, while those on mainmasts were frequently set flying. A number of contemporary marine paintings, however, give the impression that mainmast topsails were also rigged standing. Both standing and flying main topsails therefore seem to have been possible. The Bowen journals show main topsails only on the *Peter & Mary* and on the schooner *Sally* (later the *Grenville* and James Cook's first independent command).

A temporary anomaly in eighteenth century schooner rig was the spritsail yard. Schooners in the Royal Navy between 1764 and the mid-1780s were rigged with spritsail yards, and one is also shown in the 1747 Bowen sketch of the schooner *Peter & Mary*.

While a fore-and-aft schooner's squaresail was bent directly to the crossjack yard and was hoisted when needed on a horse forward of the foremast, the squaresail of a topsail schooner was rigged differently. The crossjack yard of the latter was a slung spread yard for the topsail sheets and could therefore not be raised or lowered. An extra yard for the squaresail thus became necessary. This was either short and spread only the centre half of the sail, or was of a similar length to the crossjack yard and was hoisted with four halyards (placed at a quarter of the sail's

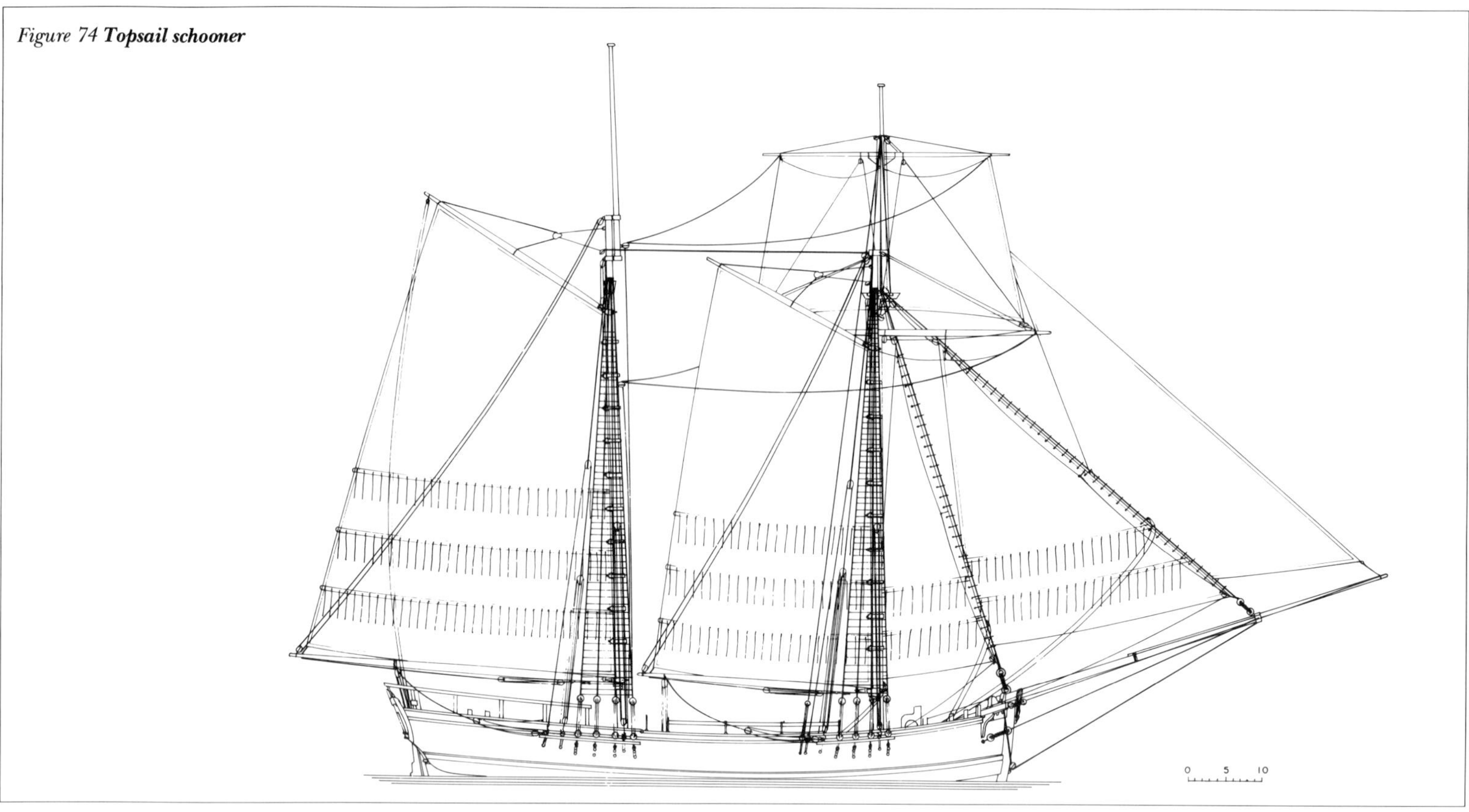

Figure 74 ***Topsail schooner***

width from the centre and at the earrings) into position beneath the crossjack yard (see Figure 88c below). In some instances the centre two halyards were replaced by one yard tackle at the slings. The shorter yard was mainly used with a deeply roached topsail, and was hoisted right up to the trestle trees; the squaresail was then cut with an extra bib to fill the deep roach.

The horse which guided the squaresail yard on a topsail schooner was set up forward of the crossjack yard, so as not to be obstructed by it.

A squaresail's clews (in schooners or in any single-masted vessel) were usually spread with the help of a spread boom, resting athwartships forward of the foremast. This squaresail boom fulfilled the same function in fore-and-aft rigged vessels as did bumkins in square-rigged ships.

Characteristic of all schooners was the schooner stay (main stay), which had to be rigged to allow clearance to the schooner sail. Initially it led from the foremast head through blocks at the fore side of the mainmast head to be set up to a tackle on deck (this arrangement could still be found during the later part of the century in Portuguese schooner-rigged vessels). This lead was reversed in the first few years of the eighteenth century, and the stay was laid round the mainmast head with an eye, rove through a block at the aft of the foremast head to be set up similarly upon deck. This direct link by a stay between the two mastheads meant that a stronger fore stay was necessary, since it effectively had to support both masts.

In some American schooners a variation from the usual schooner stay developed during the 1770s. The stay in these vessels was similar to the mizzen stay for ketches, galliots and galeases as advocated by Chapman; effectively a shroud displaced forward, the stay was set up to tackles at both sides of the vessel, just abaft the foremast backstays. This idea was pursued further after 1790, especially in American schooners, and the main stay was led past the foremast on both sides and made fast to eyebolts in the breast hook; the same practice began to appear in European schooners after 1810. The lee stay was frequently slackened off, so as neither to chafe nor hinder the movement of the schooner sail.

Releasing the pressure of the schooner stay from the foremast made it possible to rig a lighter fore stay and to move it, like that of a ship, forward to the middle of the bowsprit. The fore staysail could thus be enlarged.

Lugger

Lugger, Logger; Lougre, chasse-marée

'A lugger is a small vessel with two masts' was the general consensus among contemporary writers, though many luggers had a third mast stepped right at the stern with a jigger bumkin protruding aft. The bowsprit (Röding called it an outrigger) was nearly horizontal and could be taken in.

Other vessels with a lugger rig were single-masted (like the Ewer, a fishing vessel from the Elbe estuary), or two-masted, of which the Breton chasse-marée is a good example. During the eighteenth century the latter had a round stern, was decked or open, lean built and not longer than 45 feet; it was thus considerably smaller than two- or three-masted luggers, which had a length of up to 75 feet. It was a swift vessel, as the name (translated) 'tide (or fish) chaser' indicates. The eighteenth century chasse-marée had only one lugsail to each mast. The foremast stood close to the stem and the mainmast, stepped in the centre of the keel, was raked very sharply aft.

The larger and mostly three-masted lugger was built with a flat stern and decked. The larger sail space also allowed a greater number of sails. Naval luggers sometimes carried topgallant sails in addition to their normal lug and lug topsails; these were set flying to a temporary spar. While a chasse-marèe had no bowsprit, luggers were fitted with one and set between one and three head sails. Falconer's *Dictionary* of 1815 noted 'some luggers are mere open boats, and have only a small mast and a ring-sail set to it over the stern, and the foot spread by a small boom'.

Lugsails were quadrilateral and bent to yards which were slung at one third of their length, rather than at the centre like

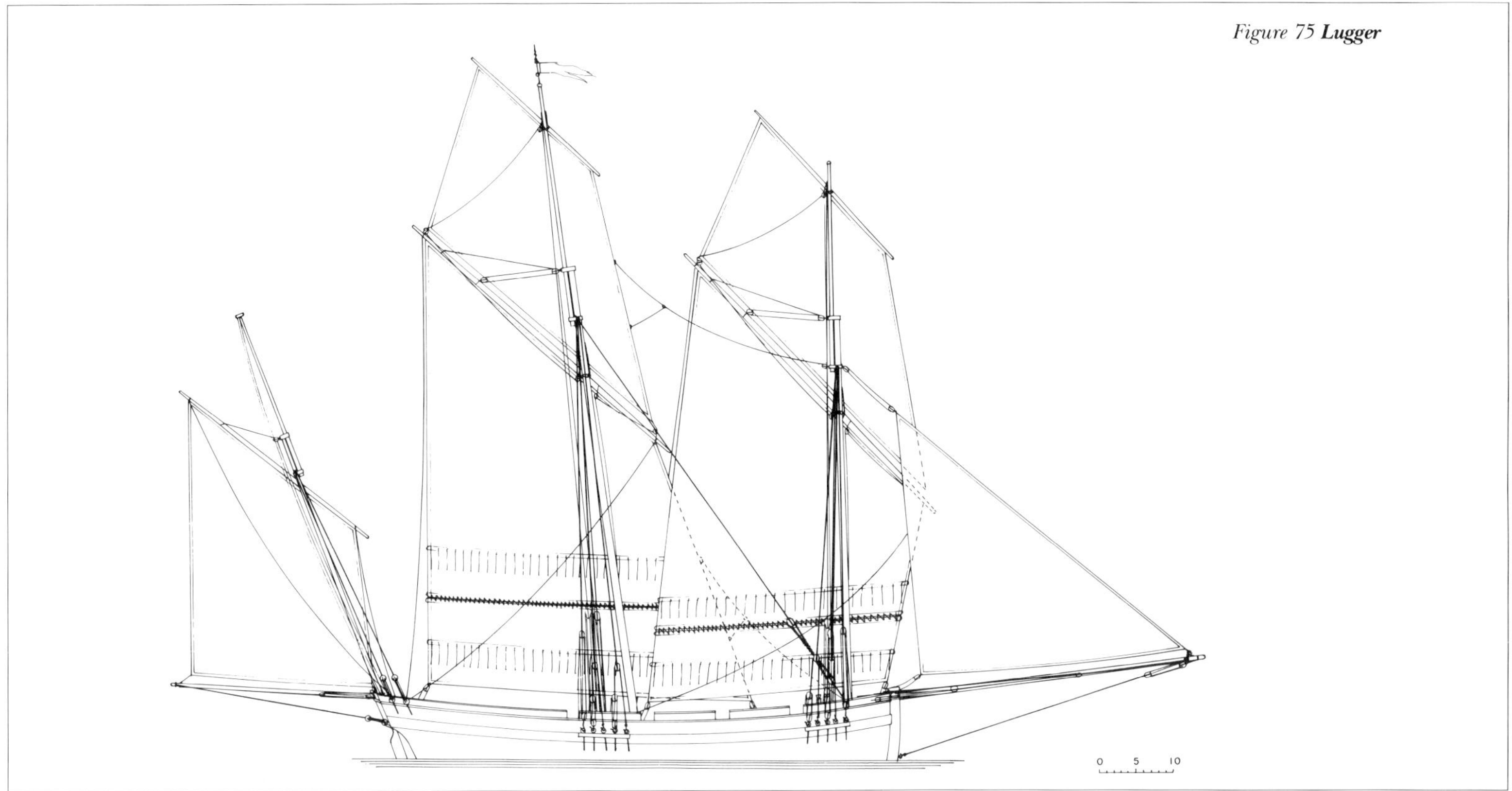

Figure 75 **Lugger**

squaresail yards. They had bonnets or reef-points, or both. The rigging of lugsails was very simple, as Steel explained:

> The yards have haliards, lifts and braces. To the lee-clue of the sail is a sheet, and to the windward-clue a tack, which is occasionally shifted as the vessel goes about. When this is often repeated, they loose ground in stays.

Each mast had a topmast fixed with iron bands abaft the mast. As noted above, temporary topgallant yards were sometimes lashed to these topmasts to accommodate topgallant lugsails. The sail on the third mast was called a ringsail, and its sheet was led to a jigger bumkin, or outrigger, over the stern. Shrouds in these vessels were set up with tackles.

In certain situations, to keep the lower lee leech of a lugsail steady, a loose spreader was employed; this device had its origin in a boom used on the squaresails of Viking ships. It was a forked boom set into a reef cringle at the lee-leech. The inner end stood on deck in a specially formed shoe, or was lashed either to the mast or to an eyebolt at the side. These spreaders were employed in smaller vessels generally and were not a particular part of the lugger rig.

While the lugsail is one of the oldest types of sail, the lugger as a vessel type was a development of the eighteenth century. Blanckley, Chapman and Falconer made no mention of the lugger; Ozanne's drawings included a number of impressions, and later writers like Steel, Röding and Falconer (1815) provided reasonable descriptions. Luggers were fast sailing vessels, sailed well close hauled and very near the wind. Mainly used around the English Channel, the lugger is usually regarded as originating on the English coast as a fishing or transport vessel; the chasse-marèe, by contrast, belonged to the Breton coast.

Röding called the lugger 'a fast sailing vessel fitted out for war, especially used by the English', but Falconer's 1815 edition noted that 'the French employ luggers as privateers but there are not many vessels of this description in our service'. Luggers came into French service in 1773; the first, *l'Espiegle*, was built to lines taken off an English lugger. Four more were named in the French Navy list of 1780, and as noted above, Ozanne provided very good pictures of these vessels. A lugger was first mentioned in the Royal Navy list of 1783, and lists from that date to 1801 include no more than one vessel. Falconer's statement on luggers in naval service was therefore probably nearer to the point than Röding's.

Dogger

Fischhuker; Dogre

The dogger's main and mizzen mast were very reminiscent of a ketch or howker. The mainmast and topmast were scarfed and bolted together in such a way that they appeared to be one mast. The short mizzen mast sometimes had a topmast extension with trestle trees and a cap, and larger vessels may also have carried a topsail.

Originally used on fishing vessels only, the dogger rig could also be found from the late seventeenth century onwards on Dutch merchant howkers. Some had the long bowsprit of a true dogger, while others had the bowsprit stepped forward of the windlass. The smaller fishing vessel usually had a mainsail and a topsail, while the larger merchant type sometimes also carried a topgallant sail. The latter carried a mizzen yard and later a gaff, while the fishing vessel originally had a squaresail rigged to the mizzen mast, and later a gaff sail.

The mainmast shrouds were set up very far aft to give the main yard a greater bracing circle, and the main yard jeers were set up to a ramshead and knighthead abaft the main hatch. In addition to the sails noted above, two to three head sails were rigged. In his description of the rig Röding noted that the mainsail had to be lowered upon deck to be reefed. The bowsprit (or as it was sometimes named, the outrigger) was loose and could be taken in; a jib-boom was sometimes lashed to it.

The dogger was basically a fishing vessel of the Continental North Sea coast and a parallel development to the howker and ketch from the sixteenth century hoekers and catches; for more details see Ketch or howker above.

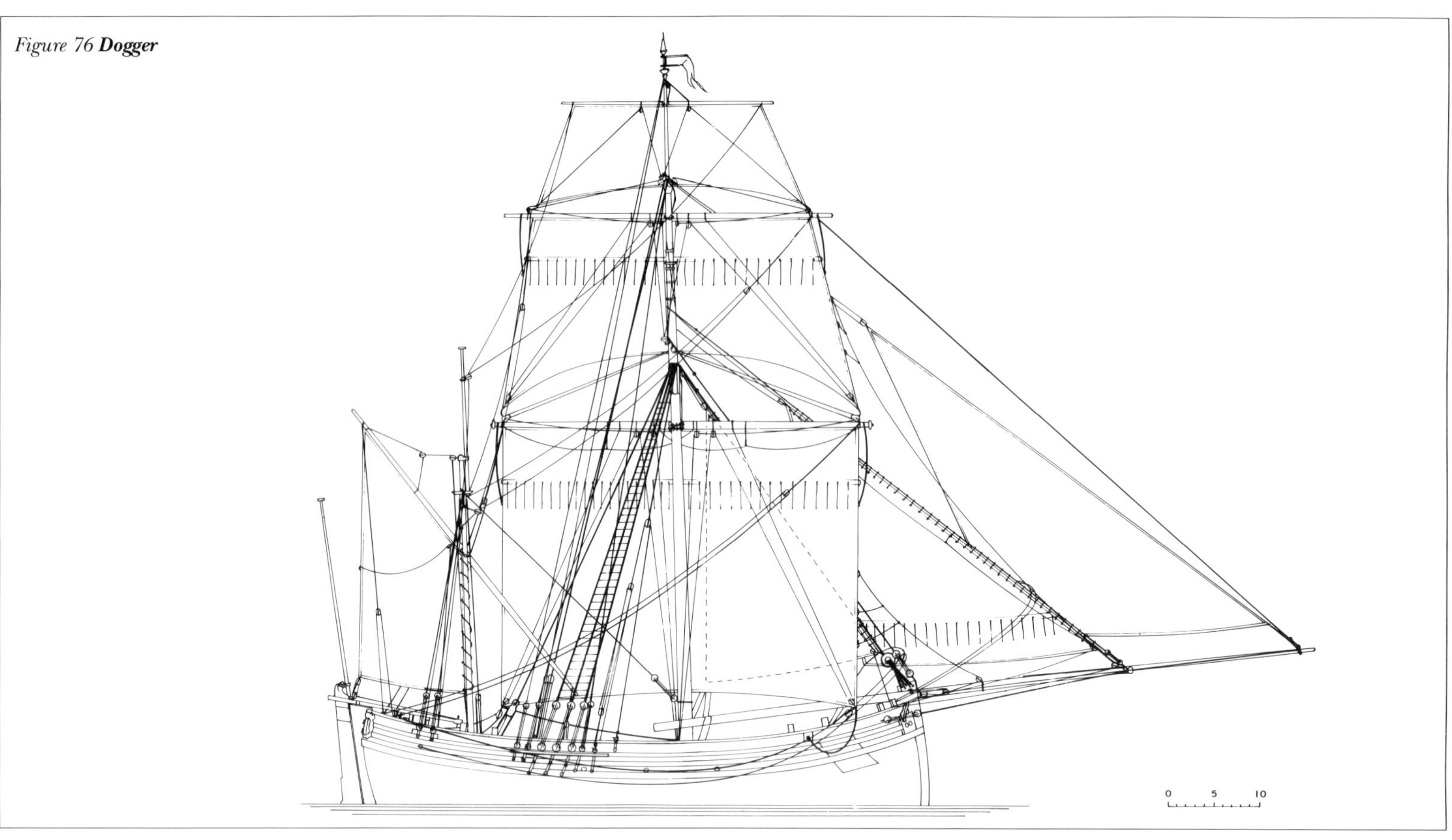

Figure 76 **Dogger**

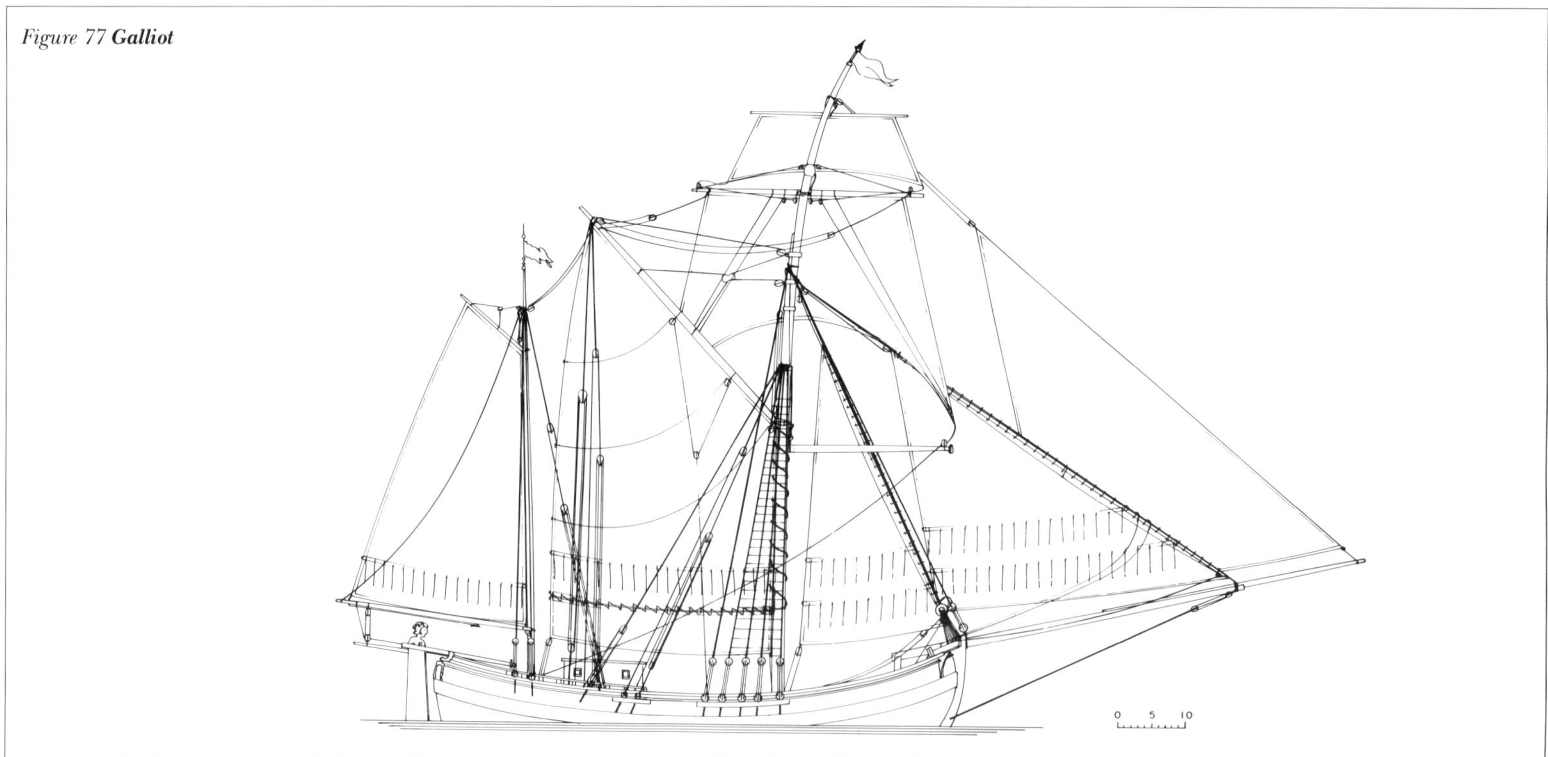

Figure 77 **Galliot**

Galliot

Galliot; Galliote hollandaise

The galliot, like the dogger, was a vessel of the Continental North Sea coast, and showed considerable Dutch influence. The origin of the rig can be traced to the *oorloge convoyers* and later Watt convoyers of the sixteenth and seventeenth centuries. In the second half of the seventeenth century its use had already spread to the Baltic Sea: the earliest mentioned galliot in a Swedish fleet list was the *Fortuna* of eight guns, bought in 1657 from Holland. Röding described the galliot as a vessel employed in Denmark and Sweden.

The galliot was usually a 1½-masted fore-and-aft rigged vessel, but one and three-masted galliots were also in existence; these were, however, galliots by construction but not by rig (for instance, a typical combination was a howker-galliot). The following notes relate to the basic galliot rig.

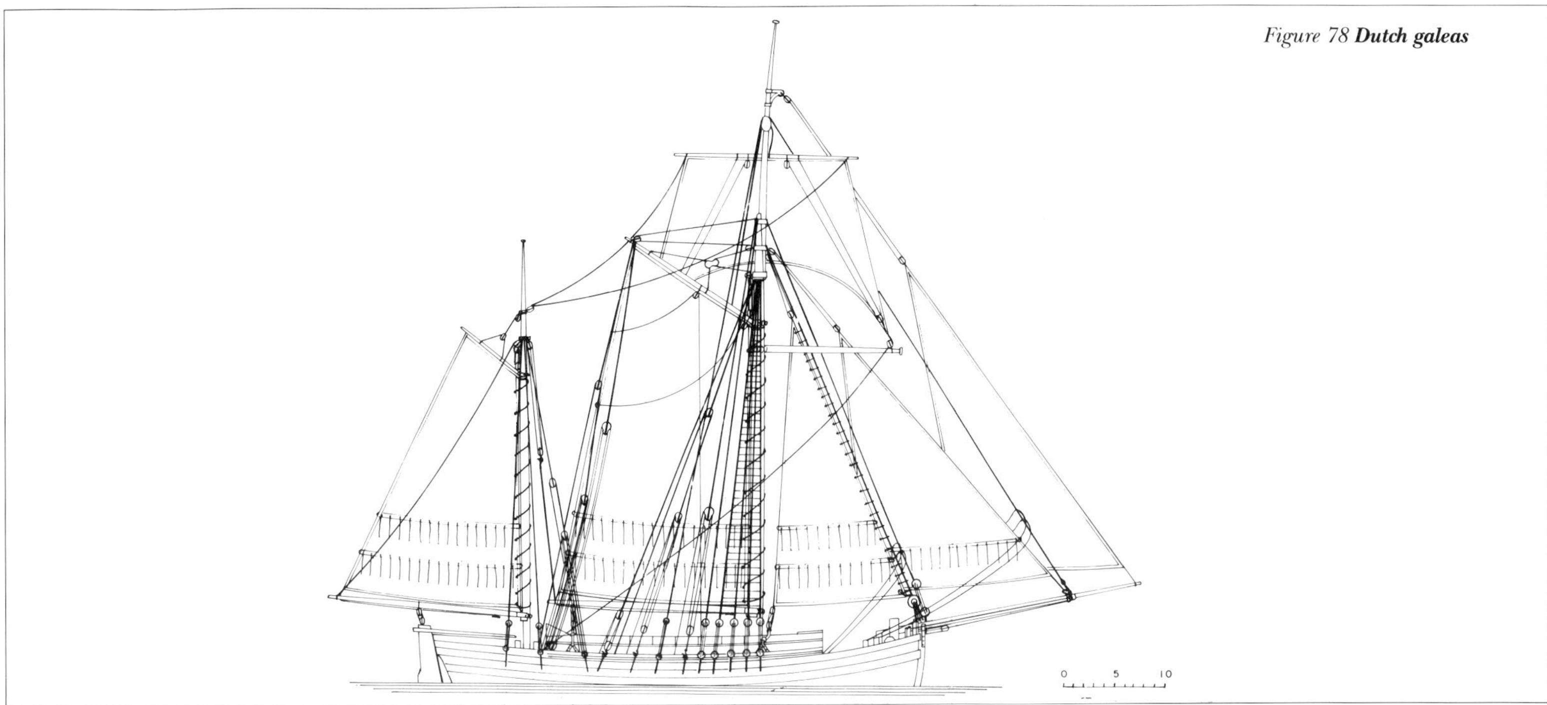

Figure 78 **Dutch galeas**

The main feature of the galliot rig was a mainmast with a scarfed topmast, the upper part of which was often curved forward in Dutch fashion. Besides a large sprit or gaff sail, the mainmast carried a topsail (set flying, according to Falconer) and sometimes a small topgallant sail. The mast had four or five pairs of shrouds, and the stays ran to the stemhead and the bowsprit peak.

A jib-boom was often lashed to the bowsprit, and three head sails were rigged. The fore stay sail and the inner jib were seized to hanks on the stays, while the outer jib was set flying on the jib-boom.

The smaller mizzen mast had a gaff and boom rigged sail of smaller proportions. For control of the boom sheet the round sterned galliot had an outrigger extended over the stern. The mizzen mast stay was more like a pendant, very short and with a single block spliced into its end with a runner reeving through it. This runner was made fast with a thimble and hook to the port side of the vessel, and the other end was set up to a long tackle hooked to an eyebolt in the waterway abreast of the mainmast mast tackle and backstay on the starboard side.

Dutch galeas

Galeasse; Galeasse hollandaise

'A small vessel employed by the Danes, Swedes, Hamburgers and Dutch which has a mainmast and a small mizzen mast fitted. It is similarly shaped to a single-masted galliot...' began Röding's short chapter on the galeas, a vessel which developed around the middle of the eighteenth century in the Baltic Sea. It was a flat sterned, broad-beamed craft well suited as a freighter.

The rig had much in common with a that of a galliot. Differences included the length of the bowsprit and jib-boom and the fore stay set up. Where both stays and the preventer stay led from the masthead (scarfed section) in a galliot, the inner jib stay became the galeas' topmast stay and was made fast above the topsail yard. Only the fore stay sail was seized to hanks; the two jibs were set flying.

The main topmast polehead was not as distinctly curved forward as was that of the galliot. Usually only one topsail was set, and the mainsail was set not simply to a gaff as on the galliot, but to a gaff and boom. The gaff on a galeas was therefore hoisting, in contrast to the standing gaff of the galliot. Another characteristic feature of the galeas was the main boom topping lift, which led to the mizzen masthead. In contrast to the galliot's loose-footed mainsail with a bonnet, the galeas' mainsail had reef points and was of greater height than a normal gaff sail.

The mizzen gaff sail's foot was also spread by a boom, but an outrigger was not needed. The boom sheet tackle was set up to the taffrail of a flat stern.

Ketch-yacht

Hukerjacht; Hourque yacht

Yachts came in the seventeenth century from Holland to Denmark and Sweden, and the original lines were soon transformed into those of a vessel suitable for freight. In the middle of the eighteenth century the yacht had already become one of the most common fast sailing merchant vessels in the Baltic. A distinction can again be made between 1½ and single-masted yachts. As with other types, 1½-masted yachts were classified according to the type of rig they used: ketch-yacht, yacht-galliot and yacht-galeas, for example.

The ketch-yacht, very much like a small ketch, was fitted with a long bowsprit and jib-boom, a mainmast with a topmast and topgallant mast and a mizzen mast with a short topmast.

The head and mainsails were ketch-rigged, and the wing sail had a boom with a topping lift set up like that of a galeas. The single sail on the mizzen mast was gaff-rigged and its foot was also spread by a boom. The wingsail and the spanker boom represented the main points of distinction between a ketch-yacht and a ketch.

Illustrations in H Szymanski's *Deutsche Segelschiffe* indicate that the rigging characteristics of 1½ masters very often blend into each other. The galliot *Emanuel* of 1770, for instance, had a lower yard sail which does not look like a temporary squaresail, the galeas *Die Lerche* of 1800 was rigged like a ketch-yacht but had a galeas mast and a very high wing sail, cut like a galeas mainsail, and the howker-galliot *Themis* of 1806 again had a ketch-yacht rig, but with a temporary squaresail on a long

Figure 79 **Ketch-yacht**

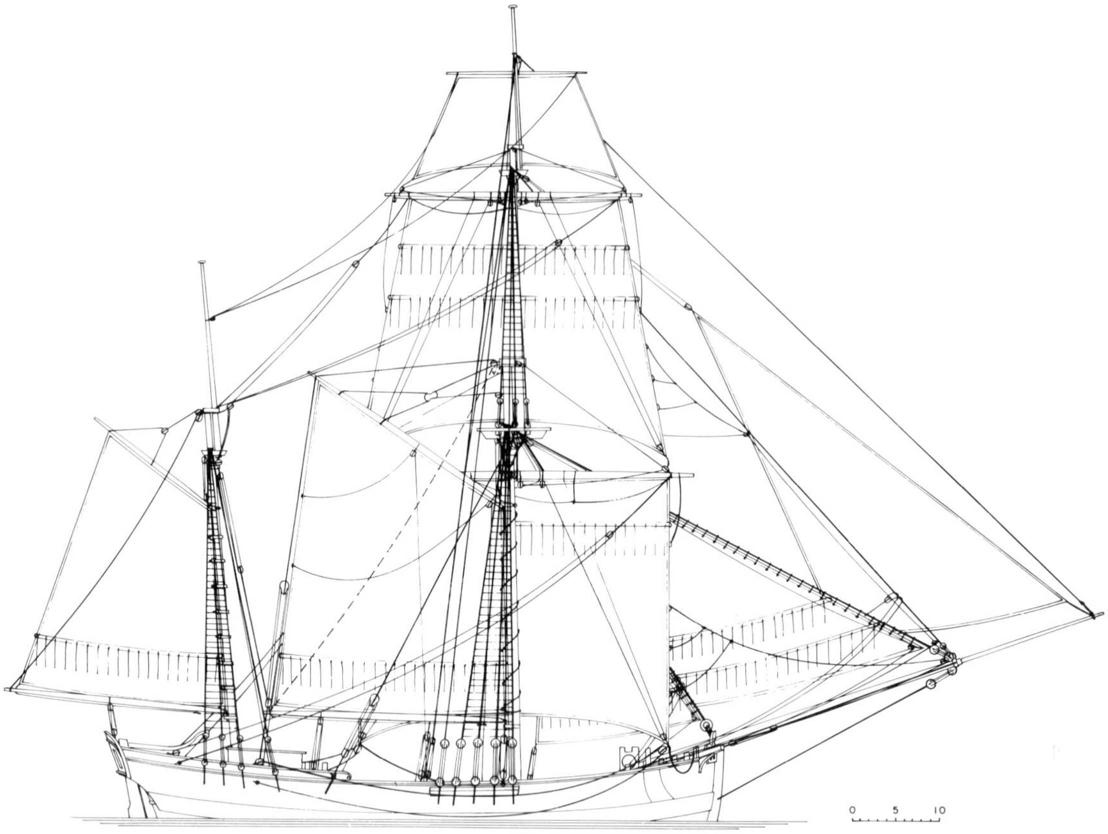

Figure 80 **Yacht**

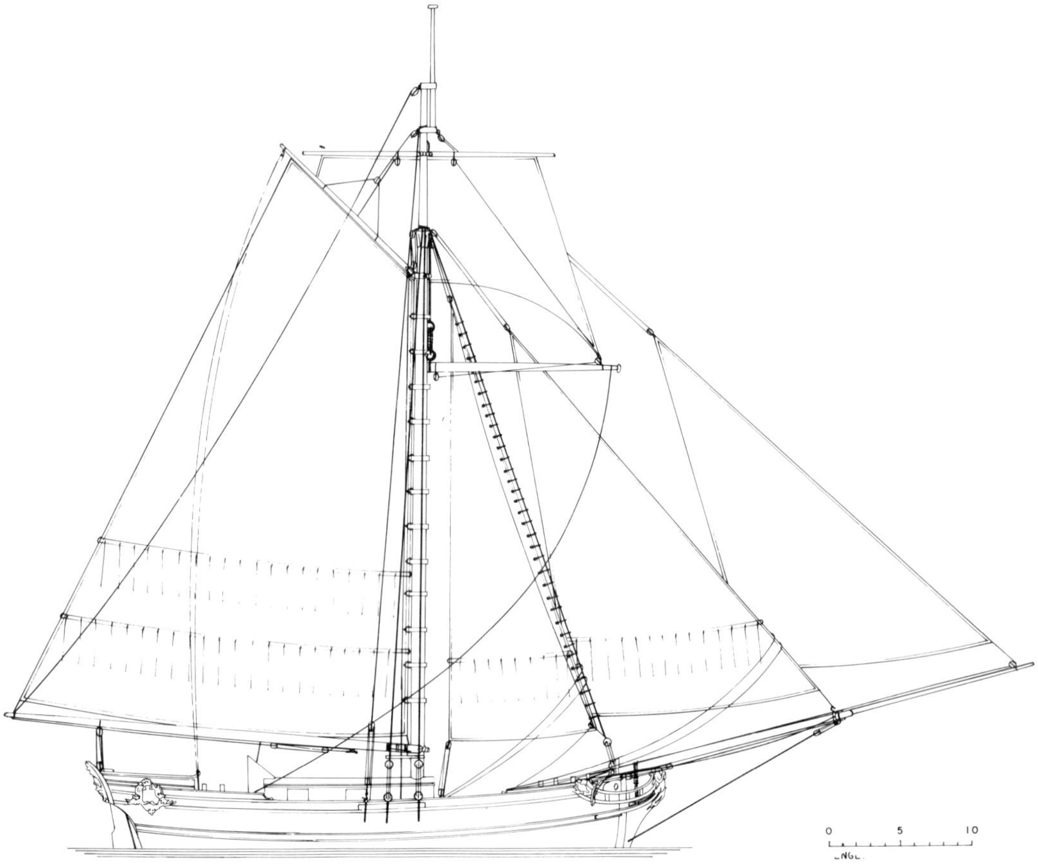

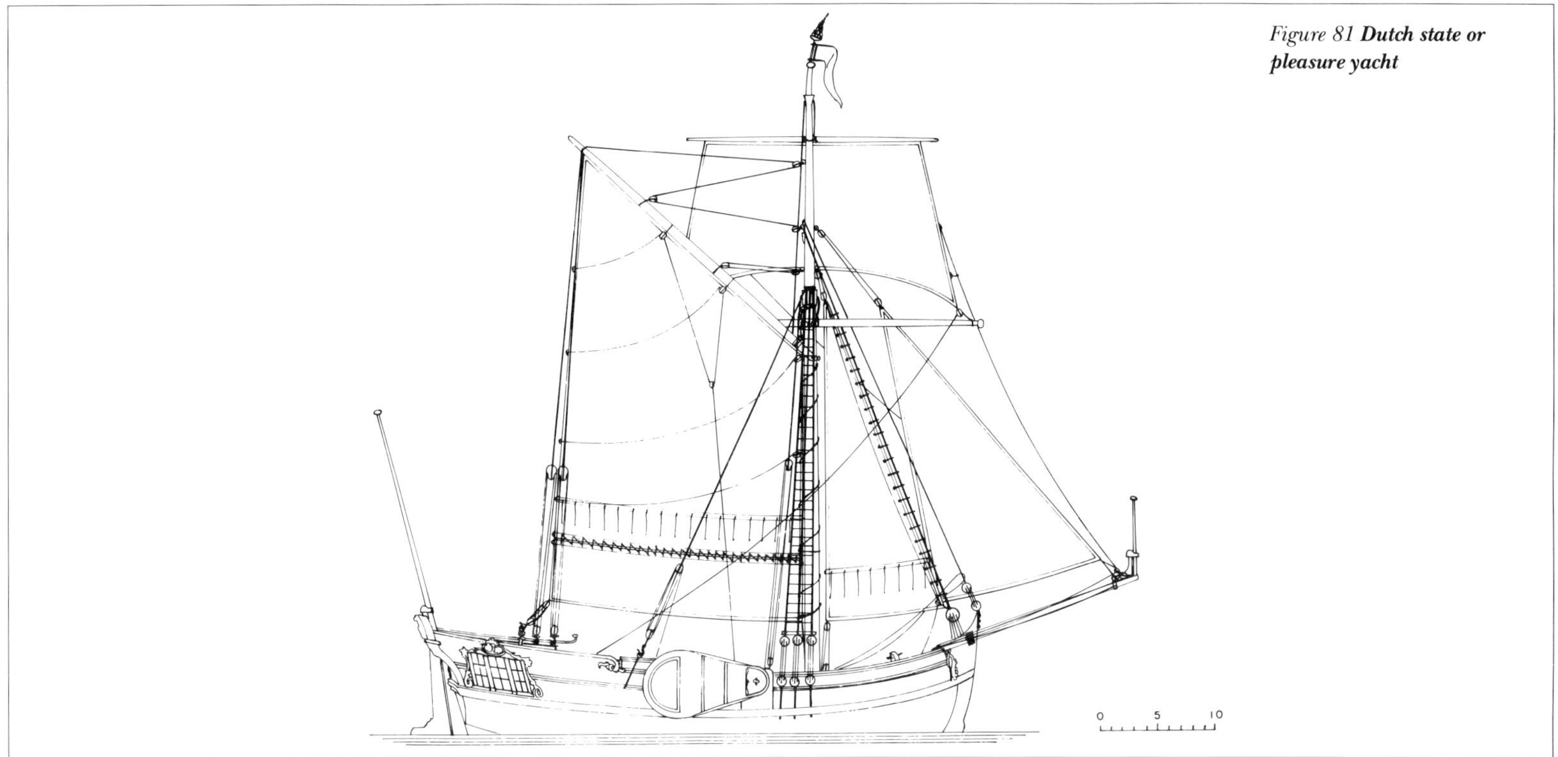

Figure 81 ***Dutch state or pleasure yacht***

squaresail yard, and her tall mizzen mast carried a higher then usual gaff sail with only a short boom, with no need of an outrigger.

Yacht

Jacht; Yacht

A single-masted yacht had a very long mast without an extra topmast. The required topmast length was an integral part of the mast, marked only by the hounds of the mast and the topmast.

Szymanski noted above the topmast hounds a short, forward curved flagpole extension, a *Trommelstock* (drumstick), sitting in a sheet metal sleeve. Chapman showed this short flagstaff as straight, fastened to the fore side of the masthead with figure-of-eight iron bands. Such minor variations indicate that regional peculiarities or the preferences of the owner always played a great part in determining the individual appearance of each vessel.

An average sized yacht mast had two shroud pairs. The gaff peak halyard was set up with a tackle to the channel abaft the shrouds on the starboard side, and the boom topping lift on port side. The stay was made fast beside the stemhead to a ringbolt on deck. A staysail was set on the stay with hanks.

The bowsprit was often extended by a jib-boom, and two jibs were set, both flying. The tack of the inner jib was hooked to a traveller, and the tack of the outer rove through a block at the jib-boom peak.

The sails set on the mast consisted of a large gaff and boom rigged mainsail and a small topsail. As with all other fore-and-aft rigged vessels, a temporary squaresail was also hoisted when necessary. The ratio of gaff length to boom length, 0.45, was somewhat smaller, shorter than in other gaff-rigged vessels. Chapman's pleasure yacht rig drawing showed a gaff and a boom lacing, and wooden hoops for the mast. This boom lacing is unsupported by other contemporary evidence, in which lacing was only mentioned in connection with head and mast, but not with the boom. Boom lacing is usually found on racing yachts of the later nineteenth century, and it may be that Chapman's illustration both provides us with a very early example of boom lacing and also underlines his genius as a ship designer.

As noted above, yachts were of Dutch origin and came to Denmark and Sweden as pleasure craft. The first Swedish royal yacht, *Lejonet*, was built in 1657 (her place of construction is unknown), but a second yacht of the same name was built for Karl XI in 1669 in Amsterdam. Another yacht, the *Mary*, presented in 1660 to the exiled King Charles II on his return to England by the burghers of Amsterdam, introduced the type to the British Isles. Sir Anthony Deane took the craft as a pattern for several yachts he constructed in the following years for his king, and he also built two yachts in 1674-5 for Louis XIV, King of France. Thought of in the beginning as pleasure and transport craft only, these vessels found their way back into the merchant service as well as serving very successfully as fleet reconnaissance vessels.

Röding noted the frequent use of yachts in England and America, but his definitions of yacht and sloop were identical (see Sloop below). He also mentioned a topgallant sail, and commented that the yacht was not only a good sailer close hauled, but also had excellent tacking abilities; he added that yachts depended on a squaresail when sailing before the wind. 'Yachts are also employed, because they are fast sailers, to send news, packets and letters, and these are called despatch yachts or packet yachts'.

Dutch state or pleasure yacht

Holländische Herrenjacht, Staatenjacht; Yacht d'amirauté hollandaise

Derived from Dutch small shallow water trading craft, the state yacht was a direct descendant of the fighting vessels of the Sea Beggars, combined with the desire of the ruling class in the newly founded United Netherlands for a suitable transport and representative vessel. As merchant craft, often armed with guns, they had already proved their worth in the long Dutch independence

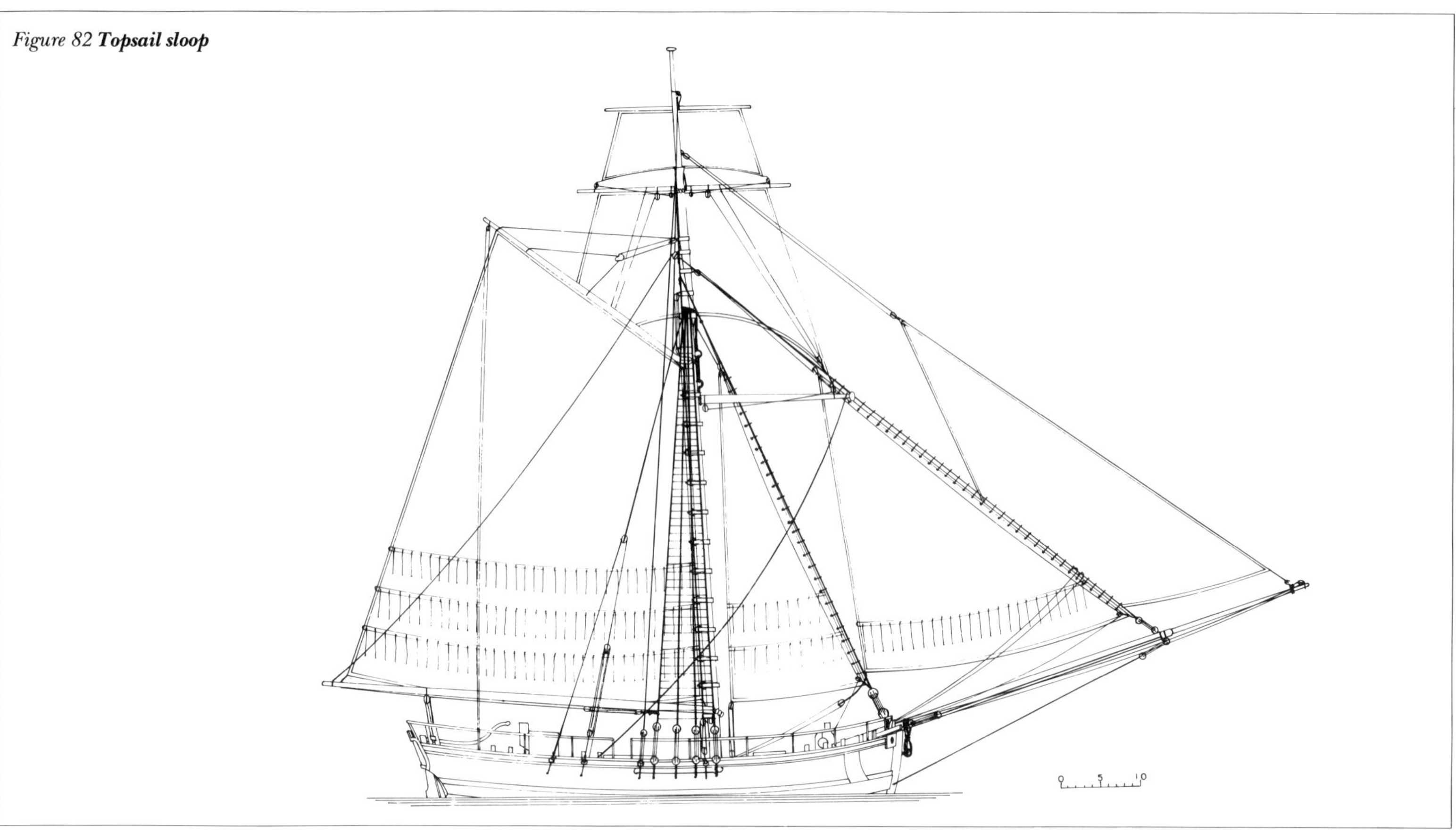

Figure 82 ***Topsail sloop***

struggle under Prince William of Orange against Spain.

Originally the term 'jacht' was not a synonym for pleasure craft, but was applied to a small and swift, usually three-masted, merchant vessel. An early departure from this was Prince Maurice's yacht *Neptunus* of 1595, known as an *oorlogsjacht* (war, or armed yacht). She had two masts stepped, with the foremast sprit and the mizzen mast lateen-rigged. The rig was therefore not like a that of a ketch, with a main and mizzen mast, but rather similar to that of the then newly developed *oorlog convoyers* or *cromsters*; the rig could be traced back to 1564 (see Smack in Chapter VI).

Another yacht owned by the Prince and built in 1615 had a single-masted sprit rig. Serving as naval command vessels for the ruling Prince, these vessels set the trend towards the seventeenth and eighteenth century Dutch State and pleasure yachts, which in turn influenced greatly the later yachts described above.

Of modest depth and flat bottomed, these yachts were fitted with lee-boards and had a long, half-height deck superstructure for the state rooms, the so-called pavilion. A number of smaller vessels of various hull shapes were developed during the eighteenth century in parallel to these larger yachts, derived from the one and two-masted speeljachts of the early seventeenth century. They were bezan rigged, a gaff and boom rig, and until the beginning of the eighteenth century were mostly without a bowsprit.

State yachts, also known as transom yachts, were rigged with a spritsail until about 1660, and after that with a standing gaff (halfsprit) rig. Developed parallel to the bezan rig, the halfsprit rig was nevertheless completely unrelated to it; its origin lay some 35 years earlier in the replacement of the sprit of a boeier rig by a halfsprit, and it retained all the other features of a sprit rig except for the long and cumbersome sprit. The term 'half-sprit', given originally to a long standing gaff, was still used in eighteenth century literature such as Blanckley's *Naval Expositor* of 1750: 'one Mast with an half Spreet...'

In addition to reef points, the mainsail, like most loose footed gaff sails, sometimes also had a bonnet. The mast was fitted with a stay and preventer stay. Turned into the stay's lower end was a large five-hole deadeye, set up with a lanyard directly to an equal number of holes in the stemhead. The preventer stay, with an ordinary deadeye turned in, was set up to an iron-bound deadeye bolted to the face of the stem. A fore stay sail was seized to hanks and a jib was set flying to a bowsprit which usually carried a jackstaff. After the gaff's introduction to the transom yacht, a square topsail, another important feature of the boeier rig, also became part of the rig of a state yacht.

Sloop

Schlup, Schaluppe; Chaloupe, sloop

In merchant shipping the name 'sloop' was not clearly defined. Röding noted that in German waters a sloop might also be called a yacht, while both sloop and yacht-rigged vessels of war were frequently referred to as cutters in England. Falconer, examining the English terminology more closely, specified the difference between cutter and sloop as the type of bowsprit; the former's was running and the latter's standing. In other words, the cutter's bowsprit rested beside the stem and could be retracted inboard when necessary, while the sloop's was set permanently above the stem; the sloop's bowsprit was also rigged with a jibstay. A sloop's sails were generally not as large as a cutter's in proportion to the vessel. There were also other differences between the two types, mainly in hull shape and masting.

Szymanski noted differences in mast construction between yachts and sloops, with the latter, in contrast to the former, having a topmast stepped, This observation does not reflect a general rule; the topmast of a sloop might equally be scarfed and hooped to the masthead. A sloop's gaff was of normal length, again in contrast to that of a yacht.

As with schooners, two types of sloop existed, the topsail and the fore-and-aft rigged sloop.

How imprecise was the eighteenth century term 'sloop' can be seen from Blanckley's comment that 'sloops are sailed and masted as Men's Fancies lead them, sometimes with one Mast, with two, and with three, with Bermudoes, Shoulder of Mutton, Square, Lugg, and Smack Sails; they are in Figure either square or round Stern'd', and from both editions of Falconer. Falconer's 1769 sloop illustration is very similar to his illustration of a yacht in 1815, with a square topsail and a spritsail beneath the bowsprit. His 1815 sloop was fore-and-aft rigged only, and was also a pleasure boat or yacht. Blanckley's illustration was of a fore-and-aft rigged sloop. Furthermore, the term 'sloop of war' has no connection with the rigs described here. These were naval vessels below the Sixth Rate, carrying between eighteen and six guns and with either a ship or snow rig. They were better known in Continental navies as corvettes.

Sloops in the modern sense with a gaff and boom rig and a bowsprit were a development of the first decades in the eighteenth century. The Prince of Orange's small bezan yacht, which appears in L P Verschuir's (1630-1686) painting of Charles II arriving at Rotterdam on the way to his restoration to the British throne in 1660, was a clear forerunner of this type; it differed from other bezan yachts by carrying a bowsprit. The bezan rig became better known in England as a Bermudoes rig, and it consisted of a sharply raking mast, a short gaff, a long boom and two head sails. Early pictorial evidence of these vessels is W Burgis' *View of New York* of 1717, and the *Ferret* class of sloops in the Royal Navy from 1710-11 were probably the first naval sloops rigged in this fashion.

Topsail sloop

Rahschlup; Chaloupe de hunier, sloop

The topsail sloop was developed slightly later then the fore-and-aft version, and can be observed in Burgis' 1725 *View of Boston.* 'Topsail sloop' is a term used here only to distinguish between the two differing rigs; the topsail sloop's rig was essentially that of a yacht. In addition to the sails shown in Figure 82, the topsail sloop might also carry a squaresail when sailing before the wind. One of the differentiating points between sloop and yacht rig was that the head of a sloop's mainsail was wider than that of a yacht. Falconer's 1769 sloop and 1815 yacht also had a spritsail yard rigged to the bowsprit, and their crossjack yard braces led forward (in 1769 to the middle of the jib stay and then down to the bowsprit peak, and in 1815 without the stay block direct to the bowsprit peak; both versions were double). Where the braces led via the jib stay, jibs could not be hanked and had to be set flying.

The use of squaresails in general was firstly mentioned by Falconer in 1769. 'in sloops and schooners, and other small vessels, the sail employed for this purpose [scudding] is called the square-sail, (*voile de fortune* in French)'. For the rigging of a squaresail see Topsail schooner above.

Fore-and-aft rigged sloop

Gaffelschlup; Bateau bermudien

The fore-and-aft rigged sloop, as shown by Falconer (1815 edition) and in several late eighteenth century English marine paintings, did not carry any square rigged sails, except for the temporary squaresail, but it had a gaff topsail set above the mainsail; the rig of pleasure yachts was often similar. These sloops did not have the extreme rake of the mainmast which characterized sloop rig early in the century and was retained later in the Bermuda sloop.

Fore-and-aft sloops did not always carry a jib-boom; where none was fitted, a fore staysail only was seized to hanks on the stay and an additional jib was set flying. A sloop mainsail was generally fitted with three or four reef bands, with the uppermost halfway up the fore leech. 'Sometimes the reefs are fitted without bands' noted Steel, and '[the mainsail] also frequently has a balance-reef from the nock to the upper reef-cringle'. Reef bands were regularly left without points. Falconer explained as follows:

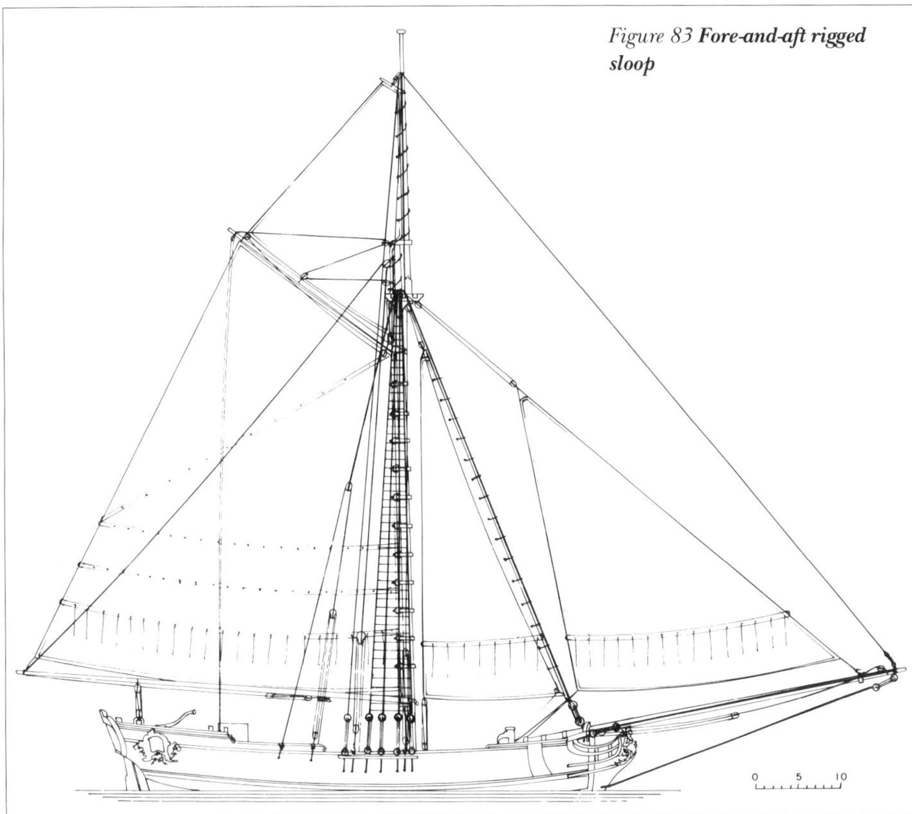

Figure 83 **Fore-and-aft rigged sloop**

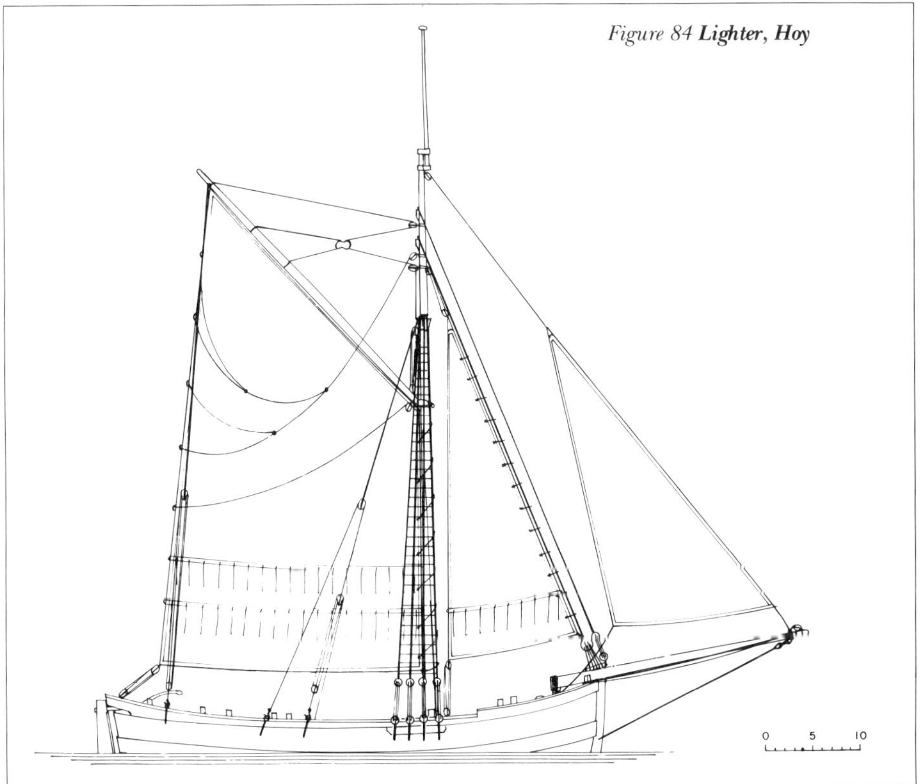

Figure 84 **Lighter, Hoy**

> When a sail is reefed at the bottom, it is done by knittles, which being thrust through the eyelet-holes thereof, are tied firmly about the space of canvas of which the reef is com-

posed, and knotted on the lower side of the bolt-rope. These knittles are accordingly removed as soon as the reef is let out.

Lighter, hoy

Leichter; Bateau lesteur, allège

Lighters, barges, hoys and similar small craft usually had a light gaff or sloop rig. Some transport vessels, however, were still sprit rigged, such as the Dutch *kaag* and others described below in Chapter VI.

Norwegian jekta (yacht)

Norwegische Jacht; Bateau marchand norvège

The Norwegian yacht, or *jekta* as the type was called in Norway itself, was an open transport vessel with a decked-in bow and a small cabin at the stern. Jektas varied in size and could carry up to 60 lasts. In contrast to the yachts of the North and Baltic Sea, jektas were square rather than gaff-rigged. They usually carried one squaresail, and Röding's Danish index included the following note under *Top-jaegt*: 'a Norwegian yacht, which also carried a topsail above the mainsail; they are no longer in use'.

Such a topsail, not shown in Röding's illustration of a Norwegian yacht, can be seen in a drawing of a *Nordfjordjekti* of 1881 in Færoyvik's *Inshore Craft of Norway*; the author noted that older jekts did not carry topsails. Except for the matter of the topsail, the two illustrations, 150 years apart, have enough in common to allow us to identify Röding's Norwegian yacht as a Nordfjord jekta.

A slight variation in the bow had also occurred between the two illustrations, with the older vessel having two extra planks fitted as a washboard to keep the room below the bow decking drier (it was here that perishable goods were stored). A washboard was, according to Falconer:

> ... a broad thin plank fixed occasionally on the top of a boat's side, so as to continue the height thereof, and be removed at pleasure. It is used to prevent the sea from breaking into the vessel, particularly when the surface is rough, as in tempestuous weather.

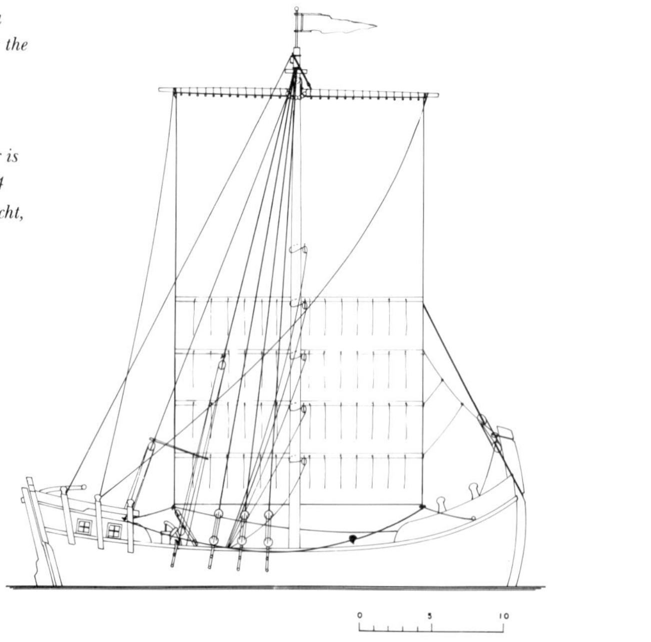

Figure 85 ***Norwegian jekta***
This jekta rigging plan follows the lines of an 1881 Norwegian drawing, but incorporates the different features from a 1794 illustration by Röding. The rig is basically that shown in a 1794 illustration of a Norwegian yacht, without a topsail.

The rear quarter windows and ornamental eighteenth century stern in the Röding illustration were also replaced in that by Færoyvik with a more austere taffrail and windows in the stern.

Four reef bands shown by Röding, in contrast to the three bonnets shown by Færoyvik, represent the only rigging differences. The latter author also provided a clear description of the midship superstructure (actually the cargo) shown in the Röding illustration; this structure consisted of planks fitted inside light vertical beams and secured within board to keep a high deck load of hay or similar together and protect it from getting wet.

The sail plan produced in Figure 85 is based on Röding's illustration, taking into account the 1881 lines drawing.

Cutter

Kutter; Cotre

Cutters evolved during the second quarter of the eighteenth century in Southeast England as swift Channel craft. They were soon employed by smugglers as well as in the British Customs Service. First noted in Royal Navy lists in 1762 and in French in 1771, cutters were relatively large single-masted vessels of up to 200 tons burthen and an armament of six to eighteen guns. The French Navy list of 1780 had six entries for cutters, and a Royal Navy list from 1764 included thirty-eight cutters. In addition to these two major eighteenth century naval powers, the navies of Sweden and Holland also included a number of cutters.

Falconer described a cutter as 'a small vessel commonly navigated in the English Channel, furnished with one mast and a straight running bowsprit that can be run in on the deck occasionally, except which, and the largeness of the sails, they are rigged much like sloops'.

Röding's description of a cutter confirmed Falconer's:

> [A cutter is] a vessel, carrying nearly the same rigging as a yacht or sloop, except that the mast rakes slightly aft and has larger sails. The cutter's main or boom sail can additionally be enlarged by a ringtail sail. Cutters differ in shape from yachts mainly in being larger and leaner built, also in having a greater hold and less dead work; therefore they draw more water and are able to carry more sail, and thus to sail much closer hauled. Because of their extraordinary sailing speed they are often employed in England's illicit trade. The government therefore also maintains cutters as Revenue vessels to arrest these smugglers. The latter have a crew of some thirty men, six to eight guns and some swivel guns. Cutters intended for trade in England are not allowed by law to carry as high a rig as that on the Revenue vessels, so as to render the former unable to escape.
>
> Cutters serve as adviso yachts and packet boats and during wartime also as privateers. They are now also being built in France and especially in Dunkirk.

Steel described the rig of cutters and other small vessels in great detail. His notes, slightly altered and enlarged, form the basis of Chapter V below. He combined his description of the rigging of large naval cutters with that of sloops, 1½ masters and other small vessels, thus giving an insight into the rigging of smaller vessels as well. The rigging detailed in Chapter V is there-fore not applicable in its entirety to all vessels, particularly with regard to the matter of slung or hoisting gaffs and crossjack yards.

V

DETAILS OF SINGLE-MAST RIG

Mast

Mast; Mât

Girtline blocks

Jolltaublöcke; Poulies de cartahu

Girtline blocks on cutters were lashed to the masthead like those on ships, and were used to hoist the rigging up to the masthead. They were removed when rigging was completed.

Mast tackle

Masttakel; Caliorne de mât

The mast tackle pendants were wormed, parcelled and served over their whole length. Each was doubled, and the bight was seized to create an eye which fitted over the masthead. The ends were then spliced together, and a single block was seized in the lower bight. The ends of all splices were tapered, marled down and served over with spun-yarn.

The tackle runners had a hook and thimble spliced into one end and were served over. They rove through the pendant blocks and were spliced round the strops of long tackle blocks.

The tackle fall was bent to a becket at the lower end of a long stropped single block, with the ends seized. The long strops, with hooks and thimbles spliced in, were hooked to eyebolts in the sides.

Shrouds

Hoofdtaue; Haubans

Four pairs of shrouds, or on smaller craft sometimes fewer, were prepared and placed over the masthead like those in ships. The full length of the after shroud on both sides was protected against chafing by being wormed, parcelled and served with spun yarn.

Stay

Stag; Étai

The stay, like that in a ship, was wormed over its full length. It went over the masthead and had a deadeye set into the lower end with a running or Flemish eye. The stay was then hauled taught with a lanyard reeving through the deadeye's holes and holes drilled into the stemhead. Stays for sloops and similar vessels had a stay collar with a seized-in deadeye lashed to the bowsprit.

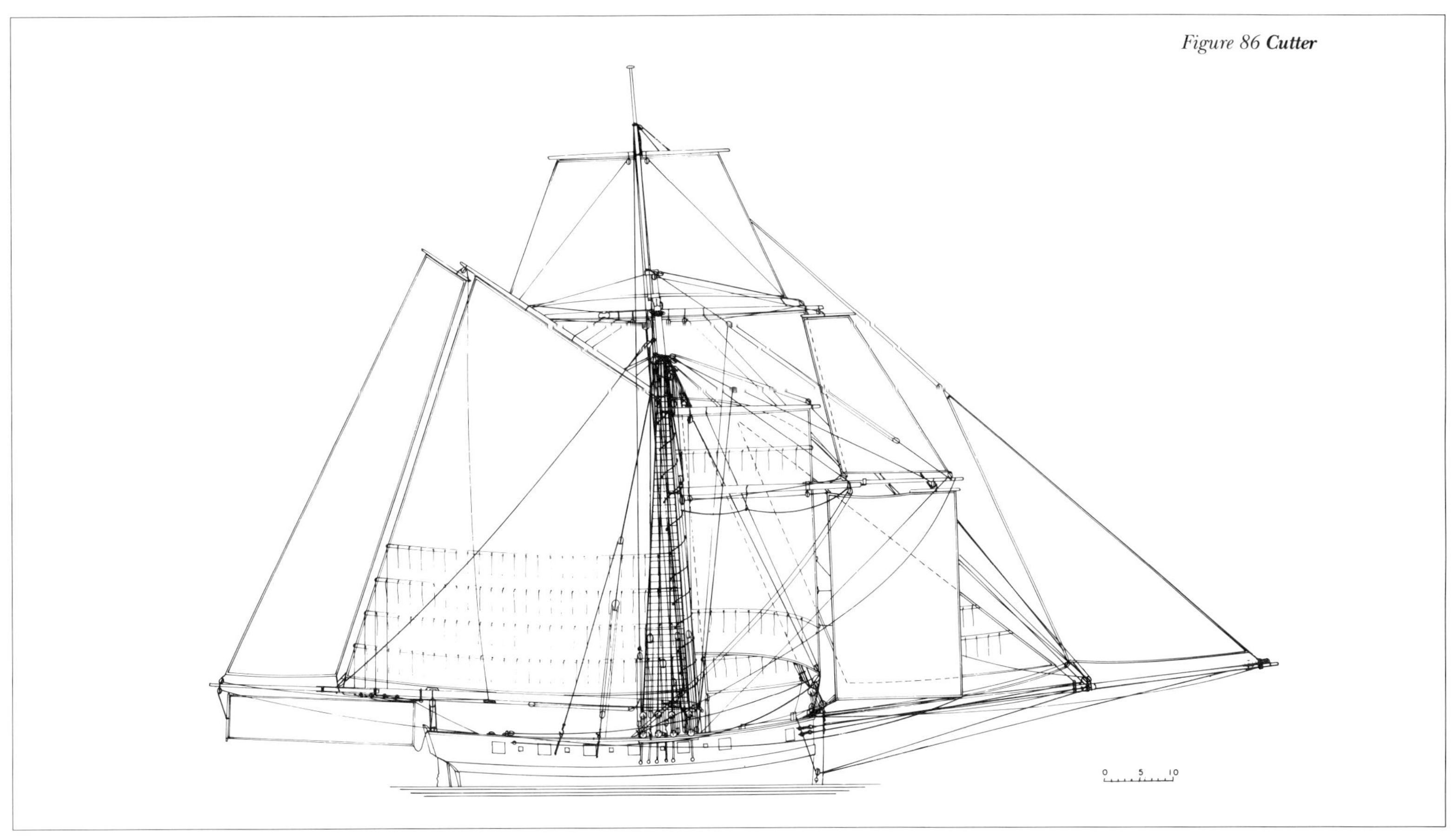

Figure 86 **Cutter**

Figure 87 ***Rigging details I***
(a) Two variations of square rig on a ship-rigged mast for small vessels. The left side of the drawing shows the squaresail fitted with a bonnet and hoisted with a short yard (club yard) below the spread yard; the right side shows an extended squaresail with a bib hoisted up to the trestle trees. Note also the differing bowlines, halyards, sheets, etc.

(b) A side elevation of a ship-rigged mast with the squaresail running on a horse.

(c) A side elevation of an entirely fore-and-aft rigged mast. Here the squaresail, rigged occasionally, is bent to a crossjack yard and runs on a horse.

(d) A horse strop fitted to a squaresail yard on a mast carrying a topsail.

(e) The iron fitting on a crossjack yard for sliding on the horse of an entirely fore-and-aft rigged mast.

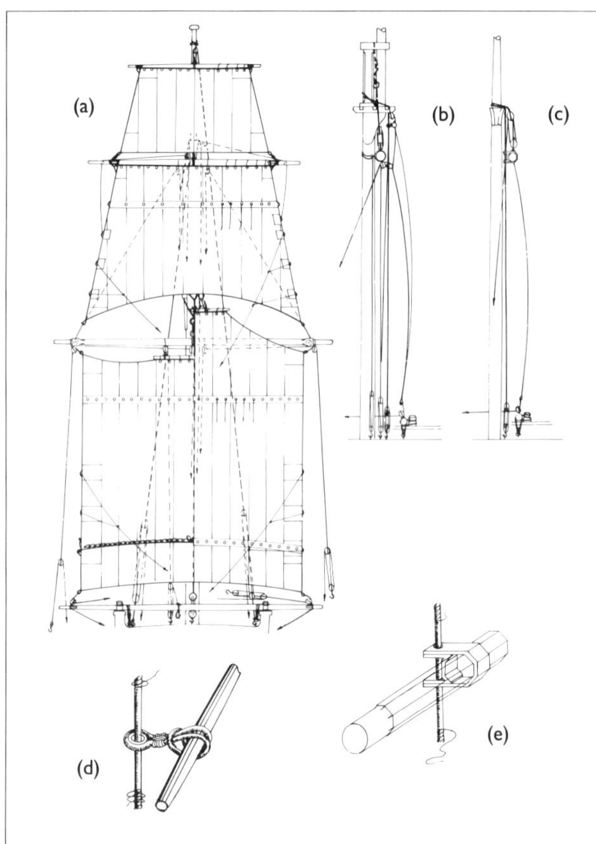

Figure 88 ***Rigging details II***
(a) Two further variations of square rig on small vessels fitted with cutter masts: studdingsails are added in both. Note the topsail's deep roach, which became necessary because of the higher positioning of the fore stay due to a larger gaff sail. The left side of the drawing shows the method in use during the eighteenth century: an overly large and narrow squaresail is hoisted up to the hounds between the spread yard (with studdingsail booms) and the topsail. The squaresail boom also has studdingsail booms fitted. The hoisting of studdingsails was only possible abaft the square-rigged sails. The method shown on the right side was more common at the end of the eighteenth century: the spread yard, rigged with slings and standing lifts, has a squaresail yard fitted with studdingsail booms hoisted to it, with a centre halyard tackle and single running lifts. The large opening between the deeply roached topsail and the lower and wider squaresail is filled with a save-all topsail. In this rig, the studdingsails are set forward of the square-rigged sails. Rigging shown in Figure 87(a) has not been duplicated in this drawing.

(b) A side elevation of a cutter mast, showing sails as carried before 1800.

(c) The rigging of a spread yard in about 1800 with slings and standing lifts, and the hoisting of a squaresail yard with a centre halyard tackle and running lifts. While the spread yard was rigged with braces which ran only aft, the squaresail yard had braces running forward as well. These details are taken from a watercolour by Roux, of 1801.

(d) The setting of a save-all-topsail. Ears were hitched to the topsail yard and the sheets to the spreadyard.

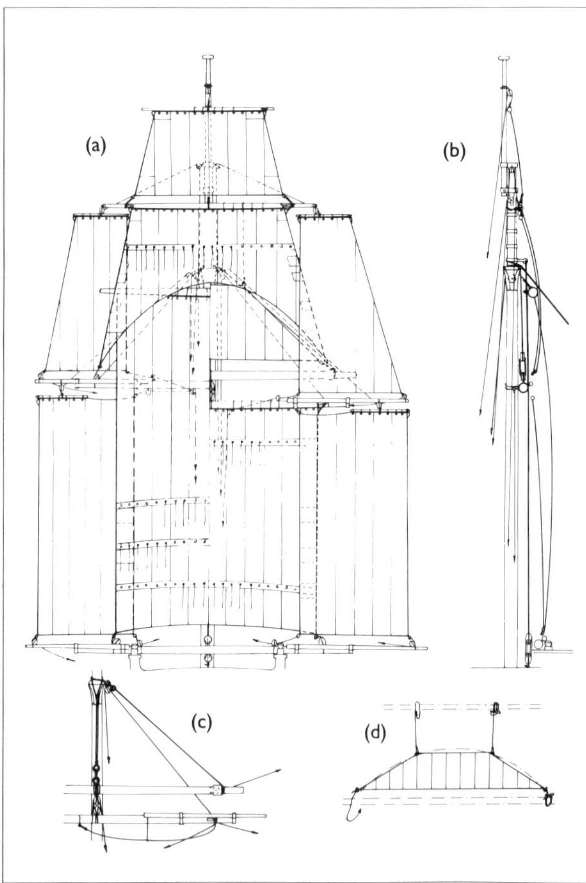

Preventer stay
Borgstag; Faux étai

Prepared similarly to the stay, the preventer stay was set up with a lanyard to another deadeye in an iron-bound strop, bolted to the stem's face. Sloops and other vessels with a standing bowsprit usually did not have a preventer stay, but they were often rigged with a jib stay.

Bowsprit and jib
Bugspriet und Klüver; Beaupré et foc

One of the main characteristics of a cutter was the running bowsprit which rested beside the stem. Sloops or yachts had a standing bowsprit above the stemhead. Cutters had also a plain stem, while those of sloops or yachts were usually built with a cutwater and a figurehead or ornaments.

Bowsprit shrouds
Backstage; Haubans de beaupré

Bowsprit shrouds were prepared with a hook and thimble and hooked to each side of a square hoop at the end of the bowsprit. With a thimble turned into the inner end, they were set up with a lanyard to an eyebolt at each side of the bow, and the ends were hitched and made fast.

On Continental vessels the shrouds were middled, with the bight round the bowsprit end and a seizing underneath. The ends led toward the bow, where they set up to deadeyes or tackles.

Horses
Fusspferde; Marche-pieds

Sloops had horses similar to the jib-boom horses in ships. They were sometimes fitted with stirrups.

Bobstay
Wasserstag; Sous-barbe

Vessels with a standing bowsprit were rigged with a bobstay similar to that in a ship; details can be found in Chapter II.

Middle jib tack
Inner Klüverhals; Point d'amure de foc

The middle jib tack was clinched through the swivel eye of the bowsprit traveller, rove through a sheave hole in the outer bowsprit and through an iron-bound block hooked and moused to an eyebolt in the side of the stem near the waterline, then led over the bow to the windlass to be hauled out; it was then stopped and belayed round a timberhead.

On sloops the middle jib was hanked to the jib stay, with the tack made fast round the stay deadeye lashing.

Jib halyard
Klüverfall; Drisse de foc

Jib halyards rove through a block lashed to the head of the sail and through others on each side of the masthead. One end had a treble block turned or spliced in and connected by a fall to a double block hooked to an eyebolt in the deck on one side. The other end was made fast to an eyebolt in the deck opposite.

When the jib was bent to jib stay hanks as on sloops, the halyard was set up like a foresail halyard. It might also be rigged single, being bent to the head of the jib and reeving through a block at the masthead upon deck, where a double block was turned or spliced into the end and a fall connected this to a single block hooked to an eyebolt.

Jib sheets

Schoten; Écoutes

Jib sheets were either single or rigged with blocks and falls. Single sheets were bent to the sail's clew and led over the bow to the windlass. Double sheets were rigged with two double blocks lashed to the clew and connected by falls to single blocks hooked near the cat heads to eyebolts in each side. Each fall led through a hole in a timberhead, or a lead block lashed to the side. Sheet blocks might also be spliced into sheet pendants, which were then bent to the clew. Sometimes, too, they were stropped round thimbles in a double-eyed hook moused to the clew.

Jib downhauler

Niederholer; Hale-bas

The jib downhauler was made fast to the head of the jib and led upon deck. The downhauler of a hanked jib led through a few hanks and a block at the lower end of the stay, or the traveller, and on to the fore deck.

Jib inhauler

Einholer; Hale-dedans

The jib inhauler was fastened to the traveller and led on to the fore deck. Lever, however, stated that 'small Ships have no In-hauler, the Down-hauler answering both purposes'.

Heel rope

Windreep; Cordage de talon

Only running bowsprits needed a heel rope. With its standing part made fast to a timberhead or eyebolt, it rove through a sheave hole in the bowsprit heel, then through a lead block hooked to an eyebolt in the bow, with the leading part going to the windlass.

Flying jib

Aussenklüver, Jager; Clinfoc

The rigging of a flying jib was similar to that of a middle or main jib set flying.

Foresail

Stagfock; Trinquette

Foresails were generally bent to hanks around the stay.

Halyard

Fall; Drisse

With the standing part made fast round the masthead, the foresail halyard rove through a block lashed to the sail's head, then through another lashed underneath the stay collar close to the masthead, and down upon deck.

Downhauler

Niederholer; Hale-bas

Bent to the head of the sail and led through a few hanks, the foresail downhauler rove through a lead block at the stay near

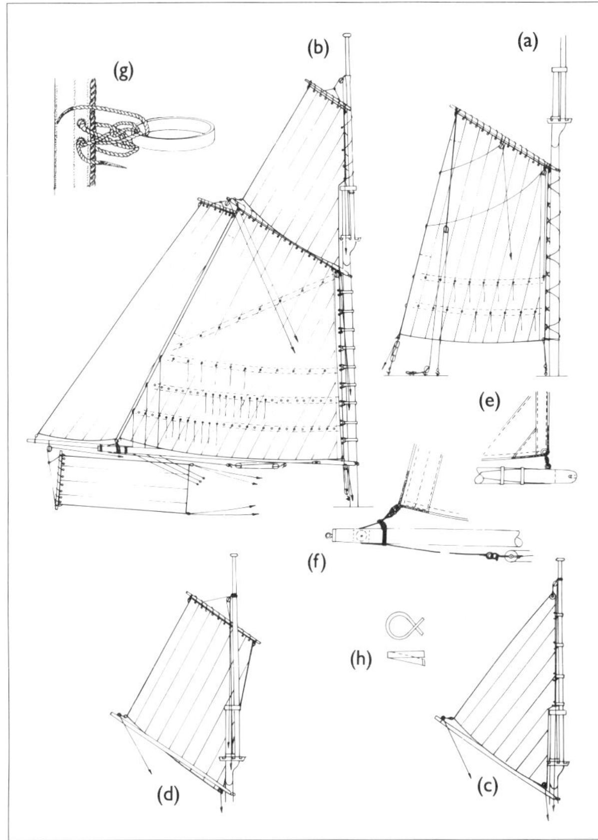

Figure 89 ***Rigging details III***

(a) A boomless gaffsail (as set, for example, on a schooner) with brails, vangs, sheet and tack, and bent to the mast with a lacing. Halyards are not shown.

(b) A mainmast with a boomsail, a gaff topsail, a ringtail sail and a watersail, with all the necessary running rigging except for the gaff halyards and the boom topping lift. The two ropes on each clew of the watersail are the tack and sheet.

(c) A fore-and-aft topsail, shoulder-of-mutton type.

(d) A fore-and-aft topsail, lugsail type.

(e) Detail showing an alternative method of seizing the tack of a main sail.

(f) Detail showing the seizing of a main sail sheet.

(g) Detail showing the sail bent to a mast hoop in fore-and-aft rigs.

(h) A wooden hank for the hoisting of a shoulder-of-mutton sail, according to Röding.

the foot of the sail and was belayed to a cleat on the gunwale.

Tack tackle

Halstalje; Palan d'amure

The foresail's tack was kept down by a tackle, with the upper block hooked to the tack and the lower to an eyebolt in the bow below the stay. The tackle fall was belayed to the crosspiece of the windlass.

Bowlines

Bulins; Boulines

A bowline was connected to the clew of the foresail by a hook spliced into its end, and it rove then through a block lashed to the shrouds at each side and through a cringle in the sail's leech, to belay round a pin in the shroud rack. Such a bowline was usually needed only on large foresails.

Sheet

Schot; Écoute

The foresail sheet rove through a block running with a thimble on an iron horse, or in some sloops instead of a block through an iron-bound deadeye; after passing a block or seized-in deadeye at the clew, the sheet returned to the horse block and back again until the whole sheet was expended, then was frapped together and hitched.

Boom

Baum; Gui

Topping lift

Dirk; Balancine de gui

The standing part of the boom topping lift was clinched round the masthead or hooked to an eyebolt at its aft side. It led down through a block at the boom end and up again through another block hooked to the same eyebolt, then came down on the starboard side of the vessel and had a double block turned or spliced in. A fall connected this block to a single block hooked to an eyebolt in the aft part of the channel, and was belayed to a pin in the shroud rack. Sometimes a runner was used and the topping lift was rigged like a ship's driver boom.

Steel described the boom topping lift of a sailing barge as follows:

> [The boom topping lift] is a long pendent, that goes over the outer end of the boom, with an eye spliced in one end, and in the other end is spliced a double or single block, that connects by its fall to a single block, hooked in an eye-bolt at the upper part of the mast-head, and the fall leads down to the shrouds at the side [see also Topping lifts under 'Rigging of masts and yards'].

Sheet

Schot; Écoute

The boom sheet rove through a double block stropped round the boom just within the taffrail and through another stropped round a thimble on a horse. Very large cutters had a treble block on the horse. The sheet was belayed to a large cleat on the taffrail or in small vessels to the pin of the block on the horse. The boom block's strop was confined by a comb cleat, or by two stop cleats on the upper side of the boom.

Tack tackle

Halstalje; Palan d'amure

The tack tackle consisted of a double block fastened to the sail's tack and a single block hooked to an eyebolt in the deck or near the partners in the mast. The connecting fall was made fast to the lower block. The sail's tack was sometimes also lashed without a tackle directly to an eyebolt in the boom.

Reef pendants

Schmierreeps; Bosses de ris

Also called earrings, the reef pendants were four in number on larger cutters. The number on a smaller vessel depended on the number of reef bands on the mainsail. They rove through holes in the outer boom end vertically below the reef cringles and had a thimble spliced in one end. The other ends rove through the respective reef cringles, and when a reef had to be taken in were catspawed to the hook of a luff tackle to haul down the leech. Afterwards they were frapped round sail and boom until extended and fastened with a hitch. Usually old hammocks were placed between the pendants and the sail for protection. Nearly the full length of the pendants was wormed and served over with spun yarn.

Guy pendant

Gei-Stander, Bullentau-Stander; Pantoire de retenue de gui

Guy pendants had a hook and thimble at one end and a thimble or eye spliced to the other. The pendant was hooked to a thimble stropped to the boom close to the boom sheet block and within the confines of the comb cleat. A luff tackle was hooked into the thimble at the other end. The luff tackle single block was hooked near the windlass to a timberhead or eyebolt, and the fall led inboard. Röding noted that the block was hooked to one of the main shroud chainplates.

Gaff

Gaffel; Corne

Tye, halyard

Piekfall; Itague, drisse de coqueron

The gaff tye rove through a block on a span clinched or spliced round the middle of the gaff. The standing part was fastened round the masthead or hooked there to an eyebolt, and the hauling part led through an iron-bound block hooked to another eyebolt in the masthead. At the lower end was a double halyard block. The gaff tye was not spliced, like other halyards, round this block, but led through the block's strop and fastened with a hitch, and the remainder was expended in turns round the block and the strop. A fall connected the block to a single or double block hooked to an eyebolt in the deck abaft the mast.

Steel described the tye or halyard of a sailing barge twice, and he noted specifically that it was rigged according to the fancy of the vessel's master. In his first description the standing part was fastened to the lower end of a block hooked to an eyebolt in the masthead just below the topping lift. The tye rove through a span block at the peak, then up to the block at the masthead, back through another span block near the middle of the gaff, then to a fourth block again hooked to an eyebolt in the masthead below the second block and upon deck. His second description had the standing part placed over the masthead with an eye splice, and the tye leading through a block at the gaff peak, passing a block at the masthead below the standing part, through a span block near the peak and finally through a block at the masthead below the second block and upon deck. The tye was normally set up with a halyard at the lower end, like the cutter's.

Inner tye, throat halyard

Klaufall; Drisse d'en dedans, drisse de mât

The standing part of the inner tye was hooked to an eyebolt in the jaws of the gaff, and it rove through an iron-bound block hooked below the rigging to the lower eyebolt in the aft side of the masthead. A double block bent to its lower end was connected with a halyard to a single block hooked to an eyebolt on deck, in the same manner as (but on the opposite side to) the other tye.

Peak downhauler

Piekniederholer; Hale-bas de coqueron

The peak downhauler rove through a small block stropped round a thimble in the eyebolt driven into the outer end of the gaff. It was belayed to a cleat on the lower side of the boom.

Throat downhauler

Klauniederholer; Hale-bas de mâchoire

Hoisting gaffs also had a throat downhauler, the upper block of which was hooked or stropped to an eyebolt in the lower side of the gaff jaws. The lower block was hooked or stropped below the boom to an eyebolt in the aft face of the mast, with the hauling part leading upon deck. Sometimes the downhauler rove through a block at the sail's nock and then upon deck.

Topmast and topgallant mast

Marsstenge und Bramstenge; Mât de hune et mât de perroquet

Shrouds

Hoofdtaue; Haubans

Topmast and topgallant mast shrouds were fitted over the masthead like the topgallant shrouds on a ship. Lead thimbles for the lifts were seized to the upper part of the shrouds just below the throat seizing. The shrouds rove through holes in the cross trees upon deck, and were set up to the lower deadeyes with thimbles and lanyards. This arrangement was, however, characteristic of the later part of the century, and then not in all regions; for further details see Topgallant shrouds in Chapter II.

Stay

Stag; Étai

An eye spliced into the upper end of the topmast stay fitted over the masthead. The lower part of the stay led over the middle sheave of a treble block lashed to the underside of the bowsprit at its outer end. A thimble turned into the end of the stay was set up with a lanyard to an eyebolt near the stem.

Standing backstays

Pardunen; Galhaubans

Standing backstays were fitted over the masthead with a cut splice if there was only one pair. When two pairs were used, eyes like these on shrouds were seized into the bights. Thimbles were turned into the lower ends and the stays were set up with a gun tackle purchase to thimbles stropped round the lower deadeyes, or were simply hauled taut with lanyards.

Top rope

Windreep; Guinderesse

The top rope was similar to that in a ship.

Crossjack yard

Bagienrah; Vergue sèche

Quarter block

Schotführungsblock; Poulie d'écoute

Quarter blocks were double stropped with long and short legs, and hung downwards within the centre cleats of the crossjack yard. The long legs came up on the aft side of the yard and went through the bights of the short legs on the fore side, to be lashed.

Strops

Stropps; Estropes

Two strops with a thimble seized in the bight were spliced, or lashed through eyes, round the middle of the yard. One thimble pointed aft and the other upwards.

Clewline blocks

Geitaublöcke; Poulies de cargue-point

Clewline blocks, as in ships, were lashed with two eyes round the yard. Some sloops and light rigged vessels did not have clewlines: lowering the yard, whether crossjack or squaresail yard, was much more the norm then bending the squaresail to a permanently fixed crossjack yard. Röding noted that 'vessels which carry a square sail have also a horse forward of the mast on which the yard of the square sail runs with a thimble up or down'.

Horses, foot ropes

Fusspferde; Marche-pieds

On a standing yard horses were set up as in ships. Temporary yards on fore-and-aft rigged vessels did not have horses.

Brace pendants

Brass-Schenkel; Pantoires de bras

Brace pendants were fitted over the yardarms with a spliced eye, and a single block was spliced into the other end, just as in ships. Continental vessels often had no brace pendants at all, but were sometimes rigged with long stropped blocks instead.

Fore braces

Vorwärtsbrassen; Bras de l'} avant

The standing parts of the fore braces were clinched round the outer bowsprit end, and the braces rove through the pendant blocks to the outer sheaves of the treble block noted above under Topmast or Topgallant mast stay, and came in upon deck.

After braces

Achterbrassen; Bras de l'arrière

The after braces led in upon the quarterdeck through a snatch block or a sheave hole in the sides and were belayed to a cleat or timberhead.

Sheet block

Schotblock; Poulie d'écoute

The sheet block was stropped into the splice of the lift block, and the eye left over beneath the seizing was served and fitted next to the brace pendant over the yardarm. Merchant vessels sometimes had a similar arrangement, especially when the topsail sheets did not lead through sheave holes in the yardarms.

Lift

Toppnant; Balancine

Lifts rove through blocks on a span round the cap or masthead and down upon deck.

Halyard or tackle

Fall oder Takel; Drisse ou caliorne

The tackle to sway up the yard was treble or double, according to the size of vessel. The upper block was hooked either to an eyebolt in the fore side of the masthead or to a strop round the mast. The lower block was hooked to the stropped thimble in the centre of the yard (see Strops above), with the tackle fall reeving through a sheave hole in the topsail sheet bitts or through a lead block hooked to an eyebolt in the deck and led aft.

Horse

Leiter; Filière d'envergure

The horse for the squaresail yard (crossjack yard on fore-and-aft rigged vessels) had an eye spliced into its upper end and was lashed to the masthead. Its lower end set up with deadeyes and a lanyard to an eyebolt in the deck forward of the mast. Topsail rigged vessels had the horse fitted forward of the crossjack yard.

The horse was led through aft facing thimble on the squaresail yard strop.

Squaresail boom

Breitfockbaum; Vergue de tréou, ou vergue de fortune;

A boom was used to spread the clews of a temporary squaresail; it was lashed across the deck of a vessel with a one or two masted fore-and-aft rig.

Topsail yard

Marsrah, Toppsegelrah; Vergue de hunier

Tye

Drehreep; Itague

Clinched round the yard's slings, the topsail yard tye rove through a sheave hole in the masthead and had a double block spliced or turned into its lower end; this was connected by a fall to a single block in the channel. The fall then led through a lead block on the gunwale and was belayed to a cleat or timberhead.

Clewline blocks, horses, brace pendants and lifts

Geitaublöcke, Fusspferde, Brass-Schenkel und Toppnanten; Poulies de cargue point, marche-pieds, pantoires de bras et balancines

All these were fitted to the yard in a similar manner to those in small ships. The lifts were led through thimbles in the topmast shrouds and upon deck.

Braces

Brassen; Bras

Topsail yard braces rove like the fore braces. The leading parts went over the sheaves of a double block at the bowsprit end and came in upon deck to be belayed where most convenient.

Bowlines

Bulins; Boulines

Topsail yard bowlines were set up to bridles like those in ships, then rove through thimbles stropped to the end of the bowsprit and upon deck.

Mainsail

Gross-Segel; Grand voile

The head of the mainsail was bent to the gaff with earrings and lacing like a ship's mizzen. The mast leech was seized to wooden hoops round the mast when a hoisting gaff was used, and otherwise laced to the mast.

Tricing line

Aufholer; Hale-breu

Spliced to the tack of the mainsail, a tricing line rove through a small block fastened to the eyebolt for the gaff's throat downhauler and led upon deck, where it was belayed to a cleat near the mast.

Sheet

Schot; Écoute

The mainsail sheet was spliced into the clew of the sail and rove through the sheave hole in the boom. It had a thimble turned into the end, into which a luff tackle hooked and was be set up to another thimble stropped near the jaws round the boom. When the sail was hove out, the clew was lashed with an earring to an eyebolt in the boom end, or round it.

Trysail or storm mainsail

Sturmsegel; Voile de tempête

Approximately half the mainsail's size, the trysail was rigged much like a ship's mizzen, but with sheets and no boom.

Squaresail or crossjack

Bagien; Voile de fortune

Bent to the yard and comparable to a main or fore course, this sail was usually known only as a squaresail, though Steel called it a crossjack. The latter name was used by Falconer only for the sail carried on the crossjack yard of a three-masted ship, and he commented 'this sail, however, has generally been found of little service, and is therefore very seldom used'.

The squaresail was only bent to a crossjack yard if the vessel did not carry a square topsail; if a topsail was set, it was bent to an extra yard known as the squaresail yard or club yard, which was hoisted up to the crossjack yard. In vessels with only a topgallant mast stepped, this yard was hoisted past the crossjack yard to below the hounds. The squaresail frequently had a bib to fill the space left by the deeply roached topsail.

Squaresails were often also fitted with a bonnet: according to Falconer, the sail was 'furnished with a large additional part called the bonnet, which is then attached to its bottom, and removed when it is necessary to scud'.

Bowlines

Bulins; Boulines

Set up to the sail's leeches with bridles, the squaresail bowlines rove through a double block at the end of the bowsprit and came in upon deck to belay on the bitts, or to cleats on the inner bowsprit.

Topsail

Marssegel, Toppsegel; Voile de hunier

A cutter topsail was rigged in the same manner as a topsail on a ship.

Topgallant sail

Bramsegel; Voile de perroquet

A cutter topgallant sail was set flying like a royal sail on a ship.

Gaff topsail

Gaffel-Toppsegel; Flèche-en-cul

A gaff topsail was laced to a small gaff at the head. Gaff topsails were commonly only rigged to masts without square topsails, since the rigging two sails performed the same functions. The use of gaff topsails on cutters can only be traced back to the second half of the eighteenth century.

Halyard

Fall; Drisse

The gaff topsail halyard was bent to the inner part of the gaff and rove through a sheave hole in the topgallant masthead upon deck.

Topping lift

Piekfall; Balancine de corne

The standing part of a gaff topsail topping lift was made fast round the topgallant masthead above the sheave holes. The lift rove through a thimble or small block at the gaff's peak, through a sheave hole or small block at the topgallant masthead and then upon deck.

Tack

Hals; Point d'amure

The gaff topsail tack was made fast a little above the hounds.

Sheet

Schot; Écoute

Bent to the clew of the sail, the gaff topsail sheet rove through a thimble seized to the mainsail peak, and led down upon deck.

Lower studdingsails

Untere Leesegeln; Bonnettes basses

Lower studdingsails were rigged like those in ships and set flying, similar to a ship's fore studdingsails. Sometimes the squaresail boom was also fitted with studdingsail booms, indicating a standing foot.

Topmast studdingsails

Stenge-Leesegel; Bonnettes de hunier

Topmast studdingsails were approximately the same as those in ships.

Halyards

Fallen; Drisses

Bent to the studdingsail yard, the topmast studdingsail halyards rove through a jewel block at the topsail yardarm, then through a block above the rigging at the masthead and down upon deck.

Sheets and tacks

Schoten und Halsen; Écoutes et points d'amure

Topmast studdingsail sheets and tacks were rigged as in ships.

Ringtail sail

Brodwinner; Bonnette de tape-cul, bonnette de baume, bonnette de brigantine

Similar in shape to a topmast studdingsail, a ringtail sail can be considered a studdingsail for a mainsail set to a gaff and boom (as in all sloops, brigs, cutters and schooners, etc). It was bent to a small yard and hoisted up by the peak downhauler. The sail's foot was extended by a spar or small boom lashed to the aft end of the main boom.

Mizzen

Besan; Artimon

A mizzen mast was stepped in those vessels commonly known as 1½-masted. The mizzen sail might be a gaff, square or spritsail. When a square mizzen was set, it was bent to a yard hoisted by a halyard running through a sheave hole in the masthead, and the clews were spread by sheets. When a sprit mizzen was set, it was bent to the mast with grommets and was peaked with a sprit; a sheet hauled the clew aft to a small boom, the outrigger. A gaff-rigged mizzen was laced to a mizzen gaff and mast, and was controlled by a sheet as for the sprit mizzen. Similar mizzen mast in larger ships were called jigger masts (see Jigger mast in Chapter III).

Water sail

Wassersegel; Bonnette de sous-gui

If the ringtail sail was a gaff mainsail's upper studdingsail, the water sail was the lower. Its head was bent to a small yard.

Ringtail and water sails were not restricted to small craft. Hutchinson's and Röding's illustrations and the latter's and Falconer's text confirm that these sails were also set on three-masted ships.

Halyard

Fall; Drisse

The watersail halyard was made fast to the middle of the yard, rove through a small block under the outer end of the main boom and was belayed to a cleat at the taffrail.

Sheets

Schoten; Écoutes

Watersail sheets led in from the clews of the sail over the quarters.

Save-all topsail

Save-all-Marssegel; Voile de tout préserver

Hoisted only in calm conditions, the save-all topsail was very rare. It was set in the bight between the topsail's foot roach and a squaresail. The clews were lashed near the lift blocks to the crossjack yard.

Halyards

Fallen; Drisses

The save-all topsail halyards were bent to the sail's earrings, and rove through blocks at each quarter of the topsail yard and upon deck.

VI

Rigs for vessels with spritsails

A large number of sprit-rigged vessels existed during the eighteenth century side by side with the gaff-rigged vessels discussed in the previous chapters. The sprit rig was characteristic of many types found particularly in the Dutch, English and German North Sea coastal regions. A further group of sprit-rigged craft was formed by the river ships on Continental rivers and canals and in the British Isles. The four types discussed below are representative only, chosen because their rigs demonstrate important differences. Further sprit rigs can be found in Chapter VIII, including the scapho, sacoleva, Turkish coaster and the small Chusan fishing boat.

Figure 90 ***Sprit-rigged koff***

Koff

Kuff; Kof

The koff was a vessel rigged in many different ways. It was a flat bottomed vessel of burthen with a broad bow and round stern, and broad wales, a short rudder and a free moving tiller. Koffs were also sometimes fitted with leeboards. W V Cannenburg noted that koffs came into use in Holland around the middle of the eighteenth century, derived from a Frisian koff built in 1735 in Lübeck. Koffs were originally coastal traders, but they were soon making extended voyages into the Baltic Sea, to Norway, the West Indies and Brazil. Dutch koffs as described by Cannenburg, Groenewegen and Le Comte were 1½-masted with the mizzen mast forward of the tiller, and gaff-rigged with the mainmast also carrying a topsail and a topgallant sail. A drawing of an Antwerp koff from 1878 in *Souvenirs de Marine* and a late eighteenth century model in the Nederlandsch Historisch Scheepvaart Museum are very much in line with this. A sail and rigging plan of such a gaff-rigged vessel has been included here for comparison with the sprit-rigged plan (see Figure 91).

Röding defined a koff as follows:

> [A koff is] a vessel employed by the Dutch, Danes and Swedes. It has a mainmast and a mizzen mast which stands on the deck rather than at the stern as in a smack. Each of these masts carries a spritsail with bonnets and also a topsail. Forward of the mast there is usually a fore sail, a middle jib and a flying jib. Koffs are sometimes fitted with lee-boards...

His description suggests spritsail rig, but the illustration of a koff given in his work is of a koff with gaff sails, as described by Cannenburg and others.

Falconer and Steel agreed on the description of a koff. They saw the type as:

> ... a Dutch vessel of burden, with a main and foremast, and a large spritsail abaft each. Thus they sail very close to the wind but when the wind is aft, they carry flying topsails and a square-sail upon the foremast, and upon the bowsprit two or three jibs.

The sprit-rigged koff illustrated here is based on the latter description and on an illustration by Steel. Similar in mast position but gaff-rigged is the Dutch koff of 1848 shown in Paris' *Souvenirs de Marine.*

Contemporaries from different countries thus provide a variety of opinions on the appearance of a koff, and each is probably correct to some extent. The average size of a koff was between 100 and 300 tons.

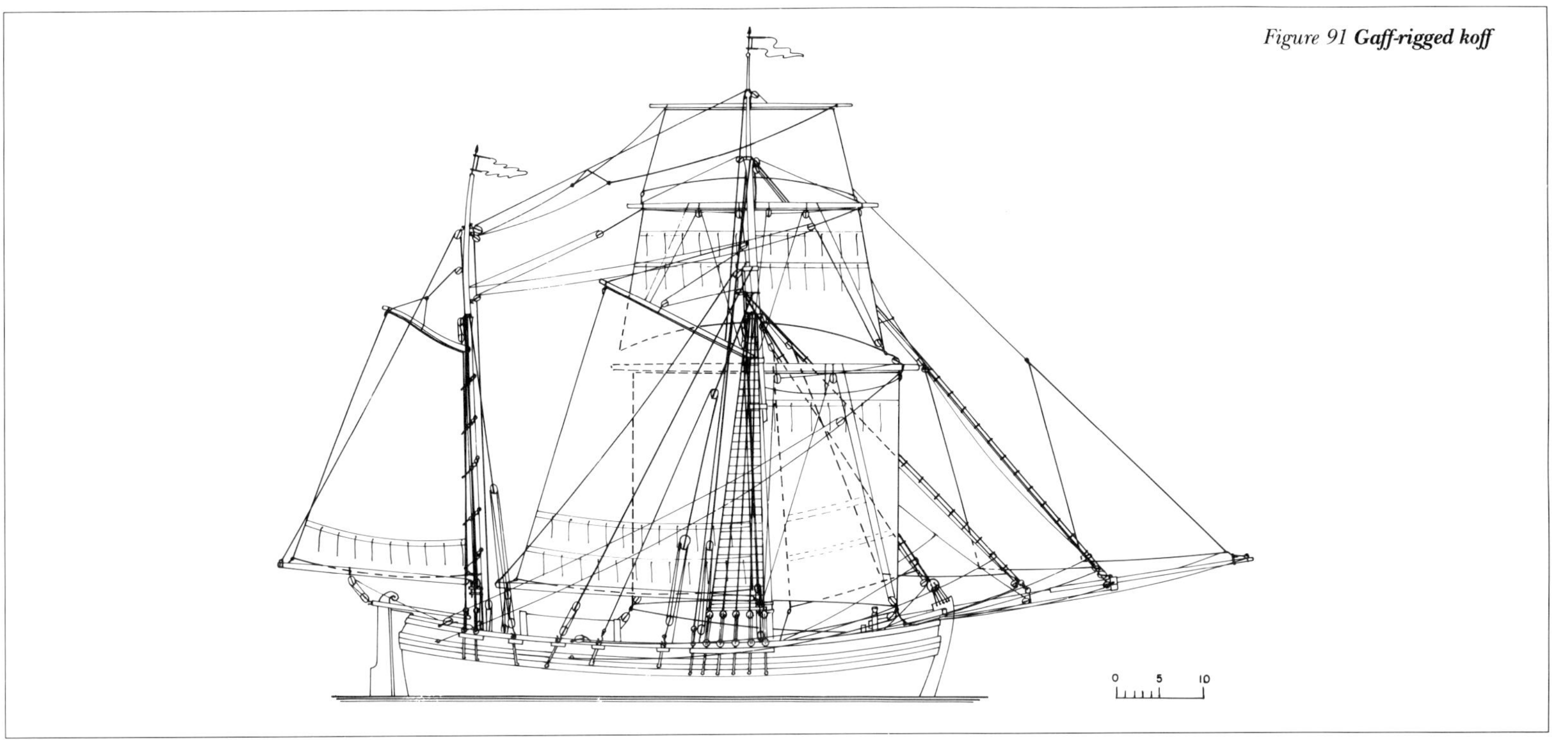

Figure 91 **Gaff-rigged koff**

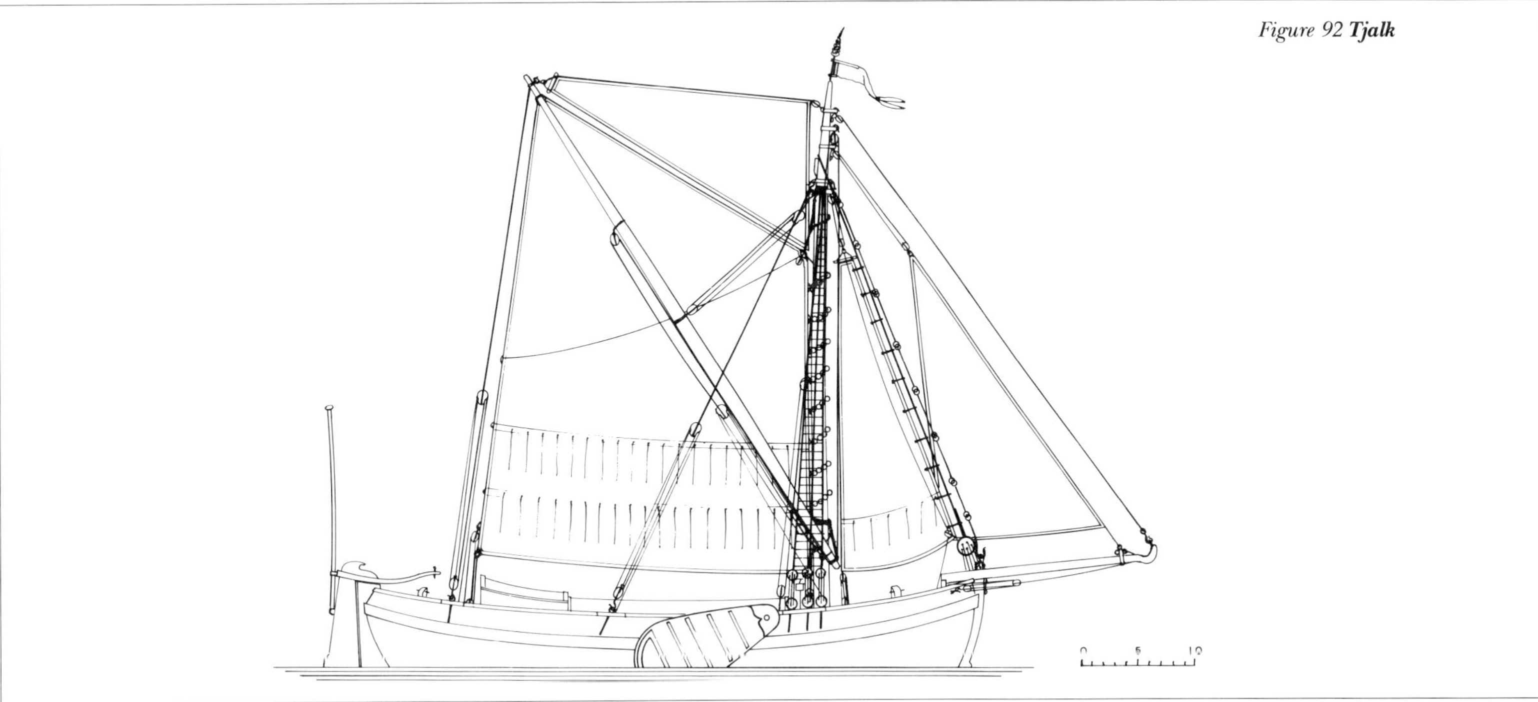

Figure 92 **Tjalk**

Tjalk

Tjalk; Tjalk

The tjalk was the most common small Dutch coastal carrier. The name is Frisian and was first mentioned by Nikolas Witsen in his *Architectura Navalis* of 1690. Cannenburg noted that 'tjalk' could be seen as a general term for many types of round-built freight vessels in use in various parts of the Netherlands. They usually differed only slightly.

If the term is used in this general sense, a distinction has to be made between small and wide vessels, and also between those vessels with turnover boards with a free moving tiller and the hektjalk, which had the stern built up with extra planks so that the tiller could not be turned over board. All tjalks had a flat bottom, were of shallow depth, and had large leeboards. They became one of the most common freight carriers on rivers and canals in Frisia and along the Watten, the tidal mud-flats extending on the Frisian coastline right up to the Elbe estuary. Single-masted during the eighteenth century, tjalks in the nineteenth sometimes also had a small mizzen mast stepped, similar to that of a smack, and their masts could be lowered. The capacity of these vessels was between 30 and 80 tons.

Röding described the tjalk's rig as consisting of a mast without a topmast, stepped relatively far forward and carrying a large spritsail. A foresail and a flying jib were also carried. Chapman added a topsail, Groenewegen indicated a squaresail as well, and a hektjalk model in Amsterdam shows a sprit topsail in addition to the normal sprit rig as shown in Figure 92.

In the nineteenth century a gaff rig superseded the sprit rig; the gaff was relatively short and the sail not very high but with a large footing and a boom.

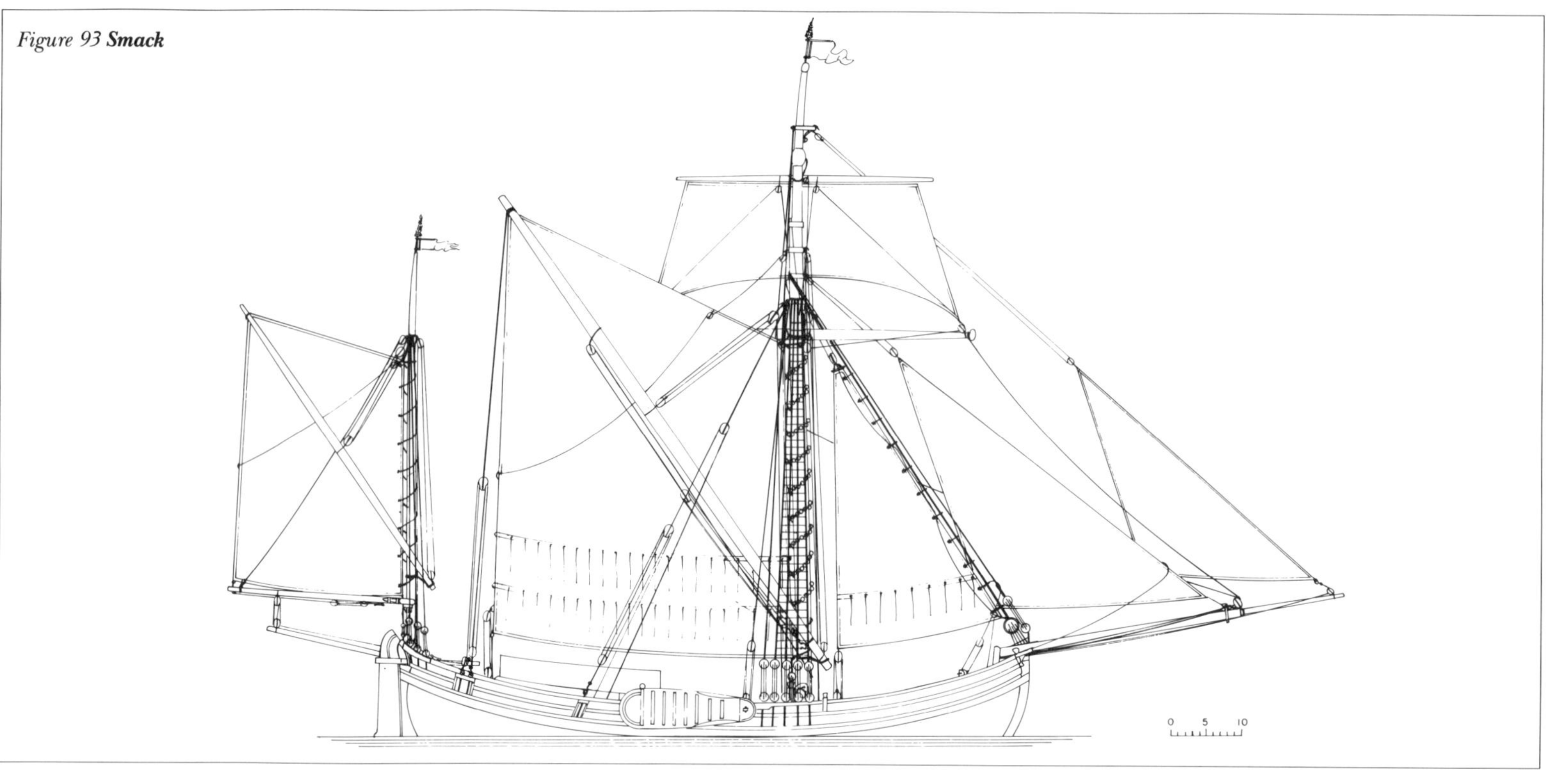
Figure 93 **Smack**

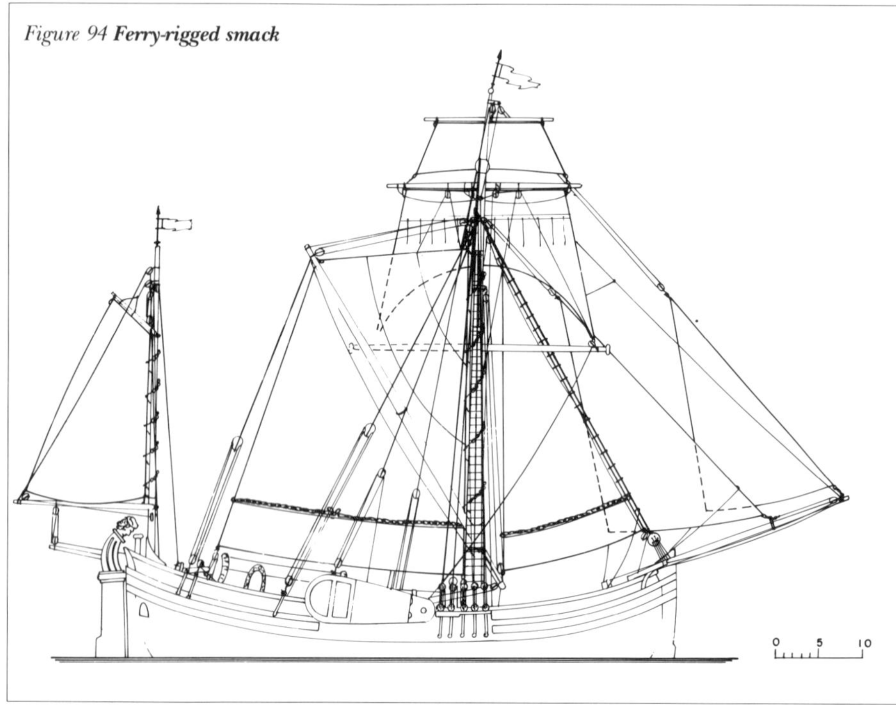
Figure 94 **Ferry-rigged smack**

Smack, Dutch hoy

Schmack; Semaque

English, Dutch and German records indicate that in England the term 'smack' was usually applied to a quite different type of vessel from the Continental vessel of that name. 'Smack: a small vessel commonly rigged as a sloop or hoy, used in the coasting or fishing trade or as a tender in the King's service' was the definition in the first edition of Falconer, while the later edition noted the type as being rigged like a cutter. English authors generally referred to the vessel described here as a Dutch hoy.

The Continental smack was a seagoing vessel of Dutch origin, also employed in the German North and Baltic Sea coastal trade in the eighteenth century. Cannenburg described smacks as strong, seagoing hektjalks. They were 1½ masted, sprit-rigged for most of the century and carried between 20 to 100 lasts (1 last equals approximately 2 tons). They ventured as far as Spain.

Round built both fore and aft, flat bottomed and fitted with leeboards like nearly all Dutch coastal vessels, a smack differed in her rig from others mainly in that her mizzen mast was stepped very close to the sternpost on a kind of transom above the tiller, which also constituted the vessel's taffrail. The side planking rose aft above the transom and created a triangular helm port opening at the stern. Like galliots and koffs, smacks had a superstructure known as the roof built on deck near the master's cabin. This was the crew's quarters and the galley.

The mainmast with its scarf-fitted topmast was frequently rigged with a topsail in addition to the spritsail. A foresail was bent to hanks on the fore stay and two jibs could be set to the bowsprit. The mizzen mast had a spritsail rigged with an outrigger to haul out the sheet. Pictures by Groenewegen and Röding indicate that smacks in the 1780s were already gaff-rigged, but without a boom at the mainmast. They sometimes carried a flying topgallant sail in addition to the topsail on the mainmast, and a squaresail was set when necessary. Groenewegen also illustrated a smack or hektjalk with what he described as a ferry rig, in which the mainmast was sprit-rigged and the mizzen mast rigged with a gaff and boom (see Figure 94).

Aak

Aak; Aak

'Aak' was a collective term for a variety of vessels employed on the River Rhine and its tributaries. One of the oldest pictorial records of the type is the *Wönsam Prospect of Cologne* of 1530. This Rhine view of the city of Cologne shows a number of river craft, among which the *Keulche Aak* (Cologne aak) is predominant. These were small, clinker-built transport vessels with a large semicircular covered hatch, flat bottomed and swimheaded. No leeboards were used, and the aak was rigged with a simple spritsail plus a triangular foresail.

Evolutions from this basic type were the *Samoreuse*, *Bönder*, *Dorsten Aak*, *Ruhr Aak*, *Neckar Aak* and others, all clinker-built and swimheaded; some of these types already had a sternpost by the eighteenth century. A short bowsprit allowed the setting of a jib, and in many other respects the rigging was similar to that of sprit-rigged coastal craft. Larger aaks like the bönder carried two masts, with the mizzen mast usually standing in the aft part of the roof (deckhouse).

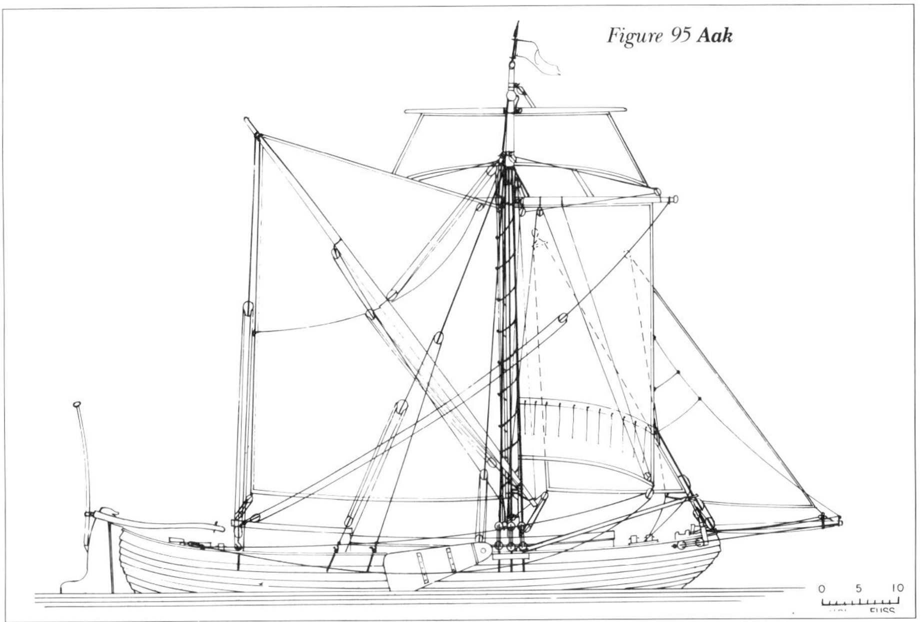

Figure 95 **Aak**

Sprit rigging

Spriettakelung; Gréement de livarde

Steel also provided a good description of a spritsail-rigged mast, and the following notes are partly based on his work.

Shrouds

Hoofdtaue; Haubans

Shrouds and runner pendants were fitted over the masthead with eyes, or bights were seized close to the mast. They rested on a grommet driven down to the masthead stop.

Stay

Stag; Étai

The stay was laid with a running eye over the masthead above the shrouds, and was set up to a three or four-fold tackle. A large thimble or the block itself was turned into the lower end of the stay for this purpose. The lower tackle block was hooked into a large strop round the stem and a fall connected the two blocks. The mast could be lowered by slackening the fall.

The stay of a standing mast was set up with a deadeye or block to holes through the stemhead, connected by a lanyard.

Topping lift, sprit yard pendant

Dirk; Palan de balancine de livarde

The topping lift was fixed to the middle of the yard. It was rigged either as a pendant or a tackle, but in both cases a grommet, with or without a thimble, was spliced round the middle of the yard. One end of the pendant was spliced to the grommet, and the other rove through a block at the masthead and downwards, with a thimble spliced into its end. In larger vessels a double-block tackle was hooked to the thimble and to an eyebolt in the deck; smaller vessels used a luff tackle (see Figure 96b-a.). Dutch vessels often had a double-block tackle hooked to the thimble on the grommet (or stropped direct to the yard) and to an eyebolt in the masthead (see Figure 96a-b.). The hauling part then led down to a cleat.

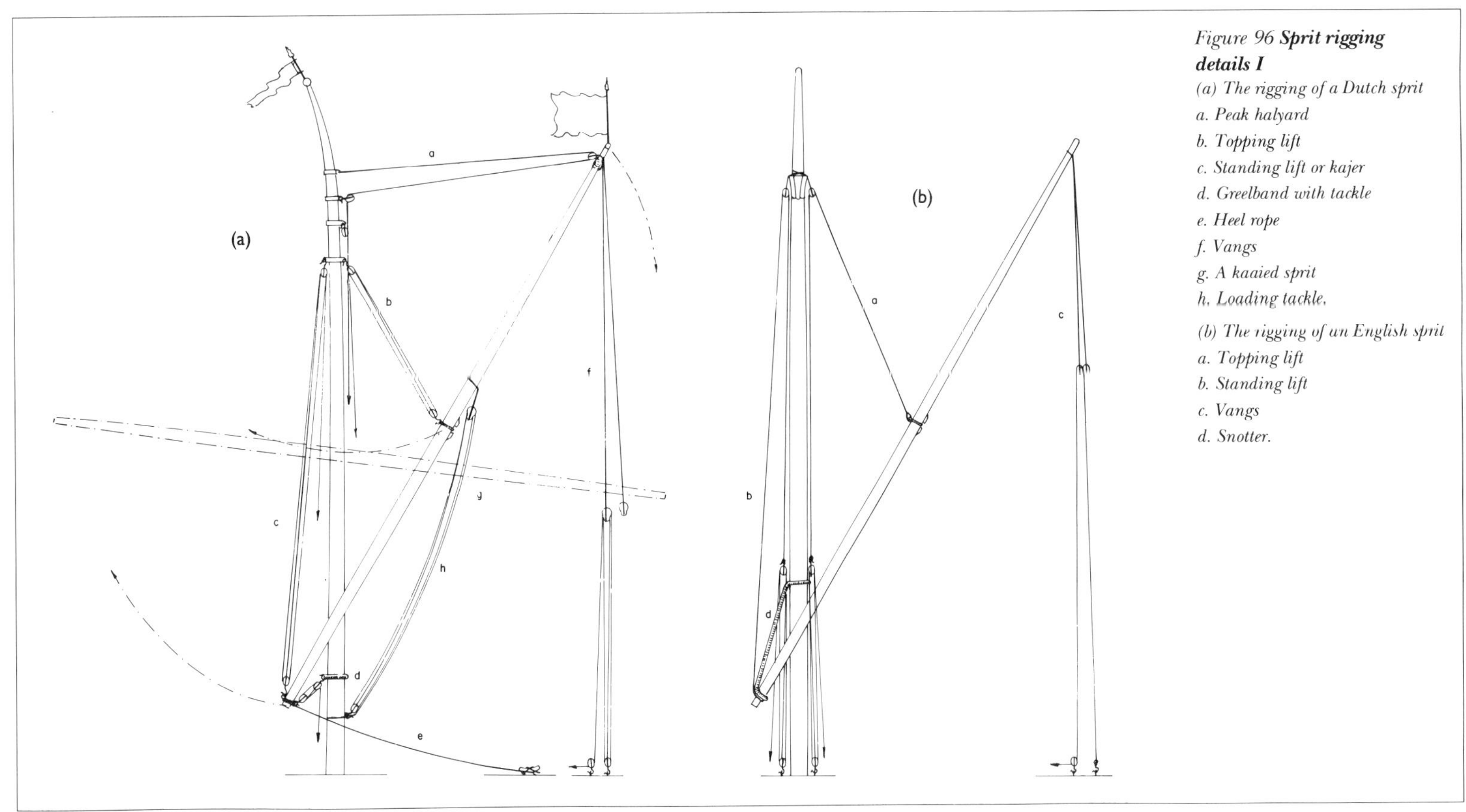

Figure 96 **Sprit rigging details I**

(a) The rigging of a Dutch sprit
a. Peak halyard
b. Topping lift
c. Standing lift or kajer
d. Greelband with tackle
e. Heel rope
f. Vangs
g. A kaaied sprit
h. Loading tackle.

(b) The rigging of an English sprit
a. Topping lift
b. Standing lift
c. Vangs
d. Snotter.

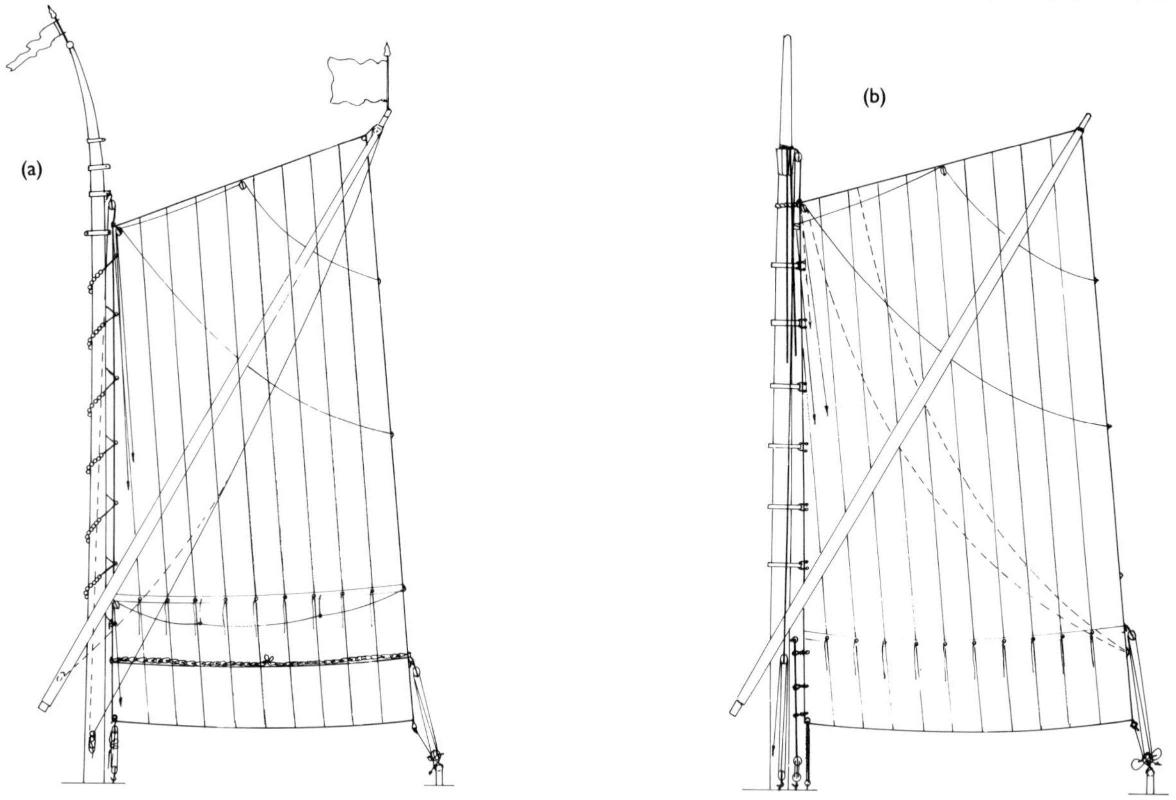

Figure 97 **Sprit rigging details II**

(a) *Dutch spritsail rigging*

(b) *English spritsail rigging*

Standing lift

Stehendes Toppnant, Kajer; Balancine de fond de livarde

The standing lift had an eye spliced into the end which went over the sprit's lower end to enable a topping of the yard. The lift then led through a block at the fore side of the masthead down again, with a thimble spliced into its end (see Figure 96b-b.). Depending on the sprit's size, a double block or luff tackle was hooked to the thimble and to an eyebolt in the deck or side. Instead of this arrangement, Dutch vessels usually had the tackle set up between the sprit's lower end and the masthead, with the hauling part of the fall coming down forward of the mast to be belayed to a cleat (see Figure 96a-c.).

Vang

Ger; Palan de garde

Steel noted that 'some large barges have vangs like a ship's mizen, and a down-hauler at the peek-end of the sprit-yard' (see Figure 96b-c.). Vangs were part of all sprit-rigged Dutch vessels (see Figure 96a-f.), as far as can be ascertained from a large number of pictures of sprit-rigged vessels.

Peak halyard

Piekfall; Drisse de coqueron

Some Dutch vessels had a peak halyard made fast to the masthead. It rove through a block at the sprit's peak then through another near the topping lift's block at the masthead and down upon deck (see Figure 96a-a.).

Heel rope

Fussreep; Guinderesse de livarde

The rigging of a Dutch sprit also featured a heel rope to bring the lower end of a topped-up sprit back towards the tackle at the rope band round the mast (see Figure 96a-e.). It was either a simple rope fastened to the sprit's heel, or consisted of a block stropped to the heel with a whip reeving through, the standing part of which was set up round a cleat in the aft part of the vessel.

Snotter

Sprietstropp; Chambrière de livarde

The snotter was made of two or more turns of rope loosely around the mast and sprit yard, with the ends spliced together. It was marled tightly and served over with spun-yarn, then covered with leather. It was seized close to the mast, and the stop of the sprit rested in the smaller eye at the other end (see Figure 96b-d.).

Instead of a snotter, Dutch vessels had a *greelband* laid round the mast and resting on a collar (see Figure 96a-d.). A short tackle connected this to the sprit's heel, and could be loosened when the sprit was *kaaied* (*kaai* means 'wharf', and as a verb it means to set the yards apeak, that is, to bring all yards in line with the keel and then top them, so as not to interfere with other vessels when lying at a wharf.) A greelband was a type of snotter without the eye for the sprit (a *greling* was a hawser, slightly thinner then a stream anchor's); no comparable English term exists.

Horse

Leiter; Filière d'envergure

A horse went over the masthead and was set up similar to that described in Chapter V.

Spritsail

Sprietsegel; Voile à livarde

'The spritsail is bent to hoops, that slide on the mast above the snotter, and to hanks below it on a horse abaft the mast,' noted Steel on the spritsail of an English barge. A Dutch spritsail was not bent to hoops, but laced, with trucks filling the lacing turns round the mast; alternatively, robands were used, marled and served and probably covered with leather, as can be ascertained from a large number of illustrations and some models. Dutch vessels did not have a snotter, and so they had no need for a horse below the snotter. The attachment of the spritsail to the mast was therefore uninterrupted, with the lacing or robands being loosely fitted to ensure the smooth hoisting and lowering of the sail.

Halyard

Fall; Drisse

The spritsail halyard was made fast to a grommet at the nock of the sail, and rove through a block at the masthead upon deck.

Tack

Hals; Point d'amure

The spritsail tack was secured with several turns round a cleat or eyebolt in the deck and through the thimble in the sail's tack, then frapped together and the end hitched.

Nock

Nock; Point de mât, point de drisse

The spritsail nock was secured in similar fashion to a grommet loosely encircling the mast.

Peak

Piek; Coqueron de livarde

The peak of the spritsail was extended by the top end of the sprit, which was fitted through an eye formed by the boltrope at the peak. Dutch models and illustrations suggest that the sail's peak was either lashed to an eyebolt or a peak halyard rove through a sheave hole in the sprit's upper end to a block attached to the nock grommet or masthead and then upon deck.

Sheet

Schot; Écoute

The standing part of the sheet was bent to the sail's clew. It rove through a block which traversed on an iron or wooden horse near the stern, then through another block hooked to a thimble in the sail's after leech about 4 feet above the clew, and was belayed to a pin in the block on the horse.

Brails

Dempgordinge; Cargue d'une voile

Spritsail brails on English vessels were set up in either of two ways. In the first method they were made fast to the cringles in the after leech of the sail and led upwards on both sides to small blocks seized into the head rope, then through blocks in the upper part of the shrouds. They then rove through trucks seized to the middle of the shrouds and were belayed to pins in the shroud rack. Alternatively, the brail consisted of two short legs, one spliced into the head rope near the nock and the other about 4 feet up the head. The leading part rove through a cringle on the after leech of the sail about 5 to 6 feet above the clew and came up on the other side to a block, seized to the nock of the sail, and was led down by the mast.

Dutch brails led from the cringles in the after leech, one about a quarter below the peak and the second slightly more than half below, to blocks at the head rope and nock, and were belayed at the shroud racks. A third brail was set approximately at the lower quarter of the after leech and rove parallel to the foot through two or three beckets in the sail, and was belayed to cleats on the mast.

Throat downhauler

Nockniederholer; Hale-bas de mâchoire de corne, hale-bas de croissant

The spritsail throat downhauler was spliced into the nock of the sail and led down by the mast.

Barges and lighters

Kähne und Leichter; Chalands et allèges

A wide variety of eighteenth century vessels fit into this category, but common to all was their use as transport craft on inland waterways and in harbour. River vessels were products of their environment; their length, width, shape and rig were completely determined by the navigational peculiarities of the waters they sailed on. Most of those vessels had a one or 1½-masted sprit rig. While small lighters often had only a simple spritsail, larger barges on Rhine, Elbe, Thames and other large rivers also carried a foresail, a jib, a squaresail and a topsail.

Strikeable masts, tabernacle masts

Umlegbare Masten; Mâts à tabernacle, mâts à bascule

Sloops, smacks, tjalks, barges, lighters and all other vessels which habitually passed beneath bridges had their masts stepped in a trunk or wooden cap on the deck, and held secured by an iron strap on the aft side (or, alternatively, had a bolt through the heel, or the heel itself hinged to the deck). The mast could be lowered aft by easing the fall of the stay tackle. To raise the mast, the fall was brought to the windlass and hauled until the mast stood upright again. The fall was then stopped to the windlass bitts.

The mast was not always hinged at the heel; occasionally a turning bolt was fitted in the upper part of the trunk, with the heel itself extending into the hold. A loose tabernacle plank covered the mast housing.

To assist in lowering the mast, strike booms were needed. These booms were set up temporarily athwart the mast in steps close to the sides, or were hinged there. Their upper ends crossed to form a fork above the upper stay tackle block, and were lashed to the stay. These booms maintained the correct angle for the fore stay as the mast was lowered; without them the stay could not have supported the mast once it was lowered beyond a certain point. The same booms were necessary when raising the mast, to overcome the dead angle at which the stay could not lift the mast.

These booms were not mentioned in contemporary sources: the only reference to sheers is in the context of stepping and unstepping lower masts, a process in which a similar mechanical principle was employed. The difference between sheer legs used for this purpose and the booms described above is that the former were used to lift a mast vertically to take it out or hoist it in; their use was probably therefore rare.

Tabernacle masts were not often fitted to vessels larger than small channel craft on which two men could raise or lower the mast. Only after the larger rivers were bridged in the nineteenth century did sheer legs became more common in illustrations of larger vessels. None of the authors except Steel mentioned mast trunks or tabernacles, but draughts of vessels exist where masts too large and heavy to be handled by a few men were set in trunks, and such masts could clearly not be lowered or stepped without sheer legs. Such sheer legs were also common among the junks of China's waterways.

VII

Rigs for boats

Launches, longboats

Grossboote, Barkassen; Grands canots, chaloupes

A ship's launch was usually rigged like a fore-and-aft sloop (see Figure 98a) or like a small schooner. A foresail was normal and a flying jib could be set to a running bowsprit. Boat masts were usually supported by a pair of shrouds set up with tackles.

Pinnaces and rowing barges

Pinassen und Labberlots; Pinasses et grands canots de vaisseau

Pinnaces and rowing barges were rigged in a variety of ways, with sloop, sliding gunter, lateen, sprit or a simple squaresail rig. These boats carried up to three masts and sometimes a bowsprit.

Sliding gunter rig

Schubstengetakelung; Gréement houari

The sliding gunter rig was described by Falconer as follows:

> Abaft the mast are set lateen-sails and sliding top-masts. The lower part of the sail is bent to hoops that encircle the lower or standing mast and the upper part of the sail is laced to the top-mast, which slides up and down the lower-mast by grommets, or iron rings, fastened to the heel of the mast. The sail is fastened at the lower part, with a tack to the mast, and, at the peek, with a small earring. The sail is hoisted by a halyard, one end fastened to the heel of the top-mast, and the other end reeved through a sheave-hole in the lower mast-head; it leads down towards the heel of the mast, and there belays. The sail is extended by a sheet, fastened to the clew, and led aft. To the heel of the top-mast is sometimes fastened a down haul rope, that leads down towards the heel of the mast.
>
> These sails furl in a close manner to the lower-mast, by lowering the top-mast and confining the sail in folds by a furling-line. On the bowsprit is spread a jib, which assists the vessel in going about. These sails are called sliding gunters, and are used in the navy's pinnaces and barges.

Nothing can be added to this comprehensive explanation (see Figure 98b).

Lateen rig

Lateintakelung; Gréement latine

The lateen rig originated in the Mediterranean Sea and found its way into boat rigs in eighteenth century Northern Europe through the use of galleys and other Mediterranean types in northern fleets, especially in the Baltic.

Lateen rig was well suited to sheltered waters and light winds, and great speeds could be reached by lateen-rigged vessels. Characteristic was the triangular sail bent to a yard hung at about 45 degrees on a short mast. The reef band reached from the tack to a point on the after leech about a quarter down from the peak (see Figure 99a).

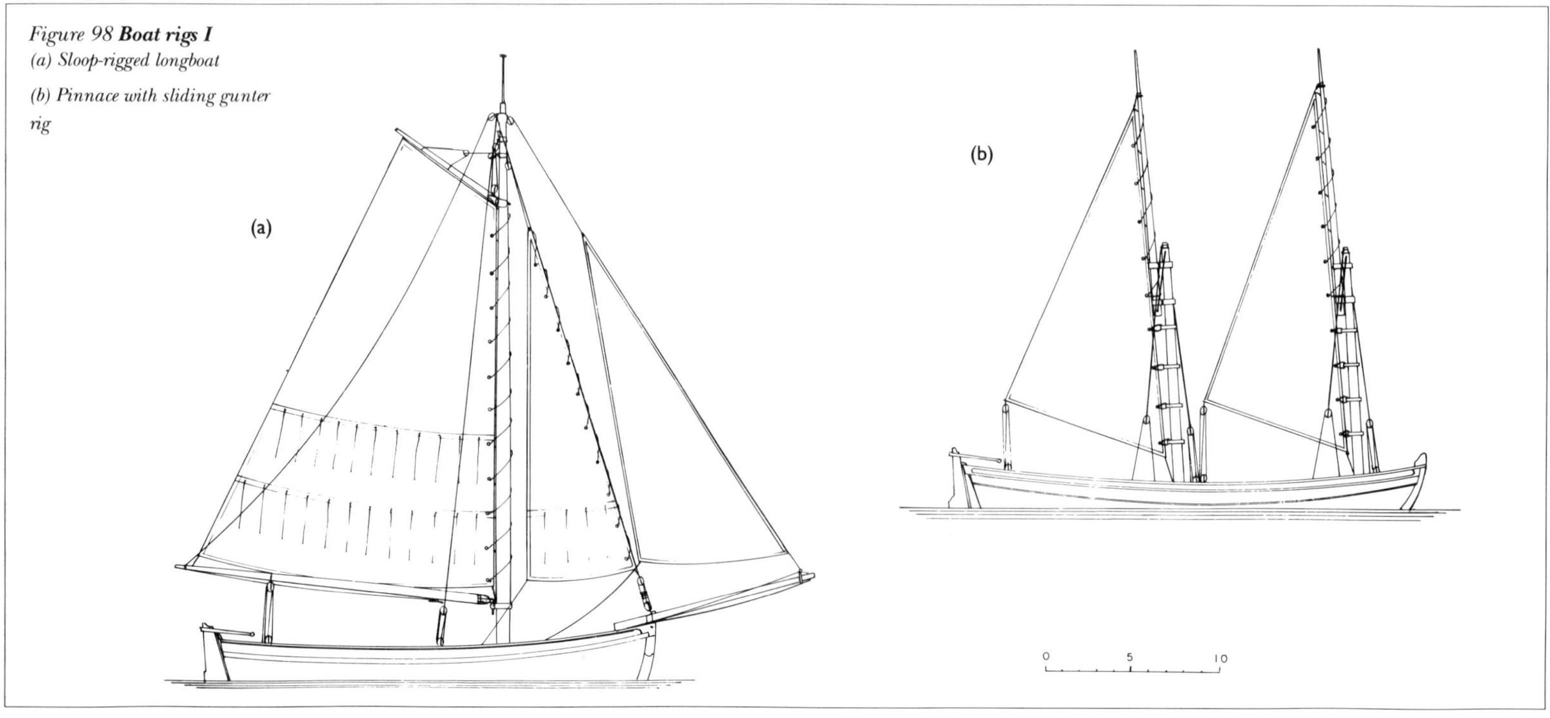

Figure 98 ***Boat rigs I***
(a) Sloop-rigged longboat
(b) Pinnace with sliding gunter rig

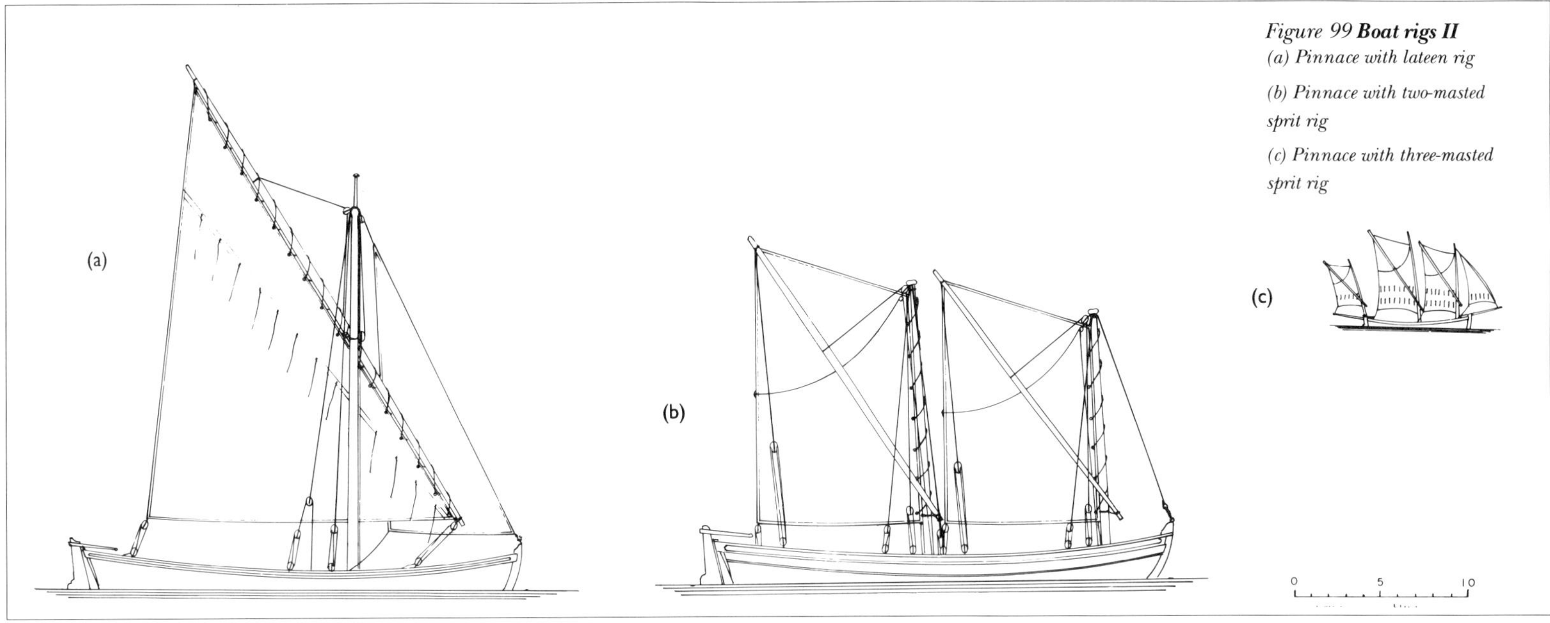

Figure 99 ***Boat rigs II***
(a) Pinnace with lateen rig
(b) Pinnace with two-masted sprit rig
(c) Pinnace with three-masted sprit rig

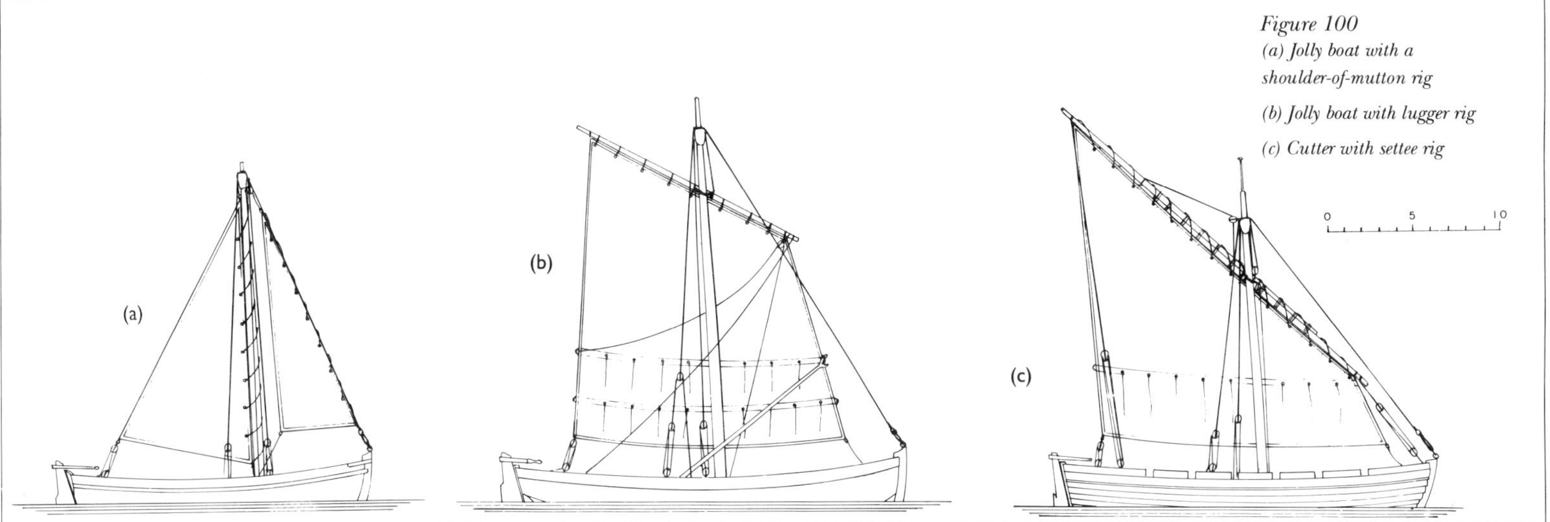

Figure 100
(a) Jolly boat with a shoulder-of-mutton rig
(b) Jolly boat with lugger rig
(c) Cutter with settee rig

Sprit rig

Sprittakelung; Gréement livarde

Boats had one, two, or three-masted sprit rigs, all of which were usually somewhat simpler than the sprit rigs used in larger vessels (see Figure 99b and c). The sprit usually stood in a snotter and a simple topping lift kept it in position. The peak of the sail was extended by the upper end of the sprit, and the nock was hoisted by a halyard. The sail was laced to the mast, and was fitted with brails, a simple tack and a tackle for the sheet. Vangs were also provided. When a pinnace or barge was rigged with three masts, the third was stepped at the stern and an outrigger for the sheet was extended over the taffrail. A bowsprit was sometimes fitted for a flying jib (see Figure 99c).

Cutters and jolly boats

Kutter und Jollen; Cotres et petit canots

Ship's cutters and jolly boats were usually single-masted. The rigs described below were also employed in other boats, and those rigs described above were also sometimes used for cutters and jolly boats.

Shoulder-of-mutton rig

Schafschinkentakelung; Gréement de voile aurique, Gréement d'oreille de lièvre

A shoulder-of-mutton rig was the simplest way to set a sail in a boat. Neither gaff nor boom, sprit or lateen yard was needed. The triangular sail was laced to the mast and rigged only with a halyard, a tack and a sheet - the barest minimum rigging, but suitable for a small boat (see Figure 100a). A foresail was often carried on the stay. Röding described the shoulder-of-mutton as an English sail, rarely found in German waters.

Settee rig

Settietakelung; Gréement de sétie

The settee sail was an Arab version of the lateen sail. In contrast to the latter, the sail's luff leech was cut with a slight forward angle, and the sail was quadrangular rather than triangular. The yard was ordinarily shorter than a lateen yard, and was hoisted at a flatter angle than 45 degrees. At about one fifth height above the clew a reef band ran parallel to the slightly gored foot leech. A settee yard was fitted with a halyard, a peak halyard, a tack tackle and vangs. The sail being bent to the yard controlled by a tack and sheet (see Figure 100b).

VIII

Foreign and exotic rigs

Just as the preceding chapters were able to deal fully only with the major northern rigs, so this chapter concentrates on major characteristic types from the wide range of sailing vessels found elsewhere in the eighteenth century world. The ships and craft discussed here were at home in southern Europe, the Middle East, the Far East, the Pacific and on the American Atlantic seaboard. They provide a colourful picture of the range of vessels a trading or exploring seaman might have encountered during the eighteenth century.

Bark

Bark; Barque

The origin of the term 'bark' can be found in the Latin word *barca,* which was applied to a range of small vessels in the Mediterranean Sea.

The bark of Northern Europe has been described in Chapter III, and needs no further discussion here. Perhaps the best known example of that type was HMS *Endeavour,* the vessel in which James Cook undertook his first voyage to the Pacific; beside Cristopher Columbus' *Santa Maria* it is today one of the most popular vessels in maritime history.

The bark's evolution began with the larger types of ship's boats and smaller merchant craft, under the names barca, barco, barge, Barke, Barkasse, barque or chaloupe. John Smith (1627) described the use of barks as fireships:

> Now betweene two Navies they use often, especially in a harbour or road where they are at anchor, to fill old Barkes with pitch, tar, traine oile, lincet oile, brimstone, rosen, reeds, with dry wood and such combastible things; sometimes they linke three or foure together in the night, and put them adrift as they find occasion.

The common 'barca' as described by Furttenbach in his *Architectura Navalis* of 1629 was a completely decked vessel with a head, a flat stern and a raised poop. She was 52 palmi (approximately 41 English feet) long and rigged like a feluca (see below) with two lateen sails. He stated that these vessels were sailed or rowed by six men, and transported all sorts of merchandise, especially citrus fruits. The sheer plan produced by Furttenbach differs little from Paris' plans of more then two centuries later.

A French *barque* from Bordeaux of 1679 was built similarly but without a head, and was square-rigged. Other barks were open boats with two masts, built double-ended. Nevertheless, all these seventeenth century barks had in common that their masts were fore and main, while the 'catches' of the same era had a main and mizzen mast.

It was the French *barque longue* or double chaloupe which was used in the Navy, and subsequently developed into the French naval corvette. These were between 50 and 70 English feet long and had a tonnage of 20 to 90 tons. Between two and ten guns

Figure 101 **Bark**

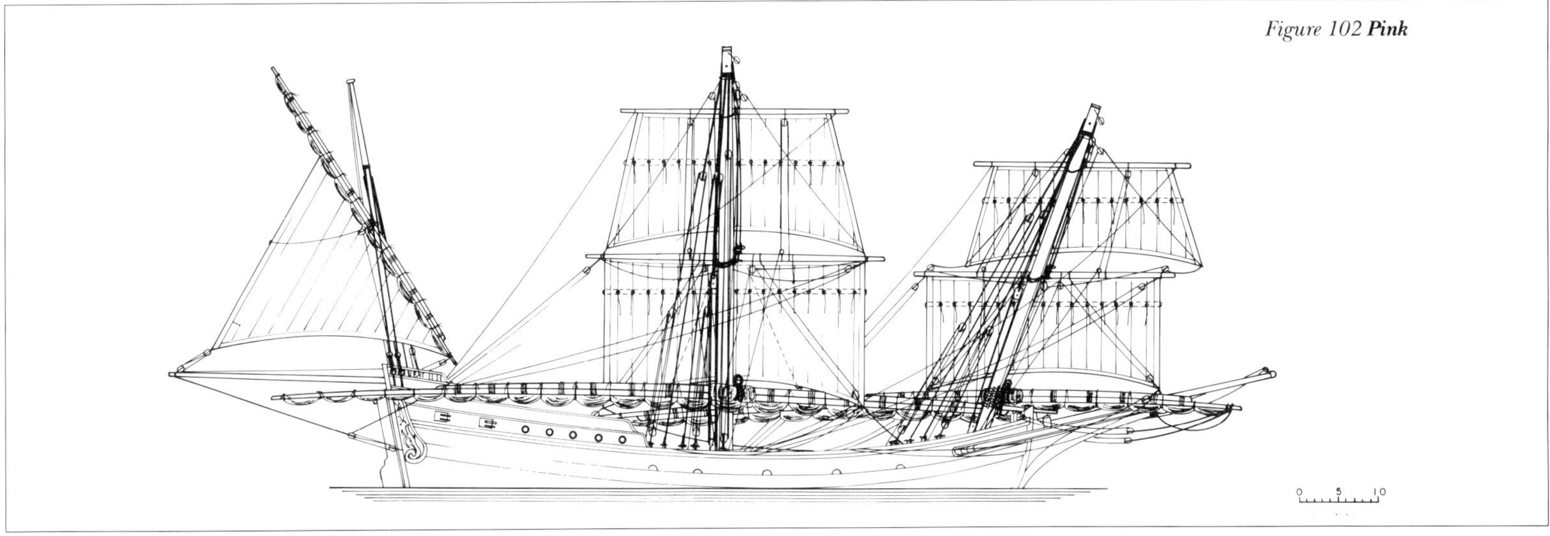

Figure 102 **Pink**

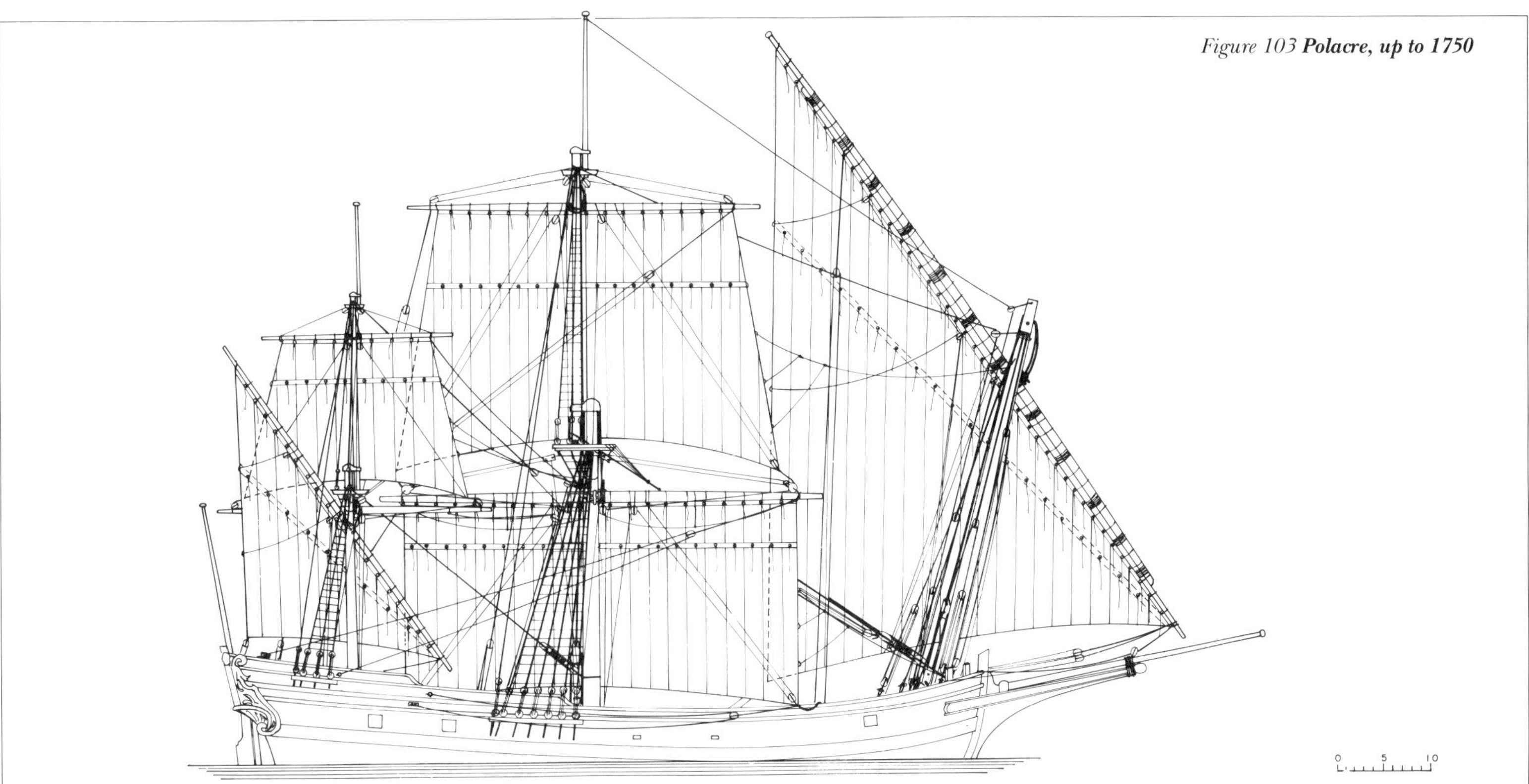

Figure 103 **Polacre, up to 1750**

were mounted, normally 3- or 4-pounders. The term 'corvette' replaced 'barque longue' in about 1677-8. The double chaloupe was an open vessel rigged only with two squaresails, whereas the barque longue was also rigged with topsails. The Danish Navy also employed *Barkalongen* in the late seventeenth century. The Spanish *barca longa* was a completely different craft, a large fishing vessel with two or three lug-rigged masts.

Röding noted that the Mediterranean bark was a vessel equipped for war or trade, short, wide and fully built with noticeable tumblehome and a fair degree of rake to the stem and the sternpost.

Barks were three-masted vessels, usually without a bowsprit (the bowsprit was replaced by a beak, a common sight on Mediterranean craft; some barks had this beak extended by an outrigger, a feature Röding considered characteristic of polacre rig). Masting consisted of foremast raked sharply forward, a vertical mainmast and a mizzen mast raked slightly aft. The fore and main were short block masts, with a squared head incorporating sheaves (the block). The mainmast, however, was sometimes a polacre mast, while the mizzen mast, normally a mast with a topmast, was sometimes fitted as a block mast or a normal pole mast.

The rigging of barks was as varied as the masting, ranging from a full lateen rig, through a rig with a mizzen topsail and small squaresails on the main and foremasts, to a complete polacre rig. These rigs came with or without a triangular foresail.

Like their predecessors, eighteenth century barks were equipped with rowing ports and carried between six and eight pairs of sweeps.

Pink

Pinke; Pinque

The pink was another Mediterranean type, but was sometimes also referred to as a bark. She was rigged similarly to a bark but built only for trade, and had a flatter floor and had no sweeps. Her stern was much smaller than a bark's.

Röding noted that pinks were mainly found in Naples and in Spain, but the pinks in Paris' *Souvenirs de Marine* were Genoese. The identification of these three areas is sufficient to suggest that

Figure 104 **Polacre (xebec), about 1780**

pinks were at home on most parts of the European coastline of the West Mediterranean.

The squaresail rig shown in Figure 102 was used when running before the wind; otherwise the lateen rig was normal (the lateen yards are shown lowered).

In northern waters a pink was a three-masted square-rigged merchant vessel with a small stern, finer fore-and-aft underwater lines than a bark's and an ornamented head. 'Englishmen call therefore any ship which is high in the rear and has a small stern "pink sterned"' (Röding).

Polacre
Polaker; Polacre

A polacre was a Mediterranean vessel chiefly employed in the transportation of merchandise. Though also fitted with three masts, a polacre was rigged differently from a bark or pink.

The term 'polacre' originally had a different meaning from our modern sense of the word. Now we refer to a polacre rig without taking any special hull type into consideration; this was not always so. In 1629 Furttenbach considered the polacre a particular vessel type, the second largest square-rigged type in the Mediterranean Sea. He wrote:

> In addition to the *nave*, the Italians employ another vessel which they call *polaca*, which is comparable with a small *nave* and which can carry a large cargo of goods, usually wine, grain, salt or timber.

He noted that the vessel had a length of 85 *palmi* (approximately 68 English feet).

By the eighteenth century, however, as noted above under Bark, polacre rig was not confined to one type of vessel; in the French Navy especially efforts were made to apply the rig also to xebecs. This proved unsuccessful, since xebecs with a polacre rig were less manoeuvrable than their lateen-rigged sisters.

Falconer noted that polacres had squaresails set to their mainmasts and lateen sails to the fore and mizzen masts, but he also referred to a variant polacre rig found in Provence, where all masts were square rigged. Steel and Röding described polacres as fitted with square-rigged pole masts; Röding described the rig in detail (see Cat in Chapter III). Steel described the rig defined by Falconer as 'Polacre-Settee', and commented 'the main-mast ever keeps the rigging of the polacre'.

The descriptions of masts and yards given by Falconer and Steel were very similar to Röding's, except that Falconer noted the absence of horses on any of the yards while Steel described horses on the lower yard for furling or loosening the sails. Steel's description of this feature corresponded to a drawing of a polacre-rigged French xebec of 1784 by Victor Jouvin. Jouvin showed trestle trees on the fore and mainmasts, and also additional topgallant masts. A watercolour by Roux from 1800 entitled *French armed chebec* confirmed Jouvin's observations, apart from a few details of the mizzen mast.

From these illustrations it can be ascertained that two backstay pairs were set up to the masts and that the bowsprit sometimes carried a spritsail. The mizzen was either a lateen sail or cut like a ship's mizzen and laced to the mizzen yard and mast.

Early nineteenth century illustrations (for example, another watercolour by Roux from 1801 of the Greek polacre *Bella Aurora* provide evidence of alterations within the rig in line with late eighteenth century innovations in ship rigging. Thus the bowsprit is shown with a cap, a jib-boom and a dolphin striker, while the mizzen yard is replaced by a gaff and boom. If before the introduction of a bowsprit and complete square rig after the middle of the century, polacres sometimes had a triangular jib only (see Bark above), then the new polacre rig had a full complement of fore-and-aft rigged staysails (fore top staysail, jib, storm jib, main top staysail, mizzen staysail and mizzen top staysail). Except that it had pole masts and lighter rigging, polacre rig was now very much like ship rig. One early example of the new polacre rig can be seen in the foreground of J Vernet's painting of 1756 entitled *View of the old Port of Toulon*.

It is worth noting, however, that polacres of the late seventeenth and early eighteenth centuries were all rigged as

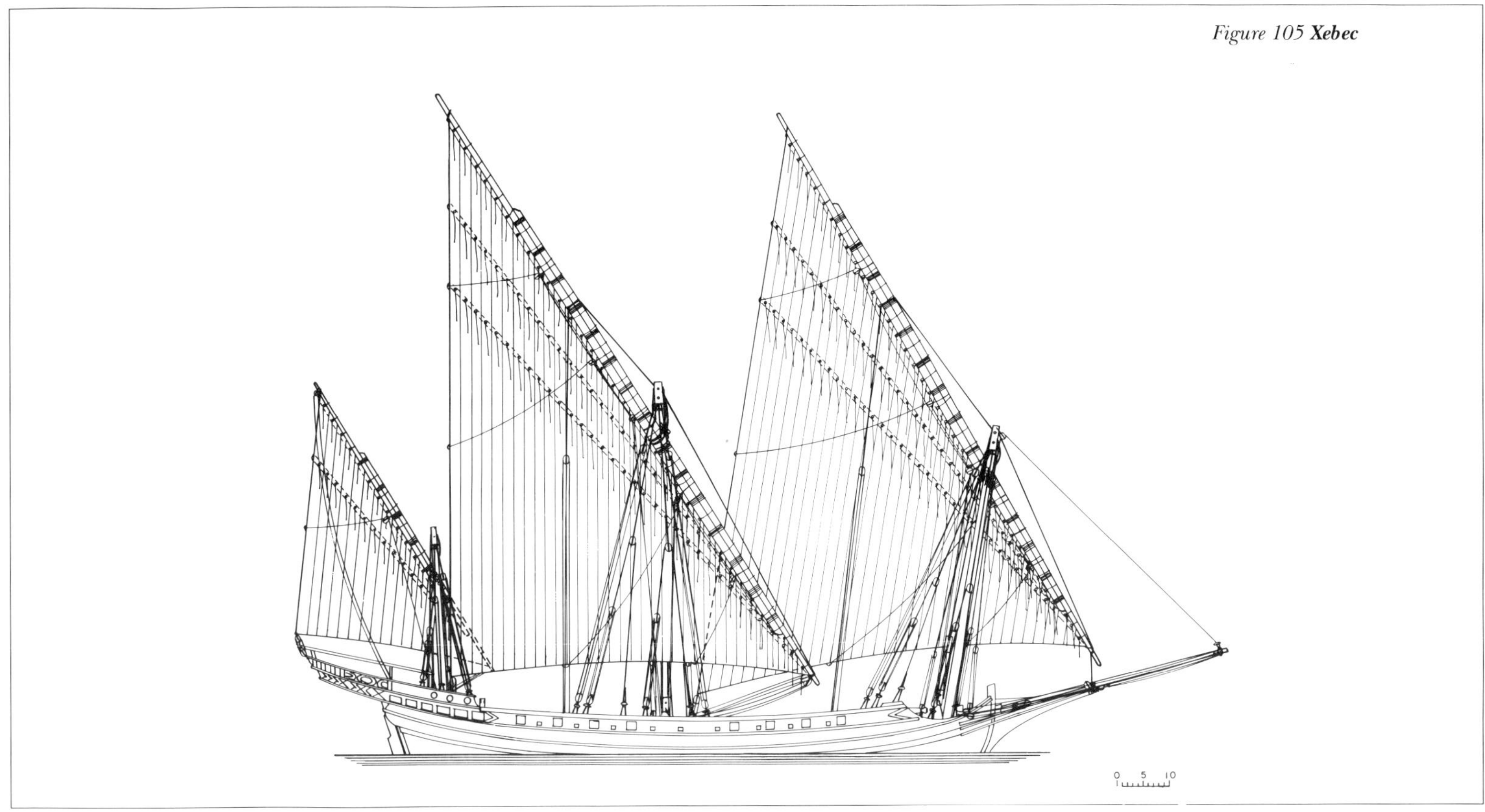

Figure 105 **Xebec**

described by Falconer. Illustrations by four different artists of that period confirm that the earliest dating of a so-called polacre mast (pole mast) cannot be established before the early years of the eighteenth century. The square-rigged mainmast of earlier polacres was fitted with a topmast and a topgallant mast, and all four artists (J Jouve in 1679, G du Pas in 1720, C Randon and D Serres) showed the mizzen mast, also fitted with a topmast, carrying a topsail yard. In Jouve's illustrations the main topmast is abaft the mainmast and the topgallant mast or flagstaff forward of the topmast. Randon's topmast and topgallant mast in an illustration from the end of the seventeenth century are normally stepped, while etchings by du Pas from about 1710 confirm Randon's masting. Both square-rigged masts had tops and topmast shrouds. The topgallant masts rested between trestle and cross trees and light shrouds supported them.

While Jouve and du Pas show only topsails, Randon and Serres also showed a topgallant yard or sail. Serres, a marine painter in England since about 1752, would probably not have painted Mediterranean vessels from life, but would have relied on sketches made in his earlier seafaring time or on second-hand information (in both cases possibly on out-of-date material). His polacre was rigged with topmast and topgallant mast, while the polacre-rigged xebec in the *Liber Nauticus* had a pole mast stepped as a mainmast.

A very early impression of a completely square-rigged polacre (fore course, main course and topsails on all three masts), but without a real bowsprit and without a spritsail, is du Pas' etching entitled *Ship-Barque* of 1710.

Xebec, chebec

Schebecke; Chébec

Besides 'galley', 'xebec' was the most widely and generally used colloquial term for vessels in the Mediterranean Sea. In fact, a xebec was a vessel 80 to 130 feet long with a distinctive hull, as Falconer noted:

> [The xebec is] very different from [the polacre] and almost every other vessel. It is furnished with a strong prow, and the extremity of the stern, which is nothing more than a sort of railed platform or gallery, projects further behind the counter and buttock than that of any European ship.
>
> Being generally equipped as a corsair, the xebec is constructed with a narrow floor, to be more swift in pursuit of the enemy, and of great breadth, to enable her to carry a great force of sail for this purpose, without danger of overturning. As these vessels are usually very low-built, their decks are formed with a great convexity from the middle of their breadth towards the sides, in order to carry off the water, which falls aboard, more readily by their scuppers. But as this extreme convexity would render it very difficult to walk thereon at sea, particularly when the vessel rocks by the agitation of the waves, there is a platform of grating extending along the deck from the sides of the vessel towards the middle, whereon the crew may walk dry-footed, whilst the water is conveyed through the grating to the scupper.

The origin of the xebec is hard to trace, but the corsairs of North Africa's Barbary Coast recognised early the type's great suitability to their trade. According to legend Ali el-Uluji (Ali the Apostate) had a strong personal influence on the development of the xebec, and it was through him that the type became the favourite vessel of the North African pirates. Born in 1508 in Calabria and enslaved by Algerian pirates, he spent fourteen years as a galley slave until converting to Islam. Then he rose rapidly through the ranks and was soon known as one of the most ruthless corsair captains. He commanded the Turkish left flank of ninety-three galleys and galliots at Lepanto, and his conduct during the battle and daring handling of his ships in the subsequent retreat to Istanbul endeared him to the Sultan, who bestowed the rank of commanding Admiral of the Ottoman Fleet and Viceroy of Algiers on him. He died in 1587.

Ali's involvement in the development of the xebec may be fictional, but only a few decades after Lepanto, Furttenbach wrote of the Turkish *caramuzzal*, 'this is nothing but a higher and strengthened Tartana, a swift-running vessel in much use by the

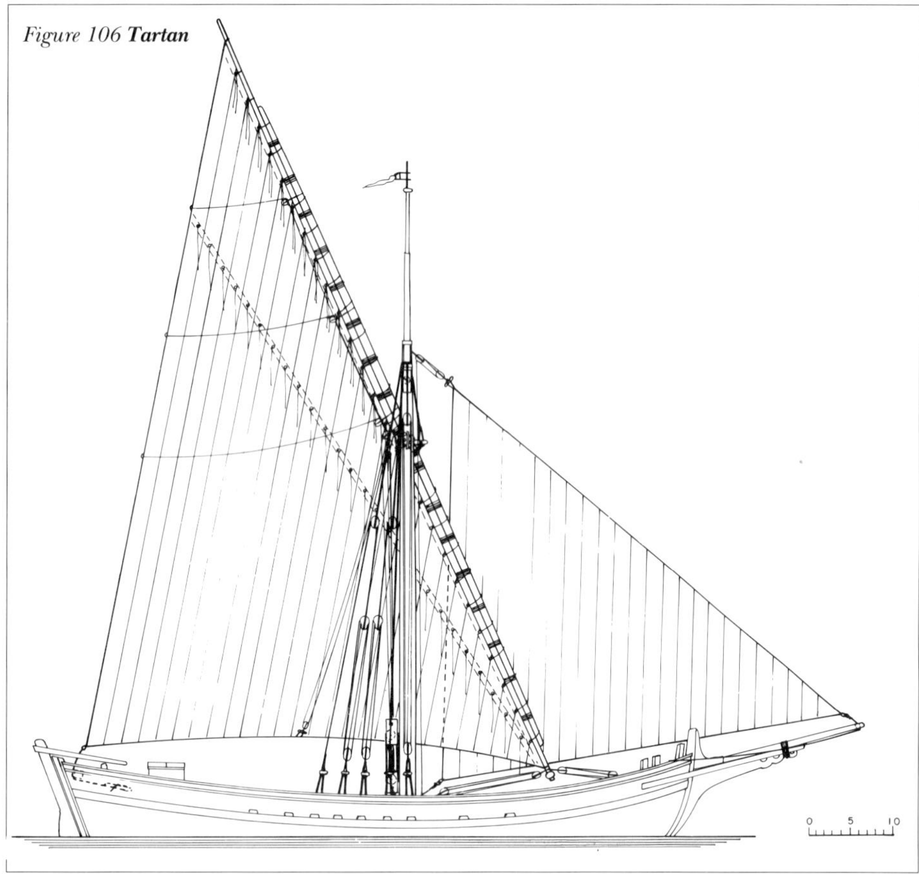

Figure 106 **Tartan**

Turks and Barbaries in the Mediterranean Sea for shipping and to do great harm to the Christians'. He explained that the Barbary pirates ambushed smaller Christian vessels, mainly during the fine weather months from May to August, from the harbours of Susa, Algiers and Tunis. These vessels were armed with eighteen or more guns and had carried fifty to eighty well-equipped corsairs. Since the tartana was closely related in construction to the xebec, the caramuzzal must have been much the same vessel by another name. Xebec itself is an eighteenth century term.

In an effort to fight fire with fire, the French and the Spanish navies in the eighteenth century also employed xebecs in the Mediterranean Sea. According to Paris these vessels were armed with between eighteen and twenty-four guns, and Falconer noted that 'the xebecs, which are generally armed as vessels of war by the Algerines, mount from sixteen to twenty-four cannon, and carry from 300 to 450 men, two-thirds of whom are generally soldiers'. Röding noted twelve guns for the smallest and forty guns for the largest xebecs. The Algerian xebec drawn by Chapman carried sixteen 6-pounders, four 12-pounders and eight 3-pounders.

Lateen-rigged xebecs had three block masts and carried sails for any change in the weather. '[The xebec] is capable of sailing with three different sets of sail, according to the force or direction of the wind' noted Falconer, and he quoted an Algerian captain of his acquaintance that 'the crew of every xebec has at least the labour of those of three square-rigged ships, in that the standing sails are reset to answer every situation of the wind'.

When the wind was fair and from near astern, it was normal to set wide squaresails to the mainmast and very often also to the foremast. If the wind was unfavourable for courses but still moderate, then the squaresail yards were taken down and large lateen yards and sails hoisted instead. When the wind increased to a storm, these yards were again lowered and on all masts replaced by shorter yards and proportional lateen sails.

As noted above under Polacre above, not all xebecs were rigged in this way, and some xebecs in the French Navy had a polacre rig. Röding explained this as an attempt to simplify the arduous procedure of sail handling for every occasion, but noted a loss in sailing qualities; Steel simply noted that 'they never sail so well as they did in their primitive arrangement'.

In *Souvenirs de Marine* Volume 2, Paris reproduced a photograph of a xebec model of 1750 with a lateen rig, and another photograph with plans of the xebec *Mistique* of 1750-86, with an old style polacre rig but with a pole mainmast and a large jib. Another xebec with twenty-two guns and a later style polacre rig (see Polacre above) can be found in Volume 1. On account of the stepped topgallant masts on this vessel, horses were rigged to the topgallant yards, and the mainmast had a flying royal yard.

Röding and Steel both noted that the mizzen mast of a xebec was fitted with a topmast and that its shrouds were set up to channels as on a ship, with Steel adding '... [this mizzen topmast] has been lately added, to keep them to the wind'. He also mentioned that the shrouds in general were 'set up, similar to the runners in English cutters or sloops, to toggles fixed in the sides. These shrouds are easily shifted when the vessels go about.'

Xebecs, like many other Mediterranean vessels, were also equipped with sweeps to ensure continued mobility in calm wind conditions. The number of sweep pairs was between eight and twelve, depending on the vessel's size, and the sweep ports were placed between the gun ports.

The rigs of xebecs, barks, pinks and polacres were so interchangeable during the eighteenth century that each of these types might in fact have been fitted with any of the rigs discussed above.

Tartan, tartana

Tartane; Tartane

The tartan was a coaster of the West Mediterranean with one mast and a short bowsprit. The rigging consisted of a large lateen sail and a similarly large flying jib. Falconer and Steel noted that tartans, like other fore-and-aft rigged vessels, used a squaresail when running before the wind. Röding noted the existence of both one and two-masted tartans, and the likelihood of a square topsail on the mainmast. A model of a French tartan rigged in this fashion can be seen in the Musée de la Marine in Paris.

As a trading or fishing vessel, the tartan was 45 to 60 feet in length and did not change much over the centuries. As early as 1629, Furttenbach wrote as follows:

> The Tartana is somewhat smaller than the aforementioned Polaca; a vessel of between 60 and 70 *palmi* is called not a Polaca, but a Tartana. [A *palmo* is an old Italian measurement based on the length of the palm of a hand; Furttenbach equated 1 *palmo* with 10 Nuremberg inches, a length of 243mm.]

Usually a tartan's mast length was about seven eighths of the deck length, and an additional flagstaff increased the height on the starboard side. However, the French tartan noted above had a topmast fitted abaft the mast and a set of triangular skysails above the topsail. The mizzen mast, about half the length of the mainmast, stood just forward of the tiller, and was used to set a lateen sail proportionate to its size. Another French tartan in an etching by Jouve from 1679 had her two masts stepped as main and foremast, with two equal size lateen sails. This arrangement also featured in Chapman's sail plan of a tartan. From the material available we can therefore conclude that rigs for these vessels varied widely, and that 'tartan' was probably considered more a hull type designation than a definition of a rig.

Like most of the Mediterranean vessels a tartan's mast had no stay, and the shrouds were set up as running tackles to toggles. Because of its large size, the yard was made from two pieces

Figure 107 **Muletta (Bean-cod)**

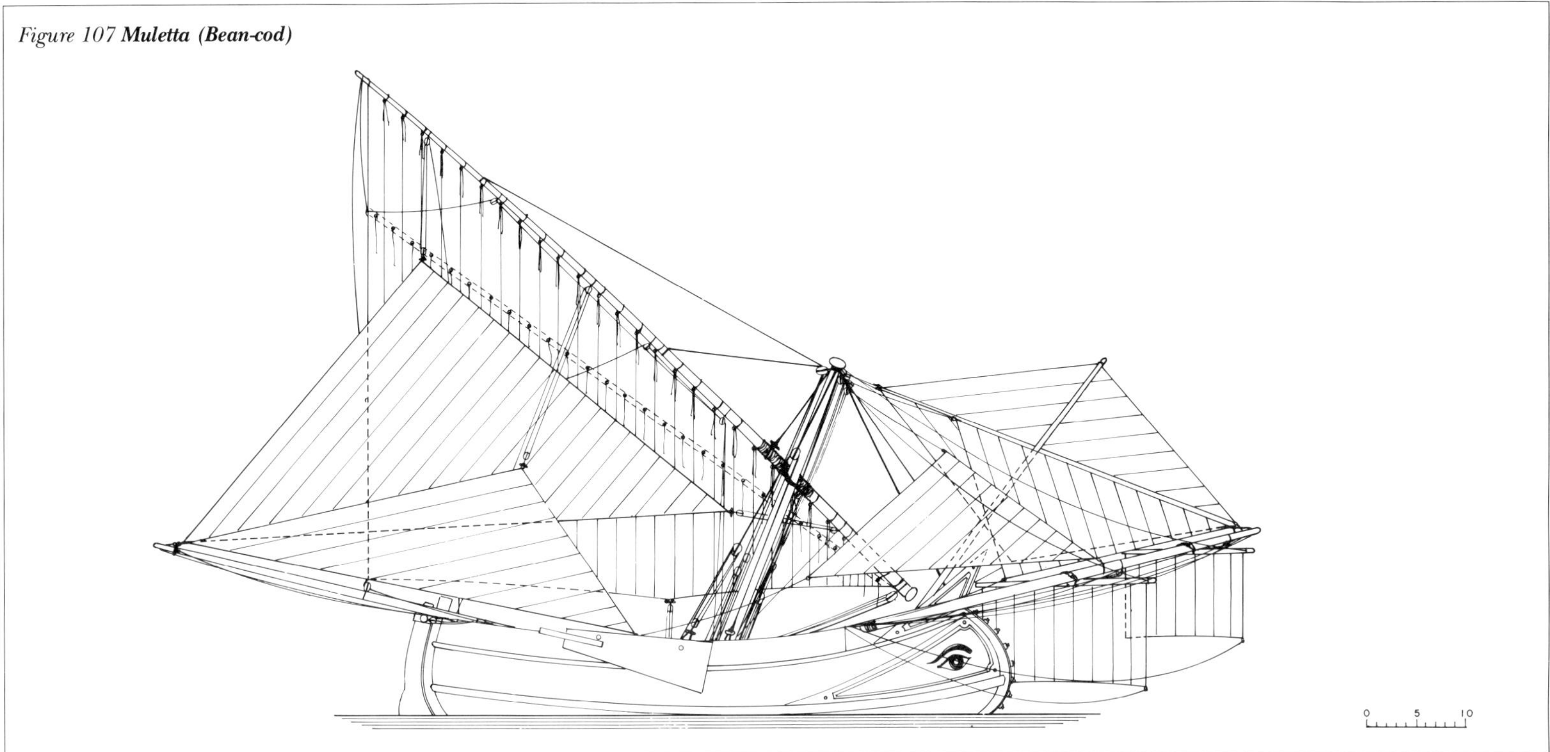

lashed together. The yard tyes were halved and a kind of ramshead block was hung in the bight; the tyes rove through cheek blocks at each side of the masthead and were spliced or seized to a toggle which fitted into an eye strop round the yard.

The basic tartan rig was adapted, with certain local variations, in many of the smaller Mediterranean and Portuguese craft. The Neapolitan Fleet had a number of gunboats rigged in this fashion and after purchase of a few of these gunboats by the US Navy, the rig found its way for a short while even into the American gunboat fleet.

Muletta, bean-cod

Muletta, Moleta; Mulet

One of these variations of the tartan rig could be found on a Portuguese fishing or pilot vessel referred to by English sailors as a bean-cod.

Falconer described the vessel thus:

> [A bean-cod is] a small fishing-vessel, or pilot-boat, common on the sea-coasts and in the rivers of Portugal. It is extremely sharp forward, having its stem bent inward above into a great curve, the stem is also plated on the fore-side with iron, into which a number of bolts are driven, to fortify it, and resist the stroke of another vessel, which may fall athwart-hawse. It is commonly navigated with a large lateen sail, which extends over the whole length of the deck and sometimes also that of an outrigger over the stern, and is accordingly well fitted to ply to windward.

Serres and Röding provided illustrations which in many ways are very similar to a model in the Science Museum in London. The only significant difference is that the model includes a multitude of sails. Serres and Röding showed the vessels sailing in normal conditions rather than under all sail. Only the large lateen sail and the jib would normally have been employed by a bean-cod on passage to the fishing grounds or working as a pilot boat.

An array of differently shaped sails was used to control the vessel when fishing. Particular head or stern sails amongst these were set or taken in to provide manoeuvrability while the vessel was drifting with a net. In addition to the jib, two spritsails beneath the bowsprit and further triangular sails could be set, with clews extended upwards or downwards by a boom which was nearly vertical. This primitive spinnaker helped to keep the bow of the drifting vessel in position.

A long outrigger spread the sheets of two further triangular sails which were hoisted by halyards to the outer part of the lateen yard. The outrigger mentioned by Falconer can also be seen in Röding's illustration.

The rudder was operated not by a normal tiller, but by a short spar set athwartships through the rudder head and controlled by tackles coming in over the sides.

This strange rig was very well suited for the tasks performed by the bean-cod, and probably resulted from the accumulated experience of generations of Tajos estuary fishermen. Spanish fishing boats from the Catalonian coast were similarly arranged with regard to mast position and lateen sail, but did not carry a bowsprit and jib.

Felucca

Felukke; Felouque

Feluccas were smaller seagoing oared sailing vessels. They were mainly employed as coastal traders, and in shape and rig were not unlike a galley, though of smaller proportions. Röding noted that a felucca was about 52 feet in length and 12 feet in breadth. Chapman noted a length of 43ft 10in and a breadth of 8ft 10in.

Röding noted twelve pairs of sweeps per craft and Chapman ten pairs, while Falconer described the felucca as a small vessel with ten to sixteen oars and a lateen rig, and Furttenbach noted six or seven oars per side. The last also indicated that a 'Filucca' was double-ended and lacked the galley's beak (*sperone*). This is confirmed by Chapman's draught, which shows the stem extended only by a short cutwater.

The vessel had a main and a foremast, both raked forward at 3 degrees according to Röding. A short outrigger over the bow helped to keep the fore yard's tack down according to both Steel and Röding.

Röding explained further:

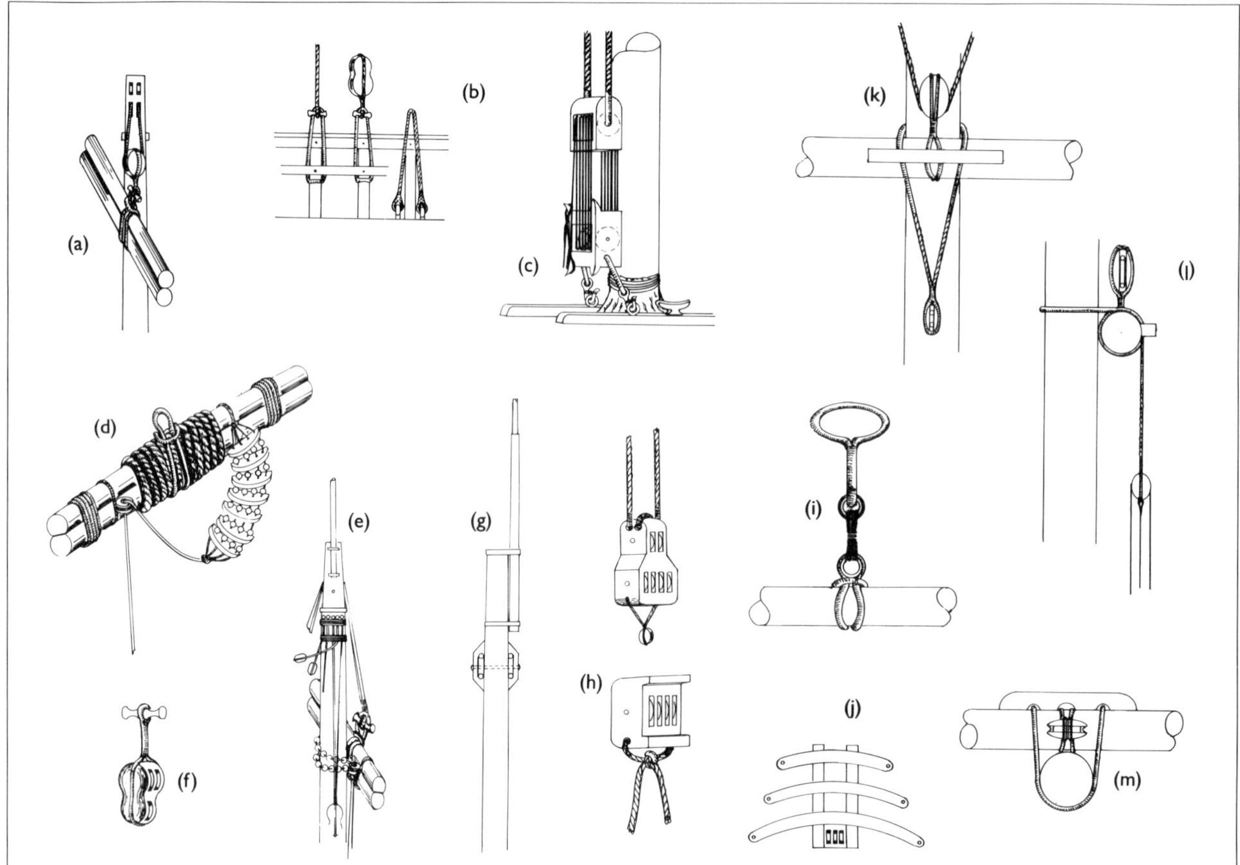

Figure 108 ***Details of rigs shown in Figures 101-109***
(a) Toggled tye block of a bark, pink or xebec, hooked to a yard strop.
(b) A toggled shroud block and runner set up to a strop. These were either laid round the frame futtocks or fastened to eyebolts in the deck.
(c) Halyard blocks of a bark or pink.
(d) The centre section of the yard of a bark, with a puddening, a strop for the tye block and a parrel. The parrel might also have been without ribs.
(e) The masthead of a xebec and similar Mediterranean craft, showing the tye, parrel and seizing of the shrouds (these were not laid with the bight over the masthead in the manner of northern shrouds, but were single with a knot in the upper end; they were laid side by side and were seized to the masthead by two or more lashings.
(f) A three-sheaved tackle block with a toggle, as used on a xebec.
(g) The masthead of a tartan, seen from aft. The tye sheaves were fitted as cheek blocks to the sides of the masthead, and the flagstaff was also fitted to the side.
(h) The halyard blocks of a xebec.
(i) Detail showing how a polacre's lower yard was set up with slings.
(j) The trestle and cross trees of a polacre with topgallant masts.
(k) A polacre's topsail yard, showing the tye and (rope) parrel, seen from forward.
(l) The same yard, side elevation.
(m) The same yard, plan view.

A felucca fitted out for war is very strongly built in proportion to its size. She is armed with 2-pounders in the bow and thirty-two guns in swivels at the sides, mounted on the gun-wale. On deck are twelve small hatches for the twelve oars. The oarsmen sit not on benches, as is common in other vessels, but on the hatch coamings. Their feet are braced against footrests at half height in the room.

Feluccas were indigenous to all countries on the Western Mediterranean seaboard. The name is also still in use for river craft on the Nile.

Galley
Galeere; Galère

The galley was one of the oldest Mediterranean ship types. Its origin lay in the oared fighting ships of antiquity, and galleys could still be found early in the nineteenth century in the maritime nations of Southern Europe and in Russia.

Two main periods of development can be identified during the six centuries in which the galley fulfilled an active role in European naval affairs. The first was between the thirteenth and sixteenth centuries, and galleys from that era are classified as galleys *alla sensile*, whereas subsequent vessels are known as galleys *al scaloggio*. Obvious main differences between the two were the number of masts (the earlier vessel had only one, the latter two) and the way in which the vessel was rowed. The galley *alla sensile* had a single professional oarsman to each oar, while the newer design had a much reduced number of oars, but with up to five men per oar; these were no longer trained men, but galley slaves.

The climax of the naval role of Mediterranean galleys came on 7 October 1571, when 200 Christian and 273 Turkish galleys faced each other at Lepanto; the victory of the Christian fleet under Don John of Austria broke the Turkish stranglehold on the Eastern Mediterranean Sea forever.

Another major encounter, this time between the dominant Baltic powers of the early eighteenth century, occurred in 1714 when Swedish ships engaged a Russian galley force near Hangö; the Russian force destroyed ten Swedish galleys for the loss of one of their own. The last Baltic action involving galleys was the battle between Swedish and Russian fleets in Svensksund in 1790.

Galleys were mainly fighting vessels. In about 1700 the powerful Venetian republic had approximately two hundred galleys and galeases in service. A century years later the French Mediterranean Fleet still listed forty galleys, and the 1815 edition of Falconer's works noted that the Russian Baltic Fleet had 41 galleys with 705 guns under sail.

Chapman's *Architectura Navalis Mercatoria* included not only the draught of the Maltese galley *La Capitana*, but also the plan of a galley constructed not for war but for pleasure – in other words, as a yacht. This vessel had sixteen pairs of oars and can be seen as a half galley (Röding noted that the smallest galleys were called half galleys and had between sixteen and twenty pairs of oars, thus differing from a true galley, designated a *galère grosse*, of between twenty-six and thirty pairs, and a second class galley, a *galère bastarde* of between twenty-one and twenty-five pairs of oars). Half galleys had two men to each oar, as Furttenbach and Chapman pointed out, while the true galley, according to the same sources, had five men.

An eighteenth century galley was between 140 and 180 feet in length, though Röding noted only 120 to 140 feet, which seems more fitting for a half galley. They usually had two masts, a main and foremast, but some had also a mizzen mast; this is indicated in paintings of Venetian galleys, the model of the Russian galley *Dvina* of 1721 in St Petersburg, the drawing of the French galley *La Ferme* from 1691 in Paris' work and by Röding. The additional mast was also explained by Coronelly in 1692 as an attempt to achieve greater speed under sail.

The mainmast was secured by seven to ten pairs of shrouds and the foremast by four or five. These were set up to tackles with toggled flat blocks, though the shroud block might alternatively be turned in, with the toggle of the lower block hooking into a strop fastened around the top of a frame futtock piece.

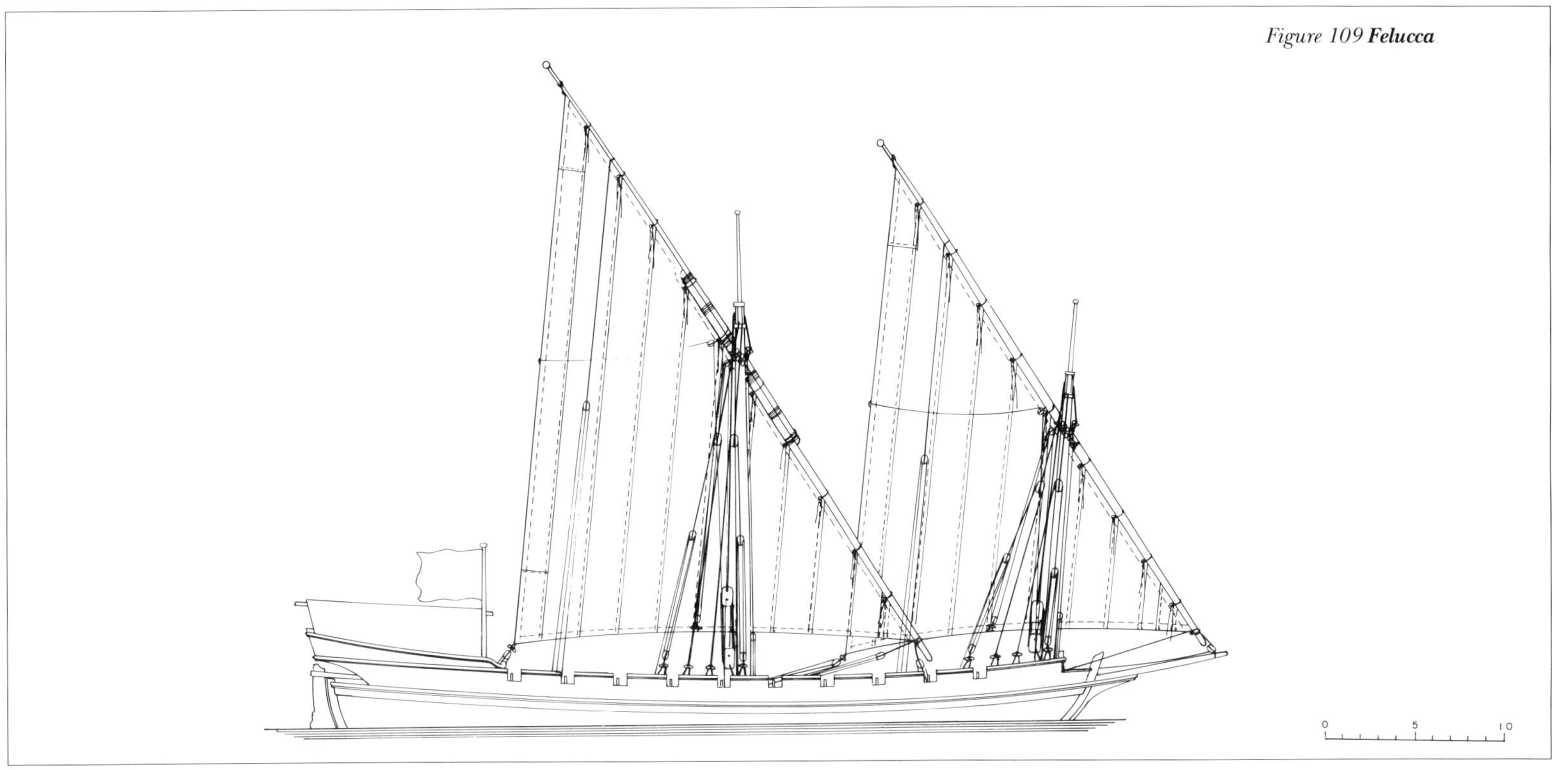

Figure 109 **Felucca**

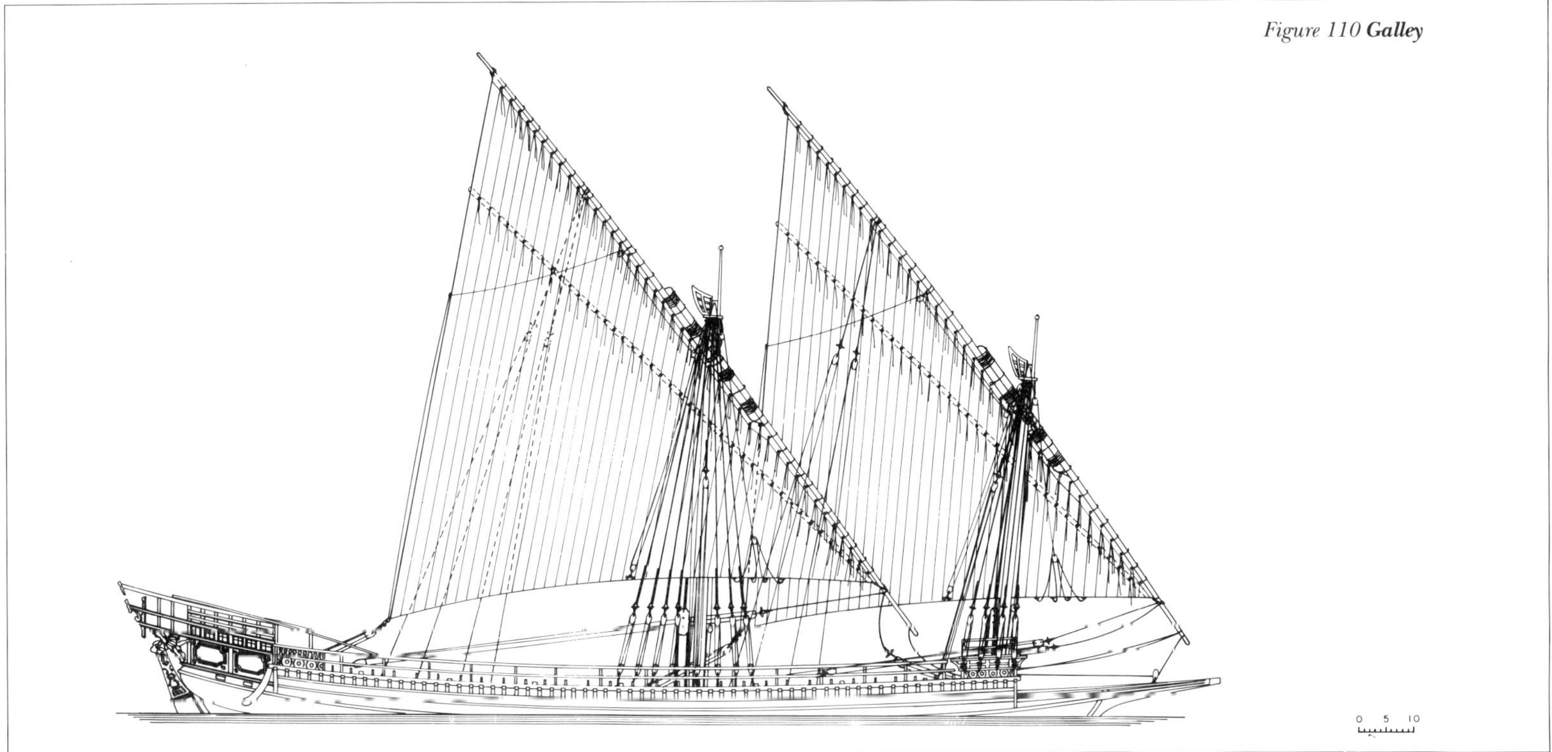

Figure 110 **Galley**

The lower block might also be laid with an eye strop round an iron-bound, chained toggle known as a *rigau*.

Galley yards and sails were similar to those in xebecs. When the vessel ran before the wind the sails let out to both sides like wings, and a squaresail was set on the foremast. Röding noted that galleys had trouble manoeuvring in bad weather and were therefore not fit for long voyages. They were useful against other vessels only in a calm sea, but were very successful along coastlines where sand banks, narrow passages and similar made it impossible for other warships to operate.

If the vessel had to be rowed against the wind, the yards were stowed amidships, and the same was usual in naval actions: galleys were always rowed into battle.

During the seventeenth and eighteenth centuries Mediterranean type galleys were also built in Holland, the Scandinavian countries, Russia and elsewhere.

Galeas

Galeasse; Galéasse

The galeas was developed as a sort of super galley in the mid sixteenth century by the Venetian engineer Francesco Bressan; it was intended to carry more armament and more fighting men, and to be more comparable with the increasing size and firepower of the man-of-war, according to Paris. Röding noted that previously the Venetians had used galeases as large oared trad-

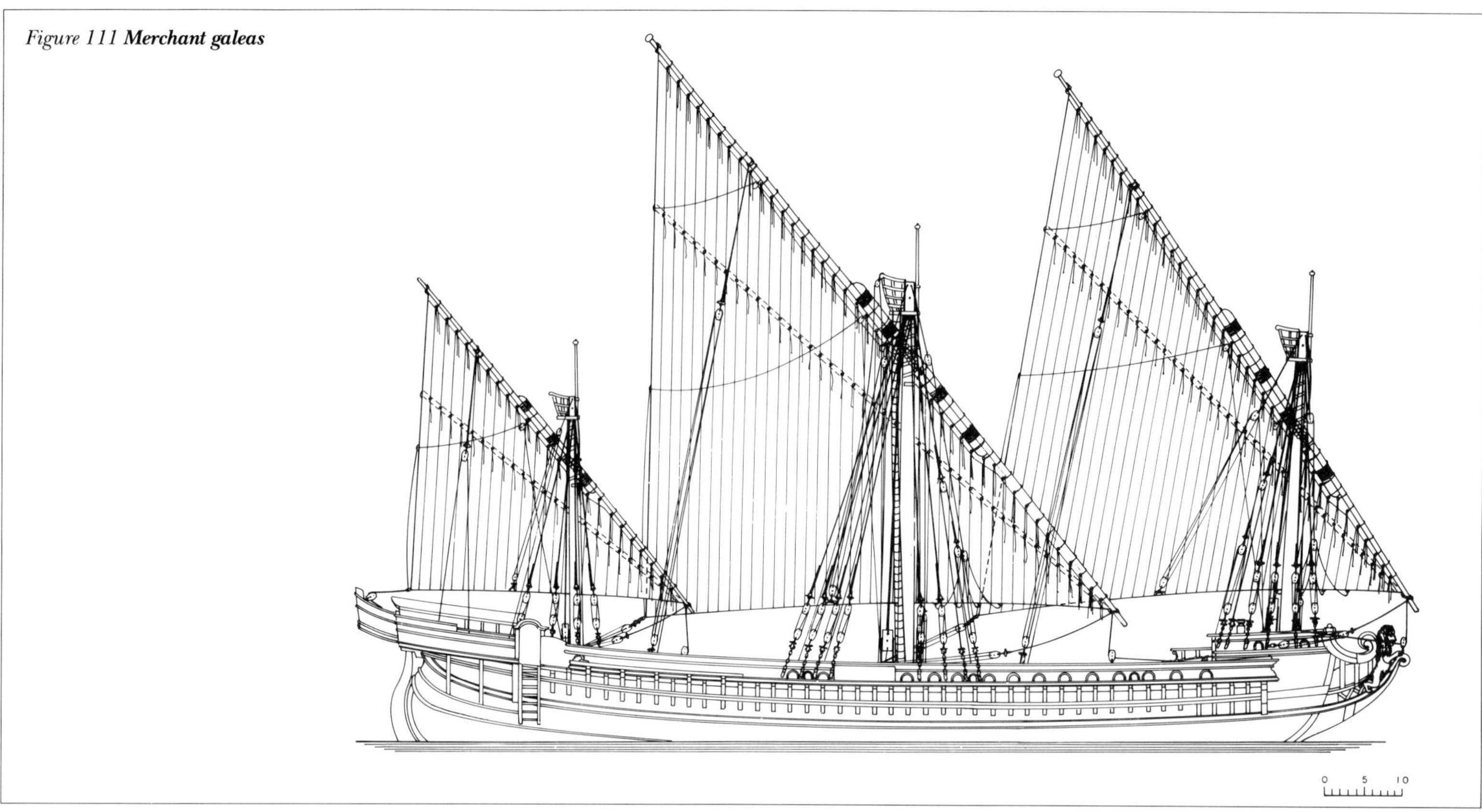

Figure 111 ***Merchant galeas***

ing vessels, and that the fighting galeas developed later, invented by Gian-Andrea Badoaro.

Merchant galeases, also known as great galleys, had been developed in the thirteenth century, and became the backbone of the republic's merchant fleet. The great time of these vessels was in the fourteenth and fifteenth centuries, and the 1726 model of a Venetian merchant galeas in the Musée de la Marine in Paris is a very late example.

Furttenbach noted that the *galeazza* was purely a naval vessel. Larger in size and far more powerfully armed than galleys, these galeases were commanded only by Venetian noblemen who had pledged their heads if they should refuse, under any circumstance, to engage fewer than twenty-five galleys.

At Lepanto the six Venetian galeases were a deciding factor in the Christian victory, but the galeases in the Spanish Armada of 1588 did not perform so well, though they were destroyed rather through bad weather than in action. The limited seaworthiness of galeases led later to the ruling that these vessels should not leave harbour during the four bad weather months of the year, under pain of severe penalties for both captain and owner.

Fighting galeases were built not only in Venice and Spain; France also built a few. The type was relatively short lived in maritime history, with the last taken out of active service between 1715 and 1720. Just as the North European fleets maintained only a few First Rate men-of-war, so only a limited number of galeases were part of the Mediterranean fleets. Venice, a maritime superpower of the period, had at no time more than seven galeases in service together, and only four large galeases went with the Spanish Armada.

These formidable ships had, according to Falconer, a complement of 1000 to 1200 men; Röding mentioned 800 to 1200, and Paris quoted a French document from 1690 which gave the crew as 1001 officers and men. They comprised:

452 oarsmen (galley-slaves) (nine per sweep abaft the mainmast and eight per sweep before the mast)	350 soldiers (marines) 60 sailors 12 helmsmen 40 guards
36 gunners	12 bowmen
4 officers (marines)	2 masters
2 master's mates	4 navigators
2 clerks	2 surgeons
2 provosts	2 carpenters
2 caulkers	2 bakers
10 captain's stewards	1 chaplain
1 lieutenant	1 captain

References to size and armament vary with the source consulted:

Furttenbach (1629)	length 216 palmi (172 English feet)
French manuscript (1690) according to Paris	length 50.02m (164 English feet)
Falconer (1769)	length 162 English feet
Röding (1793)	length between 160 and 170 feet

Oarsmen per sweep and sweeps per side were as follows:

Furttenbach	6 and 28 respectively
Paris	8 to 9 and 21 to 25
Falconer	6 to 7 and 32

And armament as follows:

Furttenbach	11 guns of differing calibre plus 24 swivel guns
Paris	30 guns, 18 swivel guns and 14 small guns
Falconer	four 36pdrs, four 24pdrs, six 18pdrs
Röding	several heavy guns fore and aft, guns below deck, swivel guns at the gunwale and mortars at the forecastle.

Galeases carried three masts and were rigged with lateen sails similar to those of galleys. These could be interchanged with squaresails when needed. Most of the known pictures show a lateen rig, but the galeases of the Spanish Armada can be seen square rigged. That later experiments with permanently rigged

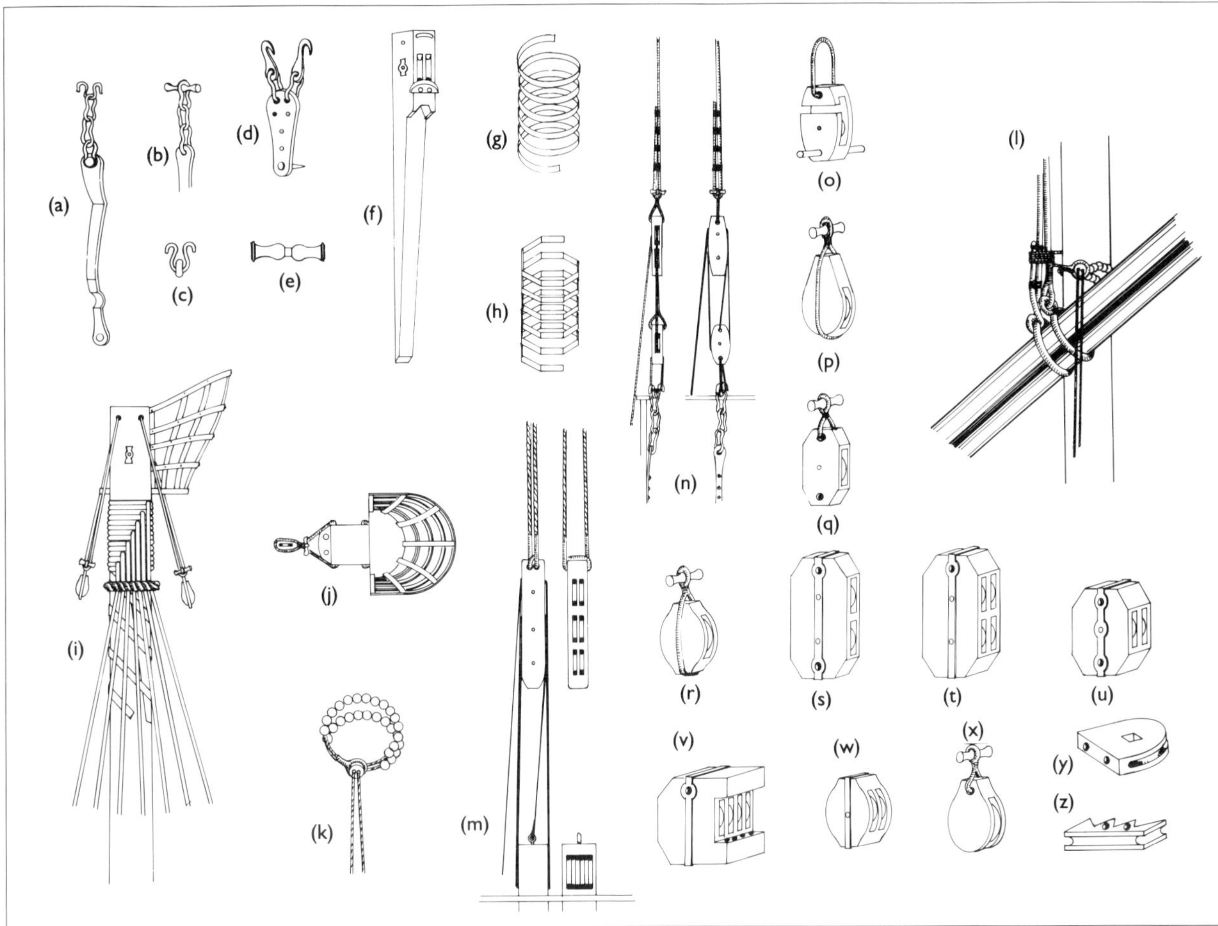

Figure 112 ***Details of Figures 110 and 111***

(a) The chains of a galley or galeas.

(b) A chain with a toggle, known as a rigau.

(c) The double hook of a chain for holding a shroud strop.

(d) The hook for an awning, known as a ganche.

(e) A toggle.

(f) The masthead fitted to a galley's mast, made from elm with two sheave-holes for the yard tye, known as a calcet.

(g) An iron spiral for securing a calcet, round.

(h) The same, octagonal.

(i) Side elevation of the masthead of a galeas, with a calcet, basket, iron spiral, shroud capelage and tackle pendants.

(j) The masthead of a galeas, plan view.

(k) The lateen yard parrel of a galeas.

(l) The rigging of tyes and a parrel to a galeas yard.

(m) A galeas tye block and knight.

(n) The setting up of shrouds in a galeas.

(o) A sheet block with a strop and belaying pin.

(p) A brail block.

(q) A single-sheaved block for the shroud tackle.

(r) A tackle and brail block.

(s) A double block, with sheaves in series.

(t) A four-sheaved tye and halyard block.

(u) A double block, with sheaves in parallel.

(v) A four-sheaved lower halyard block.

(w) A double block for tackles.

(x) A single block with a toggle.

(y) A fore tack block.

(z) A rib for the main parrel.

squaresails were made seems to follow from the comment of Paris that the damaged 1690 model of the French galeas *La Réale* had traces of a squaresail rig and round tops, and that at the time of restoration a note on the rig including the word 'project' was found in the model. Unfortunately the model was not restored to its original rig, but was rerigged in line with generally accepted ideas of the lateen rig. The restoration process was so thorough that no suggestion remains in the model photographs of round deadeyes for the fore and mainmasts, though these are indicated on the drawing. The lateen rig was executed according to known illustrations of galeases, all of which to a greater or lesser extent are plagued by artistic licence, and this resulted in a serious simplification of the model's rig. One need only look at the rigging of a galley, in comparison a much lighter vessel, drawn in 1697 by Barras de la Penne (captain of a French galley), to recognise the inadequacy of this rig for the larger vessel.

Trabaccolo

Trabaccolo; Trabaccolo

Mainly found in the Adriatic Sea and in Italian waters, the trabaccolo originated in Chioggia near Venice. She was a round-built transport ship of approximately 60 to 100 feet in length. The length to width ratio was 3:1 and the vessel was generally thought of as a very good seagoing coaster.

Commonly employed in trade, trabaccolos also carried troops, and the Mediterranean squadron of the US Navy bought two trabaccolos in the early nineteenth century to be rebuilt as bomb vessels.

A trabaccolo was two masted, with the mainmast vertical and the foremast slightly raked forward. The mast positions on deck were as follows: mainmast five nineteenths, foremast fifteen nineteenths, of the deck length from the sternpost. The length of the mainmast from keel to the masthead was three times the beam, and that of the foremast 1ft 8in less.

The trabaccolo was lug-rigged, with a loose lugsail at the foremast and a standing lugsail at the mainmast. On both sails the foot was laced to a boom.

The difference between a loose and a standing lugsail is easily recognised from the hoisting point on the yard. On a loose lugsail the halyard was fastened at about two fifths of the yard's length from the fore arm (thus not quite in the centre) and the tack was hooked to an eyebolt some distance forward of the mast, often in the bow. The sail always hung on the lee side; when the vessel went about, the lower end of the yard was swung round the aft side of the mast, and the tack tackle was hooked to the other eyebolt. A standing lugsail had the halyard fastened at one quarter to one third of the yard's length from the lower end, and the tack set up to an eyebolt in the deck level with or abaft the mast. This sail was not shifted: when the vessel tacked, the sail remained either on the lee or the weather side of the mast.

The trabaccolo carried a running bowsprit, and one or two jibs could be set flying. Like many of the vessels in the Western Mediterranean Sea, the trabaccolo lacked mast stays.

Houario

Houario; Houari

Steel described the houario as:

> Small vessels with two masts and a bowsprit, sometimes used as coasters or pleasure boats, in inlets and rivers of the Mediterranean. Abaft the masts are set lateen-sails, with sliding topmasts. The lower part of the sail is bent to hoops that encircle the lower or standing mast and the upper part of the sail is laced to the topmast, which slides up and down the lower mast, by grommets or iron rings fastened to the heel of the topmast. The sail is fastened, at the lower part, with a tack to the mast, and, at the peek, with a small earring.
>
> The sail is hoisted by a haliard, one end fastened to the

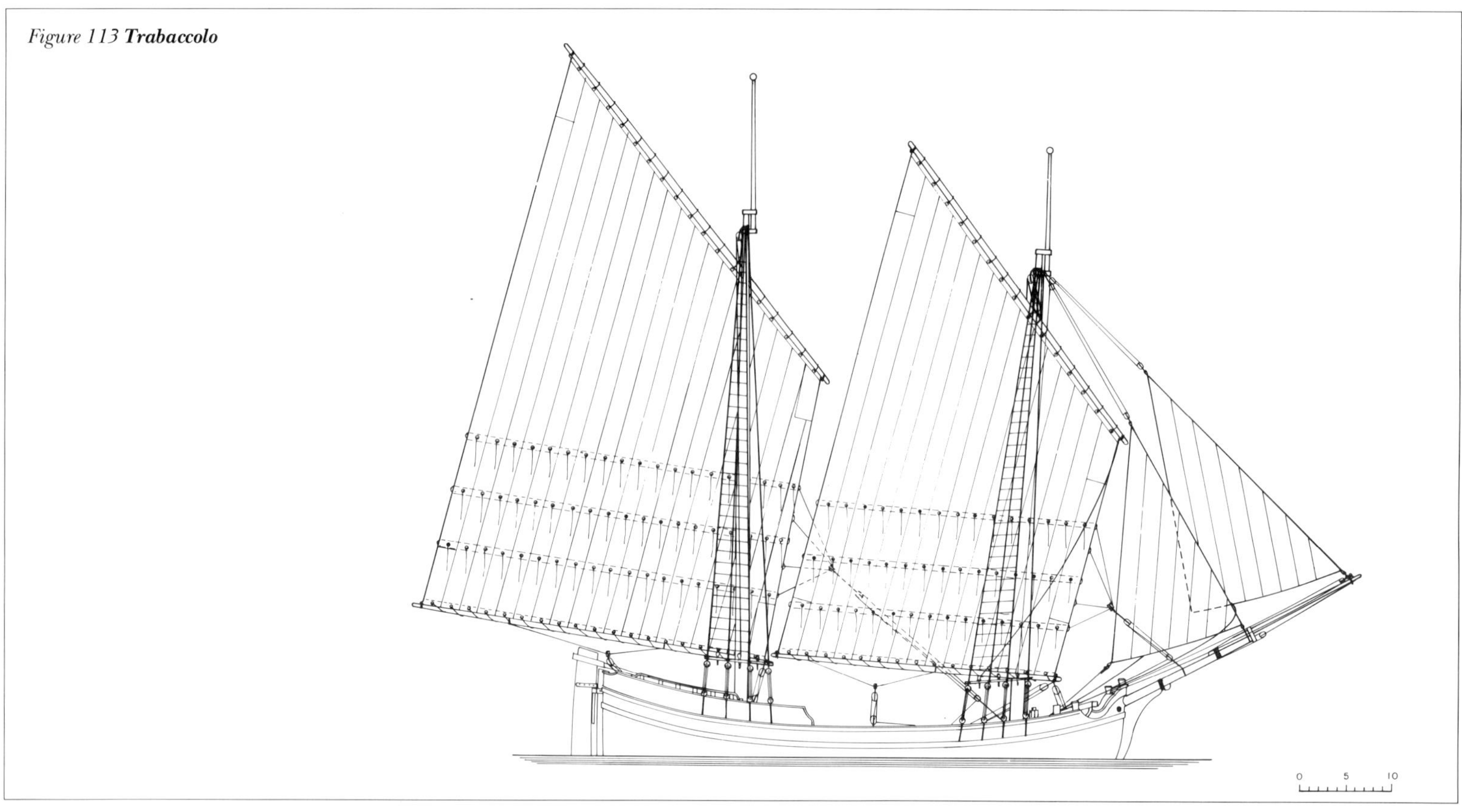

Figure 113 **Trabaccolo**

heel of the topmast, and the other end reeved through a sheave-hole in the lower mast-head. It leads down towards the heel of the mast and belays there. The sail is extended by a sheet, fastened to the clue, and led aft. A downhaul-rope is sometimes fastened to the heel of the topmast, and this leads down towards the heel of the mast. The sails furl in a close manner to the lower-mast, by lowering the topmasts, and confining the sail in folds, by a furling-line. On the bowsprit is spread a jib, which assists the vessel in going about. These sails are called sliding gunters, and used in the English Navy's pinnaces and barges.

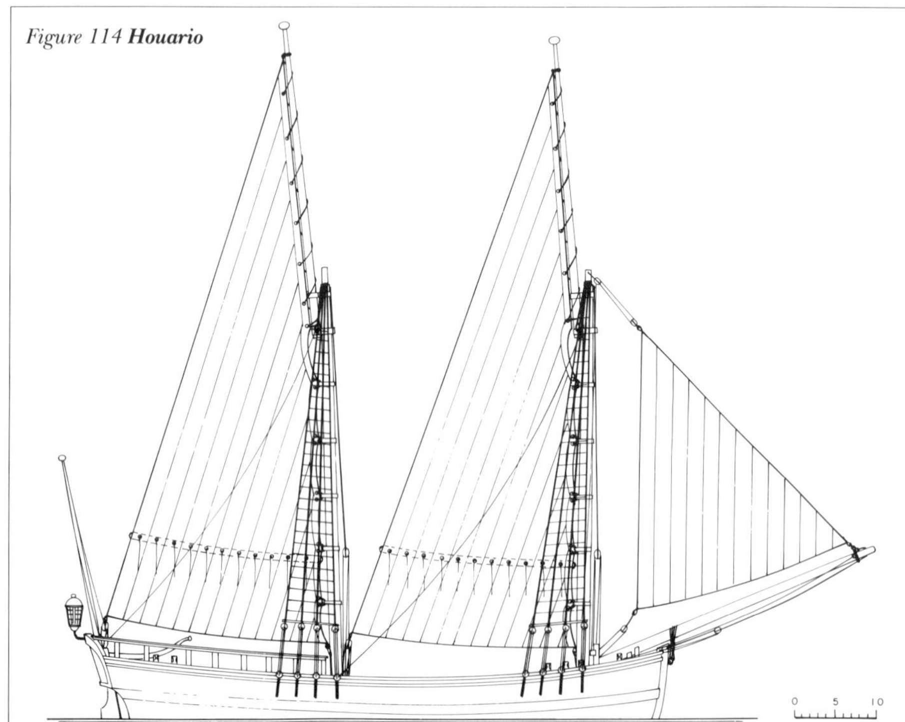

Figure 114 **Houario**

The houario rig was, in fact, not restricted to Italian coasters and Royal Navy boats: the young US Navy had also a few gunboats rigged as houarios.

According to drawings of early nineteenth century gunboats in Chapelle's *The History of the American Sailing Navy*, the foremast was sliding gunter rigged, while the mainmast had a gaff and boom rig. South East Asian vessels like the gay-bao also had a sort of sliding gunter rig.

Scapho

Scapho; Scapho

Vessels of the Eastern Mediterranean often differed markedly in their rigs from their counterparts in the western half of the sea. Whereas the rigs described above were mostly based on lateen sails, Levantine vessels usually had a squaresail or sprit rig, or a combination of the two.

Scapha, according to Röding, was the name by which the Romans referred to their small craft. The Greek scapho was also a small, clinker-built, double-ended vessel with a vertical standing short mast and a sprit as long as the vessel itself. The large spritsail was almost rectangular, with a slight gore in the top leech towards the mast. A peculiarity of the Mediterranean spritsail was the standing topping lift between the sprit's peak and the masthead, to which the sail's head was laced or hanked.

Levantine vessels, in contrast to their western cousins, were rigged with a mast stay, and it was common practice to bend a foresail to hanks on the stay. Scaphos lacked a bowsprit.

Sacoleva

Sakoleva; Sacolève

In a general comparison of Mediterranean vessels with those of Northern Europe, a sacoleva equates in rigging and freight capacity very closely to the smack of the Frisian North Sea coast. She was 1½-masted, with a forward rake of 14 degrees to the mainmast. The topmast, secured with three lashings to the fore side of the mainmast, was one seventh greater in length than the

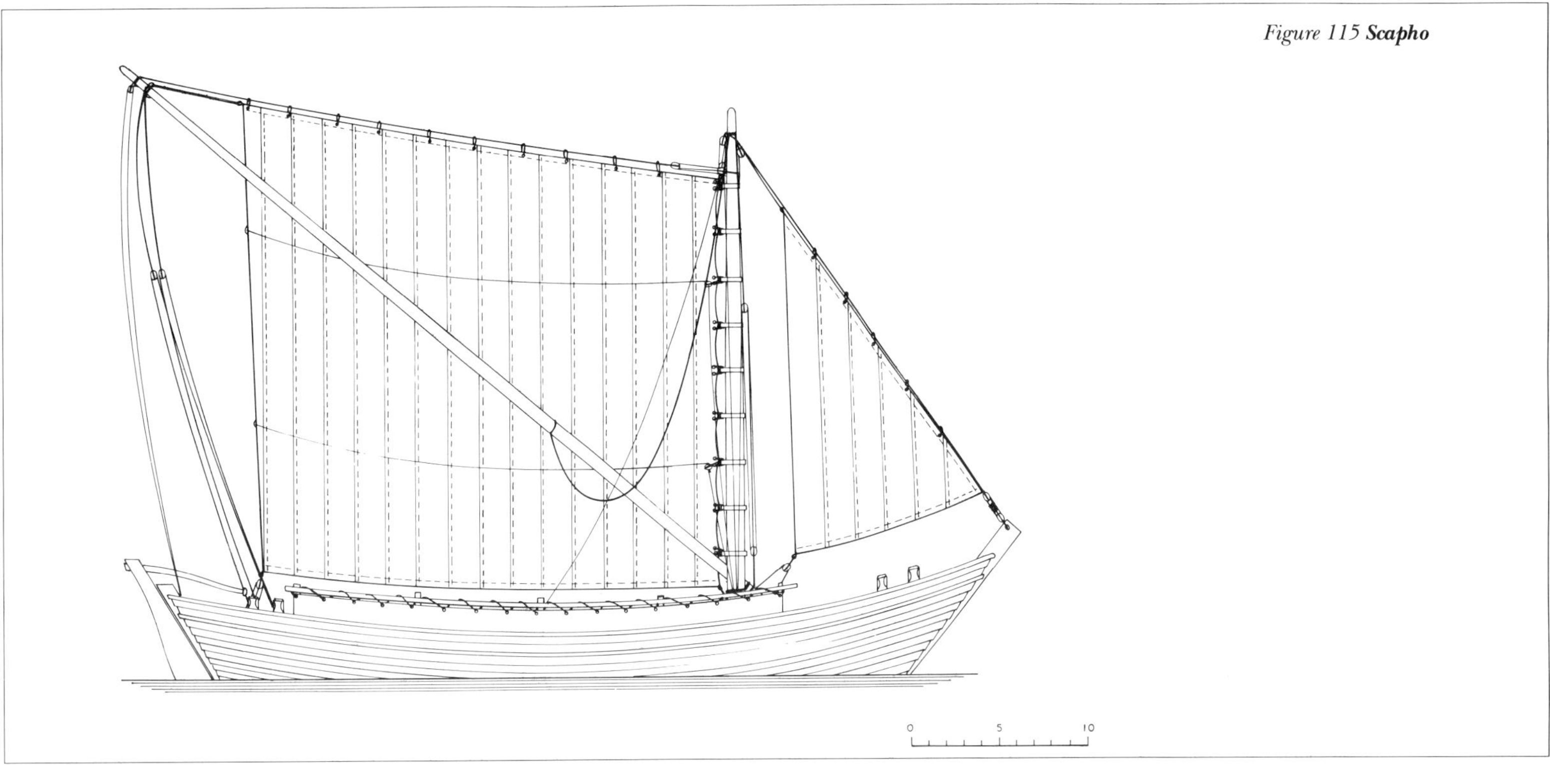

Figure 115 **Scapho**

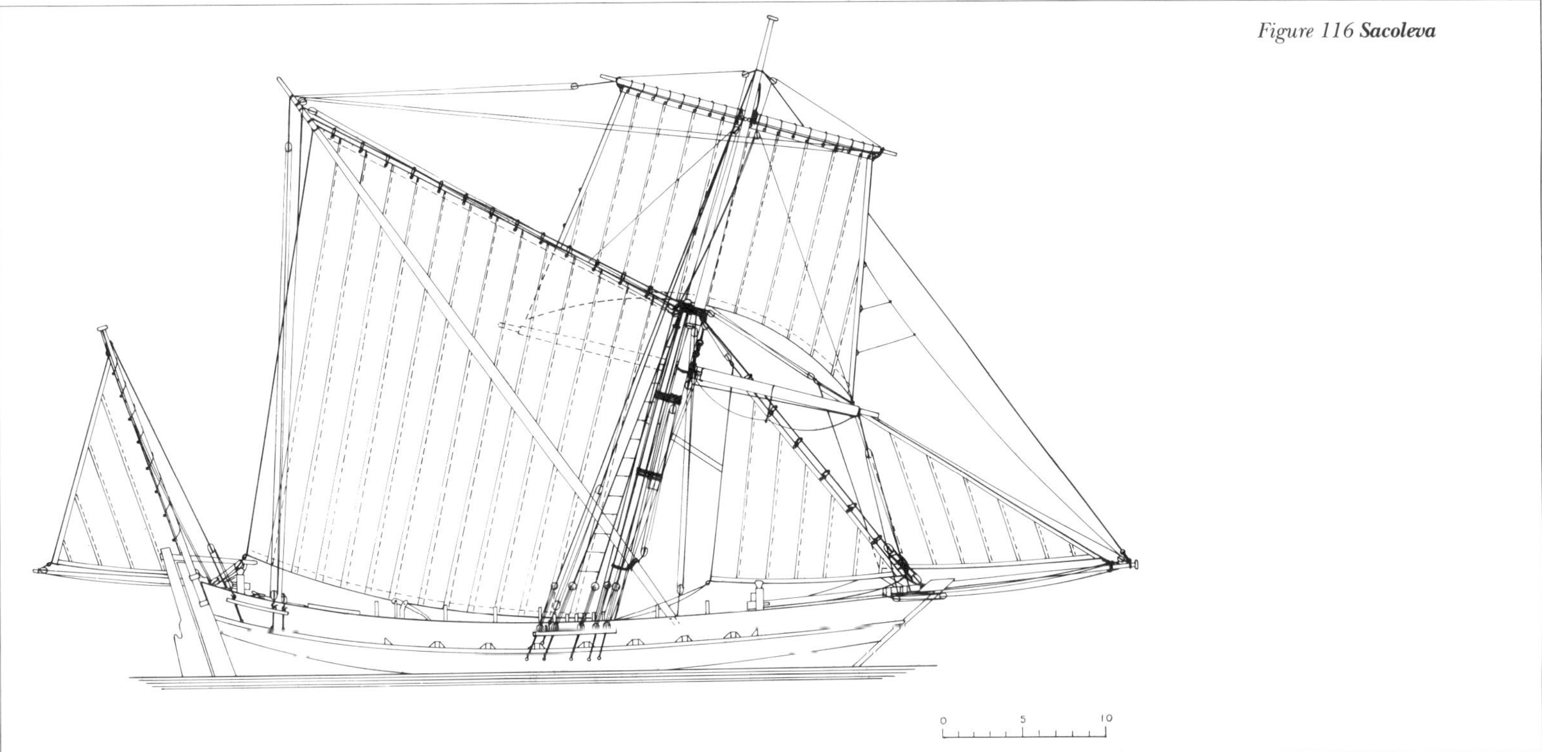

Figure 116 **Sacoleva**

standing mast, but three sevenths of this length was taken up by the part doubled and lashed to the latter.

The mizzen mast, stepped well aft, had an aft rake of approximately the same angle as the forward rake of the mainmast, and had a length of about two thirds that of the mainmast. Like the smack, the sacoleva also carried a bowsprit and an outrigger over the stern.

Further similarities between the two types can be seen in the sails. Both types carried a large spritsail with a long sprit yard, a square rigged topsail, two to three foresails and jibs, and a mizzen. There are differences in mast rake, the lashing of the topmast, some rigging and the shape of the mizzen, but the overall concept was clearly very similar.

While the smack's mizzen was a spritsail and later a gaff sail, the sacoleva had a lateen sail, or the simplest of all sails, a shoulder-of-mutton.

A Baugean etching reveals that sacolevas of the nineteenth century might also rig a flying topgallant sail, and sometimes might have a third short mast.

Paris provided a few measurements of a sacoleva, taken from a model in the Musée de la Marine in Paris:

Length	12.60m
Width	3.50m
Length of mast, including topmast	12.20m
topmast	5.10m plus 3.40m (8.50m)

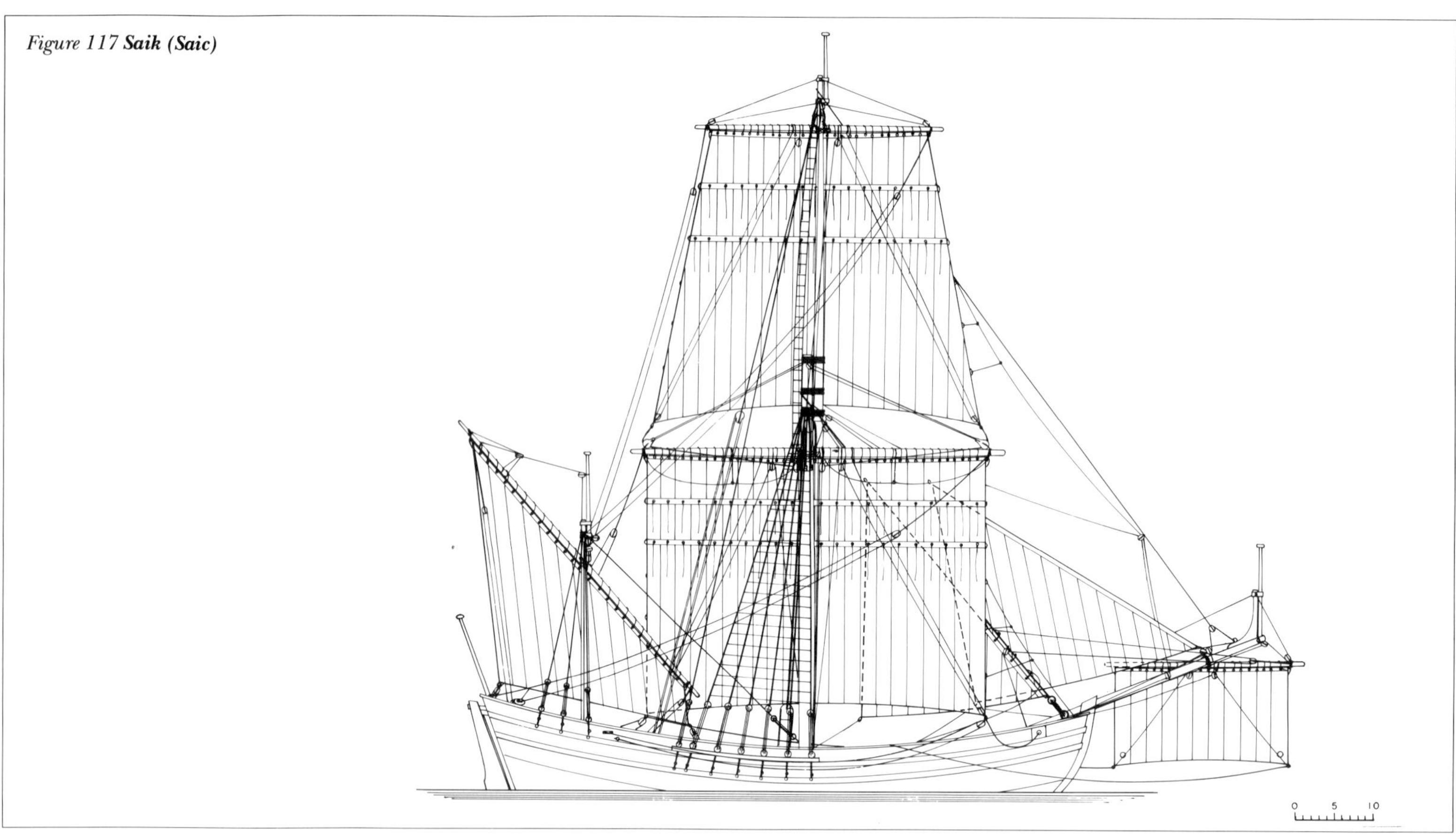

Figure 117 ***Saik (Saic)***

crossjack yard	6.80m
topsail yard	5.40m
mizzen mast	5.00m
outrigger	2.50m
jib-boom	3.50m
Sail area	84.60sq m

Of the last figure, 44.40sq m were for the spritsail, 19.40sq m for the topsail, 8.45sq m for the foresail, 7.76sq m for the jib and 4.56sq m for the mizzen.

Not included were the measurements for the bowsprit and the sprit yard. These were approximately as follows:

Bowsprit	7.40m
Sprit yard	11.80m

The sacoleva was mainly a freight carrier of the Greek islands, voyaging to Italy and to the Black Sea, but sacolevas and their crews won lasting fame through their actions during the long Greek independence struggle from Turkey in the nineteenth century.

Saik, saic

Saik, Saike; Saique

'Saic: A sort of Grecian ketch, which has no top-gallant-sail or mizen-topsail,' was the definition in the earlier edition of Falconer's works; in the 1815 edition the saik was defined as 'a Turkish vessel, very common in the Levant, used for the conveyance of merchandise. It has only one mast, which, together with its top-mast, is extremely high.'

An etching by Randon (1640-1704), shows this vessel in more detail, and the similarity to a ketch is clear, as is the overly high mainmast with its two square-rigged yard sails. The illustration also shows a short mizzen mast with a settee mizzen, and a bowsprit with a spritsail.

Foresails and jibs are not visible on the etching, but Röding, another contemporary who mentioned the saik, noted not only that the vessel was Levantine in provenance and that it had a main and mizzen mast, the former square-rigged with two sails, but also that it was rigged with a foresail and jibs.

The saik had a capacity of 200 to 300 tons and a length of between 60 and 100 feet. The flat stern and overall hull shape suggests that the saik had much in common with the Red Sea sambuk (see below).

Turkish coaster from Constantinople

Türkischer Küstenfahrer von Konstantinopel; Caboteur turc de Constantinople

The vessel illustrated in Figure 118 represents only one example from a multitude of square and sprit-rigged one or two-masted vessels employed as traders by the Turks. Like the sacoleva it had a topmast lashed to the mast, though the latter was without rake.

A crossjack yard, a topsail yard and a topgallant yard were rigged, and also a long sprit yard. The bowsprit was very steeply angled. Sails carried by the vessel were a main (sprit) sail, which according to Levantine custom was not fastened to the mast, but to hanks on a standing topping lift between the sprit peak and the masthead, and also a topsail and a topgallant sail.

In addition to these normal sails, a squaresail could temporarily be set, as shown. Since the crossjack yard was slung rather than hoisting, the squaresail was hoisted with two nock halyards, reeving through blocks at the yardarms. These halyards led aft and also acted as braces.

The fore stay led to the stem and a foresail was rigged, seized to hanks as described for the scapho. A flying jib was set between the end of the bowsprit and the masthead.

Dhow

Dhau; Boutre

By the time European traders first ventured to India and East Asia, Arab merchants had already had extensive contact with

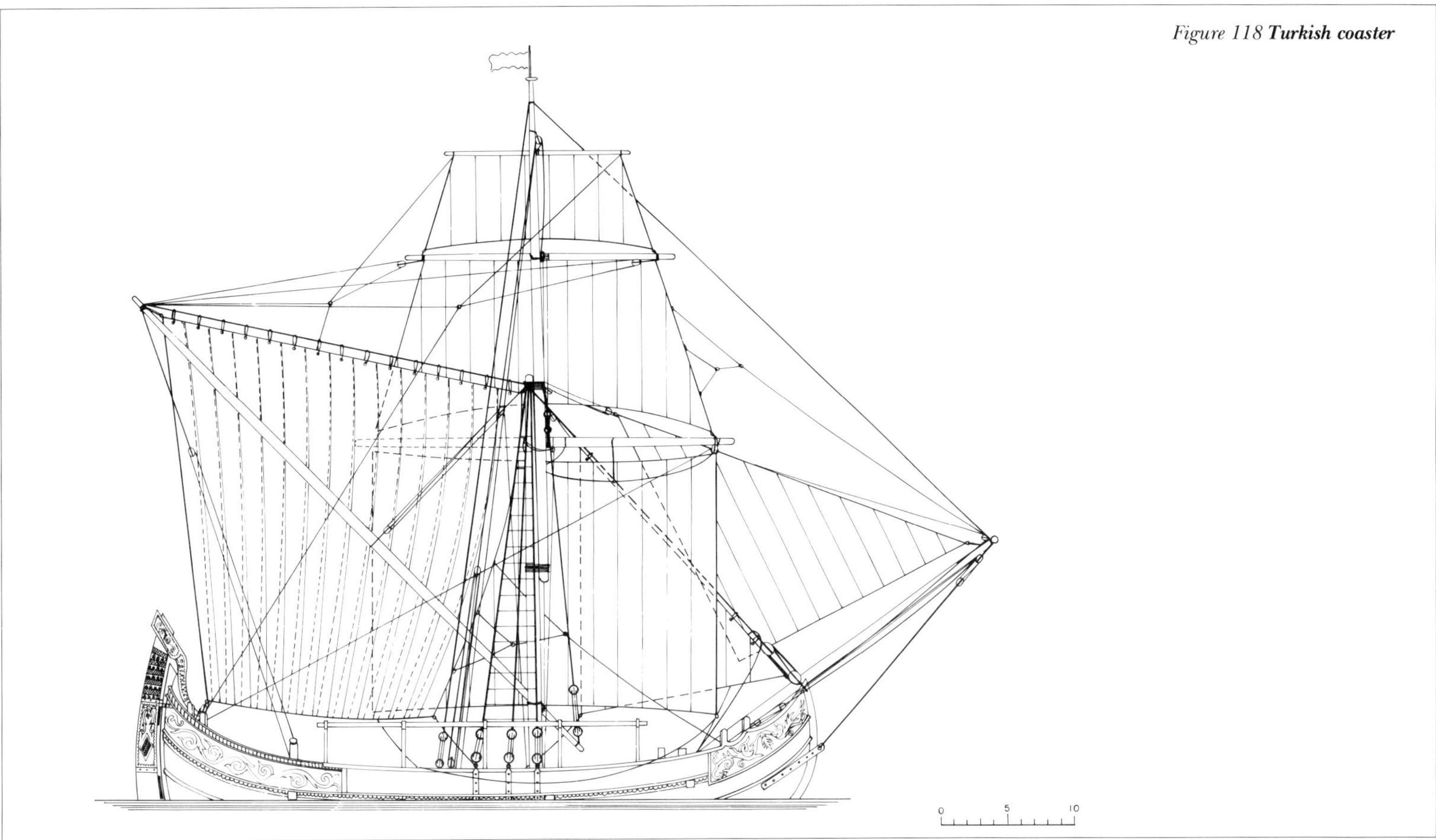

Figure 118 **Turkish coaster**

these regions for centuries. Without Arab navigational expertise and the assistance of the Sultan of Malindi in providing a pilot, Vasco da Gama's journey to India would have been much more difficult.

A flourishing shipbuilding industry was a prerequisite for any mercantile Empire, and, as with European shipping, a great number of vessel types evolved in the Arab world, differing both geographically and in purpose.

Little is now known about the great period of Arab sea trading. Shipbuilding in many places in the Arabian Peninsula is, however, still undertaken today in conditions which have changed little over the centuries. Coir and wooden pegs are often still preferred to nails; the Portuguese observation, after their first encounter with Arab vessels in 1498, that vessels of about 100 tons were very well built, of good timber, and well stitched with cords 'because they have no nails', might equally apply as recently as the early twentieth century if the reference to the non-availability of nails is not taken too literally.

Arab shipbuilding was undertaken by tradesmen who never prepared or referred to a draught; existing draughts of dhows were made by Europeans, mainly in the nineteenth and twentieth centuries. Given that Paris' draughts of Arab dhows drawn more than a century ago show vessels which are not significantly different from modern dhows, it can reasonably be assumed that his drawings are also accurate for dhows built one hundred years before his time.

All Arabian vessels were colloquially known as dhows, though the term was not used by Arabian sailors; it evolved from the Swahili word *dau* and referred specifically to a small double-ended vessel used in Zanzibar. The dhowmen called their vessels by more precise type names, and there were many possibilities; dhows were either double-ended or transom-sterned, and might have one, two, or in some cases three masts.

The rig was not the Mediterranean lateen, but a settee rig, indigenous to Arabian ships. The settee sail was a lateen sail with a luff leech; it was therefore not triangular, but had four sides.

Baghla

Baggala; Baggala

The largest of the dhows was the baghla (*baghla* is Arabic for mule). Baghlas ranged from 150 to 500 tons, and were mostly rigged with two, but sometimes three, masts.

Prominent features of the type were a full deck, a quarter-deck, and a straight, sharply raked stem which ended in a carved, turban-like head. The aft end of the vessel comprised an elaborately carved flat transom stern with a number of windows, and some imitation side galleries. European influence, exerted from the sixteenth century, on these originally double-ended vessels is clear in these features.

The baghla's rig was typical of most Arab vessels. The masts were raked forward, and the mainmast, which was stepped abaft the main beam, was lashed both above and below the deck to a similarly raked pole mounted forward of the beam in the same mast step (see Figure 123f). One or two sheave-holes were cut into the masthead according to the vessel's size, and the masts were supported laterally by two or three shroud pairs consisting of pendants set up to tackles which were toggled (or sometimes hooked) to round strops in the sides. A stay was sometimes fitted to the mainmast head and set up to the stemhead.

Small yards were made in one piece, but larger yards were lashed together from two or three. A yard cleat was dowelled and keyed in and strongly lashed to the yard near the centre. The jeers were then bent round this cleat rather than the yard itself, according to Paris (see Figure 123b); photographs of various types of modern dhows reveal, however, a lashed doubling at the centre of the yard, with the jeers laid around this.

The jeers, single or double, were served all over and rove through the masthead sheave hole (or holes) diagonally upon deck, with a multi-sheaved halyard block in the bight. A knight, strongly secured to the keel and with one sheave more than the halyard block, was placed amidships just forward of the quarter-deck at a suitable angle for the jeers, and a halyard made fast to

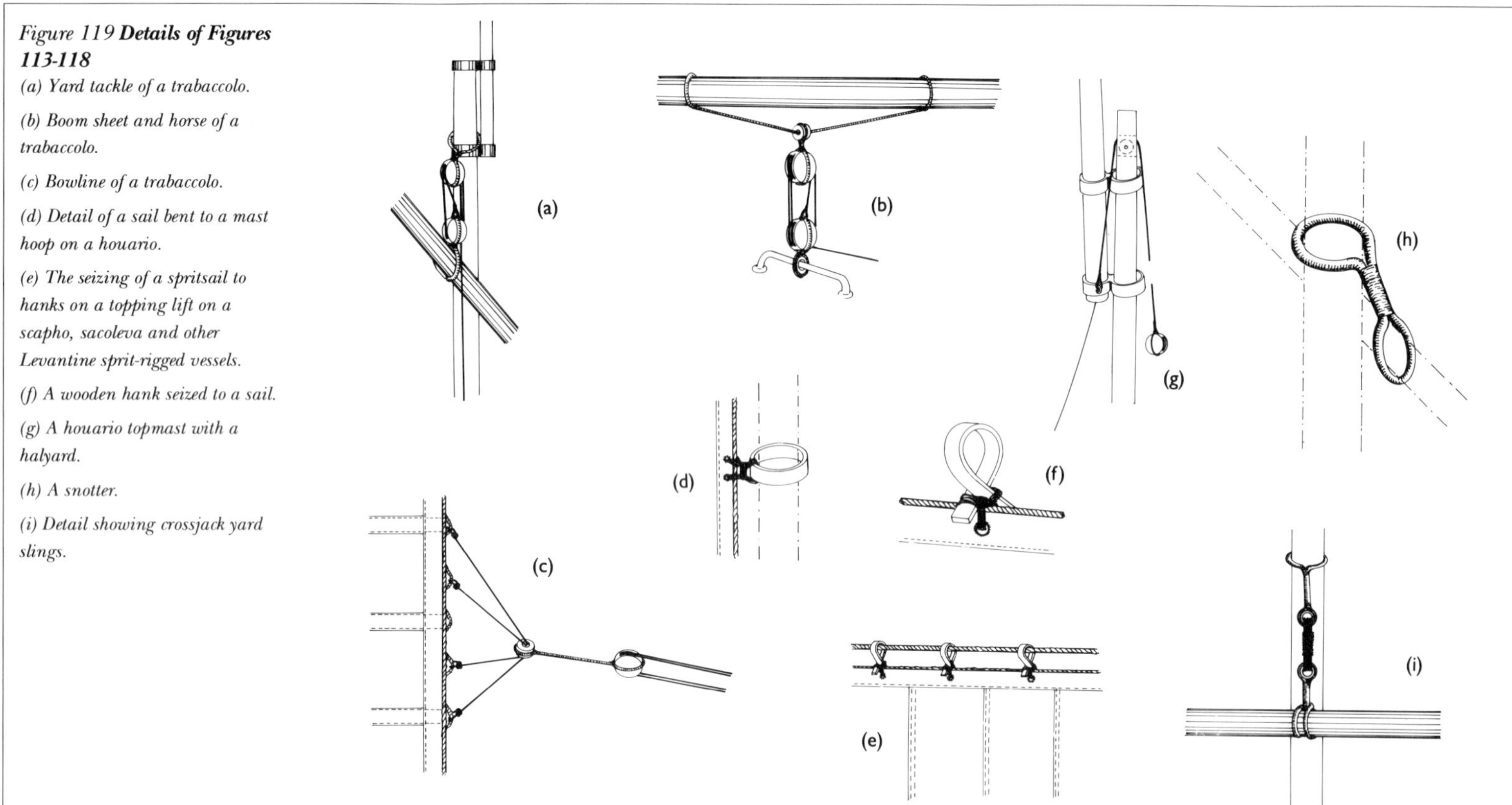

Figure 119 ***Details of Figures 113-118***

(a) Yard tackle of a trabaccolo.

(b) Boom sheet and horse of a trabaccolo.

(c) Bowline of a trabaccolo.

(d) Detail of a sail bent to a mast hoop on a houario.

(e) The seizing of a spritsail to hanks on a topping lift on a scapho, sacoleva and other Levantine sprit-rigged vessels.

(f) A wooden hank seized to a sail.

(g) A houario topmast with a halyard.

(h) A snotter.

(i) Detail showing crossjack yard slings.

the knight rove through all the sheave holes in the block and the knight (see Figure 123d).

Larger vessels had two tiers of sheaves in the knight, the lower tier for the throat halyard, and the upper for a peak tye halyard. A single peak tye was fastened to the upper quarter of the yard, rove through a block at the masthead and led parallel to the jeers or throat tyes upon deck, where it went over the upper tier of sheaves and was belayed.

Figure 120 **Baghla**

On larger yards a pendant with a single block was fastened to the yard's upper quarter, and a long stropped single block to the masthead. The tye, fastened to the masthead, rove through the pendant block and the long stropped mast block, and upon deck as before.

A single or double open truck parrel (according to Paris) held the yard close to the mast; modern photographs show ribs but no trucks on these parrels. The hauling part of the parrel was led through a foot block near the knight. Large yards had double vangs and also a tack tackle, which acted as a lower brace. The weather vang, the upper, was set up fore-and-aft to the anchor beam across the bow, but could be shifted when need arose.

Sails were cut settee fashion, as mentioned before, and had one or two brails, a tack and a sheet.

A vessel similar in many ways to a baghla was a ganja, but these were rather smaller, with considerably less transom ornamentation and a flowery scroll at the stemhead.

Ganjas ranged from 70 to 200 tons in capacity. The kotia or 'buggalow' also belonged to the same group (the latter name was almost certainly an English corruption of *baghla* or *baggalah*). Similar in size and appearance to a ganja, the kotia was in fact an Indian dhow.

All the above types were square sterned. The bhum was the largest Arabian double-ended vessel, and probably represented the original Arabian dhow before European influence introduced the more elaborate transom stern.

The bhum had no carved ornamentation (indeed, no decoration at all other than painted bands), a straight stem and an unbroken sheer line. It ranged between 60 and 200 tons.

Despite their smaller size, ganjas, kotias and bhums made the same long voyages as baghlas, and served as the tramps of the Indian Ocean. Their provenance was the Persian Gulf and the Indian Ocean coast to the east of it.

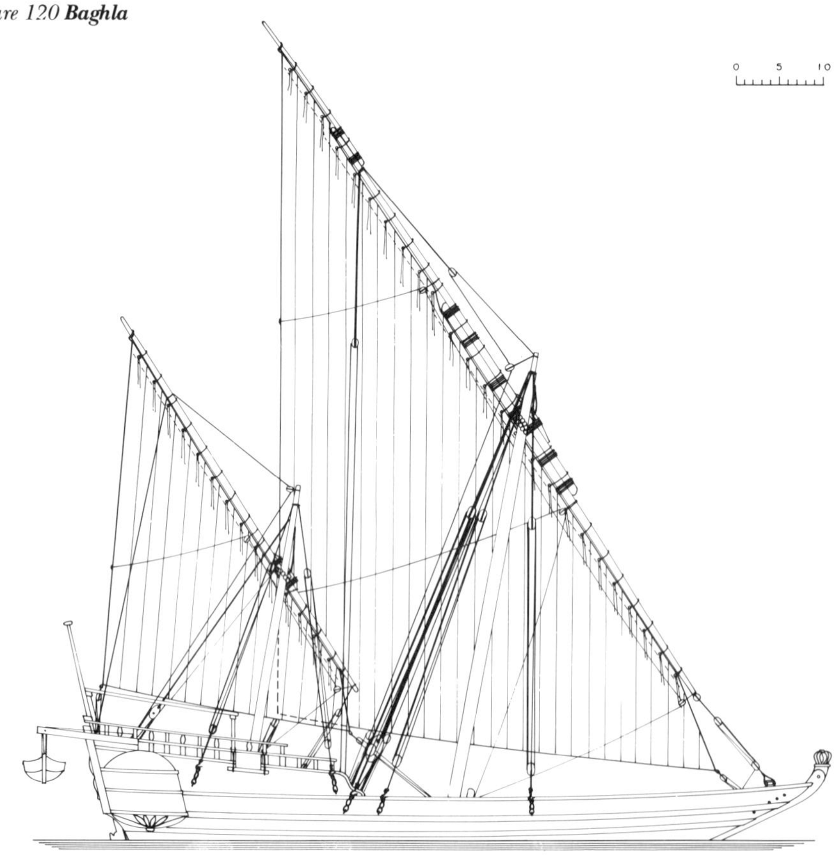

Sambuk

Sambuk; Sambouck

While the baghla, ganja and bhum were indigenous to the Persian Gulf, the sambuk was the typical larger craft of the Red Sea. It was similar in construction to the ganja, but lacked carved ornamentation; instead, it was richly painted with geometrical patterns.

Sambuks ranged from 30 to 200 tons in size. The larger vessels were fully decked, while the smaller were open with only a covered poop deck and sometimes also a small platform in the bow.

Large and medium sambuks were always rigged with two masts, but smaller vessels often had only one. With minor variations, the rig was the same as that of a baghla. Mainsails of three different sizes were set according to weather conditions, as in Mediterranean lateen-rigged vessels.

Sambuks traded mainly between the Arabian ports on the Red Sea and the African coast or Zanzibar, but also between the Red Sea ports and Bombay.

The zarook, a double-ended cousin of the sambuk, can in fact be considered its forerunner. These vessels originated in Southern Arabia and in Yemen. They were open, with one or two masts and a capacity usually below 100 tons.

Jahazi, jehazi, gehazi

Jehazi; Gehazi

The jahazi was a local Zanzibarian variation of the sambuk. These vessels were generally much smaller then their cousins and had only one mast, vertical or raked slightly forward and stepped similarly to the mast of a sambuk. Because of its relatively small size, the settee sail of a jahazi was rigged in a simpler manner.

These small sailing vessels were very seaworthy and traded not only between Zanzibar and the African continent but also with Madagascar and Aden.

Gay-bao

Gay-Bao; Gay-Bao

Paris described the gay-bao as an Annamese coaster. A timeless design like those of the dhow, prao or junk, the gay-bao was still being built in the second half of the twentieth century as Paris had recorded during the nineteenth; an eighteenth century seafarer would not have encountered anything greatly dissimilar.

Spoon shaped and with its underwater part woven from bamboo, the gay-bao was a basket craft made watertight with a caulking mixture of resin, chalk from sea-shells and coconut oil. This mixture also proved resistant to the teredo worm, but the main reason for its use was probably the ready availability of the ingredients in the region. Timber bulkheads divided the vessel's hull and the deck beams, fitted into timber freeboard strakes, extended up to 1 foot outboard.

Gay-baos ranged from 40 to 80 feet in length, with a length to breadth ratio of about 4.5:1. Two or three masts were stepped, and rigged with a sort of sliding gunter rig. The relatively short lower mast carried a long yard (or topmast), of which the lower end was fastened with a strop round the mast. The mainmast, in this case the aftmost, carried a boom set to the foot of the sail (according to Paris); a more modern illustration shows booms to all sails.

Shrouds were set up to elongated wooden trucks (see Figure 127c) and lashed to holes in the sides, though curved timbers were fitted to the sides as a kind of channel for the main shrouds (see Figure 127b). The halyard rove over a sheave in the masthead and was made fast near the centre of the yard. A single topping lift kept the boom in position. The sails were triangular and controlled by sheets; the mainsail only also had two brails.

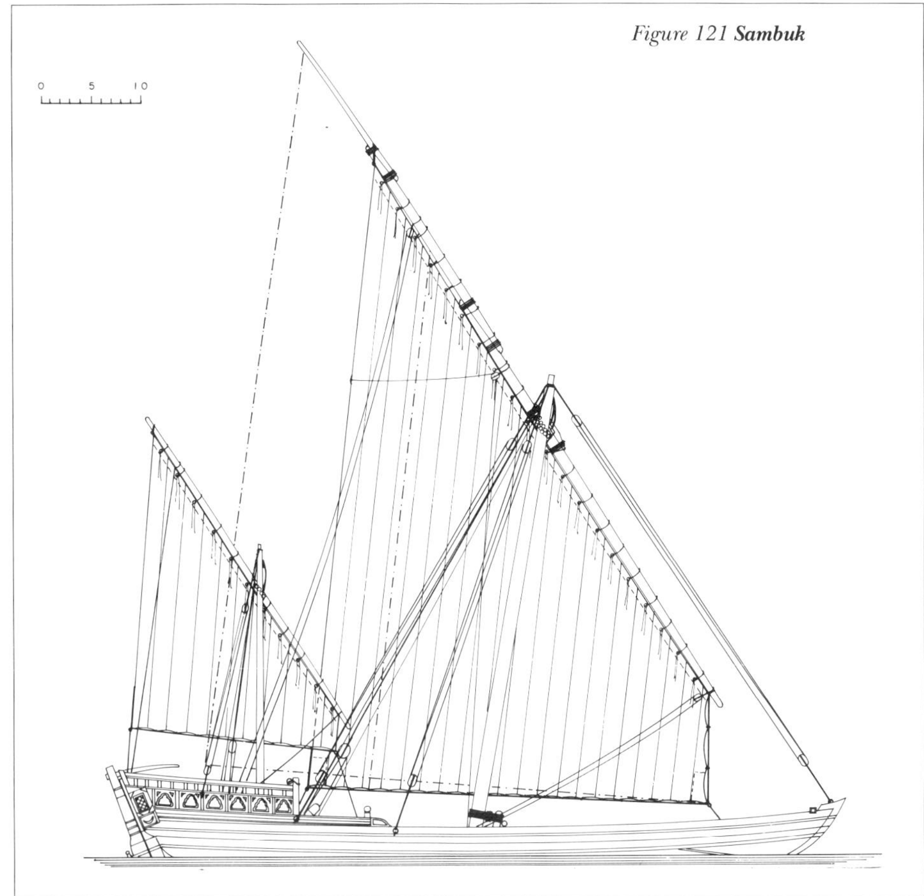

Figure 121 **Sambuk**

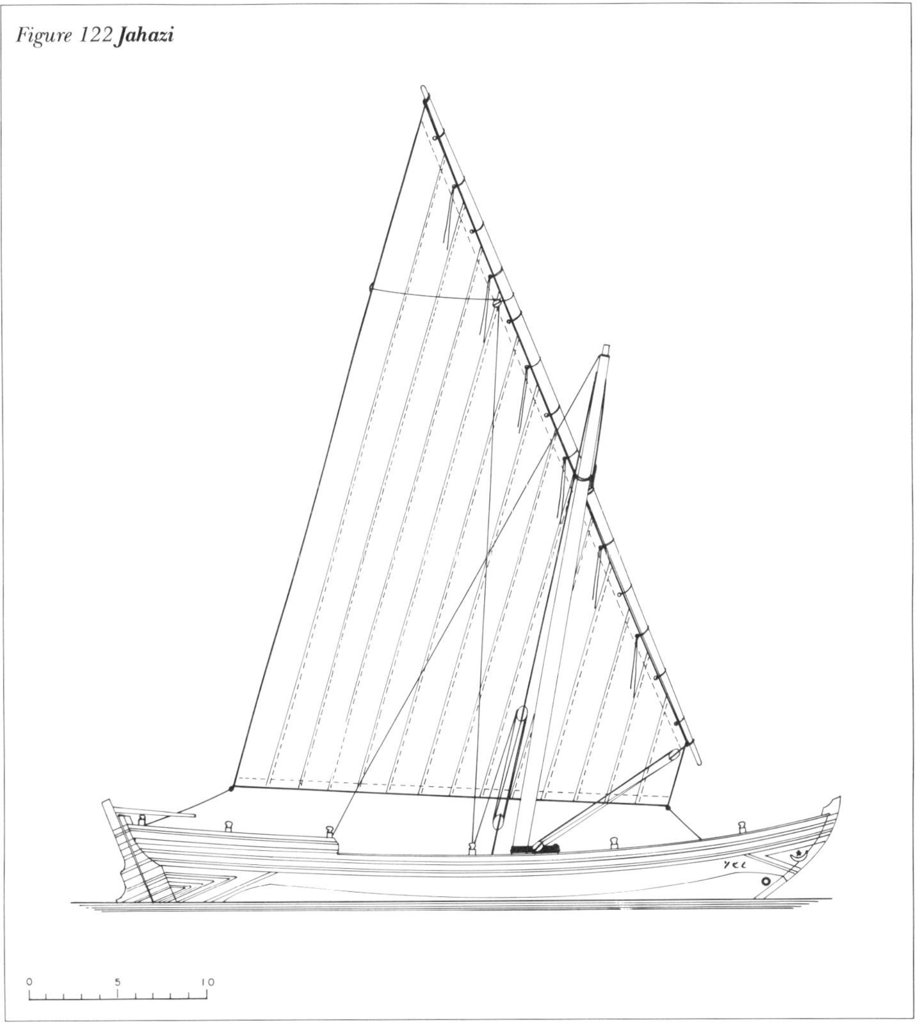

Figure 122 ***Jahazi***

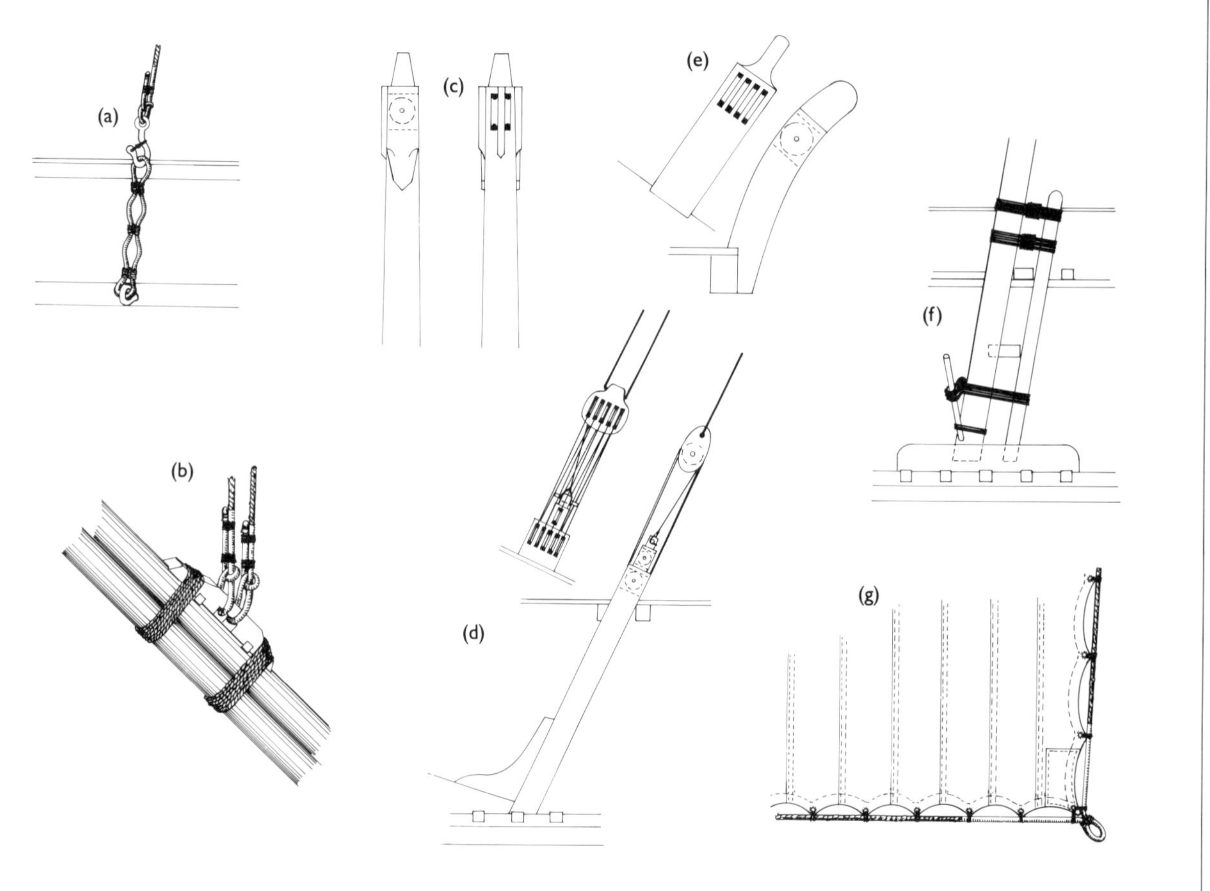

Figure 123 ***Details of Figures 120-122.***
(a) A dhow chain-strop with a shroud block strop hooked in.
(b) Yard tye of a baghla or sambuk.
(c) Masthead of a baghla or sambuk.
(d) Jeer or tye block and knight of a baghla.
(e) Knight of a sambuk.
(f) Mainmast step and lashing on a baghla and sambuk.
(g) Boltrope of a sambuk's sail, fitted to the luff and foot leech.

Some gay-baos, however, were rigged with a simple type of lugsail.

Prao mayang

Prau mayang; Prahu mayang

Just as 'dhow' was a European colloquial term for all Arabian vessels, so 'prao' was a collective name for a large variety of Malay and Indonesian vessel types.

The characteristic rig for a prao was a long, four-cornered sail made from bast, set to a yard with grommets and to a boom by a lacing. A throat halyard, attached to the yard at one third of its length from the lower (forward) end, rove through a sheave-hole in the masthead and was belayed to a cleat at the aft side of the mast. The sail itself was not rectangular, but narrowed at its upper end away from the mast. In case of a prao bedang this narrowing resulted in an overlapping of yard and boom, which were then lashed together; the sail of this type was therefore triangular.

Yards and booms are usually of complementary thickness, but the boom of the prao mayang, as an excellent contemporary model (restored and drawn up by the author) in the Deutsche Museum in Munich reveals, was a heavy object, constructed from four pieces; the extra weight probably helped to hold the somewhat unruly large sail close to the mast.

The short mast had a sheave for the throat halyard at its head, but this was the only sheave noted on the Munich model; all blocks (trucks) were flat, egg-shaped pieces of timber with a hole through the flat side, but without a sheave (see Figure 127d). A peak halyard was bent to the yard near the second third of its length, and led through a long stropped block at the masthead upon deck, where it was belayed to a grommet at the side.

Three shroud pairs provided lateral stability to the mast. These were not set up with blocks, but were hitched direct to grommets around timberheads at the sides (see Figure 127h). A double stay was passed round the beam which supported the bowsprit beam and belayed to bow bollards or knightheads (see Figure 127g). None of the ropes which supported the mast were hauled taut with tackles or similar: all were set up and belayed by hand alone.

A brace or tack was fastened to the lower (forward) end of the yard, and a vang was hitched between the end of the third and beginning of the fifth section of the yard's length with a span. A two-holed truck connected the vang to this span, and the vang was belayed to one of the grommets at the poop rail, while the tack was belayed to a similar grommet at the bow. A kind of topping lift or brail, hitched to the middle of the yard at one side of the sail, led beneath the boom and up on the other side to a truck at the masthead, again belaying to a grommet at the side. The boom itself was controlled by a single tack and sheet.

The supporting beam for the bowsprit in the bow section was set athwartships into timber brackets about 1 to 2 feet above deck (see Figure 127g). The bowsprit, constructed from two pieces, rested in notches at the upper side of the bow bulkhead, with its two inner ends locked underneath and lashed to the beam. The shorter port side piece was lashed outside the stem to the full-length starboard piece, which formed the actual bowsprit. A simple bobstay led from the lashing to a cleat on the stem and up to a bow bollard on the starboard side.

The outer end of the bowsprit had a hole for the lower end of a halyard for the triangular flying jib, and a truck for the running end of the halyard was placed at the fore side of the masthead. The halyard was belayed to the foremost grommet at the side, while the jib sheet was belayed to a grommet in the bow, where a rack with four grommets was placed on each side.

A mizzen mast stepped temporarily at the end of the poop carried a rectangular gaff sail, which was also rigged in the simplest manner possible.

The European-looking jib and gaff-rigged mizzen can without doubt be attributed to European colonial influence in the

Figure 124 **Gay-bao**

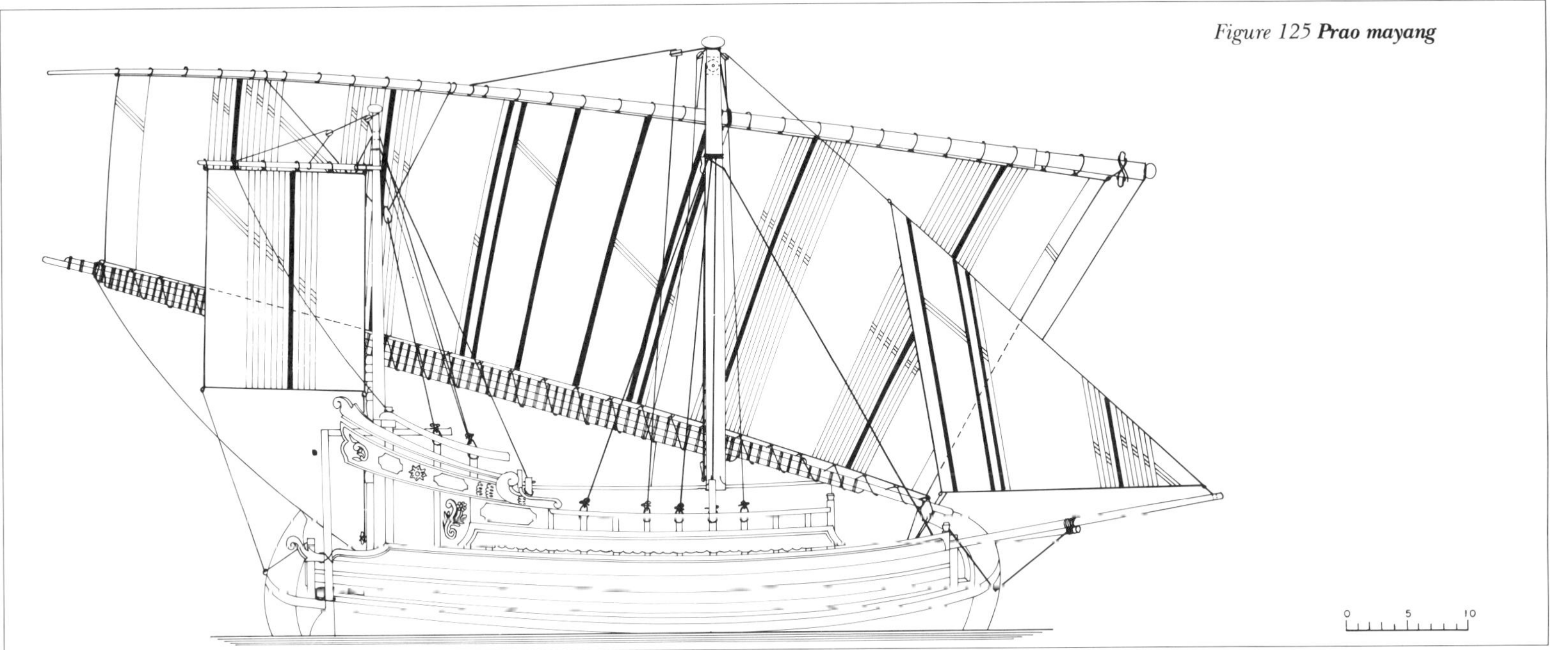

Figure 125 **Prao mayang**

region. It was impossible to set both together when the mainsail was set, but they were almost certainly used when the wind became too strong for the mainsail.

Fijian outrigger canoe

Ausleger-Kanu von Fidschi; Canot fidjien

Amid a multitude of varying catamarans and canoes, the Melanesian outrigger canoes of Fiji and Tonga had a special place. They were of numerous sizes, from 20 to 100 feet long, only a few feet wide, and capable of travelling long distances. The largest of these fast vessels carried up to fifty people.

In eighteenth century maritime literature these vessels are considered as prows. Steel called them 'flying prows' and Röding 'proas'. The notes of these two authors are, although in different languages, so similar that they must both have been drawn from a third source. The original source was a drawing of a 'flying proa', done in the Marshall Islands by a young officer during Lord Anson's circumnavigation of 1742, the later Admiral Sir Percy Brett.

Illustrations provided by Steel and Röding of such canoes were identical including even their smallest mistakes. Both show deadeye lashings on stays and shrouds, and a topping lift leading from the boom end through a block at the yard's peak to the mast; all these details existed only in the imagination of the artist. Brett showed no such rigging, and the canoes of modern times

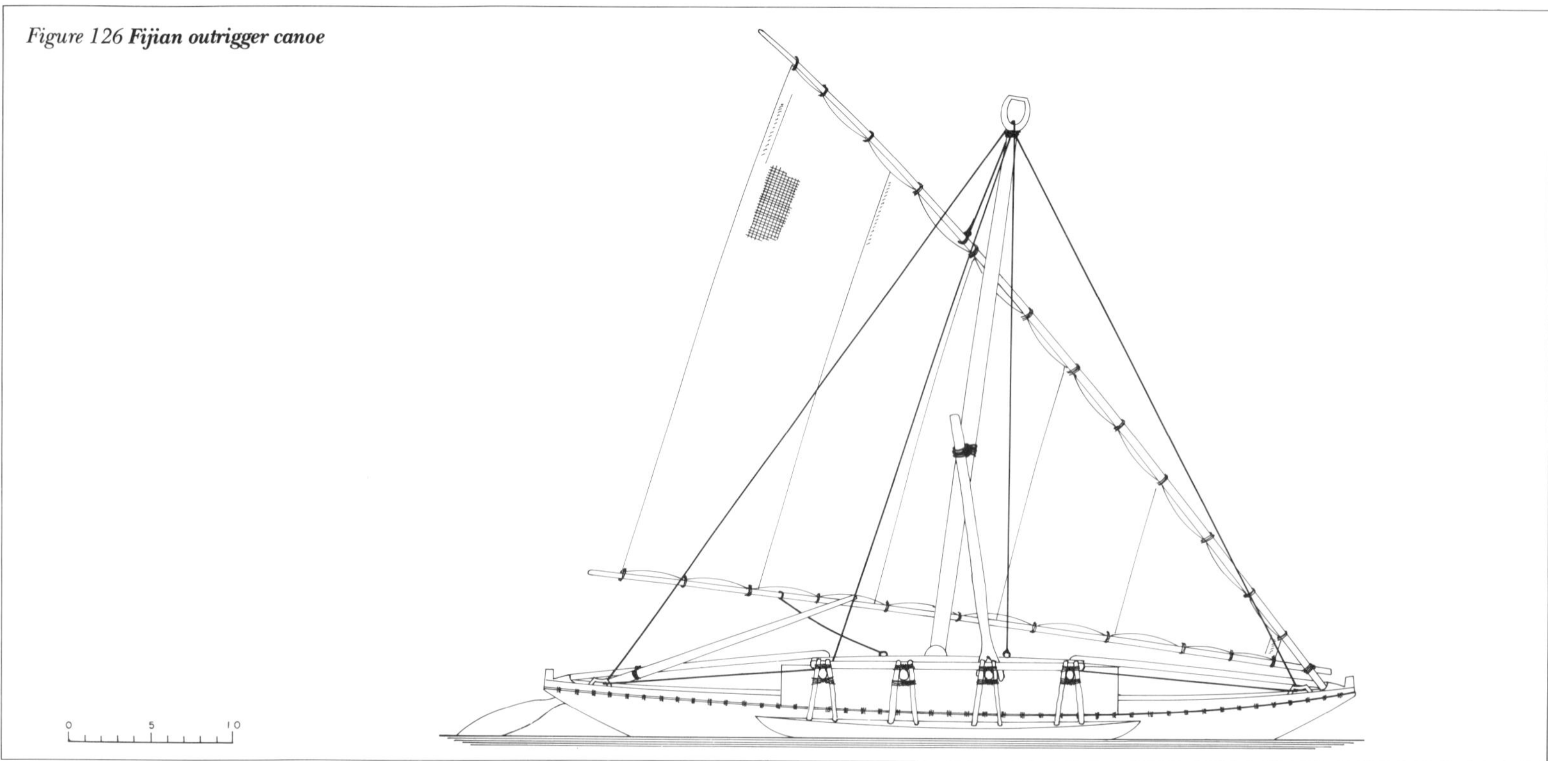

Figure 126 ***Fijian outrigger canoe***

are not rigged that way.

Steel's description of the vessel was very detailed, and is therefore worth repeating here:

> Flying-prow: A sort of narrow canoe, about 2 feet broad and 36 long, used about the Ladrone islands. Their lee side is flat, and the weather side round. A mast is stepped in the weather gunwale and to the same side is fixed a frame composed of bamboo, projecting out about 11 feet under the extremity of which, and parallel to the vessel, is suspended an oblong block of wood, formed and hollowed like a canoe, and thus a balance is produced, which prevents the prow's upsetting, for the weight of the frame (which may be and sometimes is increased by men running out upon it, according to the exigency) prevents [the vessel] falling over to leeward, while the floating properties of the hollowed block of wood at the extremity of the frame, resist the tendency of rolling over to windward. This construction is so extremely light, that she seems to feel no resistance in her passage through the water.
>
> Their rigging consists of two stays, that set up at the ends of the prow, and four shrouds that set up at the four corners of the frame. The mast-yard and boom are of bamboo. The sail is made of mat, shaped like a settee-sail: the lower end of the yard is confined forward in a shoe. In going about, they keep her way so, that the stern becomes the head and, to shift the sail, the yard is raised, and the lower end taken along the gunwale, and fixed in a shoe as before; the boom is shifted at the same time, by slackening the sheet, and peeking the boom up along the mast, then by hauling upon another sheet, the boom is brought to where the lower yard-arm was before, and is hauled aft at the other end. They are steered by paddles at each end.

In contrast to Brett's flying proa, the Fijian canoe was symmetrically shaped and double ended. Such canoes were fore-and-aft decked and had a boxed-in superstructure amidships, to give shelter to transported goods or passengers. The number of outrigger beams varied from two to six and a deck was laid over these beams from the weather side to a few feet over the lee side. Beams connected the lee corners of this deck with the bow and stern, which were identically fitted with a yard shoe (see Figure 127j), so that the stern could become the bow when the vessel went about.

The mast was loose and raked forward from its step, which was cut out of the centre deck plank (see Figure 127l); like the sail, the mast could easily be removed. The mast was supported by a forked pole reaching up to nearly half the mast's height, at which point the pole and mast were lashed together. The pole's fork rested on one of the outrigger beams.

The two stays led not to the a central point at the bow and stern, but through weather side cleats a short distance from both ends of the canoe. They were belayed to the foremost and aftmost outrigger beams. Shrouds led from the masthead to the outriggers on the weather side only. No rigging was set up on the lee side; only the steering sweep was on the lee side of the vessel, stropped to a cleat similar to those for the stays.

The yard and boom were lashed together at the forward end, and assisted by the forward rake of the mast, the yardarm fitted firmly into the semicircular shoe inside the bow of the canoe. The sail was triangular.The yard was hoisted with a single tye through the forked masthead, belayed before the mast to a cleat on deck. A similar cleat was fitted at the same distance abaft the mast for the sheet. These cleats were simply bent pieces of tree branch lashed to the deck planks through holes drilled into them (see Figure 127k). No blocks or trucks were used, nor were nails used in any part of the construction; all parts were lashed or stitched with coir.

Foochow pole junk

Futschou-Pfahl-Dschunke; Jonque de fukién

If we look for illustrations of junks in eighteenth century European books, we usually encounter only one type, the Fukien junk, or, as she was classified in G R G Worcester's work *The Junks and Sampans of the Yangtze*, the 'Footchow pole junk'. Unchanged for hundreds of years, these vessels were the large seagoing ships of China, trading with ports as far afield as the Persian Gulf and the Red Sea.

The vessel received its name from its place of construction, and the cargo usually transported in it. In the timber-rich mountainous regions of Fukien, the felled trees were formed into rafts

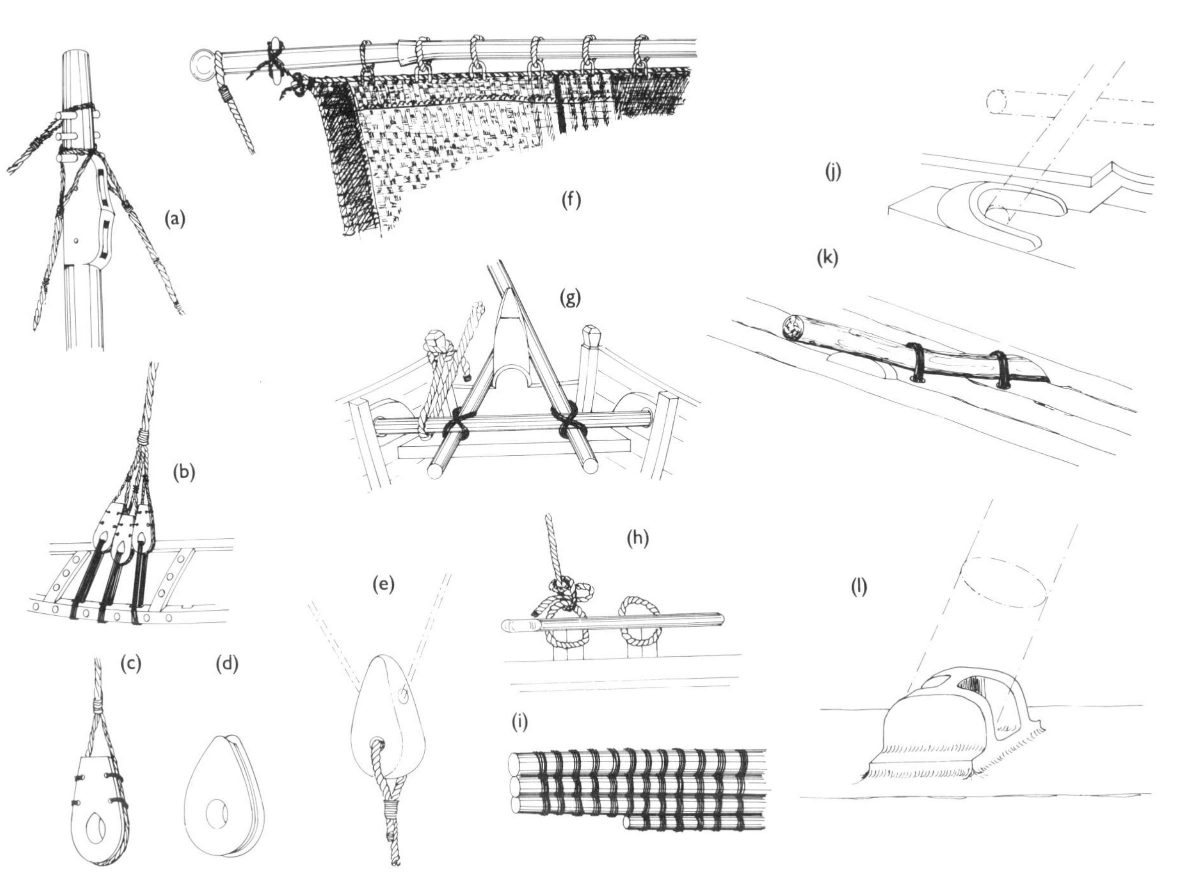

Figure 127 ***Details of Figures 124 -126***

(a) Masthead of a gay-bao.

(b) Detail of a gay-bao shroud, showing three trucks seized to the side of the vessel.

(c) Truck (block) of a gay-bao.

(d) Truck (block) of a prao mayang.

(e) Vang block (truck) of a prao mayang.

(f) The bending of a prao mayang sail to the yard with grommets.

(g) Bowsprit lashing in the bow of a prao mayang. The port stay leads round the bit and is belayed to a bow bollard. The starboard stay is not shown for clarity.

(h) A shroud hitched to a grommet, on a prao mayang.

(i) Detail of the lashing of a prao mayang boom.

(j) Detail of the yard shoe of an outrigger canoe.

(k) Belaying cleat on deck of a Fijian outrigger canoe.

(l) The mast step on a Fijian outrigger canoe, formed as part of a deck plank.

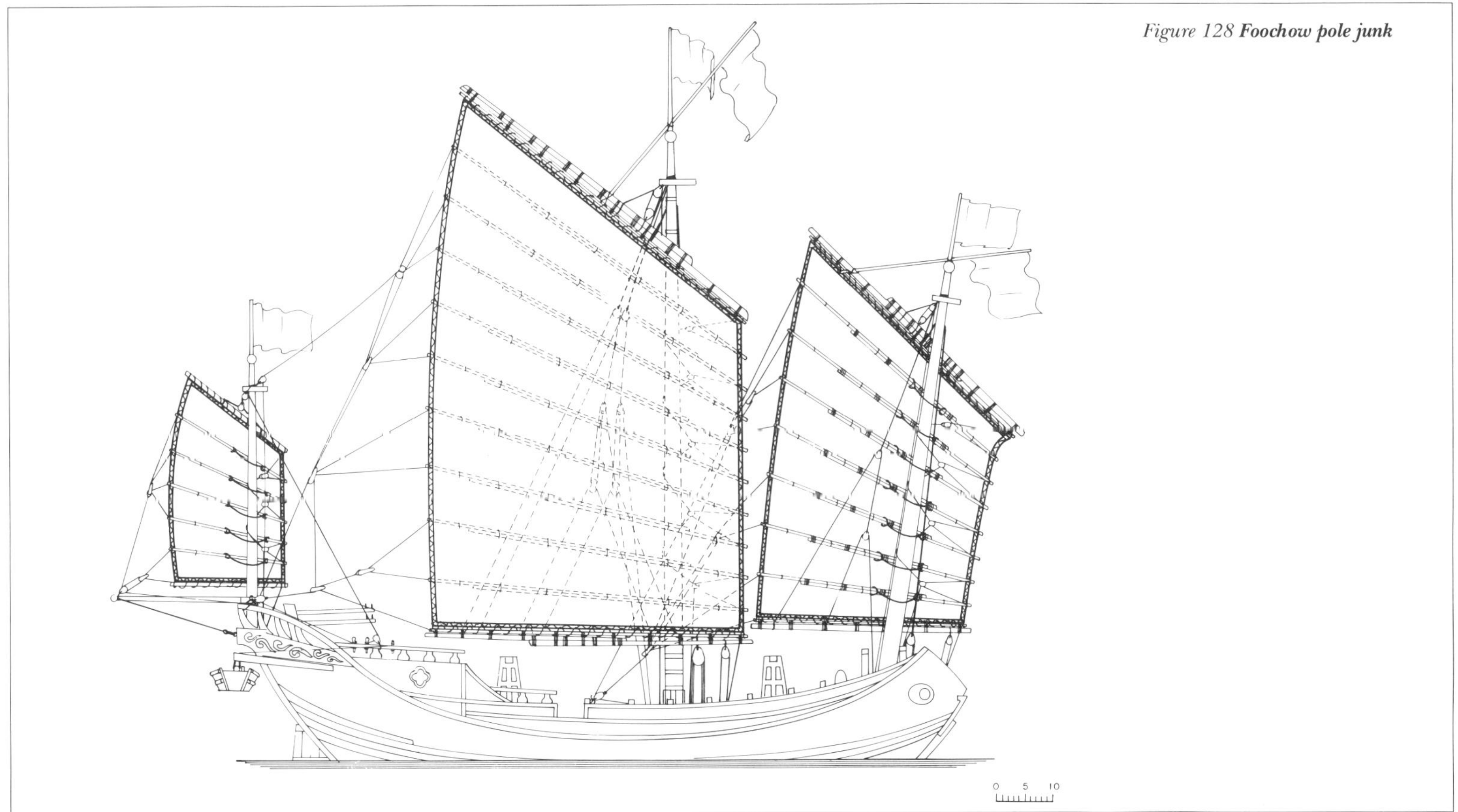

Figure 128 ***Foochow pole junk***

and floated down river to the port of Foochow. There the long timbers (poles) were bundled and fastened outboard, and the shorter logs stored in the hold, of the junks which were built in the city's shipyards. Fukien timber was shipped in this manner to Shanghai and many other places in China.

A typical Chinese lug rig was carried on the three masts. The

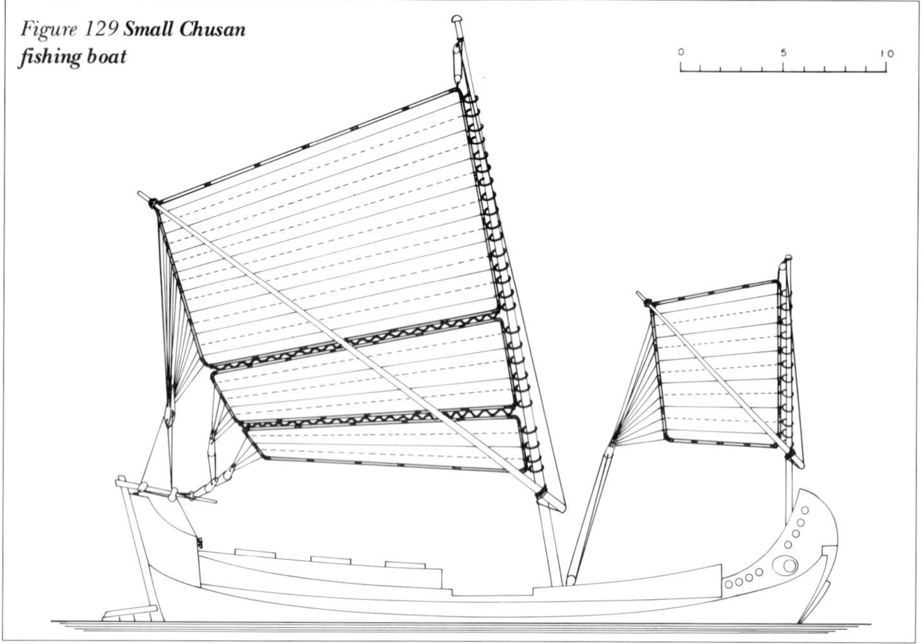

Figure 129 **Small Chusan fishing boat**

foremast was raked slightly forward, and the main and the mizzen masts were vertical, with the mizzen mast positioned at the very aft end of the poop, abaft the rudder.

Fabricated from two pieces, the larger yards for the main and foremasts were joined together with a number of rope lashings. Some booms were also built in a similar fashion. Sails were made from sailcloth of half the width of European material and usually sewn together with a smaller number of stitches per foot. A boltrope, stitched to the sail's tabling, was called the inner rope.

An outer boltrope was laid out in the sail's shape, usually together with the battens as well as the upper and lower secondary boom for bending the sail to, and the sail was fitted into the resulting frame. The inner boltrope was then seized, at irregular distances of about 6 inches, with spun-yarn to the outer, and the battens were sewn to every sailcloth. The sail's shape or cut varied with every shipping district and at least half a dozen or more differing cuts were known.

As implied above, the sail was not directly laced to the yard or boom, but to a secondary boom, usually a bamboo pole of batten thickness (see Figure 131a and b). These bending booms were seized every 1½ to 2 feet to yard or boom. Only smaller mizzens were bent direct and without these secondary booms.

Junk rig differed considerably from all other rigs. Stays and shrouds did not exist and the masts were either pole masts or made masts, in either case very strong and hooped. In contrast to European masts, those of the Foochow pole junk were made from hardwood, though Worcester noted that junk masts were otherwise mainly made from Fukien pine. The masts usually stood in tabernacles, so that they could be lowered (using much the same method as described in Chapter VI above).

Large yards had two halyards, with the throat halyard bent to a yard cleat (see Figure 131c) one third away from the lower yard arm. It rove from the rear through a sheave-hole in the masthead downwards upon deck, were it was set up to a treble and double block tackle (see Figure 131k). The secondary halyard was a tackle, set up to the yard a few feet behind the throat halyard, with its other block at the masthead. The tackle fall led upon deck through a foot block aft of the mast.

Each batten had its own loose rope parrel laid round the mast. Spans connected the fore ends of all battens with each other. A hauling parrel line was fastened to the uppermost span block and it was possible to pull the luff leech in towards the mast by taking this line round the mast and through the next span block, then repeating this for each span.

According to the sail's size, the topping lift was either single or double, and it was usually double whipped. Large booms employed a runner tackle, the lower block of which was of the leg and fall type, with a span rove through the lower opening and bent at both ends to the boom.

Rather than a sheet, there was a whole sheeting system, since it not only controlled the boom end, but acted also as a sort of vang, exercising control over the whole aft leech. Each batten-

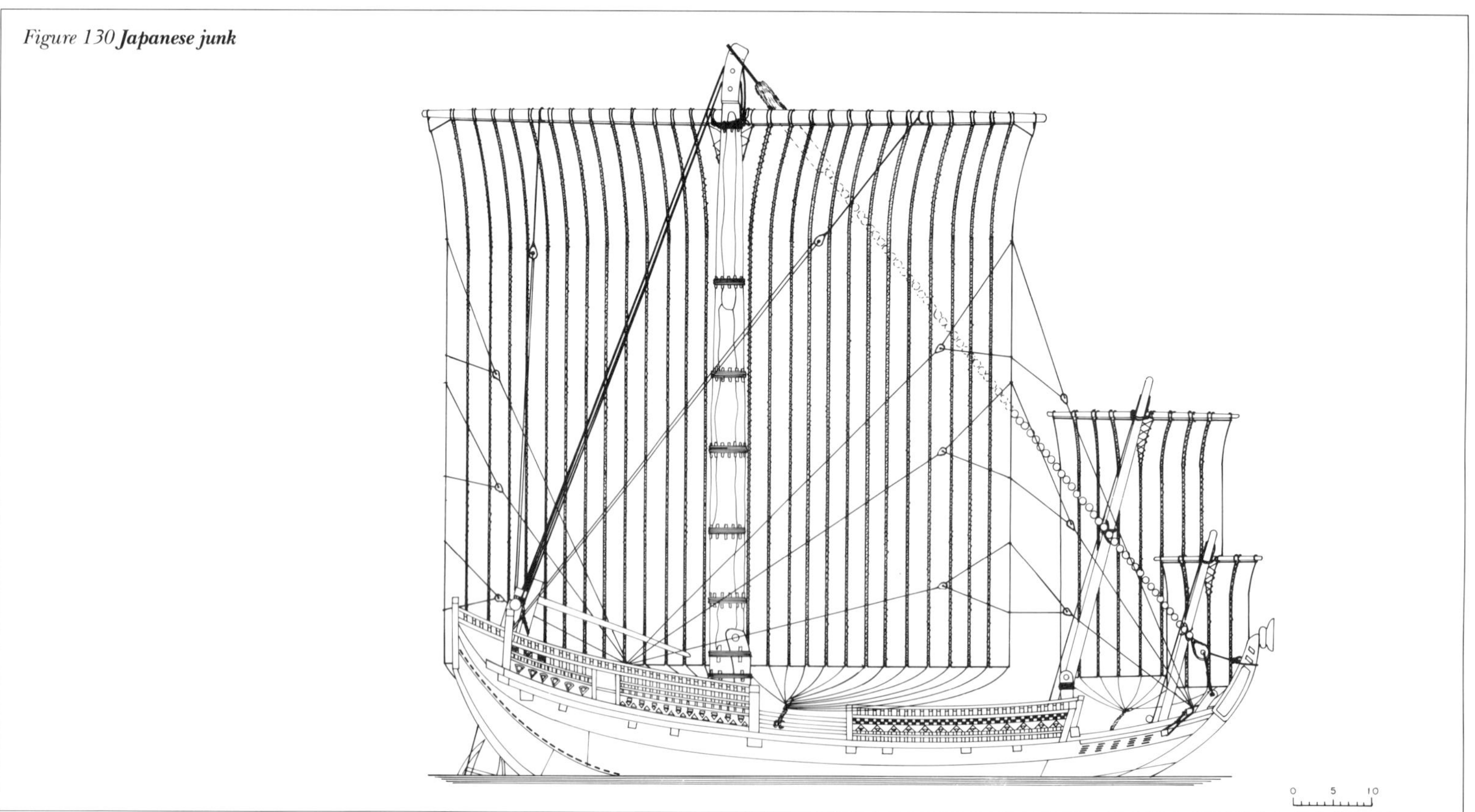

Figure 130 **Japanese junk**

end had a sheet attached to it, which, with the help of blocks, formed a sort of intricate bridle system, ending finally in one rope to be belayed. As Worcester explained, this multiple sheet could differ from one junk type to another and was also determined by the number of sail battens.

A Chinese junk's ingenious rigging required a lot less ropework than a similar sized European sailing vessel.

Small Chusan fishing boat

Chusan Fischerboot; Bateau de pêche de Chusan

Hundreds of types of vessels were indigenous to China, and their hull and sail shapes varied considerably according to region; in rigging, however, only two main lines are recognisable. One, the lug rig, has been described in connection with the Foochow pole junk, and the other was a sprit rig.

A sprit rig was very simple and used on small vessels in harbours and on rivers, as well as on nearly every fishing vessel on the island of Chusan.

Chinese spritsails were nearly square and their sailcloths were sewn horizontally together, with the seams sometimes covered with a rope. A pole was used as a mast and a bamboo sprit held the sail's peak up. The spritsail was controlled only by a multiple sheet; harbour and small river craft usually had no more rigging than this.

Chusan fishing craft carried a slightly more complex variation of this rig. The sail's horizontal cloths run not parallel to each other, but converged towards the aft leech of the sail, to give the sail a tapered look. A boltrope enclosed the sail and a second, doubled outer rope was fitted at a set distance from the inner (see Figure 132a). Fastened to the outer rope's mast side were hoops to connect the sail to the mast as well as for hoisting and lowering. These sails frequently had one or two bonnets, which were prepared similarly.

The mast was not very thick and had a slight aft rake. The sprit, a bamboo pole, was sometimes made from two parts, one of which slipped into the other. If such a connection was not possible, a hardwood dowel linked the bamboo parts together. The lower end rested in a kind of snotter, while the upper end was set into an outer rope eye at the sail's peak. A grommet pushed on to the upper end of the sprit acted as a stopper and prevented the sprit from slipping further than required through the peak eye.

The halyard was a tackle, set up to the sail's nock and the masthead. The snotter was not of the European type but a rope laid round the mast and over a notch in the sprit's lower end, and belayed round both mast and sprit. Often the hauling part of the halyard was used for this task, and thereby belayed.

The sheet was bent with crowsfeet to each sailcloth, and each bonnet had its own sheet also with crowsfeet, with all of these linked together with a few blocks to a main sheet rope.

Finally, it is worth noting that neither stay nor shrouds were rigged to the mast.

Japanese junk

Japanische Dschunke; Jonque japonaise

Japanese junks had one, two or three masts according to their size, but only one type of rigging. In contrast to Chinese junks, these vessels were square rigged. The bulky, made mainmast was nearly square, with only the fore side rounded (see Figure 132i). Fitted into the masthead, one above the other, were two sets of double sheaves for the main yard's halyards. Above these sheaves the masthead was notched to form two ears over which the fore stay was slipped (see Figure 132h).

The fore stay was hooked with two eyes over these protruding

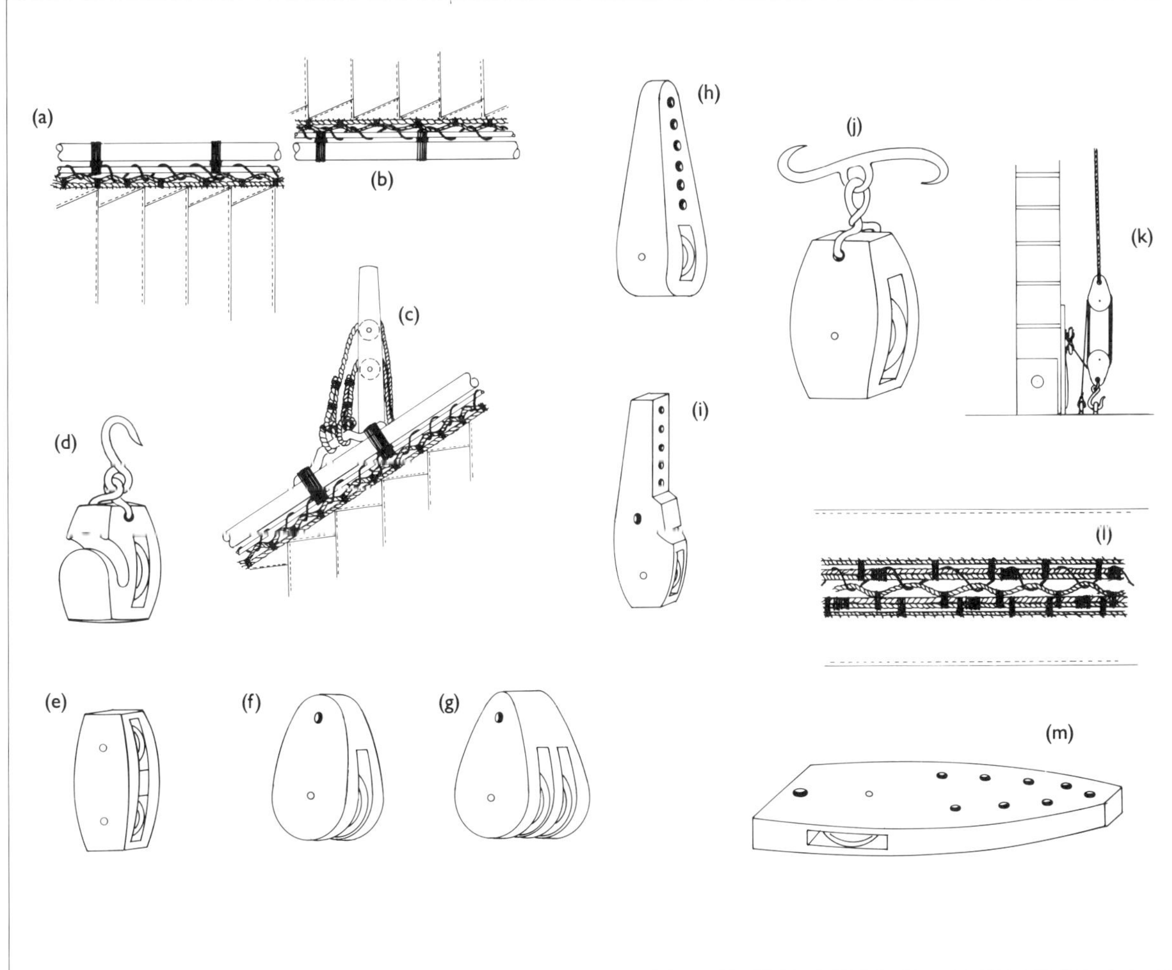

Figure 131 ***Details of Figures 128 and 129***

(a) Detail of the lacing of a Chinese junk lugsail to the yard and boom.

(b) In both cases the sail was laced not direct to the yard or the boom, but to an extra lacing boom, lashed to the spars. Note the inner and outer boltrope and the end preparation of the sailcloths with diagonally cut or folded ends, sewn accordingly.

(c) The tye on a Chinese yard, seized to a yard cleat.

(d) Detail of a Chinese cleat block with a hook.

(e) A Chinese sisterblock.

(f) A Chinese single split block.

(g) A Chinese double split block.

(h) A Chinese euphroe with a sheave.

(i) A Chinese split euphroe with a notch and sheave.

(j) A Chinese sheet block with a hook.

(k) The set-up of a halyard on deck in Chinese rigging.

(l) Detail of the bonnet lacing on a Chusan fishing boat spritsail (note that the sailcloths are horizontal). On the upper side of each bonnet an additional rope is seized in shallow loops to the outer (double) boltrope.

(m) A euphroe with a sheave, used in the sprit rig of a Chusan fishing boat.

Figure 132 ***Details of Figures 129 and 130***

(a) Detail of the boltropes of a Chinese spritsail. The outer is double and forms an eye at the peak corner to take the upper end of the sprit. The sailcloths run horizontally and overlap by half their width. The inner boltrope is sewn to the cloths, and seized to the outer.

(b) A treble block as used for rigging the spritsail's sheet.

(c) The clew of a Japanese square sail. Each sailcloth has its own boltrope, with a sheet bent to the outer clew of each cloth. Sheets were bundled and belayed as one unit. The boltropes of the cloths are laced together (for clarity the boltropes are shown here drawn apart; in fact, they ran practically side by side).

(d) The centre section of a large Japanese yard, with four halyards and a standing parrel. In this section the sail opens up in a V-shape over the width of two cloths. Each sail-cloth is bent to the yard with two grommets. The halyards were also set up to yard grommets.

(e) The yardarm of a Japanese junk. The top edge of the outer sailcloth was cut diagonally.

(f) Detail of a grommet with a halyard bent to it.

(g) A Japanese main stay, hooked with two eyes over the masthead ears, leather covered and with trucks.

(h) A Japanese mainmast, seen from forward.

(i) A section through a half-round Japanese mainmast.

(j) A bowline block from a Japanese junk.

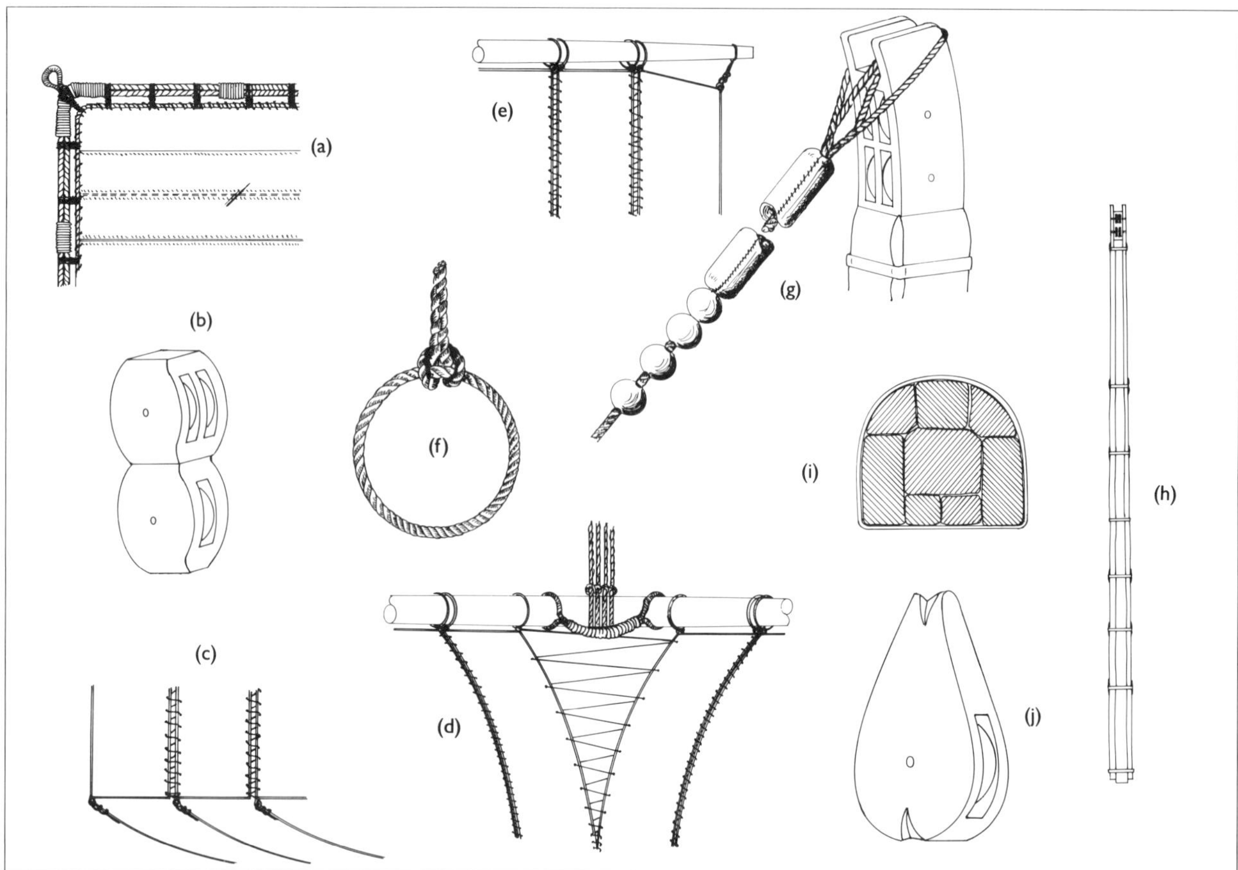

masthead ears, and the stay's upper part, in the section where the yard could chafe, was leather covered (see Figure 132g). A drawing by Paris shows that the remaining part of the stay below the leather protection was completely covered with wooden trucks. A larger, flat single-hole truck (block) was turned into the lower end, with a span reeving through; one leg of this span was made fast round the stemhead and the other to a bow bollard (knighthead).

Four halyards were needed to hoist the main yard. They rove through the sheave-holes in the masthead and led to the stern. A rope parrel kept the yard close to the mast and brace pendants were spliced round the yards, about one sixth of the yard length inside the yardarms.

Sails were square cut and made from a number of vertical sailcloths, each of which was bordered with a rope, sewn together without overlapping (see Figure 132c). In the upper fifth of the sail the two centre cloths diverged, so that at the yard the centre seam of the sail opened up over a width of two sailcloths, to prevent the sail from chafing in the yard section where the halyard and parrel were fastened (see Figure 132d). Only a loosely fitted line prevented this triangular opening from splitting the sail in the middle.

Three bowlines with single bridles were attached to the mainsail's side leech in the first, second and third quarter, from below, with similar lines going aft. Sheets were bent to each sailcloth's outside clew and bundled into two on the mainsail, and into one on other sails.

The forward-raked foremast was stepped on the fore part of the keel, and was only half the length of the mainmast. It was a pole mast of a much smaller diameter than the main, and it had no stay. Since the foresail's size was only a fraction of the mainsail's, braces were not used and the halyard was single.

When such a vessel had a third mast stepped, this stood on the lower stem, forward of the foremast, and was again only half the length of that mast; its sail was only a quarter the size of the foresail. Besides halyards, both foresails were rigged only with sheets to each sailcloth.

Balsa and jangada

Balsa und Jangada; Balsa et jangada

Built from the soft wood of :he balsa tree, the balsa was for centuries one of the dominant vessels of the South American west coast.

The epic voyage from Peru to the Tuamotu Islands in 1947 of Thor Heyerdahl and his raft *Kon-Tiki* proved the seaworthiness of these apparently primitive craft.

Steel's was the most succinct of the contemporary descriptions of the raft:

> Balsa, or Catamaran: A raft made of the trunks of the balsa, an extremely light wood, lashed together, and used by the Indians and Spaniards in South America. The largest have nine trunks of 70 or 80 feet in length, are from 20 to 24 feet wide, and from 20 to 25 tons burthen. There is always an odd log, longer then the rest, placed in the middle projecting aft. They have but one mast, in form of sheers, whose heels rest on each side the raft on which is hoisted a large square-sail. When a fore-staysail is set, a pair of sheers is rigged forward. These rafts run with foul winds, and steer, as well as any other kind of vessel, by means of an invention similar to, and perhaps the original of, that called a Sliding Keel. They have for this purpose planks about 10 feet long. and 15 or 18 inches wide, which slide vertically in the spaces between the trunks which form the raft. It is only necessary to immerge them more or less, and place a greater or less number at the head or stern of the raft, to make them either luff to, or keep from the wind, tack, wear, lie-to, and perform every necessary manoeuvre. If one of these planks be drawn up forward, the raft will keep away and, if one is raised abaft, she will come to the wind. The number of these planks is five or six and their use is so easy, that, being once underway, they work but one of them, drawing it up, and immerging it one or two feet as may be necessary. The demonstration of the theory of working of ships will confirm the effects of this

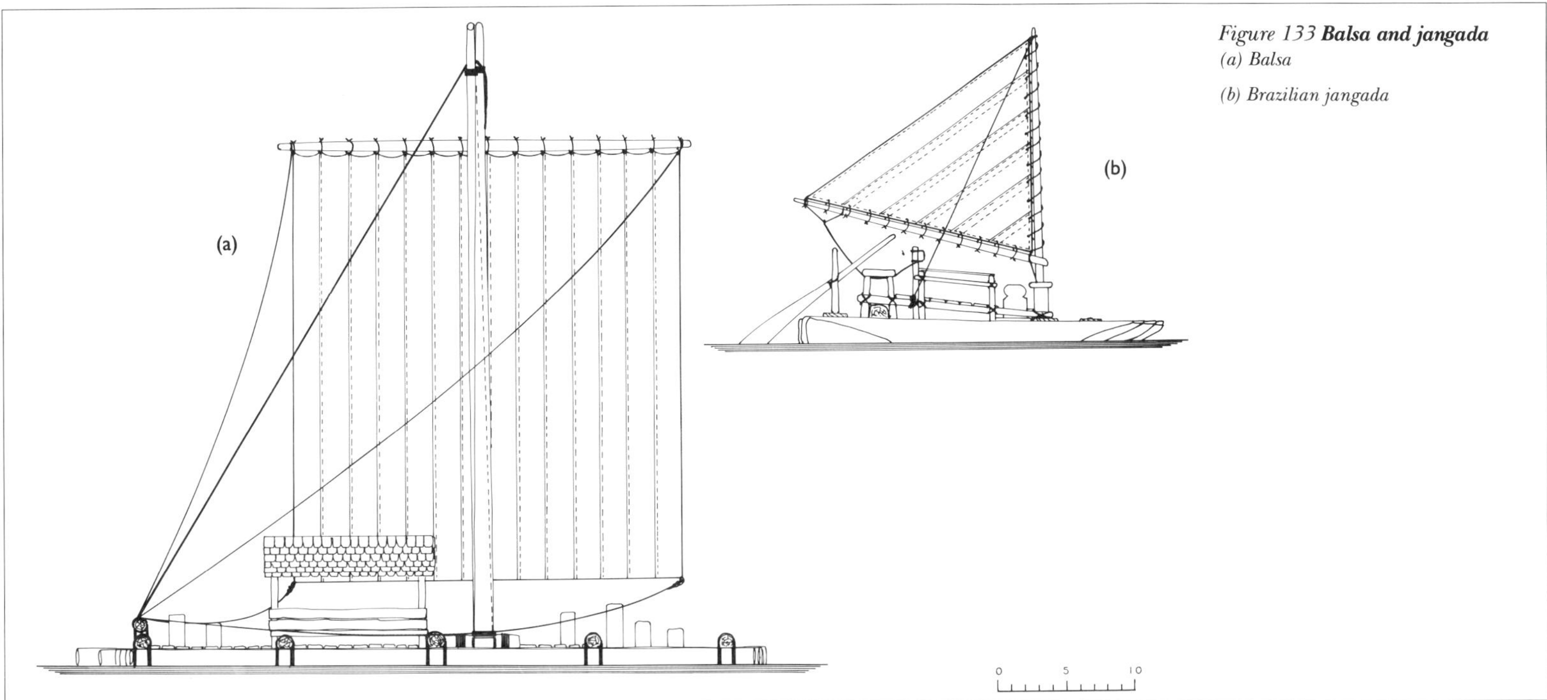

Figure 133 **Balsa and jangada**
(a) Balsa
(b) Brazilian jangada

construction which might perhaps be well adapted to many cases of emergency, after shipwreck upon coasts, destitute of all other materials for ship-building.

As the balsa was typical of the Pacific coast of South America, so the jangada was a prominent native craft of the Brazilian Atlantic coast.

The jangada was rigged with a light mast, a boom and a triangular sail. The mast was removable and could be repositioned in five different locations, according to the wind direction. A flying backstay was its only lateral stabiliser, and a boom sheet controlled the sail.

Jangadas had only one centreboard, but had a sweep at the stern for steering. These vessels were between 20 and 26 feet long and 6 to 8 feet wide; they were fast sailers, and are still in use today for coastal fishing.

Bermudian sloop

Bermudische Schlup; Sloop bermudien

The Bermudian sloop rig has been reconstructed often. The results have usually been questionable, and the reconstruction considered here makes no claim to have solved all the problems either. It takes account of all the research to date, but incorporates also material not previously used.

A shortage of contemporary illustrations meant that the Bermudian was usually depicted as a European topsail sloop with an overly large boom sail, a relatively short gaff and a mast with a fair rake aft. The main influence behind this interpretation was an illustration by Serres in his *Liber Nauticus*, but his Bermudian sloop, which had the characteristic hull shape, mast rake and long boom, had been rerigged in Royal Navy style. Evidence of this rerigging includes the spritsail yard on the bowsprit, a feature of Royal Navy schooners between 1764 and the 1780s and a feature of Royal Navy yachts and sloops also, as indicated by Falconer. The rigging shown by Serres was nearly identical with the sloop rig described in the earlier edition of Falconer, and is marked as a naval vessel by the fact that it flies a common pendant. The Bermudian sloop rig, in short, was transformed into that of a European sloop in Serres' picture.

Bermudian sloops were basically fore-and-aft rigged sloops, with all the attributes described in Chapters IV and V above. Early pictorial evidence is available in Burgis' *View of New York* from about 1717, his *Sloop off Boston Light* from 1720 and his *View of Boston* from 1725. The last named also shows a few topsail-rigged sloops, all of them marked as naval vessels.

Chapman, who provided a draught of a Bermudian sloop as well as listing all the spars, hinted at the type's original topsail rig. His notes were considered in previous reconstructions only in conjunction with the Serres illustration and other similar pictures, and his hints have largely been missed. Chapelle made one such reconstruction in *The Search for Speed under Sail*, and included for comparison another by M F Edson Jr; a third reconstruction was by R Napier in his article 'The Bermuda sloop 1740'.

If Chapman had found nothing to distinguish the Bermuda sloop's spars from those of a normal sloop, I doubt that he would have recorded them. In his whole work, he recorded the masts and spars of only three vessels, all three being unusual and not widely known. The reconstructions noted above were all based on the belief that Chapman's crossjack yard must have been the spread yard for the topsail sheets. This was not the case.

This yard, with a length of only 40 percent of the mast length (compared to 63 to 65 percent in normal sloops), was in no way a spread yard, but was the squaresail yard. By omitting a crossjack yard, in the sense of a yard employed to spread the topsail clews, Chapman confirms Röding's suggestion (supported by an illustration) that a Bermudian sloop rig had no need for a topsail spread yard.

An annotation to Röding's description of the sloop read as follows:

> The Bermudian sloop or Bermudian yacht, often used by Frenchmen and Englishmen for trading, differs from the common [yacht] in that the lowest topsail, which they carry direct above the boomsail, has a pair of very long ears [that is, clews], reaching down to be belayed to the gunwales.

This was a curious looking topsail, but it explains Chapman's short crossjack yard fully. This yard had very much in common with a cutter's squaresail yard, and careful examination of both rigs reveals a clear developmental step: the Bermudian topsail's long ears were shortened in eighteenth century cutter rig and were spread by a yard at half mast height. The studdingsail booms on Chapman's squaresail yard can also be found on some other squaresail yards, and were a necessary extension of this

Figure 134 ***Bermudian sloop***
The rig shown follows Chapman's 1768 spar dimensions and Röding's 1794 description and illustration.

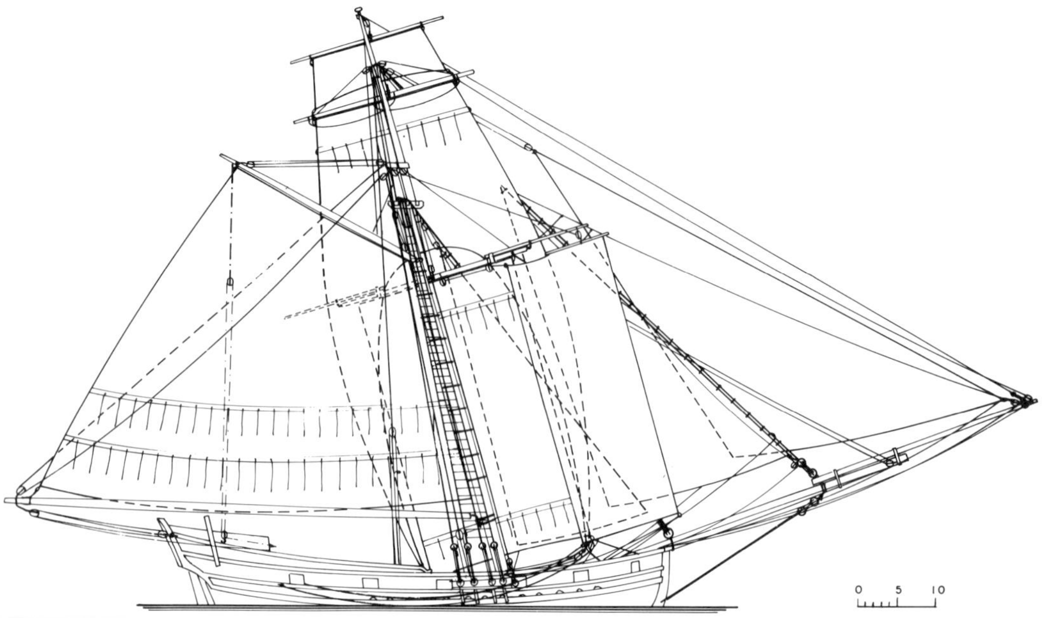

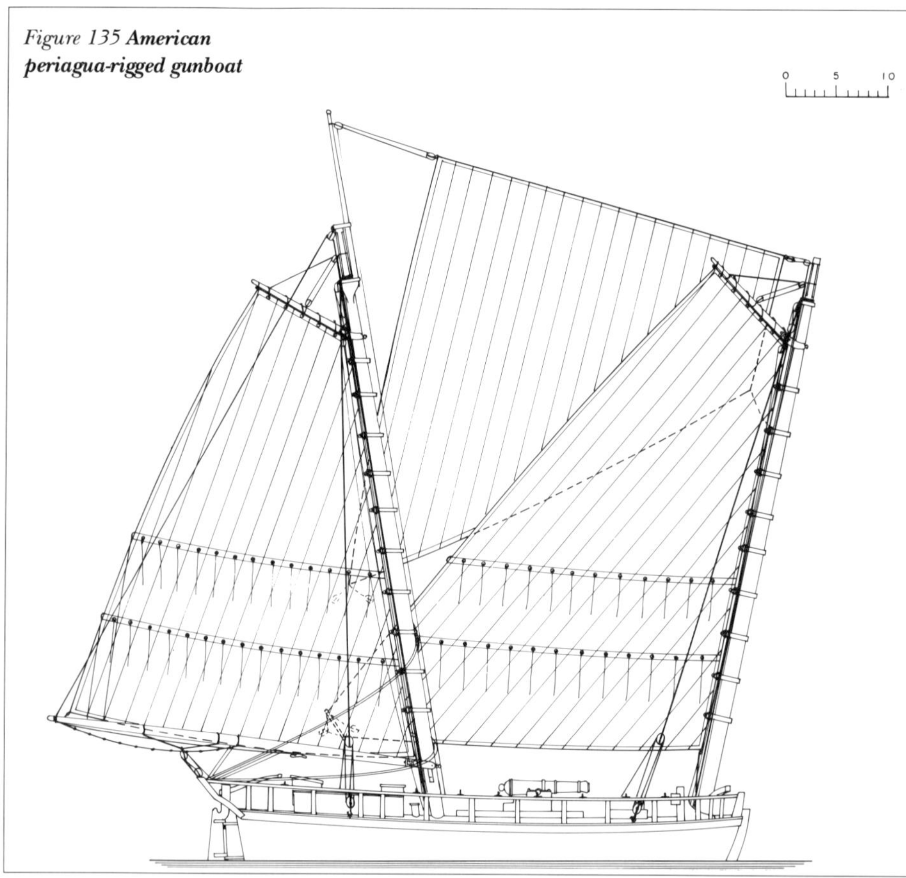

Figure 135 ***American periagua-rigged gunboat***

relatively narrow squaresail when the wind was light and the vessel sailed before the wind. Here again the squaresail was only a temporary sail, used in those conditions.

Periagua

Piroge; Pirogue

Falconer described the periagua as follows:

> [A periagua is] a sort of large canoe, used in the Leeward Islands, South America, and the gulf of Mexico. It differs from the common vessels of that name, as being composed of the trunks of two trees, hollowed and united into one fabric whereas those which are properly called canoes, are formed of the body of one tree.

The rigging of these South and Central American vessels was reminiscent in some respects of a schooner without a bowsprit. This rig spread during the eighteenth century along the North American east coast, including the Mississippi and Hudson estuaries. The lighters and ferry boats of New York harbour used it also, and the name became 'perry-auger' among the rivermen.

In North America the native periaguas of the south became flat-bottomed barges without keels, with bow and stern sections decked, and with a capacity of 25 tons maximum. They were propelled by sweeps or by two gaff sails, which were rigged to strikeable masts. Since they had no keel, lee-boards had to be fitted.

The nascent US Navy also used this rig in the early nineteenth century for fourteen gunboats. These were designed by Christian Bergh in 1806 and had a simple rig, with the foremast very far forward in the bow and with a marked forward rake (see Figure 135). The mainmast was raked aft at a similar angle. A small gaff at the foremast carried a loose-footed sail, and the mainsail was rigged schooner fashion with a boom. A flying staysail could be set between the mastheads. The ropework for this rig was restricted to the bare essentials.

As noted at the outset, the purpose of this chapter is to provide the reader with a general view of the vessel types and rigs an eighteenth century seaman might have encountered.

It must be emphasised that it has been impossible, in this chapter, to remain strictly within the confines of the eighteenth century in the descriptions of various rigs and types of vessel. Some of the Levantine and Far Eastern vessels have been described from drawings made in the nineteenth century, or from drawings which I made during the restoration of contemporary models for a number of museums, especially the Deutsche Museum in Munich.

The vessels and rigs discussed here generally did not undergo such rapid evolution as those of Northern Europe. Some vessel types, in fact, were still being built in the second half of the twentieth century in much the same traditional way as that indicated by these drawings more than a century old. The inference therefore is not unreasonable that these vessels were already built and rigged in the same way well before their details were recorded.

IX

SAILS

Sails were needed to harness the wind, a ship's motive power, and consisted of varying numbers of pieces of canvas made to a particular width, and known as sailcloth. There were two types of sail: the squaresails, which were bent to yards and hoisted athwartship, and the fore-and-aft sails, which might be staysails, gaffsails or fore-and-aft yard sails. All of these sails were either triangular or quadrangular (usually not square; the term 'squaresail' might suggest that this was the theoretical distinction between the two types, but in fact the name only indicates the way the former type of sail was rigged, square to the ship).

The four sides of a squaresail were the head, the foot and the leeches. A triangular sail had a stay, a foot and an aft leech, usually simply called the 'leech', and a quadrangular fore-and-aft sail had an additional fore or mast leech, also called the 'bunt'.

Each corner of a sail had its particular designation as well. On a squaresail the upper corners were called the 'earrings', and the lower the 'clews'. 'Peek' or 'peak' was the name for the upper corner of a triangular sail, a 'clew' was the lower aft corner, and a 'tack' the lower forward corner. On a four-cornered sail the forward upper corner was termed the 'nock'.

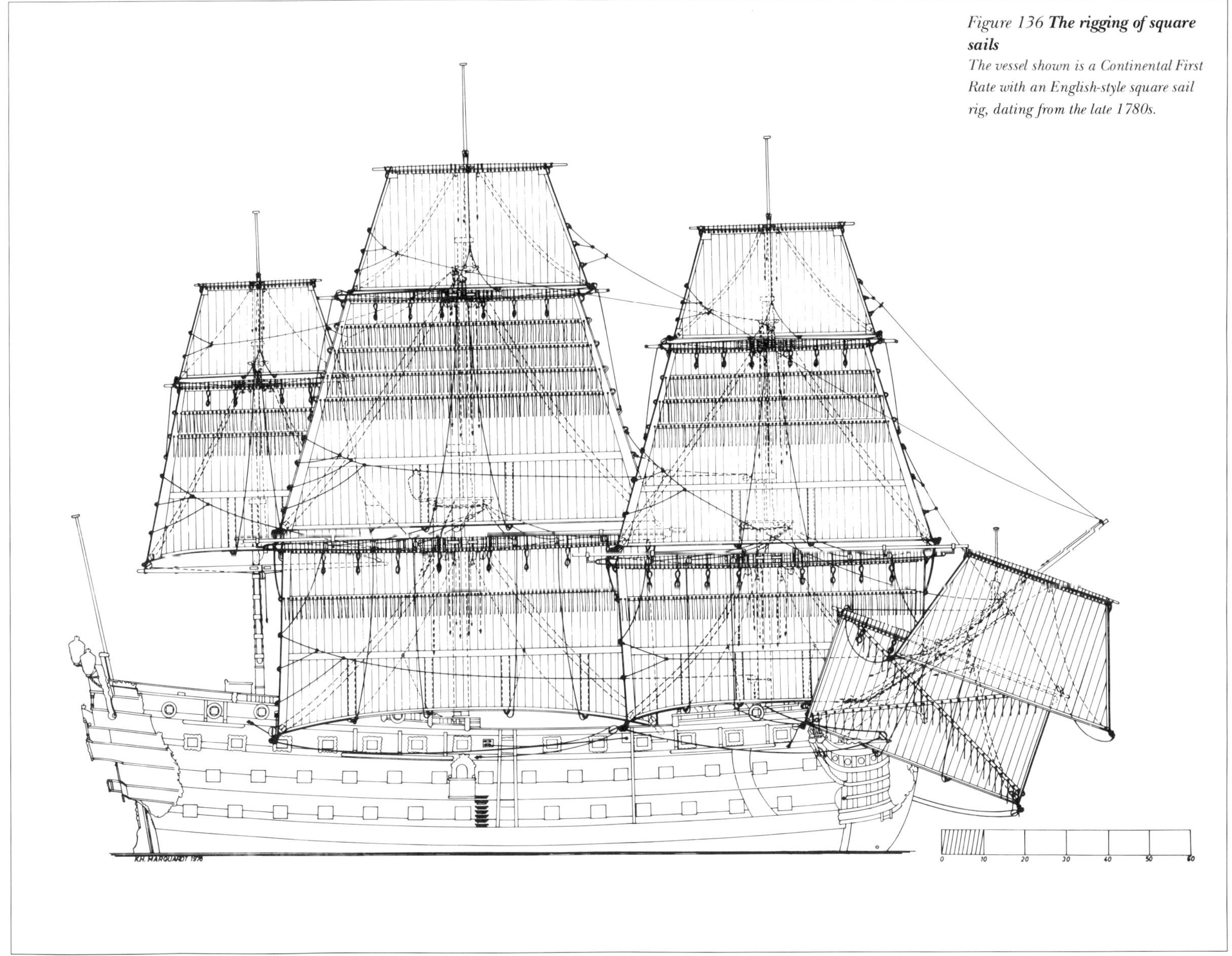

Figure 136 **The rigging of square sails**
The vessel shown is a Continental First Rate with an English-style square sail rig, dating from the late 1780s.

During the eighteenth century a full-rigged three-masted ship carried the following sails. The sails only introduced during the eighteenth century are shown with the year of introduction in brackets:

Square sails	
bowsprit	spritsail, spritsail topsail
foremast	fore course, fore topsail, fore topgallant sail
mainmast	main course, main topsail, main topgallant sail
mizzen mast	mizzen topsail, mizzen topgallant sail
Fore-and-aft sails	
bowsprit	fore staysail (English 1773, only on men-of-war), fore topmast staysail, jib (1705), flying jib (1794, English), Continental standing jib, inner jib (1715?) and storm jib (1715?)
between fore and mainmast	main staysail (only on ships with 50 guns and fewer), main topmast staysail, middle staysail (1773), main topgallant staysail (1709), main royal staysail (1719)
between main and mizzen mast	mizzen staysail, mizzen topmast staysail (1709), mizzen topgallant staysail (1760)
mizzen mast	mizzen

Until about 1760 all staysails were cut triangular; only after this were the staysails between masts quadrangular.

Besides a fore staysail and a fore topmost staysail, Continental head gear consisted of a standing jib, a second or inner jib and a storm jib. While the standing jib, like the fore and fore topmast staysails, was hanked to a stay, the inner jib was usually set flying. The storm jib, as the name indicated, replaced the former two in bad weather and was set closer to the bowsprit.

The mizzen was a lateen or settee sail, bent to the mizzen yard in large ships up until the 1780s. The shortening of this sail in conjunction with the lacing of the fore leech to the mast began on smaller ships in about 1730 and was completed by 1745 in all classes except for large ships of the line and East Indiamen.

The general use of royals above the topgallant sails did not occur until the last few decades of the century. Steel mentioned royals and Röding listed them for all masts, but royals were known long before the end of the eighteenth century. It is generally assumed that the royal sails of the *Sovereign of the Seas* (1637) were a unique fitting for this large ship, and that these sails were not used again in large or smaller ships until the late eighteenth century. However, a listing in Davis' 'Boatswain's Sea Stores for the East Indiaman *Degrave*', dated 2 October 1676, notes, in addition to an unusual sprit topgallant sail, also a fore topgallant royal and a main topgallant royal. This request gives us a new insight into the use of royals during the seventeenth century. They were probably also very well known to the ships of the East India Company. Davis also noted in the same list a fore topsail staysail, a fore topgallant staysail, and a mizzen topsail staysail; these also predate the normally accepted introductions of such sails.

Fore and main masts had lower and topsail studdingsails. French ships occasionally still also had studdingsail booms on their crossjack yards up to the 1780s, and Röding noted mizzen topmast studdingsails in 1794. Davis (1711) indicated mizzen studdingsails in English ships by noting 'two Cross-Jack Studding-sail Halyards...' and 'a pair of Cross-Jack Studding-sail Halyards'. These were only used on English ships early in the century and probably more on East Indiamen than on men-of-war, since Davis' list only indirectly suggests the former, and in the *Degrave* list of 1676 he requested '2 Cross Jack Studding-sails' without mentioning these in his rigging lists for naval ships.

After the general introduction of royal sails, topgallant studdingsails also came into being. They date from about 1775.

Regulations for the manufacture of sails

Vorschriften für die Herstellung der Segeln; Réglementations sur le fabrication des voiles

Sailcloth, made from hemp, sometimes with the addition of cotton or flax, was the fabric from which sails were made. The varying applications of sails on a mast demanded a different strength or weight of material, and therefore a short general survey of the main quality grades of various maritime nations, especially the terms used for these grades, is important.

Netherlands

The best cloth in the Netherlands was called *kanefas*, followed by *karral* or *karrel*, *karreldoek*, and *klaverdoek*. The latter, being only half as good as *karreldoek*, was used for topgallant sails. Similar light cloth grades were the *eeversdoek* and the *lightdoek*.

Baltic Sea

Sailcloth from Sweden was only known as 'Swedish cloth', and it was said of the best Russian sailcloth that it was even better then the Dutch *kanefas*. It was known as 'best blue mark', and was not only better, but also about one third cheaper. Further grades were 'small blue mark', 'black mark', 'green mark' and 'red mark'.

France

> The best and heaviest sail-cloth made in various centres in France is the *toile à trois fils* (22 to 26 ounces per ell) [a French ell equalled 1188mm], followed by the *toile à deux fils* (17 to 19 ounces per ell), and *toile à un fil*, which is the lightest. *Toile de doublage* is only ordinary and used in the doubling of sails. *Toile à prélart* is the poorest of all and is only used for tarpaulins. Besides these there are still two grades known under the names *noyale* and *melis*, which are the names of their places of production. There is *noyale à trois fils*, which serves for the lower sails of ships of the line, and *noyale à deux fils*, not quite as heavy, and used for topsails. *Melis double* is similar in quality to *toile à un fil* and *melis simple à un fil* is lighter still and only used for topgallant sails.

This is Röding's outline of the main quality grades for French sails. He also noted *cotonine à trois fils*, *cotonine double* and *cotonine simple* as sailcloth for galleys, xebecs and other lateen vessels. *Cotonine* was made from a mixture of hemp and cotton. *Cotonine à carreaux* was blue and white chequered and used for awnings and the sails of smaller boats. *Toile écrue de 7/8* was a light cloth of 7/8 ell width used for topgallant and studdingsails.

French sailcloth usually had a width of 21 French inches and came in bolts of 50 ells of length. A bolt of French sailcloth was therefore about 40mm narrower, but much longer than the English equivalent.

Other Continental countries

Spain and Portugal imported the heavier grades and called them *lona*, the lighter *loneta* and the lightest or topgallant cloth *brim*. *Bitre* in Spain was a slightly rougher cloth than *brim*. In Italy, *cottonina* was the sailcloth for lateen sails, and for vessels of other rigs Italian sailmakers used *lona*.

For topgallant sails and other light sails flax-based canvas was also in use on the Continent. This material was known as 'Flemish canvas', 'Russian linen', 'raven cloth' or 'Westphalian linen'.

England

Davis noted that sailcloth for courses and topsails went under

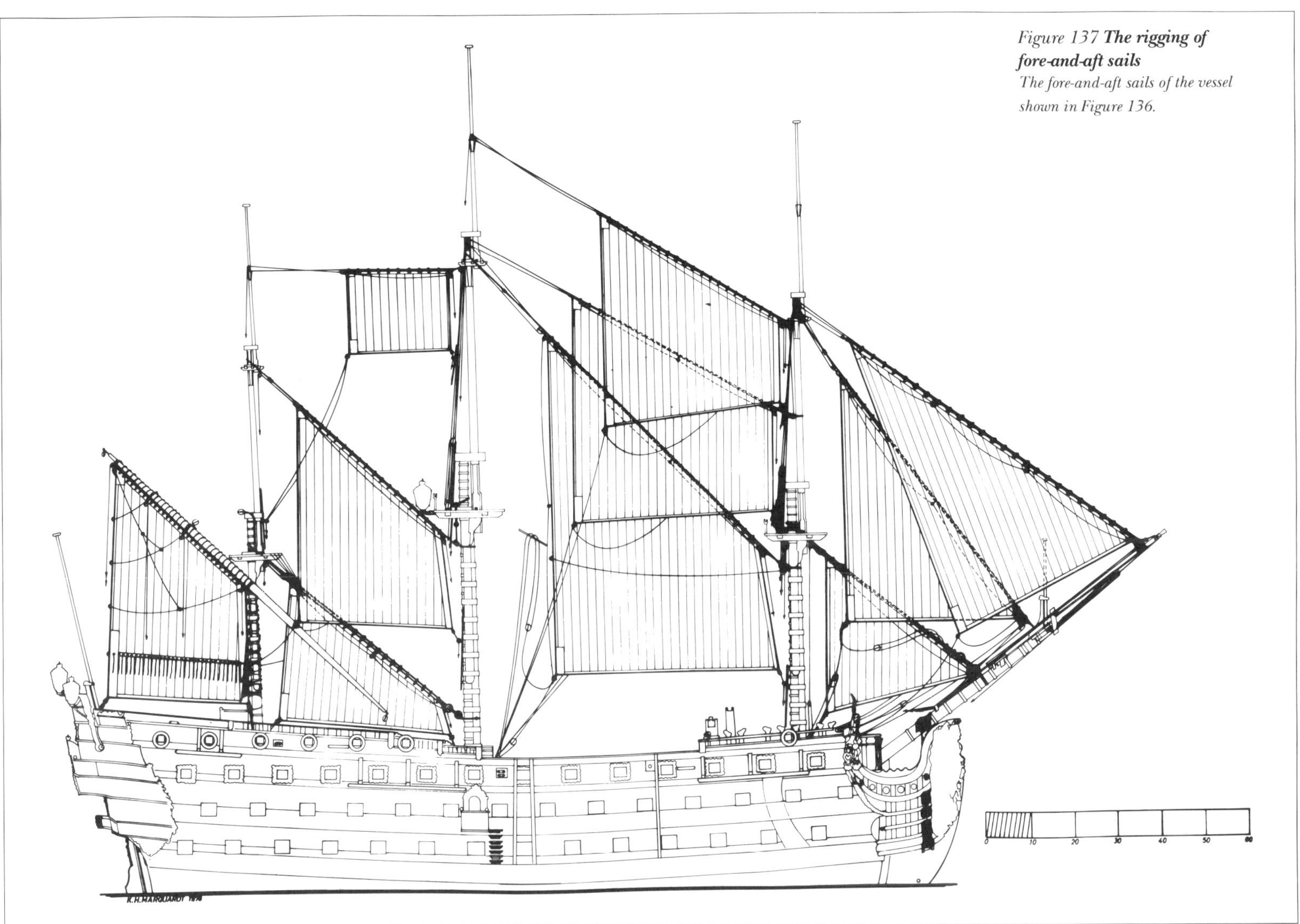

Figure 137 ***The rigging of fore-and-aft sails***
The fore-and-aft sails of the vessel shown in Figure 136.

the name 'duck', fore and main courses might also be made of 'Suffolk' and the topsails and lower staysails of 'great Noyals'. 'Vittrey', 'German duck' and 'English peartree' were the names for lighter cloth used to make topgallant sails, upper staysails and studdingsails. For the lower studdingsails he specified 'Ipswich', and for the spritsail topsail, the mizzen topsail and the upper studdingsails 'small Noyals'.

Strict rules were laid down by most shipping nations for quality maintenance. Among the strictest were the British Acts of Parliament during the reigns of William III, George II and George III, which ordered the captain of an English ship under penalty of £50 to carry only British-made sails on board when first setting out to sea – £50 was at that time about two months' salary for a captain. These laws, passed for the protection of British industry, also ensured the maintenance of a consistent material quality and therefore provided production guidelines for the different types of sailcloth.

A bolt of British sailcloth was 24 inches wide and 38 yards in length. Ten quality grades were differentiated, with the weight of the bolt determining the cloth strength.

Number 1	44lb	Number 6	29lb
Number 2	41lb	Number 7	24lb
Number 3	38lb	Number 8	21lb
Number 4	35lb	Number 9	18lb
Number 5	32lb	Number 10	15lb

The first six grades were known as double and the others as single cloth. Every alteration in a bolt's width or length had to be proportionally correct by standard measures and weights, and every bolt had to carry the correct number. The regulations set out in detail how many yarns were to be worked into certain widths, specified which flax not to use, and laid down that no flax-yarn should be whitened in English cloth on forfeiture of sixpence per yard so made or sold. These regulations had to be posted in all the places where sails were manufactured. A violation carried a penalty of 40 shillings.

In sailmaking the sailmaker, a most respected tradesman, also had many other regulated items to contend with besides the quality of sailcloth.

The yarn was to be only the best English twine of three threads, spun at 360 fathoms to the pound. The twine in the Royal Navy was waxed by hand, for the courses, topsails and staysails, with a mixture of ⅚ of genuine beeswax and ⅙ of turpentine, and for the smaller sails with a mixture of ⅖ beeswax, ½ of hog's lard and ⅒ of clear turpentine.

In the English merchant service the use of oil-thinned tar was common. The strict regulations laid down for the building and maintenance of vessels of war did not apply here.

There were also attempts to give the sails a longer life by giving them different finishes. Sails of smaller vessels were treated with a mixture of horse fat, tar and red or yellow ochre; others were wetted with seawater and had a paste of ochre and seawater rubbed into the cloth, with linseed oil brushed on to both sides of the sail. These treatments were common for small merchant vessels, but were not employed in the Navy.

When sailcloth was joined, the double flat seams were supposed to be sewn together with 108 to 116 stitches per yard of length. The widths of these flat seams were 1½in for the fore, main and topsails in ships with more than 50 guns, and under 1¼in for these sails in ships of 44 guns . All other sails had flat seams of 1 inch width. The middle of the seams of courses and topsails were also stitched over the whole length with double seaming twine with sixty-eight to seventy-two stitches per yard. In merchant ships it was common to give each seam two rows of stitches when the sail was half worn, which would last until the sail wore out.

Flat seams in the Navy ran parallel, whereas in merchant ships they sometimes tapered towards the head. Steel also noted that boom mainsails and the sails of sloops generally had the seams broader at the foot than at the head. Tapering seams were not allowed in the Royal Navy, and instead the goring of leeches was adapted. French men-of-war, however, had the seams of their lower courses tapered. Each cloth overlap was 1 inch at the foot and 3 inches at the head on main courses, and on fore courses slightly less in the opposite direction.

Tabling

Umschlag, Saum; Gaine

'A sort of broad hem formed on the skirts and bottoms of a ship's sail, to strengthen them in that part which is attached to the bolt-rope,' was Falconer's definition of tabling. The tablings of all sails were sewn with sixty-eight to seventy-two stitches per yard. The widths on each end of the sails are provided in the following table.

Table 53: Width of tablings in English inches

Name of sail	*head*	*leech*	*mast leech*	*foot*
Spritsail	3-4	3		3
Spritsail topsail	3	2½	-	2½
Fore sail	4-6	3-5	-	3-5
Fore topsail	3-4½	3	-	3
Fore topgallant sail	3	2½	-	2½
Fore royal sail	2½	2	-	2
Main sail	4-6	3-5	-	3-5
Main topsail	3-4½	3	-	3
Main topgallant sail	3	2½	-	2½
Main royal sail	3	2	-	2
Mizzen sail	3-4	3	3½-4	2-3
Mizzen topsail	3-4½	3	-	3
Mizzen topgallant sail	3	2½	-	2½
Fore staysail	3-4½	2-3	-	2-2½
Fore topmast staysail	3-4½	2-3	-	2-2½
Jib	3-4½	2-3	-	2-2½
Main staysail	3-4½	2-3	-	2-2½
Main topmast staysail	3-4½	2-3	3½-4	2-2½
Middle staysail	3-4½	2-3	3½-4	2-2½
Main topgallant staysail	3-4½	2-3	3½-4	2-2½
Mizzen staysail	3-4½	2-3	-	2-2½
Mizzen topmast staysail	3-4½	2-3	3½-4	2-2½
Mizzen topgallant staysail	3-4½	2-3	3½-4	2-2½
Studdingsails	3-4	1½-2½	-	1-2

Most of the following information is based on Steel's *Elements of Mastmaking, Sailmaking and Rigging* of 1794, which is among the oldest detailed descriptions of sailmaking.

Linings

Verstärkungen, Dopplungen; Renfort

Leeches were strengthened by sewing another layer of sailcloth on top of them. The number of stitches specified for these linings was sixty-eight to seventy-two per yard.

Fore and main courses had leech linings of a full cloth width from the earrings to the clews. This applied to late eighteenth century Continental sails as much as to British. French fore courses, with a gore reducing towards the clews, had a stepped leech lining over three or more cloths. Head linings were usually non-existent and Röding noted and illustrated a foot lining of a full width for Continental ships. This was corroborated by Paris, but not by Boudriot, who did not mention a foot lining for courses. These linings were placed on the inner side (rear) of the sail, though Paris appears to specify the outer side, probably because of a printing error. Steel stated for English ships that all linings except top linings and mast cloth were on the outer side of a sail.

Since Continental sails had leech linings on the inner side, additional patches of cloth, one width wide and 1½ widths long, were sewn across to the outer, where bunt, leech or bowline cringles spliced into the boltropes. Röding and Paris positioned these patches, termed by the former as *Bolten* (no equivalent English term exists), on the lower courses at the sides, with no buntline doubling. Boudriot noted six buntline doublings, two widths long and one ell wide, but on the sail's inner side. English ships had buntline linings on the outer side, where they were needed not only for increasing the strength at the point of fastening, but also as a protection of the actual sail against chafing.

Main and fore courses also had one or two reef bands, which, in the case of two bands, were placed one sixth of the sail depth away from the head and from each other. Röding noted only one reef band for lower sails, which was the accepted rule for Continental vessels. Nothing is recorded about the distance of a single reef band from the sail's head, but contemporary models and Boudriot suggest one quarter of the sail's depth, while the sail drawing by Röding indicates one fifth of that depth. In either case, a single band sat approximately midway between the two bands as noted above.

The ends of these bands ocurred about 4 inches below the leech linings in English sailmaking, and the width of a reef band was one third that of a cloth, that is 8 inches. For French ships Boudriot stated a width of half a sailcloth, with the reef band placed on the inner side of the sail and over, not under, the leech lining. This wider reef band is also visible in Röding's illustration, but it is placed on the outer side.

English sails were further strengthened with a full width middle band halfway between the lower reef band and the foot. Steel stated that this band should be folded to two thirds of its width (16 inches) and Lever and Steel took the view that it should only be sewn to the sail when the sail was half worn. This again was a merchant ship sail and the middle band was usually not part of a Continental sail. Röding, without identifying it with a particular style of sailmaking, noted that very large sails sometimes had a full width of cloth sewn across the others, simply known as a 'band'.

Four buntline cloths can only be found on English-styled sails. These were full width cloths spaced equally, sewn vertically to the outer side, with their upper ends below the middle band. Buntline cloths on merchant ship sails were usually about a third shorter.

English topsails had leech linings, a top lining, a mast cloth, buntline cloths, a middle band and reef bands. In the early decades of the century fore and main topsails had two, during the greater part of the century three, and in the later years four reef bands, with the mizzen topsail having always one fewer. Continental topsails, according to Röding, had only three rows of reef bands, and the mizzen topsail also one fewer. After 1800 French ships had the same arrangement as British ships adopted after 1790.

The width of topsail leech linings on Royal Navy topsails was 1½ cloths at the foot and half a cloth at the earrings. For the merchant service Steel noted that 'the leech-linings are but nine inches broad at the head and 15 inches broad at the foot'. Röding's illustration of a Continental topsail does not show a full lining, but only that the outside cloth was doubled, which amounted to about one quarter of the leech from the clew up. Instead of an additional lining on the inner side, six patches were placed on the outer side at the cringles. He also mentioned these doublings under the heading *Bolten*. The leech lining in Boudriot's topsail drawing of a 74-gun ship covered the same lower quarter but an additional lining was sewn on in steps up to the lower reef band, with the lower edge of each lining cloth matching the lower edge of the clew lining at the leech. Beneath the middle and upper reef band, as well as the earring, additional patches were sewn.

Boudriot showed no lining on the outer side of sails. Paris considered the stepped leech lining as belonging to the late seventeenth and early eighteenth centuries, and that illustrated by Röding as characteristic of the late eighteenth century. Patches were also used on British merchant ships. Lever commented that 'in the Merchant Service, Patches are frequently clapped on in the Wake of the Reef-tackle and Bowline Cringles, the strain on the first being very powerful when hauled out to reef the Sail, and on the last, when going by the Wind'.

Mast doublings featured on English topsails only, and were not used on Continental sails. A mast doubling consisted of two cloths, sewn to the middle inner part of the sail between the foot and the middle band, and after about 1790 up to the lower reef band, to protect the sail from damage by chafing on the mast. Further protection was provided by the top lining, an inner sail doubling confined by the buntline cloths on the outer and the mast cloths on the inner side, which were of the same material and became an integrated part of that lining. A topsail's top lining covered one fifth of the foot on Royal Navy sails and reached from the foot up to the middle band. Steel noted for merchant ships that 'the top lining and buntline-cloths cover one-third of the cloths in the foot, and are carried up one-third of the depth of the sail'. The height of a Continental top lining, reported Röding, was one third of the topsail's depth, and its width was, according to the illustration supplied, one third of the foot. This agrees with the data provided by Steel for British merchant ships. Boudriot's measurements for this lining were similar for height, but he considered the width to be one third that of the sail's head.

'In the Merchant Service, these Linings are generally objected to for it has been found by experience, that rain water lodging between the two parts of the Canvas, is apt to rot the Sail' was Lever's comment.

Stitched to the outer side of English topsails were a middle band, similar to that of lower courses, and two buntline cloths. Buntline cloths on Royal Navy sails extended below the middle band, while those on merchant ship sails were half a yard shorter than the top lining. Continental sails did not have a middle band or buntline cloths. They had a full-width foot lining for strengthening the lower part of the sail.

The reef bands were of half a cloth in width, put on double, and in the Royal Navy they were placed one eighth of the depth of the sail from the head, and the same distance apart. Mizzen topsails on ships of 44 guns and fewer had one reef band fewer than larger ships, one seventh of the sail's depth apart and below the head. On merchant ships reef bands were not always equally spaced; some masters chose to have a greater distance between the second and the third reef bands, Lever noted. The distances for Continental sails were equal to those in the Royal Navy, according to Röding, but Boudriot noted 4 feet below the head for the first reef band, one third of the sail's depth below the head for the third and a distance between the two for the second. Boudriot's reef bands were stitched to the inner side, while reef bands on Continental sails in general were on the outside, with only leech, foot and top lining on the inside.

While Continental topgallant sails had no lining, as can be seen from illustrations by Röding, Paris and others, Royal Navy main topgallant sails had (according to Lees) a full width leech lining from clew to earring, and fore and mizzen topgallant sails had a lining extending halfway up, with an extra patch at the earrings. Steel, describing topgallant sail linings for a 20-gun ship, provided the main topgallant sail with a triangular lining covering the outer cloth at the clew, and an earring patch. Fore and mizzen topgallant sails had only clew and earring patches. These patches appeared also on the topgallant sails of smaller vessels such as sloops, and can be considered common practice.

Royals usually had no lining at all. Since these sails were only hoisted under moderate conditions, the tabling gave enough strength to the sail.

Mizzen sails were lined at the after leech with a single width of cloth to a height of 5 yards from the clew. The peak and nock each had a lining patch 1 yard deep. French ships had a similar clew lining, but of half a cloth breadth, and no peak patch is mentioned. A mast leech lining was not referred to by Steel but was described less than 20 years later by Lever: '...it is made with a Mast Leech, which is lined with an additional Cloth'. Lees described this lining for the whole eighteenth century, and Boudriot noted half a cloth as the mast leech lining on French ships.

Parallel to, and one fifth of the mast leech depth above, the foot of a mizzen sail, a reef band was sewn across the sail. For French sails this band was one quarter that height. These reef band positions are for the shortened mizzen, which was introduced to English ships between 1730 and 1745 and to French ships in approximately 1750. The settee or lateen mizzen, carried until then and in large ships up to 1780, had a reef band which extended from the sail nock (settee sails) or tack (lateen sails) to about two thirds of the way up the after leech.

Spritsails and sprit topsails had no lining, but spritsails had two reef bands sewn on crosswise. These bands were of one quarter cloth width and reached from the leeches, 27 inches above the clews, to the first or second seam on the head inside the earrings. On Continental sails, they reached from three eighths or three sevenths down the leech to the opposite earring. They had the same width as the lower course reef bands. In the early years of the century English ships, as a 1719 Establishment rigging plan indicates, had only one reef band parallel to, and one fifth the depth below, the head. That this practice did not disappear completely during the eighteenth century is confirmed by Steel's note that 'sometimes a reef band is put on from leech to leech, at one-fifth of the depth of the sail from the head'.

Triangular staysails usually had a lining of 2 yards length from the clew up; the tack was cut in such a way that the remaining piece formed the lining and the peak lining patch was either 18 inches or 1 yard long. Merchant ship sails generally had a clew piece of only 1 yard long, and a peak lining 18 inches long. French triangular staysails had the tack lined with two or three cloths and the aft leech similar to English sails.

On quadrangular staysails the clew and peak pieces were similar to those of triangular sails, but in the Royal Navy the mast leech or bunt was usually lined with half a width of cloth. Sails in the merchant navy mostly had nock and tack pieces only, of 27 inches in length, except for the middle and the mizzen topmast staysail. This was also so for the bunts of French Navy sails. Linings on staysails were sewn to the port side.

The lower studdingsails sometimes had linings 9 or 18 inches long on all four corners. Main lower studdingsails in English ships also had a 6 inch wide reef band at one eighth depth below, and parallel to, the head.

For details of linings on all sails see Figures 139 to 147.

Holes

Löcher; Trous

Each sail was provided with a number of holes, to be used for bending and reefing the sail and other purposes. We can differentiate between head tabling holes, reef band holes, marling

Figure 138
(a) Hole with sewn-on grommet.
(b) A standard English and Continental clew up to 1796.
(c) A Continental (Russian) clew with a thimble for the tack block attached, 1750.
(d) A clew with seized-in tack block, used after 1796 in the Royal Navy. The sheet block and clew garnet block were fitted with a round strop over the clew.
(e) An English earing.
(f) A French earing, also used elsewhere on the Continent during the second half of the eighteenth century.
(g) Detail showing the sewing and marling of an English clew to the sail.
(h) A bowline and buntline cringle.
(i) An English reef and reef-tackle cringle.
(j) and (k) Details showing the bonnet lacing on the square sails of a small vessel, according to Steel.
(l) The bonnet lacing on courses (seventeenth century rigging).

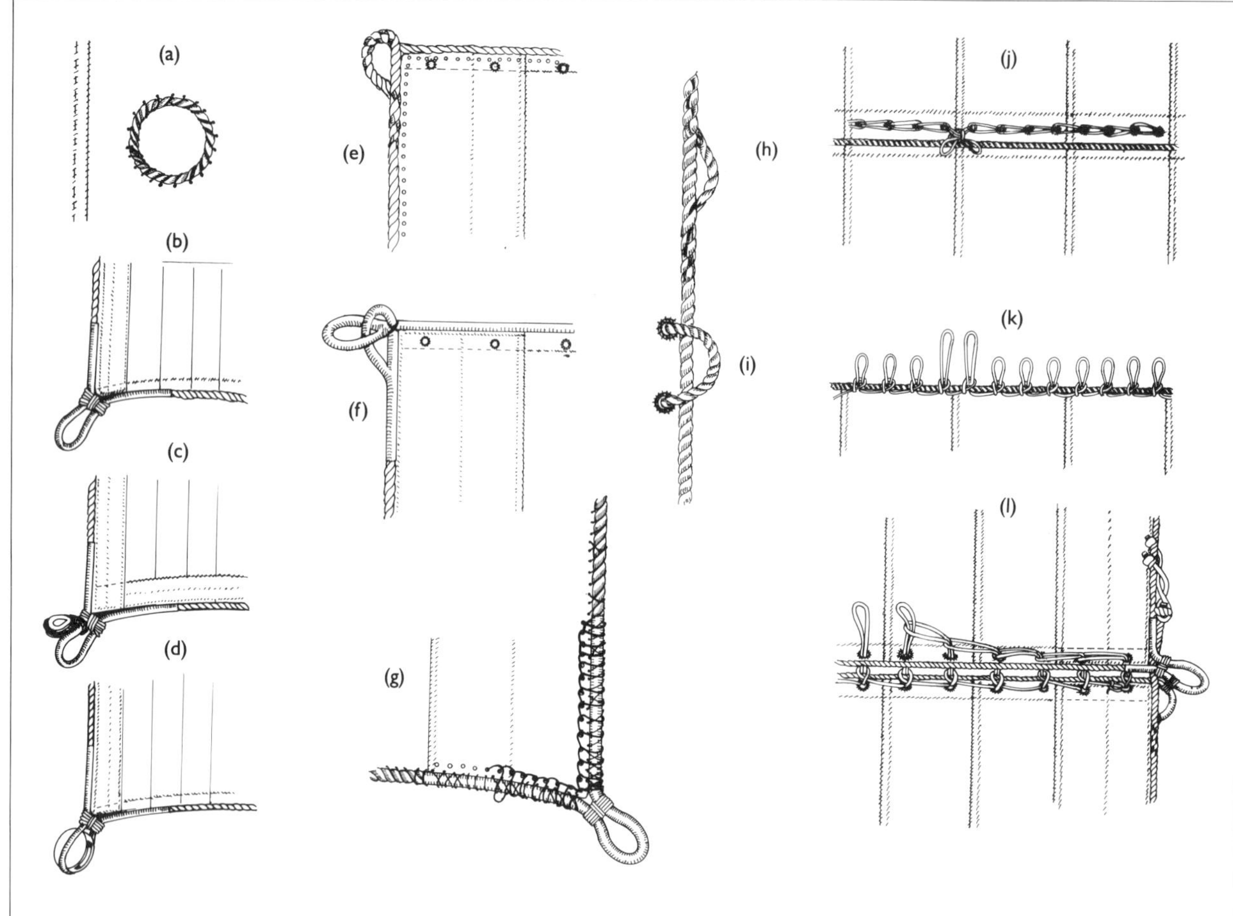

holes and water holes. All these were made with a pegging awl and framed with a grommet (see Figure 138a). Reef and head holes of large sails had grommets formed from 12-thread line, and on smaller sails 9-thread line was used. The larger grommets were sewn on with eighteen to twenty-one stitches, while for the smaller only sixteen to eighteen stitches were needed. Two holes were made in each cloth for reef and head holes in naval vessels; Lever noted that some sailmakers made one hole in the middle of the cloth and the next in the seam. The number of head holes on Continental sails, according to Röding and Boudriot, was one hole per sailcloth. Reef holes corresponded to this (according to Röding), but for the lower courses only, while for topsails Boudriot prescribed one or two.

Steel made a distinction in his portrayal of reef bands between those for square and loose-footed sails, and those for boom sails, where instead of reef points, reef hanks or knittles were used. The purpose of knittles was described by Falconer:

> When a sail is reefed at the bottom, it is done by knittles, which being thrust through the eyelet-hole thereof, are tied firmly about the space of canvas of which the reef is composed, and knotted on the lower side of the bolt-rope. These knittles are accordingly removed as soon as the reef is let out.

Knittle-reefed boom sails had only one reef hole in the seam of each cloth, since the overlapping of a seam, in conjunction with the reef band, was the strongest part of a sail.

Steel also made following observation:

> In order to strengthen sails, it has been recommended to have the holes in the heads and reefs placed thus: one hole to be made in the seam, another in the middle of the canvas, and so on alternately the hole in the seam to be half an inch lower than the hole in the middle of the canvas. By this the strain would lie upon the holes in the seam, which are more capable of bearing it than those holes which are in the single canvas. It is likewise recommended to cut these holes with a hollow punch, instead of making them with a stabber or pricker.

For all staysails except the jib, the distance between holes in the stay tabling was given as 2ft 3in and for the jib itself 3 feet.

In addition, all sails, with the exception of topgallant and royal sails and their staysails, had marling holes. These served to fasten the boltrope securely round the clew, and on topsails also at the top brims. The term 'top brim' is explained by Falconer thus:

> In sail-making, [a top brim is] a space in the middle of the foot of a top-sail, containing one-fifth of the number of its cloths. It is so called from its situation, being near the fore-part of the top, or platform on the mast, when the sail is extended.

The need for marling was clearly explained by Lever:

> The clews are wormed, parcelled and served; and Holes being made in the Tabling, the Sail is marled down to it, because the Service is too strong for the sail Needle to enter: the two parts are seized together with a round seizing.

Röding remarked that the foot rope was usually marled and Boudriot noted only three marling holes in the clew.

Marling holes were much smaller than the roband and reef holes and had grommets of logline fastened with nine to eleven stitches; twelve holes were worked in each cloth. In courses marling holes were three quarters of the tabling depth away from the rope, and in topsails half.

Main courses had marling holes from the clew up to the lower bowline cringle, and in the foot to the first buntline cringle. Fore courses had the same in the foot, but the holes in the leech reached only up to one eighth of the leech's depth. Main and fore topsails had marling holes 3 feet in both directions and at the top brims. Spritsails, mizzen topsails and all staysails, with the exception of those noted above, had marling holes at 2 feet each way.

For French ships, as illustrated by Boudriot, there were only three marling holes at the clews of the main and fore course, and only one each for most of the other sails. These holes were not marling holes in the sense of contemporary English and German definitions. The connecting line did not marl, but rather seized the clew to the throat seizing and to the leech and foot ropes. French sailmaking practice was in this respect, as in many other ways, different from that on the other side of the Channel.

In English practice the clew was sewn to a thinner, reinforcing rope, which in turn was marled to the boltrope.

Water holes were cut into the lower part of a spritsail. English spritsails had two and Continental spritsails often three, though both the author of *Der geöffnete See-Hafen* and Röding mentioned two. Those on English spritsails had a diameter of 4 to 6 inches and were placed in the second outside cloth, between the reef band and the foot. For French spritsails Boudriot reported three with a diameter of close to a cloth width (18 to 20 French inches). They were placed between the first and second cloths, or in the second and middle cloths, about a cloth width above the foot.

Boltrope

Liektau; Ralingue de chute

After the sails had been prepared up to this point, a rope was sewn around them for further stability. This was called a 'boltrope', with the part attached to the foot being the 'footrope', the clew section the 'clewrope', the leech part 'leechrope', and at the head, the 'headrope'. The headrope, usually, was approximately half as thick as the leech and footropes; Boudriot specified two thirds of the leech or footrope for French ships, while the clewrope was thicker than the other two. As Lever pointed out:

> The Clews, or lower corners of the Sails, are made of larger Rope in the Royal Navy, but in the Merchant Service it is generally omitted as unnecessary, being heavy and unhandy. The Clew is now a continuation of the Foot Rope; and if it be thought necessary to strengthen it, a Strand, of the same sized Rope, is opened and laid round it.

The splicing of the clews into foot and leechropes was explained by Steel as a short splice, with the large strands (clew) regularly tapered, going three times through and the smaller foot or leechrope going twice through, the clewrope. All splices were cross-stitched to the sail over their full length

> [The] Bolt Rope should be well made, of fine yarn, spun from the best Riga Rhine hemp well topped, and sewed on with good English-made twine of three threads, spun 200 fathom to the pound. The twine in the Royal Navy is dipped in a composition made with bee-wax, 4lb; hog lard, 5lb; and clean turpentine, 1lb; and, in the Merchant Service, in tar softened with oil.

This was Steel's description of boltropes, and he noted that boltropes on any sail should be neatly sewn on through every contline (a contline is the spiralling space between each strand of a rope), and that the rope should be kept tightly twisted, to avoid stretching during the process of sewing. Care should be taken that the right amount of slack was taken up.

Lever commented as follows:

> Bolt Ropes in the Merchant Service, are generally one third less than those formerly used and were they still less, it might be found to answer the purpose. The Sails would be thus light and handy, which is a matter of great consequence where Ships are so lightly manned: and to show the insufficiency of those very large Bolt Ropes, we need only observe a Dutchman's Jib, which is in the opposite extreme, being frequently not much stronger than Hambro' Line [Hambro' line was thick housing or house-line].

Boltropes were warmed in a stove by the heat of a flue and tarred with the best Stockholm tar. The heating and drying out of the rope made it more pliable and the tar could penetrate better. Choosing the right amount of tar to make the rope most workable was part of the sailmaker's trade. Röding mentioned that the best tar came from Norway and Sweden.

Boltrope sizes varied in circumference from 6 inches to 1 inch and the actual sizes for each English sail are given in Chapter X. French boltropes were measured as a fraction of the ship's beam, taking inches for feet. The following fractions are for foot and leech ropes; head rope measurements were two thirds of these:

Main course, fore course, mizzen staysail and fore topmast staysail	one ninth
Standing jib, inner jib, middle staysail and main topmast staysail	one tenth
Main topsail, fore topsail, mizzen course	one twelfth
Main staysail, main topmast staysail	one sixteenth
Topmast studdingsails, topgallant studdingsails	one eighteenth
Lower studdingsails	one twentieth
Main topgallant sail, fore topgallant sail, mizzen topsail, spritsail course, spritsail topsail and mizzen topgallant sail	one twenty-fourth

Clew, clue

Schothorn; Point d'écoute

Fourteen turns of an English clewrope were left at the lower corners of all sails to allow the clew loops to be formed. The length of clewropes for each sail was similar to the distances of marling holes from the clew, as noted above.

An additional inner clewrope was marled to the served length of a French boltrope's clew section and the sail was sewn through this. The reinforcing rope was thinner than the outer boltrope.

Earring cringles

Nocklegel; Pattes d'empoiture

Earring cringles in English sailmaking were made from a 14-turn or strand twist extension of the leechrope above the point where the headrope was spliced into it. This extension was turned back on the outside and its end was spliced into the leechrope (see Figure 138e). Within the splice, the sail was cross-stitched to the rope, with the first cross stitch double and the last treble.

The French method of connecting leech and headropes differed from the English. In French practice the headrope was thrust through an opened-up leechrope, with both ropes forming an earring cringle, and each cringle end led through the eye formed by the other (see Figure 138f). Röding's illustration showed the returned leechrope not spliced directly to itself but thrust through the rope and forming a second cringle before it was spliced. How far this was the practice and for what purpose is not clear.

Cringles

Legel; Pattes

Bowline and buntline cringles were spliced into the boltrope (see Figure 138h). 'Cringles should be made of the strands of new boltrope, half an inch smaller than the bolt-rope of the sail,' wrote Steel. For the making of cringles a strand of sufficient length of the boltrope was taken out and one end was pushed through two strands (one on each side) of the boltrope to one third of its length; the longer end, forming the cringle loop, also went through two rope strands and was laid over the first, with the shorter end doing the same, until the cringle resembled a rope, the ends being pushed under the boltrope strands like a splice. The ends of the outer buntline cringles, next to the serving of the clew on a course had to be long enough to be worked under the serving and to reach the end of the clew rope. The openings of buntline and bowline cringles were four turns of the boltrope long.

At the end of the century, thimbles came into use in connection with these cringles. Steel made no reference to this, but Röding noted their use for bowline cringles, because of the great stress upon these. Two decades later Lever commented that 'all these Cringles are now generally worked round Thimbles'. Steel occasionally referred to thimbles worked in the staysail tack and peak only, but only in the context of the fore-and-aft sails of cutters, sloops and similar vessels.

Reef tackle cringles

Refftakellegel; Pattes de palanquin de ris

Reef and reef-tackle cringles on French sails were worked like the other cringles, but on English sails they were led additionally through holes in the tabling, with the lower end passed once more than the upper through the boltrope, since it was the more likely to be pulled out (see Figure 138i).

Slack of sail

Bauchiges Segel; Mou de voile

The belly of a sail was determined by the amount of slack sewn into the boltrope. Main and fore courses had 2 inches per yard foot or head length, and 1½ inches per yard at the leeches. In other words, 36 inches of sail was sewn to 34 inches of head or footrope, and to 34½ inches of leechrope.

Topsails were allowed 3 inches for every cloth in the foot, 2 inches per cloth in the top-brim section and 1½ inches per yard on the leeches.

Mizzen courses had only 2 inches of slack in the mast leech, but the aft leech, head and foot were sewn without any. Spritsail courses similarly had no slack sewn in.

Jibs had 4 inches per yard at the stay, 1 inch per cloth at the foot and none in the leech. All other staysails had 3 inches per yard at the stay, with foot and leech like the jib.

Two inches per cloth in the foot and 1 inch per yard at the leeches was correct for topgallant sails.

Studdingsails had 1½ inches slack per yard in goring leeches, but none in square leeches, and 1 inch slack per cloth in head and foot.

Bonnets

Bonnetten; Bonnettes

A bonnet was an additional part of a lower sail, which could be fastened with lacings to the foot of the sail. It had the shape and

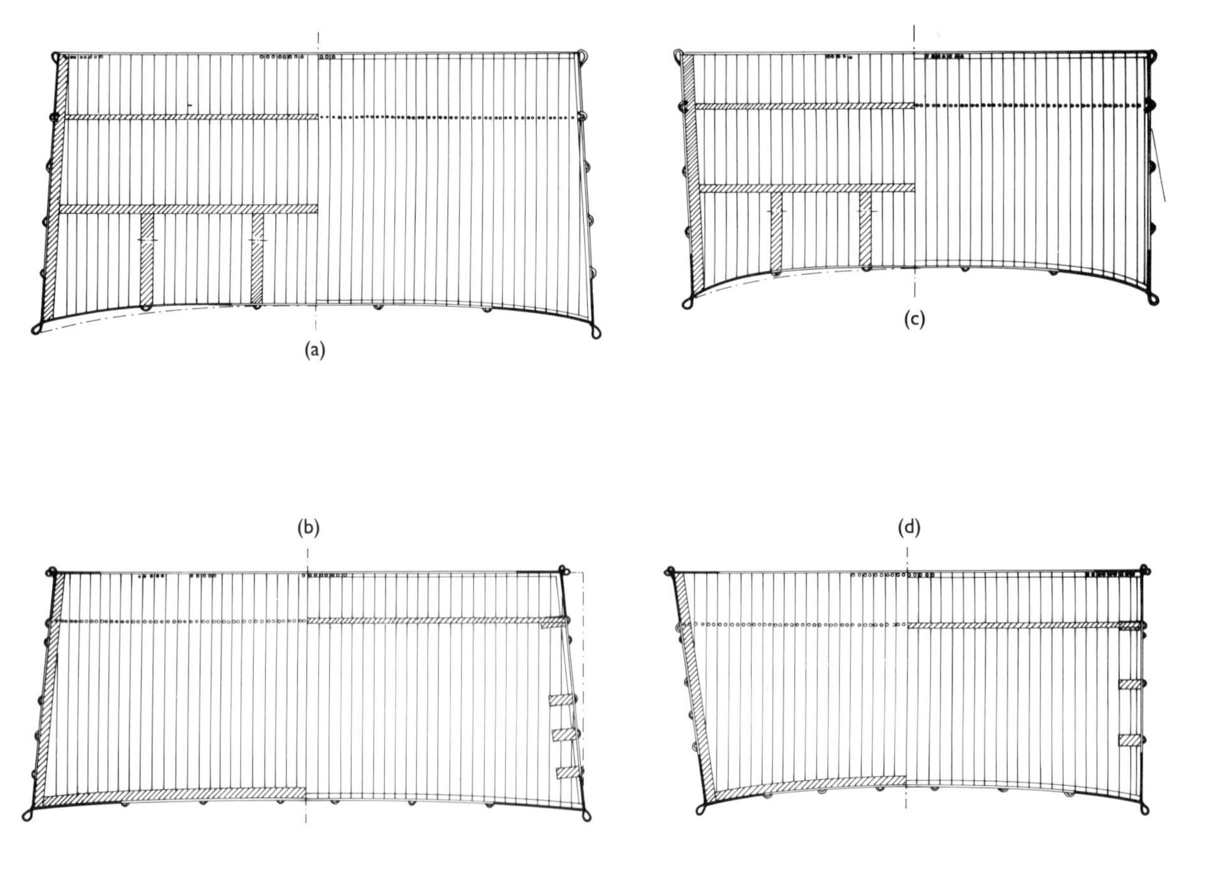

Figure 139
(a) An English main course, showing (left) the fore or outer side with the linings shown hatched, and (right) the aft or inner side with tablings. The holes for the reef band and head tabling are only partly marked. The parts of the boltrope shown as a solid line are served. The dot-and-dash lines at the foot and the buntlines indicate the shape of a merchantman's sail.

(b) A Continental main course, shown as above; the linings (hatched) are found on both sides of the sail. The dot-and-dash line indicates the sail's shape according to Korth.

(c) An English fore course, as above; the gore shown at the left-hand leech is typical of the period after 1790, and the right-hand (straight) leech is typical of the period before 1790. The dot-and-dash lines again indicate the shape of a merchantman's sail.

(d) A Continental fore course; the left side is French, from early in the century, with extra clew lining not marked. The right side is according to Korth.

width of the foot and was about one third of the sail deep. Still being used in larger ships in the seventeenth century, it was only employed in smaller vessels during the eighteenth century. Falconer and Steel noted the use of bonnets only in ships with one mast, Blanckley on the mainsails, foresails and jibs of sloops, yachts and hoys, and Röding as follows:

> Only koffs, smacks and one-masted galiots still have [bonnets] on their sails. Koffs usually carry double bonnets on their lower or main sails. The bonnet, laced to the sail itself is called the storm bonnet and is the smaller, but the other, which is similarly fastened to the storm bonnet is named the drabbler or lower bonnet. Smacks now have a bonnet not only on the gaff sail, but also on the fore staysail. Bonnets are especially useful when the wind becomes too strong, as they can be loosened in an instant to reduce the sail size.

Steel also noted that bonnets had a head tabling, to which a line was sewn in bights, 6 inches apart, forming the latchings. The leeches and foot of a bonnet were tabled in the same way as the sail it was attached to. Under 'Latchings' he noted:

> These loops are 6 inches asunder and 6 inches long, except the two middle ones, which are 12 inches long, to fasten off with. The loops are alternately reeved through holes in the foot of the sail and through each other, and fasten by the two long loops in the middle with two half-hitches, by loosing of which they unreeve themselves.

Röding illustrated the bonnet line fastened to the course and locked at one side. He described the lacing or latching as marled to the course and thrust through the holes of the bonnet; other authorities describe the lacing as marled to the bonnet and not the course. In his description the bonnet lines run from the outside towards the mast and were locked there, only squaresails were locked in the middle. He did not mention the number of latching holes necessary, but the illustration shows only one per cloth and many modern authors have taken this as the norm, which would mean latchings 2 feet long.

In 1627 Captain John Smith referred frequently to bonnets and drabblers and the sewing of 'latchets' in the bonnet, but he did not give their length and number. These were provided by Anderson, who researched seventeenth century rigging at length. He found two latchings per cloth used in English ships until 1680. The photographs by Heinrich Winter of a contemporary model destroyed in 1944 of a Dutch man-of-war from 1660-70, with its original sails, show that Dutch ships, like English vessels of the 1700s, had four latchings, with the mizzen even showing five latching holes per cloth. Anderson's findings and the Dutch model's bonnet holes are corroborated by a number of contemporary paintings.

When two bonnets were carried, the lower was named a drabbler, as noted above; while Röding wrote of it in connection with koffs, Falconer defined this extra bonnet as follows: 'Drabler: an additional part of a sail, sometimes laced to the bottom of the bonnet of a square-sail, in sloops and schooners.' The German term for the lower bonnet was *Fatzen* and the French *bonnette de sous gui*. For details of latchings see Figure 138j, k, l.

X

CUT AND SHAPE OF SAILS

Chapter IX dealt with the manufacture of sails in general. The following chapter will give details of the cut and shape of each sail of a full-rigged ship. Most of the details provided are from men-of-war, since they were built and rigged to more uniform regulations. Lever commented as follows in regard to merchant ships:

> The shape of the [fore-]Sail must be regulated by the height of the mast, and the squareness of the Yard. Ships in the Merchant Service vary as to these proportions, some having taunt lower Masts and narrow Yards, others short Masts and square Yards, and others again, in a medium between the two former.

Some of the sails of smaller vessels are also considered below in similar detail.

Main course

Gross-Segel; Grand-voile

The main course was quadrilateral and was bent to the main yard. English sails were made from canvas Number 1 or 2, French from *toile à trois fils*.

Width Head: 18in each side less than the distance between the yardarm cleats for English sails, the length of the yard less the yardarms for French; Foot: the length of the yard including the yardarms.

Height The foot's bunt had to clear the boats carried on booms in the waist.

Leeches One cloth was gored on each leech for Royal Navy ships, while merchant ships sometimes had two cloths gored. The leeches of French sails were tapered towards the head by increasing the seams of each cloth by up to 2 inches. Continental sails at the beginning of the century had parallel leeches, and later were like the sails of English merchantman. Korth still noted parallel leeches, and sail plans of several Danish warships from 1789 and 1812 indicate either the same, or leeches gored by one or two cloths.

Head Straight.

Foot Roached. On Continental sails the curve was minimal and gradual, one fifteenth of the drop or less. English merchant sails had a goring of 2 inches per cloth. The foot of Royal Navy sails was parallel to the head in the middle and was gored from the second cloth within the nearest buntline cringle, first by 1 inch per cloth and increasing after every second cloth by another inch, so that the cloth at the clew had 5 to 6 inch goring; this produced an overall curve. The foot of French sails (according to Boudriot) was divided into three equal sections. The centre bunt was parallel to the head, with the outer thirds gored in a straight line to give approximately one seventh to one eighth more depth at the leeches than at the bunt.

Reef band Two for ships with 38 guns and fewer and one for ships with 44 guns and more, in English ships. French courses usually had one only, and Röding noted one as normal for Continental vessels. Korth noted two for a Continental Third Rate.

Lining The leech lining on Continental and English courses was one cloth. A foot lining was fitted only on Continental courses, except those of French ships, where for most of the century only patches at the buntline cringles were used. Four buntline cloths and a middle band were used on the main courses of Royal Navy ships. The courses of English merchant ships were rarely fitted with a middle band, and then only when the sail was half worn. Buntline cloths on naval ships extended up to (and under) the middle band, and had a length of one quarter of the sail's depth in merchant ships. French courses had buntline cringle patches only. Patches were also used on Continental sails for strengthening the bowline and reef tackle cringles; English merchant courses sometimes had these too. On English courses the linings were fitted to the outside of the sail, whereas Continental sails had the lining on the inner side, but the reef band and patches on the outer. French ships had lining, reef band and patches on the inner side.

Boltrope The thickness of ropes was normally measured as circumference. A distinction was made between the foot and leech ropes on one hand, and the head rope on the other; on English courses the latter was only about 40 percent of the circumference of the former, while on French courses the figure was about 65 percent. On Royal Navy courses the clew rope was 2in larger for ships from 50 guns upwards, and for ships with fewer guns 1½in larger. The clew was wormed with a ¾in ratline, parcelled with old canvas, well tarred, and served with spun yarn. It was then marled to the sail with marline or houseline and seized with several turns of inchline, pulled tight with three cross turns. The courses of ships of the highest Rates had leech and foot ropes of 6in circumference and head ropes of 2½in. This measurement decreased for ships of 20 guns to 4in and 1½in respectively. For further measurements see the table at the end of this chapter.

Cringles A reef cringle was fitted at each end of the reef bands, and three bowline cringles at equal distances between the lower reef band and the clew. Continental courses had three bowline cringles, with the first at half height and the others equally spaced between half height and the clew. Four buntline cringles were normal for English courses, and six for Continental sails.

Fore course

Fock; Misaine

In general the fore course was identical to the main course; it was bent to the fore yard. Differences between for and main course were as follows:

Height The bunt of an English fore course was just above the main stay; on French sails it was few feet above the beakhead rail.

Leeches Royal Navy fore courses up to about 1790 had leeches perpendicular to the head; later they were gored by one cloth towards the clew on each side. The goring was sometimes two cloths on merchant ships. Lever explained this practice as follows:

> When the [fore-] Mast stands well aft, [the fore course] is sometimes of equal breadth at the Head and the Foot but more frequently, particularly if the Mast be forward, it is a Cloth broader at the Head than at the Foot: that is, a Cloth on each side is gored, or cut sloping, from the Head to the Foot, so that it is half a Cloth broader on each side.

He also noted that such a small goring could also be achieved by an increase in the seams. Continental fore courses sometimes followed English practice (as late as the early 1800s Korth was still advocating equal width at head and foot), and sometimes French, by which a goring of five to six cloths each side was normal in the early eighteenth century. By about 1780 a goring of three cloths (as in the *Royal Louis*, according to Paris) was usual, while Boudriot noted for both 1759 and 1780 a reduction of one cloth through a tapered seam toward the foot. The reason for this reduction in the foot of a fore course has to be sought in the use of a bumkin, which had to be a reasonable length to be workable, or, before the introduction of the bumkin in the need for the weather tack to be as close as possible to a deadblock or the cutwater's lead holes when the ship was close-hauled.

Foot The goring of the foot was greater than that on a main course, and again the reason would seem to be the need to ensure that the clew was as close as possible to the bumkin lead block or the tack lead holes. The goring of an English fore course began two cloths within the clew nearest the buntline cloth, with each cloth gored by 5 to 6 inches. The foot of a French fore course was similar to that of a main courses, and other Continental fore courses followed the French lead.

Lining English and Continental fore courses were lined like main courses. French fore courses had additional clew linings: beside the full leech lining, an extra half length and an extra quarter length cloth were sewn from the clew up.

Cringles As for the main course; English fore courses had only one bowline cringle fewer.

Boltrope First Rate ships: 5½in and 2¼in; Sixth Rate ships: 3½in and 1½in.

Mizzen course

Besan; Artimon

The mizzen course was quadrilateral and bent to the mizzen yard or gaff. The Continental full mizzen was triangular, a lateen sail, while the English was a settee sail (that is, a lateen sail with a luff leech, thus with four sides). English mizzen courses were made of canvas Number 2 or 3, and French sails of *toile à deux fils*.

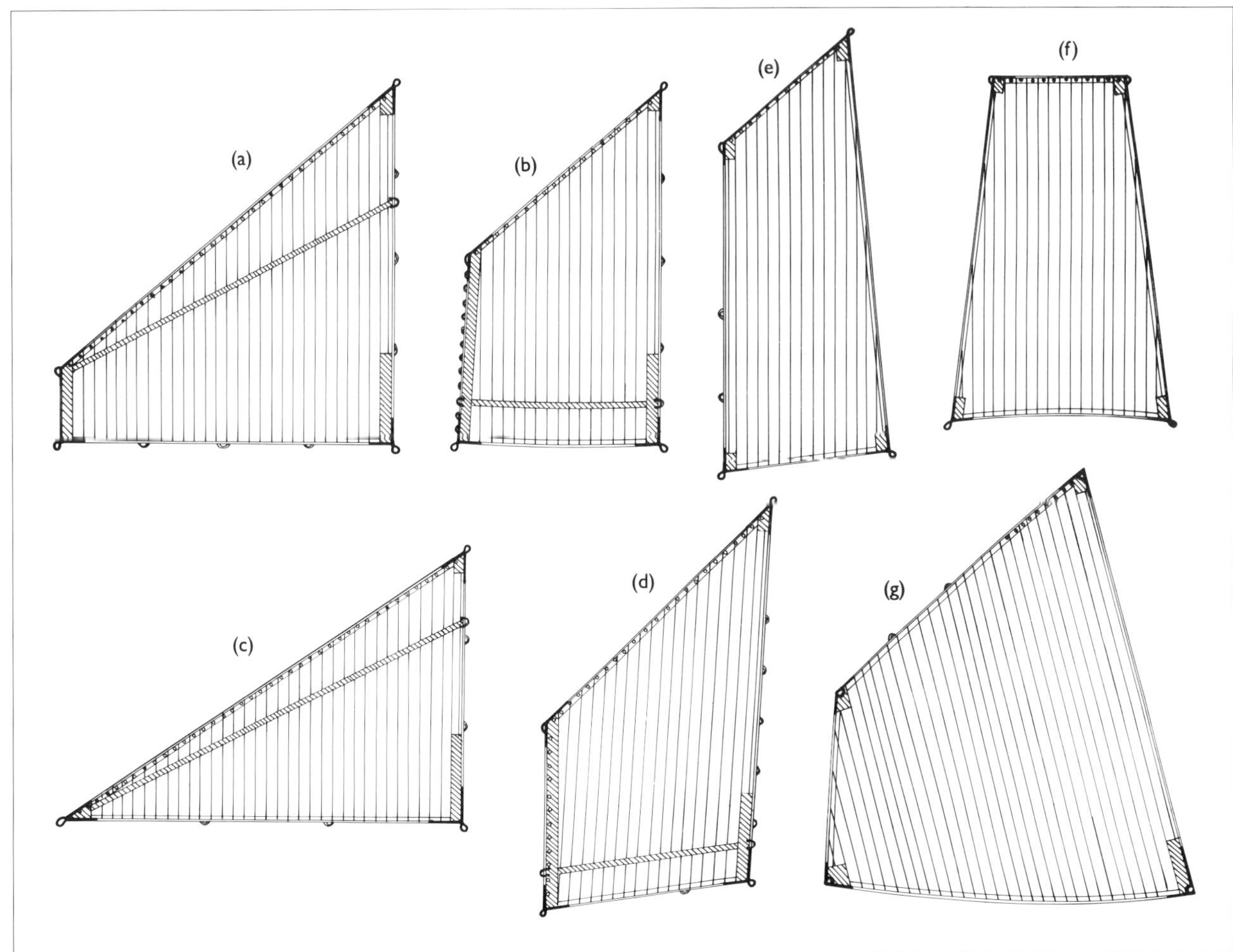

Figure 140
(a) An English settee mizzen, still part of the rig of a First Rate until 1780. Linings are shown hatched and holes in the reef band are not marked. The tablings and linings were on the port side. The parts of the boltrope shown as a solid line were served.

(b) An English mizzen from about 1730 onwards. The mast leech cringles might also be holes, as shown in (d).

(c) A French full mizzen; in contrast to the English mizzen, this was a lateen sail.

(d) A French mizzen from about 1750 onwards.

(e) A driver after 1780; this sail was mainly used on merchantmen, as an aft extension of the mizzen.

(f) A driver before 1780, as a square sail hoisted to the mizzen yard or gaff peak. This type of driver was still in use on the Continent in the early nineteenth century. The sheets were fastened to a boom extended overboard on the lee side.

(g) A driver in the Royal Navy after 1780. This sail was hoisted instead of a mizzen and had, in addition to a short yard, two head halyards and a throat halyard. The foot was spread by a boom, extended temporarily over the taffrail.

Width	From 9 inches within the cleats to the mast for English mizzen courses; a French short or 'English' mizzen had a head length of seven twelfths the length of the mizzen yard.
Height	The foot was 6 to 7 feet above the deck on English ships. On French, the mast leech was approximately 2½ times the mizzen masthead length shorter than the mizzen mast above the quarterdeck.
After leech	The after leech was straight on English mizzen courses. French mizzen courses, depending on the position and rake of the mast and the angle of the yard, were either gored towards the peak by 1 to 3½ cloths, or towards the clew by 1 to 1½ cloths.
Mast leech	In the Royal Navy the mast leech of a mizzen course was usually gored by one cloth towards the nock, and in merchant vessels sometimes by two cloths. French courses usually had an ungored mast leech.
Head	The head was gored by 16 to 22 inches per cloth on English mizzen courses. French mizzen courses had heads gored at between 45 and 50 degrees.
Foot	English: square for the central two cloths, and gored by 1 inch per cloth gored towards the ends. French: each cloth gored by about 4 inches towards the clew.
Reef band	On English mizzen courses the reef band was 6 to 8 inches wide and about one fifth of the mast leech in height from the foot. Reef bands on French mizzen courses were one quarter of mast leech height up and ran parallel to the foot; they were half a cloth wide. The reef band of a full mizzen was of the same width, but was placed one third of the after leech height down from the peak, and extended either to the tack (lateen sail) or the nock (settee sail).
Lining	On English mizzen courses the mast leech, and the bunt of the earlier settee sail, were usually fully lined with one cloth, and the after leech was lined 5 yards up from the clew. If the mast leech was not fully lined, the nock was lined to a depth of 1 yard. Lining was also necessary at the sail's peak. A French mizzen's mast leech was lined with half a cloth, and the after leech with the same width, but from the clew to slightly above the reef band only.
Cringles	English: cringles were fitted at both ends of the reef band, and a cringle for the throat brail was fitted 5 yards above the clew. Larger ships had a further two cringles for middle and peak brails at equal distances towards the peak. Settee sails also had three foot brail cringles. If the gaff was slung, a cringle was spliced into the mast leech rope every 27 inches for the lacing, but if the gaff could be hoisted eyelet holes were worked into the leech to fasten the hoops which went around the mast. Continental: the only difference was in the number of brail cringles, which was between two and five, and foot brail cringles, which was between two and four.
Thimbles	French: the clew and tack were laid around iron thimbles, with the clew eye, but not the tack eye, being wormed.
Boltrope	First Rate: 4½in and 1¾in; Sixth Rate: 2½in and 1½in. On French mizzen courses the foot rope was the largest, and it extended up to the reef band; the leech ropes were slightly smaller, and the head rope smaller still.

Driver, spanker

Treiber, Brotwinner; Tappe-cul, paille en cul

The driver was quadrilateral, and bent or laced to the driver yard. It was set only in very light winds. An English driver was made from canvas Number 6, and a permanent spanker from canvas Number 4, 5 or 6; French drivers were made from *toile de mélis simple*.

The driver was a mizzen (studding-) sail, and in England until the 1770s (and on the Continent up to the early 1800s), it was hoisted as a squaresail to the mizzen yard or gaff peak. After this it became a fore-and-aft extension of the mizzen course, which in merchantmen was set abaft it like the ringtail sail of a boom sail; in men-of-war it became a large sail set instead of the mizzen, so that the temporary rigging of a boom was necessary.

The sail plans of the Danish 80-gun ship *Neptunus* and of a 48-gun and a 40-gun ship from 1789 confirm that both the boom and the long gaff of a large driver (spanker) were permanent fittings on these ships. The 40-gun frigate *Havfruen* had a further ringtail sail hoisted behind the spanker. *Christian VII*, a ship of 90 guns of 1803, and *Venus*, a frigate of 36 guns of 1812, show another novelty for ships: a triangular gaff topsail. From this evidence it is clear that Danish ships of the late eighteenth century were in the avant-garde of rigging improvements. English drivers were not rigged permanently before 1806.

Width	The width of a driver until 1780 was similar to that of the mizzen, or two to three cloths more at the foot. That of the merchant driver after about 1780 was about two to three cloths less than the width of the mizzen, and that of a full naval driver was equal to the mizzen's width plus the width of the extension. The head of a Danish spanker was between two thirds and half the foot wide.
Height	The foot was 6 or 7 feet above deck.
Leeches	The leeches of the square-rigged driver before 1780 were gored towards the head by three to four cloths.
After leech	From around 1780 the after leech of the merchant driver was gored towards the head by three to four cloths. The after leech of a man-of-war driver was not gored.
Mast leech	Boom drivers had a roached mast leech with a gore of four to six cloths towards the nock. Danish spankers had a straight mast leech.
Fore leech	The fore leech of a merchant driver was straight and had a length of eleven twelfths that of the mizzen course's after leech.
Head	The driver head was square before 1780; the head of a merchant driver after 1780 was straight, but with a similar gore to that of the mizzen course. The driver boom sail, as the full driver was also called after 1780, had a slightly roached head with a gore of 9 to 12 inches per cloth. Danish spanker heads were gored at between 30 and 40 degrees.
Foot	Until 1780 the foot of a driver was gored like a that of a topsail. The foot of a merchant driver was gored by 3 to 4 inches per cloth towards the clew. For the boom driver no strict rules applied; it was roached and the gore of each cloth depended on the required curve. Danish sails were either roached or straight.
Lining	All four corners of a driver had a full-width lining 1 yard in length, except the clew of the boom driver, where a length of 2 to 3 yards was required. A spanker was lined like a mizzen course.
Reef band	Drivers had from one to three reef bands, depending on the size. They were one quarter cloth wide and one sixth the height of the mast apart, parallel with the foot.
Cringles	Two bowline cringles were fitted at one quarter and half way up on the fore leech rope of a merchant driver. Two cringles in the head rope at one quarter and half length from the nock were usual for a boom driver when it was temporarily hoisted to a mizzen gaff. When it was laced permanently to a gaff, no head cringles were fitted, but the mast leech rope was similar to that of the mizzen course (laced through cringles 30 inches apart or through holes in the leech). Spankers also had brail cringles as for the mizzen course, and reef band cringles at each end of the bands.
Thimbles	On the driver boom sail thimbles were generally fitted to the clew, tack, nock and peak.
Boltrope	First Rate: 2¼in and 1¼in; Sixth Rate: 1¾in and 1¼in.

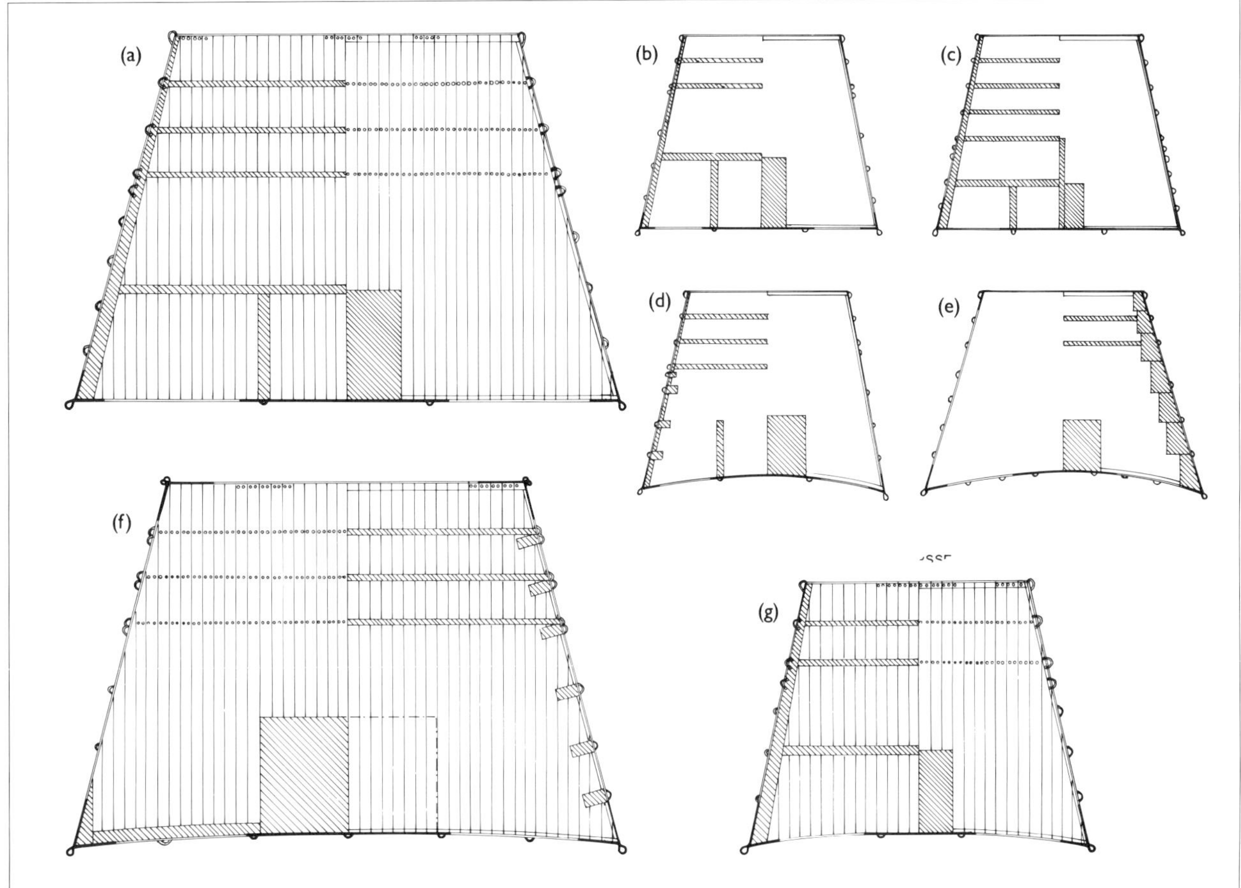

Figure 141
(a) English fore and main topsails, 1710 to 1788. The fore side is shown left, with linings hatcheds, and the after side right, showing the top lining. Those parts of the bolt rope shown as solid lines are served.
(b) English fore and main topsails before 1710, simplified.
(c) English fore and main topsails after 1788, simplified.
(d) Fore and main topsails on an English merchantman, simplified.
(e) Continental fore and main topsails from the first half of the century, simplified.
(f) Continental fore and main topsails from the later part of the century. The sail is drawn as shown by Paris; it is likely, however, that the sides as shown were accidentally reversed in Paris' work. According to Röding the side indicated by Paris as the fore side (left) is actually the aft side; the top lining would certainly have been fitted to the inner side of a sail.
(g) An English mizzen topsail up to 1788.

Try-sail, spencer

Schnausegel; Voile de senau

The quadrilateral try-sail or spencer was laced to a gaff on the spencer mast (try-sail mast). It was cut similar to, and made from the same material as, the mizzen course. A try-sail was not only a feature of snow rig, but was also sometimes used in three-masted ships (see Jackass bark in Chapter III). Cutters and sloops also carried a try-sail. It occasionally replaced the mainsail in stormy weather.

Width The head was about two thirds the width of the foot in three-masted ships. The head of cutter or sloop try-sails was two fifths the width of the foot.

Leeches While the mast leech in three-masted ships was usually similar to the depth of the mainsail, the mast leech of a cutter or sloop try-sail was three quarters the length of the mainsail, gored by eight to ten cloths. The after leech was one sixth deeper than the mast leech. The foot was gored by 5 to 7 inches per cloth, leaving two or three square cloths at the clew.

Lining Try-sails were lined like the mizzen course, but with one to three extra strengthening bands, like middle bands, running across.

Reef bands Try-sails had three reef bands, like the spanker.

The sail was similar to a mizzen course in all other aspects.

Main topsail

Grossmarssegel; Grand hunier

The main topsail was quadrilateral, and its head was bent to the main topsail yard. It was made from canvas Number 2 or 3 (English), or *toile à deux fils* (French).

Width English: head, to 18 inches inside the yardarm cleats; foot, equal to the head of the main course plus one to two cloths. French: head, to within the yardarm cleats; foot, equal to the head of the main course minus 1/48 of the main yard's length.

Length English: when the yard was hoisted to the hounds, the clews extended down to the main yard. French: the length equalled the main topmast's length minus the masthead and half the length of the hounds.

Leeches English: gored as necessary between the given widths in straight line. French: similarly gored but slightly roached from the lower reef band to the head.

Head Straight

Foot English: Royal Navy, straight and parallel to the head; merchant ships, gored from 2 to 4 inches per cloth at the outer thirds. French: Navy (according to Boudriot), as English; (according to Paris) gored by one tenth of the height in a curve from clew to clew. Continental: gored by ½in per cloth in a curve, though Danish main topsails were gored at the foot like those of English merchant ships.

Reef bands Two until about 1710 (1730 in France), then until about 1790 (1800 in France) three and later four, one eighth of the sail's height apart. French reef bands were 4 feet from the head for the first and one third of the sail's height below this for the third, with the second halfway between the other two.

Lining The main topsail leech lining in English men-of-war was half a cloth wide at the head and 1½ cloths at the clew,

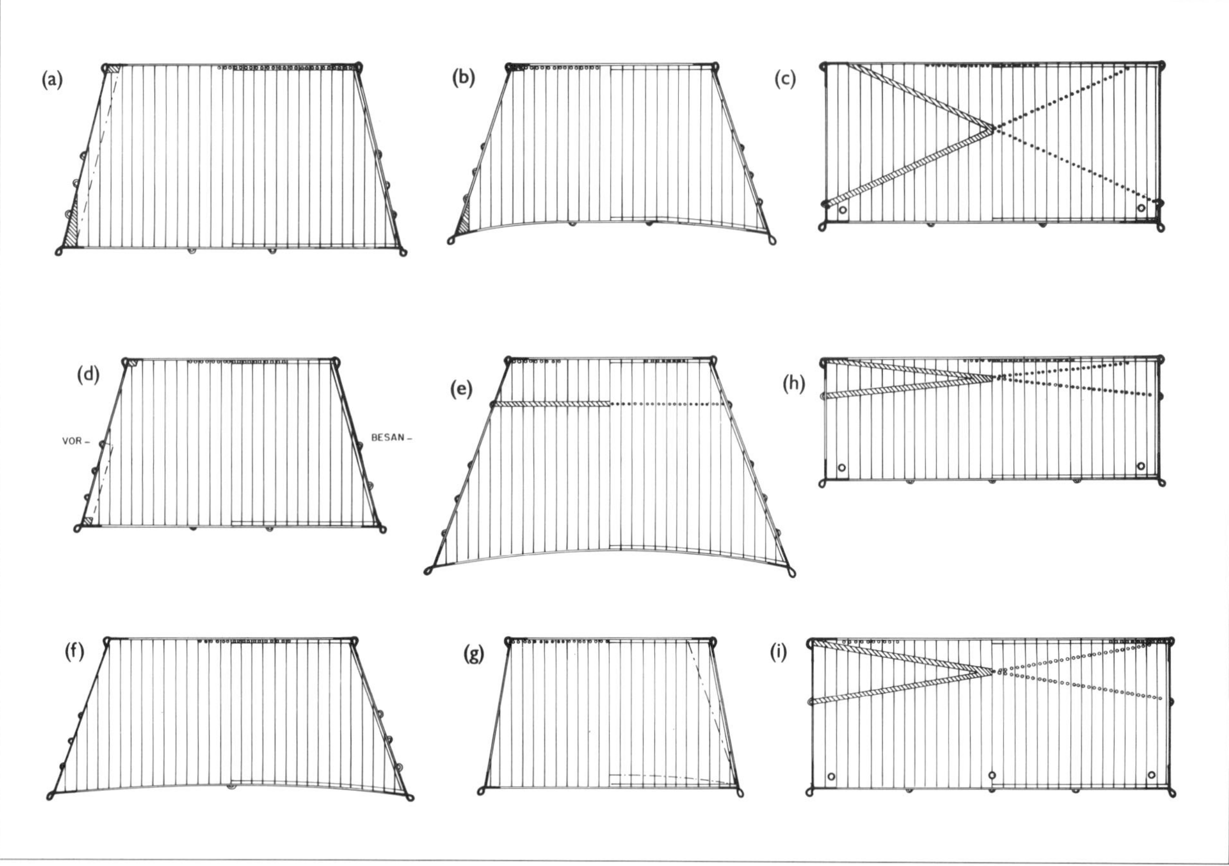

Figure 142
(a) English man-of-war main topgallant sail. The hatched patches at clews and earings were sometimes combined into full leech linings. The parts of the boltrope shown as a solid line were served; at the earings and clews this illustration follows most modern authors. According to Steel (1794) the boltropes of topgallant and royal sails were not served, and Lever (1819) noted serving only at the clews. According to these authors the boltrope of the spritsail was also served only at the clews.

(b) The main topgallant sail of an English merchantman.

(c) The fore and mizzen topgallant sails of an English man-of-war. The linings coverred the corners, with the clew lining sometimes going up to the upper bowline cringle (shown here by the dot-and-dash line). The fore topgallant sail had three, and the mizzen topgallant sail two, bowline cringles.

(d) The main topgallant sail of a French man-of-war (First Rate) of the early eighteenth century. Danish men-of-war of 40 guns and above were similar in the later part of the century. The reef-band shown here on the fore side might also be fitted to the aft side, as the fore topgallant sail of the early Royal Louis indicates; contemporary models show both alternatives.

(e) A Continental main topgallant sail from the middle of the century.

(f) An English main royal sail. The dot-and-dash line shows the shape of a Continental sail.

(g) An English spritsail. The sail had more depth and a more open reef band cross than a Continental sail. Until the early 1720s only one reef-band was fitted, parallel to the head and one quarter of the depth down.

(h) A Continental spritsail of the beginning of the century.

(i) A Continental spritsail of the middle of the century. The depth increased from 0.35 times the width to 0.4 and later to 0.5 times the width. A third water-hole was also frequently made. For the placing of reefbands see (d).

and in merchantmen 9 inches wide at the head and 15 inches at the clew. On French main topsails a triangle at the clew of up to about one quarter the height of the leech was doubled, and additional cloths were sewn on in steps as a lining up to the lower reef band. Under each reef band and the earrings were extra patches. Continental sails (according to Röding and Paris) had these clew triangles doubled only. The head tabling was about 10 to 12 inches wide, and the foot tabling 6 inches wide, though Continental main topsails had a foot lining one cloth in width. English ships had a middle band and two buntline cloths sewn to the sail's fore side, with the buntline cloths extending under the middle band. Merchant main topsails were often without this middle band, and the buntline cloths ended 18 inches short of the top lining. The position of a middle band depended on the number of reef bands, since it was sewn halfway between the lower reef band and the foot. The top lining and mast cloths were sewn to the inner or rear side of the sail. The mast lining of an English main topsail had the two centre cloths up as far as the middle band until 1788, and after that up to the lower reef band. A Royal Navy main topsail top lining was made from canvas Number 6 or 7 and covered the area between the buntline cloths. It extended to the lower edge of the middle band. On merchant main topsails the buntline cloths and the top lining covered one third of the foot. Lever explained that in the merchant service top linings and mast cloths were often not used because they collected rainwater and thus hastened the rotting of the sail. Continental top linings were one third of the foot wide and one third of the sail's height high; no mast linings were used. The width of a French top lining was one third of the head. Patches on the fore side of the Continental sails (or on the inner side according to Paris) were doublings for all boltrope cringles; they were one cloth wide and about two long, and sewn on at right angles to the leech. These were not, however, noted by Boudriot for French topsails. Lever noted that such patches were also used on the topsails of English merchantmen, but only for reef tackle and bowline cringles.

Cringles English: reef cringles at the ends of all reef bands, and four bowline cringles, with the uppermost in the middle of the leech rope and the other three equally spaced between that and the clew. One reef tackle cringle was fitted between the upper bowline and the lower reef cringle. Two buntline cringles were also fitted. Continental: all reef band cringles were double, and the bowline cringles single. Reef tackle cringles were generally not used on Continental topsails, but Paris indicated these in his sailplan of the *Royal Louis* of 1690, though this plan was a reconstruction and no further evidence can be found. Three or four bowline cringles were fitted, similar to those on English sails. The buntline cringles, however, contrasted with English practice. Because buntlines were led differently on Continental ships, four to six cringles were fitted in the foot rope.

Boltrope First Rate: 5¾in for the foot rope, 5in for the leech ropes and 2½in for the head rope. Sixth Rate: 3¾in, 3in, and 1½in. The boltrope against the top brim was wormed, parcelled and served like the clews, and was marled to the sail. The marling holes extended 3 feet each way from the clew, and along the top brim to match the top lining reached.

Fore topsail

Vormarssegel; Petit hunier

The measurements and details of the main topsail width, height, leeches, head and foot tabling, the reef bands and lining are applicable, reduced in proportion, to the fore topsail.

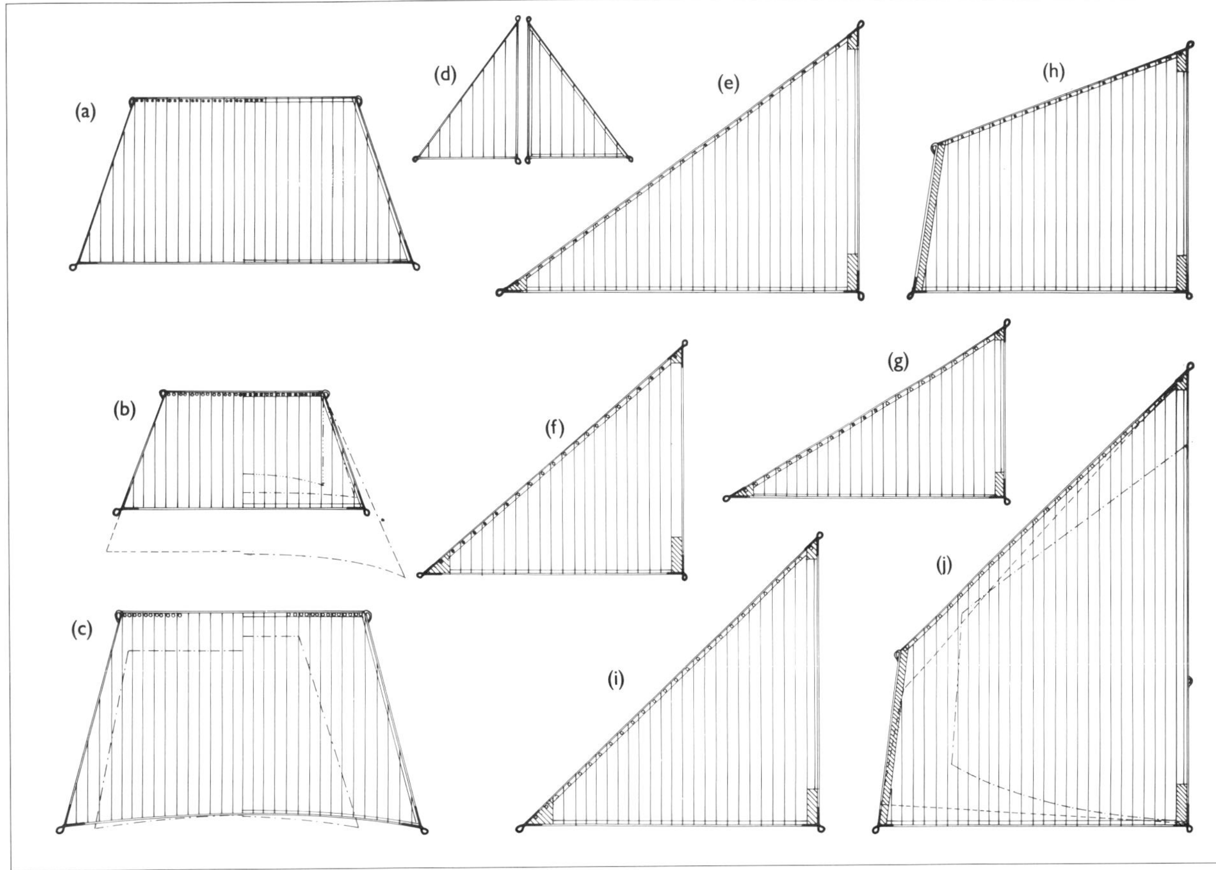

Figure 143
(a) An English spritsail topsail (jib boom); fore side (left) and aft side (right).
(b) An English spritsail topsail (spritsail topmast). The shape marked with a dashed line is also English, while the dot-and-dash and the dash/four-point lines indicate French sails, and the dot-and-double dash line is a Dutch spritsail topsail.
(c) The spritsail topsail (jib boom) of a French First Rate. On the left the dot-and-dash line outlines the spritsail topsail of a Danish Third Rate and on the right that of a Danish frigate (Fifth Rate).
(d) Sky-sails; these were rigged above the royal sails, and only came into use after the introduction of the latter.
(e) A main staysail, used for smaller ships and very rarely used on larger vessels.
(f) A fore staysail, introduced in the fourth decade of the century and only rigged on men-of-war.
(g) A mizzen staysail before 1760.
(h) A mizzen staysail after 1760.
(i) An English main topmast staysail before 1760.
(j) An English main topmast staysail after 1760. The dashed line indicates a French sail, and the dot-and-dash line a Danish sail of about 1780.

Cringles The only difference between main and fore topsail cringles was in the number of bowline cringles. Fore topsails usually had three. Only the sail shown by Röding, very similar to that shown by Paris, suggests two bowline cringles.

Boltrope First Rate: 5¼in for the foot rope, 4½in for the leech ropes and 2in for the head rope; Sixth Rate: 3¼in for the foot, 2¾in for the leeches, 1½in for the head.

Mizzen topsail

Besanmarssegel; Perroquet de fougue

The mizzen topsail was quadrilateral and bent to the mizzen topsail yard. It was made from canvas Number 4, 5 or 6, and French sails were made from *toile à mélis double* or *toile de mélis simple.*

Width English: head, to 12 inches within the yardarm cleats; foot, reaching the sheet blocks of the crossjack yard. French: head, within the yardarms; foot, within the crossjack yardarms, less 1/48 the yard's length.

Height Similar to that of the other topsails.

Leeches Similar to those of other topsails.

Head Straight.

Foot English: in contrast to the other topsails in the Royal Navy the foot was gored from two cloths outside the buntline cringle towards the clews by 27in, while a merchant sail had a straight foot. French: Navy (according to Boudriot) straight; Navy (according to Paris) slightly roached. Danish: gored as in the Royal Navy.

Reef band Until approximately 1710 only one reef band was fitted; two bands were fitted until the late 1780s in all ships. Merchantmen had two reef bands throughout the century. Royal Navy ships with more than 50 guns had three reef bands after the late 1780s. In the Danish fleet even smaller frigates were fitted with three reef bands from the late 1780s on. With three bands the distance between them was one eighth of the sail's depth; for a lesser number one seventh of the sail's depth.

Lining Like the other topsails.

Cringles The cringle arrangement was similar to that of the fore topsail, but only ships with 44 guns and more had a reef tackle cringle between the lower reef cringle and the upper bowline cringle.

Boltrope First Rate: 3½in at the foot, 2¾in at the leeches, and 1½in at the head. The clews and the top brim were wormed, parcelled and served as on other topsails, and, as noted above, marled to the sail.

Main topgallant sail

Grossbramsegel; Grand perroquet

The main topgallant sail was quadrilateral and bent to the main topgallant yard. It was made from canvas Number 6 or 7, or for French sails from *toile de mélis simple* of the second type.

Width Head, to 6 inches inside the yardarm cleats, with the clews reaching the main topsail yardarms.

Height Extending from the fully hoisted topgallant yard to the sheet blocks of the main topsail yard.

Head Straight.

Leech Gored to the necessary angle.

Foot Straight in the Royal Navy, and roached in the merchant service. The outer thirds of the foot were gored by 2 to 3 inches per cloth. French sails were straight or slightly roached. Those of other Continental ships were more like the English merchant service.

Reef bands No reef band on English sails. Early French sails had one, and Danish ships of 40 guns and above carried one later in the century (1789).

Lining	Earring pieces of 9 inches were fitted on each corner at the head. The cloth at the clews was cut to form its own lining. Continental main topgallant sails had no lining.
Cringles	English: three bowline cringles, the uppermost at the middle of the leech and the others equally spaced beneath; two buntline cringle at the foot, each one eighth the length of the foot away from the centre. Continental: bowline cringles similar, one to three buntline cringles.
Boltrope	First Rate: 2¾in, and 1½in at the head. Sixth Rate: 1¾ and 1¼in.

Fore topgallant sail

Vorbramsegel; Petit perroquet

The fore topgallant sail was identical in every respect to the main topgallant sail. Steel, however, noted that in the merchant service no bowline cringles were fitted to this sail. It was bent to the fore topgallant yard.

Mizzen topgallant sail

Besanbramsegel; Perruche

The mizzen topgallant sail was identical to the other topgallant sails. It was made from canvas Number 7 or 8, or *toile de mélis simple.* According to Steel it had no bowline cringles, but Lever noted that only smaller vessels carried no bowlines on this sail.

Reef bands	Danish ships of 40 guns and more had a reef band at one fifth of the depth below the head.
Boltrope	First Rate: 2 inches, 1¼in at the head. Sixth Rate: 1¼in, 1 inch at the head.

Main royal sail

Grossroyalsegel; Grand cacatois

The main royal sail was quadrilateral, made from canvas Number 8 and bent to the main royal yard.

Width	At the head, to 4 inches inside the yardarm cleats and at the clews equal to the head of the main topgallant sail.
Head	Straight.
Leech	Gored as necessary.
Foot	In the Royal Navy straight and on Continental ships like the other sails, roached to some extent.
Lining	None.
Cringles	None.
Boltrope	Like that of the mizzen topgallant sail.

Fore royal sail

Vorroyalsegel; Petit cacatois

The fore royal sail was similar to the main royal sail.

Mizzen royal sail

Besanroyalsegel; Cacatois de perruche

The mizzen royal sail was similar to the other royal sails, but was not often carried.

Boltrope	First Rate: 1¾in, 1 inch at the head. Sixth Rate: 1 inch, ¾in at the head.

Sprit course

Blinde; Civadière

The sprit course was quadrilateral and bent to the spritsail yard. It was made from canvas Number 2 or 3 in English ships, and from *toile à deux fils* or *toile de mélis double* in French.

Width	To 9 inches inside the yardarm cleats.
Head	Straight.
Leeches	Usually straight; the Danish ship of the line *Christian VII* of 1803 had spritsail leeches gored by approximately three cloths.
Foot	Straight.
Reef bands	English: until 1730 a single band was fitted parallel to the head, one quarter of the sail's height from the head; this arrangement was noted by Steel as still sometimes in use in 1794, though he noted a position one fifth of the sail's height below the head. Thereafter two bands 8 inches wide were fitted crossing diagonally from the first or second seam at the head to the leech 27 inches above the clew. Continental: From the earring to the leech, about two fifths of the depth (three eighths in French practice) from the head.
Lining	None.
Waterholes	English: two of 4 to 6 inches diameter, placed between the reef band and the foot in the second cloth. Continental: three, with the second in the middle cloth; their diameter was nearly equal to the width of a cloth.
Cringles	A reef cringle was fitted at the leech ends of the bands, and two buntline cringles at the foot, one third of the sail's width apart.
Boltrope	First Rate: 3¼in for foot and leeches, and 1¾in for the head. Sixth Rate: 2 inches, and 1½in for the head.

Sprit topsail

Bovenblinde, Schiebeblinde; Contre civadière

The sprit topsail was quadrilateral and bent to the sprit topsail yard. It was made from canvas Number 6 or 7, or *toile de mélis simple.*

Width	At the head the same number of cloths as the fore topgallant sail, at the clews equal to the width of the spritsail.
Height	The same height as the main topgallant sail. Merchantmen usually had sprit topsails 1 to 2 yards deeper. The larger sail was also preferred on Continental ships.
Leech	Gored over four to five cloths.
Head	Straight.
Foot	English: straight. Continental: roached to some extent.
Lining	None.
Cringles	None.
Boltrope	As on the main topgallant sail.

Sprit topgallant sail

Obere Schiebeblinde; Perroquet de civadière

A sprit topgallant sail was quadrilateral and similar to the sprit topsail. It was made from canvas Number 8 and bent to the sprit topgallant yard, which hung at a right angle under the outer end of the jib-boom (according to Steel), or was 'set upon the flying

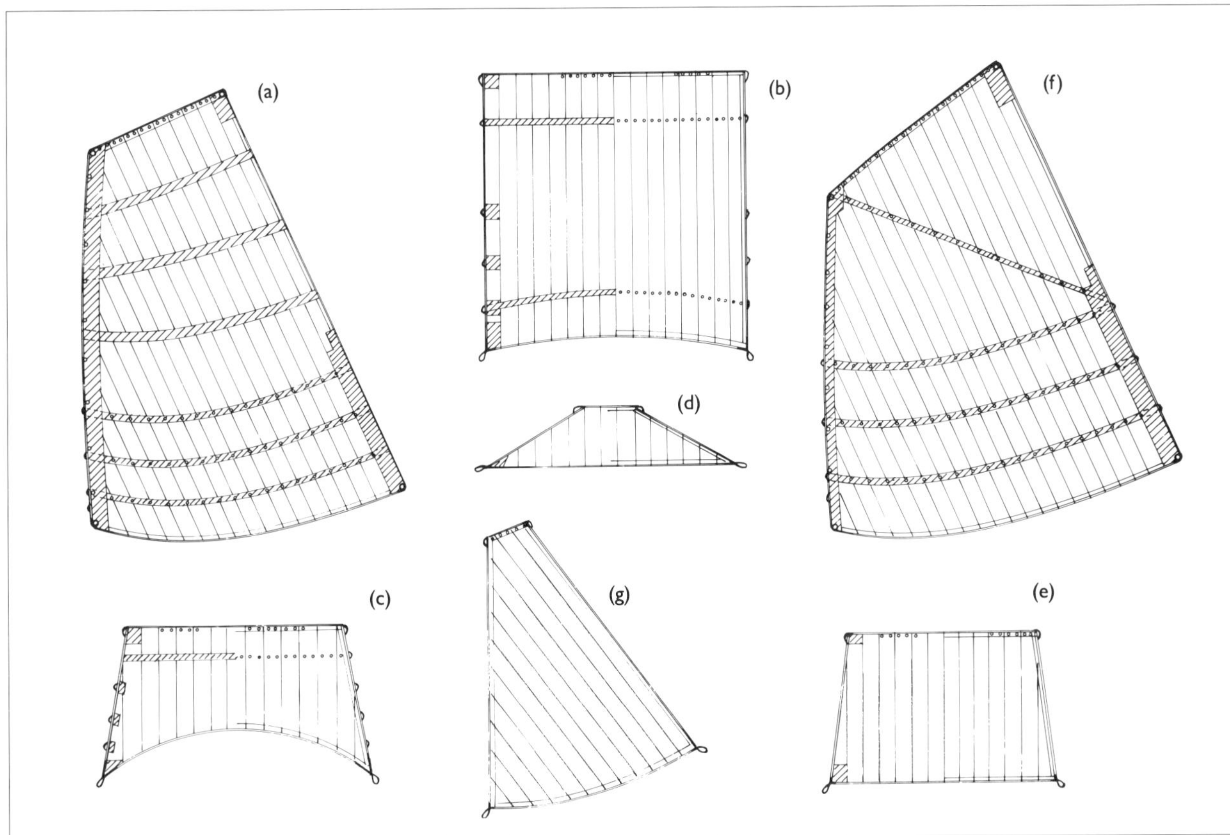

Figure 144
(a) A cutter's trysail or storm main sail.
(b) A sloop squaresail or crossjack; fore side (left) and aft side (right).
(c) A sloop topsail.
(d) A save-all topsail.
(e) A sloop topgallant sail.
(f) A sloop main sail.
(g) A sloop gaff topsail.

jib-boom, in the same manner that the sprit-sail top-sail is set upon the inner jib-boom' (Falconer, 1815). Both sources agreed that this sail was very seldom used, and was not part of a usual set of sails.

Head Two to four cloths less in width than the foot.
Foot As many cloths as were in the head of the sprit topsail.
Leech Gored by one or two cloths towards the head.

Sky-scraper, sky-sail

Skeisel; Grecque, contre cacatois, aile de pigeon

A sky-scraper was a triangular sail sometimes (and on some ships only) set above the royal sails. It was made from canvas Number 8, or *toile de mélis simple*. Steel noted that this sail was very seldom used and not usually made in general practice.

Mast leech Equal to the length between the hoisted royal yard and the truck of the relevant mast.
Foot Half the number of cloths in the head of the royal sail set below.

A pair of these sails were hitched to the royal yard with a tack and sheet, and the peak halyard rove through the truck sheave hole.

Squaresail, crossjack

Breitfock, Voile de fortune, voile quarrée

The squaresail or crossjack was quadrilateral and made of canvas Number 6 or 7, or *toile de mélis simple*. The head was bent to the crossjack yard on fore-and-aft rigged vessels, or to a spread-yard on topsail rigged fore-and-aft vessels. It was only used when the wind was astern, and otherwise not hoisted.

Width Extending to within 6 inches of the yardarm cleats.
Head Straight.
Height Four fifths the depth of the main sail's mast leech.
Foot Two to three square cloths in the middle, then gored by 1 inch per cloth toward the clews.
Reef bands Two reef bands 4 inches wide, one at the top and one at the bottom of the sail, one sixth of the sail's depth away from head and foot respectively.
Bonnet Instead of the lower reef band, a bonnet was sometimes laced to the sail.
Lining 1 yard at each clew, 18 inches at each earring, and patches of 18 inches at each cringle; these linings were on the rear (inner) side of the sail.
Cringles One cringle at each end of the upper reef band, and three bowline cringles, the uppermost in the middle of the leech, with the others equally spaced towards the clew.
Boltrope Foot and leeches 2 inches to $1\frac{1}{2}$in, at the head $1\frac{1}{2}$in to 1 inch, and when a clew rope was used this was $2\frac{1}{2}$in.

Topsail

Toppsegel; Hunier

The topsail was a quadrilateral sail, bent to the topsail yard in fore-and-aft rigged vessels. It was made of canvas Number 6 or 7, or *toile de mélis simple*.

Width To 18 inches within the yardarm cleats at the head, and extending to the crossjack yardarm cleats at the foot.
Height In the middle about one third of the depth of the squaresail.
Head Straight.
Leeches Gored by one to two cloths towards the head.
Foot The roach depended on the position of the jibstay and was normally between one third and half the depth, or at a rate of 10 to 12 inches per cloth, with the middle cloth square.

Reef bands One reef band, 4 inches wide and positioned one third of the depth of the middle cloth below the head.
Lining 18 inches at each earring, and small patches at each cringle.
Cringles A reef cringle at each end of the reef band and three bowline cringles, the uppermost in the middle of the leech, and the others equally spaced.
Boltrope As for the squaresail.

Save-all topsail

Save-all Toppsegel; Bonnette de sous-gui d'hunier

@FIRST PAR = A save-all topsail was quadrilateral and made from canvas Number 8. It was seldom used, except in very calm weather.

Head Two or three cloths wide and straight.
Foot Extending to the yardarm cleats of the crossjack yard.
Leeches Gored between head and foot.
Boltrope 1 inch all round.

Topgallant sail

Bramsegel; Perroquet

A topgallant sail was quadrilateral and made from canvas Number 8. It was bent to the topgallant yard of fore-and-aft rigged vessels.

Height This sail was between 3 and 5 yards deep, or equal to the leech of the topsail.
Head Straight and extending to within 6 inches of the yardarm cleats.
Foot Straight and extending to within between the cleats of the topsail yard.
Leeches Gored by one cloths or more.
Lining Sometimes patches 18 inches in length were placed on the rear (inner) side of the sail at the clews and earrings.
Boltrope 1 inch all round, or ¾in at the head.

Gaff mainsail

Gaffelgross-Segel; Grande voile à corne

A gaff mainsail was quadrilateral and used on sloops, cutters, English smacks and other such vessels. It was made from canvas Number 1 or 2, (*toile à trois fils*) for these types, and for a brig from Number 5 or 6, (*toile de mélis simple*). The sail was bent to the gaff, with its tack and clew extended by the boom, and the mast leech was fastened to hoops which encircled the mast.

Width The head extended from the jaws of the gaff to within 9 inches (brig), 12 inches (sloop and smack) or 18 inches (cutter) of the peak cleats. The foot was extended from the mast to within 18 inches (brig and smack), 1 to 2 feet (sloop), or 2 to 3 feet (cutter) of the boom's sheave hole at the outer end.
Height The mast leech, or bunt, was nearly the length of the mast between the lower part of the hounds and the boom, and the after leech was one third (one fifth in smacks) greater than the depth of the mast leech.
Bunt The mast leech was gored by five or six cloths with a slight curve in brigs, six to eight cloths in cutters and sloops (with a slight curve in cutters and sometimes in sloops), and by up to ten cloths in smacks.
Head Brig and smack: gored with a slight curve, 4 to 5 inches per cloth. Sloop: gored at a rate of 3 to 6 inches per cloth and sometimes cut with a curve. Cutter: gored with a slight curve, 5 to 7 inches per cloth. The head was generally wider and peaked less in merchant vessels than in naval vessels.
Foot Gored on brigs by 5 to 7 inches per cloth with a slight curve, leaving the last four to five cloths before the clew square, or at a rate of 14 to 18 inches per cloth for every cloth in the bunt. The foot of a cutter's sail was gored with a curve of 5 to 7 inches from the tack to the middle of the foot, then two cloths were left square and the remaining cloths gored by 1 inch per cloth. The gaff mainsail of a sloop was gored in a curve similarly to that of a brig (5 inches to 6½in per cloth, or 12 to 14 inches for every cloth in the bunt). On a smack's sail, the same number of cloths as in the bunt were gored by 12 to 14 inches per cloth, and the last five or six cloths before the clew at a rate of 3 to 4 inches per cloth.
Reef bands Brig: three in number, 6 inches wide and parallel to the foot, the uppermost nearly halfway up the bunt and the others equally spaced below; a brig mainsail sometimes also had a balance reef from the nock to the upper reef cringle. Cutter and smack: four reef bands parallel to the foot, 8 inches wide, the uppermost three sevenths of the bunt's depth from the tack, the others at equal distances between that and the foot; a balance reef as above was fitted on the smack. Sloop: three or four reef bands parallel to the foot, 4 to 6 inches wide, the uppermost nearly at the middle of the bunt, and the others spaced equally between that and the foot; a balance reef was frequently fitted.
Lining Brig: bunt, half a cloth, or 1 yard for the nock and tack with small triangular pieces at each bunt hole; aft leech, one cloth from the clew to 1 yard above the upper reef band, 18 inches of this cut down at the upper end and doubled under or cut off; peak piece 1 yard. Cutter: the bunt was lined with a full cloth, otherwise like a brig. Sloop: the bunt like the brig's, peak piece, 4ft 6in, and the clew lining 2 feet above the reef band, cut down as noted above. Smack: a full cloth at the bunt and otherwise like a sloop.
Cringles Reef cringles at each end of the reef bands, and a luff cringle at the bunt, halfway between the lower reef band and the tack.
Thimbles Large iron thimbles were fitted in the nock, peak, clew and tack, and normal thimbles in each reef and luff cringle.
Boltrope Brig: head, leech and foot 1½in, bunt 3½in. Cutter and smack: head, leech and foot 1½in, bunt 3 inches. Sloop: head, leech and foot 1½in, bunt 3½in. Clewropes for all 4 inches.

Gaff topsail

Gaffeltoppsegel; Voile de flèche, flèche-en-cul

A gaff topsail was quadrilateral or triangular and was made of canvas Number 8. It was bent to a small gaff or yard above the main gaff, or was hoisted as a shoulder-of-mutton sail; it was introduced in the 1770s. Originally the gaff topsail was considered appropriate only in fore-and-aft rigged vessels, but Danish men-of-war also carried a gaff topsail of the shoulder-of-mutton type above the spanker from about 1800 onwards, and Röding noted it as a gaffsail at the mizzen topmast which was rarely used. It can also be seen in Hutchinson's *Treatise on Naval Architecture* of 1794.

Fore leech On small craft the depth of the fore leech was four fifths of the depth of the mainsail, and on Danish ships it was

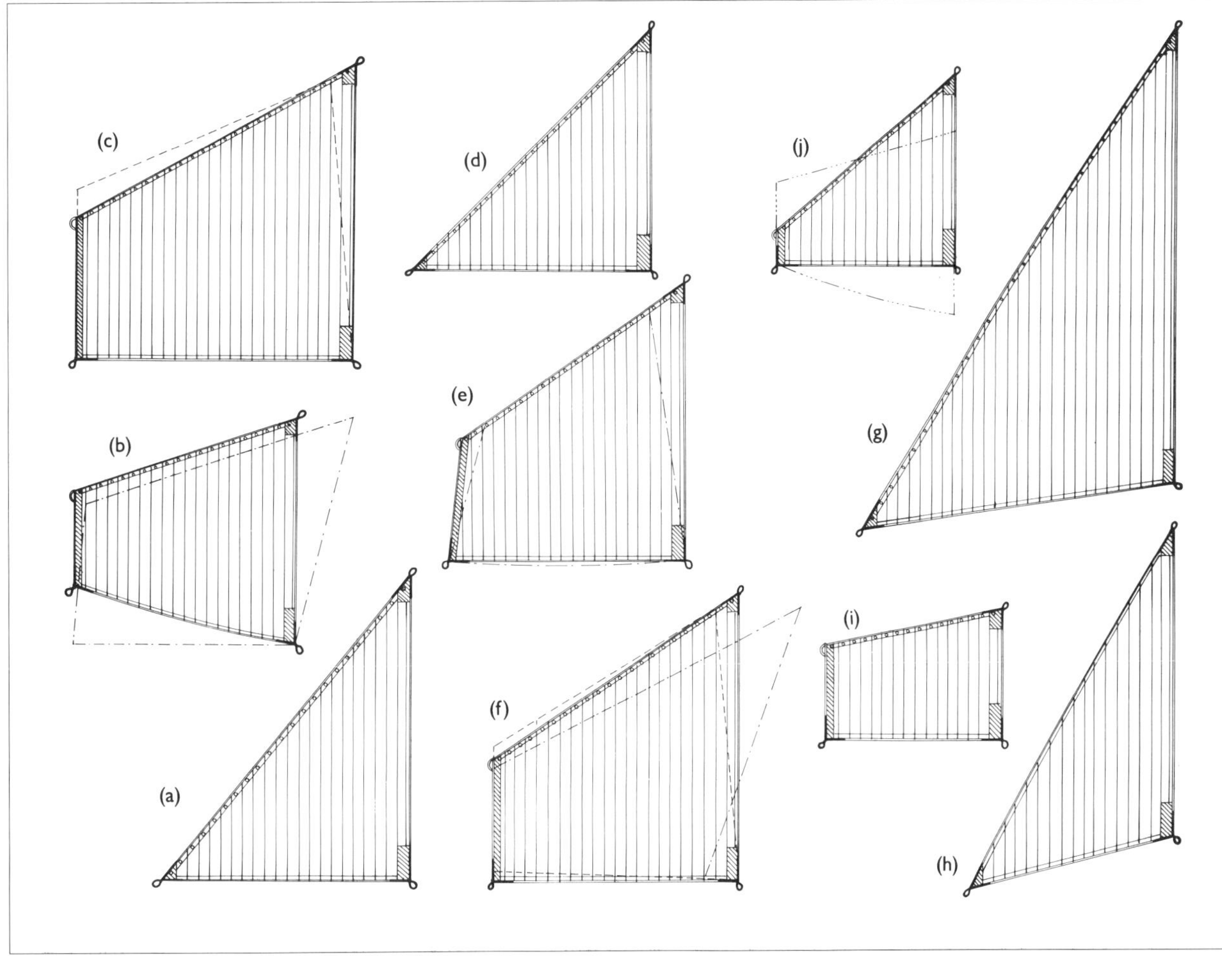

Figure 145
(a) A fore topmast staysail.
(b) An American middle staysail; the dot-and-dash line shows the shape of a Danish middle staysail.
(c) An English middle staysail; the dashed line indicates the shape of a French sail.
(d) The shape and size of a mizzen topmast staysail and a main topgallant staysail; before 1760 these were similar.
(e) An English mizzen topmast staysail after 1760. The shape of a Danish sail is marked with a dot-and-dash line.
(f) An English main topgallant staysail after 1760. A French sail is indicated by the dashed line, and a Danish sail with a dot-and-dash line.
(g) A jib; when set flying, the sail would have been without holes in the stay.
(h) A flying jib; the earliest appearance of such sails dates from the middle of the 1770s, and the sail was officially introduced in the early 1790s.
(i) An English mizzen topgallant staysail, not found before 1760.
(j) A French mizzen topgallant staysail; the dash-and-four dot line shows the shape of an American sail.

nearly the depth of the mizzen topsail and topgallant sail combined. The fore leech was gored according to the number of cloths in head and foot.

Head Only two or three cloths wide, with a gore of 6 to 8 inches per cloth.
Foot Extending from the aft side of the topmast to the peak cleat of the gaff, with a gore of 6 to 8 inches per cloth to match the angle of the spanker gaff.
Aft leech Parallel to the cloths.
Boltrope Fore leech 1½in, head, foot and aft leech 1 inch, subsequently clew rope of 2 inches.

Main staysail

Grosstagsegel; Grand-voile d'etai

The main staysail was triangular and was hoisted to a staysail stay or the main preventer stay. It was seldom used on large English vessels, but common on Continental ships of all sizes. It was made from canvas Number 1 to 3, or from *toile à trois fils* or *toile de mélis double.*

Stay The sail's stay was gored by 17 to 19 inches per cloth. The holes were 27 inches apart.
Leech Straight.
Foot Straight in the Royal Navy, but roached in merchant ships. The foot of a Continental main staysail was usually gored by 1 to 3 inches per cloth.
Depth Sufficient just to clear the boats stored above the reserve spars on deck.
Reef bands In merchant service the sail frequently had a reef band at about 4 feet above the foot.
Bonnet The reef band was sometimes replaced by a bonnet.
Lining The cloth at the tack was cut to form its own lining. The clew piece extended 2 yards up the leech, and the peak piece was 1 yard in length.
Thimbles Sometimes fitted at the tack and peak; when thimbles were not used here, the tack and peak were formed like the clew.
Boltrope First Rate: 3¼in; Sixth Rate 2¼in.

Fore staysail

Vorstagsegel; Trinquette, tourmentin

The fore staysail was triangular and made from the same material as the main staysail. It was hoisted on the fore preventer stay, or on a staysail stay. Introduced into the Royal Navy in 1773, it was mainly used on men-of-war. Continental use of the sail goes back to the beginning of the century.

Stay Gored by 21 to 23 inches per cloth, and the holes were 27 inches apart.
Leech Straight.
Lining As the main staysail.
Thimbles As the main staysail.
Boltrope As the main staysail.

Foresail

Vorsegel, Stagfock; Trinquette

The foresail was triangular, made of canvas Number 1 or 2 (*toile à trois fils* or *toile à deux fils*), and bent to the fore stay of a sloop, schooner or similar vessel. Foresails were only used on small craft.

Width	Enough cloths to fill the distance between the stay and the mast, clearing the mast.
Height	The depth of the leech was about the same as the depth of the mainsail.
Foot	Gore from 1 inch toward the clew.
Reef bands	Two reef bands 4 inches wide were generally fitted one eighth of the leech's depth above the foot and one eighth of the depth apart.
Bonnet	Sometimes a bonnet was used instead of the reef bands.
Lining	At the clew, one cloth to 18 inches above the upper reef band, where it was cut to half the width and continued for another yard. The peak had a lining of 27 inches, and the tack cloth was cut and doubled.
Head stick	A head stick of one cloth width was fitted across the peak.
Cringles	One on each end of the reef bands; one a sixteenth of the leech depth below the lower reef band, and one a sixteenth of the depth above the upper reef band, at the leech.
Thimbles	Usually fitted to tack and clew.
Boltrope	Stay: 2½ to 3 inches; foot and leech: 1 to 2 inches; clew rope 3 inches.

Mizzen staysail

Besanstagsegel; Foc d'artimon

The mizzen staysail was triangular until 1760 and quadrilateral after that in England. Continental ships up to the end of the eighteenth century had either triangular or quadrilateral mizzen staysails. The sail was set on the mizzen staysail stay, and was made from canvas Number 2 or 3 (French sails were from *toile à deux fils* or *toile de mélis double*).

Stay	Gored at 10 to 12 inches per cloth, with holes 27 inches apart.
Leech	Straight.
Foot	Straight, 6 to 7 feet above deck.
Mast leech	After 1760: three fifths of the leech in the Royal Navy, none or half to two thirds of the leech in the Danish Navy, and none or three eighths of the leech in the French Navy. Merchant ships had mizzen staysails without mast leeches, or one quarter to one third the depth of the aft leech. The mast leech was gored over two cloths towards the nock.
Lining	Half a cloth at the mast leech, extending 2 yards up the leech at the clew and one yard at the peak. A mast lining was not used on merchant ships. The peak, clew and nock pieces there were 27 inches long.
Thimbles	As for the main staysail.
Boltrope	As for the main staysail.

Storm mizzen

Sturmbesan; Artimon de fortune

A storm mizzen was triangular and was similar to a fore topmast staysail. Made from canvas Number 2 or 3 (or *toile à deux fils*), it was bent at its fore leech to a horse abaft the mizzen mast, with the foot extended towards the taffrail.

Main topmast staysail

Grossmarsstengestagsegel, Gross-Stengestagsegel; Voile d'étai de grand hunier, grande voile d'étai de hune

Main topmast staysails were triangular until about 1760 and then quadrilateral. Hoisted on the main topmast preventer stay, the sail was made from canvas Number 5 or 6 (or from *toile de mélis double* or *simple*).

Stay	The stay was gored by 22 inches per cloth, with the holes 27 inches apart.
Leech	In the Royal Navy and on large merchant ships the aft leech was 4 to 5 yards longer than the depth of the main topsail. On smaller ships it was 1 to 2 yards longer. According to a French source it was 1¼ times the length of the main topmast.
Mast leech	Usually two fifths of the length of the leech, but in merchantmen sometimes half; it was also gored by two cloths towards the nock. The bunt on French ships was twice the length of the masthead and ran parallel to the leech, while on Danish ships it was half the length of the leech. On some Continental ships it was one quarter that length.
Foot	Straight. On Royal Navy ships the foot was 1 to 2 yards greater in length than the leech. In merchant service it was two to three cloths, and in French ships $^{17}/_{24}$ of the distance between fore and mainmasts (on Danish ships it had approximately three fifths that length). The foot of Continental sails was frequently gored by about one twelfth of the stay goring.
Lining	Half a cloth at the bunt, extending 2 yards at the clew and 1 yard at the peak. Merchantmen had the same clew piece, no bunt lining, but patches at the tack, nock and peak of 27 inches in length. Continental sails generally followed the latter arrangement.
Thimbles	As for the main staysail.
Cringles	One or two brail cringles at the leech.
Boltrope	First Rate: 2 inches for leech, foot and bunt, and 3½in for the stay. Sixth Rate: 1½in and 2¼in.

Fore topmast staysail

Vormarsstengestagsegel, Vorstengestagsegel; Petit foc

The fore topmast staysail was triangular, and was hoisted on the fore topmast preventer stay. It was made from canvas Number 5, 6 or 7 (or *toile de mélis double* or *simple*).

Stay	Gored by 30 inches per cloth; the holes were 27 inches apart.
Leech	Equal in length to the depth of the fore topsail.
Foot	In the Royal Navy, two to three cloths plus one for each yard of leech depth. In merchantmen, one cloth per yard leech depth, and in Continental ships sometimes 1½ cloths per yard. In the French Navy the foot was two thirds the depth of the leech.
Lining	The tack cloth was doubled, the clew piece 2 yards long and the peak piece 1 yard. Merchantmen had linings of 1 yard for the clew and 18 inches each for peak and tack. French sails had two tack cloths doubled and a headstick fitted to the peak.
Thimbles	As for the other staysails.
Boltrope	First Rate: 2¾in; Sixth Rate: 2 inches.

Inner jib

Innenklüver, Binnenklüver; Faux foc

An inner jib was triangular and made from *toile de mélis simple* or *toile à un fil*. It was either set flying or hoisted on a staysail stay, and was hauled out on a rope traveller. Inner jibs are peculiar to Continental rigging and were not used on English ships.

Stay	The stay was gored by about 28 to 32 inches per cloth, mostly no eyelets for hanks.
Leech	Equal in length to the main topmast.

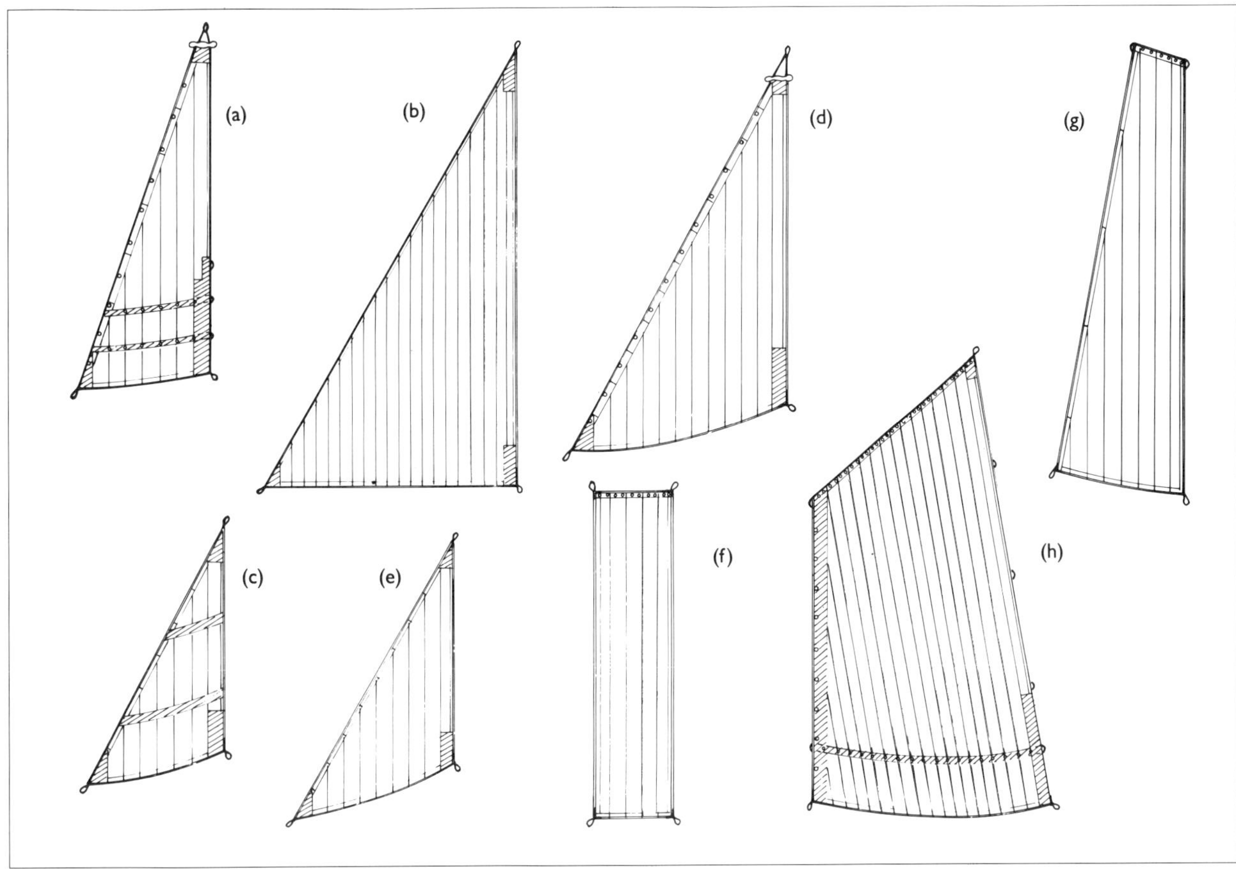

Figure 146
(a) A sloop foresail.
(b) A Continental inner jib.
(c) A sloop storm jib.
(d) A sloop jib.
(e) A sloop flying jib.
(f) A sloop watersail.
(g) A sloop ringtail sail.
(h) A French mizzen sail of about 1780.

Foot — Cut at a right angle or less towards the leech, and about half the length of the leech.

Lining — Two tack cloths doubled, clew and peak approximately 1 yard.

Thimbles — As for the other staysails.

Boltrope — The same as for the fore topmast staysail.

Storm jib

Sturmklüver; Foc de tempête, troisième foc

A storm jib was triangular and hoisted instead of the inner jib in stormy weather. Made from *toile à deux fils*, it was used on Continental ships and was about two thirds of the size of the inner jib. Fore-and-aft rigged vessels in general also had a storm jib, made of canvas Number 1 or 2.

Foot — The foot was gored by 5 or 6 inches per cloth, increasing towards the clew.

Lining — 4ft 6in at the clew, 1 yard at the peak, and the tack cloth was doubled. Two to three strengthening bands of half a cloth width were sewn on parallel to the foot, and equally spaced between the foot and peak.

Boltrope — The stay rope was 5 inches, foot and leech 2½in and the clewrope 3 inches.

Middle staysail

Mittelstagsegel; Grande voile d'étai centrale

This sail, introduced in 1773, was quadrilateral and was hoisted on the middle staysail stay. It was made from canvas Number 6 or 7, or *toile de mélis simple*.

Stay — Each cloth was gored by 13½in, and the eyelets (holes) were 27 inches apart.

Leech — In the Royal Navy the leech of a middle staysail was 4 to 7 yards longer than the depth of the main topgallant sail. On merchant ships it usually had the depth of the main topgallant sail, or 1 to 3 yards more. On French ships the leech was equal to the depth of the main topsail plus half the main topmast head, and on Danish ships it was between three quarters and eight ninths of the depth of the main topsail. While the aft leech on English ships was vertical, that on French ships was either vertical or was gored by the width of one to two cloths towards the stay; on Danish ships a goring up to a quarter of the sail's width was evident at the stay.

Foot — The foot was horizontal and square to the leech on Royal Navy ships, and six to eight cloths longer than the leech measured in yards. On sloops of war and brigs the foot was only one to three cloths more, and on merchant ships five to ten cloths more. On French ships the length of the foot was about two cloths less than that of the main topmast staysail, it also either rose slightly towards the clew, or dropped by about three eighths of the leech length. Danish middle staysails had a horizontal foot, or a drop at the clew by one seventh of the leech length. American ships also had a drop of up to one quarter of that distance.

Bunt — English: five twelfths of the depth of the leech and parallel to it. French: 1½ times the mainmast head and parallel to the leech. Danish: half to three fifths of the leech and sometimes gored toward the top by one cloth. American: similar to the English or Danish method.

Lining — Half a cloth width at the bunt, 2 yards at the clew and 1 yard at the peak.

Thimbles — Thimbles were normally fitted as on all other staysails. On those staysails without thimbles, the peak, nock and tack were prepared like the clew and marled to the sail.

Boltrope — First Rate: Stay 3¼in, otherwise 1¾in. Sixth Rate: 2 inches and 1¼in.

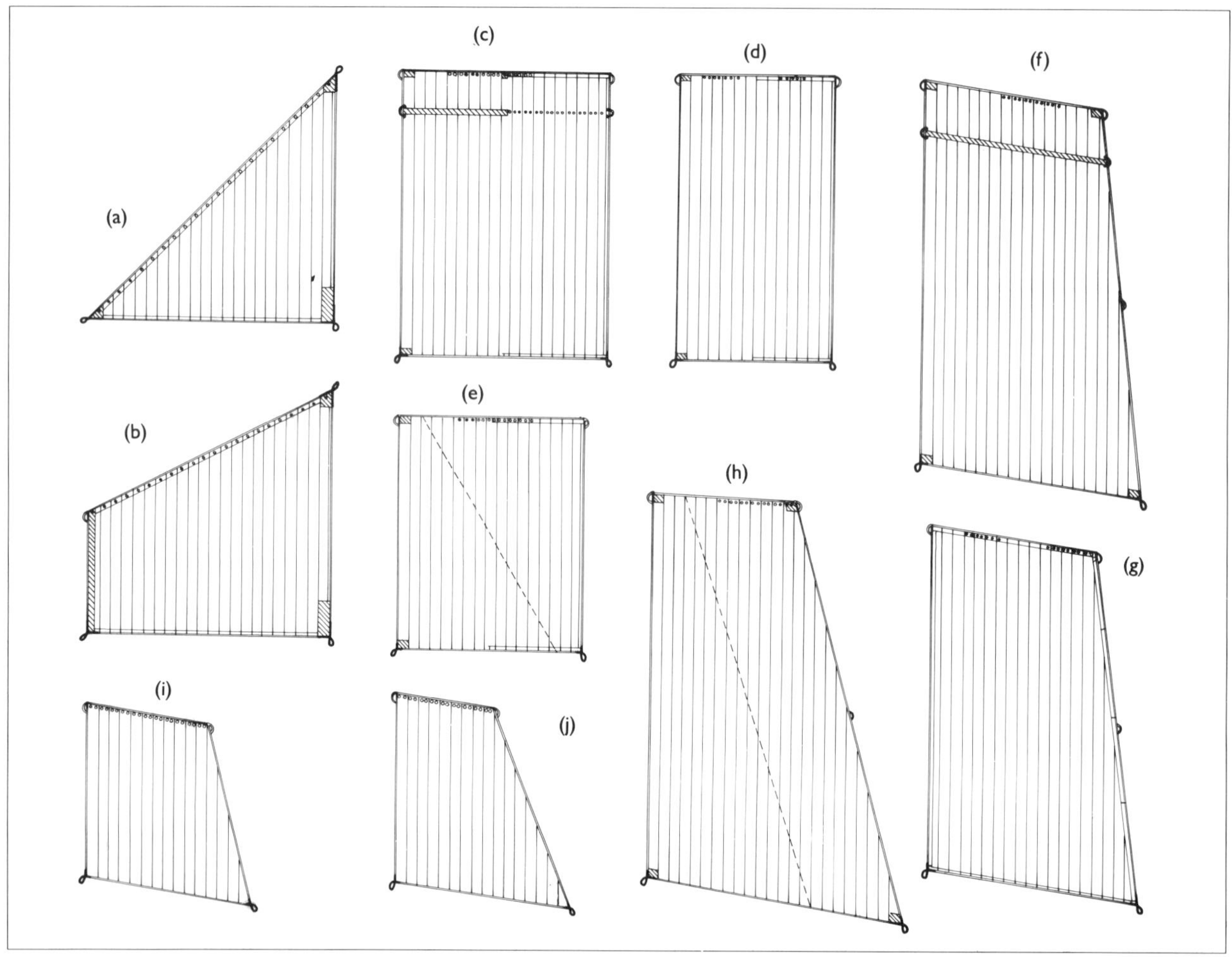

Figure 147
(a) A main royal staysail after 1719.
(b) A main royal staysail after 1760.
(c) An English lower studdingsail after 1745. The sizes of the main and fore lower studdingsails differed slightly, and only the main studdingsail had a reef-band.
(d) An English lower studdingsail before 1745.
(e) A French lower studdingsail. The dashed line indicates a fore studdingsail at the beginning of the century.
(f) An English main topmast studdingsail (fore side).
(g) An English fore topmast studdingsail (aft side). This sail had no reef-band.
(h) A French main topmast studdingsail. The dashed line indicates the shape at the beginning of the century.
(i) An English topgallant studdingsail.
(j) A French topgallant studdingsail.

Mizzen topmast staysail

Besanmarsstengestagsegel; Diablotin, foc d'artimon

Introduced around 1709, the mizzen topmast staysail was triangular until 1760 and after that quadrilateral. It was set on the mizzen topmast stay, and was made from canvas Number 7 or *toile de mélis simple.*

Stay	The stay was gored by 24 inches per cloth, with holes 27 inches apart.
Leech	One to two yards deeper than the mizzen topsail on English ships. French sails had a leech depth equal to that of the mizzen topsail.
Bunt	English: gored by one cloth towards the nock; French: parallel to the leech. The bunt length of an English sail was one third to three sevenths of the leech length, and that on a French equal to the height of the mainmast head.
Foot	Rectangular to the leech and two to five cloths longer than the aft leech length in yards on English ships. French sails had a foot length of three quarters of the distance between the masts, with the foot parallel to the deck or with a goring towards the tack similar to the gore at the stay.
Lining	As other staysails.
Thimbles	As other staysails.
Boltrope	First Rate: stay 2¾in, otherwise 1¾in. Sixth Rate: 1¾in and 1¼in.

Main topgallant staysail

Grossbramstengestagsegel; Voile d'étai de grand perroquet

Introduced in 1709, the main topgallant staysail was triangular until 1760, and thereafter quadrilateral. It was set on the main topgallant stay, and was made from canvas Number 7 or *toile de mélis simple.*

Stay	Gored by 24 inches per cloth, and the holes were 27 inches apart.
Leech	English: similar to that of the middle staysail. French: like the English or three quarters of the middle staysail; Danish leeches were like the latter.
Bunt	English and Danish: one third to three sevenths of the leech. French: equal to the length of the main topmast head. The bunt was sometimes gored by one cloth towards the nock.
Foot	English: the foot of Royal Navy main topgallant staysails was thre to six cloths longer than the number of yards in the leech, and that of merchant sails between two and eight cloths more, both square to the aft leech. French: half the distance between the masts, with the foot dropping towards the clew by the same gore as the rise at the stay, or slightly less.
Lining	Half a cloth at the bunt, one yard at the peak and 2 yards at the clew. Merchantmen had 1 yard at the clew and 18 inches at the tack, nock and peak.
Thimbles	As other staysails.
Boltrope	As for the mizzen topmast staysail.

Jib

Klüver; Grand foc

The jib was triangular, and was introduced together with the jib-boom. It was set flying at times and was hoisted on the jibstay for much of the eighteenth century. Jibs were made from canvas

Number 6 or 7 or *toile de mélis simple.*

Stay The stay was cut with a curve or roach. The length of the regular gore per cloth can be found by dividing the depth of the stay by the number of cloths. The holes were 36 inches apart.

Leech Twice the length of the fore staysail on English ships, the length of the main topmast on French.

Foot English: gored by 3 inches per cloth, straight in the Royal Navy and roached on merchant ships, and with one cloth more in the foot than the number of yards in the leech. French: rectangular to the leech and one third of the leech plus one third of the depth of the main topsail in length.

Lining English: as for other staysails. French: three cloths doubled at the tack, but clew and peak pieces as on English sails.

Head stick Jibs on sloops and other small craft had a head stick of one cloth width.

Bonnet Jibs in fore-and-aft rigged vessels were sometimes fitted with a bonnet.

Thimbles As for other staysails. The jibs of small craft jibs often had thimbles in the tack and clew.

Boltrope First Rate: stay 3½in, otherwise 1¾in. Sixth Rate: 2½i n and 1¼in respectively.

Figure 148
The sailplan of the Danish East Indiaman Danemark, dated 'Kobenhavn, den 6t Juny 1782'.

Flying jib

Aussenklüver; Clinfoc

The flying jib is not to be confused with a normal jib (which was sometimes set flying, and was then also referred to as a flying jib). The sail under discussion here came into existence only early in the last decade of the eighteenth century, together with the flying jib-boom, and was only ever set flying (without a stay), as the foremost sail on a ship. It was made from canvas Number 8 or *toile de mélis simple.*

Stay As on the jib.

Leech About seven ninths that of the jib.

Foot About eight thirteenths that of the jib.

Boltrope Two thirds the circumference of the jib boltrope.

In all other respects the flying jib was similar to the jib.

Mizzen topgallant staysail

Besanbramstengestagsegel; Voile d'étai de perruche

Introduced after 1760, the mizzen topgallant staysail was quadrilateral and was set on the mizzen topgallant stay. The sail was made from canvas Number 7 or 8, or *toile de mélis simple.* It was not often used.

Stay Gored by 5 to 10 inches per cloth. The holes were made 27 inches apart.

Leech Two fifths the length of the main topmast, measured from the cap to the hounds, or 1½ times the depth of the mizzen topgallant sail.

Bunt Two fifths the length of the leech, or sometimes more. The bunt was normally parallel to the leech, but the bunt on Danish mizzen topgallant staysails was sometimes gored towards the tack by two cloths.

Foot Ten to twelve cloths, or the depth of the mizzen topgallant sail.

Lining 18 inches at the clew and peak.

Main royal staysail or spindle staysail

Grossroyalstagsegel, Flieger über dem Bramstengestagsegel; Voile d'étai de grand cacatois, voile d'étai de perroquet volant

Introduced in 1719, the main royal staysail was triangular until 1760, then quadrilateral. It was hoisted on the main royal stay or a staysail stay and made from canvas Number 8 or 9. Röding also noted a 'mizzen spindle staysail' or *voile d'étai de la perruche volante* as the smallest of all staysails. These sails were rarely used, and Steel noted that they were not normally included in a standard suit of sails.

Stay Gored by 24 inches per cloth in triangular sails, and 10 to 12 inches in quadrilateral. The holes were 27 inches apart.

Leech 1¼ times the depth of the main royal sail.

Bunt Half the length of the leech.

Foot Equal in length to the depth of the main topgallant sail.

Lining Like the mizzen topgallant staysail.

Lower main studdingsail

Unteres grosses Leesegel; Grande bonnette

Lower main studdingsails were quadrilateral and were bent to the main studdingsail yard. They were made from canvas Number 6 or 7, or from *toile de mélis simple,* and were not often used after 1800.

Width English: on large ships two cloths more in length than the number of yards in the leech; on small ships one cloth fewer than yards in the leech length; on merchantmen two to seven cloths more than yards in the leech. French:

	one third of the mainsail's width at the head and five twelfths at the foot.
Head	Straight.
Foot	Straight.
Leech	In the Royal Navy 2 to 3 yards longer than that of the mainsail. Merchantmen and Continental ships had lower main studdingsails equal in length to the mainsail, or 1 yard more. French ships had the outer leech gored toward the head by one to two cloths.
Reef bands	One in the Royal Navy after 1745, 6 inches wide and placed one eighth of the depth below the head.
Cringles	At each side of the reef band.
Lining	9 or 18 inches on the fore (outer) side at each corner.
Boltrope	First Rate: 2¼in for leech and foot, and 1¼in for the headrope. Sixth Rate: 1¾in and 1¼in respectively.

Lower fore studdingsail

Unteres Vorleesegel; Bonnette de misaine

Lower fore studdingsails were quadrilateral and were bent to the fore studdingsail yards. They were made of canvas Number 6 or 7, or of *toile de mélis simple.*

Width	English: one cloth less then the main studdingsail. French: one third of the width of the fore course.
Head	Straight.
Foot	Straight. On a French sail slightly wider than the head, to compensate for the gore of the fore course.
Leech	The depth was the same as the main course or 1 to 2 yards more. Merchant ships and Continental ships had lower fore studdingsails with a depth equal to the depth of the fore course, or 1 yard more.
Lining	As for the main studdingsail.
Boltrope	As for the main studdingsail.

Main topmast studdingsail

Grossmarsleesegel; Bonnette de grand hunier

Main topmast studdingsails were quadrilateral and were bent to the main topmast studdingsail yard. They were made from canvas Number 6 or 7, or *toile de mélis simple.*

Width	English: the width of the foot was two cloths less than the number of yards in the leech. French: the head was 7/24 of the length of the main topsail yard and the foot was the same width as the main studdingsail head.
Leech	1 yard more than the leech of the main topsail. The outer leech was gored four cloths toward the head in the Royal Navy, seven cloths on merchant ships and between five and ten cloths on Continental ships.
Head	Gored by 4 inches per cloth on English ships, 2 to 3 inches on Continental ships.
Foot	English: gored by 4 inches per cloth. Continental: gored by 4 to 6 inches per cloth.
Reef band	One for sails in the Royal Navy, 6 inches wide and one eighth of the sail's depth from the head.
Cringles	At each end of the reef band, and one at the outer leech at half height, for the downhauler.
Lining	As for other studdingsails.
Boltrope	As for the lower studdingsails.

Fore topmast studdingsail

Vormarsleesegel; Bonnette de petit hunier

Fore topmast studdingsails were quadrilateral and were bent to the fore topmast studdingsail yard. They were made from canvas Number 6 or 7, or *toile de mélis simple.*

Width	One cloth less than the main topmast studdingsail for English sails. For French, 7/24 of the length of the fore topsail yard.
Leech	As for the main topmast studdingsail.
Head	As for the main topmast studdingsail.
Foot	As for the main topmast studdingsail.
Cringles	One for the downhauler at half height on the outer leech.
Lining	As for other studdingsails.
Boltrope	As for the lower studdingsails.

Mizzen topmast studdingsail

Besanmarsleesegel, Kreuzleesegel; Bonnette de perroquet de fougue

Mizzen topmast studdingsails were quadrilateral and bent to the respective studdingsail yards. On English vessels they were rigged only in the early years of the century, but in France and other Continental nations (according to Röding) throughout the eighteenth century. They were made from *toile de mélis simple*, and were similar in proportions to the other topmast studdingsails.

Main topgallant studdingsail

Grossbramleesegel; Bonnette de grand perroquet

Main topgallant studdingsails were quadrilateral and bent to the respective studdingsail yards. They were introduced around 1775 and were made from canvas Number 7 or 8, or from *toile de mélis simple.*

Width	English: the foot was five cloths more than the number of yards in the leech on large ships; on small ships it ranged between equal to the leech and three cloths more than the number of yards in the leech. France: at the head, one quarter the length of the topgallant yard minus one sixteenth, and at the foot equal to the head of the main topmast studdingsail.
Leech	English: 18 inches to 1 yard more than the depth of the main topgallant sail, and gored toward the head by two to four cloths. French: the depth of the main topgallant sail and gored by four to five cloths.
Head	Gored by 3 to 5 inches per cloth.
Foot	Gored by 3 to 5 inches per cloth.
Boltrope	First Rate: 2 inches and 1¼in; Sixth Rate: 1½in and 1 inch.

Fore topgallant studdingsail

Vorbramleesegel; Bonnette de petit perroquet

Fore topgallant studdingsails were similar in every respect and proportion to main topgallant studdingsails.

Watersail

Wassersegel; Bonnette de sous-gui

The watersail was quadrilateral and made from canvas Number

7 or *toile de mêlis simple.* Any sail which was set in calm conditions beneath the lower studdingsails or the ringtail boom extension was known as a 'watersail'. Ships as well as smaller fore-and-aft rigged vessels carried a sail for use under such conditions.

Depth	From half to three quarters the length of the boom.
Width	Four to five cloths.
Leech	Straight, or gored by one cloth.
Head	Square to the cloths.
Foot	Square to the cloths.
Boltrope	1½in all around.

When fore-and-aft vessels carried lower studdingsails these were similar to a watersail, with square leeches and a depth one yard greater than that of the squaresail.

Ringtail sail

Brotwinner, Brotgewinner; Bonnette de tape-cul, bonnette d'artimon

The ringtail sail was quadrilateral and made of canvas Number 7 or 8, or *toile de mêlis simple.* Occasionally it was hoisted abaft the aft leech of the mainsail in fore-and-aft rigged vessels. A sail of this kind, but more square, was sometimes extended in light winds on a small mast erected for the purpose on the upper part of the stern of some vessels; the foot of such a sail was spread by a boom projecting horizontally from the stern (see Jigger mast in Chapter III). The ringtail sail was bent to a small yard at the outer end of the gaff, with the foot spread by a boom extension lashed to the outer end of the boom proper.

Depth	Similar to the aft leech of the boom sail.
Head	Gored to match the peak of the boom sail.
Foot	Gored by 1 inch per cloth (greatest depth at the clew).
Boltrope	1 inch for head, foot and aft leech, 1½in for the fore leech.

Wingsail for a ketch

Grossgaffelsegel einer Ketsch; Grande voile à corne du quaiche

A ketch wingsail was quadrilateral and similar to a ship's mizzen course. It was made from canvas Number 6 or 7 and bent at the aft side of the mainmast to mast hoops. The head was extended by a gaff.

Smoke-sail

Rauchsegel; Masque

A smoke-sail was a small sail, hoisted against the foremast when a ship rode head to wind to prevent the smoke of the galley from blowing on to the quarterdeck. It was thus not a true sail at all.

A list given by Rees in his *Cyclopaedia* of 1819 itemises the type and number of sails in a suit supplied to a British man-of-war for 8 months service:

2 main courses	1 mizzen staysail
2 main topsails	2 main topmast staysails
1 main topgallant sail	1 middle staysail
1 main royal	1 main topgallant staysail
2 fore courses	1 fore topmast staysail
2 fore topsails	1 royal staysail
2 fore topgallant sails	2 jibs
1 fore royal	1 flying-jib
1 spritsail course	1 mizzen topmast staysail
1 spritsail topsail	2 main studdingsails
1 driver boom sail	2 main topmast studdingsails
2 mizzen courses	2 main topgallant studdingsails
2 mizzen topsails	2 fore studdingsails
1 mizzen topgallant sail	2 fore topmast studdingsails
1 main staysail	1 fore topgallant studdingsail
2 fore staysails	1 smoke-sail

A suit of sails for a First Rate man-of-war required about 16,000 yards of canvas. Additional requirements for the sails were approximately as follows:

100 yards of old canvas	574lb of twine line
56lb of beeswax	50lb of resin
14lb of turpentine	122lb of tallow
39 gallons of train oil	6½ barrels of tar
5½cwt of spun yarn	42,164 yards of waxed sewing twine
21,822 yards of dipped sewing twine	32cwt 2qr 24lb of boltrope

Rees also provides 'The Quality of Canvas of which the different Sails are made in the Merchant Service':

Canvas Number 1
Main and fore courses, and main and fore staysails of East India ships.

Canvas Number 2
Main and fore courses, and main and fore staysails of West India ships.

Canvas Number 3
Mizzen course, main and fore topsails, sprit course, and mizzen staysails of East India ships.

Canvas Number 4
Mizzen topsails of East India ships.

Canvas Number 5
Driver boom sails of large East India ships, and mizzen topsails of West India ships.

Canvas Number 6
Driver boom sails of East and West India ships; fore topmast staysails of East India ships; main topmast staysails of West India ships; spritsail topsails and main and fore topgallant sails of large East India ships.

Canvas Number 7
Main and fore topgallant sails, middle staysails, flying jibs, lower studdingsail, main topmast studdingsails and main topgallant staysails of East and West India ships, and fore topmast staysails of West India ships.

Canvas Number 8
Mizzen topgallant sails and main topgallant studdingsails of East and West India ships; mizzen topmast staysails and royal sails of East India ships; and small flying jibs of East India ships, if any.

> As there can be no fixed dimensions of merchant-ships' sails, the same number of cloths, or nearly so, of ships of the same tonnage in the royal navy may be taken, and about seven-eighths of their depth [Rees].

As noted at the beginning of this chapter, most of the sails discussed here were found on an eighteenth century fully rigged ship. As many as possible of the sails of smaller (fore-and-aft rigged) craft have been included, but not all sails could be covered. Any sail not discussed in detail here, particularly those of vessels from non-European parts of the world, can be found in the Chapters III to VIII. Sails which are not specifically referred to in these chapters fitted into one or another of the groups of sails discussed, and their precise shape and details can be inferred from those discussed in detail.

Tables

The tables below provide information about numbers and sizes of materials used in making sails, and the dimensions of sails on ships of various sizes.

Table 54: Number of reefs (1), points (2), robands (3) and gaskets (4) per sail

Sail:	*Sprit course*				*Fore course*				*Main course*			
Number of guns	*(1)*	*(2)*	*(3)*	*(4)*	*(1)*	*(2)*	*(3)*	*(4)*	*(1)*	*(2)*	*(3)*	*(4)*
100	2	64	64	6	1	90	90	8	1	102	102	8
90	2	60	60	6	1	86	86	8	1	98	98	8
80	2	58	58	6	1	82	82	8	1	94	94	8
74	2	62	62	6	1	86	86	8	1	98	99	8
70	2	58	58	6	1	82	82	8	1	92	92	8
84	2	58	58	6	1	80	80	8	1	92	92	8
60	2	54	54	6	1	78	78	8	1	88	88	8
50	2	52	52	4	1	72	72	6	1	82	82	6
44	2	46	46	4	1	66	66	6	1	74	74	6
38 & 36	2	50	50	4	2	132	66	6	2	152	76	6
32	2	50	50	4	2	128	64	6	2	148	74	6
28	2	50	50	4	2	128	64	6	2	148	74	6
24	2	40	40	4	2	112	56	4	2	128	64	4
20	2	36	36	2	2	104	52	4	2	116	58	4
Sloops	2	32	32	2	2	92	46	4	2	108	54	4

Sail:	*Fore topsail*				*Main topsail*				*Mizzen topsail*			
Number of guns	*(1)*	*(2)*	*(3)*	*(4)*	*(1)*	*(2)*	*(3)*	*(4)*	*(1)*	*(2)*	*(3)*	*(4)*
100	3	180	52	6	3	210	61	6	2	92	42	4
90	3	171	50	6	3	198	58	6	2	88	40	4
80	3	165	48	6	3	192	56	6	2	86	40	4
74	3	177	52	6	3	204	60	6	2	90	41	4
70	3	165	49	6	3	189	56	6	2	85	39	4
64	3	165	49	6	3	189	56	6	2	85	39	4
60	3	156	45	6	3	180	52	6	2	80	36	4
50	3	147	43	4	3	168	49	4	2	72	33	4
44	3	135	39	4	3	150	44	4	2	66	30	2
38 & 36	3	138	41	4	3	153	46	4	2	70	33	2
32	3	135	40	4	3	150	44	4	2	68	32	2
28	3	135	41	4	3	150	44	4	2	69	30	2
24	3	117	34	4	3	132	38	4	2	62	27	2
20	3	112	32	4	3	132	38	4	2	58	26	2
Sloops	3	99	30	4	1	111	32	4	2	48	22	2

Table 55: Sizes of all bolt ropes

Circumference in English inches: leech and foot rope (1) and head rope (2)

Guns:	*110 & 100*		*90, 84 & 74*		*80*		*70 & 64*	
Name of sail	*(1)*	*(2)*	*(1)*	*(2)*	*(1)*	*(2)*	*(1)*	*(2)*
Main course	6	2½	5¾	2¼	5½	2¼	5½	2¼
Fore course	5½	2¼	5¼	2¼	5	2	5	2
Mizzen course	4½	1¾	4¼	1¾	4	1¾	3½	1¾
leech	3½		3¼		3		3	
Main topsail	5¾	2½	5½	2¼	5¼	2	5	2
leeches	5		4¾		4½		4¼	
Fore topsail	5¼	2	5	2	4¾	1¾	4½	1¾
leeches	4½		4¼		4		3¾	
Mizzen topsail	3½	1½	3½	1½	3¼	1½	3¼	1½
leeches	2¾		2¾		2½		2½	
Sprit course	3¼	1¾	3¼	1¾	3	1	3	1½
Sprit topsail, main & fore topgallant sails	2¾	1½	2¾	1½	2½	1½	2½	1½
Main & fore staysails	3¼		3¼		3		3	
Fore top staysail	2¾		2¾		2½		2½	
Main top staysail	2	3½	2	3½	2	3¼	1¾	3
Middle staysail	1¾	3¼	1¾	3	1¾	2¾	1¾	2¾
Mizzen staysails	3¼	3¼	3¼		3		3	
Main topgallant & mizzen top staysails	1¾	2¾	1¾	2¾	1½	2½	1½	2½
Main, fore & topmast studdingsails	2¼	1¼	2¼	1¼	2	1¼	2	1¼
Topgallant studdingsails, mizzen topgallant sail & main & fore royals	2	1¼	2	1¼	1¾	1¼	1¾	1¼
Mizzen royals	1¾	1	1¾	1	1½	1	1½	1
Flying jibs	1¾	3½	1¾	3¼	1½	3	1½	3
Driver sails	3	1¾	2¾	1¾	2¾	1¾	2½	1½
Boat sails	Lateen		1½	1	Settee		1½	1

Guns:	*60*		*50*		*44,38,36, 32,28*		*24,20*	
Name of sail	*(1)*	*(2)*	*(1)*	*(2)*	*(1)*	*(2)*	*(1)*	*(2)*
Main course	5	2	4¾	1¾	4½	1¾	4	1½
Fore course	4½	1¾	4¼	1¾	4	1¾	3½	1½
Mizzen course	3¼	1½	3	1½	2¾	1½	2½	1½
leech	2¾		2½		2¼		2	
Main topsails	4¾	1¾	4½	1¾	4¼	1¾	3¾	1½
leeches	4		3¾		3½		3	
Fore topsails	4¼	1¾	4	1½	3¾	1½	3¼	1¼
leeches	3½		3¼		3		2¾	
Mizzen topsail	3	1½	3	1½	3	1½	2¾	1¼
leeches	2½		2½		2½		2¼	
Sprit course	2¾	1½	2½	1½	2¼	1½	2	1¼
Sprit topsail, main & fore topgallant sails	2¼	1¼	2¼	1¼	2	1¼	1¾	1¼
Main & fore staysails	2¾		2¾		2½		2¼	
Fore top staysail	2½		2¼		2¼		2	
Main top staysail	1¾	2¾	1¾	2¾	1½	2½	1½	2¼
Middle staysail	1½	2½	1½	2½	1½	2¼	1¼	2
Mizzen staysail	3		2¾		2½		2¼	
Main topgallant & mizzen top staysails	1½	2¼	1½	2¼	1¼	2	1¼	1¾
Main, fore and topmast studdingsails	2	1¼	2	1¼	1¾	1¼	1¾	1¼
Topgallant studdingsails, mizzen topgallant sail & main & fore royals	1½	1	1½	1	1½	1	1¼	1
Mizzen royals	1½	1	1¼	1	1¼	1	1	¾
Flying jibs	1½	2¾	1½	2¾	1½	2¾	1¼	2½
Driver sails	2½	1½	2½	1½	2½	1½	2¼	1¼
Boat sails:	Lug sail		1½	1	Sprit sail		1¼	1

Vessels:	*Sloop*		*Brig*		*Cutter*		
Name of sail	*(1)*	*(2)*	*(1)*	*(2)*		*(1)*	*(2)*
Main course	3½	1½	3½	1½	head	3	1½
					leech		1¾
Fore course	3	1½	3	1½	Try or storm mainsail		1½
Mizzen course	2¼	1½					
leech	1¾					3	1¾
Main topsail	3	1½	3	1¼		2¼	1½
leeches	2½		2½			2¼	
Fore topsail	2¼	1½	2¾	1¼	Save-all topsail		
leeches	2¼		2¼			1½	1
Mizzen topsail	2¼	1¼			Main topgallant sail	1¾	1
leeches	2¼		1¾				
Sprit course	1¾	1¼			Squaresail	2¼	1½
Sprit topsail, main & fore topgallant sails	1½	1	1½	1	Gaff topsail		1
Fore top staysail	2		1¾		head	2¼	
					leech		1¾
Main top staysail	1¼	2	1¼	2	Fore sail	2	3
Middle staysail	1¼	1¾	1¼	1¾			
Mizzen staysail	2				Storm foresail		3½
					leech	1¾	
					foot	2¾	
Main topgallant & mizzen top staysails	1¼	1¾					
Main, fore & topmast studdingsails	1½	1	1½	1	Ringtail sail	1½	½
Topgallant studdingsails, mizzen topgallant sail & main & fore royals	1¼	1	1¼	1	Water sail	2¼	1¾
Mizzen royals	1	¾			First jib		6
					leech	1½	
					foot	4	

Flying jib	1¼ 2¼	1¼ 2¼	Second jib		6
			leech	1½	
			foot	4	
Driver sails	2 1¼		Third jib		6
			leech	1½	
			foot	4	
Boat sails	Jib	1¼ 1	Storm jib		5
			leech	1½	
			foot	3½	

Merchantmen:	*1200 tons*		*700 tons*		*500 tons*		*400 tons*	
Name of sail	*(1)*	*(2)*	*(1)*	*(2)*	*(1)*	*(2)*	*(1)*	*(2)*
Main course	5½	2¼	4½	1¾	4	1½	3½	1½
Fore course	5	2	4	1¾	3½	1½	3	1½
Mizzen course	3½	1¾	2¾	1½	2½	1½	2¼	1½
leech	3		2¼		2		1¾	
Main topsail	5	2	4¼	1¾	3¾	1½	3¼	1½
leeches	4¼		3½		3		2½	
Fore topsail	4½	1¾	3¾	1½	3¼	1½	2¾	1¼
leeches	3¾		3		2¾		2½	
Mizzen topsail	3¼	1½	3	1½	2¾	1¼	2¼	1¼
leeches	2½		2½		2¼		1¾	
Sprit course	3	1½	2¼	1½	2	1½	1¾	1¼
Sprit topsail, main & fore topgallant sails	2½	1½	2	1¼	1¾	1¼	1½	1
Main & fore staysails	3		2½		2¼		2	
Fore top staysail	2½		2¼		2		2	
Main top staysail	1¾	3	1½	2½	1½	2¼	1¼	2
Middle staysail	1¾	2¾	1½	2¼	1½	2	1¼	1¾
Mizzen staysail	3		2½		2¼		2	
Main topgallant & mizzen top staysails	1½	2½	1¼	2	1¼	1¾	1¼	1¾
Main, fore & topmast studdingsails	2	1¼	1¾	1¼	1¾	1¼	1½	1
Topgallant studdingsails, mizzen-topgallant sail & main & fore royals	1¾	1¼	1½	1	1¼	1	1¼	1
Mizzen royal	1½	1	1¼	1	1	¾	1	¾
Flying jib	1½	3	1½	2¾	1¼	2½	1¼	2¼
Driver sails	2½	1½	2¼	1¼	2¼	1¼	2	1¼

Clew ropes for the main and fore course were 2 inches larger from 50 guns upwards, and 1½ inches larger below that size. Clew ropes for the mizzen course were half an inch larger and 3 fathoms long. Cringles were half an inch smaller than the rope. Clew ropes to flying jibs (the flying jib, in Steel's use of the term, was the normal jib) and drivers were half an inch larger and 2 fathoms long. All staysails had clew ropes half an inch larger and 2 fathoms long.

Table 56: Dimensions of sails for ships of each class in the Royal Navy and merchant service (after Steel)

(1) Number of sail cloths in the head (cloth = 24 inches wide)
(2) Number of sail cloths in the foot
(3) Depth of each sail in yards (1 yard = 914.4mm)

Number of guns:	*100*			*90*		
Name of sail	*(1)*	*(2)*	*(3)*	*(1)*	*(2)*	*(3)*
Main course	47½	50	14½	46	48	14
Fore course	42	40	12¼	40	38	11½
fore leech			10½			9½
Mizzen course	17	18		16	17	
aft leech			20½			19½
Main top	30½	48½	21	29½	47	20
Fore top	26½	43	19	25½	41	18
middle			14¼			13¾
Mizzen top	20½	31		20½	29	
clew			15			14½
Main topgallant	22½	30½	10½	21	30	10
Fore topgallant	19½	27	9½	18½	26	8¾
Mizzen topgallant	15	21	7	15	22	7
Main royal	18	23	8½	17	22	7½
Fore royal	16	20	7½	15	19	6¾
Mizzen royal	11	15	5½	11	15	5½
Main stay		32	15		31	14
Fore stay		23	13		22	12
bunt			8½			8
Mizzen stay	23	25		21	23	
aft leech			14			13
bunt			10			9
Main top stay	26	28		25	27	
aft leech			26			24
Fore top stay		22	19		21	18
bunt			7			6½
Mizzen top stay	20	21		19	20	
aft leech			16			14½
bunt			8			7½
Middle stay	25	25		24	24	
aft leech			17			16½
bunt			7			6
Main topgallant stay	22	22		21	21	
aft leech			16½			15½
Main studding	20	20	17½	19	19	17
Fore studding	19	19	14½	18	18	14
Main top studding	16	20	22	15	19	21
Fore top studding	15	19	20	14	18	19
Main topgallant studding	12	16	11	11	15	10½
Fore topgallant studding	11	15	10	10	14	9
Flying jib (jib)		27	26		26	25
Sprit course	30	30	9	28	28	9
Sprit top	20	30½	9½	18½	28½	10
bunt			10			10
Driver	23	29		22	28	
aft leech			22½			22½

Number of guns:	*80*			*74*		
Name of sail	*(1)*	*(2)*	*(3)*	*(1)*	*(2)*	*(3)*
Main course	44	46	14	46	48	15¾
Fore course	39	37	12	40	38	13
fore leech			9			11
Mizzen course	16	17		16	17	
aft leech			19			21
Main top	28	45	19½	30	47	19¾
Fore top	24½	39½	18	26	41	17½
middle			13¼			13¾
Mizzen top	19	28		20½	30	
leech			14			14
Main topgallant	21½	29	9½	21½	30½	10
Fore topgallant	18½	25	8½	18½	26½	8¾
Mizzen topgallant	13½	20½	6¾	15	21 7¼	
Main royal	16	21	7	17	22	7½
Fore royal	14	18	6¼	15	19	6¾
Mizzen royal	10	14	5¼	11	15	5½
Main stay		30	14		31	15¾
Fore stay		21	12		22	13
bunt			8			9
Mizzen stay	21	23		20½	22½	
aft leech			13			13½
Bunt			9			9½
Main top stay	24	26		25	27	
aft leech			24			24
Fore top stay		20	17½		21	17½
bunt			6			6½
Mizzen top stay	18	19		19	20	
aft leech			15			14½
bunt			7			7½
Middle stay	23	23		24	24	
aft leech			15½			16½
bunt			6			6
Mizzen topgallant stay	20	20		21	21	
aft leech			15½			15½
Main studding	18	18	17	19	19	18½

Number of guns:	*80*			*74*	*(continued)*	
Name of sail	*(1)*	*(2)*	*(3)*	*(1)*	*(2)*	*(3)*
Fore studding	17	17	14½	18	18	15
Main top studding	14	18	20½	15	19	23¾
Fore top studding	13	17	18½	14	18	18½
Main topgallant studding	10	14	10	11	15	10¼
Fore topgallant studding	9	13	9	10	14	9¼
Flying jib (jib)		25	24		26	25
Sprit course	27	27	8½	29	29	8
Sprit top	18	27½	9½	18½	29½	10
bunt			11			11
Driver	22½	28½		25½	32½	
aft leech			23½			24

Number of guns:	*64*			*50*		
Name of sail	*(1)*	*(2)*	*(3)*	*(1)*	*(2)*	*(3)*
Main course	41½	44	14	39	41	13
Fore course	36	34	12	34½	33½	11½
fore leech			9¼			8½
Mizzen course	14½	15½		13	14	
aft leech			17¼			16¼
Main top	27	43	17½	26	40	16¼
Fore top	23	37	15¾	22	35	14¾
middle			12			11¼
Mizzen top	18½	27		17	25½	
leech			12¾			12
Main topgallant	19½	27½	9	17½	26½	8
Fore topgallant	17	24	8¼	14½	23	7¼
Mizzen topgallant	13½	19½	6¾	12½	17½	6¼
Main royal	15	20	7	13	18	6
Fore royal	13	17½	6¼	11	15	5¼
Mizzen royal	10	14	5	9	13	4¾
Main stay		28	14		26	13
Fore stay		19	11½		18	11
bunt			7½			7
Mizzen stay	19½	21		18	20	
aft leech			11½			11
bunt			9			8¾
Main top stay	22	24		21	23	
aft leech			22¼			21¾
Fore top stay		18	16		17	15
bunt			5			4½
Mizzen top stay	16	17		15	16	
aft leech			13			13
bunt			6			5½
Middle stay	21	21		20	20	
aft leech			13½			13
bunt			5			4½
Main topgallant stay	18	18		17	17	
aft leech			14			13½
Main studding	16	16	16½	16	16	15½
Fore studding	15	15	13	15	15	13
Main top studding	12	16	18½	11	16	17¼
Fore top studding	11	15	17	10	15	15¾
Main topgallant studding	9	12	9½	8	11	8½
Fore topgallant studding	8	11	8½	7	10	7¾
Flying jib (jib)		24	23		23	22
Sprit course	26	26	7½	25	25	7
Sprit top	17	27	9	14½	25	8
bunt			9½			8½
Driver	22½	29		19	24	
aft leech			20			19

Number of guns:	*44*			*38*		
Name of sail	*(1)*	*(2)*	*(3)*	*(1)*	*(2)*	*(3)*
Main course	37½	40	12	38½	40½	13¼
Fore course	33	32	10	33½	32½	11
fore leech			9½			11¼
Mizzen course	12	12		13	14	
aft leech			17			18¼
Main top	24	38½	16½	24½	39½	16
Fore top	21½	33½	14½	22	34	14½
middle			11½			11¾
Mizzen top	17	25		17½	25½	
leech			12¼			12½
Main topgallant	17	25	8	17½	25½	8¼
Fore topgallant	15	22	7	15	23	7¾
Mizzen topgallant	12½	18	6	13	18	6¾
Main royal	12	17	6	13	18	6¼
Fore royal	11	15	5	11	15	5¾
Mizzen royal	9	13	4½	9	13	4¾
Main stay		24	12		24	13¼
Fore stay		17	10		17	11
bunt			6½			7
Mizzen stay	16	18		16	18	
aft leech			10½			12
bunt			8			8
Main top stay	19½	21½		19½	21½	
aft leech			20½		20½	
Fore top stay	16	14		16	14½	
bunt			4			4
Mizzen top stay	13	14		13	14	
aft leech			12			12
bunt			5			5
Middle stay	18	18		18	18	
aft leech			12			12
bunt			4½			4
Main topgallant stay	15	15		15	15	
aft leech			12			12
Main studding	15	15	14½	15	15	14½
Fore studding	14	14	12	14	14	13
Main top studding	11	15	17½	11	15	17¼
Fore top studding	10	14	15¼	10	14	15½
Main topgallant studding	8	11	8½	8	11	8¾
Fore topgallant studding	7	10	7½	7	10	7¾
Flying jib (jib)		22	21		22	21
Sprit course	24	24	6	25	25	6
Sprit top	14	24½	8	15	25	8¼
bunt			8			11
Driver	18	24		19	25	
aft leech			19			22½

Number of guns:	*36*			*32*		
Name of sail	*(1)*	*(2)*	*(3)*	*(1)*	*(2)*	*(3)*
Main course	37	39	13¼	35	37	13
Fore course	32½	31½	11¾	30½	29½	11
fore leech			10¼			10
Mizzen course	13	14		11	11	
aft leech			18¾			18
Main top	24	38	16¼	23½	35½	15½
Fore top	21½	33	14¼	19½	31	13¾
middle			11¼			11
Mizzen top	17	25		16	23	
leech			12			11¾
Main topgallant	16	25	7¾	15½	23½	7¾
Fore topgallant	14½	22½	6¾	13½	21	6¾
Mizzen topgallant	12½	18	6¼	11½	17	6
Main royal	11	16	6¼	11	16	5¾
Fore royal	11	15	5½	10	14	5
Mizzen royal	9	13	4¾	8	12	4½

Main stay		24	14		23	13
Fore stay		17	12		16	11
bunt			7			6½
Mizzen stay	16	18		16	18	
aft leech			12			11½
bunt			8			8
Main top stay	19½	21½		19	21	
aft leech			20½			20½
Fore top stay		16	14		15	13¾
bunt			4			4
Mizzen top stay	13	14		13	14	
aft leech			12			12
bunt			5			5
Middle stay	18	18		18	18	
aft leech			12			12
bunt			4			4
Main topgallant stay	15	15		15	15	
aft leech			12			12
Main studding	15	15	14½	14	14	15
Fore studding	14	14	13¾	13	13	13
Main top studding	11	15	17¼	10	14	16¼
Fore top studding	10	14	15¼	9	13	14¼
Main topgallant studding	8	11	8¼	7	10	8¼
Fore topgallant studding	7	10	7¼	6	9	7¼
Flying jib (jib)		21	20		20	19
Sprit course	24	24	6	23	23	6
Sprit top	14½	24½	7¾	13½	23	7¾
bunt			11			10
Driver	19	25		18	24	
aft leech			22			21

Number of guns:	*28*			*24*		
Name of sail	*(1)*	*(2)*	*(3)*	*(1)*	*(2)*	*(3)*
Main course	33	35	12	30½	32½	11
Fore course	29½	28½	10¼	27	26	9
fore leech			9½			9
Mizzen course	9½	10½		10	10	
aft leech			17			15½
Main top	22	34	14¾	20	31½	13¾
Fore top	19	30	13	17	27½	12¼
middle			10½			9¼
Mizzen top	15	22		13½	22	
leech			11¼			10
Main topgallant	15	23	7½	14	20½	6¾
Fore topgallant	13	20	6½	12	18	6
Mizzen topgallant	11	16	5¾	10½	14½	5
Main royal	10	15	5½	9	14	4¾
Fore royal	9½	13½	4½	8½	12½	4
Mizzen royal	7	11	4	7	11	3¼
Main stay		23	12		22	11½
Fore stay		16	10¼		15	9½
bunt			5½			5½
Mizzen stay	15	17		14	16	
aft leech			10½			9½
bunt			8			7
Main top stay	19	21		17	19	
aft leech			20½			18
Fore top stay		15	13		14	12½
bunt			4			3
Mizzen top stay	13	14		11	12	
aft leech			12			10½
bunt			5			4¾
Middle stay	18	18		16	16	
aft leech			12			10¼
bunt			4			2¾
Main topgallant stay	15	15		13	13	
aft leech			12			10¾
Main studding	13	13	14	12	12	13
Fore studding	12	12	12	11	11	11
Main top studding	9	13	15¾	8	12	14½
Fore top studding	8	12	14	7	11	13
Main topgallant studding	6	9	8	6	8	7¼
Fore topgallant studding	5	8	7	5	7	6½
Flying jib (jib)		20	29		19	18
Sprit course	21	21	6	19	19	5½
Sprit top	13	21	7½	12	20	6¾
bunt			9			8
Driver		16	21		15	19
aft leech			19			17

Number of guns:	*20*			*Sloop*		
Name of sail	*(1)*	*(2)*	*(3)*	*(1)*	*(2)*	*(3)*
Main course	29	31	10	25½	28½	9½
Fore course	26	25	9	22½	21½	7¾
fore leech			8			7
Mizzen course	10	10		7	8	
aft leech			13½			13
Main top	19½	30	13¼	16	26½	11½
Fore top	16½	26½	11¾	14	23	10¼
middle			9¼			7¼
Mizzen top	13½	19½		11½	17	
leech			10			8
Main topgallant	13½	20	6½	11½	17	5¾
Fore topgallant	11½	17½	5¾	10	15	5¼
Mizzen topgallant	9½	14	5	8	12	4¾
Main royal	8	13	4½	7½	11½	3½
Fore royal	7	11	3¾	7	10½	3
Mizzen royal	6	10	3	4	8	3¼
Main stay		23	12		22	11½
Fore stay		15	9		12	8
bunt			5			5
Mizzen stay	13	15		12	13	
aft leech			9			9
bunt			7			5½
Main top stay	17	19		14	16	
aft leech			18			15½
Fore top stay		14	12		12	10¼
bunt			3			3
Mizzen top stay	11	12		9	10	
aft leech			10½			9½
bunt			4¼			4
Middle stay	16	16		13	13	
aft leech			10¼			10
bunt			2¼			2½
Main topgallant stay	13	13		12	12	
aft leech			10¾			9½
Main studding	12	12	13	11	11	11
Fore studding	11	11	11	10	10	9
Main top studding	8	12	14	8	11	12
Fore top studding	7	11	12½	7	10	11
Main topgallant studding	6	8	7	5	8	6¼
Fore topgallant studding	5	7	6¼	4	7	5¾
Flying jib (jib)		19	18		16	15
Sprit course	19	19	5½	16	16	5
Sprit top	11½	19½	6	10	17	5¾
bunt			8			7
Driver	14	18		12	16	
aft leech			16			15

Number of guns:	*Brig*			*Cutter of 200 tons*		
Name of sail	*(1)*	*(2)*	*(3)*	*(1)*	*(2)*	*(3)*
Main course	13½	19		22	30	
fore leech			9			18
aft leech			13½			24
Fore course	18	18	6½			
fore leech						14
Try-sail or						
storm main sail				9	19	
aft leech						17
middle			9¾			13¾
Main top	12½	19		23	27	
leech			10¼			15¼
middle			9½	Save all topsail		
Fore top	12½	19		8	24	5
leech			10			
Main topgallant	10	14	5	7½	16½	6½
Fore topgallant	9	13	4½	Square sail		
Main royal	6	10	3½	30	30	20½
Fore royal	5	9	3			
bunt			1	Gaff topsail		
Main stay		15		2½	23	16
aft leech			8½			
Fore stay		11	7	Fore sail		
bunt			4	1	16	16
Main top stay	12	14				
aft leech			12	Storm fore sail		
Fore top stay		11	9½	1	15	12
bunt			3	Ringtail sail		
Middle stay	10	10		4½	10½	25
aft leech			9	Water sail		
bunt			2	5	5	15
Main topgallant stay	9	9		First jib		
aft leech			9	1	22	20
Main studding	8	8	9	Second jib		
Fore studding	7	7	7	1	20	19
Main top studding	4	8	10½	Third jib		
Fore top studding	3	7	10	1	16	16
Flying jib (jib)		13	12	Storm jib		
				1	10	10

Merchantman:	*1200 tons*			*700 tons*		
Name of sail	*(1)*	*(2)*	*(3)*	*(1)*	*(2)*	*(3)*
Main course	40	42	13	36	38	11½
Fore course	37	37	11½	34	34	10
fore leech			9			7½
Mizzen course	16	17		15	17	
aft leech			17½			16½
Main top	25	39	16½	22	36	15½
Fore top	24	38	16	20	34	14
middle			11½			11
Mizzen top	18	28		16	23	
leech			12½			12
Main topgallant	18	28	9¾	15	23	7½
Fore topgallant	17	27	9¼	14	21	6½
Mizzen topgallant	12	19	7½	10	16	5½
Main royal	12	18	7¾	9	15	5½
Fore royal	11	17	6¾	8	14	4½
Mizzen royal	7	12	5½	5	10	3½
Main stay	26	28	13¼		22	11
Fore stay	18	20	11¾		16	10
bunt			5½			5
Mizzen stay	19	19		15	17	
aft leech			11			11
bunt			9			7
Main top stay	21	23		20	22	
aft leech			23			19
Fore top stay	17	17			14	13½
bunt		3			3½	
Mizzen top stay	14	14		12	13	
aft leech			8			9
bunt			6			4½
Middle stay	20	20		17	17	
aft leech			10			10½
bunt			5			5
Main topgallant stay	18	18		13	14	
aft leech			10			9
Main studding	20	20	13	14	14	12½
Fore studding	19	19	11	13	13	10½
Main top studding	10	17	16½	10	14	16½
Fore top studding	9	16	15	9	13	15
Main topgallant studding	7	10	10½	6	10	8½
Fore topgallant studding	6	9	9	5	9	7½
Flying jib (jib)		22	23		17	18
Sprit course	25	25	6½	23	23	6
Sprit top	17	26	12	16	24	10
bunt			9			7½
Driver	22	28		15	26	
aft leech			18			20¾

Merchantmen:	*500 tons*			*350 to 360 tons*		
Name of sail	*(1)*	*(2)*	*(3)*	*(1)*	*(2)*	*(3)*
Main course	24	24	10	18	20	9½
Fore course	22	22	9	16	16	8½
fore leech			7			6½
Mizzen course	9½	11		8	9	
aft leech			14			12½
Main top	16	22	12	15	22	11½
Fore top	14½	21½	12	14	20	10
middle						9
Mizzen top	13	18½	8¼	12	17	
leech						9½
Main topgallant	14	18	8½	13	17	7
Fore topgallant	12	16	7½	12	16	6½
Mizzen topgallant	10	12	5	9	11	6
Main royal	8	14	6½	7	13	5
Fore royal	7	12	5½	7	12	4½
Mizzen royal	5	10	3	4	9	4
bunt						1½
Main stay		12	12	13	14	
aft leech						9½
Fore stay		12	9		9	8
bunt			3			4½
Mizzen stay	12	12		9	10	
aft leech			8½			10
bunt			6½			6
Main top stay	13	13		12	13	
aft leech			12			14
Fore top stay		10	9		10	10
bunt			2½			3½
Mizzen top stay	10	11		9	10	
aft leech			9			10
bunt			5			3
Middle stay	10	10		12	12	
aft leech			9			7
bunt			2			2
Main topgallant stay	9	9		8	8	
aft leech			6			6
Main studding	13	13	10½	12	12	10½
Fore studding	11	11	9	11	11	8½
Main top studding	7	9	13½	5	11	13
Fore top studding	6	8	12	4	10	11
Main topgallant studding	4	6	7½	4	8	8
Fore topgallant studding	3	5	6½	3	7	6½

Flying jib (Jib)		15	16		12	12
Sprit course	20	20	5½	18	18	4½
Sprit top	13	21	9	11	19	8
bunt			7			6
Driver	11	15		12	16	
aft leech			13			15
Boats: Lateen		7	6½	Fore sail	3	4½
bunt			1			3¼
Settee & Lug	10	11		5	7	
aft leech			7			5
Sprit (bunt & aft)	5½	6½	5¼			7¼

Table 57: Dimensions of the sails of HMS Royal Sovereign (100 guns) 1701 (after Davis, 1711)

(1) Number of sails
(2) Number of cloths per sail
(3) Depth of sail in yards

Name of sail	*(1)*	*(2)*	*(3)*
Spritsail course	1	30	8½
Spritsail topsail	1	21½	12
Fore course	1	47	13
Fore topsail	1	33	18
Fore topgallant sail	1	16½	8½
Main course	1	54	15
Main topsail	1	38½	20
Main topgallant sail	1	20	10
Mizzen course	1	16	21
Mizzen topsail	1	27	21
Fore top staysail	1	11½	14
Main top staysail	1	13	16
Main staysail	1	15	15
Mizzen staysail	1	13	13
Main studdingsails	2	10	17
Main top studdingsails	2	9	22

Table 58: Depth and width of sails according to ship's Rate (after Korth, 1826)

(1) Depth of the sail in Prussian ells (670mm)
(2) Number of cloths in the foot
(3) Number of cloths in the head
(4) Sail's total length of cloths in Prussian ells

Ship:	*First Rate*				*Second Rate*			
Name of sail	*(1)*	*(2)*	*(3)*	*(4)*	*(1)*	*(2)*	*(3)*	*(4)*
Main course	19	40	40	760	17	34	34	578
Main top	28	39	24	914	26	33	21	702
Main topgallant	13	23	16	247	11	21	15	198
Main stay	20	20		200	19	20		190
Main top stay	19	20		190	18	17		153
Main topgallant stay	16	15		120	14	13		91
Fore course	18	36	36	648	17	30	30	510
Fore top	25	35	21	700	23	29	19	552
Fore topgallant	10	21	14	175	9	19	13	144
Fore stay	20	17		170	17	16		136
Fore top stay	18	17		153	16	15		120
Fore topgallant stay	27	15	1	216	25	14	1	187
Mizzen course	25	24		300	24	23		276
Mizzen top	14	23	16	273	13	21	15	234
Mizzen topgallant	7	16	10	91	6	15	9	72
Mizzen stay	19	18		171	18	17		153
Mizzen top stay	18	17		153	16	15		120
Spritsail course	10	23	23	230	9	21	21	189
Spritsail top	14	23	16	273	14	21	15	252
Length of cloths in a suit				5958				4857

Ship:	*Third Rate*				*Fourth Rate*			
Name of sail	*(1)*	*(2)*	*(3)*	*(4)*	*(1)*	*(2)*	*(3)*	*(4)*
Main course	17	32	32	544	16	30	30	480
Main top	25	31	21	650	24	29	19	576
Main topgallant	10	20	14	170	10	18	12	150
Main stay	18	19		171	17	18		153
Main top stay	17	16		136	16	16		120
Main topgallant stay	13	12		78	12	11		66
Fore course	16	28	28	448	15	26	26	390
Fore top	22	28	18	506	20	24	16	400
Fore topgallant	9	18	12	135	9	17	11	126
Fore stay	16	16		128	15	14		105
Fore top stay	15	14		105	14	13		91
Fore topgallant stay	24	13		168	22	12	1	143
Mizzen course	23	22		253	22	21		231
Mizzen top	12	20	14	204	12	19	13	192
Mizzen topgallant	6	14	8	66	5	13	7	50
Mizzen stay	17	16		136	16	15		120
Mizzen top stay	15	14		105	14	13		91
Spritsail course	8	20	20	160	8	19	19	152
Spritsail top	13	20	14	221	13	19	19	208
Length of cloths in a suit				4384				3844

Ship:	*Fifth Rate*		*Sixth Rate*					
Name of sail	*(1)*	*(2)*	*(3)*	*(4)*	*(1)*	*(2)*	*(3)*	*(4)*
Main course	15	28	28	420	12	23	23	276
Main top	22	27	17	484	17	22	14	306
Main topgallant	9	15	10	127	8	14	9	92
Main stay	16	16		128	14	14		98
Main top stay	15	14		105	13	12		78
Main topgallant stay	11	10		55	9	8		36
Fore course	12	21	21	252	11	20	20	220
Fore top	17	20	14	289	15	19	11	225
Fore topgallant	8	13	9	88	7	12	8	70
Fore stay	14	13		91	12	11		66
Fore top stay	13	12		78	11	10		55
Fore topgallant stay	20	12	1	130	18	10	1	99
Mizzen course	20	19		190	19	18		171
Mizzen top	10	16	11	135	9	14	10	108
Mizzen topgallant	5	10	6	40	4	9	5	28
Mizzen stay	15	14		105	13	12		78
Mizzen top stay	13	12		78	11	10		55
Spritsail course	7	15	15	105	6	14	14	84
Spritsail top	11	14	12	143	10	13	9	110
Length of cloths in a suit				3043				2250

Table 59: Dimensions of sails on a French ship of 110 guns in the latter part of the eighteenth century (after Paris)

Measurements are in metres:
(1) Length of head (length of after leech for main staysail and following sails)
(2) Depth of sail (length of stay or hoist for main staysail and following sails)
(3) Length of foot
(4) Sail area in square metres (length of bunt for main staysail and following sails)

Name of sail	*(1)*	*(2)*	*(3)*	*(4)*
Main course	32.48	13.97	35.73	476
Fore course	28.26	12.36	25.34	320
Main topsail	21.65	20.32	31.28	547
Fore topsail	19.50	18.68	27.01	435
Spritsail course	21.11	10.72	21.11	227
Mizzen topsail	16.24	13.48	21.61	255
Main topgallant sail	16.24	10.72	21.00	200
Fore topgallant sail	14.79	9.58	18.95	161
Main royal sail	10.44	5.30	15.43	68
Fore royal sail	10.07	4.55	13.97	52
Mizzen topgallant sail	12.99	7.80	15.43	111
Spritsail topsail	14.78	12.67	20.46	210
Standing jib	37.35	35.34	13.33	236
Second jib	33.13	34.44	11.37	182
Inner jib	15.91	13.98	13.32	88
Main top studdingsails	7.63	21.92	15.28	251
Fore top studdingsails	7.15	20.14	14.63	243
Main staysail	21.48	16.29	15.75	
Mizzen staysail	19.81	15.42	14.62	
Mizzen course	17.86		12.51	16.58
Main topmast staysail	11.79	24.69	13.33	9.76
Middle staysail	16.84	19.49	11.79	7.79
Main topgallant staysail	14.62	16.89	9.42	5.85

Figure 149
The complete rigging plan of an English Second Rate man-of-war of about 1700. The rigging shown is in accordance with the spar and rigging plan of the St. George by Anderson, with additional information from a large number of contemporary pictures and model photographs.

The sprit course's horizontal reef band can be found on Establishment rigging plans from 1719. Up to the end of the seventeenth century no reef-bands were carried on sprit-courses. The exact date of their introduction in the Royal Navy cannot be established, but was close in time to their first apearance on French sails; it should be noted that these reef bands were shown by Paris on rigging plans of the end of the seventeenth century. Studdingsail booms on the lower yards and staysails are omitted in this drawing for reasons of clarity.

XI

Rigging of sails

After the rigging of masts and yards had been completed, the task of bending the sails to the yards followed. Squaresails were bent to with robands, or fastened with lacings (the latter mostly to smaller yards or to gaffs) while staysails were generally seized to hanks or grommets on the respective stay to make possible the controlled hoisting of the sail.

Robands, rope bands

Anschlagbändsel, Rahbanden; Garcettes

At the beginning of the eighteenth century a roband was a single rope, or braided cordage; it was middled and its bight was pushed through an eyelet in the head tabling, then one end was led through the bight to the aft side of the yard and the other around the fore side, and the ends were bent tightly together above the yard (see Figure 152a). As an alternative, the roband was led singly through the eyelet and secured above the headrope with a throat seizing (see Figure 152b).

According to Falconer, robands of either type were of sufficient length in the 1750s to pass twice or three times around the yard (see Figure 152c).

Continental robands were neither middled nor seized, according to Röding and Boudriot. The former indicated that robands were long enough to pass through the head eyelet and twice around the yard, to be secured in the end above the yard

with a clove hitch. The latter described the roband as pushed through the eyelet and laid around the yard, with the ends then led back through the eyelet from the other side and bent tightly to the yard.

In English rigging the last decades of the eighteenth century saw a change to a roband of two parts, described by Lever as follows:

> There is a long and short Rope-band, with an Eye in each, to every Eyelet hole in the Head of the Sail. The end of the short Leg goes through the hole in the Sail, the Eye being before it: the end of the long Leg, is reeved through the Eye of the short one, and the end of the short one, through the Eye of the long one.

These robands had a total length of 7 to 9 feet (see Figure 152d).

Gaskets, furlinglines

Beschlagzeisinge, Beschlagbindsel; Rabans de ferlage, lignes de ferlage

Gaskets were made from braided cordage (plaiting) and served for the fastening (furling) of clewed-up sails. Furlinglines served the same purpose, but were made of ordinary rope. Falconer explained that the latter were only in use on smaller sails.

These gaskets or furlinglines, with an eye in one end and a reducing tail in the other, were usually bundled up and hung on the fore side of the sail. Four to eight gaskets were fitted, depending on the sail size; Röding noted eight to twelve for the lower yards of large ships. Gaskets for large sails were made from nine, and those for smaller sails from seven, foxes (foxes were three or more rope-yarns twisted tightly together). The length of these gaskets was 5 to 7 fathoms (30 to 42 feet) according to Steel, and Falconer described them as long enough to wind equally six or seven times around the yard and sail. He also indicated specific quarter, yardarm and bunt gaskets.

> [The] Quarter-Gasket is used only for large sails, and fastened about half way out upon the yard, which part is called the quarter.
>
> [The] Yard-arm Gasket is made fast to the yard-arm, and serves to bind the sail as far as the quarter-gasket on large yards, but extends quite into the bunt of small sails.
>
> [The] Bunt-Gasket is that which supports or ties up the bunt of the sail, and should consequently be the strongest, as having the greatest weight to support; it is sometimes made in a peculiar manner.

Bunt gaskets were mentioned as early as Smith's *Sea Grammar* of 1627, but only as being longer then the normal gaskets. Steel and Röding did not regard bunt gaskets as different from others, but noted that they were applied crossed like an X, to take in the bunt. Only Lever in 1811 and Falconer in 1815 gave

Figure 150
The complete rigging plan of a French Second Rate man-of-war from about 1700, based on the sail plan of Le Fendant of 1701. This plan, published in Paris' Souvenirs de Marine, is of a ship of the Second Rate, second order. The plan, in some areas sparse on details, was supplemented by other information from the same work, and allows a comparison with Figure 149, illustrating the differences between an English and French (Continental) rigging. Note the different cut of the sails and the differences in the arrangement of stays, buntlines, jeers and caps, to name but a few. Yard-tackles, studdingsails and staysails are not shown for reasons of clarity.

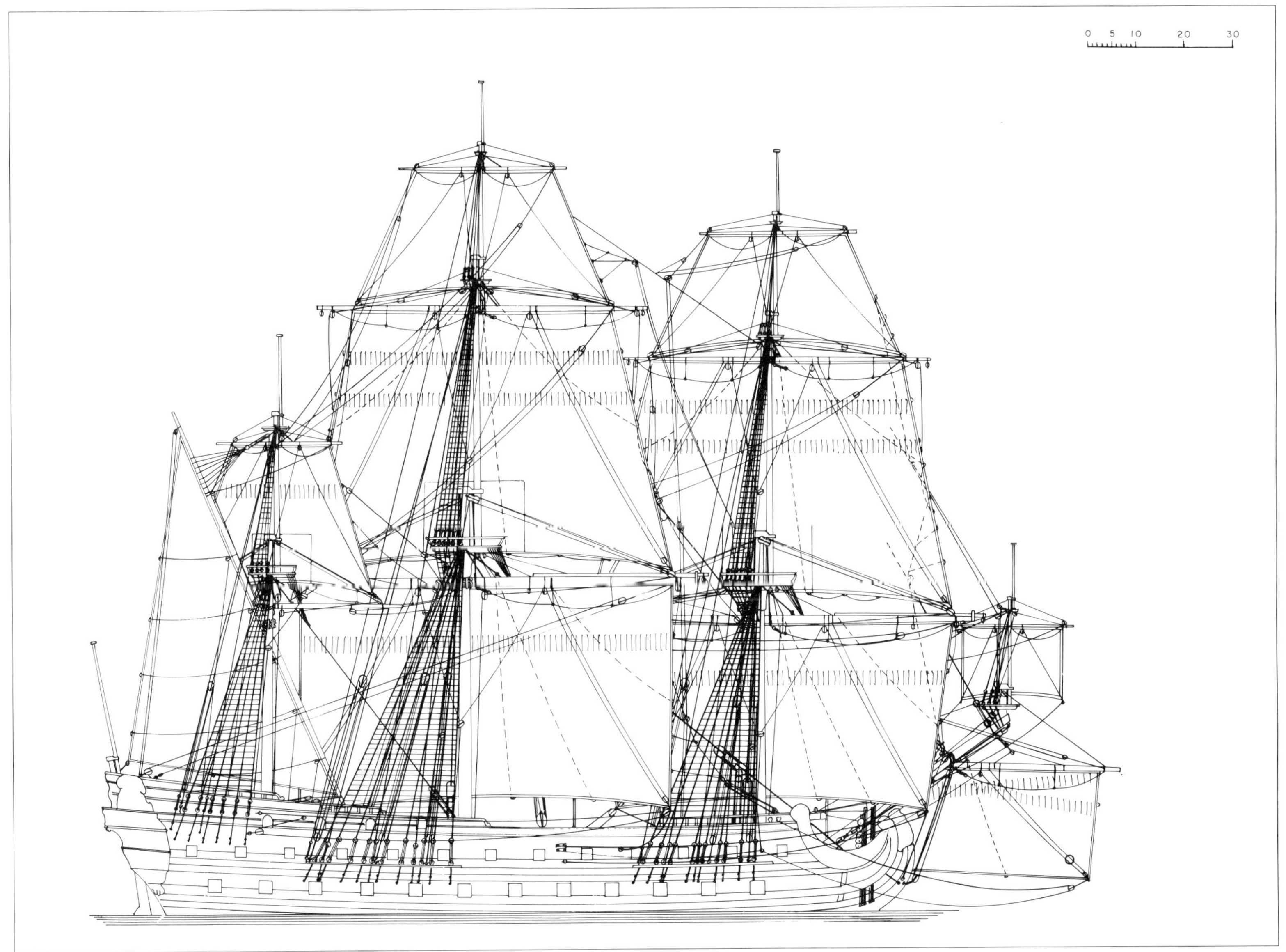

an alternative; the V arrangement they described therefore seems to have been an innovation of the end of the century.

The fastening of gaskets to the yard in English ships was done with grommets at the beginning the century (see Figure 152k). Blanckley noted 'Gromets are small Rings formerly fastened with Staples to the Yards, to make fast the Gaskets, but now never used'. Falconer (1769) mentioned only fastening to the yard. They were then probably stapled direct to the yard, with the gasket's eye at the other side of the stapled part, to enable the end of the next gasket to be secured to it.

Earrings

Nockbändsel; Rabans de pointure

Small ropes known as earrings were spliced into the earring cringles to extend the upper corners of a sail to its respective yard or gaff. During the process of bending a sail, these earrings were led twice around the yardarms and through the cringles, outside the yardarm cleats and their rigging, to stretch the sail (see Figure 152i). Then another five to seven turns were laid direct at the cleats around the yardarms (that is, within the yardarm rigging) to fix the sail's position; the two turns were known as the outer turns and the others the inner turns. Earrings had a length of about 5 to 6 fathoms.

Each reef band cringle had its own reef earring. The only difference between the nock and the reef earring was that the former was spliced into the cringle and the latter had an eye spliced into one end to make it easier to reeve or unreeve when the earring was chafed.

As with many parts of the rigging of a ship, the Continental method of using earrings was slightly different from the English. Several turns of the earring were laid like the outer turns and, according to Röding, another earring, the head earring (*Binnenbindsel* in German; *raban de croisée* in French), was set up within the yardarm square, more or less vertical.

Hanks, grommets

Stagreiter, Staglägel, Sügers; Cercles, anneaux

Hanks or grommets were used for the connection of a triangular or quadrilateral sail to a stay. During the earlier part of the century simple rope rings (grommets) were used, but the mid-century saw the introduction of wooden hanks (according to Falconer), which in later years were partly replaced by iron hanks (see Figure 152l and m). Steel mentioned both types in 1794.

While Falconer mentioned only wooden hanks as a new invention in 1769, iron hanks were already fitted to the original staysails of the contemporary model of the Russian First Rate *Zacharii i Elisaveth*, built in 1745-7, and in Continental sources references to iron hanks go back to the very beginning of the century. The author of *Der geöffnete See-Hafen* referred to the use of hanks of iron, timber or rope in connection with the hoisting of a main staysail. This suggests that iron and wooden hanks had been introduced in Continental shipping at a time when English ships still had only grommets fitted. Röding explained the use of these devices:

> Iron hanks are used thus on heavy staysails, but despite their easier running compared with wooden hanks they are not often used, because of their tendency to chafe the ropes badly. One can still find them in some koffs and smacks....

His comment suggests that iron hanks did not meet with the same approval during the eighteenth century as they later did with the iron and steel wire rigging of the nineteenth century.

Where a rope was led through the stay eyelet hole and over the stay to form a grommet, wooden hanks rode on the stay,

Figure 151
The rigging plan of a Continental ship according to Der geöffnete See-Hafen of 1705 and to several Dutch and German prints of that time. The fore topsail yard halyards might alternatively have been rigged like the mizzen topsail yard.

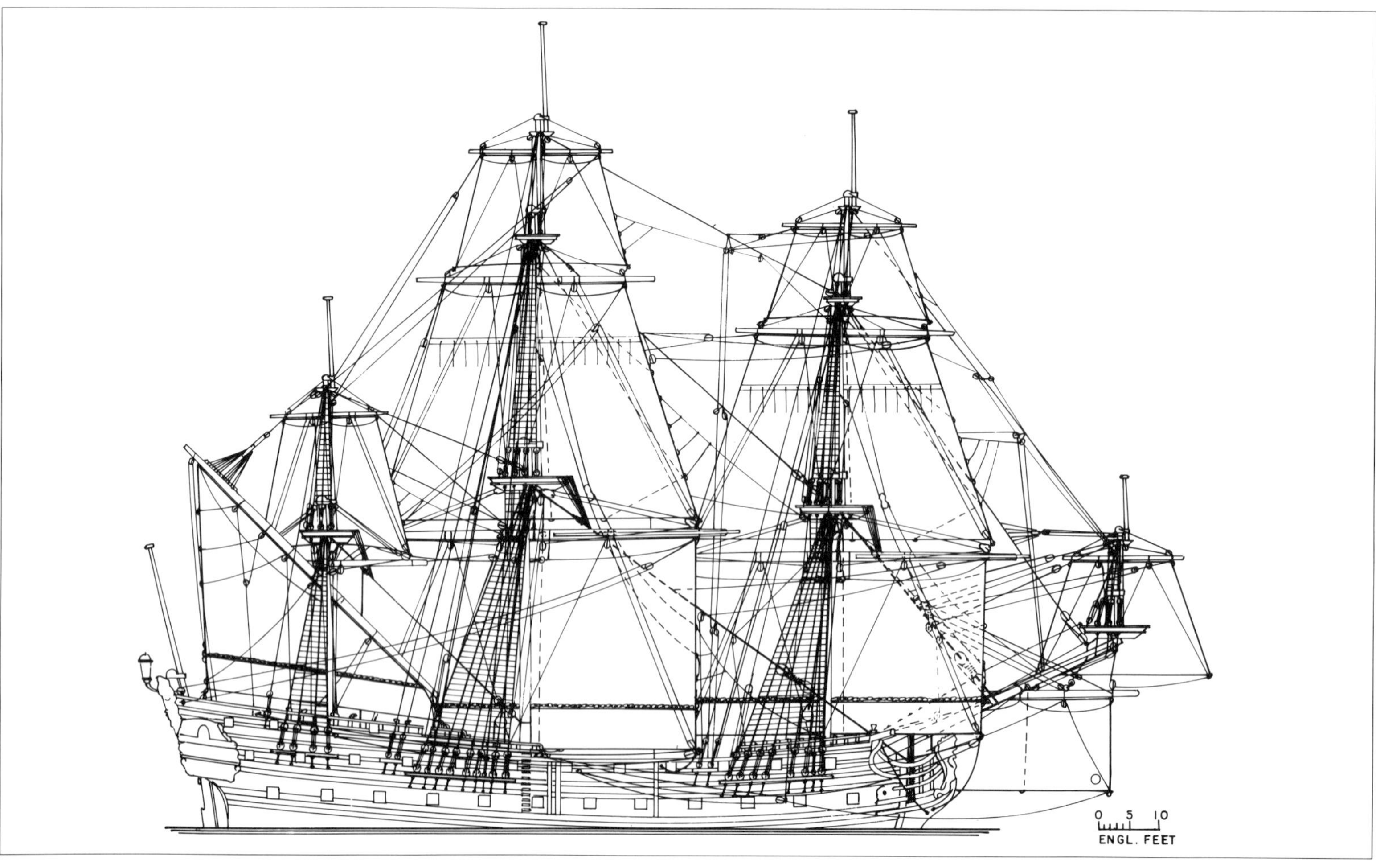

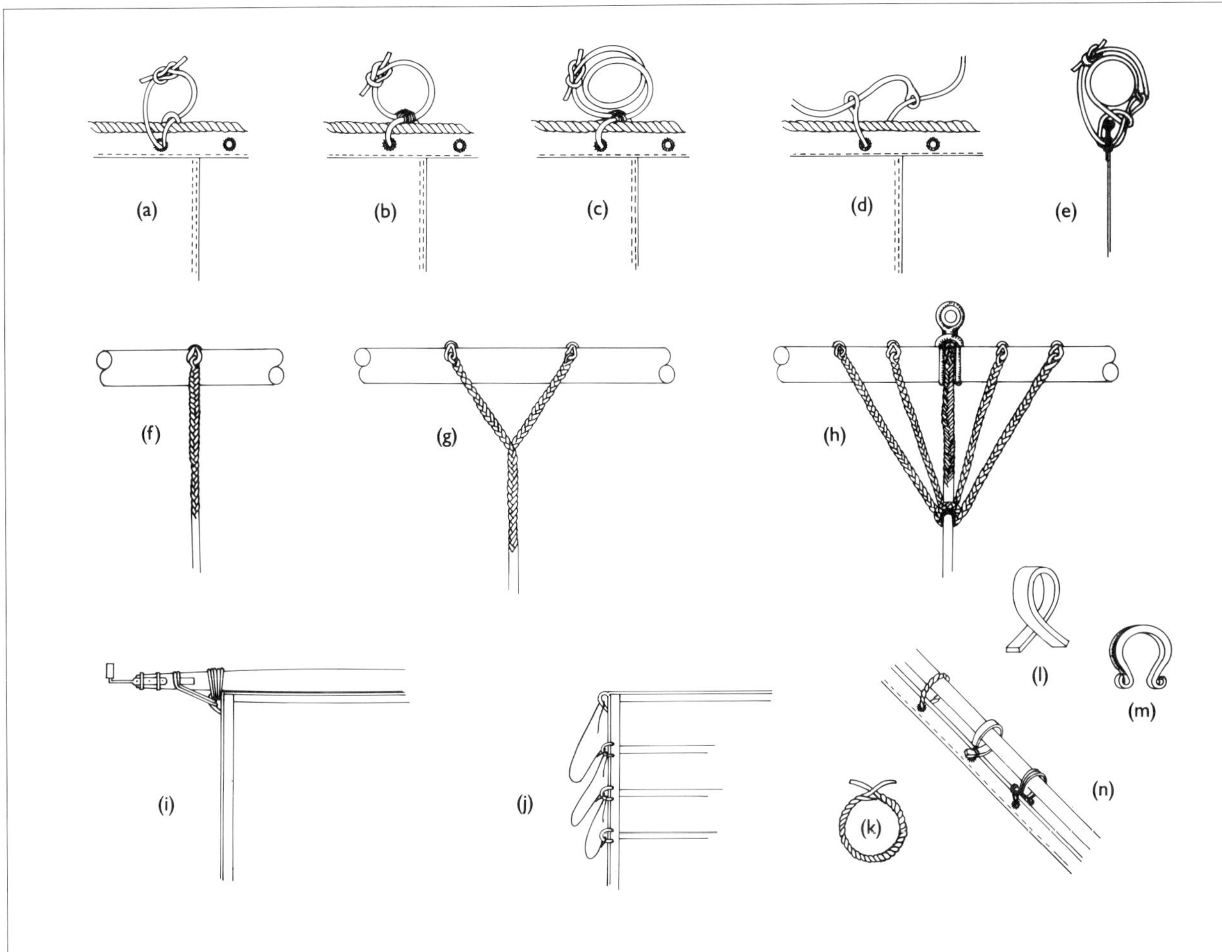

Figure 152
(a) A single roband of the seventeenth and eighteenth centuries. The roband was not simply led through the eyelet; instead, a bight has been put through, and the longer part led through the bight. Both ends were then clove hitched at the upper fore-side of the yard.
(b) A single roband with a throat seizing.
(c) A seized roband with a double turn about the yard.
(d) A roband in the last decades of the century, according to Steel and Lever.
(e) A roband middled and doubled about the yard.
(f) A gasket.
(g) A bunt gasket, V-shaped the better to take in the bunt of a sail.
(h) A bunt gasket according to Lever. The diagonal strops took the bunt in, while the vertical strop, running freely through a thimble, operated as a reeving gasket to bring the bunt even closer to the yard.
(i) An earring.
(j) Reef earrings.
(k) A grommet for the fastening of a staysail in the earlier part of the century.
(l) A wooden hank of about 1750.
(m) An iron hank of the second half of the century.
(n) The various methods of bending a staysail to its stay.

with their crossed ends seized together through the eyelet holes and over the hoist rope. Iron hanks were horseshoe shaped, with eyes turned outward at their ends and a stabilising channel along their back. Spun yarn or houseline was spliced round one eye, the line was thrust through the eyelet, through the other eye, than back through the hole and the first eye and so on until extended (see Figure 152n).

Lacing

Reihleine, Lissung; Raban d'envergure

Lacing was 'the rope or line used to confine the heads of sails to their yards or gaffs' according to the 1815 edition of Falconer; Steel elaborated somewhat:

> [Lacing is the] fastening [of] the head of a sail to a mast, yard, gaff, etc, by a line turned spiralling round them, and reeved through the eyelet-holes in the sail. When a sail is laced to a mast, it is best to take cross turns, backwards and forwards, on the fore-side of the mast only, so that the sail may slide up or down.

He also mentioned for a the mast lacing of a mizzen course either through eyelet holes or through cringles. In both cases the lacing was spliced into the throat (nock) earring. During the period in which a cut-off mizzen course was still carried on a full mizzen yard, eyelet holes were used for the mast lacing.

Lever explained the differing use of holes or cringles when the sail was laced to a gaff:

> If the Gaff be slung, Cringles are worked in the Mast Leech at equal distances, but if the Gaff traverse, then Eyelet Holes are worked in... The Lacing for the Mast Leech is spliced into the throat Cringle, goes round the Mast, and through the Cringles in the Mast Leech, backwards and forwards, hitching to the Tack. When the Gaff traverses up and down, the Eyelet Holes in the Mast Leech are fastened to Hoops which go round the Mast [see figure 153c].

This method was already used earlier in the century; the earlier edition of Falconer had noted, in connection with smaller vessels, '...sails whose foremost edges are joined to the mast by hoops or laceings'. The lacing of a mizzen course with mast leech holes was put on using the spiral method, the same as was used for the lacing of a gaff or yard.

The length of a mast lacing was one quarter of the mizzen mast's length, taking fathoms for feet, and that of a mizzen yard was equal to a mast lacing plus 7 fathoms; the respective circumferences were 3 inches and 1½in (see figure 153b).

Reef point, knittle

Reffbändsel; Garcette de ris

Reef points were usually longer than robands, and for most of the eighteenth century they were in one piece (see Figure 153d). An overhand knot secured the point at each side of the reef band eyelet hole. About two thirds of the reef point hung at the rear (inner) side of the sail, and one third at the fore (outer) side.

Reef points were plaited with seven foxes in the middle, with the foxes thinned out towards the ends, and the ends were whipped. Their length was about twice the circumference of the yard or, according to Steel, 6 to 9 feet.

Alternative reef points were introduced near the end of the century, made in what was then called 'rope-band' fashion. These came in two parts, one long and one short, with an eye in one end of each leg like a gasket; the eyes were oblong. Both parts were put through the reef band hole, one eye on the fore

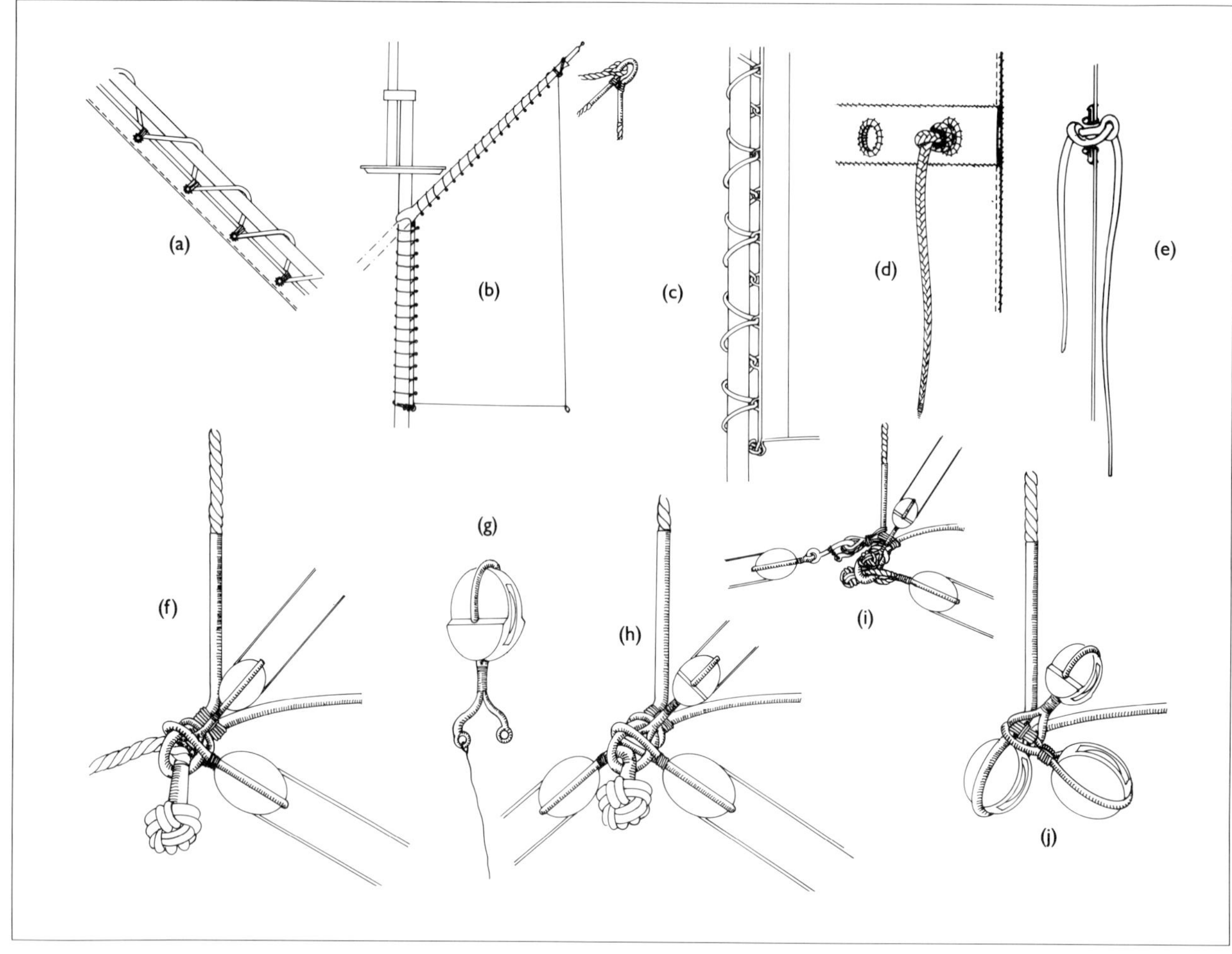

Figure 153

(a) The lacing of a staysail to the stay, as shown frequently in modern works. This is an assumed method of fastening during the later part of the seventeenth century, as noted by Anderson and Höckel (perhaps based on Anderson) and followed up by other modern authors. As far as can be ascertained by studying original sails in contemporary models (for instance the Dutch two-decker of 1660-70 examined by Winter), staysails were fastened with grommets, an observation corroborated by all eighteenth and early nineteenth century literature. All contemporary observers noted grommets and hanks, which suggests that lacing is unlikely for this period.

Lacing was first shown in Nares' Seamanship *of 1862, and then only for jibs and flying jibs; he noted moreover that even for these sails the uppermost stay hole was fitted with a grommet. For all other staysails hanks were used.*

(b) The lacing on a gaff or mizzen yard (spliced into the peak earring), and a mast lacing (spliced into the nock earring) led in a spiral through the bunt eyelet holes, or bunt cringles, and hitched to the tack. The lacing was seized to each eyelet hole to prevent a complete unreeving when it was damaged.

(c) Mast lacing through bunt cringles, after Lever.

(d) A braided reef point, fastened with an overhand knot to both sides of a reef band hole.

(e) A two-part reef point from the end of the century.

(f) A detail showing the connection of a tack with the sheet and clewgarnet blocks before 1720. The last was a shoulder block after that date. This arrangement was followed in English ships until the 1790s, and in Continental ships up to the middle of the century or in some instances even later.

(g) A stropped clewgarnet block with spliced-in eyes. In the later years of the century this was used in English rigging.

(h) The Continental method of fitting a tack block instead of a single tack, used from about mid-century onwards.

(i) Another Continental clew arrangement of about 1750. The tack block was hooked into a thimble attached to the clew, and the sheet block had two strops ending in stopper knots.

(j) The arrangement of an English clew from about 1795, with a tack block seized into the clew. The tack block strop might also have had stopper knots in both ends, which were thrust through the clew. The clewgarnet block was sometimes already without a shoulder at this time.

and one on the rear side, with the thin end of the fore part going through the eye on the rear, and vice versa (see Figure 153e).

A knittle was a small line, plaited or twisted, to be used for many purposes at sea. They were used in all sails with reef bands in the bottom section. Such reef bands did not have permanent reef points and knittles took their place temporarily when a reef had to be taken in.

For the operation of the sails a number of ropes and blocks were needed, and the following Chapter will describe these in the same sequence as was used in Chapter X for the sails themselves.

XII

Running rigging to sails

Main course

Großsegel; Grand-voile

We begin with the preparation of the clews, the lower corners of the sail, to which the clewgarnet blocks, the sheet blocks and the tacks were attached.

Clewgarnet block, cluegarnet block

Geitaublock; Poulie de cargue-point

In the Royal Navy the clewgarnet block was a single block without any special feature until about 1720, and only after that time was a specialised clewgarnet block with a shoulder used. It remained a single block throughout the century in the Continental rigging, and returned to this status again in English rigging from the early 1780s onwards. However, the shoulder block was still shown by Lever and by Rees in his work of 1819-20, which confirms the longevity of theoretically superseded items in rigging.

At the beginning the block was stropped with a closed strop and fitted with an eye over the loop formed by the clewrope. In its later form an open strop was occasionally used, with eyes spliced into both ends (see Figure 153g). This strop was put through the clew from the after side, and both ends were brought round the clewrope and seized together through the eyes. In some cases in Continental rigging stopper knots at the ends of the lower clewgarnet blocks were pushed through the clew together with the tack block strops (see Figure 153f).

Sheet block

Schotblock; Poulie d'écoute

Next a normal eye-stropped sheet block was fitted over the clew. Especially when the sail carried a bonnet, sheet blocks also sometimes had an open strop with stopper knots, for easy removal and fitting (see Figure 153i).

Tack, tack block

Hals, Halsblock; Point d'amure, poulie d'amure

The third and final fitting to the clew was the tack, or tack block

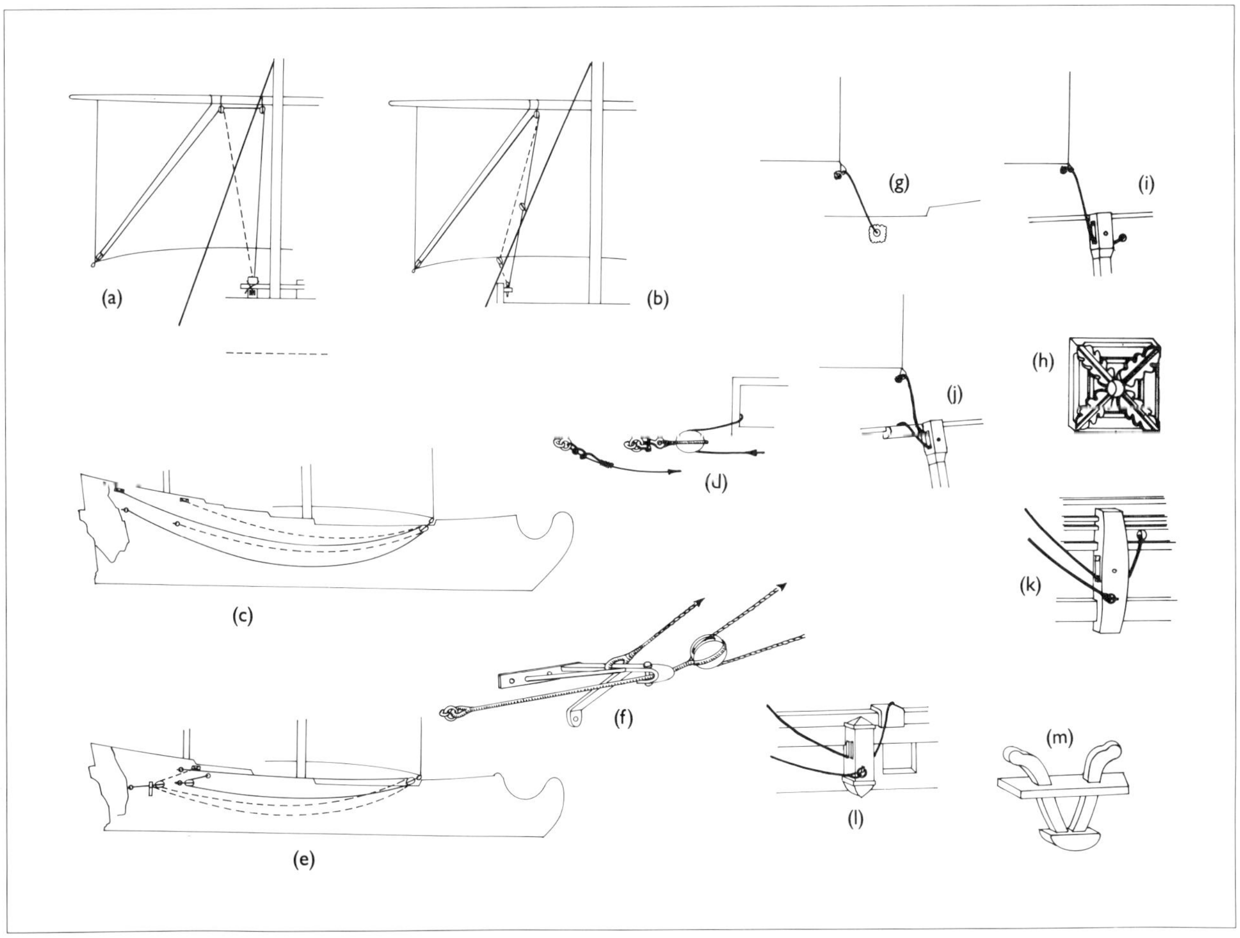

Figure 154

(a) The clewgarnet on an English course. The dashed line shows the lead during the last few decades of the century, without a lead block.

(b) The clewgarnet on a French course, with the dashed line showing again the revised lead in the last few decades of the century.

(c) An English main course sheet. The solid line indicates the lead until about 1735, with the dashed line showing that after this date.

(d) An alternative lead during the first four decades of the century in English ships.

(e) A French main course sheet. The solid line shows the lead for the first few decades, and the dashed line that for the later part of the century.

(f) An iron spread-bracket for main course sheets in French ships after about 1780. English ships did not carry these until the beginning of the nineteenth century.

(g) The lead of a main tack, about 1700.

(h) An embellished lead hole for a main tack, typical of English ships until about 1706 and of Continental ships until mid-century.

(i) An English main tack after 1706, led through a chess tree.

(j) A Continental main tack about 1750, led through a chess tree and over a bolster.

(k) A French main tack lead of about 1780, showing the standing part made fast to an eyebolt in the chesstree.

(l) A Continental main tack lead of about 1790.

(m) A kevel, used as a belaying cleat for heavy ropes such as the tack and sheet.

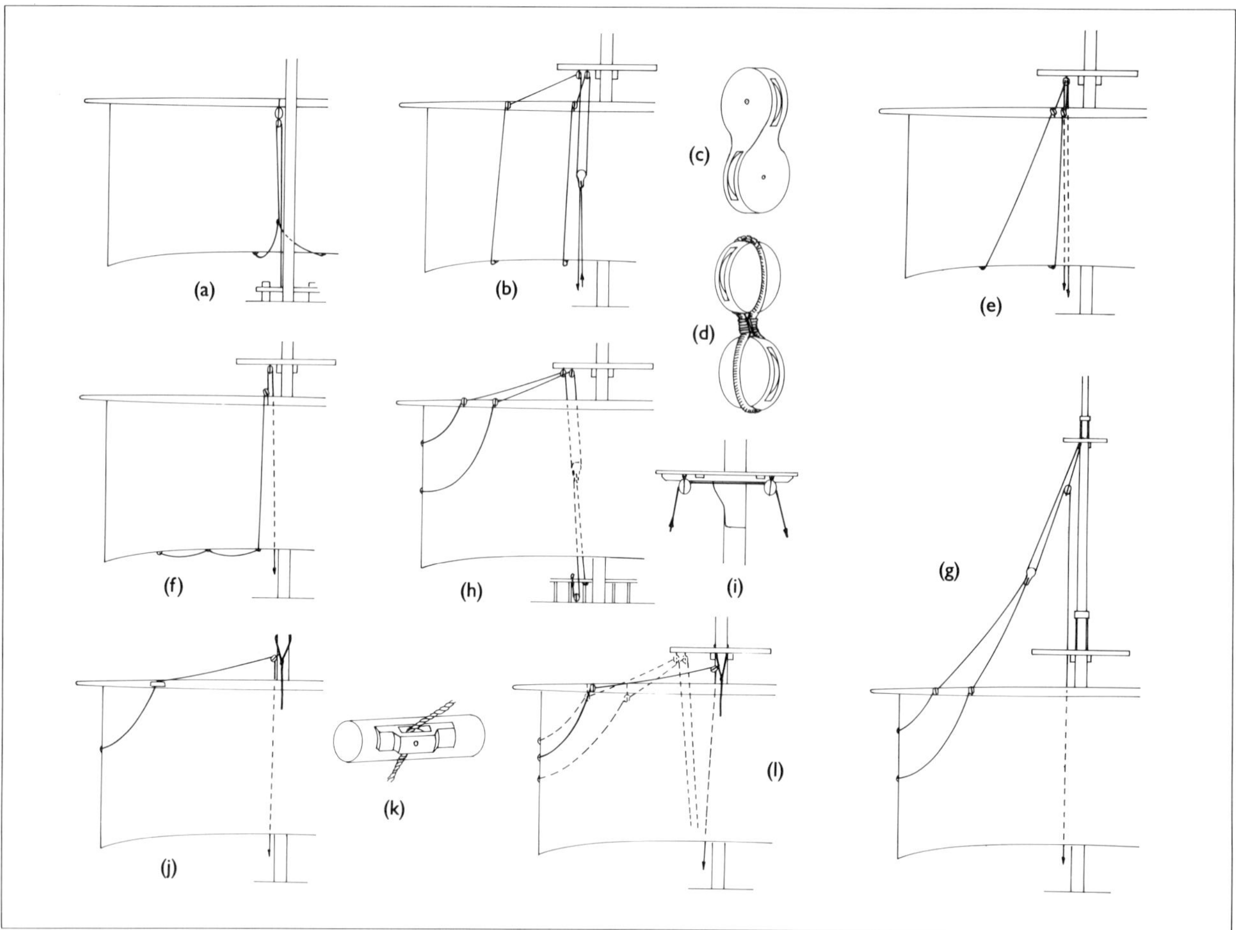

Figure 155
(a) A slab line.
(b) The buntline of a main course on an English man-of-war. The buntlines on merchantmen were mostly rigged in the Dutch fashion, single and without a shoe block, leading aft like those of the fore course.
(c) A shoe block, used in the Royal Navy until about 1775.
(d) Single blocks, stropped one on top of the other, replaced the shoe block.
(e) Main course buntlines of French origin during the first few decades of the century.
(f) Buntlines on a French main course, about 1780.
(g) English main course leechlines, up to about 1770.
(h) English main course leechlines from about 1720 onwards (initially used only in small ships).
(i) Leechline lead blocks beneath the top.
(j) A Dutch leechline with a type of cheek block.
(k) A cheek block.
(l) The French method of rigging a main course leechline; the solid line shows the lead at the beginning of the century, and the dashed line that in the last decades.

(see below under Tack). A single tack was pushed through the clew with a double wall-knot with a crown at the end (see Figure 153f). Around the mid-1790s in English rigging the single tack was replaced with a tack block, seized into the clewrope (see Figure 153j); in Continental rigging such a tack block could be found before mid-century. Continental blocks were stropped with a knot similar to that of a single tack (see Figure 153h) or sometimes with a toggle, or with a hook through a thimble stropped to the clew.

Clewgarnet
Geitau; Cargue-point

The difference between the terms 'clewgarnet' and 'clewline' is not a matter of double or single rig, but rather that the former is solely applied to the main and fore courses, while the latter pertains to all other squaresails.

In English rigging a clewgarnet was made fast to the rear of the yard about one third of the yard length outside the slings, using a timber hitch. It then rove through the clewgarnet block at the clew and through another block at the yard, slightly inside the standing part. Throughout the century this block was a shoulder block, except in the Royal Navy, where during the last quarter century a normal block was used. The clewgarnet then passed through a lead block near the slings and down upon deck, over a sheave in the bitts before the mast, and was belayed (see Figure 154a). The lead block disappeared during the last few decades of the century.

In the French Navy the clewgarnet, after passing the block on the yard, had a different lead (see Figure 154b). The lead block was not placed on the yard, but lashed to the foremost shroud at one third of its height at the beginning of the century, and at one tenth of the height in the last decades.

In Continental rigging around 1700 the clewgarnet was made fast to the yard at about one quarter of the yards's length, and the block was approximately 6 inches inside that position. The clewgarnet was then led down to the foremost shroud. A similar description to that given in *Der geöffnete See-Hafen* was provided by Röding. He had the clewgarnets made fast to the yard blocks, which were seized to the yard at one quarter or one third of its length inside the yardarms. After passing the clew-garnet block and the block on the yard, the clewgarnet rove 'through another [block] in the lower shrouds on to the ship, where it belayed to a cleat'. Boudriot's explanation reflects Röding's, with the exception that in his description the clew-garnet was made fast to the yard close to the yard block, and after passing the lead block, it went through a foot block before belaying to a pinrail. The foot block was found only on larger vessels.

Sheet
Shot; Écoute

The sheet was led through the sheet block. In English rigging up to the mid-1730s the standing part was bent to an outboard ringbolt near the quarter gallery, at the height of the quarter-deck. After that time the standing part was made fast at upper deck level and closer to the mizzen mast (see Figure 154c).

French ships at the beginning of the century, and also German vessels according to the author of *Der geöffnete See-Hafen*, had sheet standing parts made fast at quarterdeck level abreast of the mizzen mast, and near the end of the century at upper deck level close to the gallery. Around 1780, according to Boudriot, the standing part was bent to a ringbolt below the aftmost gunport near the middle deck gallery (see Figure 154e).

Up to the mid-1730s the hauling part of the sheet in English rigging rove either through a lead block hooked to another ringbolt just forward of the standing part, and from there inboard through the nearest gunport, or through a sheave hole near the ringbolt for the standing part. When the standing part was made fast abreast of the mizzen mast, this sheave hole was placed just forward of the last step in the quarter sheer. Inboard, the sheet was belayed to a kevel. Steel noted that the standing part was seized to an eyebolt with a thimble on the quarters, and that the leading part rove through a sheave hole on the same side under the quarterdeck, and was belayed to a range cleat in the waist.

Continental arrangements differed from the English. In French rigging early in the century a block was hooked to a ringbolt just below the standing part at quarterdeck beam height, and this led the hauling part to a sheave hole near the last step in the quarter sheer, where it was belayed on the inside to a kevel. Later, in the second half of the century, this block was fitted to an iron sheet bracket or outrigger (see Figure 154f). Alternatively, according to Boudriot, a long-stropped block was made fast to an eyebolt above the standing part at the forward edge of the gallery and was kept near horizontal by this bracket. In another Continental alternative the standing part was bent to a ringbolt forward of the mizzen channel, with the sheet coming inboard via a sheave hole as described above. In another early Continental description the sheet was made fast abreast of the mizzen mast to a ringbolt and came inboard through a sheave hole just forward of the standing part.

Röding noted of the sheets of lower courses in general that they ran 'from that point outboard, where the sail's clew can be found aft of the mast, when it is at its most extreme position while sailing close to the wind... The hauling part leads through a sheave hole in the same area, from which the sheet started', The hauling part was led inside and belayed at a kevel.

Tack

Hals; Point d'amure

A single tack was pushed into the clew of an English main course until the 1790s, tapering to half its largest circumference after about 10 yards; in the 1790s tacks were rigged double. French and other Continental ships had been rigged with double tacks some three to four decades earlier.

A single tack rove either directly through a hole (embellished by carvings - see Figure 154h) aft of the fore backstays, or over a sheave in the chess tree (a vertical piece of timber, fitted like a fender outboard), and then through a plain hole nearby (see Figure 154i). Steel noted the later use of a sheave hole instead of the latter plain hole: 'and through a sheave hole in the side, and belays to a large range-cleat in the aft part of the waist'. Continental rigging sometimes dispensed with the hole and led the tack inboard over a softwood bolster forward of the chess tree (see Figure 154j). The author of *Der geöffnete See-Hafen* supplemented his description of a main tack by noting that 'in storms a fourfold tackle is also set up, running through two blocks and over three sheaves, to support the tack and keep the sail more securely in place'.

The standing part of a double tack was clinched to an eyebolt forward of the chess tree (according to Steel), or at the chess tree (according to Röding and Paris), rove through the tack block, and came in as noted above (see Figure 154k). French men-of-war of the last few decades of the eighteenth century were sometimes constructed without chess trees. The tack in this case was clinched to an eyebolt aft of the fore channels, and the hauling part, after passing the tack block, went through a lead block in the spirketting above the gangway and was belayed to a kevel at the forecastle.

Slabline

Schlappleine, Kerkedortjen; Cargue à vue

A span was fastened to the inner buntline cringles at the rear of the sail, and this span was connected to a line which led through the slabline block, stropped beneath the quarter block, and downwards to the centre of the bitts forward of the mast (see Figure 155a). The rig of slablines outlined by Röding differed from this (English) method only in that he specified a slabline block at the yard rather than beneath the quarter block.

Slablines came into use around 1740 and, according to Falconer, were used 'to truss up the sail as occasion requires, but more particularly for the convenience of the pilot or steersman, that they may look forward beneath it, as the ship advances'.

Buntlines and leechlines

Bauchgordingen; Cargue-fondes et cargue-boulines

Buntlines and leechlines lines were fastened on the fore side of the sail to the bunt and leechline cringles in the boltrope, and were used for taking in the sails. The difference between the buntline and the leechline is that the former was fastened to the footrope and the latter to the leechrope.

The average English ship had four buntlines, which were one sixth of the length of the foot of the sail apart. First and Second Rates had six, and small ships only two. Buntlines were rigged in pairs, with each pair, after passing the blocks on the fore side of the yard and these hanging beneath the top, forming a bight with a shoe block (see Figure 155c) forward of the sail (see Figure 155b). The standing part of the buntline fall was made fast to the forecastle waist rail, and the line rove over the second sheave in the shoe block, to be belayed beside the standing part. The last quarter of the eighteenth century saw the shoe block replaced by two single blocks stropped together head to tail (see Figure 155d).

The rigging of buntlines in French ships differed from English practice. Early in the century two buntlines were still in use, not running parallel, but converging at an angle through blocks on the yard and led through double blocks beneath the top aft. In the later years this changed into one buntline at each side. For these, three equally spaced cringles spliced into each side of the footrope. The buntline was made fast to the outer cringle and led through the other two to a block at the yard slings, through another beneath the top and from there to the bitts before the mast (see Figure 155f); this can be seen in the drawings in Paris' *Souvenirs de Marine*. By contrast, the models (restored in the nineteenth century) of *Le Protecteur* and the *Sans Pareil* show three parallel running pairs.

Boudriot provided a further insight into this. He noted two buntline pairs and the English rig as an alternative. The outer was made fast to the outer cringle and rove through the second vertically up to a yard block similar in shape to an English clewgarnet block, then through the outer sheave of a double block hanging beneath the fore cross tree, and was belayed to the ninepin bitts. The second was made fast to the inner cringle, then led up to a lead block at the strop of the yard jeer block and through the inner sheave of the double block noted above, to be belayed as before (see Figure 156a).

In Continental rigging at the beginning of the century, according to the author of *Der geöffnete See-Hafen*, the buntline was made fast towards the outer part of the footrope and led through an iron thimble near the middle of the footrope up to a block at the middle of the yard, and a further block at the stay collar, then ran down upon deck directly abaft the sail (see Figure 156c).

Leechlines rigged in English fashion also differed from normal Continental practice. Larger English ships usually had two leechlines leading to each leechrope, with the lower made fast to the upper bowline cringle and the upper to the reef band cringle. They rove through blocks at approximately the same distance within the yardarm cleats as the cringles were down from the head. These blocks were stropped to the fore side of the yard, and in similar fashion to the buntlines, the upper and

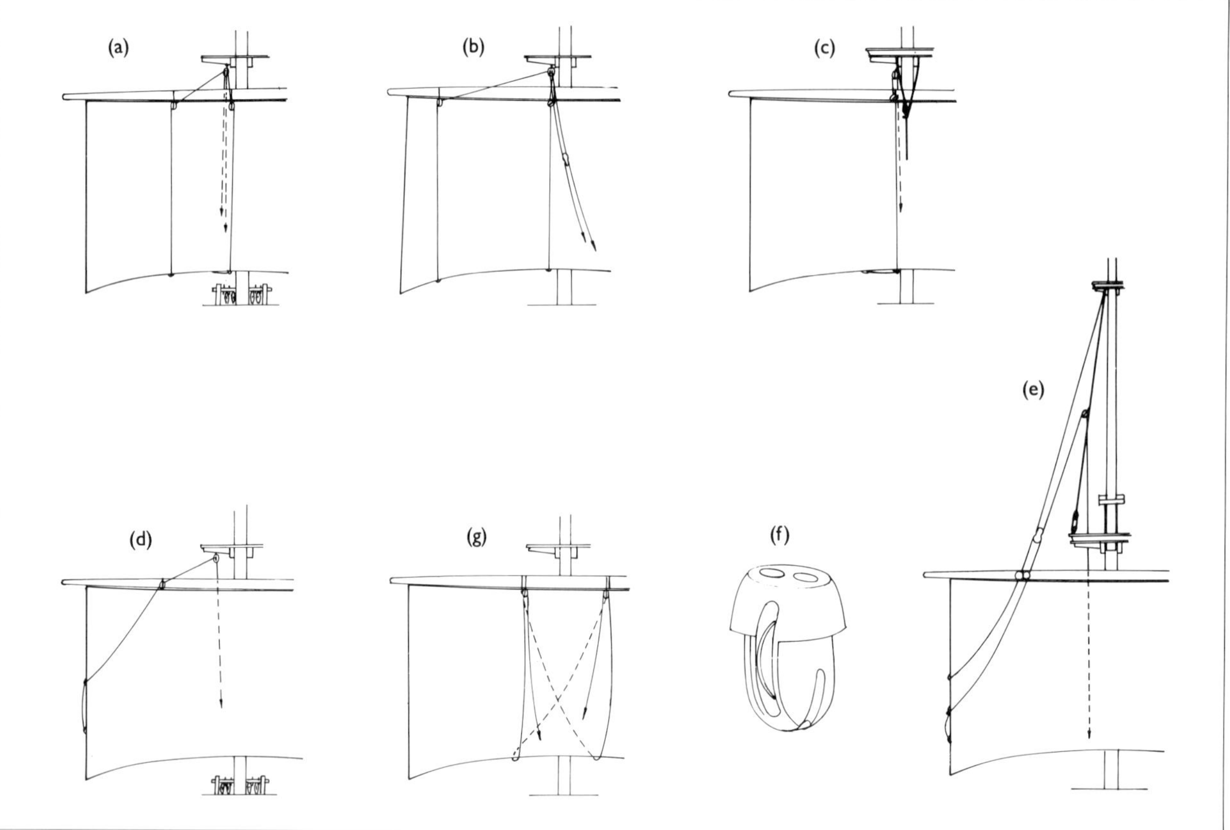

Figure 156
(a) Buntlines on a French man of war in 1780, as reconstructed by Boudriot.
(b) An alternative reconstruction of main buntlines on a French man-of-war in 1780.
(c) A Continental buntline of about 1700, as described in Der geöffnete See-Hafen.
(d) A leechline on a French man-of-war in 1780, according to Boudriot.
(e) A Continental leechline, as described in 1705.
(f) A Frech buntline block.
(g) Temporary spilling lines as described by Röding in 1794.

lower leechlines formed pairs with a shoe block in the bight. The standing part of the leechline fall was made fast round the topmast head, and after passing through the second sheave in the shoe block, rove through a long-stropped block beneath the topmast trestle trees and led down to the bitts (see Figure 155g). This method was used in large ships until about 1770, but on smaller ships an alternative lead was introduced from about 1720 onwards. The leechlines, after passing the yard blocks, rove through double blocks fore-and-aft beneath the top and again had a shoe block running in their bight (see Figure 155h). The standing part of the leechline fall was then made fast to the quarterdeck waist rail, with the hauling part first passing through a foot block before belaying at the same rail. This set-up was sometimes also used early in the century in other ships as well. Instead of foot blocks, the outer sheaves of the main jeer bitts were sometimes substituted, and the hauling part was belayed to the bitt head.

Smaller ships with only one buntline and one leechline on each side usually had the leechline leading through the outer and the buntline through the inner sheave of the double blocks beneath the top.

At the beginning of the century, Continental ships with Dutch orientated rigging still had the leechline running through a cheek block at the yard (in French rigging normal blocks were preferred), then through a block hanging beneath the top, or stropped to the stay collar, and down upon deck (see Figure 155j). German ships around the turn of the century had the leechline fall made fast to the uppermost part of the topmast shrouds and a leg and fall block hanging in the bight above the yard. The fall led up to a block above half height on the topmast shrouds and then upon deck (see Figure 156e). The leechline leg rove through the lower sheave opening of the leg and fall block, with both ends leading through two blocks on the yard (positioned at about one quarter of the yard length) and the outer leg made fast to the uppermost bridle and the inner led first through an iron thimble at the middle bridle then hitched to the lower. Fore courses had only two bridles and the outer leg led through a thimble on the upper bridle to be made fast on the lower. The inner leg was made fast direct to a cringle in the leechrope between the lower bridle and the clew.

French ships of the line also had two leechline pairs during the second half-century, with the second line made fast to the second bowline cringle from the clew. It then rove through a thimble at the upper bowline cringle, a block placed about one quarter to one third of the yard length within the yardarms, a further block beneath the top and a foot block, hooked near the mast to an eyebolt in the partners, to be belayed to a cleat on the mast side.

Bowlines

Bulinen; Boulines

Bowlines were rigged with bridles to the bowline cringles in the lower half of the leechrope. One end of the upper bridle was clinched to the upper bowline cringle, and the second bridle, with a thimble spliced into one end, was slipped on to it, then the other end of the first span was clinched to the second cringle. The second span rove through a thimble in the bowline itself and was clinched round the lower cringle (see Figure 157a and b).

In English ships the bowline ran to a double block lashed to the aft side of the foremast, about 5 feet above deck. Until about 1730 the port side bowline rove through one of the sheave holes of this block (fitted horizontally) and was belayed on the starboard side, with the starboard bowline going the opposite way. In the 1730s the block was rotated so that the sheaves ran vertically, and from then on the bowlines were belayed to the first pin outside the centre pin, at their respective sides, in the bitts (see Figure 157c, d and e).

French bowlines had bridles fitted in a slightly different way, also observed in English ships in the seventeenth century. Bowline tackles were hooked into the bridles (see Figure 157f), and the lower blocks of these tackles were hooked at each side of the foremast to an eyebolt in the partners. The tackle falls were usually belayed to deck cleats (see Figure 157f, g, h and i). French bowlines were also sometimes rigged with the weather

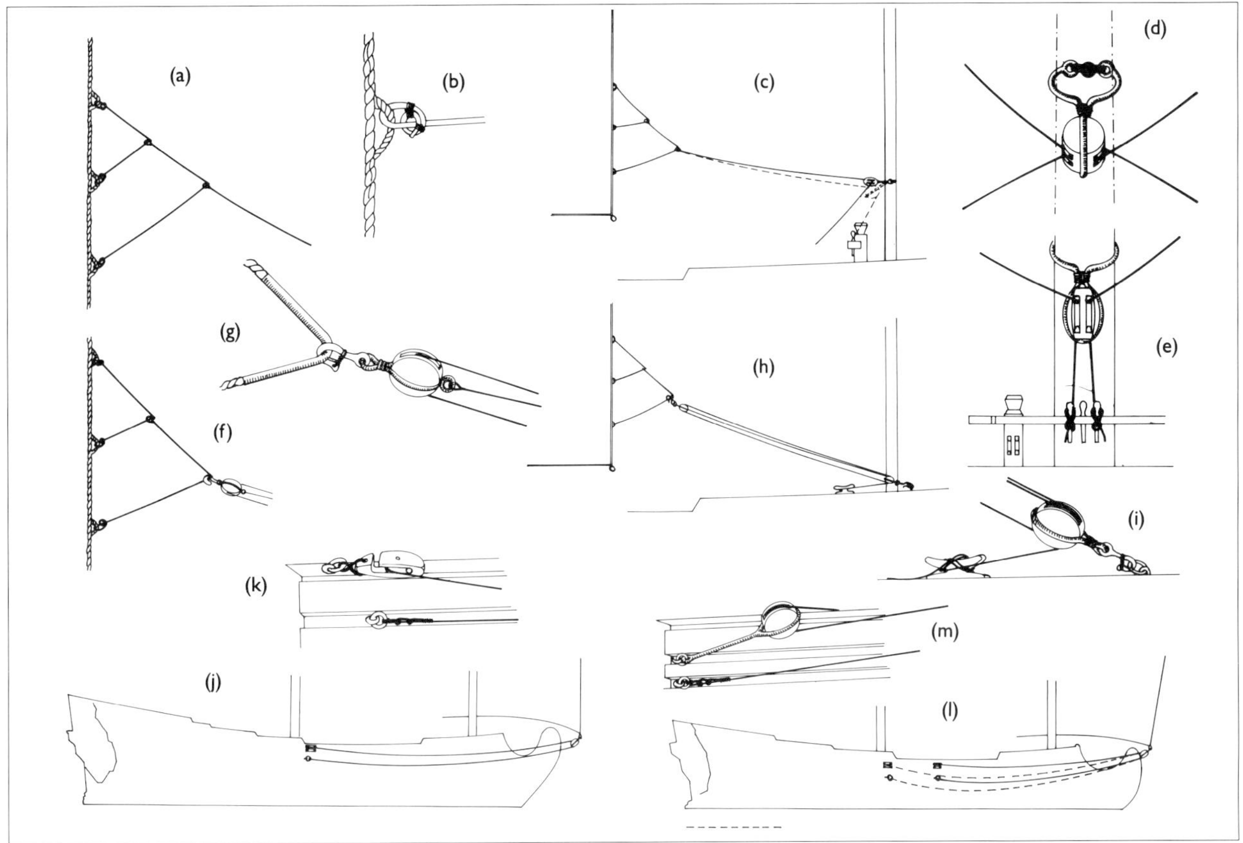

Figure 157

(a) English main course bowline bridles.

(b) A bowline bridle clinched to a cringle in the leechrope.

(c) An English main bowline until about 1730. The dashed line indicates the lead after that time.

(d) An English bowline block lashed to the foremast until about 1730. The sheaves were kept horizontal to take the bowline in from one side and to belay on the other.

(e) An English bowline block after 1730, with the sheaves vertical and the bowlines belayed to the first off-centre belaying pin.

(f) French main course bowline bridles.

(g) Bowline tackle detail of a French main course (upper block). The bridle was served or leathered within the reach of the hook.

(h) A French main bowline; the set up of bridles with tackles was also a common sight in English seventeenth-century vessels.

(i) Detail of French bowline tackle (lower block), hooked to an eyebolt and belaying at a cleat on deck.

(j) The lead of an English fore course sheet.

(k) Coastal vessels often had no sheave hole, and used instead a snatch block to lead the incoming sheet.

(l) The lead of French fore course sheets. The dashed line shows the revised lead in the second half of the century.

(m) The method of leading the incoming sheet inboard via a long-stropped block; this was used in small ships.

bowline running through a snatch block at the aft side of the foremast, and belayed on the lee side to a kevel. The lee bowline hung loosely and was also belayed to a cleat on the same side. A similar rig was described in *Der geöffnete See-Hafen* for Continental ships.

Reef tackles

Refftaljen; Palanquins de ris

Reef tackles were not rigged in the eighteenth century; they were introduced during the early nineteenth century. Lower sails were rigged instead with additional earrings, the reef earrings, which hung loosely in the reef band cringles. This was also the practice in Continental rigging.

Fore course

Fock; Misaine

Rigging to the fore course differed only slightly from that to the main course, largely because of the foremast's position.

Sheet

Schot; Écoute

In English rigging fore course sheets were usually made fast to an eyebolt in the channel wales abreast of the mainmast, with the hauling part led through a sheave hole set above this eyebolt (see Figure 157j). Fore sheets were belayed to the after kevel in the waist.

At the beginning of the century French ships had a similar arrangement at about half the distance between the mainmast and the forecastle. the eyebolt and sheave hole were later moved nearer the mainmast, slightly forward of the gangway steps (see Figure 157l). This latter position was described for Continental ships around 1705.

When no sheave holes were fitted into the sides, a long-stropped block was hooked or spliced to a second eyebolt to take the place of the sheave hole (see Figure 157m). The sheet was led inboard via this block.

Tack

Hals; Point d'amure

Early in the eighteenth century the fore course tack in English rigging led through the so called 'deadblock' in the fore part of the head rails and was belayed to the outer opposite head rail timberhead; this was the case in larger ships until mid-century (see Figure 158a). French and other Continental ships at that time still had lead holes in the cutwater (see Figure 158b). The deadblock, an ornamented hole like the hole for the main tack, was somewhat rare on Continental ships.

With the introduction of the bumkin (from 1710 in ships of the lowest Rates, and by 1745 in vessels, including the largest), the tack rove through a block attached to the outer end of the bumkin, and then through the deadblock, with the end of the tack belaying to the outer nearside timberhead (see Figure 158c).

After the adoption of the double tack, the standing part was made fast to the outer end of the bumkin, then rove through the tack block at the clew, through the block on the bumkin and belayed as before (see figure 158e), except in French rigging, where it led through a foot block near the foremast and was made fast to a cleat nailed to the foremast.

At all times, the single tack, after a length of about 10 yards at normal circumference, tapered to about half its original circumference.

Buntlines and leechlines

Bauchgordinge; Cargue-fondes et cargue-boulines

English fore course buntlines led, like those of a main course, through the blocks beneath the top, then through a second set of blocks at the aft of the top to form a bight with a shoe block

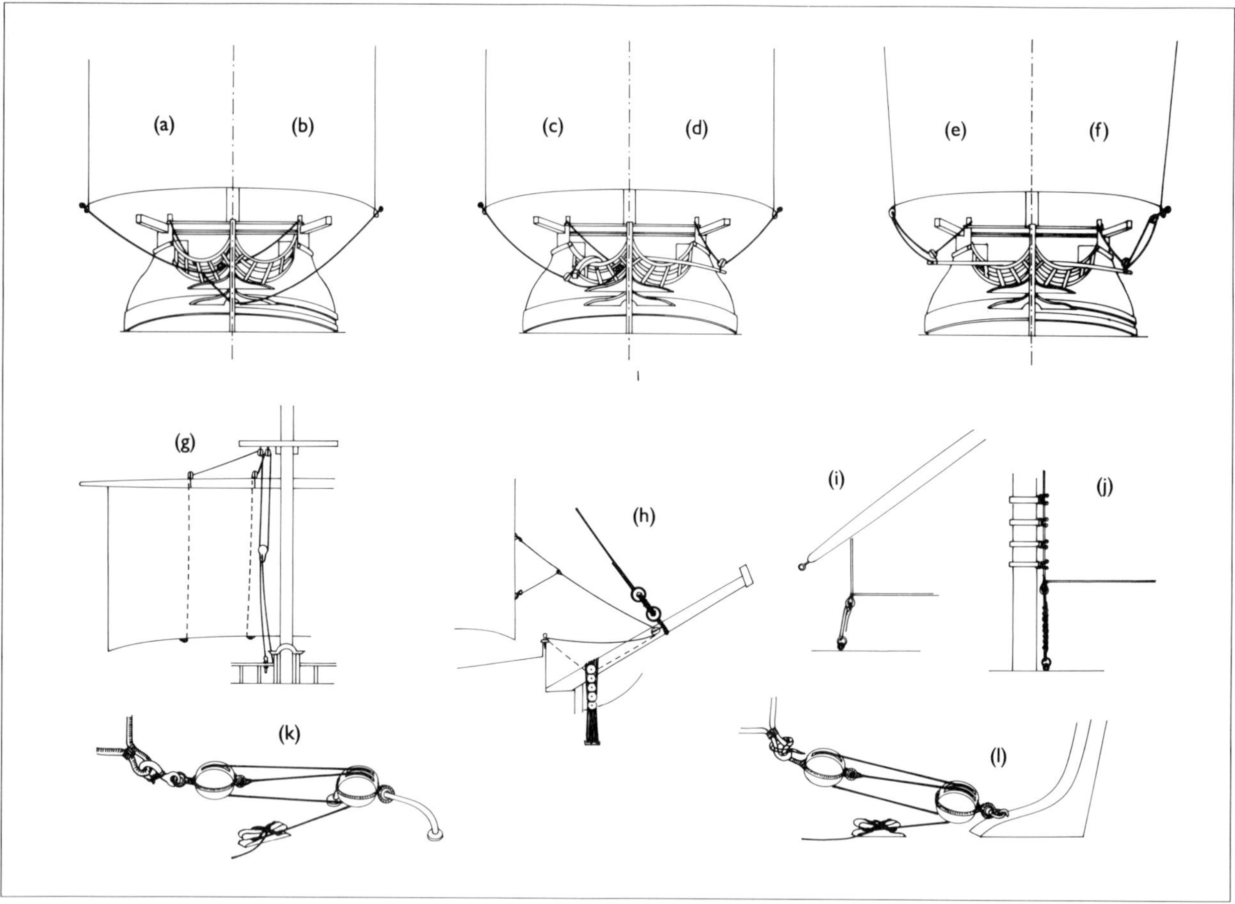

Figure 158
(a) An English fore tack, leading through a deadblock, until about 1710.
(b) A Continental fore tack, leading through the cutwater holes, until about 1735.
(c) An English fore tack, leading through a bumkin block and the deadblock, until about 1725 (in small ships).
(d) A single fore tack leading through a bumkin block (English until about 1790, and Continental until mid-century).
(e) An English double fore tack, leading through a seized tack and bumkin block, in the last decade of the century.
(f) A Continental double fore tack, already used in some ships before 1750.
(g) The lead of the fore buntlines in an English man-of-war.
(h) The lead of a fore bowline; the dashed line shows an alternative lead.
(i) The tack of an English full mizzen course (settee sail).
(j) The tack of an English shortened mizzen course.
(k) The arrangement of a sheet with an iron horse.
(l) An alternative set up of a sheet on a flagstaff step.

hanging in it. The standing part of the buntline fall was hitched to the forecastle waist rail, with the hauling part belaying there too (see Figure 158g).

French fore buntlines followed the lead of main course buntlines in principle, and as an alternative lead that of English buntlines, but with the slight difference that the shoe block was replaced by a leg and fall block.

Leechlines followed the same leads as those of the main course during the same time periods, and were belayed to the forecastle waist rail. (The inner sheave of the double blocks beneath the top was used for the spritsail yard braces). Fore course leechlines in Continental rigging also followed the lead of those of the main course; the slight difference at the beginning of the century is noted above in the description of main course leechlines.

In addition to the lines described above, Röding noted the use of preventer buntlines, which were rigged in stormy weather:

> Preventer buntlines are rigged beside the buntlines in stormy weather, round the main and fore courses, to enable the bunt of these sails to be hauled up. They are made fast to the aft side of the yard, and usually lead crosswise down the aft side of the sail and round the bunt, to reeve through blocks on the fore side of the yard and belay upon deck.

Falconer described these temporary spilling lines in very much the same way, but omitted mention of the crossing of the lines.

Bowlines
Bulinen; Boulines

Fore course bowlines in smaller ships had only one bridle span; two were used only on larger vessels. A single block, lashed to both sides of the fore stay collar on the bowsprit, led the bowline either directly to the forecastle head rail, or via the rack block on the gammoning (see Figure 158h).

Continental bowlines, according to Röding, rove through a block at the head of the bowsprit, and were belayed at the forecastle. Boudriot and Paris followed the English method, while Ozanne's illustrations show both methods. At the beginning of the century single bridled bowlines were led through blocks on the fore stay, attached at about one third of the stay height, direct to the forecastle.

Slablines and clewgarnets
Schlappleinen und Geitaue; Cargue à vue et cargue-points

Fore course slablines and clewgarnets were identical to those of the main course.

Mizzen course
Besan; Artimon

A distinction has to be made in a description of mizzen course rigging between the full lateen or settee course and the short mizzen set only aft of the mizzen mast.

Tack
Hals; Point d'amure

A tack was necessary on both types of mizzen course. On the short course it rove through an eyebolt in the partners behind the mizzen mast and through the eye in the tack of the sail, and was then frapped round itself until extended (see Figure 158j). In Continental rigging the tack was often led round the mast itself instead of through an eyebolt.

On a settee sail (or full course in English ships) a tack was

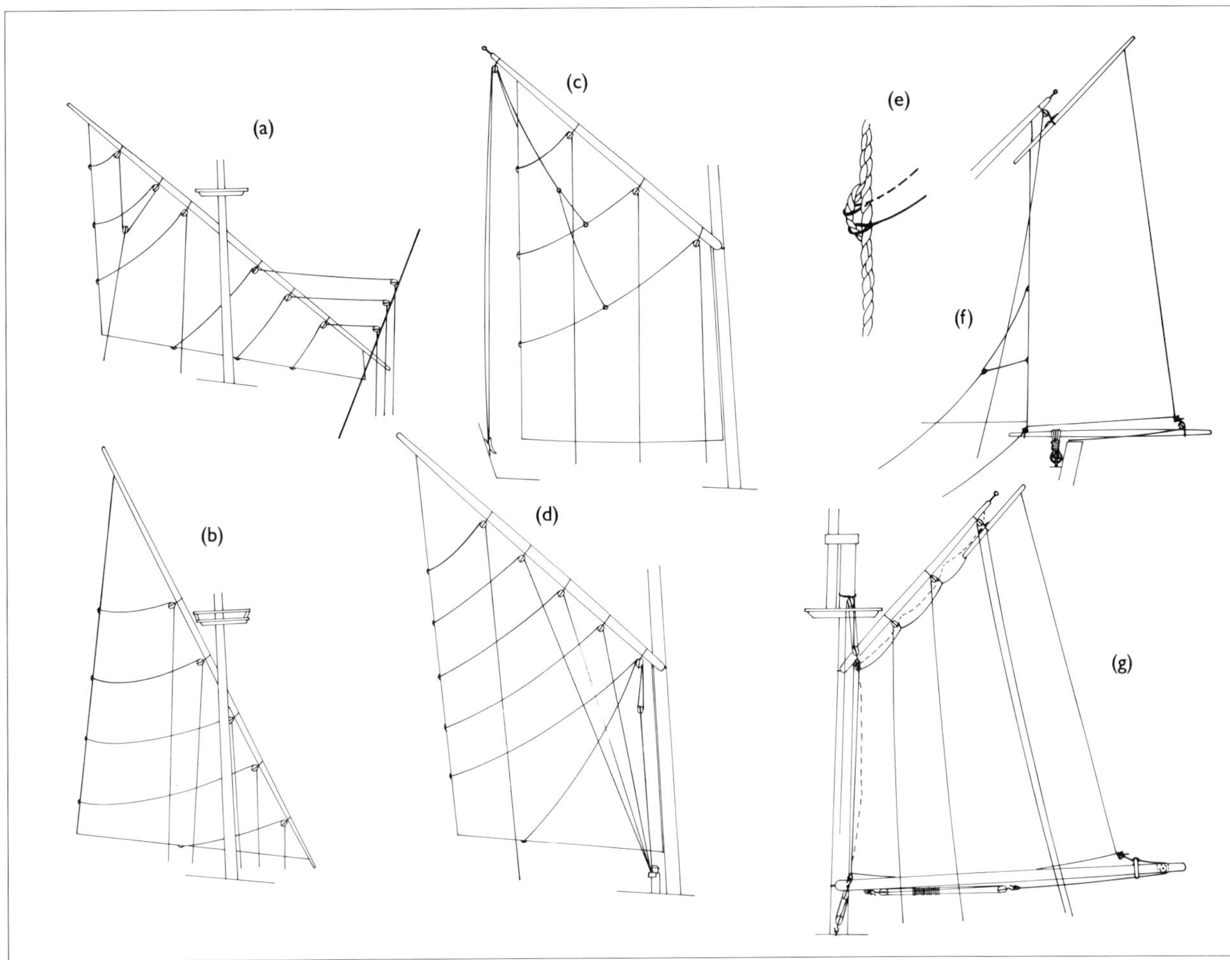

Figure 159
(a) The brails of an English full mizzen (settee type) up to about 1745.
(b) The brails of a French full mizzen (lateen type) until the 1750s.
(c) An English short mizzen with a gaff, rigged with three brails and a fancy line.
(d) A French short mizzen with a gaff, rigged with up to six brails.
(e) Brails spliced into a cringle.
(f) An English driver after about 1780, mostly used in merchantmen.
(g) An English temporary driver after 1780, as preferred in the Royal Navy.

also rigged; normally this was a single line made fast, according to the yard's position, either amidships or to the port or starboard side to an eyebolt in the deck. A tack tackle for a lateen mizzen (on Continental ships) was mentioned in *Der geöffnete See-Hafen*. In this rig an iron thimble was seized into the sail's tack, with the double block of a tackle hooked into it. The single block was hooked to another thimble stropped aft of the main shrouds to a timberhead (see Figure 158i).

Sheet
Schot; Écoute

The mizzen sheet was a tackle, the single block of which was hooked into the clew while the double block either ran on an iron horse or was hooked to an eyebolt at the step of the ensign staff (see Figure 158k and l). The standing part of the tackle fall was hitched or spliced to a becket on the single block's strop, and the hauling part, after being led through the blocks, was belayed to a cleat nearby. While in English rigging both the clew and the tack of a mizzen were simple boltrope eyes even at the end of the century, the author of *Der geöffnete See-Hafen* described both corners of the mizzen as fitted with an iron thimble. He also described the splicing of another strop with a thimble into the clewrope, to enable a relief tackle for the sheet to be hooked to it in bad weather.

Brails
Dempgordinge; Cargues

Full courses had either three (English and some Continental ships), or four or five (French) after leech brails, and one to four foot brails. The after leech brails were named, from top to bottom, 'peak brails', 'middle brails' and 'throat brails'. French ships had two or three middle brails.

In the English rigging the peak and middle brails formed a bight, in which the block of the peak brail fall was placed. This fall was belayed to the aftmost mizzen shroud. The throat brails were belayed either to belaying pins on the mast, or abreast of the mast to the quarterdeck rails; as an alternative to the shrouds the former might also be belayed to the rails (see Figure 159a). Continental peak brails were also sometimes double and rigged like the English peak and middle brails with a block in the bight. However, the two ends counted only as one brail and another middle brail was also rigged (see Figure 160a).

French peak brails were belayed to the rails near the mizzen topmast backstay, and the others either to the rails or to the mast (see Figure 159b).

The foot brails rove through blocks on the lower half of the mizzen yard, equally spaced between the jeers and the foot of the yard according to the number of brails. While French brails were led to the rails, English brails ran through further blocks seized to the aftmost main shrouds at the same height as the corresponding yard blocks, and then ran down to the rails (see Figure 159a). A Continental description from the beginning of the century listed four brails altogether, with the fourth, the foot brail, leading like the others to the rails (see Figure 160a), except in smaller ships, where all four brails were made fast to a batten fixed for that purpose into the mizzen shrouds.

The shortened English mizzen also had three after leech brails, and French mizzen might have up to five. All brails were made fast to the brail cringles at the leech and footropes, and were carried on both sides of the sail. The blocks on the gaff or yard were therefore fitted in pairs, and were also equally spaced between the peak cleats and the jaws or jeer cleats. In French ships the distance between the peak cleats and the peak brails was twice the distance between all the other blocks.

The shortened mizzen's after leech brails were belayed as follows:

English: the peak brails to a cleat on the spirketting (inboard plank), the middle brails directly or via a lead block to the aftmost mizzen shroud or the rails, and the throat brails to the mast, or abreast of the mast to the rails.

French: (after Paris) the peak brails as in English rigging, the two or three middle brails direct to the mast, or to bitts abaft the mast, and the throat and foot brails, led through a double block on the jaws and forming a bight through which a block for the brail fall ran, were belayed by this fall at or near the second mizzen shroud.

French alternative: (after Ozanne and Boudriot) the peak brails were double and formed a bight with a block (Ozanne), or crowsfeet (Boudriot), and a single fall was belayed to a pinrail. This was similar to Continental procedure at the beginning of the century. The peak and upper middle brails might also be led single and directly to a pinrail (Ozanne). The second and third middle brails rove through a treble block close to the mast, and the third sheave of this block was taken by the throat brail (Boudriot); all three were belayed to mast cleats (see Figure 160c). Only two middle brails can generally be seen in Ozanne's etchings, the lower of which lead to a block in the same position as the treble block noted above. His throat brails usually run to a point halfway up the mast leech.

Fancy line

Aufholer der Dempgordinge; Hale-breu des cargue-boulines

In English-style rigging a fancy line was frequently used in connection with the brails. This line consisted of a span with thimbles in both ends, set up to the middle and throat brails, with a further thimble running in its bight. The actual line was then spliced round this last thimble and rove through a block at the peak cleats on a gaff or yard and downwards to the taffrail, where it was belayed to a cleat (see Figure 159c).

The function of the fancy line was it to keep the lee brails far enough away from the sail to prevent them from chafing the sail cloth.

Early driver

Früher Treiber; Tape-cul ancien

The early driver, a squaresail hoisted abaft the mizzen, was bent with robands to the driver yard. In English ships this type of sail was in use until about 1780, and in Continental ships it survived until early in the next century. A painting, dated 1811, of the French ship of the line *Le Wagram* still shows one of these drivers.

Halyard

Fall; Drisse

The square driver's halyard was made fast to the middle of a span set up from one yardarm to the other on the driver yard, and rove through a block at the peak of the gaff or yard, to be belayed to a cleat at the taffrail.

Sheet

Schot; Écoute

Both clews were fitted with single sheets; these either were made fast to the driver boom (a spar temporarily pushed over the lee side), or they rove through blocks on the spar and were belayed at the taffrail.

New driver (merchantmen)

Neuer Treiber (Handelsschiff); Tape-cul nouveau (navires marchands)

After 1780 two new types of driver came into use; in general terms these can be classified as 'merchantman driver' and 'man-of-war driver'.

The former was shaped like a topsail studdingsail with its head gored to the same angle as the gaff sail, and was bent to a driver yard (see Figure 159f)

Halyard

Fall; Drisse

Hitched to the lower third of the yard, the halyard rove through a block at the gaff's peak and was belayed to the weather rail.

Tack

Hals; Point d'amure

The tack led as a single line from the sail to a belaying point (fiferail) on the weather side.

Sheet

Schot; Écoute

Made fast to the clew, the sheet rove through a block on the outer end of the loose driver boom extended over the taffrail. It was belayed to a cleat at the taffrail.

Bowline

Bulin; Bouline

A bowline with a single bridle was attached to the fore leech and belayed forward of the tack, also on the weather side.

New driver (men-of-war)

Neuer Treiber (Kriegsschiffe); Tape-cul nouveau (navires de guerre)

Men-of-war were normally provided with a driver with a total sail area equal to that of the mizzen plus a driver of the merchantman type: it was set not in addition to but instead of the mizzen, and the mizzen had therefore to be brailed up to the gaff and mast (see Figure 159g). This sail had a driver yard normally bent only to the outer third of the head. In addition to this a boom was needed, but even at the end of the century in most cases this remained a temporary fixture.

Halyard

Fall; Drisse

Four halyards were used for this driver, with the inner one made fast to the nock, the outer to the inner third of the driver yard and the others at equal distances to the head. The nock halyard was a double and single block tackle in large ships, or rove simply through a single block in smaller vessels (the tackle's double block, or the single block for the nock halyard, was lashed to the mizzen masthead and the tackle's single block was hooked to the nock earring). The blocks for the other halyards were lashed to the gaff, equally spaced, with the halyards themselves belayed to the weather side fiferail.

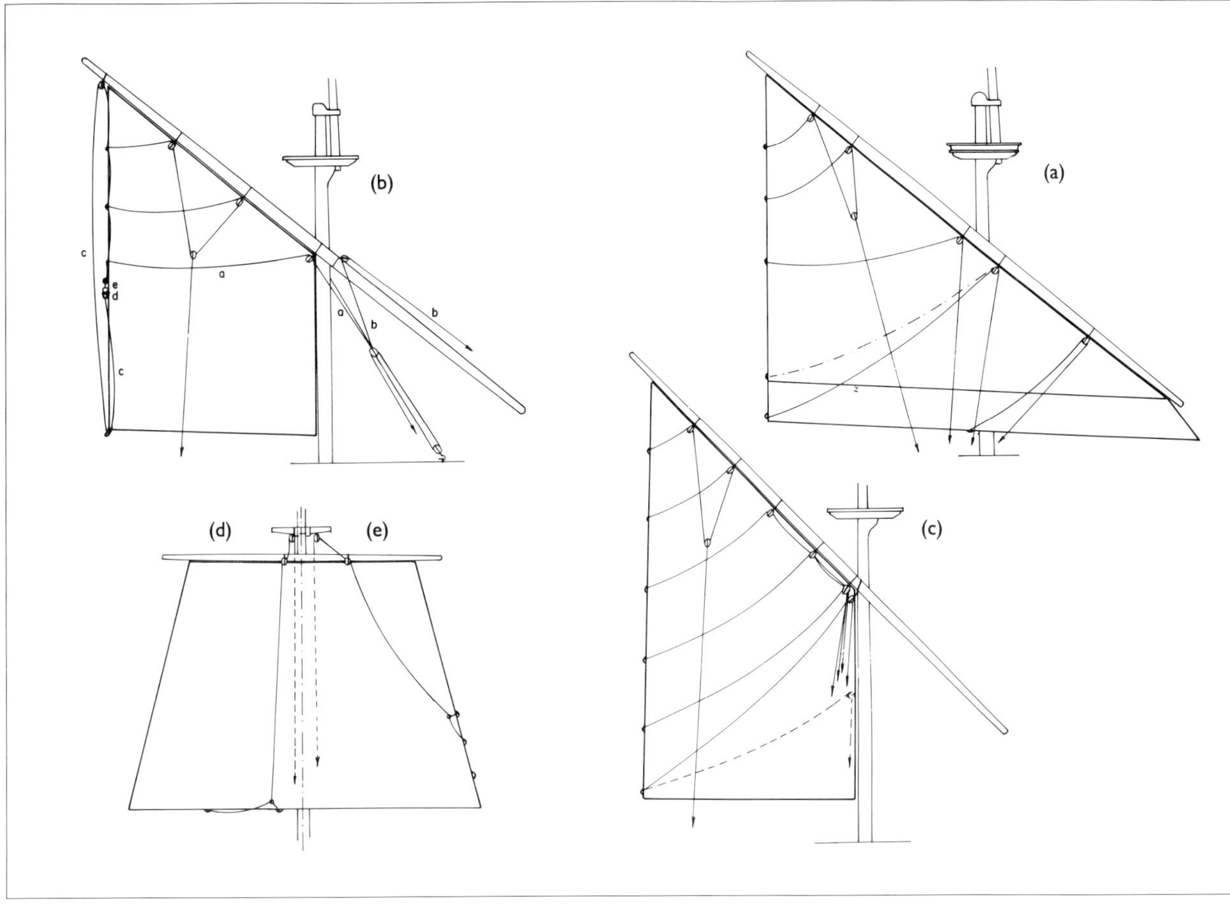

Figure 160
(a) The rigging of Continental mizzen brails according to the author of Der geöffnete See-Hafen of 1705.

(b) An illustration of Röding's description of Continental mizzen brails, with peak and middle brails joined into one. The throat brails (a.) are stropped to a block forward of the mast which forms the upper block of a tackle, the lower block of which hooks to an eyebolt on deck. (b.) is the throat brail tackle uphauler, a device to loosen the tackle by hauling the upper block close to mast and yard when the sail had to be set. (c.) is the throat brail uphauler, an endless rope passing through a peak block and the clew's eye, then through a fairlead truck (d.) below the throat brail cringle, then through the cringle itself and the other brail cringles and up to the peak block again. A large stopper knot (e.) is set upon the uphauler between the truck and the throat brail cringle. When the free side of the uphauler is hauled, the knot stops below the cringle and thus lifts the throat and other brails up to the peak.

(c) French mizzen brails of about 1780, as described by Boudriot. The dashed line shows the lead shown in Ozanne's sketches.

(d) A French topsail buntline of 1780.

(e) A French topsail leechline of 1780.

Tack
Hals; Point d'amure

The tack was also set up as a tackle, with the double block hooked into the tack cringle and the single block to an eyebolt in the deck. When the boom was fitted with jaws and rested on a mast shoulder, the lower block might alternatively be hooked to a boom eyebolt near the jaws.

Sheet
Schot; Écoute

The driver's sheet was rigged in various ways. Steel described it as follows:

> The sheet reeves through a block or sheave hole at the outer end of the boom, and is bent to the clue of the sail; a luff-tackle is cats-pawed to the other end of the sheet. The inner block hooks to the taffarel, and the fall leads in upon the quarter-deck. When this sail is bent to the mast, yard, or gaff, instead of the mizen, it bends exactly the same, only the foot of the sail is extended on the boom as above.

Lever described the sheet as 'reeved through a Sheave hole in the Boom, and clinched to an iron Traveller: in the other end a Thimble is spliced: the outer block of a Luff Tackle is hooked to it, and the inner one, to a bolt on the Boom'.

Downhauler
Niederholer; Hale-bas

A downhauler was bent to the middle of the yard and also belayed to the weather side on the quarterdeck.

Main topsail
Grossmarssegel; Grand hunier

Sheet
Schot; Écoute

The main topsail sheets were secured to the clews with a wall knot (English), or a sheet bend (French). They rove through a shoulder block (English) or a pear-shaped block (Continental) at the main yardarms, then through a lead block (quarter block) abaft the slings, and came down to the main topsail sheet bitts; there they led either through one of the ninepin sheaves (French), or a bitt post sheave (English), to be belayed (see Figure 161a, b and c).

Clewlines
Geitaue; Cargue-points

The rigging of topsail clewlines during the eighteenth century was subject to several small alterations. In English ships during the first decade the standing part was made fast three eighths of the yard's length inside the yardarms. After passing the block at the clew, the line returned to a similar block, seized to the yard just inside the standing part, and was led through the lubber's hole in the top direct upon deck (see Figure 161d). Between approximately 1705 and 1720 an additional lead block was fitted at half height on the masthead (see Figure 161e).

Around 1720 the standing part moved slightly further outboard, with the fastening point now one third of the yard length inside the yardarms. Instead of the masthead lead block, a block was seized to the topsail yard outside the slings, and the clewline run freely from the yard through the lubber's hole upon deck (see Figure 161f).

These lead blocks were abandoned in about 1735 and the clewline was rigged in the same fashion as in the early years (see

Figure 161g). Around 1790 the clewline blocks were moved to only 3 feet outboard of the centre of the yard, just outside the slings (see Figure 161h).

In French ships throughout the century the clewlines were made fast two thirds of the yard length inside the yardarms (one third of the total yard length), according to Paris and the models of *Le Fendant* of 1700 and the *Royal Louis* of 1780; after passing the blocks on the yard, the clewlines rove through lead blocks on the foremost topmast shrouds at one third of their height (1700), or one eighth of their height (1780). Illustrations suggest that the clewlines were belayed on deck in 1700, while in 1780 they were led to belaying pins in the topmast shroud spreader (see Figure 161i and j respectively). The clewlines of the *Royal Louis* were drawn single in the rigging plan.

Röding also noted single clewlines for topsails as normal. His clewlines rove through the blocks on the yard then through others on the topmast before they came upon deck.

According to Boudriot the clewlines for a 74-gun ship of 1780 were double and the lead blocks made fast to the lower mast hounds. The lines then came down inside the shrouds, passed foot blocks at the spirketting and were belayed to the pinrail or to cleats in the shrouds. One alternative to this was the placing of these lead blocks on the top rim near the second shroud; a second was the replacing of the foot block at the spirketting by an additional lead block, two thirds of the shroud's length down at the third shroud. The clewlines were then belayed direct to the pinrails.

Contemporary models of the *Sans Pareil* of 1760 and *Le Protecteur* of1793-4 were different again; the clewlines were made fast to the yard nine tenths of the distance inside the yardarms, just outside the slings, and the main topsail yard drawing of *Le Protecteur* suggests a two-sheave clewline block. The topsail yard lift lost its additional function as a topgallant sail sheet after 1790, when additional sheets were rigged, so the second sheave in the double clewline block may indicate its extra function as a sheet lead block.

Up to 1750 Dutch ships had the yard blocks and the standing part of the clewlines made fast approximately halfway between the yardarms and the slings, at one quarter of the yard's length (see Figure 161l). The author of *Der geöffnete See-Hafen* implied the same position for the standing part, with the clewline yard block 6 inches inside that point and the hauling part of the clewline running down to the third shroud to be belayed.

Buntlines and leechlines

Gordinge; Cargue-fondes et cargue-boulines

There were usually two topsail buntlines, leading on the fore side of the sail from the footrope cringles vertically up to single blocks at the yard, then reeving through long-stropped blocks at the trestle trees upon deck (see Figure 161m). It appears that buntlines at the beginning of the century were fitted like the lower buntlines: Davis' mention of a pair of buntline legs and a buntline fall for all ships from a Sixth Rate to the *Royal Sovereign*, whereas he noted only leechlines rather than leechline legs as in the lower courses, suggests that a block was fitted in the bight of the two buntline legs for a fall to reeve through. His contemporary Sutherland also listed two long-tackle blocks in addition to two single blocks for the leading of main topsail buntlines, which supports Davis' account. After 1745, instead of the long-stropped blocks, the outer sheave of a double block was used, placed beneath the top cross trees (see Figure 161n). After 1790 some larger ships had two buntlines on each side. By this time the buntlines rove through yard blocks, then through double blocks seized to the tye block's strop, through other double blocks made fast to the trestle trees and upon deck (see Figure 162a).

Single buntlines, without yard blocks, were also led directly through single blocks at the tye block strop in this period, passing through the trestle tree blocks upon deck (see Figure 162b). Buntlines were belayed to cleats or a pinrail in the lower shrouds.

Leechlines were carried until the last decade of the century, disappearing after 1790. They were generally made fast to the uppermost bowline cringle and, up to 1745, rove through the blocks at the tye block upon deck (see Figure 161m). After 1745 leechlines rove first through blocks at the yard (one quarter of the yard's length from the yardarms), then through those at the tye block strop and finally through the inner sheave hole of the double blocks beneath the top noted above and upon deck (see Figure 161n). Leechlines were also belayed in the shrouds.

Again certain notes in the works of Davis and Sutherland differ from later sources. The length of a leechline according to Davis (1½ times the topmast length) and Sutherland (7/17 of the buntline length or 14 fathoms) was only long enough to be belayed in the top; only for ships from the First to Third Rates did they specify leechlines long enough to go down to deck level. Davis noted '2 Leech-lines to the Leech, and foot of the Sail, and so the other ends to the Top Masthead and down upon the Deck'. Early eighteenth century English leechlines in larger ships were therefore led through leech cringles and made fast to or near the clew (foot of the sail) before leading to the topmast head and upon deck.

Continental rigging of topsail buntlines and leechlines at the beginning of the century differed again. The author of *Der geöffnete See-Hafen* described buntlines not as legs but as single running buntlines, made fast to the foot of the sail at its centre and at one third width. The centre buntline rove through a block above the topsail yard (probably made fast to the tye) up to the topmast trestle trees, then through a block there upon deck (see Figure 163a). Those made fast at one third of the footrope led through a thimble on the lowest bowline cringle and upwards and, after passing blocks on the yard, went through further lead blocks on the trestle trees upon deck (see Figure 163b). The author notes that this buntline lead was only followed when the topsail was hoisted with a tackle, and that when it was rigged with a tye and fall, blocks were lashed to the tye only; yard and trestle tree blocks were not necessary and the buntline led direct upon deck. Leechlines were made fast to the leech above the uppermost bowline bridle, went through a block in the middle of the yard and, without any other lead block, ran down the rear of the sail to the top to be belayed (see Figure 163b). These early leechlines were therefore generally belayed at the top in Continental rigging as well as English, rather than on deck as often indicated later.

Differences between English and French rigging of topsail buntlines and leechlines were considerable. The outline of French practice given below is based on drawings and photographs in *Souvenirs de Marine*, cross checked against Boudriot's work.

Buntlines were single and made fast to the outermost of the three footrope cringles. They rove through the other cringles, the inner of which had a thimble fitted, and led up to a double block seized to each side of the tye block strop above the yard (this is shown on the yard drawing *Le Protecteur*). From there the buntline passed through a block beneath the trestle trees to lead upon deck. Except for the spreading of the buntline along the footrope is this very similar to the short description given in *Der geöffnete See-Hafen* (see Figure 162e). Boudriot placed the trestle tree block at the topmast head on the second shroud and the buntline, after leading through a ninepin block, was belayed at the bitts. As an alternative to the blocks at the tye block, Boudriot specified a double block stropped to the topmast stay collar. After passing this block the buntline led down aft of the mast through a hole in the top decking then a truck at the rear of the catharpins and through the ninepin block as before. By contrast, the restored models of *Le Protecteur* and the *Sans Pareil* have two or three buntlines on each side, leading from each

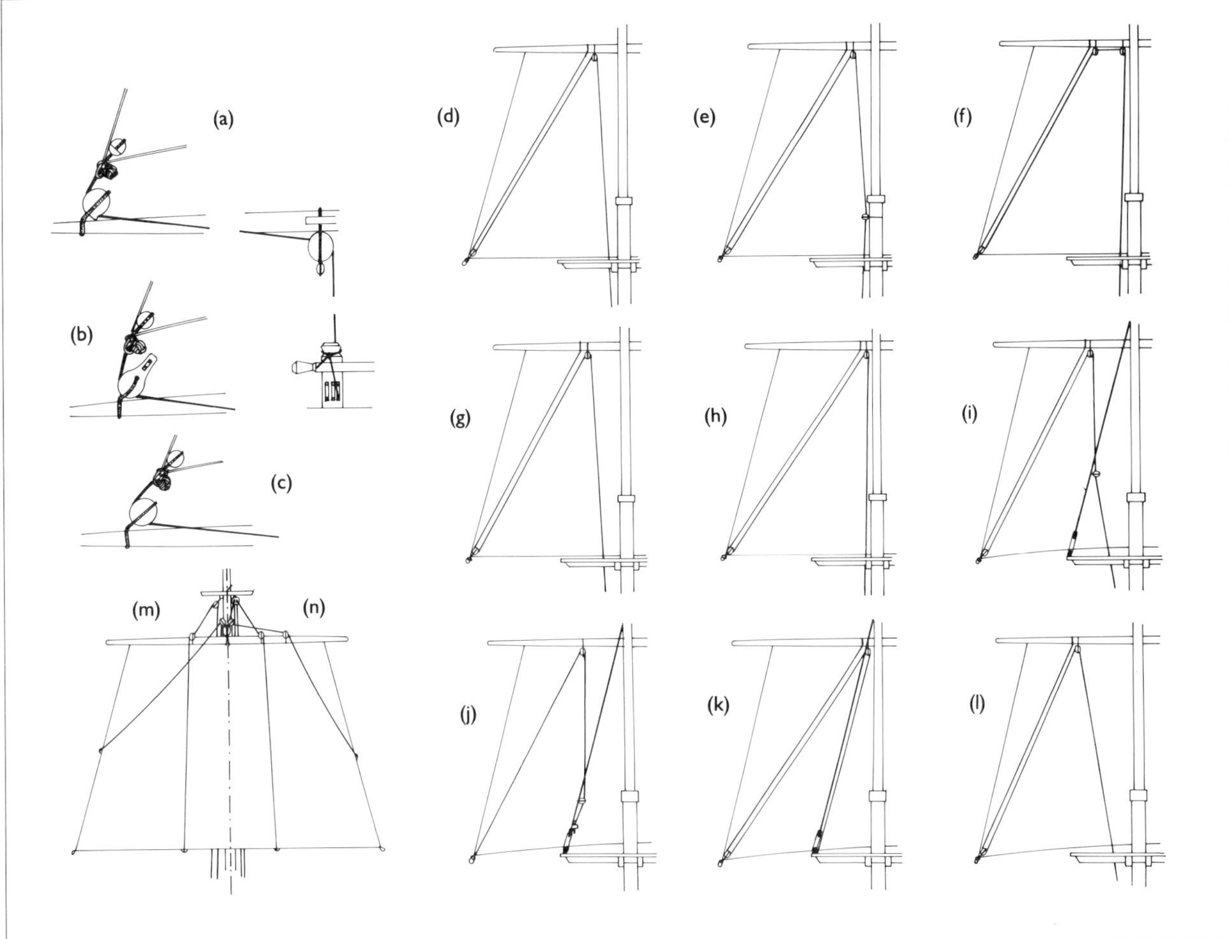

Figure 161
(a) An English topsail sheet.
(b) A Continental topsail sheet.
(c) A Continental topsail sheet at the end of the century.
(d) Topsail clewlines in English ships until about 1705.
(e) Topsail clewlines in English ships, 1705 to about 1720.
(f) Topsail clewlines in English ships, 1720 to about 1735.
(g) Topsail clewlines in English ships, 1735 to about 1790.
(h) Topsail clewlines in English ships after 1790.
(i) Topsail clewlines in French ships at the beginning of the century.
(j) Topsail clewlines in French ships in about 1780.
(k) Topsail clewlines in French ships in about 1790.
(l) Topsail clewlines in Dutch ships in the first half of the century.
(m) English topsail buntlines and leechlines until about 1745.
(n) English topsail buntlines and leechlines between 1745 and 1790.

cringle up to the yard; this is probably a restoration mistake.

Made fast to an extra cringle between the two upper bowline cringles, the leechline was led through another cringle below the reef band but above the bowline and from there to a block at the tye block (see Figure 162e). The rigging plan of *Le Fendant* provides two approaches: on the main topsail the leechline stops on the topsail yard and probably leads as described in *Der geöffnete See-Hafen*, while the fore topsail leechline leads up to the topmast trestle trees, and then most likely to the top.

In 1780 the leechline was made fast as before, but the second cringle was not used and the leechline rove direct to the block on the tye block strop. A second leechline was made fast to the upper reef cringle, rove through a block one quarter of the yard length inside the yardarm, then through another block seized to the stay collar directly beneath the trestle trees (see Figure 162f). Boudriot's description of a 74-gun ship of 1780 followed closely the rigging illustrated in this Figure. The fore topsail leechline was made fast to the middle bowline cringle, rove through a thimble in the lower reef cringle, through the yard block one quarter of the yard length within the yardarms, and then as noted above. For the main topsail the leechline was made fast to the second bowline cringle from the clew, and rove through a thimble in the third cringle, then as above. Leechlines were belayed in a similar way to buntlines.

The end of the eighteenth century saw monkey blocks fitted for the leading of buntlines in merchantmen. They were nailed to the top of the yard, either as normal blocks with a half-moon shaped foot, or iron-bound with a swivel (see Figure 162c). Röding mentioned these blocks in the appendix to his English index, and in his volume of plates he stressed the English origin of these blocks. Lever commented as follows:

> In the Merchant Service, Monkey-Blocks are often nailed on the yard: the Bunt-lines in this case are not taken to the Mast Head but they are reeved through these Blocks, and bent to the Cringles in the Foot of the Sail... In the former Method it is thought they prevent the Yard coming down readily but in the latter Mode, they act as down-haul Tackles, and have equally the Effect of spilling the Sail. The principal Objection to the Monkey-Blocks is that the Bunt-lines do not overhaul so well, for when the Yard is at the Mast-Head, their Weight lies directly from the Yard to the Deck [see Figure 162d].

Topsail spillinglines

Schmiergordinge; Étrangloirs, égorgeoirs

In the later years of the century topsails had spillinglines in addition to normal buntlines and leechlines. Steel gave the following description:

> Spilling-lines of topsails have two legs, which are each made fast with a timber-hitch round the quarters of the topsail yards, then lead down on the aft-side, return upwards under the foot of the sail, and reeve through a block on the foreside, lashed to the tye-block on the yard, and then lead upon deck abaft the mast [see Figure 163c].

Falconer referred to 'harbour leechlines'. These were made fast to the middle of the topsail yards, passed round the leeches of the topsail and rove through the blocks at the tye block, to truss the sail very close up to the yard before it was furled.

Boudriot mentioned the same type of spillingline for French ships and remarked that these were not generally fitted to English ships, which flatly contradicts Steel. Spillinglines for Continental ships, as described by Röding, differed from these detailed above. Röding's description was as follows:

> Because of their depth, topsails are also rigged with

Figure 162
(a) Topsail buntlines on a large English ship after 1790.
(b) Topsail buntlines on a small English ship after 1790.
(c) The rig of buntlines and leechlines on a merchantman in the last decade of the century. Buntlines and leechlines rove through monkey blocks nailed to the top of the yard, and led direct upon deck; they also acted as downhaulers.
(d) Topsail buntlines on a merchantman according to Lever. The buntline here rove only half the distance up the face of the sail, and after passing through a leather patch, led up the upper half of the sail on the aft side, through a yard block, and down upon deck. The advantage of this arrangement was supposed to be that it minimised chafing.
(e) Topsail buntlines and leechlines on a French ship in about 1700.
(f) Topsail buntlines and leechlines on a French ship in about 1780.
(g) Topsail buntlines and leechlines on a French ship in about 1790.
(h) Topsail buntlines and leechlines on a Dutch ship at the beginning of the century.
(i) Main topsail bowlines in an English ship until 1705 and between 1740 and 1775.
(j) Main topsail bowlines in an English ship between 1705 and 1740 and after 1775. The dashed line shows an alternative lead.
(k) Main topsail bowlines in a French ship.
(l) Main topsail bowlines in a Dutch ship.

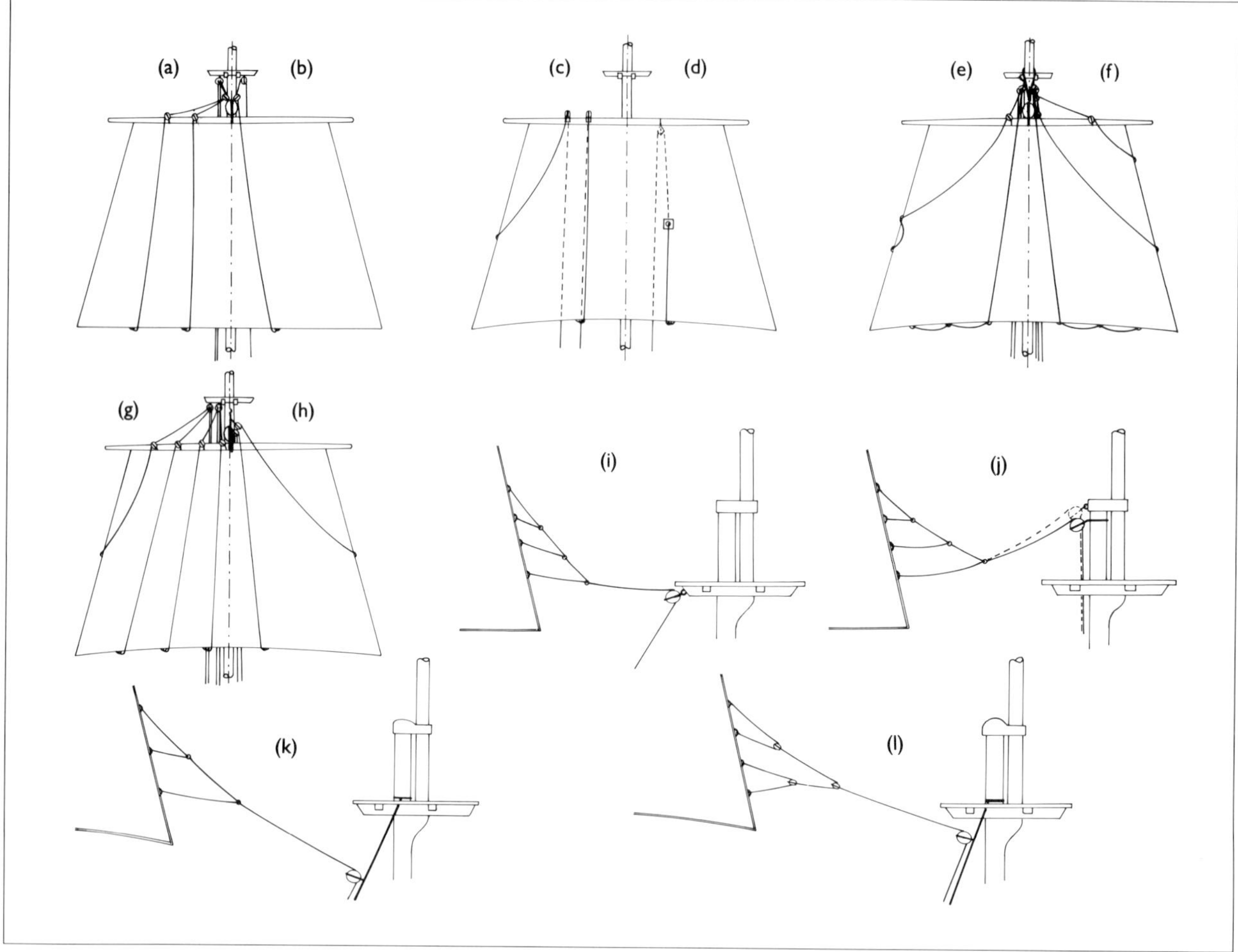

preventer leechlines, which, like the leechlines but lower, are made fast to the leechrope and reeve similarly. They are normally only fitted when the vessel lies in a roadstead; at sea they are unreeved and used as buntlines [see Figure 163d].

Spillinglines for use in heavy weather were mentioned by Röding only for main and fore courses.

Reef tackle

Refftalje; Palanquin de ris

Various methods of fitting the reef tackle are considered in Chapter II. A further description of the function of a Continental reef tackle was given by Röding:

> [A] Reef-Tackle is a tackle with a pendant to lift the outer end of a reef under the yard when a sail is reefed. The tackle itself is placed above the yard, with the end of its fall passing through a block at the masthead to the deck. The pendant leads through a hole or sheave hole at the end of the yard and is hitched to the cringle of the lowest reef. When one of the reefs is to be taken in, a seizing is made on to the pendant and the respective cringle, so that the pendant hauls only that particular reef.

Reef earrings

Stechbolzen, Steekbolten; Bosses de ris, rabans de pointure

On the use of reef earrings Röding commented as follows:

> Reef earrings [are] ropes which are double at one end, or have a long eye. In reefing they help to secure the reef cringle close to the yard, when the cringle is brought home with the reef tackle. The reef earring is slung round the yard and the single end is reeved through the reef cringle, then through the eye of the reef earring, back through the cringle and round the yard again until the earring is filled, whereupon it is fixed to the yard.

Bowlines

Bulinen; Boulines

Bowlines for the main topsail were made fast with bridles to either three or four cringles (normally four in English rigging). At the beginning of the century bowlines rove through blocks below the rear of the fore top trestle trees and from there to the forecastle waist rail near the belfry (see Figure 162i). Between 1705 and 1740 these lead blocks were moved to the aft face of either the masthead or the cap, and the bowlines came down to the bitts abaft the foremast (see Figure 162j). They passed there through sheave holes in the bitt posts and were belayed to the bitt heads.

The period from 1740 to 1775 saw the lead blocks again on the aft edge of the fore top and after that period they were once more on the aft face of the masthead beneath the cap, sometimes as a double block.

As an alternative at this time, especially in French ships, lead blocks were lashed to the aftmost foremast shrouds about 6 feet below the catharpins, and the bowlines were then belayed to pinrails in the sides (see Figure 162k).

Röding recommended running the bowlines through single blocks beneath the fore top, then through lead blocks in the aftmost shrouds and belaying them to cleats in the sides. A modern French description again gives the pinrail as the belaying point, and also suggests an alternative rig, similar to Figure 162 (j) but with the bowline, after passing through the lubber's hole, leading through a foot block in the waterway before belaying to a shroud cleat.

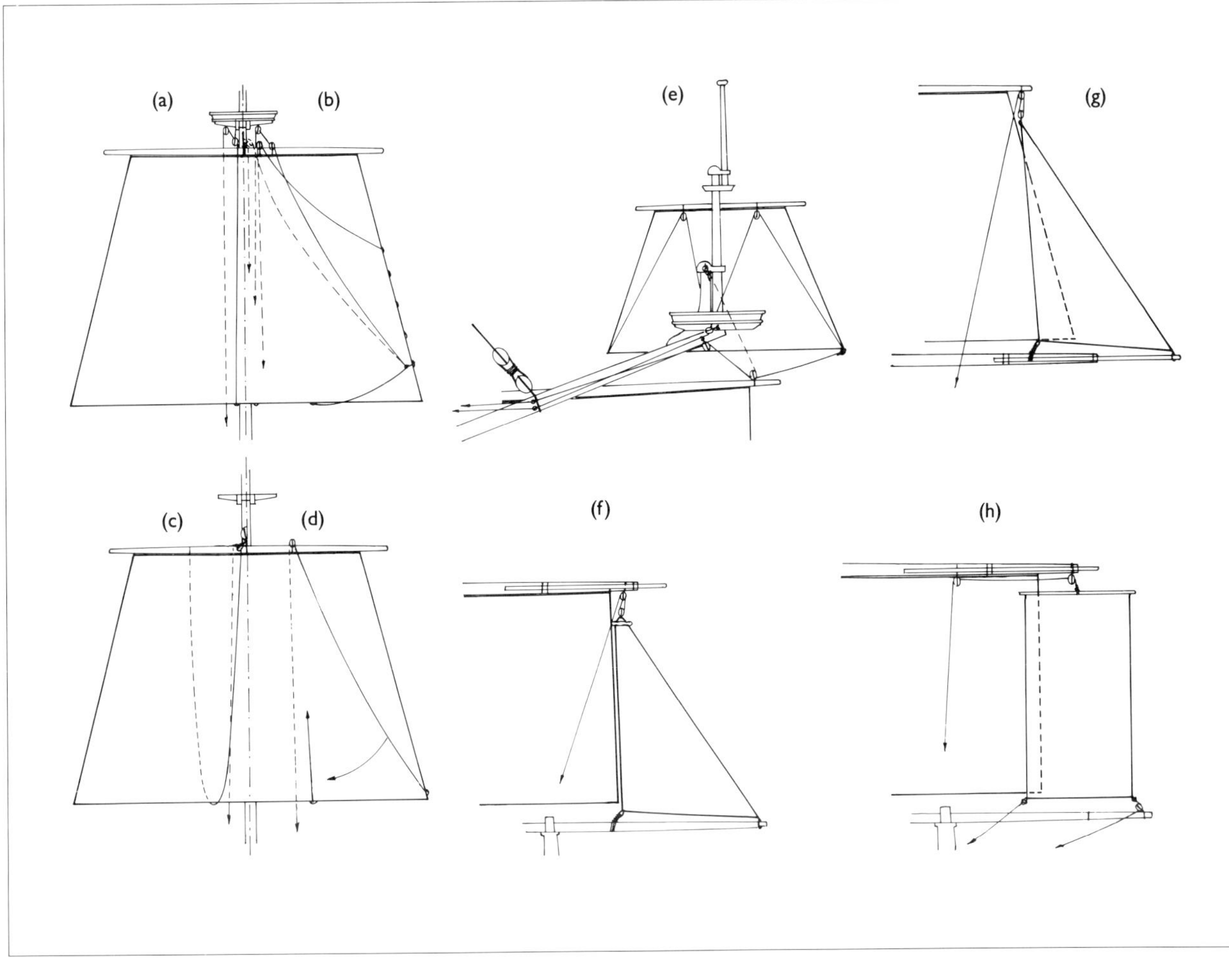

Figure 163
(a) Continental main topsail buntlines as described in 1705.
(b) Continental main topsail leechlines as described in 1705. The lower leechline shown dashed was used with a yard hoisted by a tye and fall.
(c) Temporary spilling lines according to Steel in 1794.
(d) A preventer leechline, also in use as a buntline (according to Röding in 1794).
(e) Continental sprit topsail clewlines and sheets as described in Der geöffnete See-Hafen in 1705. Note that these ropes passed through thimbles on the forestay collar instead of going through the rack block.
(f) Continental fore and main studdingsails as described in Der geöffnete See-Hafen in 1705. The clew eye went over the boom's end and the tack was seized to the boom.
(g) Continental fore and main topsail studdingsails in 1705.
(h) A recommended rig for a lower studdingsail in the early days of the century according to notes in Davis' The Seaman's Speculum of 1711.

Fore topsail

Vormarssegel; Petit hunier

With only minor differences, the rigging described for the main topsail was also used for the fore topsail.

Sheet, clewline, buntline, leechline and reef tackle

All these items were rigged as for the main topsail.

Bowlines

Bulinen; Boulines

The number of bridle spans for the fore topsail in English ships was usually one fewer than for the main topsail. French vessels had three bowline cringles for each sail, and Boudriot followed English rigging for his 74-gun ship, but he gave his East Indiaman *Le Boullogne* four cringles for each sail.

The rigging of fore topsail bowlines during the sprit topmast period was not at all uniform. Bowlines might pass through blocks on the fore topmast stay to others lashed to the bowsprit near the fore stay, then through the rack block on the gammoning to belay on the beakhead rail (see Figure 164a). They also sometimes led from the fore topmast stay blocks to a treble block on the bowsprit near the sprit topmast standard, where the outer sheaves of that block took the bowlines and the middle one the fore topmast stay tackle fall (see Figure 164b). The German author of *Der geöffnete See-Hafen* noted two bridles instead of three, as for the main topsail bowlines, with the bowlines passing through blocks stropped to the fore topmast stay at one third height, then through a double block on the bowsprit abaft the sprit top. From there the bowlines were led to blocks underneath the fore top and then vertically down to belay on a foremast cleat.

Two single blocks, seized or hooked to eyebolts at the sides of the bowsprit cap, were substituted for the bowsprit lead block around 1730, but the bowlines still rove through the stay lead blocks (see Figure 164c). The latter became obsolete in the 1770s.

In Continental ships from about 1760 onwards the fore topsail bowlines led direct to the jib-boom end near the standing jib, rove through the outer sheaves of a treble block there, passed one or more lead blocks along the bowsprit and belayed at the forecastle (according to Röding). The bowline in French ships followed a slightly different lead after passing the treble block.

There the lead block was seized to an eyebolt in the bumkin just above the upper head rail. Alternative noted by Boudriot were a block at the bees instead of the treble block together with the head rail block, or the block at the bees and the rack block.

Chapman's rigging plans of a frigate, a snow and a bilander (all dated 1768) show the fore topsail bowlines similarly rigged to Röding's description (see Figure 164d).

Mizzen topsail

Besanmarssegel; Perroquet de fougue

Except for a few details, the mizzen topsail was rigged like the other topsails.

Sheet, clewline, reef tackle

All these were rigged as detailed under Main topsail.

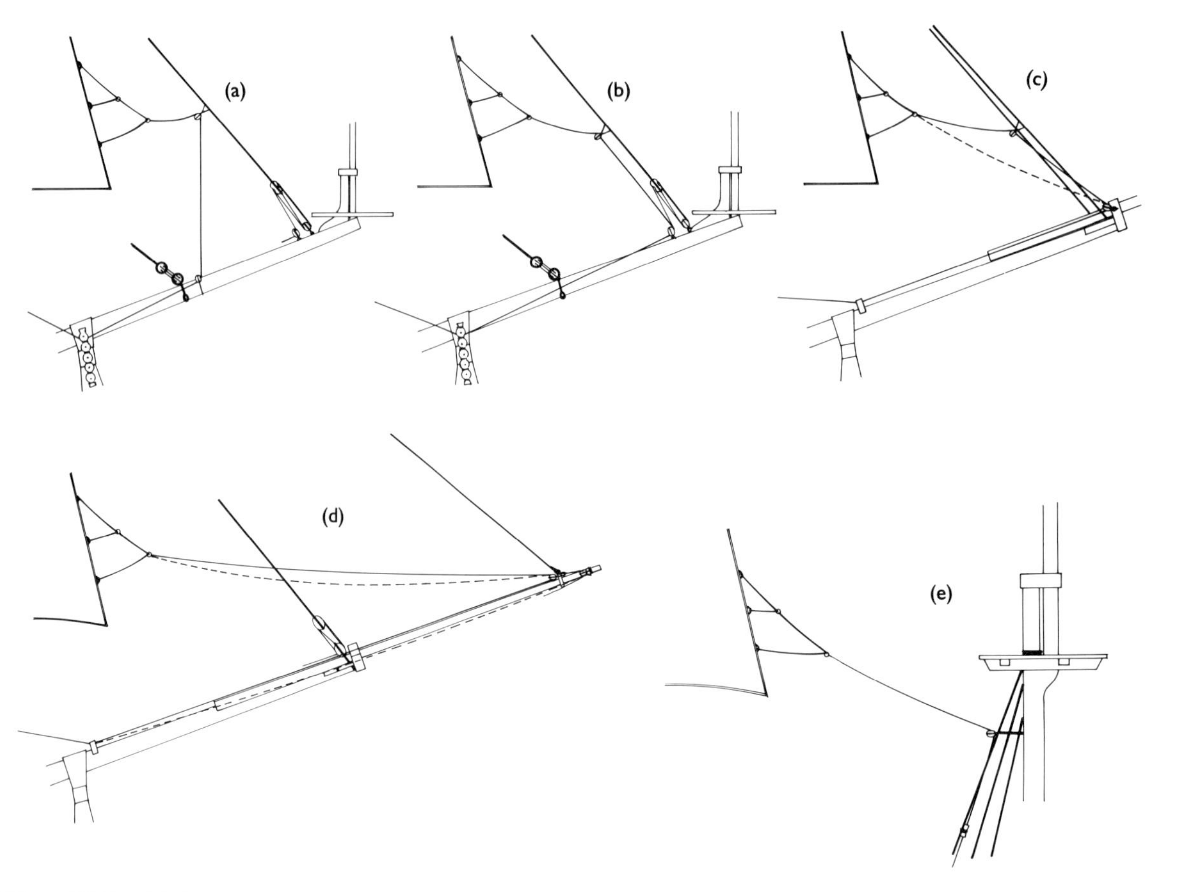

Figure 164
(a) A fore topsail bowline during the sprit topmast period.
(b) A fore topsail bowline during the sprit topmast period, with a treble lead block.
(c) A fore topsail bowline in an English ship between 1730 and 1770. The dashed line shows the lead after 1770.
(d) A fore topsail bowline in Continental ships after the sprit topmast period. The solid line shows the French lead, and the dashed line the lead shown in Chapman's sail plans of 1768.
(e) A mizzen topsail bowline.

Bowlines
Bulinen; Boulines

No significant differences between English and Continental mizzen topsail bowlines can be found. Set up to the sail with bridles like the fore topsail, the mizzen topsail bowlines led crosswise through blocks seized to the after main shrouds just below the catharpins (that is, the starboard bowline led to the port side shroud, and vice versa) then downwards through a lead truck, to be belayed at the pinrails (see Figure 164e). At the beginning of the century the author of *Der geöffnete See-Hafen* noted that the lead block in the shrouds was situated between these for the crossjack yard braces and the mizzen topsail yard braces, to be operated together, but he made no mention of these being rigged crosswise, either in his description of the braces or of the bowlines. Earlier English sources, like Sutherland or Davis did not discuss this point either; it is mentioned only in late eighteenth century works by Röding, Steel, Lever and others. From his study of seventeenth century rigging, Anderson concluded that the crossing of braces and bowlines was mainly a feature of English rigging, but it was certainly not a fixed rule, or so Davis and Sutherland would seem to suggest. Crossed mizzen topsail bowlines and braces were probably introduced to Continental rigging during the eighteenth century.

Buntlines, leechlines
Bukgordinge, Nockgordinge; Cargue-fondes, cargue-boulines

Mizzen topsail buntlines and leechlines were normally rigged like those of the fore topsail, but when a long pole topmast was stepped instead of a topgallant mast, the mizzen topsail rig was similar to that of a main topgallant sail, with neither leechlines nor reef tackle, nor sometimes even buntlines. Neither *The Seaman's Speculum* and *The Ship-builders Assistant* of 1711, nor *Der geöffnete See-Hafen* of 1705 gave any hint of mizzen topsail buntlines and leechlines for the early eighteenth century, but the first specified reeflines for all six Rates (that is, reeflines as on lower courses, not reef tackles).

Main topgallant sail
Grossbramsegel; Grand perroquet

Sheet
Schot; Écoute

The main topgallant sheet was made fast with a sheet bend to the clew of the sail and led, like the sheet of the main topsail, upon deck. According to Lever the sheet was fixed either with a double wall knot (Royal Navy), with a clinch like the topsail, or with a sheet bend (merchantmen).

Topgallant sheets were not used before 1715 or between 1735 and 1790 in Royal Navy rigging; the topsail yard lifts took their place. Topgallant sheets were noted, however, in *Der geöffnete See-Hafen*, and they were likewise well established in Continental merchant shipping. Davis (1711) made no mention of topgallant sheets on merchantmen, and Falconer (1815 edition) noted that 'in some merchant vessels, the lifts of the top-sail-yards, called the top-sail-lifts, are also used as sheets, to extend the clews of the top-gallant-sail'. This confirms that the dual use of topsail yard lifts was still customary in merchant vessels 25 years after the Royal Navy discarded the practice.

Continental ships followed very much the same pattern throughout the century, except for the last decade when the new trend of separating lift and sheet took hold. Röding explained as follows:

> The topsail lifts sometimes also serve as sheets, and then consist of a single rope, of which one end is toggled to the

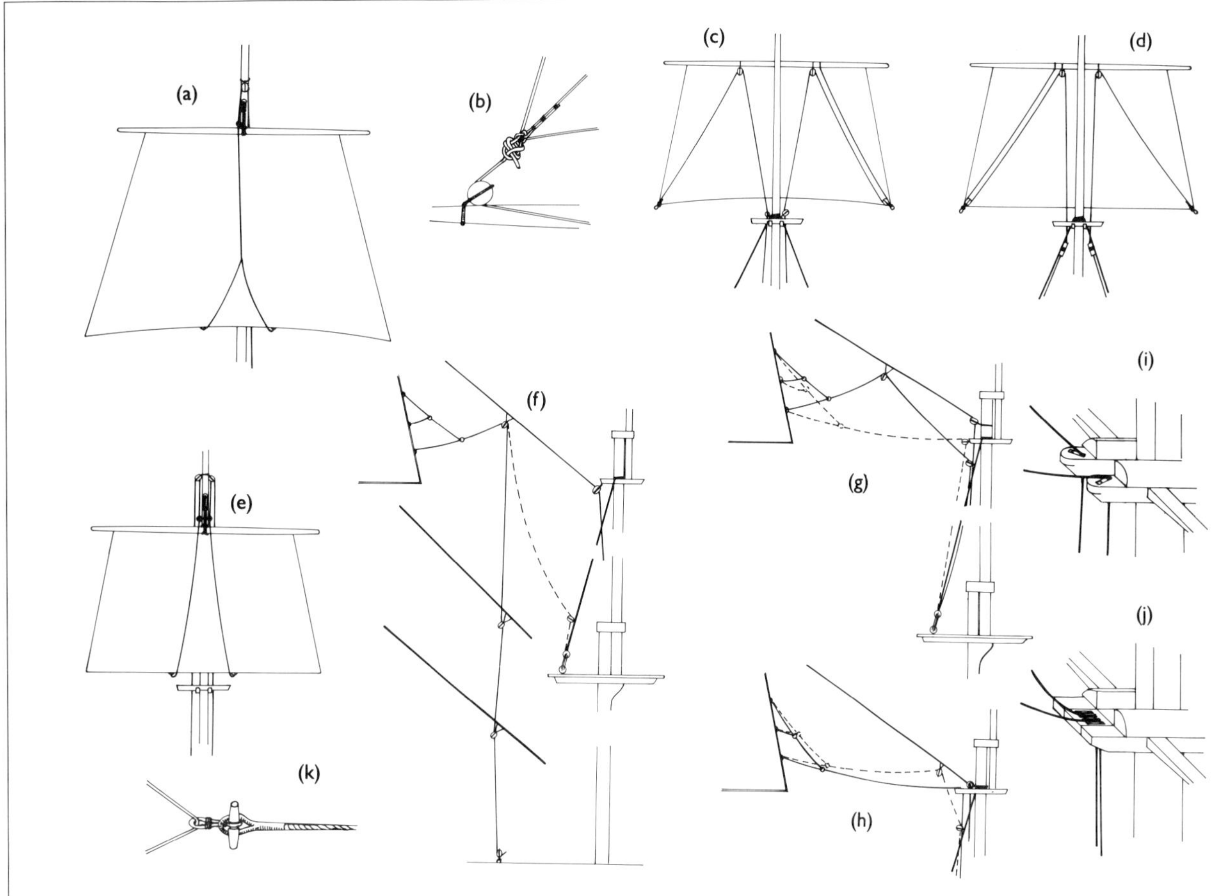

Figure 165
(a) The mizzen topsail buntline in small ships; sometimes rigged as in (e).
(b) Detail of a topgallant sheet bend and single clewline.
(c) A topgallant clewline, single and double, in Continental ships.
(d) A topgallant clewline, single and double, in English ships.
(e) The topgallant buntline in large English ships; sometimes rigged as in (a).
(f) Two Continental main topgallant bowline leads at the beginning of the century.
(g) An English main topgallant bowline until about 1750. The dashed line shows the lead between 1750 and about 1775.
(h) An English main topgallant bowline after 1775, and (dashed) the French lead in about 1780.
(i) An English main topgallant bowline after 1775 (detail).
(j) Main topgallant bowlines in merchantmen according to Lever.
(k) A toggled topgallant bowline after about 1760.

> clew of the topgallant sail. It reeves from there through a block at the topsail yardarm and another block beneath the topgallant cap upon the ship. When the topgallant sails are not set, or lowered down, the end of this lift is released from the clew and toggled instead into donkey ears placed for the purpose near the block below the topgallant cap through which the lift reeves down upon deck. When, on the contrary, the topsail is lowered, then the part of the lift which serves as a topgallant sheet must be toggled into the donkey ears.

In another connection Röding noted that by his time (around 1790) the topsail lifts were single and therefore extra topgallant sheets had to be used. Boudriot noted the use of lifts as sheets in French ships around 1780, and mentioned as an alternative arrangement what he described as the 'old' method of rigging separate sheets, rove through extra blocks at the yardarms, lead blocks near the slings, lead blocks on deck and belayed to the pinrails. He noted that the more 'modern' (actually, it would seem, nearly a century old) method of combining lifts and sheets was generally preferred. In both arrangements the belaying points were either shroud cleats or pinrails.

Clewlines

Geitaue; Cargue-points

As a rule, main topgallant clewlines were single and were hitched to the clews. They rove through blocks on both sides of the yard, then through lead trucks in the upper part of the topmast shrouds and were belayed in the top (see Figure 165d, righthand side). In some cases lead blocks were lashed to the topmast trestle trees. Measurements provided early in the century, giving a clewline length far longer than needed for belaying in the top, suggest they may have been belayed on deck.

From 1740 to the 1780 large ships were sometimes rigged with double clewlines, necessitating extra blocks in the clews and the bending of the standing parts to the yard outboard of the clewline blocks (see Figure 165c and d). In English rigging these were also belayed at the top.

Continental main topgallant clewlines came down upon deck and were belayed either to shroud cleats or to pinrails (see Figure 165c). The author of *Der geöffnete See-Hafen* noted that they came down on the third main shroud, and Boudriot described these as toggled to the clews.

Buntlines

Gordinge; Cargue-fondes

Neither leechlines nor buntlines were rigged to topgallant sails for the greater part of the eighteenth century. Only the largest ships carried topgallant buntlines at the end of the century; In English rigging these led through thimbles near the tye and through single blocks at the topgallant masthead, to be made fast in the top (see Figure 165e).

Another arrangement, mainly found in the French Navy and generally in merchant ships, consisted of a single buntline attached to a span covering the inner quarter of the footrope. After reeving through a thimble on the tye and a single block at the topgallant masthead, the buntline was belayed as above.

Although most of the larger French ships were not rigged with topgallant buntlines, Röding was quite clear that 'all square sails have buntlines,' and he also mentioned additional leechlines for the lower and topsails.

Bowlines

Bulinen; Boulines

The rigging of main topgallant bowlines was subject to variation over both time and place during the eighteenth century.

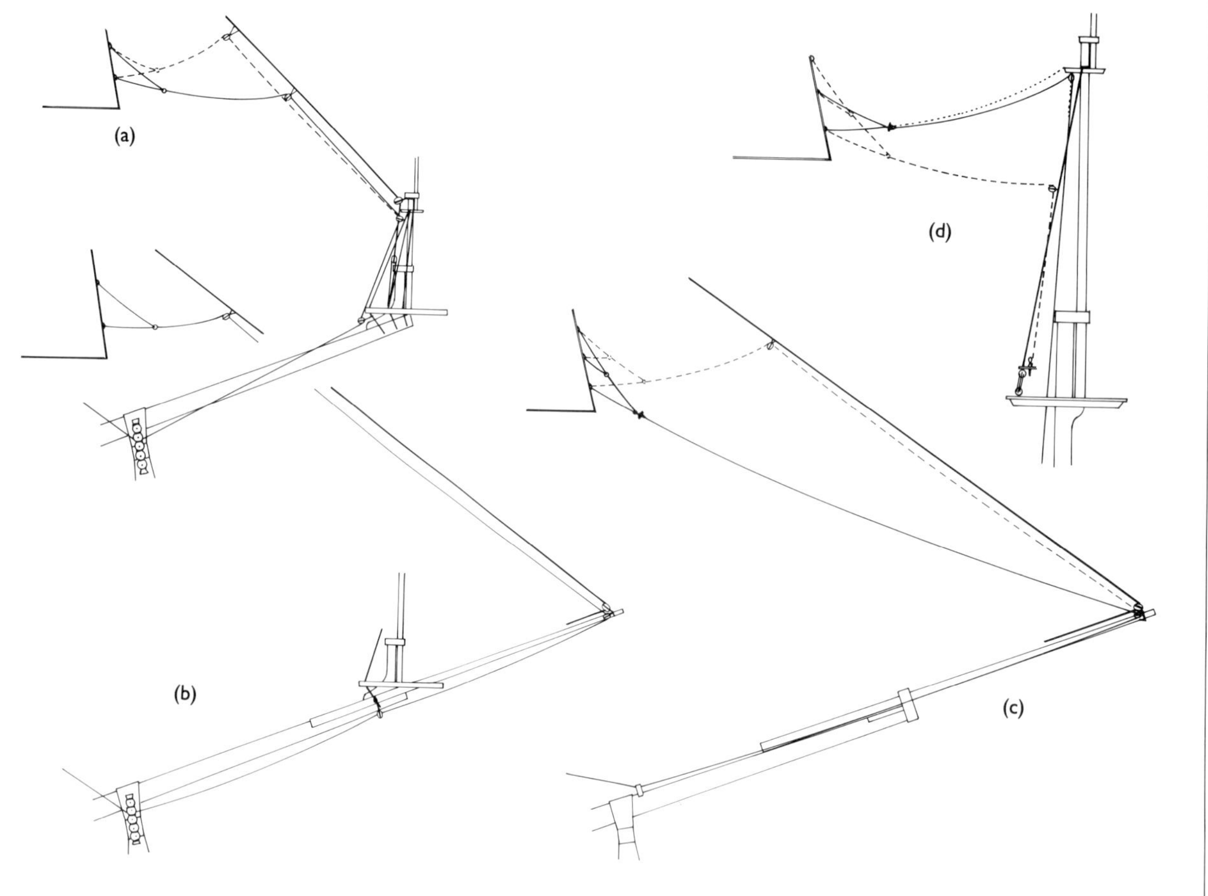

Figure 166

(a) A fore topgallant bowline as rigged during the sprit topmast period. The solid line shows the lead up to 1706, and the dashed line that thereafter.

(b) A fore topgallant bowline as rigged during the transition period from sprit topmast to jib-boom.

(c) A fore topgallant bowline up to 1760 (dashed line), and thereafter (solid line).

(d) A mizzen topgallant bowline: normal English lead (solid line); alternative lead after 1775, according to Steel and Lever (dotted line); and French lead (dashed line).

Until the 1740s bowlines generally rove through lead blocks or thimbles hanging on a short span in the middle of the main topgallant stay (described in 1705 as one quarter of the stay's length from the fore topmast). On English and early Continental ships the bowlines then led through further blocks close to the futtock stave on the aftmost fore topmast shrouds, and were belayed in the fore top (see Figure 165g); the author of *Der geöffnete See-Hafen* noted that they were led through the top and belayed to a cleat in the second aftmost main shroud.

French ships had either additional lead blocks in the upper third of the aftmost fore topmast shrouds, and the bowline made fast in the fore top, or lead blocks positioned vertically below the topgallant stay blocks on the main topmast stay and on the mainstay, with an additional foot block on deck and the bowlines belayed to a nearby cleat.

In the second half of the century the topgallant stay blocks were shifted close to the fore topmast trestle trees and further lead blocks were made fast to the upper fifth of the aftmost fore topmast shrouds, with the bowline again belaying in the fore top. From 1750 to 1775 English ships had the lead blocks made fast to the aft face of the fore topmast trestle trees, and the bowlines rove direct to the fore top (see Figure 165g, dashed line).

Röding noted lead trucks attached to the lower part of the main topgallant stay, and the bowlines reeved through blocks in the upper part of the aftmost fore topmast shrouds and then led through the lubber's hole, to be made fast to cleats on the aftmost fore shrouds.

Dutch ships around 1700 had main topgallant bowlines rigged as in the second arrangement given above for French ships, then from around 1720 followed the English arrangement. French rigging around 1780 included the block at the aftmost fore topmast shroud and the bowlines led through holes in the top, lead blocks or trucks at the aftmost fore shrouds, and belayed as described by Röding or to pinrails.

In English rigging during the last quarter century the lead blocks were integrated into the topmast trestle trees, with the sheave fitted at the aft end of the trees (see Figure 165i). The belaying point remained the same for a longer time, but at the end of the century the bowline was frequently led down to the foremast jeer bitts.

The number of bridle spans varied between two and three; at the end of the century main topgallant sails had one bridle more then the other topgallant sails in English rigging.

Lever noted an alternative to the trestle tree sheaves: a block with four sheaves completely filling the gap between the aft ends of both trees (see Figure 165j). Two of the sheaves were for the bowlines, and the other two for braces, if they were led forward. He also noted the continuing use of lead blocks stropped to the trestle trees; their survival even into the early nineteenth century can therefore still be ascertained. According to his notes these blocks were stropped not to the trestle, but to the aftmost cross tree.

Fore topgallant sail

Vorbramsegel; Petit perroquet

The fore topgallant sail was rigged similarly to the main topgallant sail, except for the rigging of bowlines. Like the main topgallant bowlines, these underwent various changes during the century.

Bowlines

Bulinen; Boulines

Until about 1706 lead blocks for the fore topgallant bowlines were fitted to a span at the middle of the fore topgallant stay.

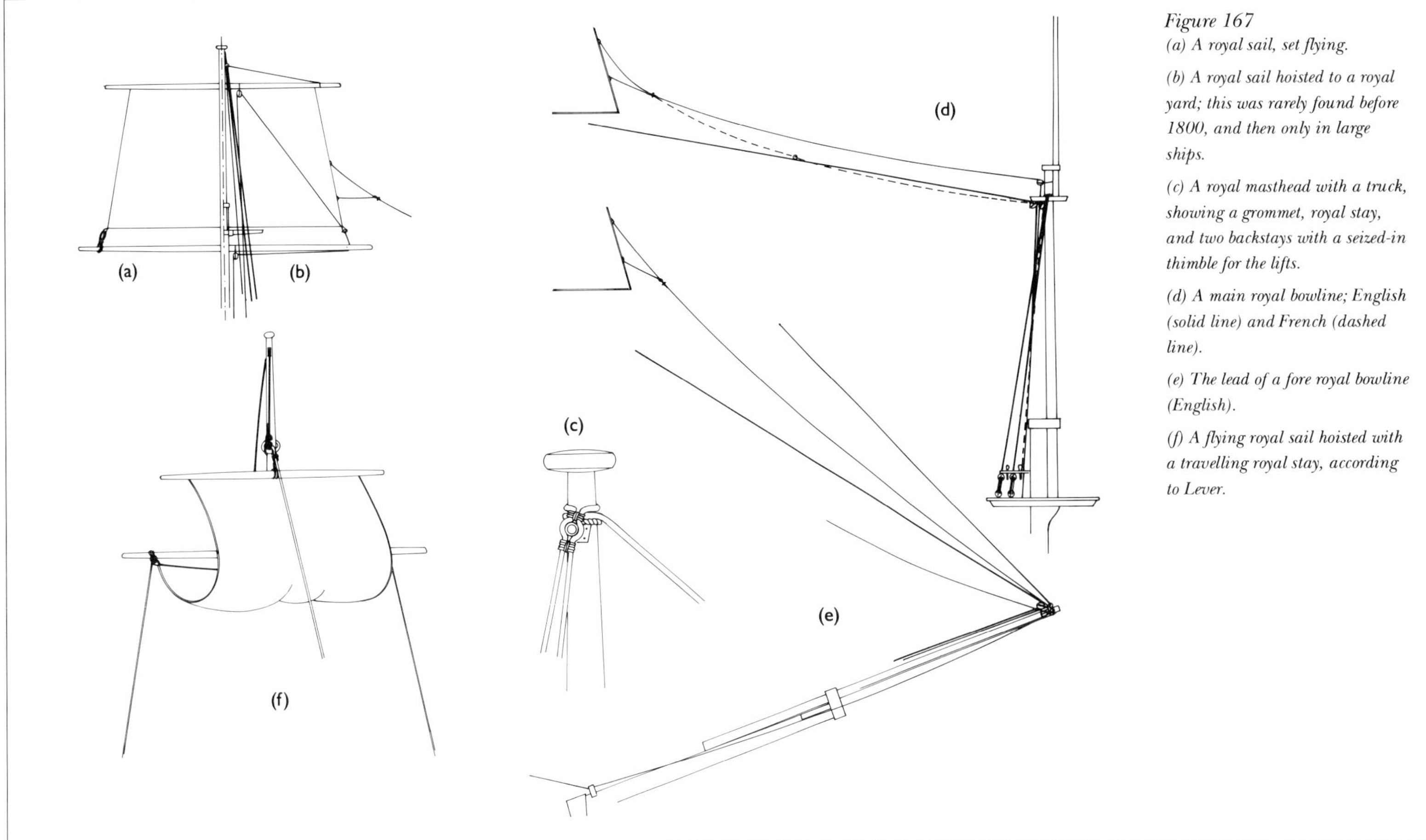

Figure 167
(a) A royal sail, set flying.
(b) A royal sail hoisted to a royal yard; this was rarely found before 1800, and then only in large ships.
(c) A royal masthead with a truck, showing a grommet, royal stay, and two backstays with a seized-in thimble for the lifts.
(d) A main royal bowline; English (solid line) and French (dashed line).
(e) The lead of a fore royal bowline (English).
(f) A flying royal sail hoisted with a travelling royal stay, according to Lever.

After that time, these blocks were moved upwards to three quarter height on the stay. The bowlines also rove through blocks on the aftmost sprit topmast shrouds, from there through blocks seized to the futtock irons, along the bowsprit, through the rack block and were belayed at the beakhead rail (see Figure 166a).

The author of *Der geöffnete See-Hafen* added a few more details. The blocks in the sprit topmast shrouds fastened about 2 feet below the sprit topmast trees, and the bowlines were led through the sprit top, then through blocks beneath it on the bowsprit and through thimbles on the fore stay collar.

After the abolition of the sprit topmast the fore topgallant bowlines were led through trucks at three quarters of the stay's height downward to lead blocks at the outer end of the jib-boom. These were either one treble or two single blocks, with the third single block, or the centre sheave, taking the fore topgallant stay.

During the transition period, when both sprit topmast and jib-boom were fitted, the bowlines often rove to the jib boom peak (see Figure 166b).

Lead trucks or blocks at the stay were discontinued in the 1760s and the bowline was led direct to the jib-boom peak blocks or thimbles. The toggling of bridles to the bowlines also began in about 1760, with the latter having an eye spliced into the upper end (see Figure 166c).

Finally, it should be noted that English fore topgallant bowlines were rigged in many ways similarly to Continental fore topsail bowlines; this can lead to some confusion for modern observers.

Continental fore topgallant bowlines rove either through thimbles seized to the strop of the treble block, or through trucks at the lower part of the fore topgallant stay, then through blocks on the outer jib-boom, thimbles at the sides of the fore stay collar, a sheave hole in the rack block, and were belayed to the beakhead rail (see Figure 164d above).

Mizzen topgallant sail

Besanbramsegel; Perruche

The only point at which the rigging of a mizzen topgallant sail differed from that of the other topgallant sails was in the arrangement of the bowlines.

Bowlines

Bulinen; Boulines

In English rigging the mizzen topgallant bowline lead blocks were made fast to the aftmost main topmast cross tree (see Figure 166d). These blocks were frequently replaced by sheaves at the aft end of the trestle trees during the last quarter century. The bowlines were belayed either in the main top, or in later years to the bitts abaft the mainmast. Small ships did not carry bowlines on this sail.

French ships had the lead blocks made fast at half height on the aftmost main topmast shrouds (see Figure 166d, dashed line), though Röding did not specify the height and noted that these bowlines were crossed like those of the mizzen topsail (and, following the same lead, were belayed beside the latter).

Royal sails

Royalsegel; Cacatois

Royal sails came into general use only during the last two decades of the eighteenth century, but had been a recognised part of rigging since the early seventeenth century (see Royal yards in Chapter III). They were officially introduced in the Royal Navy in 1779, but sail plans of Continental ships also show more common use of royal sails after that time. They were referred to by Röding as 'upper' or 'flying' topgallant sails, and he commented that they were very rarely rigged to mizzen masts, and if so, then only to the largest ships.

During the eighteenth century royal sails were usually set flying (see Figure 167a). After taking in the sail, the yard was lowered and sometimes lashed in the top to the topmast shrouds, not brought down upon deck.

A number of East Indiamen stepped royal masts, and in this case the royal sails were rigged like topgallant sails: bowlines and clewlines were therefore included.

Lever provided a clear account of royal sails from the turn of the century:

> The Royal Yards are seldom rigged across. When they are, they have a Royal Mast fidded on the Trestle trees at the Top Gallant Mast Heads: the Masts and Yards are rigged like the Top Gallant ones - Royal Masts are sometimes stepped abaft the Top Gallant Masts, on the Topmast Cap.
>
> When these Masts are not stepped, the Royals are set flying; that is, they are not rigged across, having neither Lifts, nor Braces (though sometimes the latter), but the Sail being bent to the Yard with Rope-bands made of Sennit, the Halliards which are reeved through the Sheave hole in the Pole Head of the Top-gallant Mast, are overhauled down on Deck, hitched to the Slings of the Yard, and stopped to the starboard Yard Arm (if the Yard be sent up on the larboard side), like the Top-Gallant Yard, and it is hoisted up by them. The Boy at the Mast Head having cut the Stops, secures it to the Top-Gallant Yard by a Becket for that purpose, and the Clews are lashed to the Top-Gallant Yard Arms. If it be not set at the Time it goes up, the Halliards are unbent and made fast to the Top-Gallant Stay, that they may not impede the Top-Gallant Yard, when lowering down; but if it be set, and the Fore Top Gallant Stay go with a Traveller, the Stay is let go and the Halliards being hoisted on, it traverses up with the Royal Yard by the Traveller, to which it is spliced. When the Sail is up, the Stay is set hand taught. If the Stay do not go with a Traveller, the Royal Yard and of course one of the Sheets must be shifted over it [see Figure 167f).

Sheet
Schot; Écoute

When royal sails were set flying, the clews were lashed to the yard below (see Figure 167a); in sails hoisted to a royal mast, however, a sheet bent to the clew and rove through a sheave hole in the topgallant yardarm, passed another lead block near the slings and was led upon deck (see Figure 167b).

Clewlines
Geitaue; Cargue-points

Royal clewlines were single and made fast to the clews, but only when the sail was not rigged flying. They rove through blocks near the slings upon deck (see Figure 167b).

Buntlines
Gordinge; Cargue-fondes

Royal sails had no buntlines.

Bowlines
Bulinen; Boulines

Main royal bowlines were rigged with single bridles and, in English rigging, led from the sail to blocks on the fore topgallant mast and were belayed in the fore top (see Figure 167d, solid line). French and Continental main royal bowlines rove through trucks at the upper part of the main topgallant stay, and through blocks in the upper section of the aftmost fore topgallant shrouds. They run down through the lubber's hole and via trucks to a belaying cleat in the aftmost fore shrouds (see Figure 167d, dashed line).

English fore royal bowlines ran through thimbles at the jib-boom peak and led to the forecastle. In Continental rigging the lead was again first through trucks at the upper sector of the fore topgallant stay, than down to thimbles or blocks at the jib-boom peak, following the same lead as the fore topgallant bowlines to the forecastle.

Spritsail course
Blinde; Civadière

A spritsail course was rigged with sheets, clewlines and buntlines.

Sheet
Schot; Écoute

Spritsail sheets were rigged either single or double. The single sheet was pushed through the clew with a crowned double wall knot in its end, or was made fast to the clew with a sheet bend and the end stopped. Until the end of the 1780s the sheet rove over the lower sheave in the fore course sheet channel, or during the last decade direct on to the forecastle, to belay to a cleat (according to Lever). Steel mentioned only that the sheet led inboard.

A double sheet was set up to a sheet block fitted to the clew. Until mid-century this was a special spritsail sheet block - a block rounder in shape than a normal block, and fitted with a saucer-like shoulder to protect the block's strop, which was led through holes in that shoulder. The strop was longer than a normal block strop and ended in a spritsail sheet knot, which was pushed through the spritsail clew (see Figure 168c). Röding, by continuing to show this block in his later work (1798), demonstrated its longevity in Continental ships.

In the second half of the century this block was frequently replaced by a shoulder block, like those used for leading the clewlines (see Figure 168d). The strop was shortened. Sprit course sheet blocks in French naval vessels were similar.

The position of the sheet's standing part varied throughout the century (see Figure 168a). It was made fast in the early years abreast of the foremast to a timberhead, then until 1745 in larger ships clinched to an eyebolt above the main tack passage, or to an eyebolt in the fore part of the fore chainwale in smaller vessels. From 1745 to approximately 1775 the standing part was made fast either to the chain plate of the foremost shroud, or to a timberhead at approximately the same place. After 1775 it was made fast to the same eyebolt as the fore course sheet, and during the last ten years to an eyebolt near the cathead. The lead of the hauling part was similar to that of a single sheet.

From about 1780 in French ships the standing part was shackled to an eyebolt in the lower wale, slightly abaft the cathead, with the hauling part coming in and belaying to one of the outer beakhead stanchions. The *Royal Louis*, a First Rate from this period, was rigged like an English ship, and an East Indiaman from 1760 had both parts set up to beakhead stanchions. French ships at the beginning of the century, such as *Le Fendant* of 1701, had the standing part made fast to a beakhead stanchion, while the hauling part rove first through a long-stropped lead block at the second fore shroud and then through the lower sheave hole in the fore course sheet channel; the spritsail sheet blocks were long-stropped. Another ship, probably French, flying the white ensigns and pendants of the

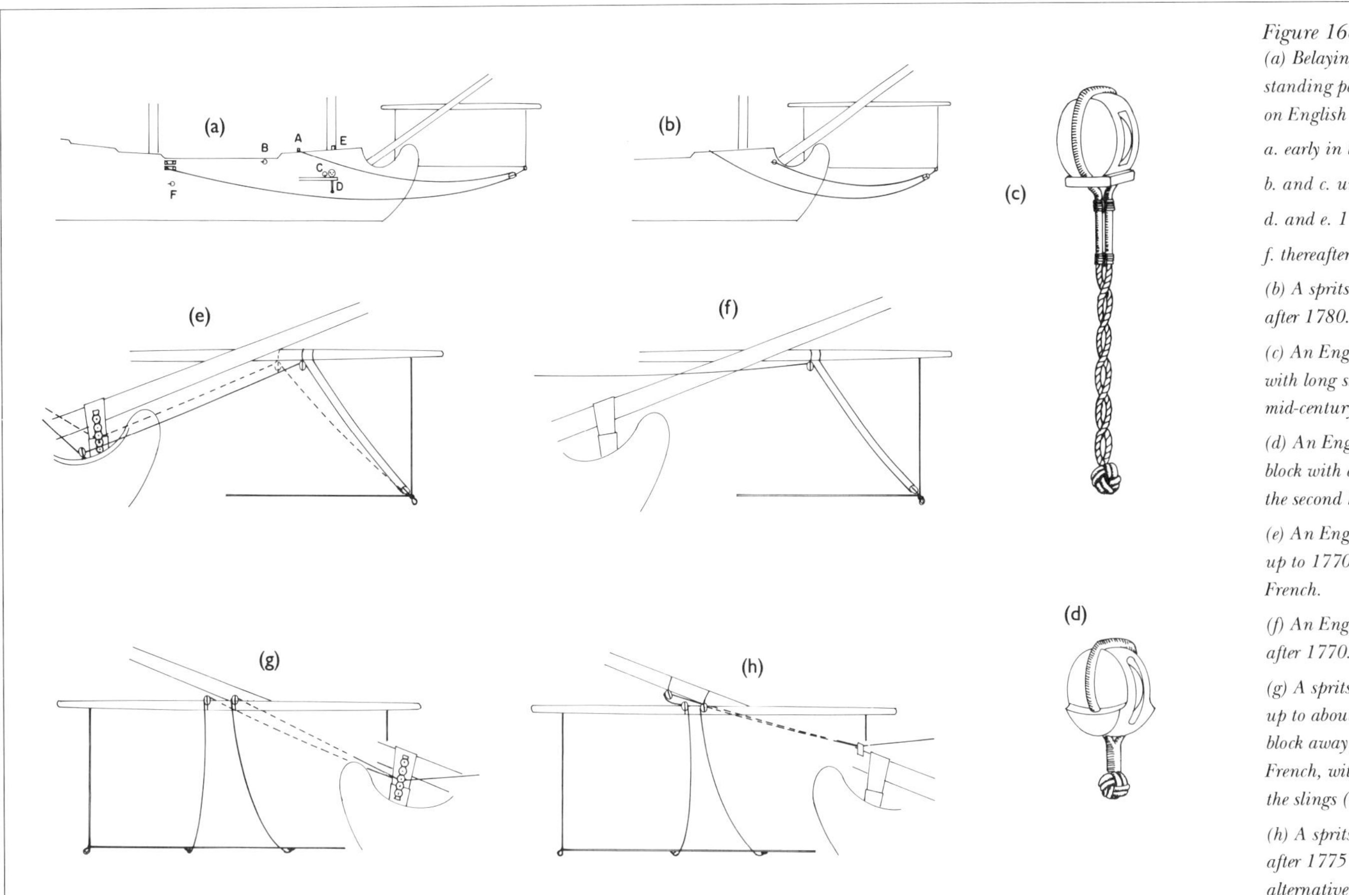

Figure 168
(a) Belaying points for the standing part of a spritsail sheet on English ships up to 1780:
a. early in the century
b. and c. until 1745
d. and e. 1745 to 1775
f. thereafter.
(b) A spritsail sheet standing part after 1780.
(c) An English spritsail sheet block with long strop, used only until mid-century.
(d) An English spritsail sheet block with a short strop, used in the second half-century.
(e) An English spritsail clewline, up to 1770; the dashed line is French.
(f) An English spritsail clewline after 1770.
(g) A spritsail buntline: English up to about 1775, with the yard block away from the slings (left); French, with the yard block close to the slings (right).
(h) A spritsail buntline: English, after 1775 (left); French alternative (right).

Bourbon era, is shown in Van IJk's *De Nederlandsche sheepsbouwkonst open gestelt* of 1697; here again long-stropped sheet blocks are evident. The standing part was made fast beneath the fore course sheet channel to an eyebolt, rove through a double block hooked or stropped to the third wale beneath the aft end of the fore channel, then through the spritsail sheet block, back through the second sheave hole of the double block and inboard through the lower opening of the fore course sheet channel.

Up to the beginning of the eighteenth century, Dutch and some other Continental ships had very long sheet pendants, reaching to the aft end of the fore channels, with single blocks spliced into the ends. The standing part of the sheet was made fast either above (according to Witsen) or below (according to Winter) the fore course sheet channel, with the hauling part again reeving through it. The *Gertruda* of 1720 had these pendants shortened to a length approximately equal to the sails's depth. The author of *Der geöffnete See-Hafen* described these pendants as reaching to halfway on the fore channel. A short strop of 1½ fathoms length with a small heart spliced into one end was hitched to the second or third fore shroud to hold the pendant up and prevent it constantly dipping into the water; this strop was known in German as a *Kundwächter*, in Dutch as a *kondwagter* and by Swedish seamen as *Kundvägtare*. The standing part of the sheet was made fast to the same eyebolt as the fore course sheet and the hauling end came inboard as described above.

Clewlines

Geitaue; Cargue-pointes

Sprit course clewlines were single or double. Davis noted single clewlines for Fourth to Sixth Rates in the first decade of the eighteenth century, and clewlines for the upper Rates which were probably meant to be similar to his '2 Clew-lines with Blocks in the Clews' for an East Indiaman of 84 feet keel length (that is, double). The author of *Der geöffnete See-Hafen* mentioned only single clewlines.

Single lines were hitched to the clews, and double to the yard about one third outside the slings; the latter rove through the clewline blocks at the clews and then, in similar fashion to single ones, through yard blocks slightly inboard of the standing parts (see Figure 168e). In 1705 these blocks were at one quarter of the yard's length, and in French ships they were placed just outside the slings (thus much closer to the centre of the yard than English clewline blocks). Boudriot, however, gave the same position as for English clewline blocks. French clewlines in any event were usually single.

Up to about 1770 the clewlines were led towards the forecastle through lead blocks at the headrails, or less commonly through one of the sheave holes of the rack block (see Figure 168e). After 1770 these lead blocks became obsolete, and the clewlines were belayed direct to the beakhead rail (see Figure 168f). Continental ships at the beginning of the century had these lines belayed on a pinrail in the head.

Buntlines

Gordinge; Cargue-fondes

Two buntlines were rigged about 3 feet outside the sprit yard's centre to the face of the sail. They rove through thimbles or small single blocks at the yard and after passing through a sheave hole in the rack block were belayed to the beakhead rail (see Figure 168g). In the last quarter century spritsail buntlines were led additionally through blocks at the bowsprit (or only through these blocks, without yard blocks or thimbles) and via the lead saddle towards the forecastle. At the beginning of the century Continental ships sometimes had only one buntline fitted with a lead block lashed to the bowsprit. The line led then into the head and was belayed beside the clewlines.

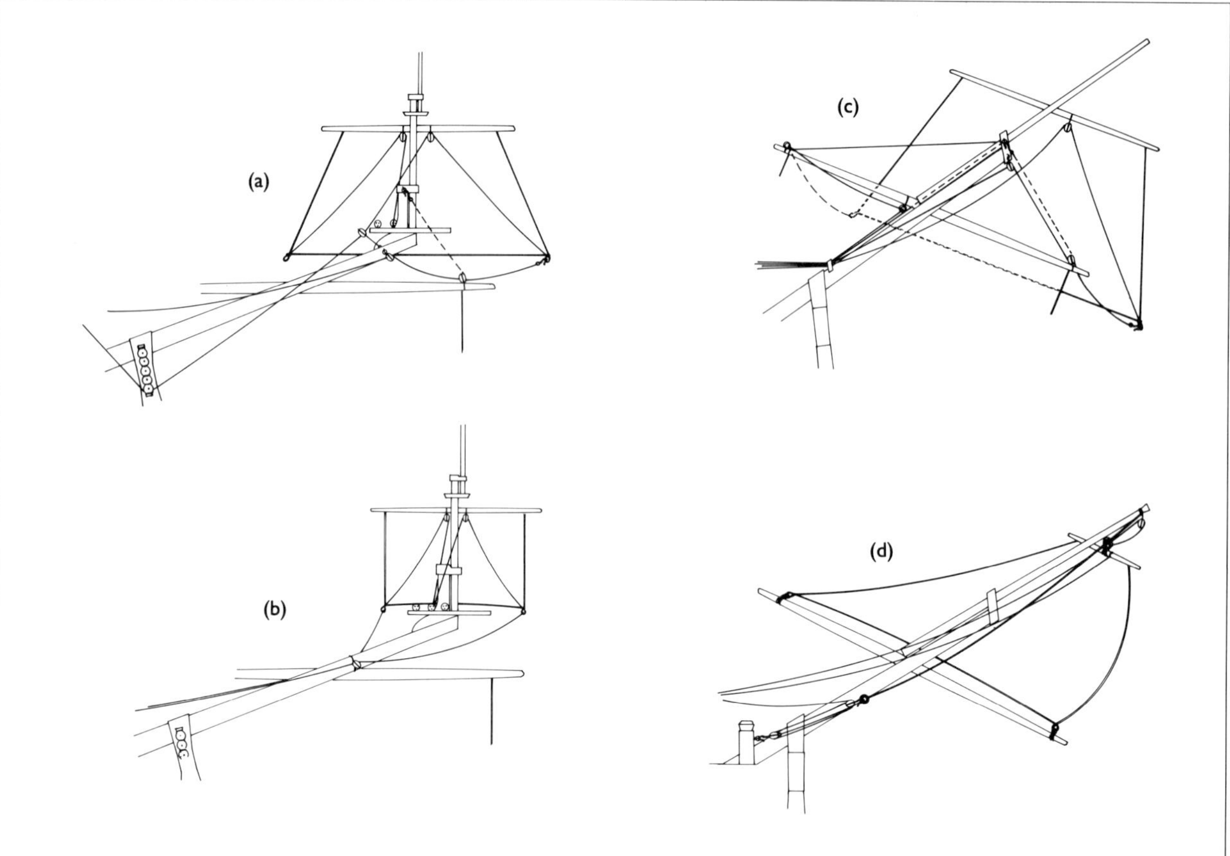

Figure 169
(a) Sprit topsail clewlines and sheets during the sprit topmast period. The dashed line indicates the lead of the spritsail lift when it was not used as a sprit topsail sheet. The port clewline is belayed in an alternative fashion, around a futtock deadeye in the bowsprit top.

(b) Sprit topsail clewlines and sheets in a French ship at the beginning of the century; these lines might also have been rigged as shown in (a).

(c) Sprit topsail clewlines and sheets on a jib-boom. On the right side of the drawing the spritsail lift acts as a sprit topsail sheet; on the left a single spritsail lift and an additional topsail sheet are rigged. The latter reeves through thimbles or small blocks at the yardarms and lead blocks beside the yard sling.

(d) A flying sprit topsail, mainly rigged in East Indiamen, as described by Lever. The sprit topsail yard is much shorter than normal and is lashed to a traveller, running on a jackstay, which is clinched round the jib-boom head and is set up with a luff tackle to the bow. The clews are lashed to the spritsail yardarms. An outhauler connected to the traveller reeves through a block at the jib-boom head to the forecastle. An inhauler hitched to the yard also leads to the forecastle, belaying at the beakhead rail.

The latter arrangement can also be seen in the rigging of *Le Fendant* of 1701, except for a different belaying point. The buntlines of a French spritsail were made fast to cringles in the footrope about one third from the centre, and rove through a block above the slings and through the fourth sheave hole in the rack block on the gammoning (see Figure 168g, right).

A different description was provided by Boudriot. The buntlines were made fast to the cringles as before, but rove through thimbles made fast with a medium length strop to a cringle in the middle of the footrope, then came up at the rear of the sail to a double block beneath the sling strop. From there they passed through the fifth sheave hole in the rack block toward the beakhead rail. It seems highly unlikely that the buntlines ran up the rear of the sail, and this may have been a slip of the pen on Boudriot's part.

Small vessels often had only one buntline, fitted to a span at the footrope and reeving through a block at the yard slings towards the beakhead.

Sprit topsail

Bovenblinde, Schiebeblinde; Contre civadière

A sprit topsail was bent to the yard with lacing and earrings according to Steel, and with robands and earrings according to Lever. Neither author provided dates or further details, so it can be assumed that both fastening methods were common practice during the same period.

Sheet

Schot; Écoute

No sprit topsail sheet was rigged, for a sail hoisted to a sprit topmast or to a jib-boom, when the spritsail was rigged with double lifts. The sheet's function was taken up by the spritsail lift, which was hooked to the topsail clew. In his rigging lists for the spritsail yard, Davis included standing and running lifts but no sheets for the sprit topsail; this indicates that this arrangement was already in use at the beginning of the century. The same indication is given in *Der geöffnete See-Hafen*, where *Spanische Toppenanten* (standing lifts) and *Toppenanten* (running lifts) are described. After being made fast to the clew of the topsail, the latter passed through the yardarm block and a block underneath the bowsprit top, ran along the bowsprit, through a thimble seized to the fore stay collar and belayed on the beakhead rail.

The introduction of single lifts made a sheet for the sprit topsail essential. It was bent to the clew with a double sheet-bend, though in Continental rigging the use of toggles in the ends of sheets, to lodge them into the clew eyes, was quite common. Single lifts rove through blocks or thimbles at the spritsail yardarms, then through lead blocks at the bowsprit forward of the spritsail yard parrel (sling), and sometimes also via the rack block, to belay as noted above (see Figure 169a).

French ships with a small sprit topsail during the sprit topmast period had no yardarm blocks or thimbles; their sheets rove direct through the bowsprit lead blocks (see Figure 169b).

Clewlines

Geitaue; Cargue-pointes

Sprit topsail clewlines were normally single and led from the clews to blocks which hung about 2 feet outside the sprit topsail yard's centre (or according to the author of *Der geöffnete See-Hafen*, at one quarter of the yard's length). Davis suggested that during the sprit topmast period these clewlines might also be rigged double: he mentioned 'two single Clew-lines 28 fathom, or as for double Braces' and 'two going double 36 fathom, or 6 times the Mast and Yard, adding 6 Foot'. A double clewline was an alternative for nearly all ships, right down to the Fifth Rate. By contrast, Röding noted in 1794 that sprit topsails had neither clewlines nor buntlines.

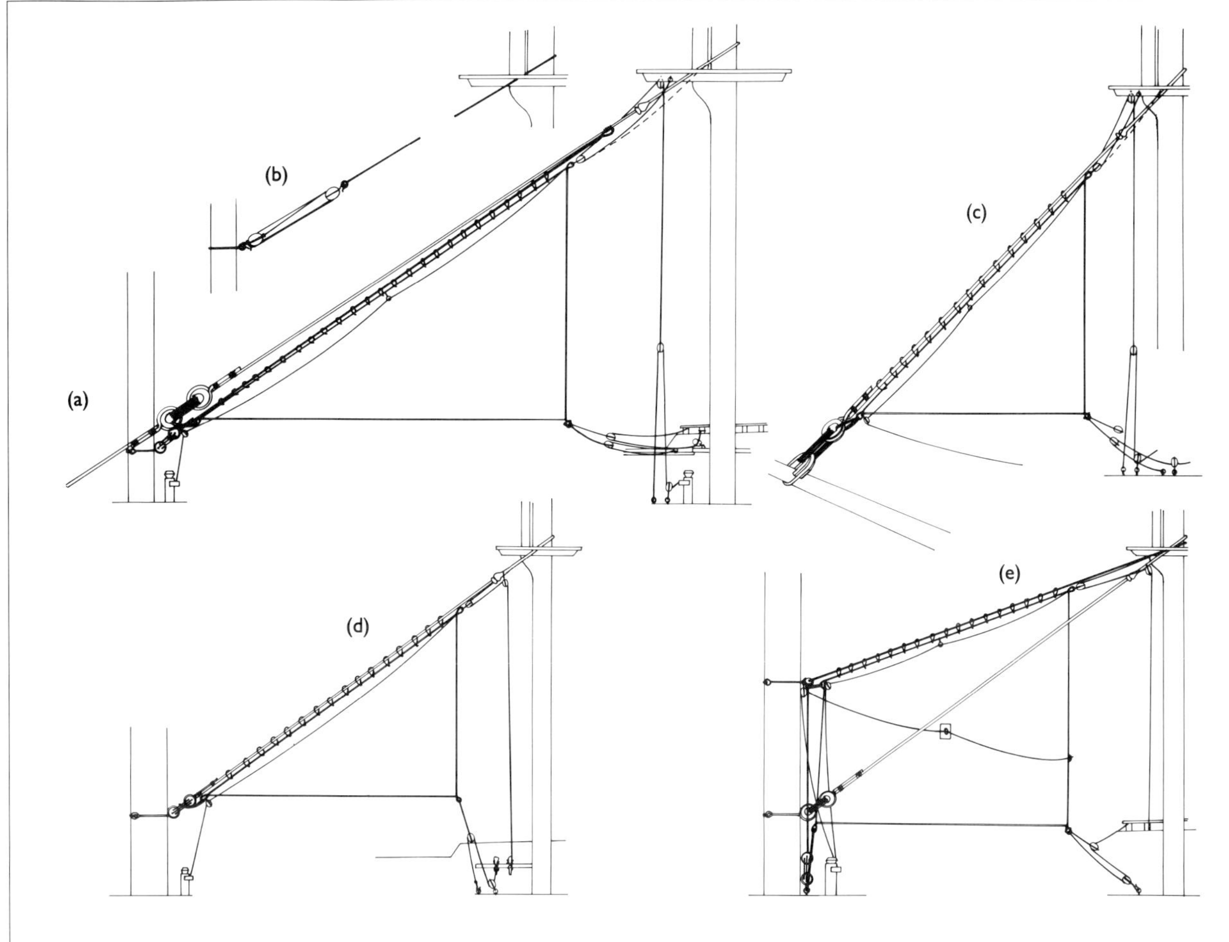

Figure 170
(a) The rigging of a main staysail. This sail was frequently hoisted to a staysail stay, but Lever noted the use of the main preventer stay for the purpose. The drawing shows the two positions, as noted by Steel and Lever.
(b) A main staysail stay set up with a luff-tackle, as described by Steel.
(c) The rigging of a fore staysail.
(d) The rigging of a mizzen staysail, carried at the beginning of the century in all, but after mid-century only in small, ships.
(e) A mizzen staysail after 1760.

Sprit topmast lead blocks were spliced to a span placed through a hole in the sprit topmast standard; the downward-leading hauling part of the clewline rove through these blocks, then through the rack block or a thimble seized to the fore stay collar, and belayed at the beakhead rails (see Figure 169a). When a topsail was rigged to the jib-boom, the incoming part was led from the yard block directly through the lead saddle or rack block, depending on what was fitted, to the beakhead rails (see Figure 169c). Sometimes an extra set of lead blocks was stropped to the bowsprit head.

Main staysail

Gross-Stagsegel; Grand-voile d'étai

As noted in Chapter X, the main staysail was not often set in large ships and was a more common sight in brigs and similar vessels. It was hoisted with hanks to the main preventer stay or a staysail stay (see Figure 170a and b). A main staysail was described in *Der geöffnete See-Hafen*, and Davis also mentioned this sail for HMS *Royal Sovereign* of 1701 (a First Rate) and HMS *Vanguard* of 1678 (a Second Rate). Boudriot noted a main staysail as part of the rigging of a French Third Rate from 1780, and Röding as part of the rigging of all ships. By contrast, Falconer, Steel and Lever all noted that a main staysail was seldom bent in ships. From these differing views it can be ascertained that the main staysail was common to all ships during the sprit topmast period, was not much in favour in the Royal Navy for ships after that period, but was still common in Continental rigging until the end of the eighteenth century.

Tack

Hals; Point d'amure

The tack of a main staysail was lashed to the stay collar.

Halyard

Fall; Drisse

Main staysail halyards were double in English ships. Until 1745 the standing part was hitched to an eyebolt beneath the port side main trestle tree; after this the standing part was laid round the mainmast head.
French ships mainly had single halyards, with the standing part hitched to the peak of the sail. Continental ships at the beginning of the century had either double or single halyards, with the standing part made fast to the lead block, or in case of single halyards to the peak of the sail.

The leading part of double halyards rove through a single block in the sail's peak, a lead thimble lashed to the stay, and through a block at the starboard main trestle tree upon deck. A single block with a whip was spliced to the lower end; the standing part of the whip was made fast to an eyebolt on deck and the hauling part led through a foot block nearby and was belayed to the mast bitts.

The author of *Der geöffnete See-Hafen* noted a lead block made fast about 4 feet above the staysail stay and a single block hooked into a thimble in the sail's peak; no whip or foot block were mentioned.

The French arrangement, as described by Boudriot, was much simpler. A single halyard passed through a block lashed to one of the cross trees and a foot block at the starboard side near the mast, and was made fast around the block.

Downhauler

Niederholer; Hale-bas

The downhauler was hitched to the peak eye of the sail, and led downwards through some of the hanks or a lead-truck sewn to the stayrope, to a block at the stay lashing, and was belayed at the forecastle.

Staysail stay

Leiter; Faux étai, contre étai

As noted above, the main staysail was frequently hoisted to a staysail stay throughout the eighteenth century. This stay was an additional rope, hitched to the main stay collar above the mouse and the stay lashing. The lower part was sometimes set up to a tackle hooked to a strop round the foremast (at the beginning of the century deadeyes were fitted to the lower end and to the foremast instead of the tackle).

Sheet

Schot; Écoute

A rope was hitched to the staysail clew to form two pendants, and into the end of each a single block was spliced. The standing part of the sheet was clinched to an eyebolt, or made fast to a timberhead, near the gangway (according to Lever; Steel noted one of the first quarterdeck timberheads only). After passing through one of the sheet pendant blocks, the hauling part rove through a block in the gangway (according to Lever), or was led through a snatch block on the side and belayed to the next timberhead. Sometimes a luff tackle was used to bowse the sheets aft.

In French rigging a thimble was seized into the clew and a tackle took the sheet's function. The other end of the tackle hooked to an eyebolt on the upper deck near the waterways. This arrangement was also described in *Der geöffnete See-Hafen* and was therefore used in Continental rigging for the whole century.

Fore staysail

Vorstagsegel; Petit foc

The fore staysail (see Figure 170c) came into use in the Royal Navy no earlier than 1773. It was, according to Lever, 'seldom used in any Ship but Men of War,' and was bent to hanks on the fore preventer stay. Lees noted that staysails were seldom fitted to the fore stay and one would expect only large ships in the late eighteenth century to be rigged with a fore staysail, but Steel not only provided a description and drawing of a fore staysail for a 20-gun ship, set on the fore stay, but also gave the measurements for the necessary rigging, and for all merchantmen as well as for all Rates in the Royal Navy.

French ships were not usually rigged with a fore staysail in the same sense as English vessels, and Boudriot described the fore staysail as a storm jib. Röding, however, noted that the sail was hoisted to the fore stay and was of a smaller size then the main staysail and the "ape" (mizzen staysail); he regarded the storm jib as a different sail, and his comment on this sail was 'the [storm jib] is only used during a storm, and its tack does not travel as far out as the others'. He also defined a jib as a sail hoisted on a staysail stay up to the fore topmast head. A fore staysail could therefore never have had jib status. The author of *Der geöffnete See-Hafen* provided a short note that everything regarding a fore staysail could be found under Main staysail, thus at least confirming the use of both.

In connection with Continental rigging it is worth noting that in an etching by Passebon, a captain in the French Navy who died in 1705, a fore staysail is shown furled on the fore stay of a 112-gun ship. This ship also had a jib-boom lashed to the bowsprit with a jackstaff and standard, but without the normal sprit topmast of the period. A flying jib completed the novelties of this remarkable vessel.

Given that this etching was made by a very knowledgeable, high ranking naval officer and not by an artist who lacked genuine nautical experience, it provides proof that the innovations of Peregrine Osborne, Marquis of Carmarthen, in 1695 in the headgear of HMS *Royal Transport* were quickly taken up by the more far-sighted of the naval officers of other nations. Passebon's etching may represent a visualisation of the practical application of such innovations rather than an illustration of the rigging of an existing vessel.

Halyard

Fall; Drisse

The fore staysail halyard was similarly rigged to that of the main staysail.

Tack

Hals; Point d'amure

The tack was a short rope, spliced to the eye in the boltrope at the tack of the sail and led alternately through the stay heart and the eye until extended, with the end hitched.

Downhauler

Niederholer; Hale-bas

The fore staysail downhauler was rigged in the same fashion as that of the main staysail.

Sheet

Schot; Écoute

The sheet was again set up to two pendants with blocks. The standing part was made fast to an eyebolt on the forecastle or to the inside of the bow. The hauling part was led through a foot block on deck and belayed to a cleat or a belaying pin in the rails. In Continental rigging the fore staysail sheet was rigged in the same fashion as the main staysail sheet.

Mizzen staysail

Besanstagsegel; Foc d'artimon

During the period in which a full mizzen was in use, whether lateen or settee, it must have been nearly impossible to hoist a mizzen staysail, but many contemporary pictures and the few remaining sail-rigged models of the period do show this sail. It is also included in Davis' and Sutherland's rigging lists for English men-of-war, and mentioned in *Der geöffnete See-Hafen* (see Figure 170d and e and 171a).

Staysail stay

Leiter; Faux étai, draille

At the beginning of the century in English rigging the mizzen staysail was often still bent direct to the stay. Davis and Sutherland made no mention of a separate mizzen staysail stay in 1711, but the use of such a stay must have become more frequent soon after this. The stay was made fast round the mizzen masthead and rove through a thimble stropped at half height to the aft side of the mainmast. A thimble was turned into the lower end

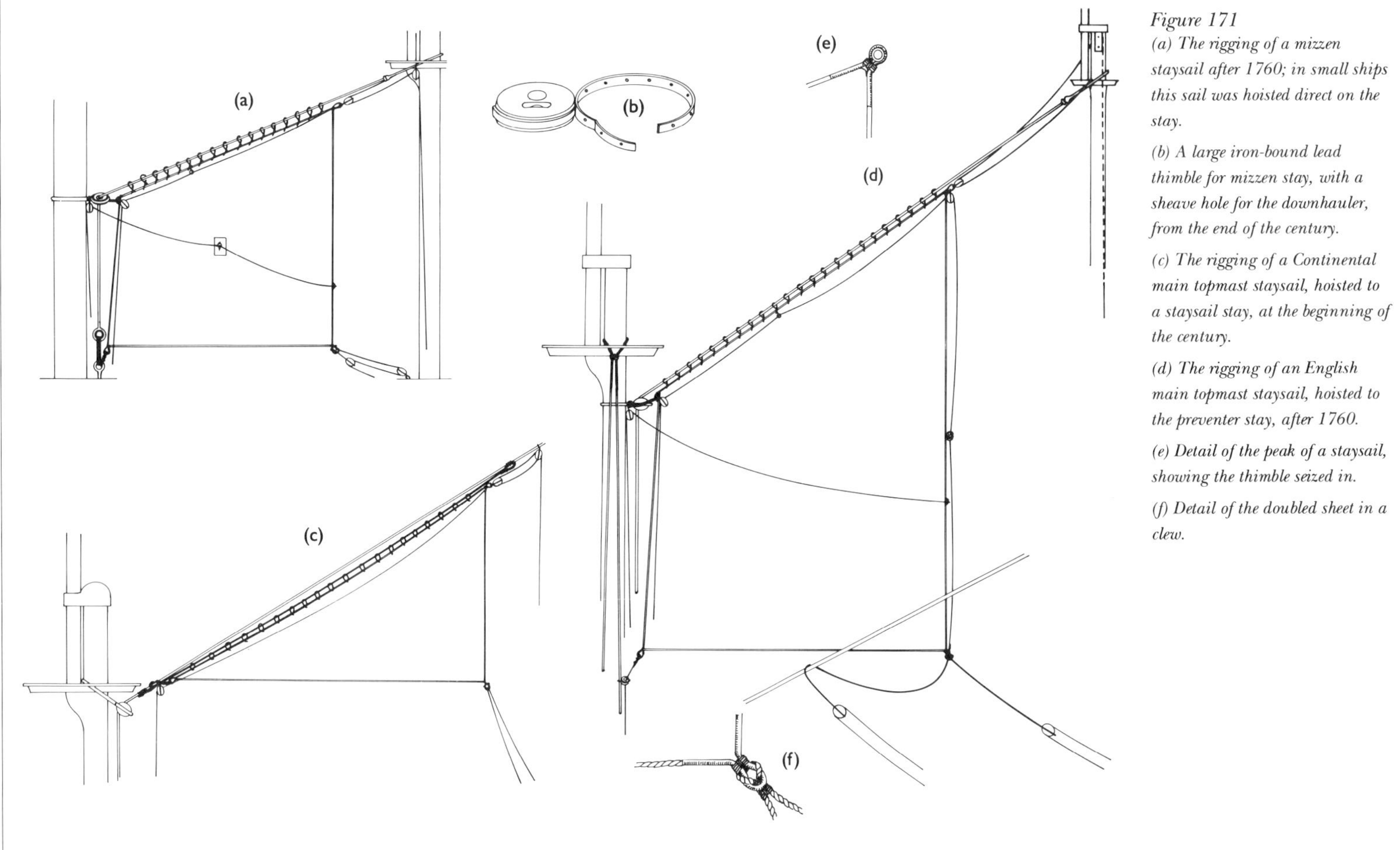

Figure 171
(a) The rigging of a mizzen staysail after 1760; in small ships this sail was hoisted direct on the stay.
(b) A large iron-bound lead thimble for mizzen stay, with a sheave hole for the downhauler, from the end of the century.
(c) The rigging of a Continental main topmast staysail, hoisted to a staysail stay, at the beginning of the century.
(d) The rigging of an English main topmast staysail, hoisted to the preventer stay, after 1760.
(e) Detail of the peak of a staysail, showing the thimble seized in.
(f) Detail of the doubled sheet in a clew.

of the stay, and this was set up to an eyebolt in the deck with a lanyard. In smaller ships after mid-century the mizzen staysail was again hoisted on the mizzen stay.

Continental ships were rigged with a mizzen staysail stay from the beginning of the century. Made fast to the stay at about three quarter height, it was set up similarly to the main staysail stay. In French rigging the mizzen staysail stay was lashed to the stay above the stay collar mouse, then it rove through the mainmast block or a thimble, and was hitched to the mizzen stay seizing.

Nock
Nock; Point de drisse

The nock of a quadrilateral mizzen staysail after 1760 was seized to the strop of the lead thimble for the staysail stay.

Tack
Hals; Point d'amure

The tack of a quadrilateral sail was made fast to an eyebolt on deck, and that of a triangular sail to the lead thimble strop noted above. Where the sail was hoisted on the mizzen stay and the stay was set up with deadeyes to the mainmast, then an extra eyebolt was fitted on deck for the tack.

Halyard
Fall; Drisse

The halyard rove through a block at the sail's peak, and its standing part was made fast around the mizzen masthead. The hauling part led through a block on the port side trestle tree or mizzen masthead, then through a lead block or truck in the upper section of the mizzen shrouds down to a pinrail or timberhead. A single halyard was hitched to the sail's peak and led the same way, to be belayed similarly on the port side.

Downhauler
Niederholer; Hale-bas

The downhauler led from the sail's peak down to the bitts abaft the mainmast via a truck sewn to the stayrope and a lead block seized to the nock. In some cases the downhauler rove through a sheave hole in the stay collar heart instead of the lead block, and was lashed round the mast.

Sheet
Schot; Écoute

In English rigging the mizzen staysail sheet consisted of a short and a long part. A block was seized to the short part, while the long end led through a block on one side of the deck, then via the block in the short leg to a timberhead or cleat at the same side, where it was belayed.

A similar foot block and cleat was positioned on the opposite side of the deck so that only the long leg had to be unreeved when shifting the sheet, and not the block.

In Continental rigging a single sheet might also be used, belayed to pinrails on either side of the ship, or a tackle as for the main staysail.

Brails
Dempgordinge; Cargues

In some instances brails were rigged to a mizzen staysail. Clinched to a cringle in the lower after leechrope, the brails passed on both sides through thimbles sewn to the centre of the sail (which was strengthened with a patch) to two single or one double block seized to the staysail stay collar round the mainmast. They were belayed to the bitts abaft the mainmast.

Main topmast staysail

Grossmarsstengestagsegel; Voile d'étai de grand hunier

At the beginning of the century, before the introduction of the preventer stay, the main topmast staysail was hoisted on a staysail stay.

Staysail stay

Leiter; Faux étai, draille

The main topmast staysail stay was hitched and seized, near the mouse, at three quarters height, to the lower side of the main topmast stay and led down to the rear of the fore stay collar, where it was set up with deadeyes.

An early Continental staysail stay (see Figure 171c) was set up similarly, but later it was led through a stropped block hanging at the aft side of the cheeks down from the foremast head, and was then set up to deadeyes on deck or to the catharpins.

Halyard

Fall; Drisse

When the sail was hoisted on a staysail stay, a block was seized above it to the stay, with the standing part of the halyard bent to a becket in its tail. The hauling part passed through a block in the sail's peak and returned through the first block upon deck.

The introduction of the quadrilateral staysail also resulted in an alteration in the leading of the halyard. The standing part was now laid round the topmast head, and the halyard rove through the sail's peak block as before, but then over the upper sheave in the starboard cheek block to the aft side of the top, coming down to a foot block in the starboard spirketting to be belayed to a pinrail (see Figure 171d).

Boudriot noted a single halyard for a 74-gun ship of about 1780, while Ozanne and Roux generally showed the sail hoisted with double halyards, except on smaller ships like corvettes; this is also confirmed in the work of another contemporary, Röding. Boudriot also noted that the halyard was led through a long-stropped block or a cheek block on the starboard side of the main topmast head, via a hole in the top, upon deck, belaying to a foot block near the mainmast or to the bitts. All contemporary pictures of Continental ships show the halyard leading to a point below the topmast trees rather than above, so the long-stropped block is more probable.

Lever confirmed this: 'If there be no Cheek-blocks at the Mast-head, a Span with a double Block at each end, is put over in the same manner... These double Blocks are for the Jib, Halliards, Stay, &c.'

Röding noted a block made fast to the cap in Continental rigging, and he made no reference to the cheek block in his German dictionary, nor was there a *joue de vache* or *demi-joue* in his French index, but his entry 'Cheek block' in the English index explained the use of this type of block in the English rigging.

The use of a cheek block for the rigging of a Continental main topmast staysail halyard seems therefore conjectural, based on English sources only. In fact, what Boudriot considered to be a cheek block (the only illustration he provided can be seen in his drawing of a main topmast head with all its rigging) was a block shaped like a Continental upper lift block and lashed to the topmast head immediately above the masthead rigging. His use of the term 'cheek block' for this block is entirely misleading.

A triangular staysail hoisted to the preventer stay had a single halyard only, which was hitched to the sail's peak and rove through a block near the mast top to be led upon deck.

Nock

Nock; Point de drisse

The nock of a main topmast staysail was seized to the preventer stay lead block collar.

Tack

Hals; Point d'amure

The tack of a triangular sail was set up similarly to the nock. A quadrilateral sail had a doubled tack, with its bight put through the eye of the sail's tack.

The ends passed through lead trucks on each side in the lower shrouds and each belayed round a shroud cleat near the deck.

Downhauler

Niederholer; Hale-bas

Hitched to the peak of a triangular sail, the downhauler rove through one or more hanks and through a lead block near the tack upon deck.

On the quadrilateral sail a different arrangement was used. Instead of being hitched to the peak, the downhauler passed through another lead block there, then through a truck in the middle of the after leech, and was made fast to the clew. The extension of this line to the clew made the downhauler act also as an uplifter, used to lift the clew to the other side of the main stay when the necessity arose.

Brails

Dempgordinge; Cargues

Because of its size, the quadrilateral sail was also rigged with brails.
These were made fast to a cringle in the after leechrope about one third of the leech's length up, led through two single or one double block at the preventer stay's lead block collar, then through another set of lead blocks seized to the strop of the main bowline block and were belayed to the mast bitts.

Sheet

Schot; Écoute

The sheet was doubled and the bight was put round the clew, with the ends then passed through it (see Figure 171f). Blocks were spliced into each end. According to Steel the standing parts of the sheet falls were made fast to both ends of a boatskid near the quarterdeck, while the hauling part led through a block on the gunwale abaft the gangway and was secured around a pin in the boatskid. Lever had the standing part fixed to a timberhead abaft the gangway, and the hauling part belayed round a cleat.

A contemporary Dutch model from the later part of the seventeenth century had a relatively small triangular staysail rigged with one double sheet set up to the middle of the main stay, and the hauling end leading along the stay towards the foremast.

Boudriot described these sheets without falls. They passed through the fourth sheave hole in the fore course sheet channels at the forward edge of the quarterdeck, and were belayed round pins in the rails or led direct to the main topsail sheet bitts.

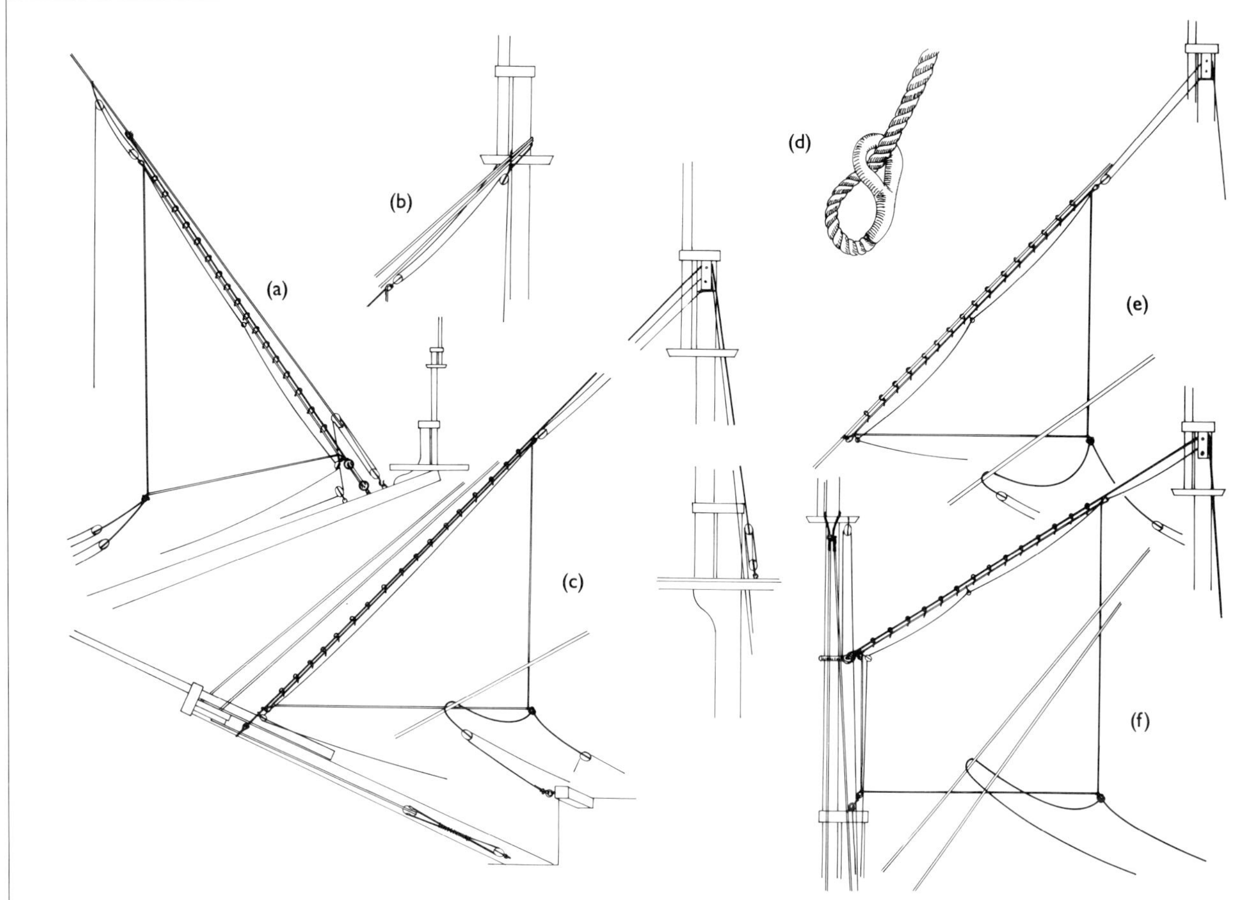

Figure 172
(a) The rigging of a fore topmast staysail during the sprit topsail period, hoisted on a staysail stay.
(b) A fore topmast staysail halyard block stropped round the topmast head after 1720.
(c) The rigging of a fore topmast staysail on a staysail stay after 1750, as found in English men-of-war.
(d) Detail of the running eye of a staysail stay.
(e) An English fore topmast staysail hoisted to a preventer stay, from the second half of the century.
(f) The rigging of a middle staysail. The staysail stay here is fastened with a grommet around the mast; an alternative arrangement was with a horse (jackstay).

Fore topmast staysail

Vormarsstengestagsegel; Second foc ou troisième foc

The fore topmast staysail was triangular, and was in use from the second half of the seventeenth century. It dated in the Royal Navy from 1660, but there is little evidence of its existence at this date in Continental rigging. A Dutch man-of-war model from 1660-70 (lost during a bombing raid on Berlin) had a fore topmast staysail, but the badly damaged rigging of this model had been restored in about 1890 and the small staysail hoisted to the fore topmast stay could have been an addition from that time. The author of *Der geöffnete See-Hafen* made no mention of such a sail, although he noted the fore staysail, then unknown in English rigging (the latter can also be seen in a painting from 1693 by the French sculptor Pierre Puget, and in an etching by Passebon from before 1705). An early example of a fore topmast staysail on a Continental ship can be seen in a painting of the Dutch 52-gun ship *Gertruda* of 1720.

Staysail stay

Leiter; Faux étai, draille

During the sprit topsail period the fore topmast staysail was hoisted on a staysail stay (see Figure 172a). Hitched and seized to the stay at approximately two thirds height, this stay was set up to deadeyes on the bowsprit slightly inboard of the fore topmast stay.

After the introduction of the preventer stay the sail was hanked to that stay (see Figure 172e). Some English men-of-war, however, were again rigged from mid-century onwards with a staysail stay (see Figure 172c). Made fast with a running eye around the bowsprit inboard of the bees (see Figure 172d), this staysail stay rove over the upper sheave in the port side cheek block and was set up to a tackle in the rear of the fore top, with the tackle fall belayed to the mast bitts.

Steel and Lever made no mention of the staysail stay; in their works the fore topmast staysail was set on the preventer stay, and this was the most common arrangement at the end of the century.

Röding noted that around 1790 the English method of rigging the preventer stay below the stay was generally accepted because it did away with an additional staysail stay, but he also mentioned a 'loose stay [that is, preventer stay] of the fore topmast, of which the latter can only be found in very large ships'. Another passage in his work made clear that the fore topmast staysail ran on the fore topmast stay.

For French ships this can also be concluded from the rigging plan of the *Royal Louis* of 1780, but Ozanne's etchings make it clear that this sail was set to an additional staysail stay. The latter can be seen too on the models of the *Sans Pareil* (1760), *Le Lion* (1780) and *Le Protecteur* (1793). Boudriot describes a preventer stay below the stay for his 74-gun ship, which does not conform with the evidence noted above. His depiction of a fore topmast preventer stay in various drawings is very inconsistent; even for the same ship it is shown either above or below the stay.

The conclusion to be drawn is that on the Continent only large men-of-war (perhaps down to a 74-gun ship) and merchantmen above 1500 tons were sometimes rigged with a fore topmast preventer stay, while all vessels of a lesser size were probably rigged with a staysail stay.

Tack

Hals; Point d'amure

The tack was a single rope, hitched to the tack eye or thimble and made fast to the deadeye of the stay or staysail stay. Where the preventer stay was led through the bees, the tack was made fast either to the preventer stay itself or to an eyebolt in the bowsprit.

Halyard

Fall; Drisse

Several changes, sometimes only small, occurred in the rigging of the fore topmast staysail halyard during the eighteenth century. It was usually rigged double, with a block hooked or hitched to the sail's peak.

At the beginning of the century a block was seized to the stay above the staysail stay, and the standing part of the halyard was made fast to a becket in the tail of the block. The halyard then rove through both blocks and down upon deck (see Figure 172a).

After 1720 the stay block was replaced by a long-stropped block around the topmast head, hanging on the port side below the trestle trees.

The standing part of the halyard was made fast on the starboard to the stay collar. Single halyards were rigged on medium-sized and smaller vessels during at this stage.

With the introduction of cheek blocks the standing part of the halyard was made fast around the topmast head and, after passing the sail's peak block, the hauling part rove over the lower sheave in the port side cheek block upon deck, where it was belayed to a deck cleat at the aft side of the forecastle (see Figure 172e). Again it should be noted that the cheek block was a feature only of English or English-style rigging.

Boudriot referred to single halyards, which led through the starboard cheek block and a ninepin block in the bitts, and were belayed to a cleat. Röding described Continental single halyards thus:

> Sometimes a single tye is also fastened to the upper end of this sail, and strapped into the other end of that tye is a single block through which a fall reeves. One end of this fall is made fast below [that is, on deck]; the other, serving as the hauling part, reeves similarly through a foot block and is belayed on deck or round a cleat.

Downhauler

Niederholer; Hale-bas

The downhauler led from the sail's peak through a thimble sewn to the middle of the hoist (stay) tabling, or through several of the hanks, to a block at the tack, and then through the rack block or the lead saddle to the beakhead rail.

Sheet

Schot; Écoute

The sheet was doubled and the bight was made fast to the clew; a single block was spliced into each end. The standing parts of the sheet falls were hitched to a ringbolt at the fore side of the cathead or to a timberhead nearby. The hauling part was led, either directly or through a foot block at the forecastle, to belay on a pin in the side rails.

Continental sheets were also doubled, but frequently did not have sheet falls. These sheets rove through lead blocks on eyebolts in or near the spirketting and were belayed to ranges nearby.

Middle staysail

Mittelstagsegel, Flieger über dem Gross-stengestagsegel; Grand-voile d'étai centrale

The middle staysail was a quadrilateral sail, introduced in the early 1770s and hoisted on a middle staysail stay specifically rigged for the purpose (see Figure 172f).

Staysail stay

Leiter; Draille

Full details of the rigging of a middle staysail stay are given under Topmasts in Chapter II. The rigging plan of the *Royal Louis* is particularly informative on the Continental method of setting up this stay; it shows the middle staysail stay leading to the main topmast trestle trees. Boudriot advocated a cheek block (though his definition of this type of block is dubious, see Main topmast staysail halyard above) on the port side of the main topmast head, with the stay itself ending on the main top. Röding commented as follows:

> The staysail stay for the middle staysail is usually fastened to a thimble running on a horse behind the fore topmast; the other end reeves through a block beneath the main topmast cap, and is set up abaft the main topmast to a burton [in the main top].

Röding's description is of a set-up very similar to the English method with a cheek block, except for that block itself.

Halyard

Fall; Drisse

The middle staysail halyard was usually single. It led from the sail's peak over the lower sheave of the main topmast's port side cheek block and followed on the port side the same lead as the main topmast staysail halyard on the starboard. The Continental halyard rove through a block or thimble stropped to the topmast head rigging and upon deck, and was sometimes led through a foot block before being belayed to the bitts.

Nock

Nock; Point de drisse

The middle staysail's nock was lashed to the stay's grommet or thimble.

Downhauler

Niederholer; Hale-bas

The downhauler was made fast to the sail's peak and rove through a lead truck or thimble on the stayrope, a block stropped to the stay's grommet or thimble, and through the top down to the bitts abaft the mast.

Tack

Hals; Point d'amure

The tack was double like that of the main topmast staysail, and rove through trucks on the after fore topmast shrouds to belay in the fore top. Boudriot noted a French arrangement with a single tack made fast to the aft face of the foremast head.

Sheet

Schot; Écoute

The sheet was doubled and came down without falls to the gangway on both sides, where it rove through foot blocks and was belayed to the aftmost beam of the boatskids. In larger French ships these sheets were also double and followed the same lead as the main topmast staysail sheet.

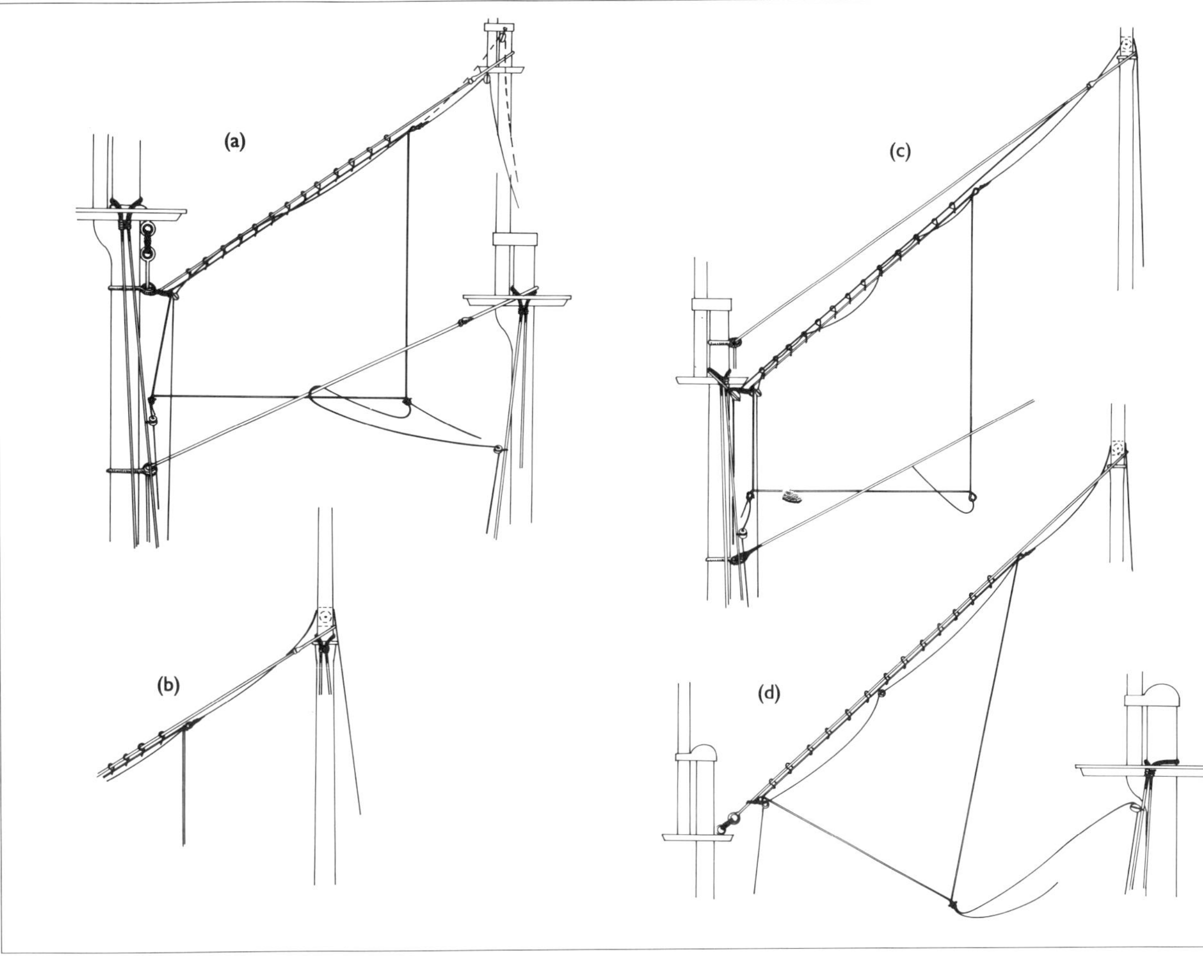

Figure 173
(a) The rigging of an English mizzen topmast staysail after 1760. The dashed halyard line is according to Steel, and the solid line according to Lever.
(b) The halyard of a mizzen topmast staysail, leading through a sheave hole in a long pole head.
(c) The rigging of a main topgallant staysail according to Lever.
(d) The rigging of a main topgallant staysail according to Röding.

Inner jib

Innenklüver, Binnenklüver; Faux foc

The inner jib was a feature of Continental rigging, and was hoisted between the fore topmast staysail and the jib. On the Continent the latter was usually known as a standing jib, in contrast to English rigging, where in the last few decades of the century the same sail was known as a 'flying jib' (not to be confused with the flying jib of the very end of the century, which was rigged to the flying jib-boom). The inner jib was either set flying or was hoisted on a staysail stay.

Staysail stay

Leiter; Faux étai, draille

Röding described a staysail stay for the inner jib as follows: 'the staysail-stay for a jib has one end secured to the traveller and the other rove through a two-sheave halyard block upon the forecastle, where it is set up to a tackle'. Lever confirmed the use of the same halyard block on English merchantmen, rather then a cheek block.

Halyard

Fall; Drisse

Depending on the size of the sail, the inner jib halyard was either single or double. Röding generally noted a double halyard, but mentioned that sometimes single halyards were used. Ozanne and Roux showed single halyards in corvettes and small frigates, while larger ships carried double halyards.

Boudriot indicated that even large French vessels were rigged with single inner jib halyards, leading over the second sheave of the port side 'cheek block' (again his misinterpretation of this type of block is a problem) down to the ninepin bitts to be belayed. The sail was set flying.

The standing part of a double halyard was made fast to the fore topmast head, and the other end led over the second sheave of the halyard block noted above towards the forecastle, where it rove through a foot block and was belayed. A single halyard was hitched to the sail's peak, rove through the double halyard block like a tye and had a block turned into its end. One end of the halyard fall was hooked into an eyebolt on deck, and the other passed through the block, through another foot block on deck, and was belayed, usually to a cleat.

Tack

Hals; Point d'amure

The tack was made fast to a traveller, which was ordinarily made of rope.

Outhauler

Ausholer; Tire-bout

Spliced to the traveller, the outhauler rove through a sheave hole in the outer end of the jib-boom or a thimble, led back through a truck on the cap or a block fitted there (according to Röding), or alternately through the rack block (according to Boudriot), and was belayed to the beakhead rail.

Inhauler

Einholer; Hale-dedans

The inhauler rove from the traveller direct to the beakhead rail and was belayed there.

Sheet
Schot; Écoute

The sheets were doubled and made fast to the clew of the sail. Both ends passed through lead blocks at the forecastle sides and were belayed to the pinrails in the sides nearby.

Mizzen topmast staysail
Besanmarsstengestagsegel; Diabloton

Like most of the staysails, the mizzen topmast staysail was quadrilateral after about 1760 (see Figure 173a and b).

Halyard
Fall; Drisse

Hitched single to the sail's peak, the halyard rove either through a block at the mizzen topmast's trestle trees (according to Lever), or through a block stropped on the starboard side of the mizzen topmast head (according to Lees), upon deck. According to Steel, where the mizzen topmast had a long pole head and no topgallant mast was carried, the halyard was led through a sheave hole in the topmast above the rigging and the sheave hole for the topsail halyards (see Figure 173b). In French rigging the halyard rove through a block lashed to the stay near the topmast hounds and was led to a port side pinrail (Boudriot).

Nock
Nock; Point de drisse

The nock was made fast to the mizzen topmast stay collar round the mainmast. Lever referred to a staysail's nock as an 'upper tack'.

Tack
Hals; Point d'amure

The tack of a triangular sail was made fast in the same manner as the nock of a quadrilateral sail. A quadrilateral sail had a double tack, leading through trucks or thimbles seized to the shrouds and belayed to shroud cleats or a pinrail.

French ships had only a single tack made fast to the mizzen stay, according to Boudriot.

Downhauler
Niederholer; Hale-bas

Hitched or spliced to the peak and led through a few hanks, the downhauler then rove through a small block or a thimble at the nock (tack) and was belayed on the breast rail.

Sheet
Schot; Écoute

The sheet was doubled and bent to the clew; each end passed through a block or thimble on the foremost mizzen shroud, and was belayed at the side to a pin in the rails. Boudriot's description of a French sheet differed from this (English) arrangement: he noted a single sheet, belayed to a cleat on the quarterdeck near the poop breastwork.

Main topgallant staysail
Grossbramstengestagsegel; Voile d'étai de grand perroquet

From its introduction at the end of the first decade of the eighteenth century until approximately 1760 the main topgallant staysail was triangular; after that, like most of the staysails, it was quadrilateral.

In English rigging the sail was originally hoisted on the main topgallant stay, but later it was hoisted on a staysail stay. Sutherland referred to the earlier rig thus: 'there are other Sails called Stay-sails used on almost every Stay as, the Main Stay-sail, Main-top-mast Stay-sail, Fore-top-mast Stay-sail, Mizon Stay-sail, and sometimes on the Mizon-top-mast Stay, and Top-gallant Stay'. Steel and Lever noted the later rig for staysails, while Röding and Boudriot indicated the bending of staysails to the stay on Continental vessels (see Figure 173c and d).

Staysail stay
Leiter; Faux étai, draille

> [The staysail-stay] is generally spliced into the Main Top Gallant Stay, a little below the Collar but sometimes it is put over the Mast Head, like the other Spring Stays. The Stay leading through the Hanks, is reeved through a Block, or Thimble, seized, or strapped, to the Fore Topmast Cross-trees, and led down upon Deck, being sufficiently long for the Bight to overhaul into the Top.

This was Lever's description. The final remark indicates the way in which this sail was taken in, with the bight being dropped down to the top so that the topmen could work on it. Steel proposed securing the staysail stay in the top and not on deck.

Halyard
Fall; Drisse

The halyard was single and rove through a sheave hole above the hounds of the topgallant mast and led upon deck, belaying to the bitts abaft the mast on the quarterdeck. Instead of the sheave hole, Continental ships had a block stropped to the stay collar.

Nock
Nock; Point de drisse

The nock was made fast to the lead block or thimble strop of the stay or staysail stay.

Tack
Hals; Point d'amure

Until 1760 the tack was rigged similarly to the nock; after that time it was rigged in the same manner as the middle staysail tack, with the trucks placed slightly above these for the latter. Boudriot noted a single tack hitched to the main topgallant stay below the stay lead block.

Sheet
Schot; Écoute

In English rigging the doubled sheet followed the same lead as that of the middle staysail. Boudriot described the same arrangement, but passed the sheet from the outside through the fixed four-sheave topsail sheet channels in the quarterdeck bulwark. A rigging plan in Röding's work showed these sheets leading to a point directly below the main top, probably to a block or thimble on the foremost main shroud, and finally belayed to the side pinrails.

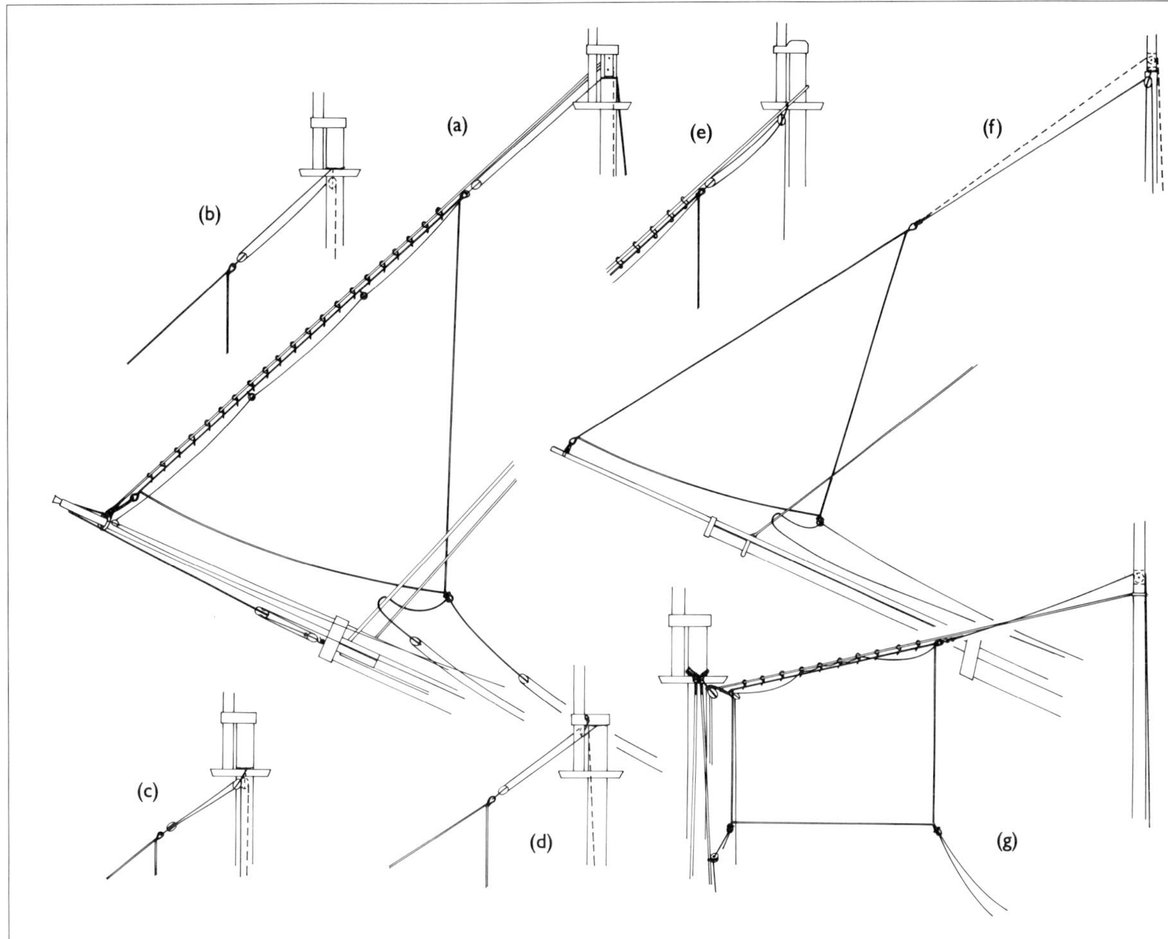

Figure 174
(a) The rigging of a jib on a jibstay after 1760.
(b) A flying jib halyard at the time of the sail's introduction.
(c) An alternative halyard lead at the same time.
(d) A jib halyard after 1720, when the jib was stayed.
(e) A Continental halyard for a standing jib.
(f) A flying jib at the end of the century.
(g) A mizzen topgallant staysail.

Downhauler

Niederholer; Hale-bas

The downhauler followed a similar lead to that of the other staysails, and was belayed at the bitts abaft the foremast.

Jib

Klüver; Grand foc

Up to 1720 and to some extent after 1775 the jib was set flying, and thus rigged only with a halyard, tack and sheet. Otherwise the jib was hoisted on a jibstay. The sail was officially introduced in the Royal Navy in 1705, but an early illustration was that of Passebon, as mentioned above under Fore staysail; Sutherland (1711) noted 'another Sail call'd a flying Gib... used with a Boom or small Mast extended at the Extremes of the Bowsprit'. Blanckley mentioned a flying jib in 1750, and Steel provided alternative rigs for nearly all the vessels in his rigging lists for a jib, standing jib or flying jib (that is, a jib set flying). Even though he might be expected to have given some hint of the new sail of the 1790s, the flying jib, as an additional sail, his descriptions of spars included only jib-booms proper: no flying jib-boom was mentioned, except in his 'Proportional lengths and diameters of booms', where the term 'flying jib-boom' took the place of 'jib-boom', which was not listed at all (the measurements were, however, these of the latter). In his detailed description of sails he included a flying jib for a sloop, but for ships only the jib bent to hanks on the jibstay. This seems to confirm that his listing of 'flying jibs' for ships in rigging lists as noted above referred only to an alternative way of rigging the jib, rather than to an additional sail.

Röding described the standing jib as hoisted on a staysail stay, and Boudriot on the jibstay; these were simply different terms for the same stay.

Halyard

Fall; Drisse

Made fast to the fore topmast head, the halyard of a flying jib led through a block in the jib's peak and another on the starboard side of the topmast head (see Figure 174b). It was then led downwards and belayed to the side pinrail. Smaller ships rigged the jib flying until the 1730s and, as noted above, even in 1750 the sail was still described with the adjective 'flying'. An alternative rig to that described was with the halyard middled and both parts reeved through topmast head blocks and belayed on deck (see Figure 174c).

From about 1720 onwards a block was stropped to the starboard side of the topmast cap, with the standing part still made fast to the topmast head (see Figure 174d). Röding noted this block also in regard to staysail halyards. Smaller ships frequently rigged only a single halyard, leading as noted above under Fore topmast staysail.

Steel noted single halyards for jibs, with large ships having a single block turned into the halyard's end with a whip reeving through, and its standing part set up to the side. Lever noted a double halyard for larger ships, reeved 'through a Block, seized to the Peak Clue'.

After the introduction of cheek blocks, the hauling part rove through the lower sheave hole in the starboard cheek block, but Lever's merchant alternative to the cheek block, a double block stropped to both sides of the topmast head and hanging beside the trestle trees, seems also to have been common. The cheek block was mainly used in the Royal Navy. Once again, Boudriot noted a cheek block [sic] for the single halyard of a

French 74-gun ship's standing jib. The halyard then led downwards abaft the foremast to belay at the ninepin bitts.

Downhauler
Niederholer; Hale-bas

A necessity for a downhauler arose when the jib was hoisted to a stay. Spliced into the peak, it rove down the stay through two thimbles, passed through a block made fast to the traveller or the tack, and was led to the beakhead rail, either direct or through the rack block.

Tack
Hals; Point d'amure

The tack was seized to an eye on the traveller, when fitted, otherwise to the jib-boom.

Sheet
Schot; Écoute

The sheets were similar to these of the fore staysail and belayed beside them. One contemporary noted that the standing part was made fast to an eyebolt in the bow, with the hauling part coming inboard through a lead block or a hole in one of the beakhead bulkhead stanchions and belaying to the side ranges.

Boudriot considered the timberhead aft of the cathead or a stanchion in the beakhead bulkhead as belaying points.

Flying jib
Aussenklüver; Clinfoc

The flying jib as a sail in its own right was introduced only at the end of the eighteenth century in ships, and initially it was confined to English rigging (see Figure 174f).

Continental ships hoisted the outer jib instead, which was a standing jib, as noted above (examples can be found in the rigging plan of the *Royal Louis* of 1780 and in many pictures by Ozanne, in Roux's watercolours of French ships and in Röding's illustrations. Continental flying jibs with their additional flying jib-booms date only from the early nineteenth century onward).

The flying jib of this period was normally set flying. In some cases, however, it was hoisted to a flying jibstay. Not only can this be seen in some of Roux's paintings and in Serres' pictures of HMS *Cambrian* and an unknown frigate in his *Liber Nauticus*, but Steel's *The Art of Rigging* of 1818 also specified a stay in the rigging lists for men-of-war, though for merchantmen his lists remained unaltered from 1794.

Flying jibstay
Aussenklüverstag; Étai de clinfoc

From Steel's 1818 lists it can be concluded that in the first decades of the nineteenth century in Royal Navy rigging, the flying jibstay was made fast round the flying jib-boom head, passed through a cheek block of smaller proportions than the jib's (about three quarters the size) and was set up to a tackle, probably in the fore top.

Halyard
Fall; Drisse

The illustrations listed above reveal that flying jib halyards were either single or double. They rove through a block on the port side of the fore topgallant masthead and led upon deck. In some ships a sheave was fitted to the topgallant masthead, like that for the main topgallant staysail halyard.

Steel's lists specify two blocks for men-of-war right down to the size of brigs, but only one for merchantmen without a flying jibstay. The additional block in naval rigging suggests double halyards.

Tack
Hals; Point d'amure

At the beginning of the century the tack was lashed around the flying jib-boom head.

Sheet
Schot; Écoute

Single sheets led directly to the forecastle.

Mizzen topgallant staysail
Besanbramstengestagsegel, Kreuzbramstengestagsegel; Voile d'étai de perruche

The mizzen topgallant staysail was introduced around 1760 and was quadrilateral (see Figure 174g). Falconer noted this sail in 1769: '...sometimes a mizen top-gallant stay-sail above the latter'. Steel described it in 1794 and Röding also commented under *Kreuzbramstengenstag-Segel* that the sail was 'hoisted to the mizzen topgallant stay'. Lees described this sail as very short-lived and placed it only in the period between 1802 and 1815.

Halyard
Fall; Drisse

The halyard was single and was spliced to the sail's peak; it rove over a mast sheave or through a block above the topgallant masthead rigging. Leading down the mast, it was belayed, according to Steel round a pin in the handrail.

Nock
Nock; Point de drisse

The nock was lashed to the stay's lead block.

Tack
Hals; Point d'amure

The tack was doubled and led on both sides through lead trucks on the after main topmast shrouds to belay in the main top.

Downhauler
Niederholer; Hale-bas

Spliced to the peak eye, the downhauler led through a few hanks and a block on the stay's lead block strop upon the main top, and was belayed there.

Sheet
Schot; Écoute

Both single sheets led to blocks or lead thimbles in the upper part of the foremost mizzen shrouds and were belayed either to a pin in a handrail or to a cleat in the shrouds.

Main royal staysail

Grossroyalstagsegel; Voile d'étai de grand cacatois

The main royal staysail was set only in ships of the largest order if these had a main royal sail rigged, and therefore also a main royal stay. Introduced in about 1719, it was triangular until 1760 and quadrilateral after that, and was rigged like the mizzen topgallant staysail. The sheets led through lead blocks or thimbles beside those of the middle and main topgallant staysails and were secured in a similar fashion.

Mizzen royal staysail, mizzen spindle staysail

Flieger über dem Kreuzbramstengestagsegel; Voile d'étai de perruche volante

Röding referred to the mizzen royal staysail as 'a staysail hoisted to a staysail stay above the mizzen topgallant staysail... the smallest of all staysails'. He also gave the English term 'spindle staysail' for royal staysails, though this name was not given in contemporary English nautical dictionaries. The spindle was an iron extension of the topgallant masthead for flying a vane and was topped with a round or conical piece of timber, the acorn. The unusual terminology used for these staysails suggests that the staysail stay's upper point of fastening must have been either on the spindle, or just below. In either case, these sails were very rarely set.

Lower studdingsails

Untere Leesegel; Bonnettes basses du grand mât et du misaine

Similarly to other square sails, the lower studdingsails in English ships were bent with robands and earings to a studdingsail yard. The narrow sails of the first half of the eighteenth century had a yard across the whole of the head rope (according to Lees), while later only half or three quarters of that length were covered, with the yard spreading the outer part of the head (see Figure 175a). Continental vessels frequently had no studdingsail yards, and the lower studdingsails were set to lower yard booms. Röding actually described these studdingsails as bent with robands and earrings to the lower yard booms, which seems unlikely.

French studdingsails until about 1760 were set without a yard (see Figure 175b), and both Paris and Boudriot noted that this was still the case for fore studdingsails in 1780, though main studdingsails then had either full yards or half yards according to Boudriot, the latter spreading the outer half of the sail's head. Ozanne's etchings indicate a similar rig; one of his pictures shows the main studdingsail apparently set over its full head length against the main yard boom, with no studdingsail yard visible, as if it were bent to the boom (see Röding's description), and in another the outer third is spread by a yard. Paris also noted in Volume 2 of his work that the then (1880) abandoned main studdingsail had a small yard which spread the inner half of the head and was hoisted in the middle to the main yardarm, while the outer earring was hoisted to the boom's end, which was contrary to the current practice.

The earlier English sources give no details of lower studdingsail yards, and it is possible that early English studdingsails were hoisted in the same fashion as Continental sails later in the eighteenth century, that is, without a yard. Blanckley confirmed this in 1750, listing under Sail 'those which are not bent to the Yards [including] Main and Main-top Studding Sails'. The use of a full yard for early eighteenth century Royal Navy lower studdingsails, as mentioned by Lees, therefore remains conjectural. Since insufficient evidence is available, however, both rigs have to be accepted as possible.

Descriptions of lower studdingsail rigging were not very detailed before Steel's 1794 *The Elements and Practice of Rigging and Seamanship*. Falconer (1769), after a lengthy excursus into the name 'studdingsail', gave only a short and very general description, but mentioned the yard and lower booms.

Not much is known about the earliest studdingsails. They are documented as early as 1620-5 in *A Treatise on Rigging*: 'studding sayles which ar set on ether side of your fore and mayne saile in the utter sides,' and Smith noted in 1627 a studdingsail 'alongst the side of the maine saile, and boomes it out with a boome or long pole'. Details of the size and shape of such sails are now lost.

A hint was given by Smith, who noted that 'studding sailes were bolts of Canvasse or any cloth that will hold wind'. He did not elaborate on the 'jibheaded' (triangular) shape, as described in *Der geöffnete See-Hafen*; this shape was probably more a feature of Continental rigging. Smith's use of the term 'bolt of canvas' suggests rather a sail with parallel sides. Another indication of the size and shape of these early sails was provided by Davis, who gave the number of cloths, their lengths and the total length of the material used for the main studdingsails of HMS *Royal Sovereign*. These figures were ten full cloths at 24 inches, 17 yards long; total for two sails 340 yards. This certainly implies a rectangular sail of half the width of a sail for an equally rated ship from about 1790.

The unknown author of *Der geöffnete See-Hafen* compared the shape of lower main studdingsails with staysails, and mentioned the fitting of a single block each to the sail's peak and to the yardarm, with a fall connecting both. Both sheet and tack were made fast to a spar extended over the side for that purpose. The sheet was probably led through a block; this was not specifically mentioned, but the note '...over which the sail's sheet runs, and the tack is also made fast to the same boom or spar' suggests it. It was likewise noted that similar studdingsails were set to the main topsails, and that the fore and fore top studdingsails were the same.

The actual number of studdingsails carried by a ship seems to be slightly obscured as well. Anderson stated that even though the fore studdingsail was mentioned in the *Treatise on Rigging* of 1620-5, these were not carried on English ships earlier then 1690, but Stork's painting *The Four Day Battle* of 1666 and others show short fore yard booms on English and Dutch ships, as if they were quite common. The evidence can be summarised as follows:

Fore and main studdingsails are listed in the *Treatise on Rigging* of 1620-5;

Main studdingsails only in Smith's *A Sea Grammar* of 1627;

Main and main top studdingsails in a list from 1670 at the National Maritime Museum in Greenwich;

Main and main top studdingsails in a list from 1677 at the British Library;

Fore and main, and fore and main top studdingsails in *Der geöffnete See-Hafen* of 1705;

Main and top studdingsails in Sutherland's *The Ship-builder's Assistant* of 1711;

Main and main top studdingsails in a list of all sails for HMS *Vanguard* (1678) and for HMS *Royal Sovereign* (1701);

'Fore Studding-sails 2, Ditto Topsail 2, Main Studding-sails 2, Ditto Topsails 2, Cross Jack Studding-sails 2' in *A Demand for the Boatswain's Sea Stores for the Degrave Frigate, dated October the 2nd 1676 by J Davis, Boatswain*;

'Fore Top-sail Studding-sail, Main Studding-sail and Main Top-sail Studding-sail' in *An Account of how many Fathoms of Rope etc will rigg HMS Bonaventure* (1683);

Rigging for fore, main, crossjack, and top studdingsails in Davis' *The Seaman's Speculum* of 1711;

Main and top studdingsails in a list dated 1720 but associated with the 1677 programme, at the British Library;

Main and main top studdingsails in Blanckley's *A Naval Expositor* of 1750;

Fore, main, fore top, and main top studdingsails in Falconer's *Universal Dictionary of the Marine* of 1769;

Lower, top and topgallant studdingsails on fore and mainmasts in Steel's *Elements of Mastmaking, Sailmaking and Rigging* of 1794.

It would appear that even though the use of studdingsails on the foremast was known since the early seventeenth century, up to the middle of the century the general practice, with several known exceptions, was to rig studdingsails only to the mainmast.

Halyard

Fall; Drisse

Studdingsails had an outer and an inner halyard in the later part of the eighteenth century, but (according to Lees) one halyard only for the smaller sails in the first half of the century. Davis noted 'Two Fore studding-sail Halyards' and 'Two Tacks to haul out upon the Yard' (in addition to 'Two Tacks for the Booms'), 'the Main Studdingsail Halyards with blocks at the quarter of the Yard', and 'Two Cross-Jack Studding-sail Halyards, with Blocks at the quarter of the Yard for an Indiaman'. He also mentioned stropping 'for Studding-sail Halyards by the Yard-arm' for 11 inch single blocks. Studdingsails in men-of-war were noted for fore and mainmasts, but not for the mizzen mast, and in some instances not for the fore course. Sutherland noted the same rig: 'there are also Sails call'd Studding Sails, made use of at the Extremes of the Main Yards and Top-sail Yards'. His rigging list included studdingsail halyards for the mainmast together with four single 10 inch blocks.

Both halyards were secured in the earrings, the outer halyard to the middle of the studdingsail yard, when the sail was bent to one. This halyard rove through the halyard block on the studdingsail boom, then through a span block at the lower cap down upon deck. The inner halyard was led through the block at the yard's quarter, another near the mast and then upon deck. The latter block probably did not exist at the beginning of the century.

Davis and Sutherland indicated that initially one halyard per sail was used, which rove through a single block at the yardarm. The halyard itself had a length some three quarters that of its 1790 equivalent, indicating to a shorter lead upon deck (perhaps not leading through a span block at the cap). The quarter blocks mentioned by Davis and the second set of blocks by Sutherland were probably lead blocks for the halyard, which may have been led down to the sides; this halyard rig would only have been possible when the sail's head was spread by a yard. If the sail was set without a yard, then the halyard (outer) rove through the yardarm block down to the side and was belayed there, and the 'Tack to haul out upon the Yard' makes sense if it is understood as the inner halyard, leading through the block at the quarter of the yard downwards. Both these reconstructions are made on the basis of the material at hand.

The rigging of a triangular studdingsail was described in *Der geöffnete See-Hafen*. A single block was lashed to the yardarm and another to the studdingsail's peak, with a fall these formed a tackle to hoist the sail.

The outer halyard on a French sail passed through the boom halyard block and, according to Boudriot, through a span block at the topmast head, while Ozanne's etchings followed the English lead. The inner halyard was rigged as noted above, bent to the middle of the yard if the studdingsail was fitted with one. The halyards were belayed to a mast cleat.

Tack

Hals; Point d'amure

Steel described the tacks as bent to the outer clew of the sail, leading through a block lashed to the outer part of the lower boom. They led fore-and-aft, with the aft-leading fore tack passing through a block in the main chains and coming inboard through a gunport to be belayed to a cleat in the waist. The forward-leading tack rove through a block on the bowsprit bees upon the forecastle. The main studdingsail tack led from the block at the end of the boom to another lashed to an eyebolt in the buttock, and came inboards via a snatch block on the quarters.

Davis noted a tack for the boom with a length equal to the beam for the fore studdingsails, and of 1½ boom lengths for main studdingsails, but if the boom came within board, then twice the boom's length for each. This was only a fraction of the lengths given by Steel, indicating a very much shorter length inboard. The mention of the boom coming inboard suggests a loose boom, as Smith and the author of *Der geöffnete See-Hafen* described, and gives the impression that swinging booms were not the norm. The latter work described the tack made fast to the boom, in similar fashion to the sheet.

According to Boudriot, French fore studdingsail tacks rove from the boom block through the fore sheet channel in the side and were belayed to a main shroud cleat, while the main studdingsail tacks passed through the swinging boom block, a further double block on the poop planksheer and were belayed fast round a mizzen shroud cleat.

Guy

Gei; Retenue

Lees noted flying lower studdingsails for all ships in the first half of the century and for small ships up to 1810; Lever commented as follows:

> The lower Studding-sail is often set flying, that is, without a Boom. When this is the case, the sail is spread at the Foot by lashing the Clews, to a small Yard: a Span, is made fast to the Yard, and the Guy bent to it.

Steel confined the flying studdingsail to the foremast and also noted that the guy led through a block lashed to the main chains, came inboard via a gunport and was belayed to a cleat in the waist.

The recommendation by Lees of flying studdingsails for the earlier years of the century is only conjectural. Davis helps to make matters clearer. He provided evidence as early as 1676 for fore and main studdingsail booms and followed this up with details for the Sixth Rate HMS *Swan* (1694), an unnamed Fifth Rate of about 1708, the Fourth Rate HMS *Bonaventure* (1699), the Third Rate HMS *Lenox* (1701), the Second Rate HMS *Vanguard* (1678), and the First Rate HMS *Royal Sovereign* (1701). His listing of all Rates and of East Indiamen with booms, and the early mention of such booms by Smith, suggests that the loose-footed studdingsail was a later innovation than has been assumed.

Sheet

Schot; Écoute

Later in the eighteenth century studdingsail sheets were doubled and the bight was put over, and the ends through, the inner clew of the sail. One end led forward and the other aft. Belaying points are not known. At the beginning of the century both Davis and Sutherland mentioned only one relatively short sheet (7 or 12 fathoms long).

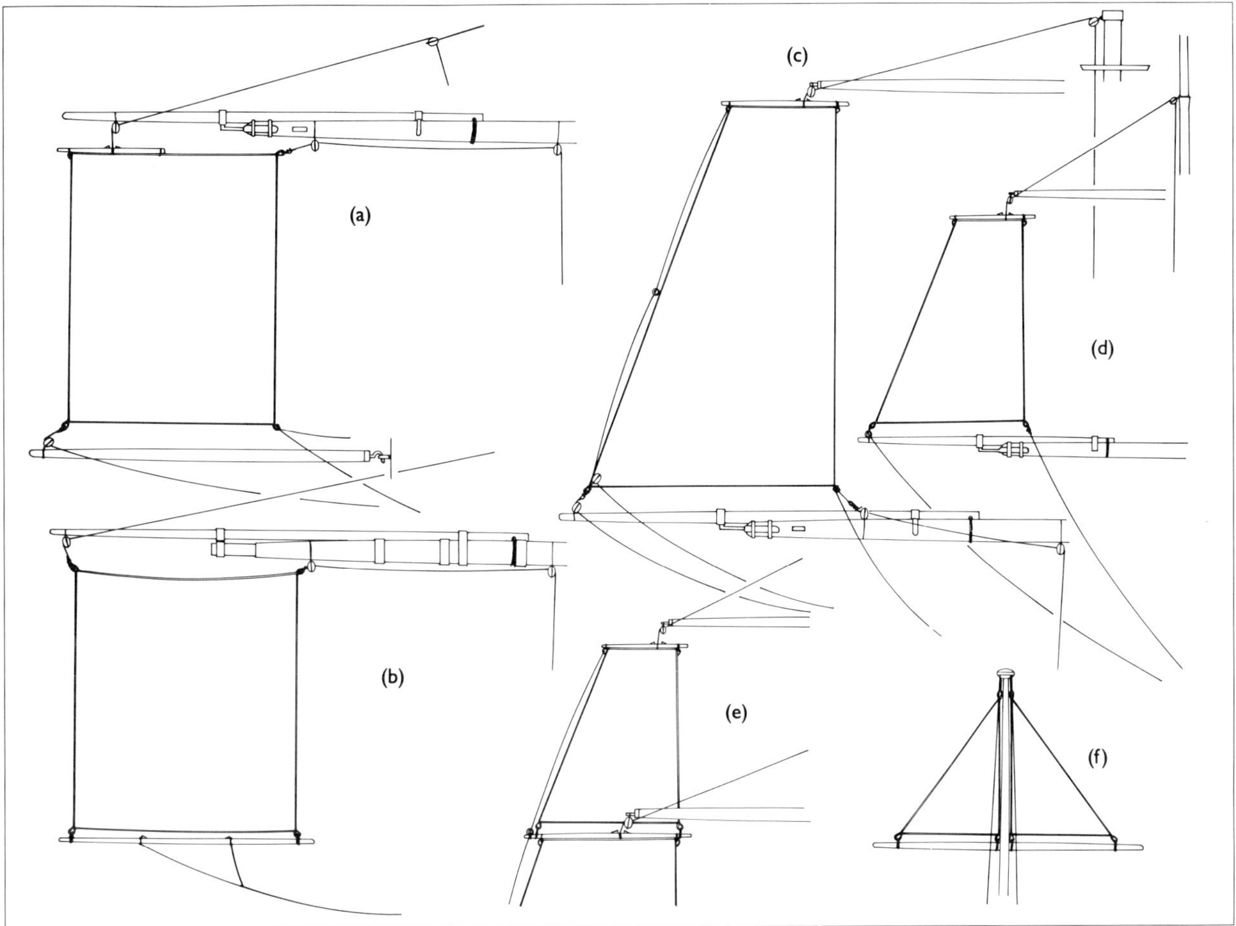

Figure 175
(a) An English lower studdingsail from the second half of the century.
(b) A French lower studdingsail, until at least 1760; this rig was also sometimes used in merchant shipping, especially the flying boom.
(c) A top studdingsail.
(d) A topgallant studdingsail.
(e) A flying topgallant studdingsail. This sail was often set flying on merchant ships, because these vessels did not carry topgallant studdingsail booms. The tack and sheet were lashed to the top studdingsail yard, and the top studdingsail downhauler rove through a thimble on the sail's yard and was bent to the topgallant studdingsail yard. Lever commented that neither hoisting the sail afore nor abaft the topsail led to satisfactory results, since the sails were frequently damaged by chafing when set afore, and the top studdingsail yard had to be lashed to the topsail yard, which had serious consequences in a sudden squall, when the studdingsails were set abaft the topsail.
(f) A skysail; both the sheet and the tack of this sail were lashed to the royal yard, and the halyards rove through sheave holes in the truck and belayed at the mast top.

Top studdingsails

Marsleesegel; Bonnettes de hune

In English rigging top studdingsails were bent to top studdingsail yards (see Figure 175c). Illustrations by Ozanne indicate that French top studdingsails were sometimes hoisted without yards, while Röding differentiated between a studdingsail bent to the yard boom and one bent to a studdingsail yard, hoisted when the yard did not carry a boom. The former sail was set forward of the topsail leech, while the latter was rigged abaft it.

Halyard

Fall; Drisse

In English rigging at the end of the century the halyard was made fast round the inner third of the studdingsail yard. After leading through a halyard block at the yardarm it rove through another block stropped round the topmast head beneath the cap, and came down through the top to belay at the bitts. At the beginning of the century Davis gave very much the same account, but did not mention the point at which the halyard was made fast to the studdingsail yard. He also described the halyard as operated either from the top or from deck, giving alternative lengths.

Continental halyards rove through the yardarm block and a lead block near the tye block downwards, passing through the lubber's hole to belay to a mast cleat. When studdingsails were set without yards, as Ozanne indicated, two halyards were needed, passing first through blocks on the topgallant studdingsail boom end and on the yardarm, then through a further, probably double, block at the topmast head beneath the cap; they were belayed, according to Boudriot, in the mast top.

Downhauler

Bekajer, Niederholer; Cargue-bas des bonnettes

A downhauler was made fast to the outer studdingsail yardarm; it came down on the outer leech, reeving through a thimble at half height and a block on the outer clew of the sail, to be belayed on the forecastle or in the waist. Röding noted the same arrangement for Continental top studdingsails hoisted to a yard, but he noted that the downhauler was then led along the main or fore yard and, after passing a block near the slings, down upon deck. Studdingsails hoisted without a yard did not have a downhauler.

Bowline

Bulin; Bouline

In a footnote, Boudriot mentioned that French main top studdingsails were sometimes rigged English fashion, with a bowline belayed in the fore top. Neither Steel nor Lever indicated a bowline for this sail, and nor did Röding; nevertheless, a bowline fitted with a single bridle span above the lower quarter of the inner leech is shown in some of Ozanne's etchings for top studdingsails hoisted without a yard. They can be seen on both fore topmast and main top studdingsails, and were very much a French rather than an English fashion. These etchings show that the bowline of the fore top studdingsail led aft, while the other followed the lead described by Boudriot.

Tack

Hals; Point d'amure

A tack was bent to the outer clew of the sail and rove through a block at the outer studdingsail boom end, another block in the gangway and was belayed to a timberhead (foremast). The tack from the mainmast led through a block lashed upon the

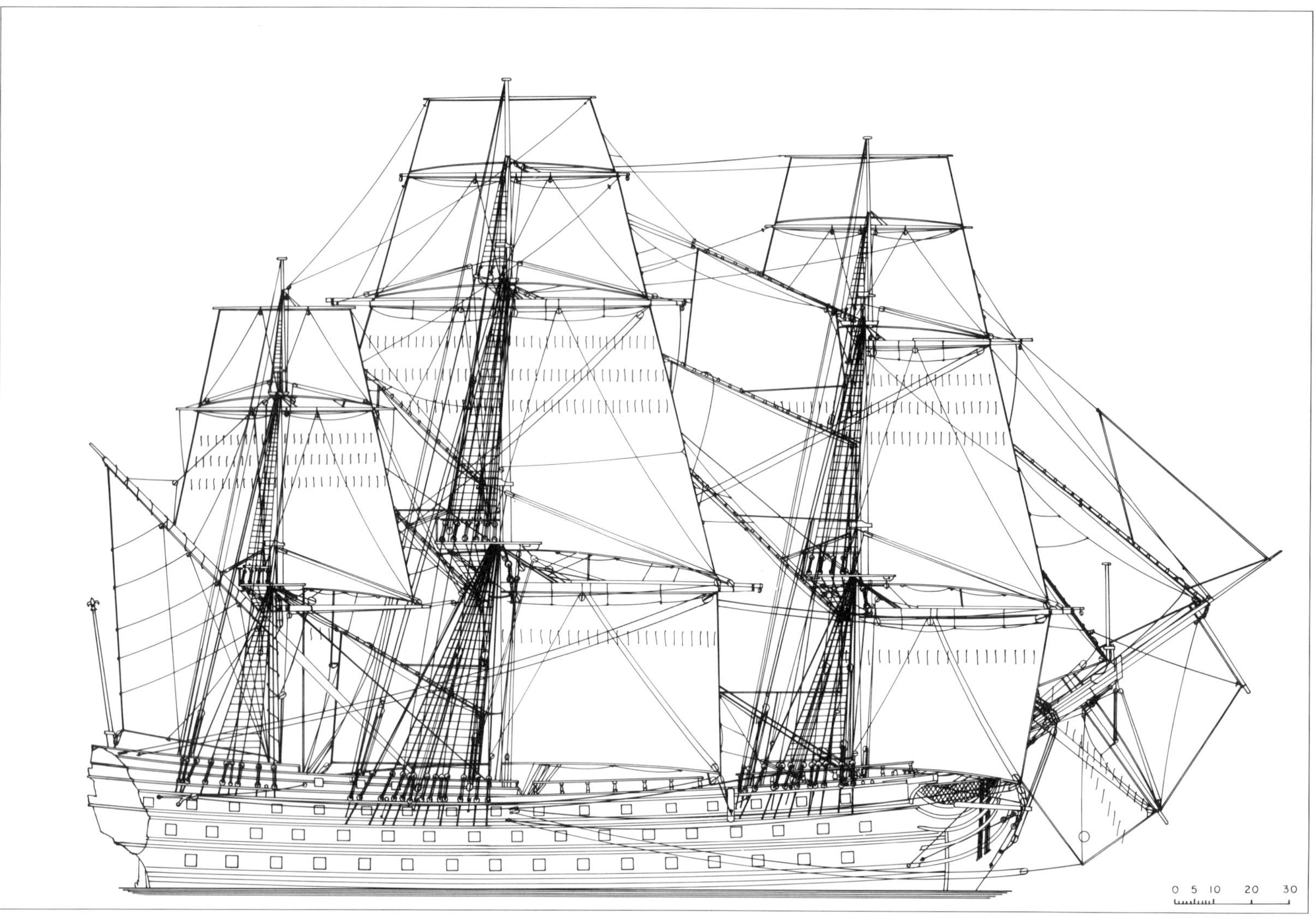

Figure 176
A sail and rigging plan of a French First Rate of about 1780, based on the drawing 'Mâture et Voilure du Vaisseau le Royal Louis de 1780' in Souvenirs de Marine. A large number of rigging details differ from English rigging practice of the time. Note, for example, that the lower lifts are double block tackles, the fore topsail has an inner and an outer leechline, and at the main topsail leech a line, possibly a counter bowline, is evident. The sail are cut differently, the stays are secured in another way, and the lower yards is carried in slings, but in the Continental manner. The main royals are not set flying, even by being rigged to the topgallant mast, and the spritsail yard does not have a parrel but is set up to deadeyes and standing lifts.

quarters. A similar lead was noted by Davis for the beginning of the century; he noted 'two Tacks into the Waste' or 'two Tacks to haul upon the Poop,' but he also suggested an alternative rig, with the tack 'laid upon the Yard'. The latter tacks were only one sixth the length of the former.

Lever described the shorter tack a century later:

> It is sometimes the Practice to lead the Topmast Studding Sail Tacks like Topsail Sheets through a Block on the lower Yard, and through another farther in, instead of its being taken to the Gangway; for in going large, (these Sails, particularly in East Indiamen, being carried when it blows fresh), should the Wind come suddenly forward, and require the weather Braces to be eased off, it is not always that the Tack can be eased in Proportion: if it be too much so, it is difficult to haul out again, and if not enough, the Boom is liable to be carried away by the Strain: therefore in Ships which go long Voyages, where this Sail is of such constant use, it is good to have a stout Brace always reeved, and the Tack to lead as before mentioned, as there will then be no risk in bracing forward, of carrying away the Boom.

Boudriot's description of fore top studdingsails indicated a tack like the short one, belayed in the top, and for the main top studdingsail a long tack leading to the block upon the poop and secured to a cleat nearby.

Sheet
Schot; Écoute

The sheet on Continental top studdingsails led like these on early British ships, where the length of the sheet was only one quarter to one seventh that of the tack. Röding described the sheet passing through a block lashed to the yard, and belayed in the top.

For the later eighteenth century Steel noted that the sheet was doubled, with the fore part leading into the forecastle and the after part of the fore top studdingsail coming down aft of the fore shrouds. The fore sheet of the main top studdingsail led into the waist, while the aft end led in abaft the main shrouds. Lever described the sheet 'bent with a long and short Leg,' with the long leg leading down forward of the lower yard. The short leg had a thimble spliced in its end and the yard sheet (boom outhauler) bent to it. The latter rove upon deck through a block at the yard's quarter and another near the slings.

Topgallant studdingsails
Bramleesegel; Bonnette de perroquet

Topgallant studdingsails appeared in the early 1770s and the topsail yards carried booms from then on. According to Lever, topgallant studdingsails were sometimes also set flying in ships without topgallant booms (see Figure 175d).

Halyard
Fall; Drisse

The halyard was bent to the yard in the same manner as that of the topsail studdingsail, and rove through a block at the yardarm, then through another stropped to the topgallant masthead above the rigging or above the hounds and led down the mast into the top, where it was belayed. Boudriot noted that the halyard of a French topgallant studdingsail led directly from the jewel block on the yardarm down to the top.

Tack
Hals; Point d'amure

The tack passed through a thimble on the end of the topsail yard boom and was belayed in the top.

Sheet
Schot; Écoute

The sheet was doubled in English rigging. One end led forward and was made fast to the quarter of the topsail yard, and the other end led into the top and was belayed to the topmast shrouds. Only a single sheet is mentioned in Continental rigging, belayed in the top.

Downhauler
Bekajer, Niederholer; Cargue-bas des bonnettes

A downhauler was only used in connection with a flying topgallant studdingsail (see Figure 175e).

Watersail
Wassersegel; Bonnette de sous-gui

A watersail was similar to a lower studdingsail, and was bent to a short yard.

Halyard
Fall; Drisse

The halyard was made fast to the middle of the yard and rove through a small block at the outer end of the driver boom to belay to a cleat on the taffrail.

Sheet
Schot; Écoute

The sheets were made fast to the sail's clews and came in over the quarters, to belay to a pinrail.

Ringtail sail
Brotwinner, Brotgewinner; Bonnette de tapecul, bonnette d'artimon

See New driver (merchantman) above.

Wingsail for a ketch
Grossgaffelsegel einer Ketsch; Grande voile à corne du quaiche

This sail was rigged similarly to a short mizzen course.

XIII

Belaying plans

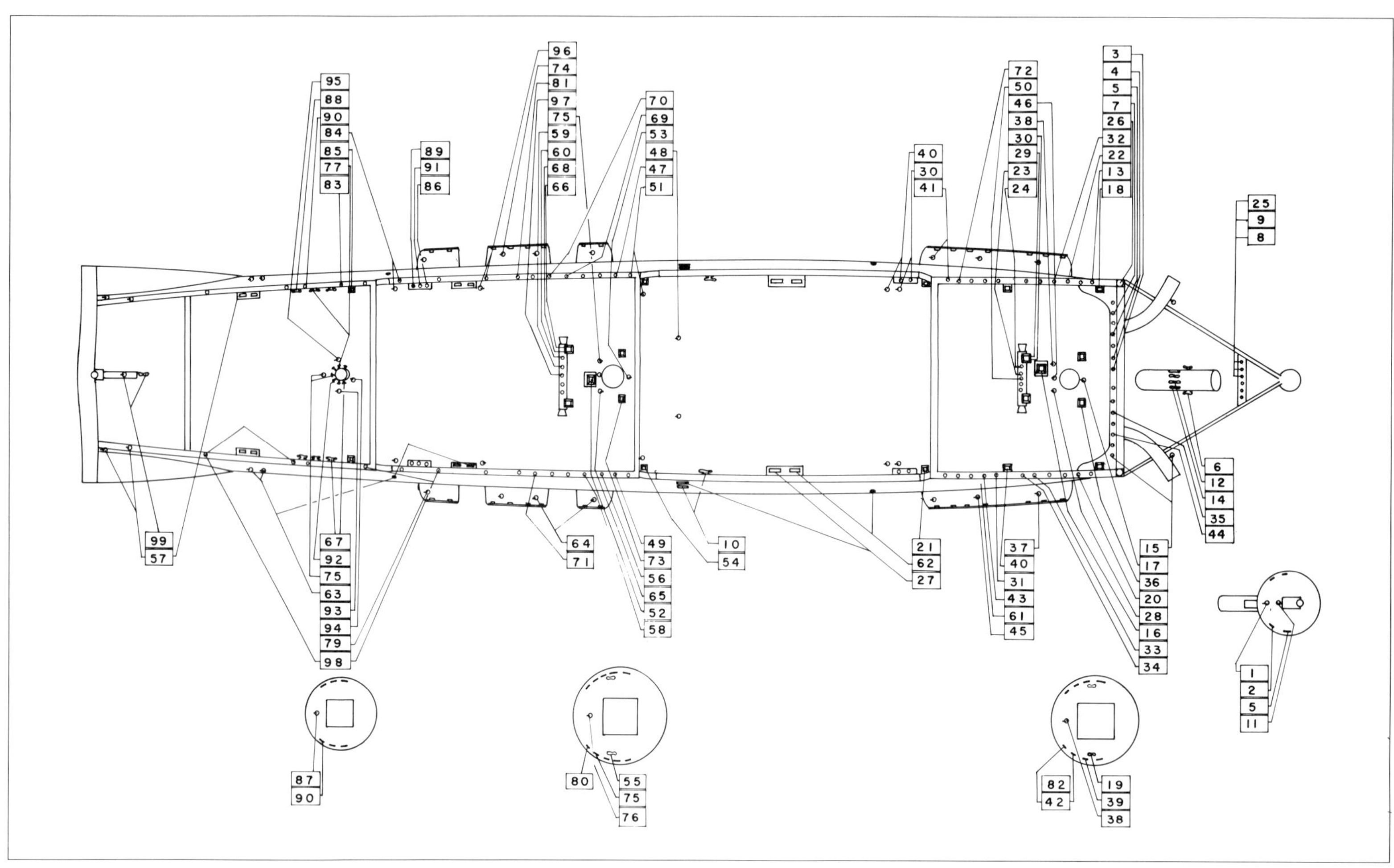

Figure 177. ***Belaying plan of a Continental ship, about 1700***

Spritsail topsail
1. Halyard tackle
2. Lifts
3. Braces, beakhead bulkhead or bowsprit cleats
4. Clewlines, beakhead bulkhead or bowsprit top
5. Sheets

Spritsail
6. Outhauler
7. Braces
5. Lifts
8. Buntline
9. Clewlines
10. Sheets

Fore topgallant stay
11. Stay tackle

Fore topmast stay
12. Stay tackle

Fore topmast staysail
13. Halyard
14. Downhauler
15. Sheets

Fore course
16. Jeers
17. Parrel tackle
18. Lifts
19. Yard tackle tricing line
20. Yard tackle tricing line
21. Braces
22. Clewgarnets
23. Buntlines
24. Leechlines
25. Bowlines
26. Tacks
27. Sheets
37. Mast tackle (side tackle)

Fore topsail
28. Tye
29. Reef tackle
30. Braces, inner sheave at the bitts, or foot block and cleat in the waist.
31. Lifts
32. Clewlines
33. Buntlines
34. Leechlines
35. Bowlines
36. Sheets
45. Shifting backstays

Fore topgallant sail
38. Tye, upon deck or in the top
39. Lifts
40. Braces, belayed with foot block in the waist, or via a lead block in the topmast shrouds to forecastle rails
41. Lifts
42. Clewlines
43. Sheets, when extra fitted
44. Bowlines

Main topmast stay
46. Stay tackle

Main staysail
Staysail stay with dead eyes to the foremast. Downhauler and tack fastened to the staysail stay.
47. Halyard
48. Sheets

Main topmast staysail
49. Halyard
50. Downhauler
51. Sheets

Main course
52. Jeers
53. Parrel tackles
54. Lifts
55. Yard tackle tricing line
56. Yard tackle tricing line
57. Braces
58. Clewgarnets
59. Buntlines
60. Leechlines
61. Bowlines
62. Tack, last forecastle timberhead or second cavil
63. Sheets
64. Mast tackle (side tackle)

Main topsail
65. Tye
66. Reef tackle
67. Braces, a cleat at the mizzen mast or a shroud cleat in the mizzen shrouds
68. Lifts
69. Clewlines
70. Buntlines
71. Leechlines
72. Bowlines
73. Sheets
74. Shifting backstays

Main topgallant sail
75. Tye, upon deck or into the top
76. Lifts
77. Braces, like 67
78. ——
79. Backstays
80. Clewlines
81. Sheets, if extra rigged
82. Bowlines

Mizzen staysail
83. Halyard
Downhauler and tack belayed to the stay
84. Sheets

Crossjack yard
85. Lifts
86. Braces

Mizzen topsail
87. Tye
88. Lifts
89. Braces
90. Clewlines, belayed in the top, or to pin rails on deck
91. Bowlines
92. Sheets

Mizzen course
93. Jeers
94. Parrel tackle
95. Topping lift
96. Tack tackle
97. Tack
98. Brails
99. Sheet

For the accurate rigging of a model it is not only important to know the run of all the ropes, their thicknesses and types, but also to know how they were belayed. A modelmaker must therefore acquaint himself with the correct belaying points. Towards this end, this chapter reconstructs the belaying plans of several full-rigged vessels from the beginning, the middle and the end of the eighteenth century, based on models, many photographs and drawings, and especially on the specifications given by Davis, Steel, Lever, Röding and others. Figure 182 shows additional detail in the form of the various types of belaying cleats, etc.

Just as rigging varied in different vessels, so belaying plans were diverse. It is important to note, however, that the belaying points for particular items of rigging were necessarily stand-

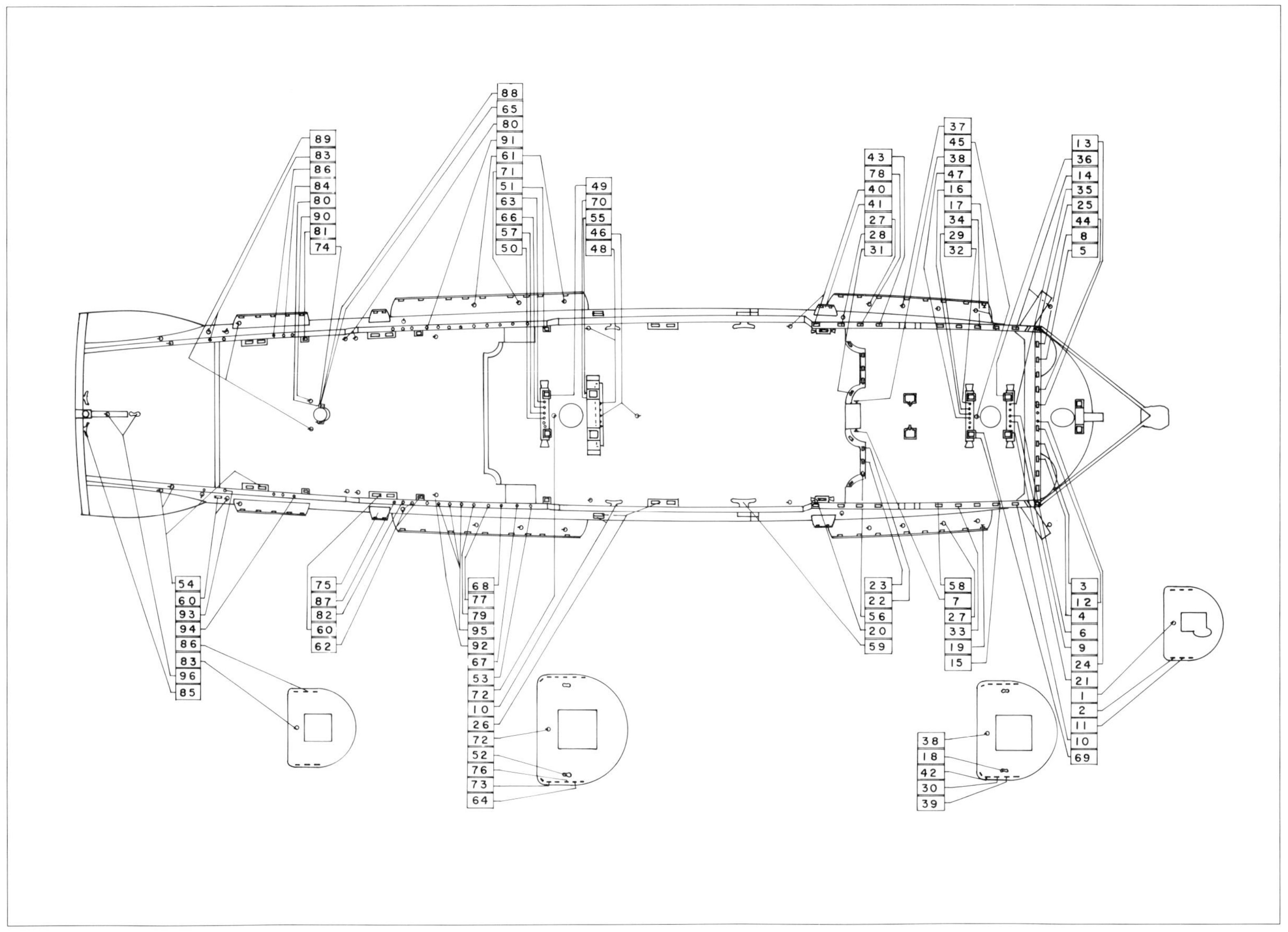

Figure 178. **Belaying plan of the running rigging in an English ship with a sprit topsail and jib-boom, about 1710**

Sprit topsail
1. Halyard
2. Lifts
3. Braces
4. Clewlines
5. Sheets

Spritsail
6. Outhauler
7. Braces
5. Lifts, also spritsail topsail sheets
8. Buntlines
9. Clewlines
10. Sheets

Fore topgallant stay
11. Stay tackle

Fore topmast stay
Staysail stay fastened to the stay
12. Stay tackle

Fore topmast staysail
13. Halyard
14. Sheets

Fore course
15. Jeers
16. Naveline
17. Lifts
18. Yard tackle tricing line
19. Yard tackle tricing line
20. Braces
21. Clewgarnets
22. Buntlines
23. Leechlines
24. Bowlines
25. Tacks
26. Sheets
27. Mast tackles (side tackles)

Fore topsail
28. Tye
29. Reef tackle
30. Lifts
31. Braces
32. Clewlines
33. Buntlines
34. Leechlines
35. Bowlines
36. Sheets
37. Shifting backstays

Fore topgallant sail
38. Tye
39. Lifts
40. Braces
41. Backstays
42. Clewlines
43. Sheets, if extra rigged
44. Bowlines

Main topmast stay
45. Stay tackle
Staysail stay fastened to the stay and set up with deadeyes to the fore stay collar

Main topmast staysail
46. Halyard
47. Downhauler
48. Sheets

Main course
49. Jeers
50. Naveline
51. Lifts
52. Yard tackle tricing line
53. Yard tackle tricing line
54. Braces
55. Clewgarnets
56. Buntlines
57. Leechlines
58. Bowlines
59. Tacks
60. Sheets
61. Mast tackle (side tackle)

Main topsail
62. Tye
63. Reef tackle
64. Lifts
65. Braces
66. Clewlines
67. Buntlines
68. Leechlines
69. Bowlines
70. Sheets
71. Shifting backstays

Main topgallant sail
72. Tye
73. Lifts
74. Braces
75. Backstays
76. Clewlines
77. Sheets, if extra rigged
78. Bowlines

Mizzen staysail
79. Halyard
80. Sheets

Crossjack yard
81. Lifts
82. Braces

Mizzen topsail
83. Tye
84. Lifts
85. Braces
86. Clewlines
87. Bowlines
88. Sheets

Mizzen course
89. Jeers
90. Topping lift
91. Tack tackle
92. Tack
93. Peak brails
94. Middle brails
95. Foot brails
96. Sheet

ardised, at least partly, across vessel types within a given time period. Seamanship was a trade, and a trained sailor usually had little time in a new environment to familiarise himself with the belaying points of ropes in his station. Taking into account also the steady influx of new crewmen on board and the fact that Navy press gangs were largely indiscriminate in their impressment of landlubbers or foreigners, the importance of a certain standardisation becomes obvious.

The qualified seaman became the somewhat reluctant tutor to his bewildered and uncomprehending new shipmates, for his own safety might depend upon the speed and thoroughness with which they learned their duties. Any uncertainty over which rope was the correct one to use might have catastrophic consequences, especially in bad weather.

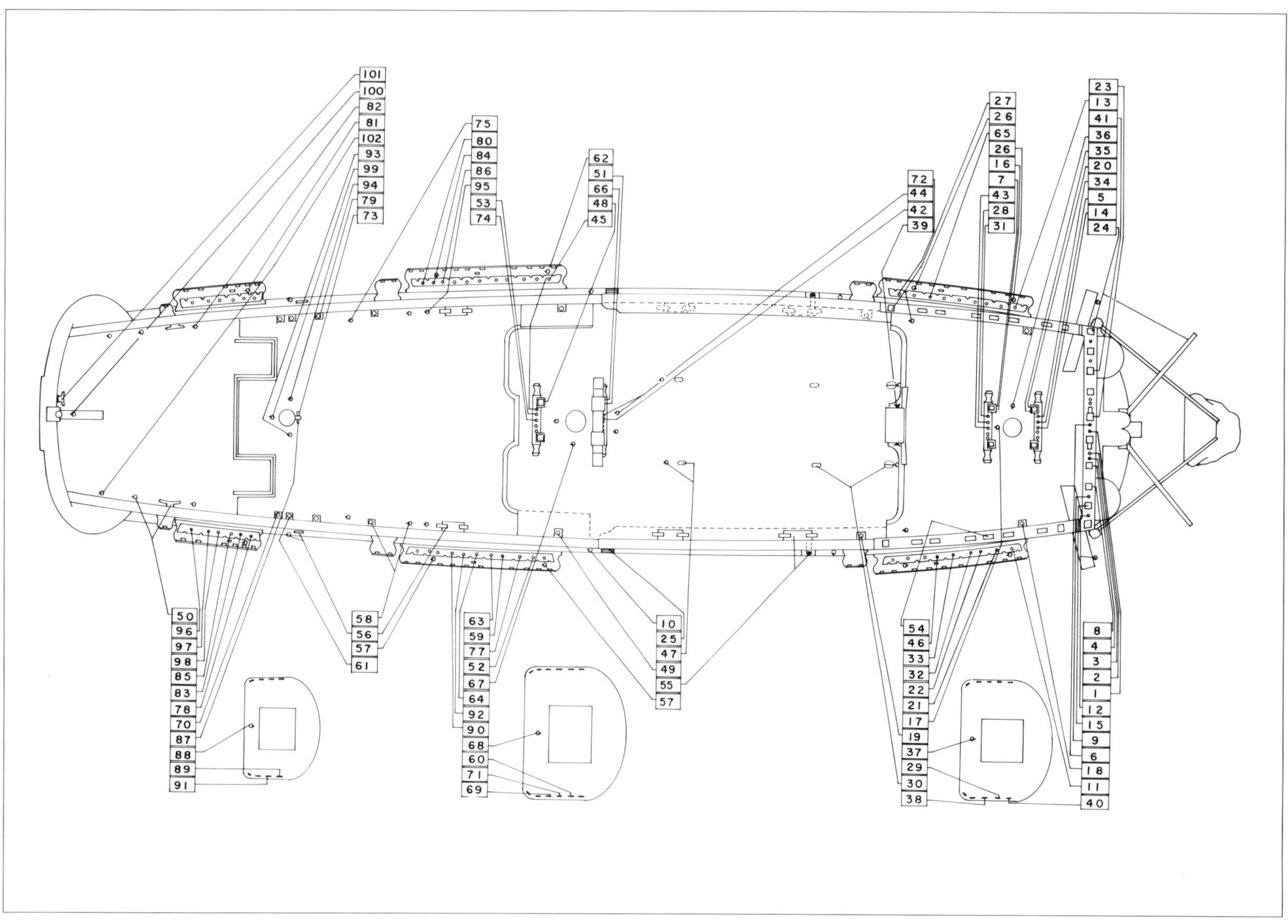

Figure 179. **Belaying plan of a Russian ship around 1750. Note in this contemporary model the stretchers in the lower shrouds, serving as pin rails.**

Sprit topsail
1. Outhauler
2. Lifts
3. Clewlines
4. Sheets, also spritsail lifts
5. Braces

Spritsail
6. Outhauler
7. Braces
4. Lifts, also sprit topsail sheets
8. Buntlines
9. Clewlines
10. Sheets

Jib
11. Halyard
12. Sheets

Fore topmast staysail
13. Halyard
14. Downhauler
15. Sheets

Fore course
16. Jeers
17. Parrel tackle
18. Lifts
19. Braces
20. Clewgarnets
21. Buntlines
22. Leechlines
23. Bowlines
24. Tacks
25. Sheets
26. Mast tackle (side tackle)

Fore topsail
27. Tye
28. Reef tackle
29. Lifts
30. Braces
31. Clewlines
32. Buntlines
33. Leechlines
34. Bowlines
35. Sheets
36. Top rope

Fore topgallant sail
37. Tye
38. Lifts
39. Braces
40. Clewlines
41. Bowlines

Main topmast staysail
42. Halyard
43. Downhauler
44. Sheets

Main topgallant staysail
45. Halyard
46. Downhauler
47. Sheets

Main course
48. Jeers
49. Lifts
50. Braces
51. Clewgarnets
52. Buntlines
53. Leechlines
54. Bowlines
55. Tacks
56. Sheets
57. Mast tackle (side tackle)

Main topsail
58. Tye
59. Reef tackle
60. Lifts
61. Braces
62. Clewlines
63. Buntlines
64. Leechlines
65. Bowlines
66. Sheets
67. Top rope

Main topgallant sail
68. Tye
69. Lifts
70. Braces
71. Clewlines
72. Bowlines

Mizzen staysail
73. Halyard
74. Downhauler
75. Sheets

Mizzen topmast staysail
76. Halyard
77. Downhauler
78. Sheets

Crossjack yard
79. Lifts
80. Braces
81. Mast tackle

Mizzen topsail
82. Tye
83. Lifts
84. Braces
85. Clewlines
86. Bowlines
87. Sheets

Mizzen topgallant sail
88. Tye
89. Lifts
90. Braces
91. Clewlines
92. Bowlines

Mizzen course
93. Jeers
94. Topping lift
95. Tack tackle
96. Peak brails
97. Middle brails
98. Throat brails
99. Tack
100. Sheet
101. Fancyline
102. Vangs

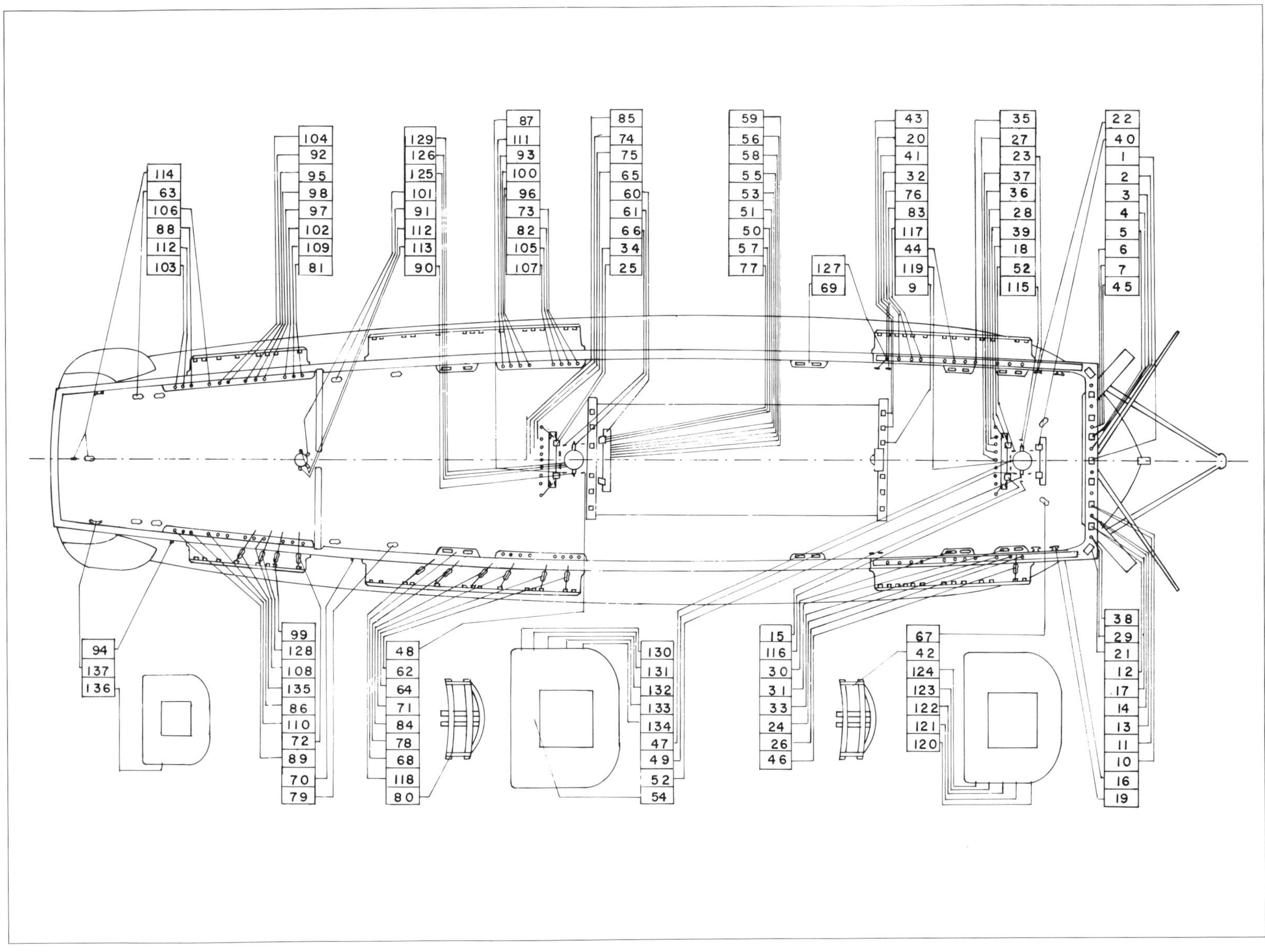

Figure 180. **Belaying plan of a French ship around 1780**

Bowsprit
1. Jib-boom guy

Sprit topsail
2. Halyard
3. Lifts
4. Braces
5. Clewlines
6. Sheets (spritsail yard lifts)

Spritsail
7. Halyard
8. Lifts (Spritsail topsail sheets)
9. Braces
10. Clewlines
11. Buntline
12. Sheets

Standing jib
13. Outhauler
14. Downhauler
15. Halyard
16. Sheets

Inner jib
17. Outhauler
18. Halyard
19. Sheets

Fore topmast staysail
20. Halyard
21. Downhauler
22. Sheets

Fore course
23. Jeers
24. Lifts
25. Braces
26. Clewgarnets
27. Buntlines
28. Leechlines
29. Bowlines
30. Sheets
31. Tacks

Fore topsail
32. Tye
33. Lifts (fore topgallant sheets)
34. Braces
35. Clewlines
36. Buntlines
37. Leechlines
38. Bowlines
39. Reef tackles
40. Sheet

Fore topgallant sail
41. Tye
42. Lifts
43. Braces
44. Clewlines
45. Bowlines
46. Sheets (fore topsail yard lifts)

Main staysail
47. Staysail stay
48. Halyard
49. Downhauler
50. Sheets

Main topmast staysail
51. Halyard
52. Downhauler
53. Sheets

Middle staysail
54. Staysail stay
55. Halyard
56. Downhauler
57. Sheets

Main topgallant staysail
58. Halyard
59. Downhauler
60. Sheets

Main course
61. Jeers
62. Lifts
63. Braces
64. Clewgarnets
65. Buntlines
66. Leechlines
67. Bowlines
68. Sheets
69. Tacks

Main topsail
70. Tye
71. Lifts (main topgallant sail sheets)
72. Braces
73. Clewlines
74. Buntlines
75. Leechlines
76. Bowlines
77. Sheets
78. Reef tackles

Main topgallant sail
79. Tye
80. Lifts
81. Braces
82. Clewlines
83. Bowlines
84. Sheets (main topsail yard lifts)

Mizzen staysail
85. Staysail stay
86. Halyard
87. Downhauler
88. Sheets

Mizzen topmast staysail
89. Halyard
90. Downhauler
91. Sheets

Crossjack yard
92. Lifts
93. Braces

Mizzen topsail
94. Tye
95. Lifts (mizzen topgallant sail sheets)
96. Braces
97. Clewlines
98. Buntlines
99. Leechlines
100. Bowlines
101. Sheets
102. Reef tackles

Mizzen topgallant sail
103. Tye
104. Lifts
105. Braces
106. Clewlines
107. Bowlines
108. Sheets (mizzen topsail yard lifts)

Mizzen course
109. Jeers
110. Lifts
111. Bowline
112. Brails
113. Tricing lines
114. Sheet

Lower fore studdingsail
115. Outer halyard
116. Inner halyard
117. Sheets
118. Tacks

Fore topmast studdingsail
119. Halyard
120. Sheets
121. Tacks

Fore topgallant studdingsail
122. Halyard
123. Sheets
124. Tacks

Lower main studdingsail
125. Outer halyard
126. Inner halyard
127. Sheets
128. Tacks

Main topmast studdingsail
129. Halyard
130. Sheets
131. Tacks

Main topgallant studdingsail
132. Halyard
133. Sheets
134. Tacks

Mizzen topmast studdingsail
135. Halyard
136. Sheets
137. Tacks

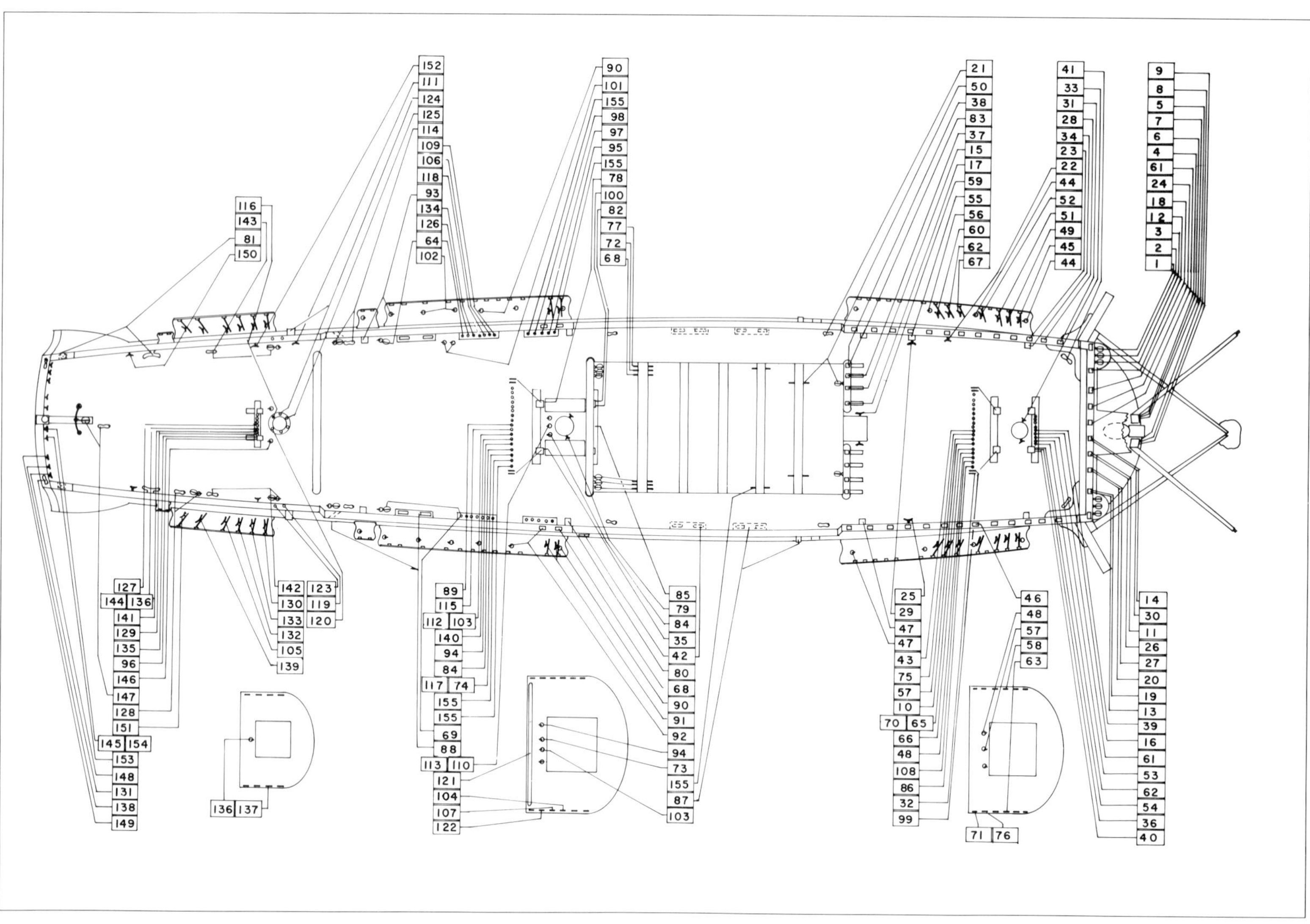

Figure 181. **Belaying plan of an English ship at the end of the eighteenth century**

Bowsprit
1. Flying jib-boom guy
2. Jib-boom guy
3. Traveller guy
4. Fore topmast preventer stay
5. Fore topmast stay
6. Fore topgallant stay
7. Inner martingale
8. Outer martingale
9. Flying jib-boom martingale
10. Jibstay tackle

Sprit topsail
11. Outhauler
12. Lifts
13. Clewlines
14. Sheets, sometimes also spritsail lifts
15. Braces

Spritsail
16. Outhauler
17. Braces
18. Lifts
19. Buntlines
20. Clewlines
21. Sheets

Flying jib
22. Halyard
23. Sheets
24. Outhauler

Jib
25. Halyard
26. Downhauler
27. Outhauler
28. Sheets

Fore topmast staysail
29. Halyard
30. Downhauler
31. Sheets

Fore course
32. Jeers
33. Truss tackle
34. Lifts
35. Braces
36. Clewgarnets
37. Buntlines
38. Leechlines
39. Slabline
40. Bowlines
41. Tacks
42. Sheets
43. Naveline
44. Mast tackle
45. Yard tackle tricing line
46. Yard tackle tricing line

Fore topsail
47. Tye
48. Reef tackle
49. Lifts
50. Braces
51. Clewlines
52. Buntlines
53. Bowlines
54. Sheets
55. Toprope
56. Shifting backstays

Fore topgallant sail
57. Tye
58. Lifts
59. Braces
60. Clewlines
61. Bowlines (alternative)
62. Sheets (alternative)
63. Buntlines

Main topmast staysail
64. Halyard
65. Downhauler
66. Brails
67. Tack
68. Sheets (alternative)

Middle staysail
69. Halyard
70. Downhauler
71. Tacks
72. Sheets
73. Stay tackle

Main topgallant staysail
74. Halyard
75. Downhauler
76. Tacks
77. Sheets

Main course
78. Jeers on upper deck
79. Truss tackle
80. Lifts
81. Braces
82. Clewgarnets
83. Buntlines
84. Leechlines
85. Slabline
86. Bowlines
87. Tacks
88. Sheets
89. Naveline
90. Mast tackle
91. Yard tackle tricing line
92. Yard tackle

Main topsail
93. Tye
94. Reef tackle
95. Lifts
96. Braces
97. Clewlines
98. Buntlines
99. Bowlines
100. Sheets
101. Toprope
102. Shifting backstays

Main topgallant sail
103. Tye
104. Lifts
105. Braces
106. Clewlines
107. Buntlines
108. Bowlines
109. Sheets

Mizzen staysail
110. Stay
111. Halyard
112. Downhauler
113. Tack
114. Sheet
115. Brails

Mizzen topmast staysail
116. Halyard
117. Downhauler
118. Tacks
119. Sheets

Mizzen topgallant staysail
120. Halyard
121. Downhauler
122. Tacks
123. Sheets

Crossjack yard
124. Lifts
125. Truss tackle
126. Braces
127. Naveline

Mizzen topsail
128. Tye
129. Reef tackle
130. Lifts
131. Braces
132. Clewlines
133. Buntlines
134. Bowlines
135. Sheets

Mizzen topgallant sail
136. Tye
137. Lifts
138. Braces
139. Clewlines
140. Bowlines
141. Sheets

Mizzen course
142. Throat tye
143. Peak tye
144. Throat downhauler
145. Peak downhauler
146. Boom topping lift
147. Boom sheet (driver), otherwise mizzen sheet
148. Boom guy
149. Vangs
150. Peak brails
151. Middle brails
152. Throat brails
153. Fancy line

Ensign staff
154. Flag line

Miscellaneous
155. Free belaying points for studdingsails, etc.

Most of the rigging came in pairs, usually belaying in the same place on both port and starboard sides; the numbers for such ropes are shown only on one side for clarity. Exceptions are only where the opposite belaying point was used for a different purpose.

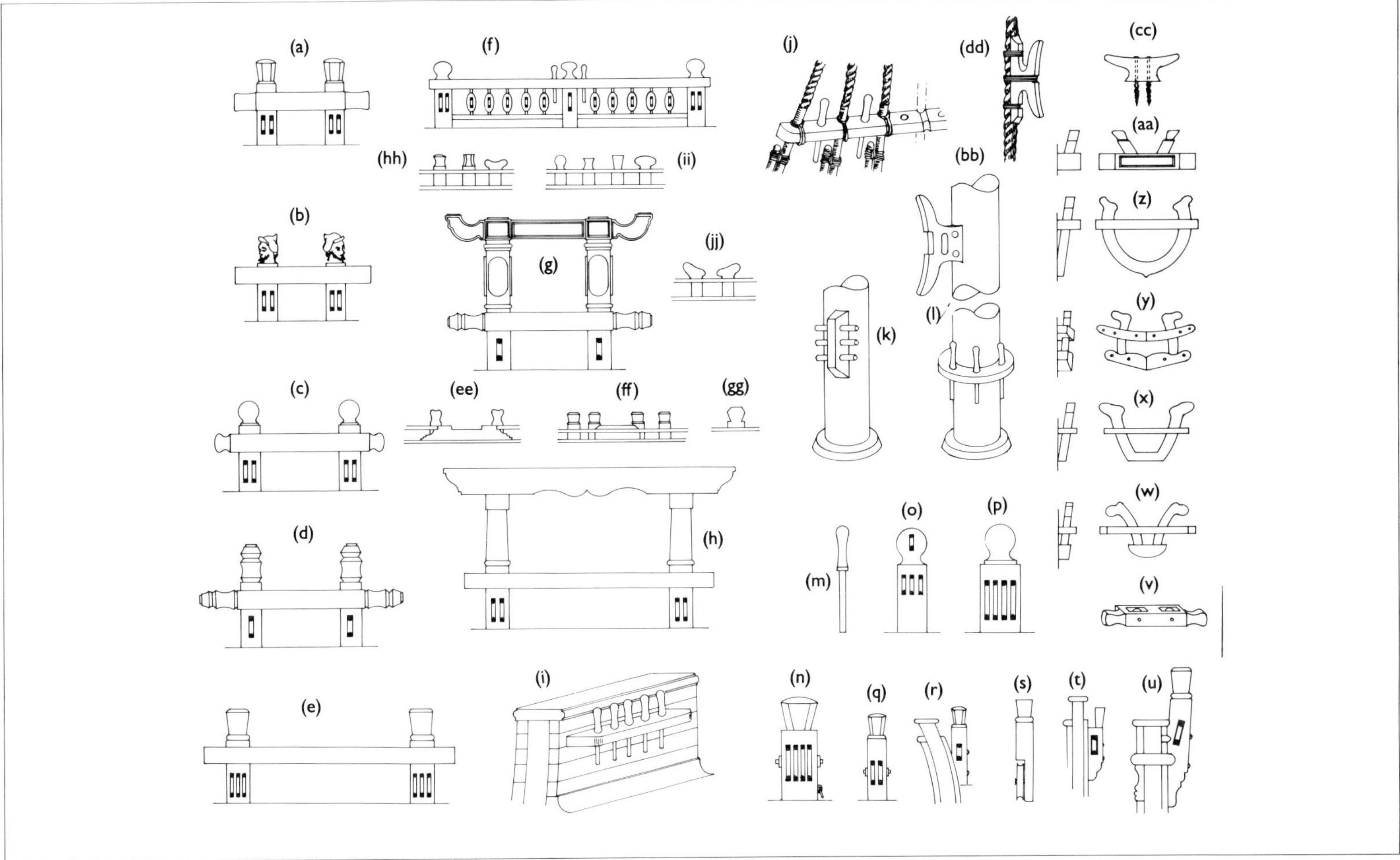

Figure 182 **Types of belaying point for running rigging**

(a) Dutch bitts of about 1700, with three pins.

(b) English bitts of about 1700, with three pins.

(c) French bitts of about 1700, with five pins.

(d) Russian bitts of about 1750, with five pins.

(e) English bitts of about 1780, with nine or seventeen pins.

(f) French bitts of about 1770, with twelve pins and ten nine-pin blocks.

(g) Russian main topsail sheet bitts with spar gallows, about 1750.

(h) English main topsail sheet bitts with gallows, about 1775.

(i) Pinrail.

(j) Shroud stretcher with belaying pins.

(k) A pin rack on both sides of a mizzen mast, where no bitts were fitted (also known as topsail sheet cleats). In smaller ships these were sometimes fitted to all masts.

(l) Belaying pins on a mizzen mast. At the end of the century these often replaced the topsail sheet cleats shown in (k). In large ships these were used in addition to mizzen mast bitts.

(m) Belaying pin; according to Falconer these were 16in long, the upper part was three sevenths of the total length, the upper diameter was 1⅜in and the lower 1⅛in, and the whole was made from ash. He also noted that belaying pins might be made from iron, in various sizes.

(n) Dutch knighthead for the jeers, about 1700.

(o) French knighthead for jeers, about 1750, according to Duhamel and Röding.

(p) French knighthead for jeers, about 1770, according to Paris. The width of the knighthead was two thirds of the mast diameter, and Duhamel gave 4ft as its height.

(q) Free-standing topsail sheet knighthead, about 1700, still sometimes found in English ships up to 1740.

(r) Dutch small knighthead fitted to the bulwarks for lower lifts, top rope, topsail halyards, etc; about 1700.

(s) Small bulwark knighthead according to Röding.

(t) Russian small bulwark knighthead, about 1750.

(u) English small bulwark knighthead, about 1750.

(v) English cavil cleat for topsail halyards and braces, also sometimes found with one sheave.

(w) French kevel of about 1750, according to Duhamel.

(x) French kevel of about 1770, according to Paris.

(y) Continental kevel of about 1790, according to Röding.

(z) English kevel.

(aa) English kevel of about 1770.

(bb) Mast cleat; the notch and slot were for the lower and upper lashings, and the holes in the foot for a cross seizing to hold the cleat closer to the mast.

(cc) Cleat for the bulwark, or to be nailed to the deck, etc.

(dd) Cleat lashed to a shroud.

(ee)) English fish davit support and kevel heads (timber heads), about 1750.

(ff) English fish davit support and timber heads, about 1770.

(gg) English kevel head, about 1790.

(hh) Three different Dutch timber or kevel heads.

(ii) Four different French timber or kevel heads.

(jj) Kevel head pair in a Russian ship, about 1750.

XIV

Blocks and tackles

Blocks

Blöcke; Poulies

The blocks of the eighteenth century were wooden casings with one or more sheaves of timber or metal. They were of various shapes and sizes, depending on their use on the ship. Connected to falls, they formed tackles for every possible purpose on board.

A block's casing or shell was made from ash or elm, and the sheaves or shivers from American hard wood (lignum vitæ), ash, brass or iron. The sheaves often had a brass coaking in the centre, to provide longer trouble-free running on the pin. The pin was made from lignum vitæ, cog-wood, greenheart, or iron. One or two scores at the ends of the shell, according to size and use, were prepared for single or double strops. A sheave had a semicircular groove cut in around its circumference; to allow the fall to run smoothly it was one third of the sheave's thickness deep.

Normal blocks were oval in shape, and the number and size of the sheaves determined the size and thickness of the shell. A single block's proportions were as follows: the sheave's thickness was one tenth more than the diameter of the rope, and its diameter was five times the thickness; the sheave hole was one sixteenth wider than the sheave's thickness; the length of a block was eight times the width of the sheave hole, the breadth six times the thickness of the sheave, and the thickness half the length. Flat thin blocks were three eighths of their own length in thickness. All blocks with more than one sheave increased in thickness by the widths of the additional sheave holes and the partitions, which had a width of five sixths of a sheave hole.

Blocks were made by hand throughout the century and the shell was usually made from one piece. Only very large blocks were made from several pieces, which were bolted together with three bolts on both ends and clinched.

Machines for the production of blocks, invented by Mark Isambard Brunel, were first installed in Portsmouth in 1804. The blockmaker's trade then became redundant in the Royal Navy. These machines produced 1420 blocks a day and provided the British fleet with a ready supply of replacement blocks at a considerable saving in cost.

The variety in size and shape of blocks was enormous; up to 200 variations were needed in the fleet. Among the different shaped blocks were the following:

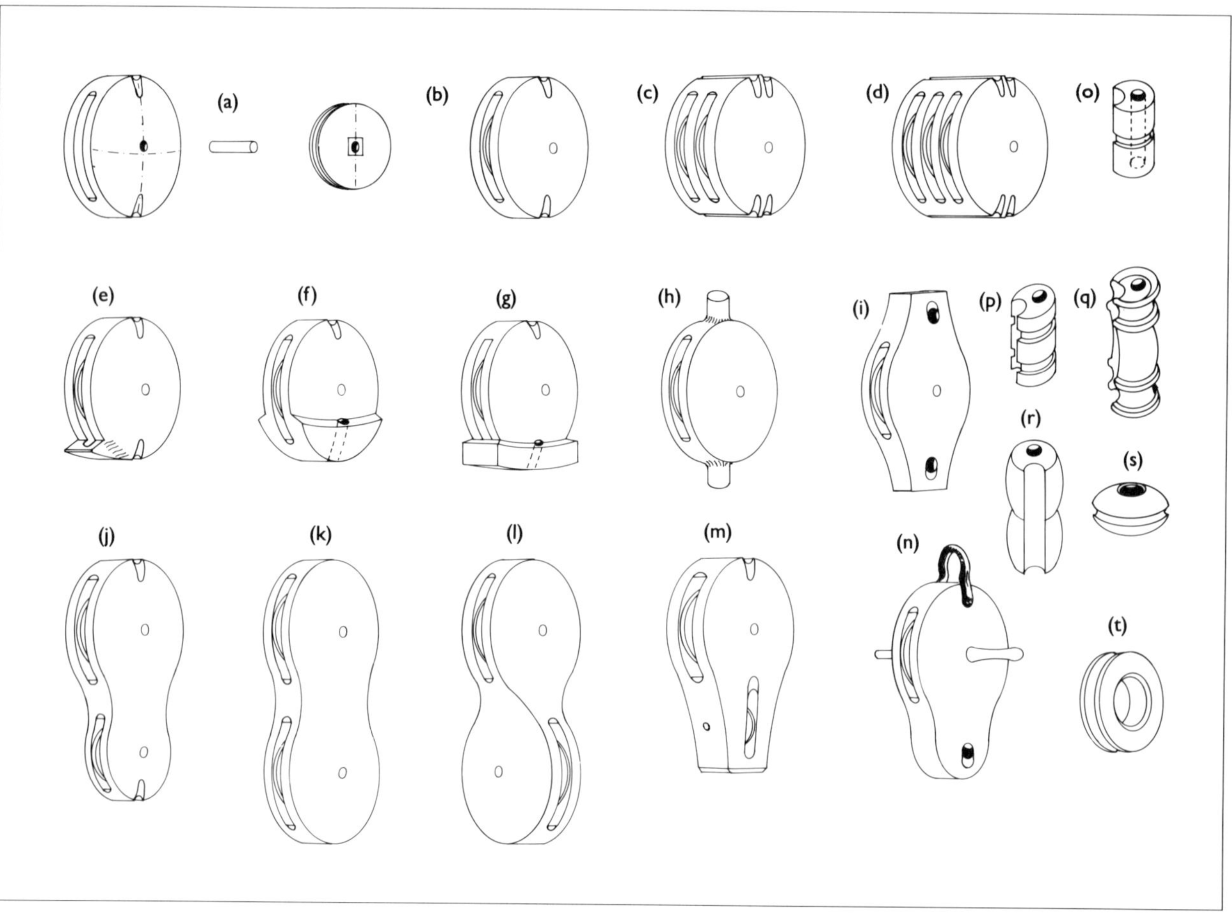

Figure 183
(a) Parts of a block: shell, pin and sheave.
(b) Single block.
(c) Double block with two scores.
(d) Treble block with two scores.
(e) Shoulder block.
(f) Clew garnet block.
(g) Spritsail sheet block.
(h) Ninepin block.
(i) Continental lift block.
(j) Fiddle block, or long tackle block.
(k) Leg and fall block.
(l) Shoe block.
(m) Continental topsail sheet and lift block.
(n) Gaff sail sheet block with belaying pin.
(o) Fairlead truck with one lashing groove.
(p) Fairlead truck with two lashing grooves.
(q) and (r) Differently shaped fairlead trucks.
(s) Wooden thimble.
(t) Wooden thimble.

Shoulder block

Marssegelschotblock, Schulterblock; Poulie à talon

A shoulder block (see Figure 183e) had a protrusion on one side of the lower end for better guidance of the topsail and topgallant sheets. These blocks were kept upright by the shoulder, thus preventing the sheets from jamming between the block and the yard.

Shoulder blocks were also in use on bumkins, to lead the fore tack inboard.

Long tackle block, fiddle block

Violinblock; Poulie à violon

The end-to-end combination of two single blocks in one shell, the lower two thirds the size of the upper, gave the impression of a fiddle (see Figure 183j).

Such blocks were used in the Royal Navy and on East Indiamen as yard tackle blocks, and on merchantmen as loading tackles (according to Rees). In earlier years they were also in use in other tackles according to Blanckley. Röding described this block as the upper block of a tackle, but by his time it had mostly been replaced by a double block, since the fall often became choked in a fiddle block.

Shoe block

Schuhblock; Poulie à olive

'Shoe block' was the term for the end-to-end combination of two equally sized single blocks in one shell, with the sheaves at right angles to one another. They served for legs and falls of buntlines (see Figure 183l).

Leg and fall block

Schenkel und Fallblock; Poulie de cargue-fond

The leg and fall block was similar to the shoe block, but with the difference that both sheaves ran in the same direction as in a fiddle block (see Figure 183k).

Snatch block

Kinnbackenblock; Poulie coupée, galoche

A snatch block was flat with its upper end opened up wide enough on one side to allow the rope or fall to be laid in or taken out (see Figure 184d, e and f). The upper end was slightly longer than that of a normal block and was fitted with a hole, to enable the block to be lashed. Navy snatch blocks were iron-bound with a swivel hook.

Snatch blocks were employed to get the deep-sea lead in, to lead the main braces and sometimes also the fore braces inboard, as a lead block for the weather main bowline abaft the foremast, as a foot block for the top rope, or as a larger block to take the bight of a cable in when a ship had to be towed.

Sister block

Puppblock, Stengewantblock; Poulie vierge

A sister block consisted of two flat single blocks arranged vertically in one shell, with a groove in both flat sides to accommodate the two foremost topmast shrouds, to which the block was lashed (see Figure 183g and h). Later in the century, these blocks guided the topsail yard lifts and reef tackle. Sister blocks on merchantmen often had only a hole instead of the lower sheave.

According to Röding, sister blocks were mainly used in English rigging. He added that 'in earlier times these blocks also had three sheaves, and that the third one was used for the studdingsail halyard'.

Clewgarnet block, clewline block

Geitaublock; Poulie de cargue-point

A clewline block or clewgarnet block was a normal single block with a shoulder. The shoulder was an extension to both flat sides, covering about one third of the block in length, with a hole in each side to lead the strop through (see Figure 183f). 'The use of the shoulder on the lower-end is to prevent the strap from being fretted or chafed by the motion of the sail, as the ship rolls and pitches' (Falconer). In French shipping this type of block was employed as a buntline block.

Spritsail sheet block

Blindeschotblock; Poulie d'écoute de civadière

A spritsail sheet block was in use up to around 1750 and had a certain similarity to a clewline block. It was slightly rounder in shape with a shoulder shaped like a saucer or a squared piece on the lower end (see Figure 183g). The scores in this block were often replaced by deeper grooves, which went all round the block, with holes in the shoulder.

Monkey block

Grenadierblock, Monkeyblock; Poulie de conduit de cargue-fond

A monkey block was a block with a saddle, and was nailed to the upper side of a yard on merchantmen, mainly for buntlines (see Figure 184a, b and c). Lever described them as 'to nail upon the Topsail Yards on Merchant Ships, for the Bunt-lines to reeve through... Sometimes [a monkey block] has a Swivel above the Saddle, to permit the Block to turn, when used for a Leech-line.' Röding described the use of these blocks slightly differently: 'both types are nailed to the upper side of the lower yards and serve, mainly on merchantmen, to lead the running rigging either to the mast, or direct upon deck'. Falconer (1815) confirmed Röding's description by noting the use of monkey blocks on lower yards on small merchantmen, and continued:

> Some are only small single blocks, attached, by a strap and iron swivel, to iron straps, which embrace and nail to the yard, the block turning to lead the small running ropes in any direction; others are nearly eight square, with a roller working in the middle, and a wooden saddle beneath, to fit and nail to the yard.

The monkey block was a product of the very end of the eighteenth century.

Topsail sheet and lift block

Marssegelschot- und Toppnantblock; Poulie de bout de vergue et balancine

The topsail sheet and lift block was a pear shaped block that had a certain similarity to a shoe block. It was used in Continental ships on the lower yardarms for leading the topsail sheets and lifts. The sheave for the sheet was placed in the thicker part of the block, and the lift's sheave was smaller and fitted at right angles to it in the neck (see Figure 183m). The lift sheave hole was enlarged to take the strop as well, and scores were cut into the block's tail. These blocks were only stropped around the thicker part.

Continental lift block

Toppnantblock, Schildpad; Poulie de balancine

The type of lift block used in Continental rigging was a flat single block with two neck-shaped ends, which were not rounded (see Figure 183i). Through each of these neck extensions a hole was drilled, perpendicular to the sheave, to take

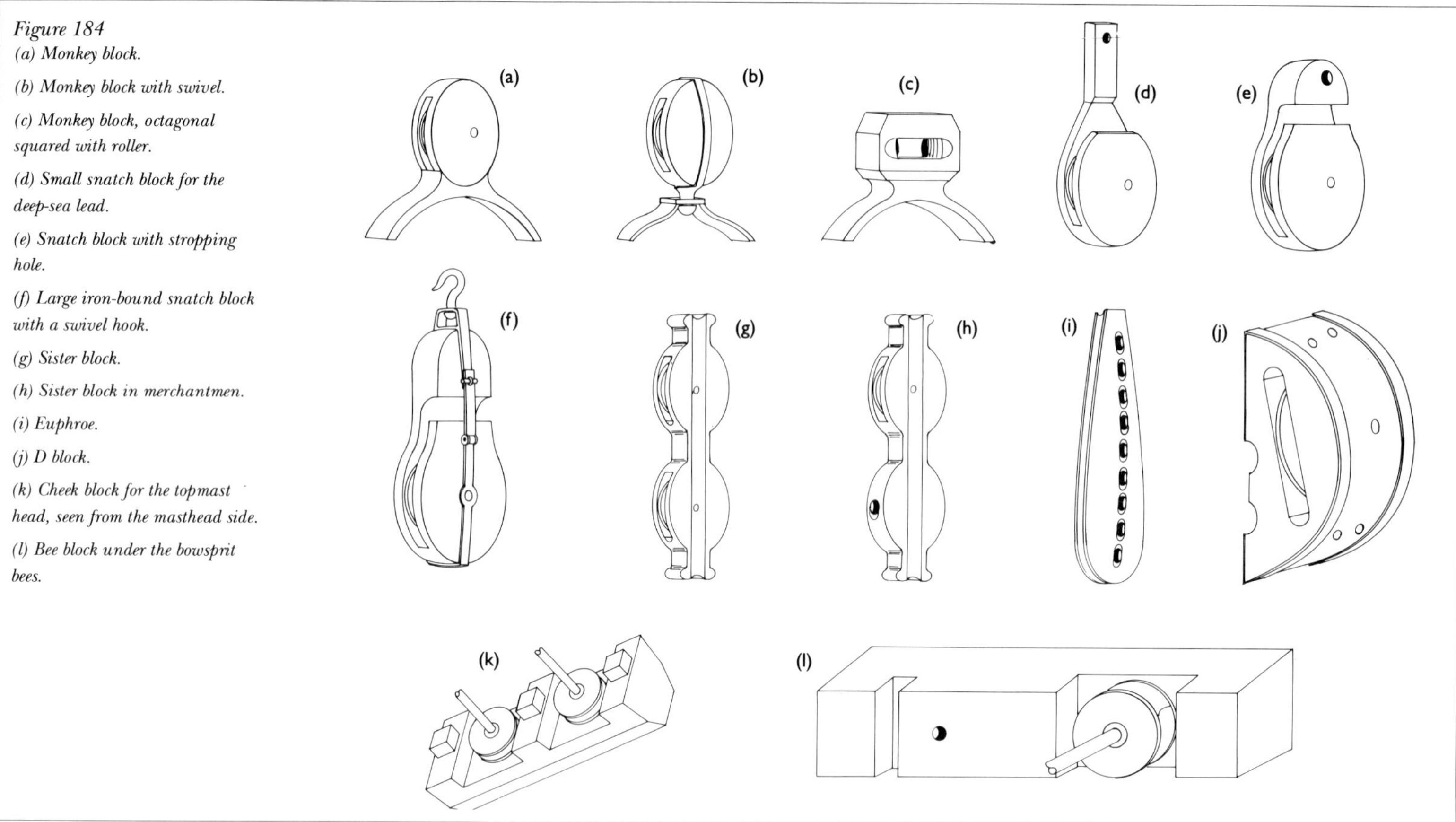

Figure 184
(a) Monkey block.
(b) Monkey block with swivel.
(c) Monkey block, octagonal squared with roller.
(d) Small snatch block for the deep-sea lead.
(e) Snatch block with stropping hole.
(f) Large iron-bound snatch block with a swivel hook.
(g) Sister block.
(h) Sister block in merchantmen.
(i) Euphroe.
(j) D block.
(k) Cheek block for the topmast head, seen from the masthead side.
(l) Bee block under the bowsprit bees.

the block's strop and for the standing part of the lift. These blocks were only used in conjunction with the pear-shaped sheet and lift block described above.

Ninepin block

Ninepin block; Poulie de pied de mât

Ninepin was an English term also taken up by Röding for a block shaped somewhat like a pin, flattened on both sides and fitted into the breastwork and under the crosspieces of the forecastle and quarterdeck bitts to lead the ropes horizontally (see Figure 183h). During the eighteenth century the use of these blocks in bitts was mainly a French fashion (they were known as *poulies tournantes* or *marionettes*), and the ninepin block is mentioned in Falconer's 1815 edition only in connection with its general use.

Rack block, gammon-lashing

Tausendbein, Wegweiser; Rateau, ratelier de poulies

'[A rack block is] lashed to each Side of the Gammoning in the Head, and has several Shives one above another, through which the Spritsail-lifts, Buntlines, Clewlines, and Sprit Topsail Sheats go' (Blanckley). According to Falconer (1815), rack blocks were:

> A range of small single blocks, made from one solid piece of wood, by the same proportion as single blocks, with ends in form of a dove's tail, for the lashing, by which they are fastened athwart the bowsprit, to lead in the running ropes.

He noted that they were seldom used. The number of sheaves in a rack block varied. Blanckley mentioned four ropes, but showed five sheaves, Falconer did not specify a number and Röding wrote of eight to ten sheaves in pairs above one another (see Figure 185a).

In addition to the rack block, and later in the century often in lieu of it, a fairlead saddle was fitted to the upper side of the bowsprit, just forward of the gammoning (though this was little known in Continental rigging).

Ramshead block

Kardeelblock; Poulie de drisse

Ships with Continental mast caps had a special halyard block hanging in the bight of the tyes, which, as the name indicates, was formed like a ram's head (see Figure 185d). One of the earliest descriptions of this block was given by Smith in *A Sea Grammar* of 1627: 'The Ramshead is a great blocke, wherein is three shivers into which are passed the halyards, and at the end of it in a hole is reved the ties, and this is only belonging to the fore and maine halyards.'

English ships abandoned the use of tyes and halyards in the third quarter of the seventeenth century, but their use in Continental ships can be followed into the next first half-century. Arab dhows used the ramshead block right into the twentieth century. A ramshead block's proportional measurements were not the same as those of normal blocks in every aspect. The length of the block was approximately twelve times the thickness of the sheaves, with the axis of the sheaves 10/21 of the height above the lower end. The hole for the tye, at right angles to the sheave holes, was situated in the middle between the upper edge of the sheave hole and the top of the block. Sometimes these blocks had a smaller block stropped to them for lifting, through which a tricing line passed.

Lead-cleat, cleat block

Scheibenklampe, Schildpad; Joue de vache

A lead-cleat or cleat block was a cleat with an inserted sheave, nailed to the fore side of the lower yards, about two fifths of the length from the yardarm to the slings. It was used to lead the leechlines of the main and fore courses on Dutch and related ships in the early decades of the eighteenth century.

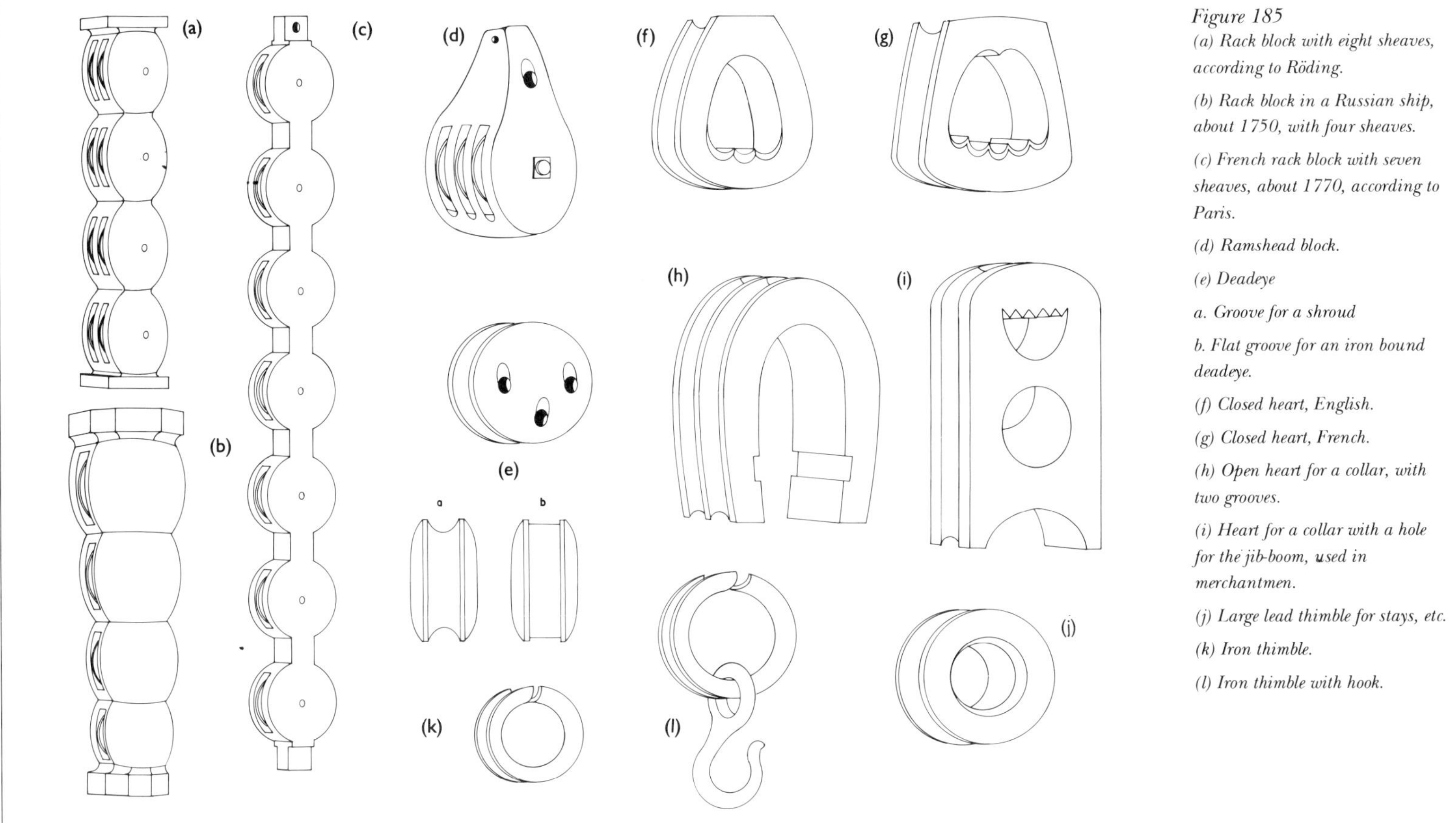

Figure 185
(a) Rack block with eight sheaves, according to Röding.
(b) Rack block in a Russian ship, about 1750, with four sheaves.
(c) French rack block with seven sheaves, about 1770, according to Paris.
(d) Ramshead block.
(e) Deadeye
a. Groove for a shroud
b. Flat groove for an iron bound deadeye.
(f) Closed heart, English.
(g) Closed heart, French.
(h) Open heart for a collar, with two grooves.
(i) Heart for a collar with a hole for the jib-boom, used in merchantmen.
(j) Large lead thimble for stays, etc.
(k) Iron thimble.
(l) Iron thimble with hook.

Deadeye

Juffer, Jungfer; Cap de mouton

A deadeye was a wooden disk with a groove around its circumference and usually three holes, set in a triangular pattern, for leading the lanyard (see Figure 185e). An egg-shaped deadeye with two holes was often in use with the parrel ropes of the mizzen yard. Deadeyes with more than three holes were often used as fore stay deadeyes on smaller craft.

Closed heart

Dodshoofdt, Doodshoofdt; Moque, poulie à moque

A closed heart was a type of deadeye (the German name means Death's head, or skull), shaped slightly like a heart, with one large hole in the middle (see Figure 185f and g). The lower side of this hole was scored four to five times for the lanyard. Around the outside of the heart a groove for the stay was cut in. The length of a heart was 1½ times the stay's circumference, its width three quarters of its length, and the thickness twice the stay's diameter. Hearts for stay and collar were of equal size.

Open heart

Offenes Dodshoofdt; Moque ouvert

The open heart came into use during the last quarter-century for the fore and fore preventer stay. It was shaped like a horseshoe and set on top of the bowsprit (see Figure 185h). It was provided with two grooves, like those of closed hearts after 1735, to enable double stropping on larger vessels only. The inner side of the opening was notched on both sides for an extra seizing of the heart to the stay or collar.

Hearts on merchant ships sometimes had an extra hole for the jib-boom, and their lower end was cut semi-circular, to fit the bowsprit surface (see Figure 185i).

Thimble, bullseye

Kausche, Klotje; Cosse

A thimble or bullseye was a wooden disk with a hole in the middle, and a groove around the circumference (see Figure 183s and t). It was often made of iron, and in later years was sometimes replaced by a more tear-shaped fitting (see Figure 185j, k and l).

Truck

Leitkausche, Leitklotje; Pomme de conduite

A truck was an elongated piece of rounded timber, with a hole drilled through its length, for leading ropes. A groove ran over the entire length one side to place the truck securely on to a shroud. One or two smaller grooves around the truck enabled it to be seized to the shroud (see Figure 183 o to r).

D-block

D Block; Poulle D

D-blocks were lumps of oak, formed in the shape of the letter D, from 12 to 16 inches long and 8 to 10 inches wide (see Figure 184j). They had a sheave fitted and were bolted to the ship's sides in the channels, to take the lifts etc. The D-block was English and came into use at the end of the century, mainly in men-of-war.

Euphroe

Spinnenjuffer, Spinnkopf; Moque d'araignée, moque de trelingage, hernier

A euphroe was an elongated hardwood batten with a number of holes through which crowsfeet from tops, awnings and so on were led (see Figure 184i). A groove was usually carved in around the outer edges for the stropping.

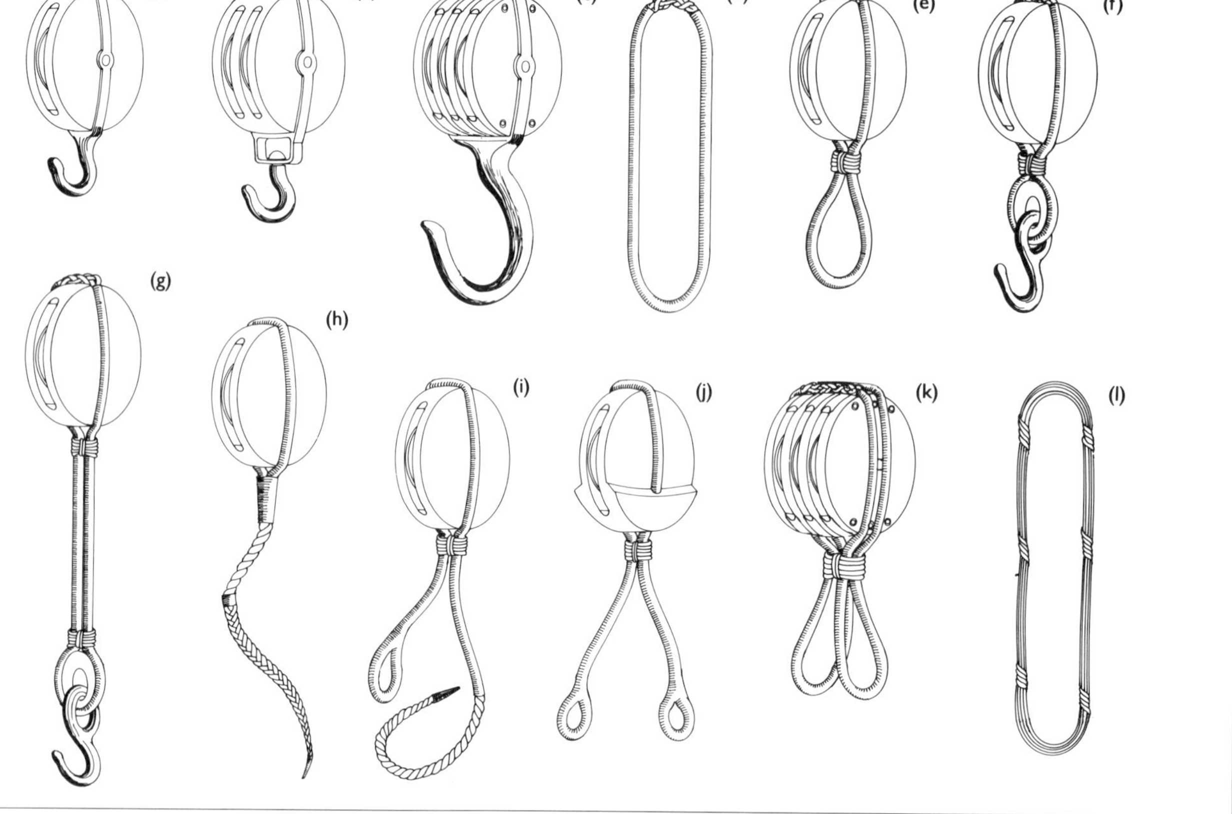

Figure 186
(a) Iron-bound block with a hook.
(b) Iron-bound block with a swivel hook.
(c) Iron-bound built cat block with a large hook.
(d) Block strop, spliced and served.
(e) Stropped single block.
(f) Stropped single block with hook and thimble.
(g) Long-stropped block with hook and thimble.
(h) Tail block.
(i) Block stropped with a short and a long leg.
(j) Stropped clewline block.
(k) Double stropped built treble block.
(l) Selvagee.

Cheek block
Wangenblock; Joue de vache

A cheek block was a squared, elongated piece of timber with two sheaves arranged vertically, bolted to the sides of the topmast heads (see Figure 184k). The cheek blocks were part of English, or English-style, rigging in the second half of the century, but were not found in Continental rigging (see Topmast in Chapter I).

Bee block
Violinblock; Poulie de violon

A bee block was a type of cheek block fitted beneath the bowsprit's bees (see figure 184l). This, too, was only found in English or English-style rigging (see Bowsprit in Chapter I).

Miscellaneous blocks
Vermischte Blöcke; Poulies diverses

Iron-bound
Eisenbeschlag; Estropé en fer

Some blocks were iron-bound and provided with a hook (see Figure 186a, b and c). A top block, for example, was single with a normal hook, an upper top tackle block had two or three sheaves with a normal hook, and a lower, with the same number of sheaves, had a ring fastened to its head for the tackle fall's standing part, and a swivel hook at its tail. A cat block was a large treble block with a hook large enough to take the anchor's ring. Some snatch blocks were also iron-bound.

Strop
Stropp; Estrope

The circumference of a rope used to strop a block varied according to the block's size.

> Straps of Blocks are generally in two Parts, and sometimes in four, which ought to be equal in Strength to the Folds of the Tackle-fall, or any other Rope. And since 4 Parts of any Rope of 6 Inches Circumference, are near equal to 2 Parts of a Rope alike in Goodness, of 8 Inches and ½ in Circumference, those two Parts will be suitable for a Strap to a Block that is used with 4 Folds, as a Tackle-fall, or any other running Rope [Sutherland].

The required length was taken and the ends were usually spliced together into a ring, which was served (see Figure 186d). Steel explained that the stropping for jeer blocks was wormed, parcelled and served, stropping of 4 inches and above was wormed and served, and all ropes below 4 inches were served only with spun yarn. The block was placed in the end with the splice, and the strop was seized tightly together close to the block, leaving a loop which could be put over the yardarm, a boom or spar (see Figure 186e). If the block had a hook, then the hook and thimble were laid into the opposite bight to the block and the part between was tightly seized together (see Figure 186f).

A tail block (see Figure 186h) had an open strop, with the block fitted into an eye splice which was served with spun-yarn. The splice, under the block, had its ends combed and marled down, and the end of the rope was whipped. Alternatively, a stop was put at a given distance from the splice, and the tail was then unlaid and the strands plaited or the yarns opened out and marled down like a selvagee.

A block strop with a short and a long leg was seized in the eye or bight with a round seizing (see Figure 186i). The short leg had an eye spliced in the end, and the eye and a larger part

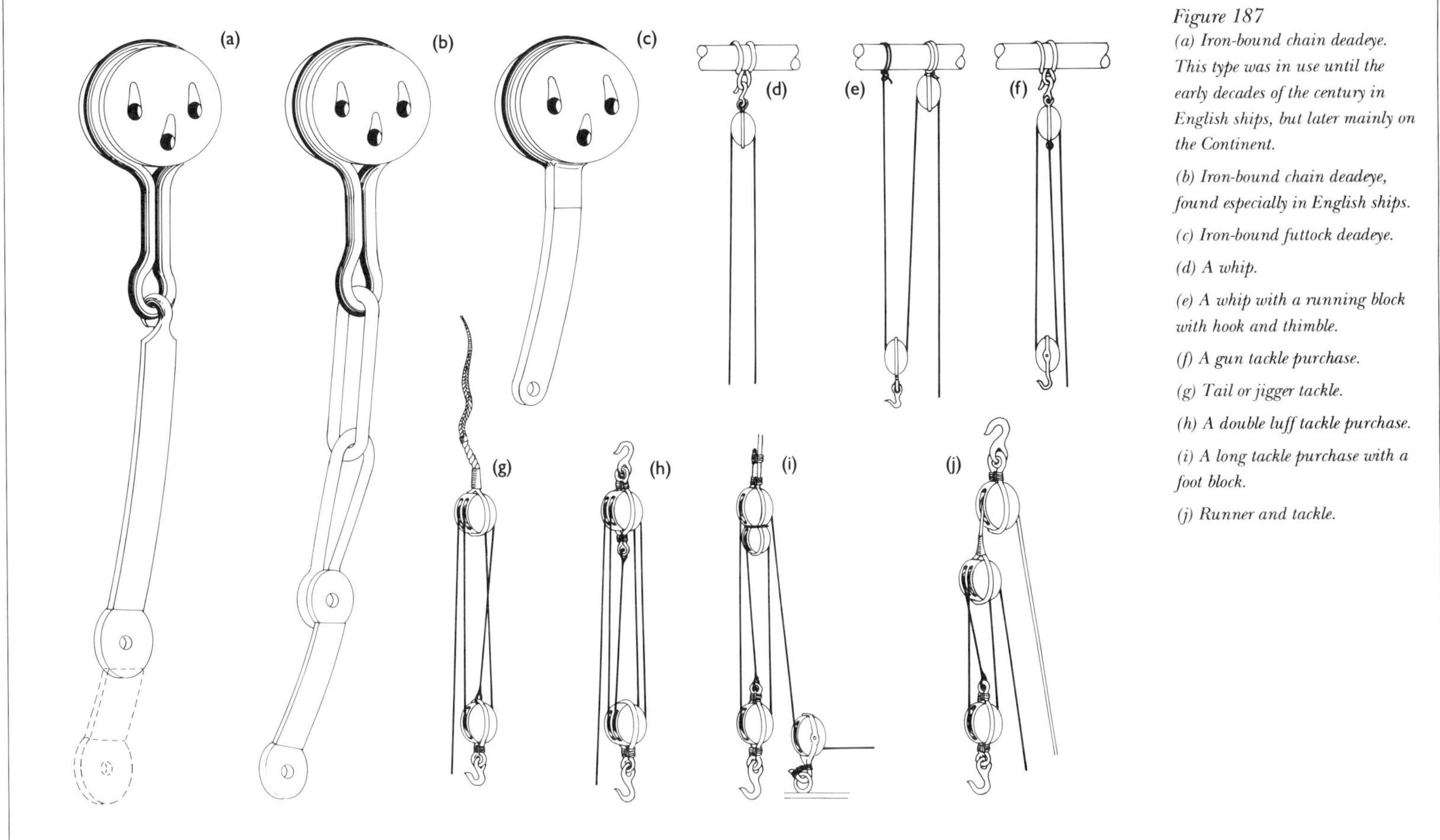

Figure 187

(a) Iron-bound chain deadeye. This type was in use until the early decades of the century in English ships, but later mainly on the Continent.

(b) Iron-bound chain deadeye, found especially in English ships.

(c) Iron-bound futtock deadeye.

(d) A whip.

(e) A whip with a running block with hook and thimble.

(f) A gun tackle purchase.

(g) Tail or jigger tackle.

(h) A double luff tackle purchase.

(i) A long tackle purchase with a foot block.

(j) Runner and tackle.

of the strop were served over, with the end opposite the eye whipped. This end was laid around the yard or similar and through the eye of the short leg, then hitched or seized to itself.

A block with two short legs was fitted in the same way, with the minor difference that such a block had an eye spliced into both ends. Clewline blocks were similarly stropped (see Figure 186j). The strop passed through the shoulders, and the second eye was spliced into the end. The seizing was put on as above.

A three or four sheaved block was stropped double. The strop was formed into a loop and served according to its size, then it was doubled, with the splice and the other bight put in the scores at the end of the block (see Figure 186k). The seizing was put on as before, except that it was crossed both ways through the double parts of the strop.

The descriptions given here applied, except for small size variations, to all stropped blocks.

A selvagee (see Figure 186l) was a very durable strop. It was made by laying rope-yarns in a loop (around two or more timber-heads) and marling them down with spun yarn. Block strops made in this way were very neat, and were used for lead blocks in the shrouds and sometimes worked with Spanish foxes.

Davis provided 'Rules for cutting out all the Stropping for any Ship', giving the proportional lengths of strops relative to the lengths of their blocks:

Bowsprit

9 times	A double Strap for the Halyard Block for the Yard
5½ times	A single Strap for ditto
6½ times	Strap for the long Tack. Block for the Boulsprit end
7 times	For Clew-lines on the Yard, with Eyes
8 times	For the Clews of the Sail
5¼ times	For Lifts for the Yard-armes
4 times	For seizing Blocks and for Pendants for Braces all the Ship over
11 times	For Lifts for the Spritsail-top at the Topmast-head
5 times	Ditto for the Yard-arms
6 times	For the Tye on the Yard, if any, with Eyes
6½ times	For the Clew-lines on the Yard
6 times	Ditto for the Clews of the Sail

Foremast and yard

4½ times	For the Pendants for Tackles at the Mast-head
3½ times	For a long Tackle Block for the Runner
17 times	A double Strap for a double Block at the Mast-head
7¼ times	Ditto for a Block for the Yard
13 times	For a double Strap for a quarter Block for the Yard
7½ times	For a Topsail Sheet for the Yard-arm
5½ times	For a Sheet for the Clews of the Sail
6 times	For Studding-sail Halyards by the Yard-arm
5¼ times	For a Clew-garnit close to the inward Cleats
7 times	Ditto for the quarter of the Yard
8 times	For Bunt-lines and Leech-lines under the Top

Fore topmast and yard

8 times	For a double Strap for a double Block for the upper Block for a Top tackle
4¾ times	For a single Strap for ditto
11 times	For a double Strap for the Lifts for the Mast-head
5 times	For the Lifts for the Yard-arms
12 times	For a double Strap for a double Block on the Yard
7 times	For a single Strap for a double Block on the Yard
6 times	For a Strap for a Runner Block at the Mast-head
6 times	Ditto with a Hitch
7½ times	A Strap for a Clew-line Block on the Yard, within 18 Inches of the Parrel
6½ times	Ditto for the Quarter of the Yard
5 times	For a Clew-line Block for the Sail
5¼ times	For a Studding-sail at the Yard-arm
4¼ times	A single Strap for the upper Block for Halyards
3½ times	A Strap for a long Tackle Block for ditto

6 times	A Strap for a single Block below
11 times	For a double Strap for the Fore Topgallant Lifts at the Mast-head
5 times	For the Yard-arms
5½ times	For a Tye at the Mast-head to lash
6 times	Ditto with one Eye to hitch
6½ times	For the Clew of the Sail
6 times	For the Clew-lines on the Yard, with Eyes
4½ times	For the upper Halyard Block
5 times	For the lower to the side

'The same Rules that serve for the Fore-mast, Top-mast, and Top-gallant-mast, will serve the Main-mast, Top-mast, and Top-gallant-mast'.

Mizzen mast

12 times	For a double Strap for a double Block for the Yard
7½ times	For a single Strap for ditto, with Eyes to lash on the Yard
9 times	A double Strap for a double Block at the Mast-head to lash
9½ times	A single Strap for ditto
4½ times	A Strap for a single Block for the Clew of the Mizon
4½ times	For a double breast below
4 times	For a long Tackle Block for the Truss-tackle
7½ times	For a single Block below
4 times	For all the seising Blocks for the Brails etc.
7½ times	For the Quarter Block for the Cross-Jack Yard, with Eyes to lash
6½ times	For the Slings to lash on the Yard
4 times	For a Topsail Sheet for the Yard-arm
11 times	For the Mizon Topsail Lifts per Mast-head
5 times	For the Yard-arms
6 times	For the Tye Block at the Mast-head to lash
5¾ times	For the Block on the Yard
6 times	A Strap for a Clew-line on the Yard to lash
6 times	Ditto for the Clew of the Sail
4¼ times	For the upper Halyard Block
6 times	Ditto for that below to the Ship's side

For necessary ropes

11 times	A double Strap for the David Block
6 times	A single Strap for ditto
3¼ times	For a long Tackle Block for the Fish
5 times	A single Block for ditto
6 times	A single Strap for the Viol Block
13 times	A double for ditto about the Mast

Tackle, purchase

Takel, Talje; Palan, caliorne

A tackle was formed by the combination of a rope with blocks. Tackles were used on a ship to raise, remove or secure heavy objects, to support the masts, to extend the sails and in the rigging. A tackle might be either standing or loose.

The simplest tackle was a whip, where the fall ran through only one block (see Figure 187d). When a block was fitted to one end of a whip and another fall passed through this, then the whole was called a whip upon whip.

A tackle with two single sheave blocks was known as a gun tackle purchase (see Figure 187f). A tackle with a double and single block was a jigger, luff or watch tackle (see Figure 187g). A combination of a whip and a luff tackle, where one end of the whip (which was then called a runner) was hooked to an eyebolt, was known as a runner and tackle. The term 'winding tackle' was given to a heavy tackle, with three or more sheaves in each block, principally employed to hoist heavy objects aboard a ship.

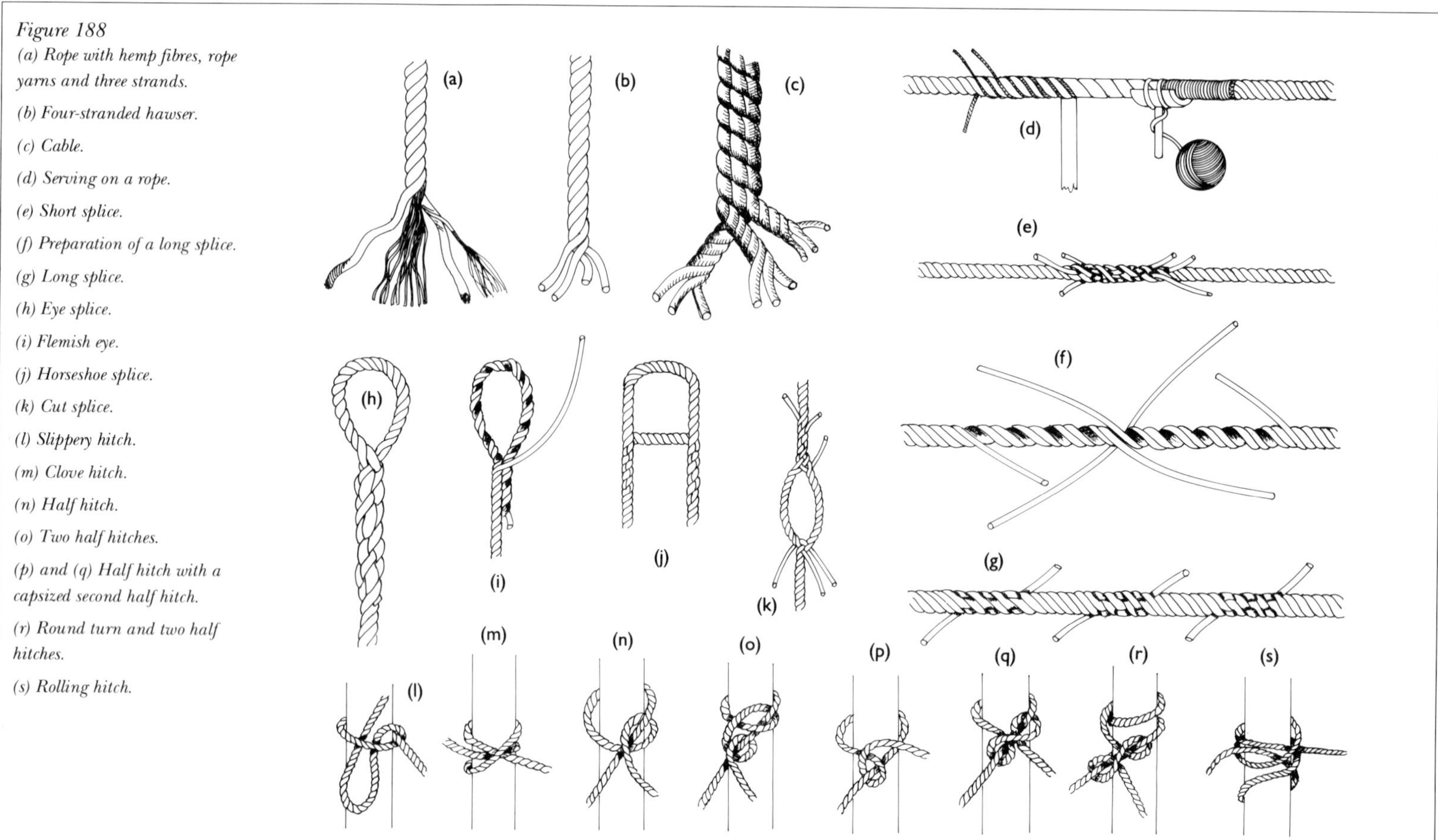

Figure 188
(a) Rope with hemp fibres, rope yarns and three strands.
(b) Four-stranded hawser.
(c) Cable.
(d) Serving on a rope.
(e) Short splice.
(f) Preparation of a long splice.
(g) Long splice.
(h) Eye splice.
(i) Flemish eye.
(j) Horseshoe splice.
(k) Cut splice.
(l) Slippery hitch.
(m) Clove hitch.
(n) Half hitch.
(o) Two half hitches.
(p) and (q) Half hitch with a capsized second half hitch.
(r) Round turn and two half hitches.
(s) Rolling hitch.

XV

Cordage, splices, hitches and knots

The term 'cordage' covers all cables, hawsers and lines used on a ship. Where cordage is still made from plant fibre today, it is produced from hemp, manila, coir or sisal. Wire ropes, which in the nineteenth century gradually replaced natural fibre in general ship use and in rigging, was first made from iron and later from steel wires and often had a hemp core or heart.

The last three fibres listed above were largely unknown during the eighteenth century, and were mentioned in contemporary documents, if at all, only as exotic materials with no real value for domestic shipbuilding. Only one material for rope-making was really available, and that was hemp.

Rees noted in his *Naval Architecture* of 1819-20 that:

> Of all the hemps yet produced at our English markets, the Russian hemp has proved to be the best; it is grown in the southernmost parts of Russia, and shipped for England from the ports of St Petersburgh and Riga. The best sort is Riga rhine hemp: the next in quality is termed Petersburgh clean hemp.

Röding also noted Russian hemp as the best, especially Moscow hemp. After that he ranked Baltic hemp, which came from Riga, Königsberg and Danzig. He listed Frankfurt am Main as a major trading centre for good hemp.

Ropes were made by the reeper or ropemaker with the help of spinning wheels. To produce ropes and anchor cables with a length of 120 fathoms or more, the ropemaker needed a narrow lane of approximately 200 fathoms in length, called a rope walk.

Cordage

Tauwerk; Cordage

Hemp

Hanf; Chanvre

Hemp was first hatchelled or combed and sprinkled with a small amount of train oil (a pint to a hundredweight) to make the hemp more pliable.

Rope yarn

Kabelgarn; Fil de caret

The hemp was then spun into rope yarn, using clockwise turns.

Strand

Kardeel; Cordon

According to the required size of the rope, a number of yarns were combined, using anticlockwise turns, into a strand.

Hawser-laid rope

Trosse; Cordage commis en haussière

Three strands laid in right (clockwise) turns formed a hawser, and this was the most common type of rope (see Figure 188a).

Shroud-laid rope

Wantenschlagtau; Aussière en quatre

Sometimes four strands were used (see Figure 188b), and the resulting rope was described as four-stranded, with the strands laid around a core of rope yarn, the heart. Four-stranded rope was known as shroud-laid, even when it was not used for shrouds.

Cable-laid rope

Kabelschlagtau; Cordage commis en grelin

Three hawser-laid ropes laid in left turns formed a cable-laid rope (see Figure 188c). The shrouds of lower masts were often cable-laid, and the majority of stays were cabled-laid with four stranded ropes. A cable-laid rope might therefore have nine or twelve strands.

Line

Leine; Ligne

If the circumference of a rope was less than one inch, the rope was known as a line. Lines were structurally similar to thicker ropes; the only difference was in the number of hemp fibres per yarn, and the number of yarns per strand. Lines can be categorised as flaglines, ratlines, lacing, marline, seizing material, houseline, twine and spun yarn.

Tarred or untarred

Geteert oder ungeteert; Goudronné ou ingoudronné

Cordage used in ships was either tarred or untarred. Untarred cordage was generally used for running rigging, whereas tarred rope was used for standing rigging, which did not run through blocks and was more difficult to replace when worn. The tar prolonged the life of the rope by protecting it from the weather. According to various contemporary writers the best material for this was Stockholm tar.

At the end of the eighteenth century William Chapman found that certain substances of tar actually encouraged dry rot in ropes, and he patented a special tar oil, which was supposed to lead to better results in cordage impregnation.

Duhamel du Monceau undertook various experiments in Rochefort in 1741, 1743 and 1746 with tarred and untarred cordage, in which he stretched prepared cordage of 3 inch circumference to breaking point. He established that untarred white cordage was 25 to 30 percent more durable than tarred ropes. In an additional test on storage it was proven that white cordage was still in excellent condition after three years in store, and actually showed higher stress qualities than on the first day of storage, while tarred cordage revealed a 20 percent decline.

Awareness of the better quality of untarred cordage must have been common among master riggers long before these experiments; Falconer reported that:

> In 1758, the shrouds and stays of the sheer-hulk at Portsmouth dock-yard were over-hauled, and when the worming and service were taken off, they were found to be of white cordage. On examining the store-keeper's books, they were found to have been formerly the shrouds and rigging of the *Royal William*, of 110 guns, built in 1715, and rigged in 1716. She was thought top-heavy, and unfit for sea, and unrigged, and her stores laid up. Some few years afterwards, her shrouds and stays were fitted on the sheer-hulk, where they remained in constant and very hard service for about thirty years, while every tarred rope about her had been repeatedly renewed.

This note confirms that shrouds and other standing rigging were not always impregnated with tar; sometimes a coating of tar was simply applied to the surface.

Seizing material
Bändselgut; Matériel d'amarrage

Seizing material was made from two or three rope yarns.

Twine and spun yarn
Segel- und Takelgarn; Fil à voile et commande

Twine and spun yarn differed only in the number of fibres in a yarn. Spun yarn (*Schiemansgarn; bitord*) was a two-stranded thin line, tarred and used for seizing and serving. It was rougher and usually thicker than houseline. Röding noted that spun yarn was sometimes also three-stranded and loosely twisted.

Houseline (*Hüsing; merlin*) was 2mm to 5mm in diameter (¼in to ⅝in in circumference), smooth, three stranded, twisted and tarred. It was used for serving and seizing. It was twisted more tightly than spun yarn.

Serving
Bekleeden; Fourrure

The serving of cordage in the rigging was part of its protection against rot and particularly against chafing. The normal procedure in preparing a rope in this way (see Figure 188d) was as follows:

(1) The rope was wormed by filling the divisions between the strands with spun yarn. This was done both to strengthen the rope and to create a smooth surface for parcelling.

(2) The rope was then parcelled by wrapping old and well tarred canvas round it, preparing it for serving and also stopping rainwater from penetrating the rope in those parts where the serving was worn.

(3) With the help of a serving mallet the serving was then laid on. This was done by laying spun yarn or houseline in tight turns around the parcelled rope.

The parcelling followed the lay of the strands, but the serving ran in the opposite direction. Standing rigging also received a coat of tar to weatherproof it further.

Mat, paunch
Matte, Stossmatte; Paillet

Another type of protection for cordage was a mat or paunch. This was, in Röding's words:

> A fabric from spun yarn or houseline, made like a mat, for covering those parts of anchor cables, shrouds, tackles, masts and yards likely to be chafed by ropes. The length and width of these mats depend on their purpose. Some are 2 or more feet wide and 6 to 8 and more feet long. There are also spiked mats, or mats with 3 to 4 inch long strands pulled through so that the ends are at one side. These ends are then opened out, so that one side of the mat remains smooth, and the other rough; the rough side protects the chafing ropes from damage. Those mats which are nailed around the yard where these come in contact with the mast are named paunches. Another mat is fitted round the forward edge of the mast tops, and this is known as the top mat; all others take their names from the positions in which they are fitted, such as yard mats, shroud mats, stay mats, etc.

Splices
Spleissen; Épissure

Splicing is the art of intertwining or interweaving two ropes to make a durable connection, or one rope with itself to produce an eye. A differentiation can be made between an ordinary mariner's splice, in which the unlaid strand was woven 'once underneath and once over' the strands of the other part, and the sailmaker's splice (or round splice), in which the strands follow round and round with the lay of the rope. Such splices were especially smooth and were used for joining together the ends of boltropes. The most common types of splice were the short splice, the long splice, the eye splice, the Flemish eye, the horseshoe splice and the cut splice.

Short splice
Kurzspleiss; Épissure courte

The short splice (see Figure 188e) connected two ropes. The strands of both ropes were opened for a convenient length (three to four turns), and then loosely interwoven so that each strand of one rope was laid beside one of the other. The middle strand of the righthand rope was then led under the opposite strand, with the others following the same way. After this, the strands of the lefthand rope were similarly led, and the whole process was repeated three to four times until the loose ends were fully expended.

Eye splice
Augspleiss; Épissure à oeil

The strands in one end of the rope were laid open for three to four turns for an eye splice, and an eye was formed by turning the end back and splicing the strands into the body of the rope (see Figure 188h). With the end of the rope pointing outwards, the first of the loose strands was laid left, the second on top, and the third right from the rope. The strand below the second was then opened up with a fid or marling spike, and the second loose strand was pushed through, the first following and going through the opened next strand as well. After that, the eye was turned around, and the last loose strand went through the remaining twisted strand. This procedure was repeated four to five times, and the number of yarns in a strand were reduced each time to give the splice a neater look.

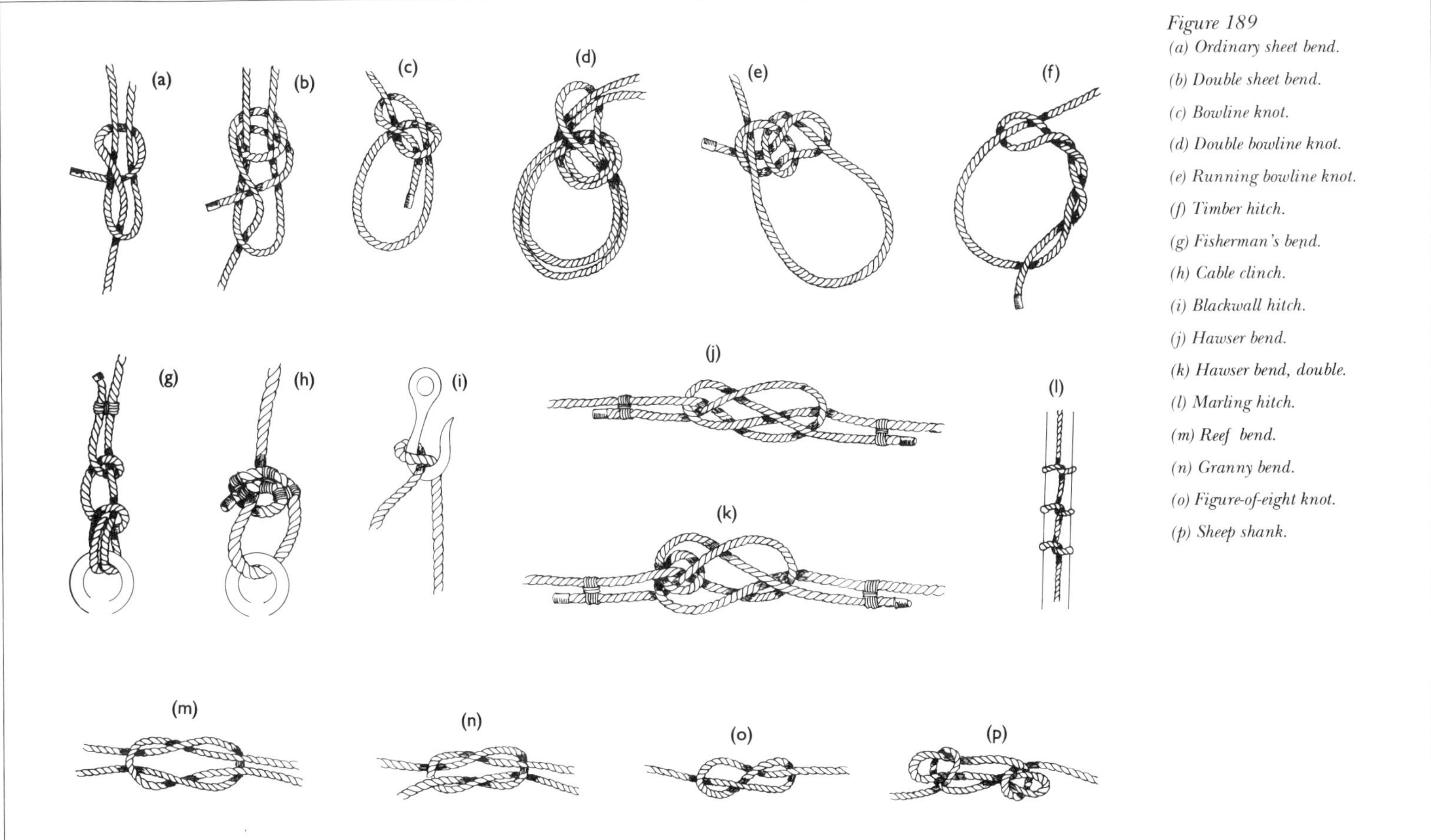

Figure 189
(a) Ordinary sheet bend.
(b) Double sheet bend.
(c) Bowline knot.
(d) Double bowline knot.
(e) Running bowline knot.
(f) Timber hitch.
(g) Fisherman's bend.
(h) Cable clinch.
(i) Blackwall hitch.
(j) Hawser bend.
(k) Hawser bend, double.
(l) Marling hitch.
(m) Reef bend.
(n) Granny bend.
(o) Figure-of-eight knot.
(p) Sheep shank.

Long splice
Langspleiss; Épissure longue

A long splice was also used to connect two ropes, especially when the joined rope was required to reeve through a block. For this splice, in contrast to a short splice, the strands were opened over a much larger distance - twelve turns from each rope. The two ropes were then placed together in the same way as for a short splice, with one strand opened for another nine turns, and the opposite strand laid into that space. Then another strand was opened in the other direction, and filled again with the opposite strand.

The two middle strands were split and an overhand knot was put on the opposing halves, and the other halves were cut off (see Figure 188f). The middle strands were then spliced together as in a short splice. The ends of the other strands were spliced with reducing yarns, so that the shape of the rope was barely altered. Such a long splice had its six strands worked into each other at three places of equal distance (see Figure 188g).

Flemish eye
Flämisches Auge; Oeil à la flamande

A Flemish eye was an eye splice frequently used on a stay collar (see Figure 188i). One strand of the rope was opened, while the other two formed the eye. The opened strand was then laid back into its place, now running in the opposite direction. The ends of the strands were then scraped down, tapered, marled and served over with spun yarn. Often the whole eye was also served over.

Horseshoe splice
Hufeisenspleiss; Épissure en greffe

A horseshoe splice took its name from its shape: a short piece of rope was spliced into the bight of a rope, so that it took the shape of a horseshoe. Such a splice was usually used for spans, so the term 'span splice' was also used.

Cut splice
Cuttspleiss; Épissure en portière de vache

A cut splice connected two ropes in such a way that the two ends formed a collar or eye in the middle of the finished rope (see Figure 188k). Cut splices were often employed to join two single backstays, mast pendants or similar.

Hitches and bends
Stecks, Stiche, Schläge; Noeuds

Seamen throughout the ages developed a number of knots to connect ropes to other objects (these are collectively known as hitches) and to join two ropes at their ends (generally known as bends). Many of those listed below are shown clearly in the drawings and need no further description.

Slippery hitch
Schlippstek; Noeud de ride

The slippery hitch is simple and easy to loosen (see Figure 188l).

Clove hitch
Webleinstek; Deux demi-clefs renversées

The clove hitch was employed to fasten ratlines in the shrouds. It is a combination of two half hitches (see Figure 188m).

Half hitch

Halber Schlag; Demi-clef

A half hitch (see Figure 188n), or more normally two half hitches (see Figure 188o), were used to fasten a rope temporarily to another part. Inverted half hitches or overhand knots followed the same principle, but the end of the rope went over the standing part instead of going around and coming up through the bight, thus constricting it.

To prevent a knot formed from two half hitches from pulling too tight, a round turn with two half hitches was used (see Figure 188r). The rope was laid in a complete turn around the part before the half hitches were made. Lever referred to two round turns over a spar with two half hitches around the standing part as a rolling hitch, and two round turns over a spar with a half hitch opposite to the second turn as a Magnus hitch.

Rolling hitch

Stopperstek, Rollstich; Noeud coulant, amarrage à fouet

A rolling hitch was similar to a clove hitch, but had an extra round turn jamming one end (see Figure 188s). This prevented the line under load from slipping along the spar to which it was hitched. For a version which was easier to release, the end could be made as a slippery hitch.

Sheet bend, common bend

Schotstek; Noeud d'écoute simple

A sheet bend (see Figure 189a) was used for the temporary connection of two ropes, particularly where the ropes were of different sizes. A bight was formed near the end of the larger rope, and the end of the other was passed through, then round the bight and underneath its standing part.

Double sheet bend

Doppelter Schotstek; Noeud d'écoute double

A double sheet bend was used when the ropes had to withstand a stronger pull. The rope passing through the bight was given an additional turn around the bight before it went underneath the standing part.

Bowline knot (hitch)

Palstek; Noeud de bouline, Noeud d'agui à élingue, noeud de chaise simple

A bowline knot formed an eye which did not pull together (see Figure 189c). The term reflects its use in rigging, while the German term is related to another use, the temporary fixing of an eye to a hawser, to slip it over a pole (*Pal* or *Pfahl*), bollard or similar.

Bowline knot upon the bight, French bowline knot

Doppelter Palstek, Doppelter Pfahlstich; Noeud d'agui à élingue double, noeud de chaise double

To make a bowline knot upon a bight the rope was doubled, and the bight was passed through a loop in the standing parts; it was then turned back on itself around the whole knot so that it could be pulled tight around the standing parts, leaving a large double loop (see Figure 19d).

Running bowline knot

Laufender Palstek; Laguis

A running bowline knot formed a noose which could be opened and shut easily. To create such a knot, the end of a rope was taken around the standing part and through the bight so formed, and a normal bowline knot was then made in it. The standing part thus ran freely through the knot (see Figure 189e).

Timber hitch

Zimmermannsstek, Zimmerstich; Noeud d'anguille, noeud de bois

The timber hitch formed a simple noose. The end part of a rope was taken around a spar or similar, then led under and over the standing part and a few times around its own end (see Figure 189f).

Fisherman's bend

Roringstek, Fischerstek; Noeud d'orin, noeud de pêcheur, noeud de filet

Used among other purposes for fastening a hawser to a light anchor such as a kedge anchor and for bending the studding-sail halyards to the yard, a fisherman's bend consisted of two turns around the anchor ring or spar, a half hitch around the standing part and under the turns, and another half hitch around the standing part alone, with the end seized (see Figure 189g).

Cable clinch

Kabellaschung; Étalingure du cable

The cable clinch was used for larger anchors, since it was impossible to form a fisherman's bend in thick cables. The cable was led through the anchor ring and in a full turn around its standing part, and secured with a number of seizings (see Figure 189h).

Blackwall hitch

Einfacher Hakenschlag, Einfacher Holländer; Gueule de loup simple

A Blackwall hitch was used for the quick securing of a load-bearing rope to a hook; it was a simple turn jammed by the standing part (see Figure 189i).

Midshipman's hitch

Hakenschlag, Maulstek; Noeud de griffe

A midshipman's hitch was simply a sheet bend applied to a hook. The term was also used for a single loop with a half hitch around the standing part and a further inside round turn below the half hitch; this formed a loop which tightened when the knot pressed against an object within it but would not shake loose when not under load.

Cat's paw

Katzenpfote, Trompete; Noeud de gueule de raie

A cat's paw was a simple method of securing a hook in the middle of a line. Two bights were taken and each was twisted three times around itself, and the two loops so formed were then put over a hook.

Hawser or carrick bend

Trossenstek; Noeud d'étalingure

An eye was formed in one rope, and the other rope was led through, passed around the throat of the eye and back through the eye again, crossing its own standing part. The end of each rope was then seized to its respective standing part (see Figure 189j).

Lever described a slightly different hawser bend. There the first eye had a half hitch before it was seized, and the incoming hawser was led only through the eye, then returned, half hitched and seized like the first.

Hawser bend, double

Doppelter Trossenstek; Noeud d'étalingure double

The double hawser bend was like a normal hawser bend except that the second rope took a full turn around the throat of the first eye before returning (see Figure 189k). Both types of hawser bends were used for the connection of mooring lines, since they were easy to undo after having been under strain.

Marling hitch

Marlschlag; Demi-clef à capeler, transfilage

A marling hitch was used to secure parcelling, to bend a sail to a yard, to secure clewropes, and similar purposes. The end of the marline was made fast, then a turn was made around the object at the necessary distance and the marline returned from below through the loop and hauled taught. This action could be repeated as often as necessary (see Figure 189l).

Reef bend

Kreuzknoten; Noeud marin, noeud de vache

The reef bend (see Figure 189m) is one of the best known knots. Formed from two overhand knots, it is easy to undo and sits flat. If it is put together with the second overhand knot reversed, it becomes a granny bend (see Figure 189n), which jams readily and is hard to open, especially when wet.

Figure of eight knot

Achtknoten, Achterstich, Sackstich; Noeud en huit, noeud d'arrêt

A figure of eight knot is similar to a simple overhand knot, but with an extra half turn around the standing part (see Figure 189o). It was used as a stopper knot to prevent a rope slipping through a block, but could also be formed in a bight to produce a useful loop which would not slip whichever standing part was under strain.

Sheep shank

Trompete; Jambe de chien

A sheep shank (see Figure 189p) was used for the temporary reduction in the length of a rope. That part of the rope not needed was taken together in bights, and a half hitch made with the parts of the rope still in use, was placed over each end. These jammed the unnecessary part of the rope when it was under stress, but allowed easy release when the full length of the rope was required.

Knots

Knoten; Noeuds

Knots, in the more precise sense of the word, are used to prevent the unlaying of a rope at its end and to provide a good finish or to prevent a rope running through a block.

Wall knot

Einfacher Taljereepknoten; Cul-de-porc simple

The wall knot (see Figure 190a and b) provided the basis of almost all other knots. The rope end was opened for three turns and the strands were laid under each other (number one under two, number two under three, and number three under one). After being pulled tight, the ends were twisted and whipped.

Wall knot with crown, double wall knot

Doppelter Taljereepknoten; Cul-de-porc double

The double wall knot was made in the same way as the simple wall knot, but the strands were laid under each other twice (see Figure 190c). The knot was finished off in the same manner.

Spanish whip

Spanischer Takling; Souliure espagnol

A Spanish whip was an ordinary wall knot reverse spliced two or three times (see Figure 190d).

Wall knot with reversed splice

Taljereepknoten with Hahnepoot; Cul-de-porc avec épissure renversé

This knot was similar to a Spanish whip, but was reverse spliced only once (see Figure 190e).

Double wall knot with crown

Fallreepsknoten; Cul-de-porc double avec tête de mort

Based on the wall knot with a reversed splice, the double wall knot with a crown had the strands led once more from below through the strands of the wall knot to double these. The ends were then cut off very short to conceal them.

Diamond knot

Diamantknoten; Pomme d'étrier

Diamond knots, single or double, were not placed at the end of a rope, but in intervals along it (see Figure 190g and h). They were used in bell ropes and foot ropes, to provide the sailor with a better foothold. To make such knots the strands of the rope had to be opened up as far as the last or lowest knot (the knot itself was similar to a wall knot) and then twisted together again up to the position of the next knot. This could be repeated as often as needed.

Shroud knot

Englischer Wantknoten; Noeud de hauban anglais

A shroud knot was used to connect the two ends of a broken or shot through shroud, and was thus technically a splice rather than a knot (see Figure 190i). Both ends were opened as for a short splice and placed close together. The strands of one end then formed a wall knot at the other and vice versa. The ends were thinned and tapered, then finally marled and served.

French shroud knot

Französischer Wantknoten; Noeud de hauban

In a French shroud knot, after the ends were opened and placed close together, the strands of one end were bent back and these of the other end formed a wall knot through these loops. The ends were finished as above (see Figure 190j). According to Lever, this knot was much smaller than the English knot, and just as secure.

Spritsail sheet knot

Blindeschotknoten; Noeud d'écoute le civadière

The spritsail sheet knot was formed at the end of a block strop (see Figure 190k); the strands at both ends of the strop were opened and all six strands together formed the knot, in similar fashion to a wall knot. The block was then placed into the strop's eye.

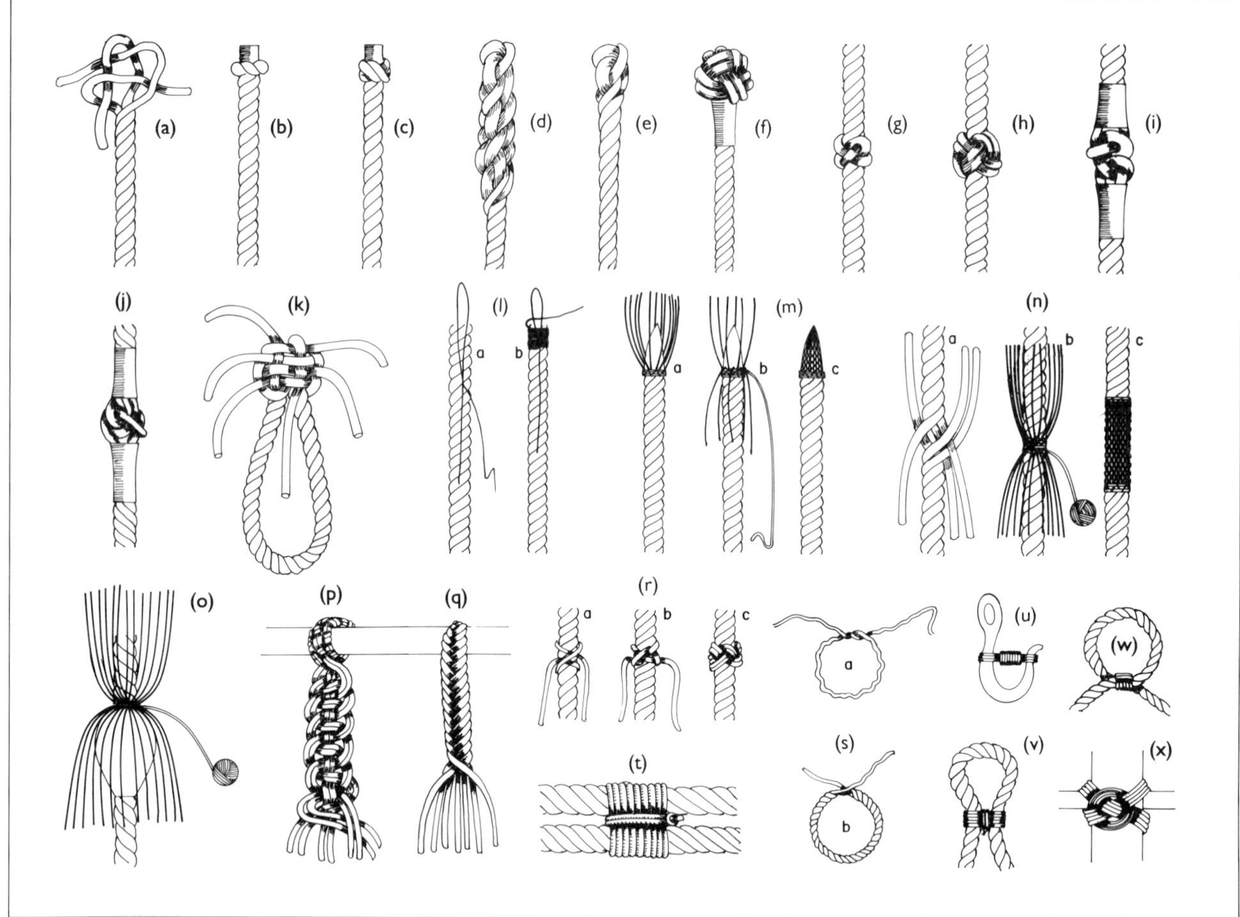

Figure 190
(a) The making of an ordinary wall knot.
(b) Wall knot.
(c) Double wall knot.
(d) Spanish whip.
(e) Wall knot with reversed splice.
(f) Double wall knot with crown.
(g) Diamond knot.
(h) Double diamond knot.
(i) Shroud knot.
(j) French shroud knot.
(k) Spritsail sheet knot.
(l) Whip
a. beginning
b. final move.
(m) Pointing a rope
a. beginning
b. work in progress
c. finished product.
(n) Grafting a rope
a. beginning
b. work in progress
a. finished product.
(o) Making a stay mouse.
(p) A flat plait.
(q) Sennit.
(r) a. b. and c. Making a Turk's head.
(s) a. and b. Making a grommet.
(t) Seizing.
(u) A moused hook.
(v) Eye seizing.
(w) Throat seizing.
(x) Rose lashing.

Whip
Takling; Souliure

A whip was the simplest way to stop a rope from unravelling. It was made by laying a loop of spun yarn on to the rope towards its end and winding the yarn in close turns round the rope, starting at a distance of about the diameter of the rope away from the end. After the whole distance to the end of the rope had been covered, the end of the spun yarn went through the loop. The loop and the inserted end were pulled under the round turns from the standing part and both ends were cut off (see Figure 190l).

Miscellaneous ropework
Vermischtes Tauwerk; Cordage divers

Pointing a rope
Hundspünt; Queue-de-rat

A finer way of securing and finishing the end of a rope was by pointing it (see Figure 190m). Pointing was the operation of tapering the rope's end and working a close netting with an even number of knittles, twisted from the same, over the reduced part to prevent unravelling and to allow it to be led better through a block or hole. It was performed in the following manner:

> Take out as many yarns as are necessary, and make knittles, which is done by taking separate parts of the yarns when split, and twisting them; comb the rest down with a knife. Make two knittles out of every yarn which is left, and lay half of them down upon the scraped part, and the other, back upon the rope. Take a length of twine, which is generally termed the warp, and pass three turns of it very taught, jambing them with a hitch. Then lay the knittles backward and forward, as before and pass the warp. The ends may be whipped and snaked with twine, or the knittles hitched over the warp, and hauled taught. The upper seizing must also be snaked [Falconer, 1815]

To finish the end off neatly, a small tapered piece of timber was inserted and woven over.

Grafting a rope
Überweben eines Taues; Garnir un cordage

The grafting of ropes, instead of splicing, was favoured by some naval captains, especially for decorating the quarterdeck. Lever commented as follows:

> Straps of blocks are often grafted instead of the short Splice, particularly on the quarter-deck: this is by no means so secure as the Splice, for if the pointing be worn by wet and friction, the Strap may give way - it is therefore better that the Straps of blocks which are to be pointed for neatness, and without a Splice, should be made Selvagee fashion, all the parts of which bear an equal strain, and if the pointing give way, the Strap will hold.

Grafting was done by opening the two ends of a rope and placing the strands within another, as for splicing, and stopping them at the joint. The yarns were then opened out, split and twisted into knittles as for pointing. The lower and upper knittles were divided and the grafting proceeded with a warp in each direction as described above under Pointing (see Figure 190n).

Stay mouses, puddening of mast or yard and the dolphin were all grafted in the same way; the only difference was that the material for the knittles was not taken from the rope itself, but applied additionally.

Sennit, plaiting

Platting; Tresse

Sennits and plaitings (see Figure 190p and q)were flat cordage, plaited from either rope yarns or foxes (foxes were a number of rope yarns, from three to nine, twisted together). Plaitings were used for many things on board: in rigging, for example, they were used for gaskets and reef points.

Turk's head

Türkischer Bund; Bonnet turc

Turk's heads were mainly used for the decoration of ropes. A Turk's head was worked with logline or similar and began with a clove hitch, the lower bight of which was brought over the upper and the end taken up through it, then another cross with the bights was made and the end taken down, after which it followed the lead. The effect was a kind of crown or turban (see Figure 190r).

Grommet

Grummet; Bague

A grommet was a ring made from a single strand which was laid three times around itself until it formed a solid rope ring (see Figure 190s).

Seizing

Bändselung; Amarrage

A seizing (see Figure 190t and v) connected two parallel or crossing ropes. The material used was spun yarn, marline, houseline or small cordage. A simple eye splice was made by leading the yarn twice through itself, and the yarn was laid round the two ropes, through the eye splice and pulled tight; then another six to ten close turns were made in the opposite direction to the first around the ropes, pushing the end through the last turn. Over these were laid another five to nine turns (one fewer than the inner turns), which were called riders. The end was then pushed up through the seizing, and two frapping turns were taken around the seizing between the ropes. the seizing line was stopped with an overhand knot when spun yarn was used and with a wall knot in the case of small cordage. A throat seizing (see Figure 190w) did not have frapping turns.

Rose lashing

Roslaschung; Amarrage de rose

A rose lashing (see Figure 190x)was a neat and attractive lashing used particularly for the fastening of puddenings and dolphins on masts and yards. The lashing lanyard was spliced into one eye of the puddening or dolphin and was taken crosswise over and under between the eyes, with the end taken around the crossing. The final lashing had an appearance reminiscent of a rose.

XVI

Netting and other accessories

In addition to the normal rigging described hitherto, most vessels carried in addition a number of rigging accessories, designed for specific tasks in action, during the weighing of the anchor, as additional rigging during bad weather, or for the manning of boats.

Some of these accessories are described below. Such items not only provide a better understanding of a sailor's working routine, but also allow a modelmaker, by using these accessories in the right way, to enhance the appearance and realism of his model.

Netting

Netze; Filets

Men-of-war carried various types of netting, for a number of purposes. They had bowsprit netting, made fast at the outer end of the bowsprit to the horses or manropes (see Figure 191a and b), in which the fore top staysail and the jibs were stowed. Hatchway netting was securely placed over the hatchways to prevent crewmen from falling into the hold and to allow ventilation which would be prevented by a solid hatch cover. Head netting in smaller vessels such as sloops was made fast to the horses in the head and the upper rail as a safety measure for crewmen working in the head. Splinter netting was nailed to the inner part of the ship's sides (extending right across the deck), for the purpose of reducing injuries through splintering timber during action. Top netting, made fast to the rail and shrouds in the top, was a further safety measure.

Boarding netting and overhead netting were two varieties rigged when the ship prepared for action. The former extended fore-and-aft from the gunwale to a certain height in the rigging to prevent the enemy from boarding during the engagement, and the latter stretched from the mainmast aft to the mizzen shrouds at about 12 feet above the quarterdeck, to protect the officers from blocks and other items falling from the masthead. An illustration from 1586 of the *Black Prynnes* showed this type

Figure 191
(a) Bowsprit netting between bowsprit cap and fore stay.
(b) Bowsprit netting.
(c) Rigged awning.
(d) Jacob's ladder.
(e) Windsail.
(f) Mast coat.
(g) Boat lashing.
(h) Port tackle.
(i) Man rope.
(j) Passing rope.

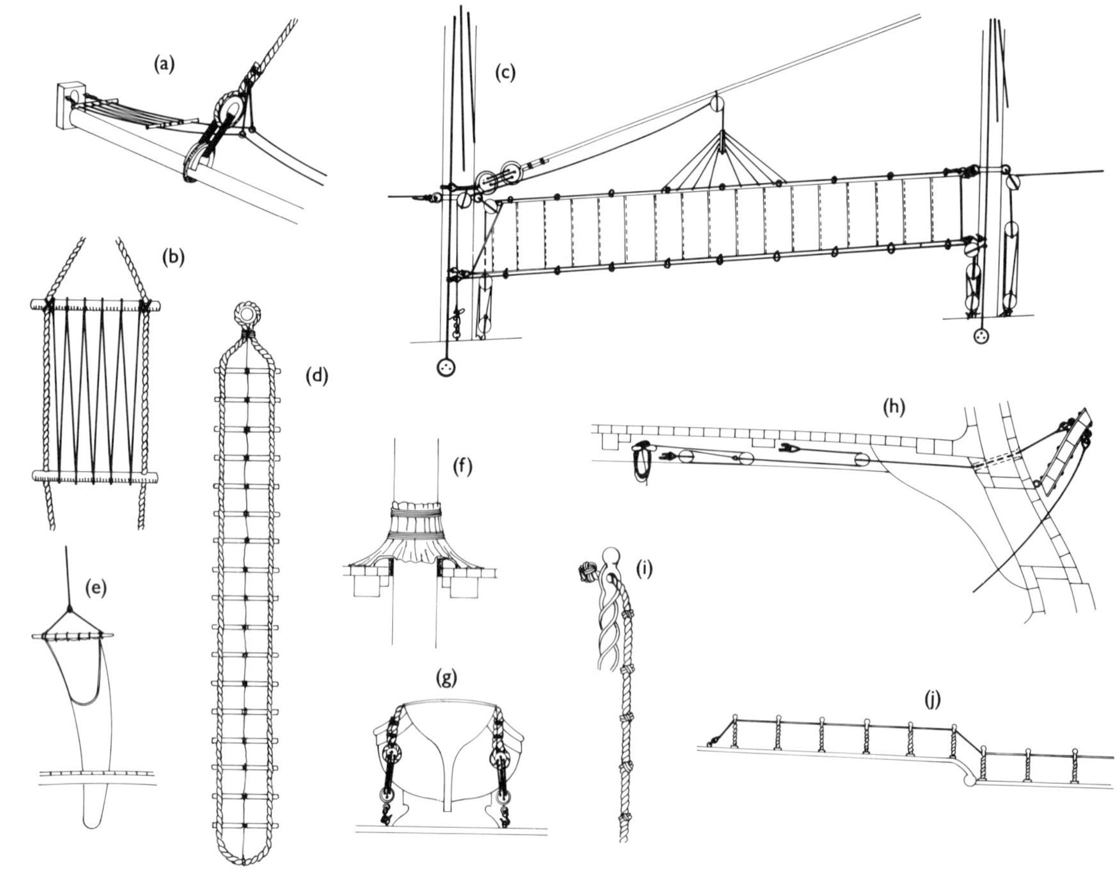

of netting rigged from the fore to the mizzen mast like an awning. The rigging of such netting was a long established way of preparing for action; for the 700 officers and crew of the *Mary Rose* this netting became a death trap when the ship suddenly capsized in a squall and sank within minutes in 1545.

> The netting, in merchant-ships, is usually stretched along the upper part of a ship's quarter, to contain some of the seamen's hammocks, and secured in this position by rails and stanchions [Falconer, 1815].

Awnings

Sonnensegel; Tentes

Awnings (see Figure 191c) were rigged in harbour, especially in the tropics, to shade the crew from the heat of the sun and to prevent the deck from splitting. They usually covered the whole deck abaft the foremast and were sewn from light canvas, shaped according to the shape of the deck. A number of trucks were made fast to the sides and along the ridge of the awning, and ropes for the rigging of the awning were led through these and made fast on deck.

The ridge rope of the poop awning passed through the midships trucks, with the standing part made fast with a clinch to the ensign staff. The leading part rove through a block at the aft side of the mizzen mast, or a roller fitted to an iron clasp-hoop on the mast, was set up with a double and single block tackle to an eyebolt in the deck. The side ropes passed through trucks on the sides of the awning, through an eye spliced into the standing part, and were seized to the fore leg of the mizzen shrouds. The leading part went through a sheave hole in a wooden stanchion fixed to each side of the stern and was also set up with a tackle.

The ridge rope of the quarterdeck awning was clinched to an eye in the clasp-hoop noted above, and was set up in a similar way to the mainmast. The side ropes were seized to the foremost main shroud, with the leading part reeving through a block at the foremost mizzen shroud, and set up with a tackle. The main deck awning was rigged in a similar way.

Awnings were spread and suspended thus: three thimbles seized in the after ends went over hooks at the mast and the shrouds. The fore part was hauled forward and laced through the eyelet holes in the awning corners to the shrouds. In harbour, awnings were spread to wooden stanchions bolted along the sides, and hauled forward with tricing lines through single blocks on uprights, instead of the masts.

In the middle of the awnings the legs of a crowsfoot were made fast at equal distances to the strands of the ridge rope, with a euphroe running in their bights. The euphroe was suspended from the stay, by a halyard which passed through a single block seized to the stay and belayed to its lower part.

Jacob's ladder

Jakobsleiter; Échelle de Jacob, échelle de tangon

While a ship was in harbour the boats were launched and made fast either to the stern or to the boat boom, extended over the forecastle. Access to the boats was by rope ladders which Falconer referred to as quarter-ladders, but which are normally known as Jacob's ladders (see Figure 191d). Falconer's description was as follows:

> [Quarter-ladders are] two ladders of rope, depending from the right and left side of a ship's stern, whereby to descend into the boats which are moored astern, in order to bring them up along-side of the ship or to use them for any other occasion.

Blanckley provided further information on the use of these ladders: 'Those [ladders] of Ropes, hung over the Stern of the Ships, are to enter out of the Boat, when the Weather is foul and the Sea high.'

For the making of a Jacob's ladder, cable-laid rope of twice the depth from the taffrail to the water level was used. This rope was spliced together into a ring, with a thimble seized into the upper end, or had an eye spliced into each end and hung with the bight downwards. The ladder was made fast to eyebolts in the taffrail.

The wooden dowels generally used as rungs were pushed between the opened up strands and had a slight groove in the middle to accommodate the middle rope. This ran from the upper thimble, was hitched around every step, and was made fast at its end to the bight; it kept the dowels 16 inches apart.

Wind sail

Windsack, Windsegel, Kühlsegel; Manche à vent

One of the great problems during the sailing ship era was the ventilation of a vessel. Besides obvious measures such as opening the hatchways, canvas tubes which tapered toward the lower end and were cut diagonally and spread by a spar at the top were employed; these were known as wind sails (see Figure 191e). With the opening turned to the wind and rigged to the top, the lower end was led through a hatchway into the ship, forcing fresh air into the hull. The more modern ventilation systems of steamships were based on the same principle.

The invention of wind sails was credited by Röding to the Danes. He also noted that this type of ventilation worked best when the hatchway used was not larger than the wind sail opening itself, or was covered outside the wind sail, since otherwise much of the fresh air escaped without first circulating through the ship. His observation was often ignored, and the sail lost much of its ventilating effect.

Mast coat, canvas cover

Segeltuchkragen am Mast; Braie de mât

Since the opening of the partners was usually larger than the mast's diameter, the subsequent gap was a constant source of water penetration every time the deck became wet. A collar was therefore formed from old canvas and lashed around the mast just above the opening (see Figure 191d). This canvas coat was then thoroughly tarred.

> A wide round or octagonal wooden hoop is nailed to the mast for additional security, and to the deck a smaller hoop, or timber strips, above which the fixed hoop round the mast rests and is able to move together with the mast; this is called a playing collar. Nailed above this is then a double coat of canvas. Instead of a wooden playing collar, sometimes a dolphin or a rope collar around the partners is used [Röding].

Yard tackle

Rahtakel, Rahnocktakel; Palan de bout de vergue

The yard tackles of the lower yards, which normally served to move heavy objects, were sometimes led aft during bad weather and hooked to eyebolts in the deck, to assist the braces.

Spar-lashing

Laschung der Bäume; Aiguillette d'éspars

The spare booms and spars carried as emergency topmasts and yards were stowed on the boatskids on each side, and were first secured individually with several lashings, then cross-lashed together and well frapped in the middle. To prevent the booms from shifting in gales of wind, several turns with a hawser were taken around all the spars and through large triangular ringbolts in the side. Sometimes the turns were passed through an opposite port, and around the side, with the turns hauled tight, frapped and belayed.

Port-tackle

Stückpfortendeckel-Takel; Palan de sabord

In the eighteenth century gunport lids usually had two ringbolts fitted to the outside. The tackle for opening these port lids consisted of a span with a single block seized in the bight and the two ends, leading through two holes above the port, spliced around the ringbolts in the lid. A runner passed through the span block with an eye in its standing part over a hook in the deck beam and another single block turned into the end of its leading part. The tackle fall was spliced to a becket in this block, and led through a second block, also single, set up to a hook in the same deck beam, and was belayed to an iron cleat near the hook (see Figure 191h).

Port lids at the quarterdeck had also a span and block, but the runner was made fast around a timberhead, or spliced to an eyebolt in the side. It passed through the span block and was hauled up by hand and belayed.

Manrope, entering rope

Fallreeptau; Tire-veille

'Manrope' was a general term for ropes used for ascending and descending a ship's side; 'they are usually covered with kersay or canvas,' noted Falconer. Röding noted the use of a red cloth cover and knots one foot apart. They hung alongside the ladder, and were secured in the upper part of the stanchions with a double wall knot with a crown (see Figure 191i). Other sources mentioned diamond knots worked into the manropes 9 inches apart.

Passing rope

Relingtau; Corde de lisse

Stopped outside the gangway stanchions with a double wall knot with a crown, the passing rope rove through all the stanchion holes either forward or aft, and had a thimble turned into the other end which was seized to a ringbolt either at the taffrail (aft) or at the knighthead (forward). It served as a handrail (see Figure 191j).

Anchor, cathead and davit accessories

Anker-, Kranbalken- und Davitzubehör; Accesoires d'ancre, de bossoir, de capon et de bossoir de traversière

Among rigging accessories for an anchor were the serving of its ring and the fastening of the buoyline and the fish pendant. The fastening of an anchor cable to the ring is detailed in the previous chapter under Fisherman's bend and clinch.

The ring was first covered with strips of tarred canvas, then a number of layers of twice-laid stuff, spun yarn or similar, cut to a length of three times the ring's diameter, were temporarily seized at the middle to the top of the ring, and wrapped around in both directions. At one eighth to each side from the temporary seizing, a permanent seizing was put on and snaked. Outside it, the puddening of the ring was continued until three quarters of the latter was served. Another seizing was put on each side, and the rest of the stuff was combed out to the same width as the seizing, and tarred (see figure 192a). The number of seizings varied from ship to ship, and in many cases the top quarter of the ring was fully seized and snaked. The puddening of the anchor protected the iron ring against chafing by the anchor cable.

The buoyrope was made fast to the crown of the anchor with two half hitches, its end seized several times to the shaft (see Figure 192b). A buoyrope had a minimum length of 18 fathoms, and in large vessels up to 50 fathoms. On smaller anchors another method of attachment was sometimes used (see Figure 192c): the end of the rope was led along the shaft, through the ring and beneath the anchor stock with a cringle around the rope and shaft. Near the crown the rope was seized twice to the shaft, and at the crown itself another cross seizing was put on.

The buoy itself (see Figure 192d, e and f) was usually made from cork, but timber-built buoys were also in existence, and were known as nun buoys. Buoys usually had the shape of a pointed egg, though French practice favoured a double cone shape.

For stability, buoys were held in several ropes known as slings and hoops, and here again various techniques existed: a Spanish method (which was also usual in English vessels), a Dutch method and a French method. In the first, two rope hoops of a slightly smaller diameter than the maximum diameter of the buoy were laid around it about one sixth of its depth from the middle; two vertical ropes in each direction (the slings) were spliced around the further hoop and led underneath the nearer to each end. The slings were fully served, and the two eyes at each end were marled together; a thimble was laid in and a seizing was put on between the buoy and the eye. The buoyrope was spliced around the lower eye. Dutch mariners frequently used a sheet bend to secure the buoyrope. 'The Buoy-Rope is often extremely useful otherwise, in drawing up the anchor when the cable is broke,' commented Falconer, and he continued: 'it should therefore be always of sufficient strength for this purpose, or else the anchor may be lost through negligence.'

A simple lanyard was spliced into the upper eye for hauling the buoy into a boat and to fasten it to the shrouds.

The *traversin de l'ancre* (the English term would be 'anchor cross rope') was not in use on English ships. The crossrope consisted of strops made fast to each anchor arm and to ringbolts in the stock's side (see Figure 192g). According to Röding these ropes were peculiar to the French and Danish Navies for anchor-fishing. He explained their use thus:

> In a number of harbours, the French use an unusual method of fishing an anchor. They lay a strap around the crown of the anchor, into which the foremast tackle (*croc de la candalette*) hooks, and wind the anchor up with that tackle; when the anchor reaches the height of the gunwale, it is then taken in on the bow with the help of the cross ropes (*traversins*).

The sheaves of a cathead and a cat block, together with the cat fall, formed the cat tackle, or cat (see Figure 192i). It was used to haul the anchor up to the cathead, so that the anchor flukes did not damage the ship's side. When the anchor was catted (that is, hauled up), a cathead stopper was passed through the ring, and went over a cleat at the cathead's side or along a vertical groove in the head's face, to be belayed to the nearest timberhead. The stopper's other end was fixed to the cathead. It was either laid around the head and secured to itself, or more often led from underneath through a vertical hole in the cathead, and stopped at the upper side by a large stopper knot. After the anchor had been hung in the stopper, the cat purchase could be unhooked.

After the catting of an anchor, a davit was used for hoisting its flukes on to the bow; this operation was known as the fishing of an anchor. The two types of davit in use during the eighteenth century are detailed in Chapter I. The short davit set up in the chainwales, employed in men-of-war, East Indiamen and large merchantmen, was rigged with three guys. The forward guy was made fast to the cathead, the aft guy to the aft end of the fore chainwale, and the third guy led from the head of the

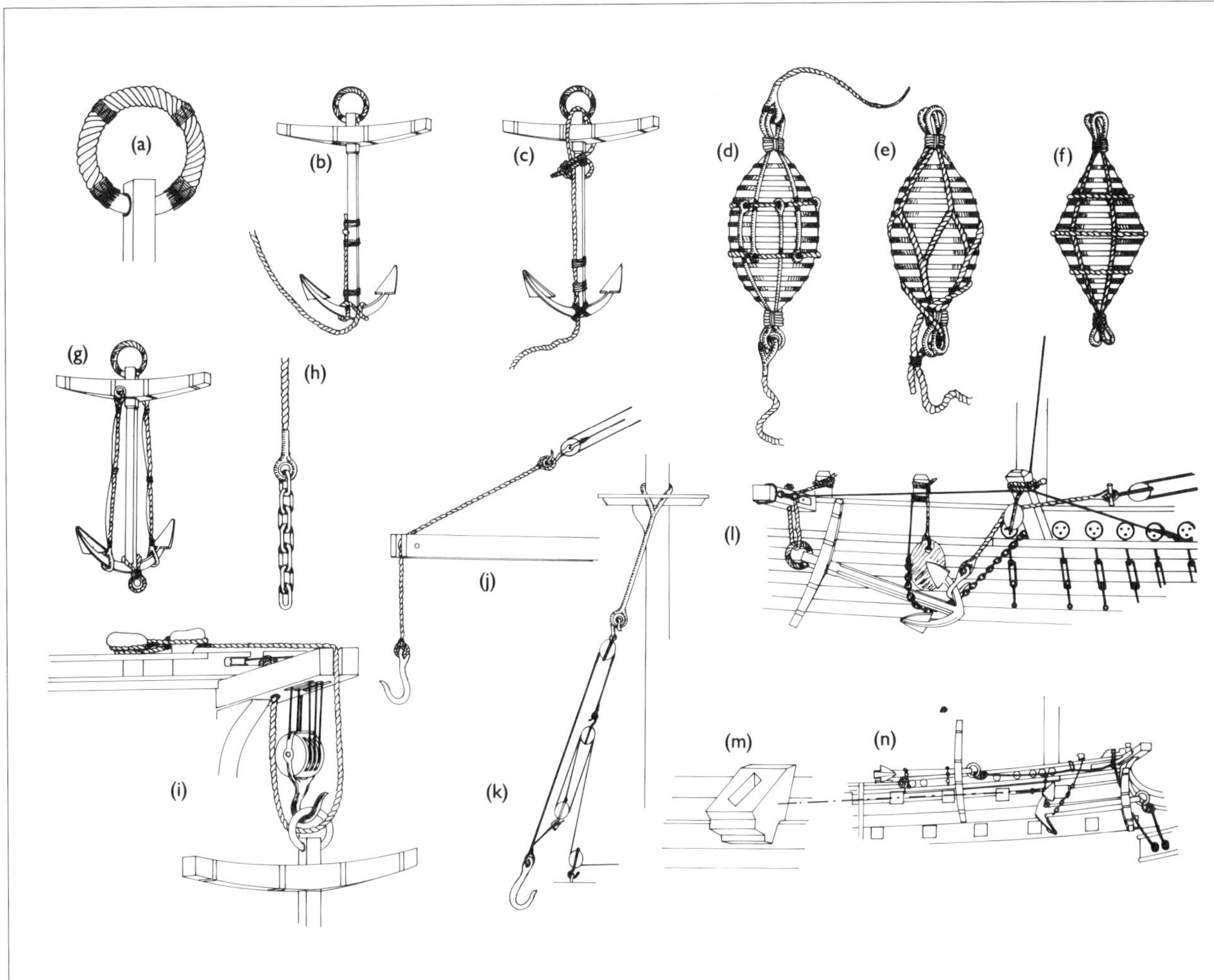

Figure 192
(a) Puddening of an anchor ring.
(b) Lashing of a buoy line.
(c) Lashing of a buoy line on a small anchor.
(d) 'Spanish fashion' anchor buoy, English.
(e) Dutch fashion anchor buoy; the buoy line is fastened with a sheet bend.
(f) French fashion anchor buoy.
(g) Anchor with cross ropes (traversins de l'ancre).
(h) Shank painter with a chain; buoy lines sometimes also had chains fitted.
(i) Cat tackle and cathead stopper.
(j) Fish tackle and long davit.
(k) Runner and tackle for the fishing of anchors in smaller ships without davits.
(l) Fishing of an anchor with a short davit. The shank painter is laid round the anchor, and the fender (shoe) between the side and the bill (fluke).
(m) Block for resting the sheet anchor in English ships at the end of the century.
(n) Rest position of sheet and bow anchor in English ships at the end of the century

davit to the masthead and was secured there.

Fishing an anchor (see Figure 192l) was done with a fish-tackle. A large single block was stropped to the davit's head, and the fish pendant rove through it, with a thimble and large hook turned into its lower end and an eye splice in the upper. The strop of a double block was put through the eye and secured by a toggle; a single block connected to this block by a fall was hooked to an eyebolt aft. Lever noted that sometimes a thimble was spliced into the inner end of the pendant, with a luff tackle hooked to it; sometimes, too, the pendant was longer and without a tackle, and the end was laid around the capstan. He explained further:

> Smaller Ships, which carry Davits, have them run out over the Gunnel athwart-Ships, the inner end resting on the Forecastle. In the outer end is a Snatch-sheave, in which the Bight of a short Pendent is placed, having the Fish Hook spliced in one end, and a Thimble in the other, to which is hooked the lower Block of the Runner Tackle. Some Ships have no Davits. The Anchors are then fished by the Runner and Tackles.

The davit as shown by Röding had a large block stropped to its end through which the pendant passed. In the outer end of the pendant the large fish hook was spliced, and stropped to the inner end was a fiddle block. The single block of this tackle was hooked to a strop set in the davit in the same way as the stopper was set in the cathead. He also pointed out that:

> The Spanish and French usually use the fish tackle without a davit; they fasten the tackle to a timberhead on the forecastle and keep the anchor away from the side with the fore yard tackle.

He described the use of a davit as mainly an English practice.

> When the anchor was brought up high enough, a wooden fender known as a shoe was hung over the side on a lanyard for the inner bill of the anchor to rest against. The shank painter, a stopper with an iron chain made fast to the chainwale (see Figure 192h), was passed around the inner arm and the shank of the anchor, and belayed on a timberhead, then hitched and stopped.

PIC 192

Futtock stave

Schwichtungslatte, Wurst; Baton de trélingage

Futtock staves were usually made from a piece of served rope. Lever commented 'the Futtock Stave is sometimes made of rope, served, and sometimes of wood: and is only seized to those Shrouds, which are to be catharpined in'. Its length was determined by the number of shrouds, of which the foremost and the aftermost were often not catharpined (see Chapter II). Röding noted that the foremost shroud only was not crossed by the stave, and that futtock staves were fitted both outside and inside the shrouds, seized securely to each shroud.

Stretcher, rack

Spreizlatte, Bockstange; Traversin des haubans

Shroud stretchers were rare during the eighteenth century, and were mainly seen on smaller ships. Steel referred to a stretcher as a rack, and described it thus: 'a short thin plank, with holes made through it, containing a number of belaying-pins, used instead of cleats: it is seized to the shrouds, and nailed over the bowsprit or windlass'. Lever noted no limitations on size or type of vessel when he mentioned stretchers in the lower shrouds, used to keep them from twisting and to make the lanyards lie fairly.

Stretchers were to be found in some topmast shrouds on large French ships, as can be seen in the rigging plan of the *Royal Louis* of 1780. An early example of a large man-of-war carrying these racks is the model of the Russian First Rate *Zacharii i Elisaveth* of 1748.

Stoppers

Stopper; Bosses

Short ropes employed to check anchor cables, suspend weighty objects, hold shrouds or similar in position when damaged, and many other purposes were known a stoppers. A distinction can be made between cathead stoppers, which suspended the anchor when catted, deck stoppers for retaining the cable when the ship was riding at anchor, dog stoppers and wing stoppers (used when the ship rode at anchor in heavy gales – these were large ropes laid around the mainmast and made fast to the cable forward of the bitts, to prevent the cable from snapping at the bitts), bitt stoppers, which prevented the cable from slipping off the bitts, shroud stoppers, which retained the shrouds when shot through, and fore tack and sheet stoppers which secured tacks and sheets until they could be belayed.

Beckets

Knebelstropps; Chambrières

Beckets were used for securing loose rigging, oars, spars or indeed anything else loose on deck, and also for hanging up the weather sheets, or lee tacks, of the main and fore sails to the foremost main or fore shroud. Basically short lengths of rope, beckets sometimes had an eye spliced in one end and a small wall knot at the other (for example, those used for retaining sheets and tacks), while others had their ends spliced together to form a wreath. The loop left in the bight of a block strop, to which the fall was hitched or spliced, was also named a becket. Steel and Falconer noted that long iron hooks, short strops and even timber pieces (brackets) were also known as beckets; the term thus was loosely used for any fitting needed to secure a vessel for sea. Falconer stated that the term 'becket' probably arose as a corruption and misapplication of the word 'bracket'.

Beckets were mentioned in Steel's tables only in connection with the topsail lifts of the fore and main masts.

APPENDIX

TABLES

The following tables show the proportional lengths of the standing and running rigging of all ships after Steel, 1794, calculated from the lengths for masts and yards given above in the tables for Chapter I.

Explanation of the Proportions
Rule: Take the number of feet contained in the length of the bowsprit, etc, then find the proportional part for any of its rigging, observing to reckon fathoms for feet. Thus, suppose it were required to know the length of a rope necessary for the woolding of a bowsprit of a twenty gun ship; the following tables direct that it should be twice the length of its bowsprit; therefore, suppose the length of the bowsprit to be 60 feet, twice that length is 120, that is 120 fathoms.

Thus began the introduction in Steel's work *The Elements and Practice of Rigging and Seamanship* to a series of tables of great variety and comprehensiveness. His tables gave not only the length of each rope, but also the circumference, and the type, size and number of blocks used for each part of the rigging, in a great variety of vessels. Besides seven men-of-war, ranging from 110 guns to 14 guns, he included a brig, a cutter, a ketch, and two sloops of contrasting sizes. Four fully rigged merchantmen from 1250 to 300 tons completed this remarkable work. An additional column of figures for a schooner was provided in his *Art of Rigging*.

These tables are today extremely valuable to modelmakers; they provide the answers to very many rigging questions, and are given in full below as they are scarcely capable of improvement.

The varyious block types are indicated by single characters, or a combination of these. A key to the abbreviations is given above Table 62.

In addition to Steel's lists, a number of tables giving information on blocks and ropes for early eighteenth century ships are also given below, taken from Davis' *The Seaman's Speculum or Compleat School Master* of 1711, and from Sutherland's *The Ship Builder's Assistant*, also of 1711. These will particularly benefit modellers and enthusiasts of ships of the sprit topmast period.

Table 60: Proportional lengths of standing and running rigging of all ships (after Steel, 1794)

Rigging of the bowsprit	
Woolding	2½ times the length of the bowsprit for large ships, and twice for small
Gammoning	2¼ times the length of the bowsprit for double gammoning, and 1½ times for single
Shrouds	half the length of the bowsprit for two pairs, and a quarter for one pair
Collars	one third the length of the shrouds
Seizing	1½ times the length of the collars
Lashings	1 fathom longer than the collars
Lanyards	twice the length of the collars
Bobstay	half the length of the bowsprit for three pairs, and one third for two pairs
Collars	one quarter the length of the bobstays
Seizing	the length of the bobstays
Lashings	one fathom longer than the collars
Lanyards	twice the length of the collars
Horses	two sevenths the length of the bowsprit
Strops	one fifth the length of the horses
Lanyards	three eighths the length of the horses

Rigging of the spritsail yard	
Horses	one fifth the length of the yard
Stirrups	half the length of the horses
Braces	1¼ times the length of the spritsail yard
Pendents	one twelfth the length of the braces
Strops	one seventh the length of the braces
Lifts	seven eighths the length of the spritsail yard
Beckets	four fathoms in ships of the line, and two in other ships
Strops	the same length as the beckets
Seizing	twice the length of the strops
Standing lifts	one sixth the length of the lifts
Strops	half the length of the standing lifts
Lanyards	three quarters the length of the standing lifts
Halliards	three fifths the length of the spritsail yard
Strops	three fathoms in ships of the line, and two in other ships
Seizing & lashing	one fifth the length of the halliard
Slings	one tenth the length of the spritsail yard
Seizing & racking	twice the length of the sling
Clewlines	two thirds the length of the spritsail yard
Strops	one eighth the length of the clewlines
Buntlines	half the length of the spritsail yard
Strops	one fathom
Earings	one quarter the length of the spritsail yard
Sheets	five eighths the length of the spritsail yard

Rigging of the jib-boom	
Horses	two fifths the length of the jib-boom
Seizing	half the length of the horses
Guy pendents & strops	five eighths the length of the jib-boom
Falls & strops	two fathoms longer than the pendents
Lashings	half the length of the pendents
Outhauler	one quarter the length of the jib-boom
Tackle fall	five eighths the length of the jib-boom
Strop	1 fathom
Stay	five eighths the length of the jib-boom
Strop	1 fathom
Tackle fall & strop	half the length of the jib-boom
Halliard & strop	1½ times the length of the jib stay
Downhauler	seven eighths the length of the halliard
Sheets	seven eighths the length of the jib-boom
Pendents	one quarter the length of the sheets
Flying jib halliard	1½ times the length of the fore topgallant stay
Sheets	equal to the length of the jib-boom
Tack	2 fathoms
Downhauler	seven eighths the length of the halliard

Rigging of the sprit topsail yard	
Horses	one fifth the length of the sprit topsail yard
Braces	1⅜ times the length of the sprit topsail yard
Lifts & strops	equal to the length of the sprit topsail yard
Halliard & strop	three quarters the length of the sprit topsail yard
Lashing	one quarter the length of the halliards
Parrel ropes	one tenth the length of the sprit topsail yard
Clewlines & strops	1¼ times the length of the sprit topsail yard
Lacing & earings	equal to the length of the sprit topsail yard

Rigging of the foremast	
Wooldings	three times the length of the foremast in large ships, and 2½ times in small
Girtlines & strops	three quarters the length of the foremast
Seizing	one twelfth the length of the girtlines
Lashings	one quarter the length of the girtlines
Pendents of tackles	one fifth the length of the foremast for two pairs, and one tenth for one pair
Strops	one third the length of the foremast for two pairs, and one fifth for one pair
Seizing	equal to the length of the pendents for two pairs, and half the length for one pair
Runners of tackles	four times the length of the pendents for two pairs, and twice the length for one pair
Strops	one fifth the length of the runners for two pairs, and one tenth for one pair
Falls of tackles	twice the length of the foremast for two pairs, and once for one pair
Strops	one tenth the length of the foremast for two pairs, and one twentieth for one pair
Seizing	four times the length of the strops
Shrouds	2¼ times the length of the foremast for ten pairs, twice the length for nine pairs, and 1⅗ times the length for seven pairs. Note that the length of the first warp is taken from the upper side of the bolster on the trestle trees to the foremost deadeye in the channel, or from the middle of the opposite side of the masthead down to the deck.
Eye seizing	one quarter the length of the shrouds
Throat seizing	half the length of the shrouds
End seizing	half the length of the shrouds
Lanyards	five eighths the length of the shrouds
Ratlines	1⅝ times the length of the shrouds
Stay	one seventh the length of the foremast, in fathoms
Seizing	1½ times the length of the stay
Lanyard	equal to the length of the stay
Stay collar	half the length of the stay
Seizing	1½ times the length of the collar
Lashing	three quarters the length of the collar
Preventer stay	equal to the length of the fore stay
Lanyard	two thirds the length of the stay
Preventer stay collar	half the length of the stay
Lashing	equal to the length of the collar
Seizing	2½ times the length of the collar
Catharpin legs	one sixth the length of the foremast for six, and one eighth for four
Seizing	four times the length of the catharpin legs
Jeers, tye & falls	in large ships with treble and double blocks, the falls are 1¼ times the length of the mast; with double blocks, the length of the mast; and with a tye, the tye is one fifth the length of the mast, and the falls equal to the length of the mast
Strops	one eighth the length of the mast
Seizing	one quarter the length of the mast
Masthead lashings	three eighths the length of the foremast
Yard lashings	one third the length of the masthead lashing
Stoppers	half the length of the yard lashing
Horses	one sixth the length of the fore yard
Stirrups	equal to the length of the horses
Seizing	equal to the length of the horses
Lanyard	one third the length of the horses
Yard tackle pendents	one tenth the length of the fore yard
Falls	equal to the length of the fore yard and pendents
Strops	equal to the length of the pendents
Seizing	twice the length of the strops
Inner tricing lines	three sevenths the length of the fore yard
Outer tricing lines	three sevenths the length of the fore yard
Strops	one sixth the length of the outer tricing lines

Rigging of the fore yard	
Braces	1⅙ times the length of the fore yard
Pendents	one eighth the length of the fore yard
Preventer	ditto one fathom longer
Strops	half the length of the pendents
Seizing	one quarter the length of the fore yard
Lashing	one quarter the length of the fore yard
Braces, preventer (war only)	seven eighths the length of the other braces
Strops	3 fathoms in large ships, and 2 in small
Seizing	10 fathoms in large ships, and 6 in small
Lifts & strops	1¼ times the length of the fore yard
Seizing	one eighth the length of the fore yard
Span for the cap	one tenth the length of the fore yard
Short span	one third the length of the span for the cap
Jigger tackle	from 20 to 30 fathoms
Strop	one eighth the length of the tackle
Truss pendents	one fifth in large ships, and one sixth the length of the fore yard in small
Falls	five eighths the length of the fore yard
Strops	one third the length of the pendents
Eye seizing	three times the length of the strops
Naveline	one fifth the length of the fore yard
Puddening of the yard	12 fathoms in large ships, and 6 in small
Clew garnets	three quarters the length of the fore yard
Strops about the yard	one eighth the length of the clew garnets
Strops	2 fathoms in large ships, and 1 fathom in small
Seizing	one eighth the length of the clew garnets
Lashing	one sixth the length of the clew garnets
Buntline legs	four sevenths the length of the fore yard
Falls	equal to the length of the legs
Strops	one quarter the length of the falls
Leech line legs	equal to the length of the buntline legs
Falls	equal to the length of the buntline falls
Strops	equal to the length of the buntline strops
Slab line & strop	three sevenths the length of the fore yard
Bowlines	two thirds the length of the fore yard
Bridles	one tenth the length of the bowlines
Strops	three quarters the length of the bridles
Seizing	twice the length of the bridles
Lashing	equal to the length of the seizing
Earings	one third the length of the fore yard
Sheets & strops	equal to the length of the fore yard
Seizing	one sixth the length of the fore yard
Stoppers	one third the length of the seizing
Tack, single	half the length of the fore yard
Tacks, double	1¾ times the length of single tacks
Strops	one eighth the length of the tacks
Seizing	one fifth the length of the tacks
Stoppers	equal to the length of the strops
Lanyards	equal to the length of the stoppers
Gammoning for the bumkin	16 fathoms in large ships, and 10 fathoms in small
Lanyard	12 fathoms in large ships, and 6 in small
Slings	one tenth the length of the fore yard
Strop	half the length of the slings
Seizing	twice the length of the slings
Lanyard	three quarters the length of the slings
Staysail halliard	twice the length of the preventer stay, plus 2 fathoms
Sheets	twice the length of the preventer stay
Strops	one third the length of the preventer stay
Tack	2 fathoms
Downhauler	twice the length of the preventer stay
Strop	one third the length of the other strops
Studdingsail	
halliards, inner	three fifths the length of the fore yard
Ditto, outer	equal to the length of the fore yard
Sheets	one sixth the length of the fore yard
Tacks	five sevenths the length of the fore yard
Strops	equal to the length of the sheets

Rigging of the fore topmast

Burton pendents	one ninth the length of the topmast
Falls	equal to the length of the topmast
Strops	three quarters the length of the pendents
Shrouds	
First warp	seven eighths the length of the topmast, in feet
Whole length	twice the length of the first warp for every pair
Eye seizing	one quarter the whole length of the shrouds
Throat seizing	1¾ times the length of the eye seizing
End seizing	1½ times the length of the eye seizing
Lanyards	two thirds the whole length of the shrouds
Ratlines	1⅜ times the whole length of the shrouds
Standing backstays	twice the length of the topmast, in feet, for the first warp, and twice that for every pair
Eye seizing	one eighth the whole length of the backstays
Throat seizing	1¾ times the length of the eye seizing
End seizing	1½ times the whole length of the backstays
Lanyards	one fifth the whole length of the backstays
Breast backstay runners	one sixth the length of the topmast, in fathoms
Falls	one third the length of the topmast
Strops	one sixth the length of the falls
Stay	two fifths the length of the topmast
Collar	one sixth the length of the stay
Tackle	equal to the length of the stay
Strop	one seventh the length of the tackle
Seizing	twice the length of the strop
Preventer stay	equal to the length of the topmast stay
Collar	one eighth the length of the stay
Tackle	equal to the length of the stay in large ships, and seven eighths of the stay in small ships
Strop	one sixth the length of the tackle
Seizing	three times the length of the strop
Lashing for the collar	one quarter the length of the stay
Shifting backstay	five eighths the length of the topmast
Tackles	three quarters the length of the topmast
Strops	one quarter the length of the stay
Futtock shrouds	three eighths the whole length of the topmast shrouds
Upper seizing	1¼ times the length of the futtock shrouds
Lower seizing	seven eighths of the length of the upper seizing
Ratlines	1¼ times the length of the shrouds
Top rope pendents	two thirds the length of the topmast
Falls	3¼ times the length of the pendents in large ships, and 2¼ times in small
Tye	three quarters the length of the topmast
Strop	one fifth the length of the tye
Seizing	1½ times the length of the strop
Masthead lashing	one fifth the length of the topmast
Yard lashing	half the length of the masthead lashing
Halliards	twice the length of the topmast
Strops	one seventh the length of the topmast
Seizing	twice the length of the strops

Rigging of the fore topsail yard

Horses	one fifth the length of the yard
Stirrups	three quarters the length of the horses
Braces	1⅔ times the length of the yard
Pendents	one sixth the length of the yard
Preventer	ditto one fathom longer than the pendents
Strops	half the length of the pendents
Lifts	1⅛ times the length of the yard
Beckets	3 fathoms in large ships, and 2 fathoms in small
Strops	three times the length of the beckets
Seizing	three times the length of the strops
Parrel ropes	one fifth the length of the yard
Racking & seizing	1⅓ times the length of the parrel rope
Clewlines	1½ times the length of the yard
Strops	one tenth the length of the clewlines
Buntlines & strops	6 fathoms more than the length of the yard
Leechlines & strops	half the length of the yard
Bowlines & strops	equal to the length of the yard
Bridles	one quarter the length of the yard
Strops	2 fathoms in large ships, and 1½ in small
Lashing	equal to the length of the bridles
Reef tackle pendents	three quarters the length of the yard
Falls	equal to the length of the yard
Strops	3 fathoms in large ships, and 1½ in small
Earings	three quarters the length of the yard
Sheets	seven eighths the length of the yard
Strops for sheet blocks	one tenth the length of the sheets
Strops for quarter blocks	twice the length of the former for large ships, and 1½ times the length of the former for small
Lashing for quarter blocks	twice the sheet block strops
Seizing	three eighths the length of the sheets
Span	one sixth the length of the sheets
Stoppers	half the length of the span
Slings	one fifth the length of the yard
Staysail stay	one fathom longer than the top stay
Tackle	two thirds the length of the stay
Halliard & strop	1¾ times the length of the stay
Sheets & strops	1½ times the length of the stay
Outhauler	seven eighths the length of the stay
Downhauler & strop	1⅓ times the length of the stay
Studdingsail halliards	1⅔ times the length of the topsail yard
Sheets	five sixths the length of the fore topsail yard
Tacks	1⅙ times the length of the fore topsail yard
Downhaulers	equal to the length of the fore topsail yard
Boom tackles	1⅖ times the length of the fore topsail yard
Tails	one third the length of the fore topsail yard
Strops	one third the length of the tails

Rigging of the fore topgallant mast

Shrouds, length of the first warp	1⅓ times the length of the fore topmast, in feet
Length of each pair	twice the length of the first warp
Lanyards	one sixth the length of the shrouds
Standing backstays, length of first warp	2⅜ times the length of the fore topmast, in feet
Length of each pair	twice the length of the first warp
Lanyards	one quarter the length of the backstays
Stay	6 fathoms longer than the topmast stays
Strop	1½ fathoms in larger ships, and 1 fathom in small
Tackle	half the length of the stay
Strop	1½ fathoms
Flagstaff stay	the same length in large ships, and five eighths the length of the topgallant stay in small
Halliards	1⅛ times the length of the fore topmast
Tye	three tenths the length of the fore topmast
Halliard	1¾ times the length of the tye
Strop	1½ fathoms in large ships and 1 in small

Rigging of the fore topgallant yard

Horses	one fifth the length of the yard
Braces	three times the length of the yard
Pendents	one tenth the length of the braces
Strops	half the length of the brace pendents
Lifts & strops	1½ times the length of the yard
Parrel ropes	one eighth the length of the yard
Clewlines & strops	twice the length of the yard
Bowlines & strops	equal to the length of the clewlines
Bridles	one eighth the length of the bowlines
Sheets	equal to the length of the topsail yard
Earings	three quarters the length of the yard
Shifting backstay	half the length of the standing backstays
Tackles & strops	half the backstay
Studdingsail halliards	1¼ times the length of the fore topmast
Sheets	half the length of the halliards
Tacks	10 fathoms more than the length of the sheets

Downhauler	equal to the length of the sheets
Strop	one quarter the length of the fore topgallant yard

Rigging of the mainmast

Wooldings	three the length of the mainmast in large ships, and 2½ times the length of the mainmast in small
Girtlines & strops	three quarters the length of the mainmast
Seizing	one twelfth the length of the girtlines
Lashings	one quarter the length of the girtlines
Pendents of tackles	one sixth the length of the mainmast for two pairs, and one twelfth for one pair
Strops	one third the length of the pendents for two pairs, and one fifth for one pair
Seizing	the length of the pendents for two pairs, and half the length for one pair
Runners of tackles	four times the length of the pendents for two pairs, and twice the length of the pendents for one pair
Strops	6 fathoms in large ships, and 3 in small
Fall of tackles	twice the length of the mainmast for two pairs, and once for one pair
Strops	one tenth the length of the mainmast for two pairs, and one twentieth for one pair
Seizing	four times the length of the strops
Shrouds	2¼ times the length of the mast for ten pairs, twice for nine pairs, and 1⅗ times for seven pairs. Note that the length of the first warp is taken from the upper side of the bolsters upon the trestle trees, to the foremost deadeye in the channel; or from the middle of the opposite side of the masthead down to the deck.
Eye seizing	one quarter the length of the shrouds
Throat seizing	twice the length of the eye seizing
End seizing	twice the length of the eye seizing
Lanyards	five eighths the length of the shrouds
Ratlines	1⅝ times the length of the shrouds
Stay	one fifth the length of the mast, in fathoms
Seizing	equal to the length of the stay
Lanyards	three quarters the length of the stay
Stay collar	half the length of the stay
Worming	five times the length of the collar
Seizing	1¾ times the length of the collar
Lashing	twice the length of the collar
Preventer stay	2 fathoms shorter than the main stay
Lanyard	half the length of the preventer stay
Preventer stay collar	one third the length of the preventer stay
Lashing	equal to the length of the collar
Seizing	equal to the length of the stay
Catharpin legs	one seventh the length of the mast when six, and one ninth when four
Seizing	four times the length of the catharpin legs
Stay tackle pendent	one quarter the length of the main stay
Falls	nine times the length of the pendent
Strops	three quarters the length of the pendent
Seizing	three times the length of the pendent
Lashing	2 fathoms shorter than the seizing
Foremost stay tackle fall	equal to the length of the stay tackle fall
Strop	equal to the length of the stay tackle strop
Seizing	equal to the length of the stay tackle seizing
Jeers, tye & falls	In large ships with treble and double blocks, the falls were 1¼ times the length of the mast, and for ships with double blocks, equal to the length of the mast; when a tye was fitted, the tye was one fifth the length of the mast, and the falls equal to the length of the mast.
Strops	one eighth the length of the mast
Seizing	one quarter the length of the mast
Masthead lashings	three eighths the length of the mast
Yard lashings	one third the length of the masthead lashings
Stoppers	half the length of the yard lashings

Rigging of the main yard

Horses	one sixth the length of the yard
Stirrups	equal to the length of the horses
Seizing	equal to the length of the horses
Lanyard	one third the length of the horses
Yard tackle	
Pendents	one tenth the length of the yard
Falls	1⅙ times the length of the yard in large ships, equal to the length of the yard in small ships
Strops	equal to the length of the pendents
Seizing	twice the length of the strops
Inner tricing lines	three sevenths the length of the yard
Outer tricing lines	three sevenths the length of the yard
Strops	one sixth the length of the outer tricing lines
Braces & strops	equal to the length of the yard
Pendents	one tenth the length of the yard
Preventer	ditto 1 fathom longer than the pendents
Strops	one third the length of the pendents
Seizing	equal to the length of the pendents
Braces, preventer (war only)	seven eighths the length of the other braces
Strops	3 fathoms in large ships, and 2 in small
Seizing	10 fathoms in large ships, and 6 in small
Lifts & strops	1¼ times the length of the yard
Seizing	one eighth the length of the yard
Span for the cap	one tenth the length of the yard
Short span	one third the length of the cap span
Jigger tackles	from 20 to 30 fathoms
Strops	one eighth the length of the tackle
Truss pendents	one fifth the length in large ships, and one sixth the length of the yard in small
Falls	five eighths the length of the yard
Strops	one third the length of the pendents
Eye seizing	three times the length of the strops
Naveline	one fifth the length of the yard
Puddening of the yard	14 fathoms in large ships, and 6 in small
Clewgarnets	three quarters the length of the yard
Strops about the yard	one eighth the length of the clewgarnets
Strops	2 fathoms in large ships, and 1 in small
Seizing	one eighth the length of the clewgarnets
Lashing	one sixth the length of the clewgarnets
Buntline legs	four sevenths the length of the yard
Falls	equal to the length of the legs
Strops	one quarter the length of the legs
Leechline legs	equal to the length of the buntline legs
Falls	equal to the length of the leechline legs
Strops	equal to the length of the buntline strops
Slab lines & strops	three sevenths the length of the yard
Bowlines	three fifths the length of the yard
Bridles	one quarter the length of the bowlines
Strops	one quarter the length of the bridles
Seizing	equal to the length of the strops
Lashing	equal to the length of the strops
Tackle	equal to the length of the bridles
Tackle strops	3 fathoms in large and 2 fathoms in small ships
Earings	one third the length of the yard
Sheets & strops	equal to the length of the yard
Seizing	one sixth the length of the yard
Lashings	6 fathoms
Stoppers	4 fathoms
Tacks, single	half the length of the yard
Tacks, double	1¾ times the length of the single tacks
Stoppers	4 fathoms
Lanyards	6 fathoms in large ships, and 5 fathoms in small
Lanyard for the puddening & dolphin	10 fathoms in large ships, and 6 in small
Slings	one tenth the length of the yard
Strop	half the length of the slings
Seizing	twice the length of the slings
Lanyard	three quarters the length of the slings

Quarter tackle	
Pendents	10 fathoms in large ships, and 7 fathoms in small
Falls	seven eighths the length of the yard
Strops	one tenth the length of the falls
Seizing	twice the length of the strops
Luff tackles	from 30 to 25 fathoms each tackle
Strops	2 fathoms to each tackle
Seizing	5 fathoms to each tackle
Staysail stay	one seventh the length of the mast
Collar	2 fathoms
Seizing	6 fathoms
Lanyard	4 fathoms
Halliard	two fifths the length of the mainmast
Sheets	equal to the length of the staysail stay
Strops	one third the length of the sheets
Tacks	one third the length of the sheets
Downhauler	1½ times the length of the staysail stay
Strop	1 fathom
Studdingsail inner	halliards, three fifths the length of the main yard
Ditto, outer	2 fathoms longer than the main yard
Sheets	three tenths the length of the main yard
Tacks	five ninths the length of the main yard
Strops	one fifth the length of the main yard

Rigging of the main topmast

Burton pendents	one ninth the length of the main topmast
Falls	equal to the length of the main topmast
Strops	two thirds the length of the pendents
Shrouds, first warp	seven eighths the length of the main topmast in feet
Whole length	twice the length of the first warp for each pair
Eye seizing	one quarter the whole length of the shrouds
Throat seizing	1¾ times the length of the eye seizing
End seizing	1½ times the length of the eye seizing
Lanyards	three fifths the whole length of the shrouds
Ratlines	1⅜ times the whole length of the shrouds
Standing backstays	twice the length of the topmast, in feet, for the first warp and twice that for each pair
Eye seizing	one eighth the whole length of the backstays
Throat seizing	1¾ times the length of the eye seizing
End seizing	1½ times the length of the eye seizing
Lanyards	one fifth the length of the backstays
Breast backstay runners	one seventh the length of the topmast, in fathoms
Falls	one third the length of the topmast
Strops	one sixth the length of the falls
Stay	two fifths the length of the topmast
Collar	one sixth the length of the stay
Tackle	equal to the length of the stay
Strop	one sixth the length of the tackle
Seizing	one third the length of the stay
Lashing	7 fathoms in large ships, and 6 in small
Preventer stay	equal to the length of the topmast stay
Collar	one sixth the length of the stay
Tackle	equal to the length of the stay
Strop	one eighth the length of the tackle
Seizing	three times the length of the strop
Lashing for the collar	one quarter the length of the stay
Shifting backstays	five eighths the length of the topmast
Tackles	10 fathoms for each tackle in small ships, and 12 fathoms in large
Strops	1½ fathom to each tackle in small, and 2 fathoms in large
Futtock shrouds	three eighths the whole length of the topmast shrouds
Seizing, upper	1⅓ times the length of the futtock shrouds
Lower	seven eighths the length of the upper seizing
Ratlines	1¼ times the length of the shrouds
Top rope pendents	two thirds the length of the topmast
Falls	3¼ times the length of the pendents in large ships, and 2¼ times in small
Tye	three quarters the length of the topmast
Strop	one fifth the length of the tye
Seizing	1½ times the length of the strop
Masthead lashings	one fifth the length of the topmast
Yard lashings	half the length of the masthead lashing
Halliards	twice the length of the topmast
Strops	one eighth the length of the topmast
Seizing	twice the length of the strops

Rigging of the main topsail yard

Horses	one fifth the length of the main topsail yard
Stirrups	three quarters the length of the horses
Braces	1⅕ times the length of the yard
Pendents	one eighth the length of the yard
Preventer	ditto 1 fathom longer than the pendents
Strops	three eighths the length of the pendents
Span about the mizzen mast	6 to 4 fathoms
Lifts	1⅛ times the length of the yard
Beckets	2 to 3 fathoms
Strops	three times the length of the beckets
Seizing	one quarter the length of the lifts
Parrel ropes	one fifth the length of the yard
Racking & seizing	1⅓ times the parrel rope
Clewlines	1½ times the length of the main topsail yard
Strops	one tenth the length of the clewlines
Buntlines & strops	6 fathoms more than the length of the yard
Leechlines & strops	half the length of the yard
Bowlines & strops	equal to the length of the yard
Bridles	one quarter the length of the yard
Frapping & lashing	one fifth the length of the yard
Seizing	6 fathoms in large ships, and 4 in small
Reef tackle pendents	three quarters the length of the main topsail yard
Falls	equal to the length of the yard
Strops	3 fathoms in large ships, and 1½ in small
Earings	two thirds the length of the main topsail yard
Sheets	seven eighths the length of the main topsail yard
Strops for sheet blocks	one tenth the length of the sheets
Strops for quarter blocks	twice in large ships, and 1½ times the length of the strops for sheet blocks in small
Lashings for quarter blocks	twice the length of the strops for sheet blocks
Seizing	three eighths the length of the sheets
Span	one sixth the length of the sheets
Stoppers	half the length of the span
Slings	one fifth the length of the main topsail yard
Staysail halliard & strop	twice the length of the preventer stay
Sheets & strops	1½ times the length of the stay
Pendents	4 fathoms
Tack	5 fathoms in large ships and 3 in small
Downhauler	three fifths the length of the halliard
Strops	1 fathom
Brails	equal to the length of the halliard
Middle staysail stay	equal to the length of the preventer stay
Tackle	2 fathoms shorter than the stay
Halliard	2 fathoms shorter than the topmast staysail halliards
Sheets	equal to the length of the halliard
Tack	5 fathoms in large ships, and 4 in small
Downhauler	two thirds the length of the halliard
Strops	8 fathoms in large ships, and 5 in small
Tricing line	three fifths the length of the downhauler
Studdingsail halliards	1½ times the length of the topsail yard
Sheets	half the length of the halliards
Downhaulers	three fifths the length of the halliards
Boom tackles	four fifths the length of the halliards
Lashing for booms	20 fathoms
Tails	one fifth the length of the halliards
Strops	one third the length of the tails

Rigging of the main topgallant mast

Shrouds, length of the first warp in feet	1⅓ times the length of the main topmast
Length of each pair	twice the length of the first warp
Lanyards	one sixth the length of the shrouds
Standing backstays, length of the first warp	2⅝ times the length of the main topmast in feet
Length of each pair	twice the length of the first warp
Lanyards	one quarter the length of the backstays
Stay	equal to the length of the topmast stay
Strop 1 fathom	
Flagstaff stay	3 fathoms longer than the topgallant stay
Halliards	1⅛ times the length of the main topmast
Tye	three tenths the length of the main topmast
Halliard	twice the length of the tye
Strop 1½ fathoms	

Rigging of the main topgallant yard

Horses	one fifth the length of the yard
Braces	1¾ times the length of the yard
Pendents	one eighth the length of the braces
Strops	3 fathoms in large ships, and 2 in small
Lifts, single & strops	1½ times the length of the yard
Clewlines & strops	twice the length of the yard
Bowlines & strops	equal to the length of the clewlines
Bridles	one eighth the length of the bowline
Sheets	equal to the length of the topsail yard
Earings	three quarters the length of the yard
Shifting backstay	half the length of the standing backstay
Tackles	20 fathons in large ships, and 14 fatoms in small ships
Strops	1½ fathoms
Staysail stay	10 fathoms longer than the topgallant stay
Halliards	10 fathoms longer than the stay
Sheets	equal to the length of the halliards
Tack	7 fathoms in large ships, and 6 in small
Downhauler	3 fathoms shorter than the staysail stay
Strop	3 fathoms in large ships, and 2 in small
Studdingsail halliards	1¼ times the length of the main topmast
Sheets	half the length of the halliards
Tacks	10 fathoms longer than the sheets
Downhauler	equal to the length of the sheets
Strops	one quarter the length of the fore topgallant yard

Rigging of the mizzen mast

Wooldings	1⅗ times the length of the mizzen mast
Girtlines & strops	three quarters the length of the mast
Seizing	6 fathoms in large ships, and 5 in small
Lashings	8 fathoms in large ships, and 7 in small
Burton pendents	8 fathoms in large ships, and 6 in small
Falls	three quarters the length of the mizzen mast
Strops	4 fathoms in large ships, and 3 in small
Shrouds	1⅛ times the length of the mast for six pairs, equal to the mast for five pairs, and three quarters for four pairs
Length of the first warp	the same rule as for the mainmast
Eye seizing	one quarter the length of the shrouds
Throat seizing	twice the length of the eye seizing
End seizing	1½ times the length of the eye seizing
Lanyards	half the length of the mizzen shrouds
Ratlines	1½ times the length of the shrouds
Stay	one sixth the length of the mizzen mast
Seizing	half the length of the stay
Lanyard	6 fathoms in large ships, and 5 fathoms in small
Collar	3½ in large ships, and 2½ fathoms in small
Seizing	3 fathoms
Lashing	3 fathoms

Rigging of the mizzen yard or gaff

Jeers	three quarters the length of the mizzen mast for treble and double, and half for double and single
Strops	8 fathoms in large ships, and 5 in small
Seizing	10 fathoms in large ships, and 8 in small
Masthead lashing	one eighth the length of the mizzen mast
Yard lashing	half the mast lashing
Derrick	half the length of the mizzen mast
Span	4 fathoms in large ships, and 3 in small
Strop	2½ fathoms in large ships, and 2 in small
Seizing	one eighth the length of the mizzen mast
Lashing	8 fathoms in large ships, and 7 in small
Vang pendents	one seventh the length of the mizzen mast
Falls	three times the length of the pendents
Strops	1 fathom
Bowlines	one quarter the length of the mizzen yard
Strops	3 fathoms in large ships, and 2 in small
Brails, peek legs	12 fathoms in large ships, and 10 in small
Falls	18 fathoms in large ships, and 16 in small
Middle	one quarter the length of the mizzen mast
Throat	1¼ times the length of the middle brails
Foot	1¾ times the length of the middle brails
Throat strops	1 fathom
Foot & middle strops	9 fathoms in large ships, and 7 in small
Lacing to mast	one quarter the length of the mizzen mast
Lacing to yard	7 fathoms longer than the mast lacing
Earings	10 fathoms in large ships, and 8 in small
Peek halliards	one quarter the length of the mizzen mast
Sheet	one third the length of the mizzen mast
Strops	3½ fathoms in large ships, and 2 fathoms in small
Seizing	twice the length of the strops
Tack	5 fathoms in large ships, and 3 in small
Slings	6 fathoms in large ships, and 4 in small
Staysail stay	one sixth the length of the mizzen mast
Collar	3½ fathoms in large ships, and 2 in small
Seizing	equal to the length of the collar
Lashing	equal to the length of the collar
Lanyard	twice the length of the collar
Halliards	twice the length of the staysail stay
Sheets	equal to the length of the staysail stay
Strops to sheets & halliard	3 fathoms in large ships, and 2 fathoms in small
Tacks	4 fathoms in large ships, and 2 in small
Downhauler	equal to the length of the staysail stay
Strop to downhauler	2 fathoms in large ships, and 1 fathom in small
Brails	2 fathoms shorter than the halliards

Rigging to the driver

Topping lifts	four sevenths the length of the driver boom, taking fathoms for feet
Span	6 fathoms in large ships, and 4 fathoms in small
Falls	46 fathoms for ships from 110 to 64 guns, 40 fathoms for those from 50 to 36 guns, 36 fathoms for those from 32 to 28 guns, 30 fathoms for those from 24 to 20 guns, and 24 fathoms for all ships with fewer guns
Guy pendents	12 fathoms for ships from 110 to 36 guns, and 10 fathoms for all ships with fewer
Falls	60 fathoms for ships from 110 to 28 guns, and 50 fathoms for all ships with fewer
Boom sheet	25 fathoms for ships from 110 to 28 guns, and 24 fathoms for all ships with fewer
Horses	14 fathoms for ships from 110 to 64 guns, 12 fathoms for ships from 50 to 28 guns, and 10 fathoms for all ships with fewer guns
Sheet pendent	14 fathoms for ships from 110 to 44 guns, 12 fathoms for ships from 38 to 36 guns, and 10 for all ships with fewer guns
Falls	20 fathoms for ships from 110 to 28 guns, and 18 fathoms for all ships with fewer

Brails	108 fathoms for ships from 110 to 28 guns, and 96 fathoms for all ships with fewer
Lacing to yard	30 fathoms for ships from 110 to 28 guns, and 24 fathoms for all ships with fewer
Tack tackle	12 fathoms for ships from 110 to 36 guns, and 10 fathoms for all ships with fewer
Downhauler	one quarter the length of the mizzen mast
Lashing	8 fathoms in large ships, and 6 in small

Rigging to the crossjack yard

Truss pendent	6 fathoms in large ships, and 4 in small
Falls	one third the length of the crossjack yard
Strops	3 fathoms in large ships, and 2 in small
Seizing	8 in large ships, and 6 in small
Span about the cap	2 fathoms
Braces	2 fathoms longer than the yard
Pendents	6 fathoms in large ships, and 5 in small
Preventer	ditto 1 fathom longer than the pendents
Strops	2 fathoms
Lifts, running	six sevenths the length of the yard
Strops	3 fathoms in large ships, and 2 in small
Slings	one tenth the length of the yard
Strops	2 fathoms in large ships, and 1½ in small
Seizing	1½ times the length of the strops
Lashing	5 fathoms in large ships, and 4 in small

Rigging to the mizzen topmast

Shrouds, length of the first warp	seven eighths the length of the topmast, in feet
Whole length	1⅙ times the length of the topmast in fathoms for four pairs, and of equal length to the topmast for three pairs
Seizing	1½ times the length of the shrouds
Lanyards	two thirds the length of the shrouds
Ratlines	1¼ times the length of the shrouds
Standing backstay	twice the length of the topmast, in feet for the first warp, and double that for every pair
Seizing	half the length of the backstays
Lanyards	16 fathoms for two pairs, and 8 for one pair
Stay	three tenths the length of the topmast
Lanyard	half the length of the stay
Collar	one quarter the length of the stay
Seizing & lashing	6 fathoms
Flagstaff stay	equal to the length of the topmast stay
Halliards	half the length of the topmast
Shifting backstay	1 fathom longer than the topmast stay
Tackle	12 fathoms in large ships, and 10 fathoms in small
Strop	4 fathoms in large ships, and 3 fathoms in small
Futtock shrouds	three tenths the length of the topmast shrouds
Seizing	twice the length of the topmast shrouds
Ratlines	1 fathom longer than the shrouds
Top rope pendents	three eighths the length of the topmast
Falls	2½ times the length of the pendents
Tye	one third the length of the topmast
Halliard	equal to the length of the topmast
Strop	4 fathoms in large ships, and 2 in small
Lashing	3 fathoms

Rigging of the mizzen topsail yard

Horses	one fifth the length of the yard
Stirrups	half the length of the horses
Braces & strops	9 fathoms longer than the yard
Pendents	4 fathoms in large ships, and 3 in small
Lifts	10 fathoms longer than the yard
Strops	3 fathoms in large ships, and 2 in small
Parrel ropes	one eighth the length of the yard
Clewlines & strops	1½ times the length of the yard
Buntlines & strops	2 fathoms longer than the yard
Leechlines & strops	three sevenths the length of the yard
Bowlines & strops	equal to the length of the yard
Bridles	one fifth the length of the bowlines
Reef tackle pendents	three quarters the length of the yard
Falls	6 fathoms longer than the pendents
Strops	1 fathom
Earings	seven eighths the length of the yard
Sheets	seven eighths the length of the yard
Strops	2 fathoms for shoulder blocks, and 3 fathoms in large ships and 2 in small for the other blocks
Seizing	14 fathoms in large ships, in 12 in small
Lashing	5 fathoms in large ships, and 4 in small
Staysail halliard	three fifths the length of the mizzen topmast
Sheets	2 fathoms longer than the halliard
Tacks	2 fathoms
Downhauler	seven tenths the length of the halliards
Strops	2 fathoms in large ships, and 1 in small

Rigging of the mizzen topgallant mast and yard

Shrouds, length of the first warp	1⅓ times the length of the mizzen topmast, in feet
Whole length	1¾ times the length of the topgallant mast, in fathoms
Lanyards	one third the length of the shrouds
Backstays	1¾ times the length of the topgallant mast
Lanyards	6 fathoms
Stay	three quarters the length of the topgallant mast
Lanyard	4 fathoms in large ships, and 3 in small
Flag halliard	equal to the length of the mizzen topmast
Tye	one third the length of the topgallant mast
Halliard	1⅓ times the length of the topgallant mast
Horses	one fifth the length of the yard
Braces	twice the length of the yard
Lifts, single	equal to the length of the yard
Parrel ropes	one seventh the length of the topgallant yard
Clewlines	5 fathoms longer than the braces
Bowlines	2 fathoms shorter than the clewlines
Bridles	3 fathoms in large ships, and 2 in small
Sheets	equal to the length of the yard
Earings	three quarters the length of the yard

'I Shall begin with the Sizes of Rigging, for all British Built Ships, whether they be Ships of War, or Merchant-Men, and shall lay down such Easy and Familiar Rules, as may be readily learn'd, even by the meanest Capacity, and shall begin with the Boulsprit.' [Davis, 1711]

In *The Seaman's Speculum or Compleat School-Master* Davis included lists showing the proportional and actual circumferences and lengths of ropes in early eighteenth century ships, and tables for all six Rates in the Royal Navy. His contemporary Sutherland also gave a list of proportional circumferences and lengths of ropes entitled 'A General Proportion for the Rigging of a Three Mast Ship' and specified in a table for a ship of nearly 600 tons the actual sizes and lengths of ropes and blocks. These lists are combined here, with the figures from Sutherland given in brackets where they differ from those of Davis.

Sutherland also made clear that certain ropes, such as lanyards, pendants and strops, were always in proportion to the ropes they served. A lanyard had half the circumference of the rope it fastened, and seizing were one sixth the circumference of the rope seized. Strops for blocks were generally in two parts, or sometimes in four, which equalled the strength of the rope they were fastened to. Ratlines were one sixth the circumference of the shrouds, and worming one eighth that of the rope wormed. Mast tackle pendants had to be as strong as the shrouds, the runners were eleven twelfths the circumference of the pendant, and the tackle falls half the circumference of the pendent.

Table 61: The circumferences of all rigging in an English ship at the beginning of the century

Bowsprit	
Horse	one ninth the bowsprit diameter; small ships ¼in more (one seventh the bowsprit diameter)
(Deadeye	twice the diameter of the horse)
(Lanyard	half the horse's circumference)
(Strops	equal to the horse's circumference)
Halliards	one ninth the bowsprit diameter; small ships ¼in more (seven twenty-fifths of the spritsail yard's diameter)
Lift	one ninth the bowsprit diameter; small ships ¼in more (six sevenths of the halliard's circumference)
(Seizing to the bowsprit	two thirds the circumference of the lift)
(Standing lifts	equal to the horses)
(Lanyards	equal to the seizing)
(Strops	three quarters the circumference of the standing lifts)
Braces	for small ships equal to, but for large ships ¼in less than, the lift circumference (five sixths of the lift circumference)
(Pendents	one fifth larger then the braces)
Sheets	one eighth the bowsprit diameter; for small ships add ¼in (cable laid, seven twenty-fifths of the spritsail yard diameter)
(Pendents, cabled	six sevenths of the sheet's diameter)
Buntlines	one fifteenth of the bowsprit diameter; for small ships ¼in more (four fifths of the clewline circumference)
Clewline	one fifteenth of the bowsprit diameter; for small ships ¼in more (five twenty-firsts of the spritsail yard diameter)
(Reeflines	half the buntline circumference)
(Slings	twice the brace circumference)
(Seizing & racking	three tenths the circumference of the slings)
Gammoning	one ninth the bowsprit diameter (one fifth the bowsprit diameter)
(Woolding	half the circumference of the gammoning)
Sprit topmast	
Shrouds	two thirds the sprit topmast diameter; for small ships ¼in more (two fifths times that diameter in the cap)
Single lifts	two thirds the sprit topmast diameter; for small ships ¼in more
Double lifts	one quarter times the sprit topmast diameter, small ships ¼in more (equal to the halliards)
Braces	equal to the double lifts (two thirds the circumference of the lifts)
(Pendents	equal to the lifts)
Clewlines	equal to the double lifts
Halliards	equal to the double lifts (one quarter the circumference of the tye)
(Tye	equal to the shrouds)
Crank lines	equal to the double lifts (equal to the shrouds); 'crank' or 'crane' lines was another name for backstay pendents
(Fall	half the backstay pendent)
(Parrel rope	equal to the brace pendents)
Foremast	
Shrouds & tackle pendents	for long voyages about one third the diameter of the mast, for other voyages six elevenths the circumference of the fore stay (two sevenths the diameter of the mast at the partners)
Tackle runners	one quarter the diameter of the mast, less ¼in for large ships (eleven twelfths the circumference of the pendent)
Tackle falls	one eighth the diameter of the mast, plus ¼in in small ships (half the circumference of the pendent)
(Lanyards, ratlines, worming	as noted above)
(Catharpin legs & falls	one quarter the circumference of the shrouds)
Fore and main stays	half the diameter of the mast, allowing about ¼in more for the foremast
(Lanyards	seven twentieths the circumference of the stay, worming as above)
(Fore stay collar	nine tenths the circumference of the stay)
Lifts	equal to the tackle falls (half the circumference of the shrouds)
(Strops for the cap	equal to the lifts)
Bowlines	equal to the tackle falls (equal to the braces)
(Bridles	equal to the bowlines)
(Lashings	five sixths the circumference of the bowlines)
Braces	¼in less than the tackle falls (equal to the lifts)
(Pendents	equal to the braces)
Clewgarnets	equal to the braces
Fore tacks at the head	¾in greater in diameter than the shrouds (equal to the sheets and tapered)
Sheets	one fifth the diameter of the mast for large ships, plus ¼in for small ships (equal to the jeers)
(Stoppers	equal to the sheets)
(Lanyards	equal to the worming of the shrouds)
Runner for hoisting the yard	1 inch greater in diameter than the shrouds
Runner fall	equal to the braces
Jeers with two double breast blocks	one quarter the diameter of the mast, ¼in less for large ships (five sixths the circumference of the shrouds)
(Lashings for the yards	equal to the shroud lanyards)
Buntlines	one twelfth the diameter of the mast (two thirds the circumference of the clewgarnets)
Leechlines	equal to the buntlines
(Reeflines	half the circumference of the leechlines)
(Robands & earings	three quarters the circumference of the leechlines)
(Parrel rope	one sixth the circumference of the mast diameter at the partners)
Single navelines	equal to the buntlines
Double navelines	(equal to the catharpins)
(Racking & seizing	two thirds the circumference of the naveline)
(Woolding	circumference ⅛in per inch of mast diameter
(Crowsfeet for the top & tackle	equal to the ratlines)
(Tackles for boats	equal to the woolding)
(Horses	nine tenths the circumference of the jeers)
(Lanyards	two thirds the circumference of the horses)
(Puddening for the yard	equal to the jeers)

Fore topmast

Shrouds	one third the diameter of the topmast, and for small ships ¼in more
(Lanyards & ratlines	as noted above)
Top rope	for one, four ninths the diameter of the topmast, ¼in more in small ships; for two, three eighths the diameter (pendents of the top rope equal to the fore shrouds)
Falls	equal to the lifts (seven twelfths the circumference of the pendent)
(Pendents of Burton tackles	one quarter of the topmast diameter)
(Falls	two thirds the circumference of the pendents)
(Futtock shrouds	equal to the top rope fall)
Backstays	in men-of-war equal to the shrouds; in merchantmen the fore topmast backstays were equal to the main topmast shrouds, and the main topmast backstays equal to the mizzen shrouds
(Lanyards	as noted above)
Topsail sheets	for men-of-war and indiamen (long voyages) four ninthsthe diameter of the topmast; for small ships half the diameter (one third of the topsail yard diameter in the slings)
(Span	two fifths the circumference of the sheet)
(Lashing	equal to the racking)
Runner for the topsail yard	for one, equal to the shrouds; for two, ½in smaller in circumference (equal to the stay)
Halliards	equal to the lifts (five eighths the circumference of the runner)
Lifts	one quarter the diameter of the topmast in small ships, and one fifth the diameter in large
(Beckets upon the cap	equal to the lifts)
Braces	equal to the lifts
(Pendents	equal to the braces)
Clewlines	equal to the lifts
Bowlines	equal to the lifts (two thirds the circumference of the fore bowline)
(Bridles	equal to the bowlines)
Topmast stay	in large ships one third the topmast diameter, and in small ¾in larger (nine twenty-fifths of the topmast diameter as circumference)
(Lanyard	as noted above)
Buntlines	in large ships one eleventh the diameter, in small ships one eleventh plus ¼in (equal to the bowlines)
Reef tackle	equal to the buntlines (tye equal to the clewlines)
Falls	equal to the buntlines (three fifths the circumference of the tye)
Leechlines	equal to the buntlines
(Slings	equal to the futtock shrouds)
(Parrel rope	one quarter of the topmast diameter)
(Racking	one quarter of the parrel rope circumference)
(Horses	equal to the parrel rope)
(Staysail stay	seven ninths the circumference of the stay)
(Cringles	equal to the lanyards)
(Lanyards	two sevenths the circumference of the stay)
(Halliards	equal to the reef tackle fall)
(Sheet	equal to the halliard)
(Tack	one third larger in circumference than the sheet)
(Reeflines	half the circumference of the staysail sheet)
(Robands & earings	one third larger in circumference than the reeflines)

Fore topgallant mast

Shrouds & tye	three eighths the diameter of the mast, plus ¼in for small ships (two fifths of the diameter)
(Lanyards	as noted above)
(Futtock shrouds	equal to the topgallant shrouds)
Halliards	one quarter the diameter of the mast in large ships, plus ¼in in small (three quarters the circumference of the tye)
Backstays	¼in smaller in circumference than the shrouds
Stay	three sevenths the diameter of the shrouds, but ¼in less for large ships (three quarters the circumference of the shrouds)
Lifts	equal to the halliards (half the circumference of the shrouds)
Braces	equal to the halliards (equal to the lifts)
(Pendents	equal to the braces)
Clewlines	equal to the halliards
Bowlines	¼in smaller in circumference than the braces (equal to the braces)
(Bridles	equal to the bowline)
(Parrel rope	equal to the brace pendents)

Mainmast

The same rules as applied for the foremast, fore topmast and fore topgallant mast served also for the mainmast, main topmast, and the main topgallant mast (Davis).

The following information on these masts was given by Sutherland.

(Tackle pendents	six twenty-seconds of the mast diameter at the partners as circumference
Runners	twelve thirteenths the circumference of the pendents
Falls	seven thirteenths the circumference of the pendents
Garnet pendent	five sixths the circumference of the tackle runner
Guy	four fifths the circumference of the pendent
Fall	equal to the main tackle fall
Shrouds	equal to the main tackle pendents
Lanyards	as noted above
Catharpin leg & falls	twice the worming
Stay	half the diameter of the mast
Lanyard	five twenty-fourths the circumference of the stay
Lashing to the foremast	half the circumference of the lanyard
Worming	one sixth the circumference of the stay
Collar	three quarters the circumference of the stay
Woolding	one ninth the diameter of the mast
Crowsfeet & tackle	equal to the worming
Jeers	eleven thirteenths the circumference of the shrouds
Boat tackle	five sevenths the circumference of the main tackle
Lifts	equal to the main tackle falls
Strop for the cap	equal to the shroud lanyards
Braces	six sevenths the circumference of the lifts
Pendents	equal to the braces
Parrel rope	one sixth the diameter of the mast
Naveline	three quarters the circumference of the catharpins
Racking & seizing	half the circumference of the parrel rope
Horses	nine elevenths the circumference of the jeers
Lanyard	half the circumference of the horse
Puddening of the yard	equal to the jeers
Sheet, cable laid	equal to the main tackle runner
Stoppers	five sixths the circumference of the sheets
Lanyards	half the circumference of the naveline
Tack, tapered	equal to the shrouds
Luff tackles	six thirteenths the circumference of the tack
Bowlines	equal to the lifts
Bridles	six sevenths the circumference of the bowlines
Tackle	five sixths the circumference of the bridles
Clewgarnets	five sixths the circumference of the braces
Buntlines	two fifths the circumference of the clewgarnets
Leechlines	equal to the buntlines
Staysail stay	one third the circumference of the main stay
Lanyard	as noted above
Cringles	five eighths the circumference of the stay
Halliards	half the circumference of the stay
Sheet	five eighths the circumference of the stay
Tack	three quarters the circumference of the stay
Studdingsail halliards	equal to the staysail tack
Sheet	five sixths the circumference of the halliard
Tack	equal to the halliard
Reeflines	three quarters the circumference of the leechlines
Robands & earings	equal to the reeflines)

Main topmast	
(Shrouds	nine twenty-ninths of the topmast's diameter at the cap
Lanyards	as noted above
Ratlines	two ninths the circumference of the shrouds
Top rope pendent	six twenty-ninths the diameter of the topmast
Fall	two thirds the circumference of the pendent
Burton pendent	seven ninths the circumference of the shrouds
Fall	four sevenths the circumference of the pendent
Futtock shrouds	eight ninths the circumference of the topmast shrouds
Standing backstays	equal to the shrouds
Lanyards	as noted above
Stay, cable laid	equal to the shrouds
Runners	half the diameter of the topmast
Halliards	three fifths the circumference of the runners
Lifts	half the circumference of the runners
Beckets at the cap	equal to the lifts
Braces	equal to the lifts
Pendents	equal to the braces
Beckets about the mizzen mast	equal to the halliards
Slings	equal to the mizzen beckets
Parrel rope	one quarter of the topmast diameter at the cap
Racking	one quarter the circumference of the parrel rope
Horses	equal to the parrel rope
Sheets	half the diameter of the topsail yard
Span	half the circumference of the sheets
Lashing	three quarters the circumference of the ratlines
Bowline	equal to the halliards
Bridles	equal to the bowlines
Clewlines	equal to the bowlines
Buntlines	two thirds the circumference of the clewline
Leechlines	equal to the buntlines
Reef tackle tye	equal to the braces
Fall	three fifths the circumference of the tye
Staysail stay	one ninth the circumference of the stay
Lanyards	as noted above
Cringles	four sevenths the circumference of the stay
Halliards	equal to the reef tackle fall
Sheet	equal to the halliard
Tack	equal to the leechlines
Studdingsail halliards	equal to the staysail stay
Sheet	four fifths the circumference of the halliards
Tack	equal to the sheets
Reeflines	half the circumference of the leechlines
Robands & earings	equal to the reeflines)

Main topgallant mast	
(Shrouds five ninths the circumference of the topmast shrouds	
Lanyards	as noted above
Futtock shrouds	equal to the topgallant shrouds
Stay	three quarters the circumference of the shrouds
Tye	equal to the shrouds
Halliards	three fifths the circumference of the tye
Lifts	equal to the halliards
Braces	two thirds the circumference of the lifts
Pendents	equal to the braces
Parrel rope	equal to the pendents
Bowline	equal to the lifts
Bridles	equal to the bowline
Clewlines	equal to the bowline)

Mizzen mast	
Shrouds	for small ships, one third the diameter of the mizzen mast; for large ships, ¼in less (nine twenty-ninths of that diameter at the partners)
(Lanyards and ratlines	as noted above)
Tackle pendents	equal to the shrouds (seven ninths the circumference of the shrouds)
(Fall	five sevenths the circumference of the pendents)
Stay	one third the diameter of the mast, plus ¼in for small ships (equal to the shrouds)
(Lanyard	as noted above)
(Collar	eight ninths the circumference of the stay)
(Crowsfeet & tackle	one eighth the circumference of the shrouds)
Jeers	for double and treble block one quarter of the diameter, for double and single block plus ¼in, and for two single blocks plus ½in (eight ninths the circumference of the shrouds)
(Lashing	five eighths the circumference of the jeers)
(Parrel rope	equal to the lashing)
Truss tackle	one ninth the diameter (like the parrel rope)
Fall	equal to the tackle
(Slings	equal to the shrouds)
Sheet	for double and single block one fifth the diameter of the mast, and for two single blocks plus ½in (seven ninths the circumference of the shrouds)
(Tack	five sevenths the circumference of the sheet)
Bowline, single	equal to the truss tackle
Bowlines	one fifth the diameter (equal to the tack)
Burton falls	equal to the truss tackle
Brails below the mast	equal to the truss tackle (four fifths the circumference of the bowline)
Double peak brails & legs	equal to the bowline (four fifths the circumference of the bowline for all brails)
(Lacing for the mizzen	half the circumference of the brails)
(Staysail halliards	equal to the brails)
(Sheets	equal to the mizzen tack)
(Tack	four fifths the circumference of the sheets)

Crossjack yard	
(Standing lifts	equal to the mizzen bowline)
(Lanyards	three fifths the circumference of the lifts)
Braces	one fifth of the topmast diameter (three fifths the circumference of the lifts)
(Pendents	four fifths the circumference of the lifts)
(Slings	equal to the mizzen sheet)

Mizzen topmast	
Topmast shrouds	two fifths the diameter of the topmast (equal to the topgallant shrouds)
(Lanyards	as noted above)
(Futtock shrouds	equal to the topgallant shrouds)
Backstays	¼in less than the shrouds
Sheets & topmast stay	one eighth the diameter of the topmast (equal to the shrouds)
(Lanyards	as noted above)
(Tye	equal to the stay)
Halliards	one fifth the diameter (four fifths the circumference of the tye)
Lifts	equal to the halliards (three fifths the circumference of the shrouds)
Braces	equal to the halliards (equal to the lifts)
(Pendents	equal to the braces)
(Parrel rope	equal to the braces)
Clewlines	equal to the halliards (equal to the braces)
Double & topping lifts for the staysail & pendent halliards	equal to the halliards
Bowlines	¼in less than the halliards (two fifths the circumference of the sheets)
(Bridles	equal to the bowlines)

Necessary ropes	
Viol	two thirds the diameter of the bowsprit (cable laid, equal to the fore stay)
(Strop	seven tenths the circumference of the viol)
(Lashings	half the circumference of the strop)
Winding tackle pendent	one third the diameter of the bowsprit (four fifths

	the circumference of the viol)
Fall	½in greater in diameter than the cat rope (half the circumference of the viol)
Fish tackle pendent	two sevenths the diameter of the bowsprit (thirteen sixteenths the circumference of the winding tackle pendent)
Fall	one seventh the diameter of the bowsprit (seven thirteenths the circumference of the pendent)
Cat ropes	equal to the fall (nine thirteenths the circumference of the fish tackle pendent)
(Lanyards	as noted above)
(Stoppers for anchors	eleven thirteenths the circumference of the fish tackle pendent)
(Shank painters	equal to the anchor stoppers)
Stoppers at the bits	½in greater in diameter than the fish tackle pendent (four fifths the circumference of the viol)
(Seizing	one twelfth the circumference of the stoppers)
(Lanyards	three eighths the circumference of the stoppers)
Buoy ropes for merchantmen	one quarter the diameter plus ¾in (cable laid, equal to the fish tackle pendent)
(Buoy slings	seven thirteenths the circumference of the buoy rope)
(Gun slings	seven eighths the circumference of the winding tackles pendents)
(Butt slings	six sevenths the circumference of the gun slings)
(Hogshead slings	two thirds the circumference of the butt slings)
(Nut slings of the guns	three quarters the circumference of the hogsheads)
(Horses in the head	seven elevenths the circumference of the gammoning)
(Lanyards	as noted above)
Fore stay collar	1 foot in length per inch of bowsprit diameter, and one third of the bowsprit's diameter in diameter
Main stay collar	at the stem for large ships, 1 fathom in length per 4 inches of bowsprit diameter, and two fifths of the diameter of the bowsprit in diameter; for small ships the same diameter plus ½in
(Poop ladders	equal to the mizzen jeers)
(Middle rope	one third the circumference of the outer)
(Lashing	two thirds the circumference of the middle rope)
(Futtock staves	equal to the mizzen shrouds)
(Cable bends	five elevenths the circumference of the shank painters)
(Enter ropes	3 inches)
(Port ropes	one twelfth of the port width in circumference)
(Puddening of anchors	equal to cable bends)
(Seizing	one third the circumference of the puddenings)

(Long boat)	
(Main stay	equal to the mizzen topmast shrouds
Burton pendent	one fifth larger than the stay
Fall	two thirds the circumference of the tye
Fore sheets	half the circumference of the Burton fall
Halliards	equal to the fore sheets
Main sheet	equal to the Burton fall
Tack	equal to the main sheet
Boat rope, cable laid	equal to the buoy rope
Guest rope	seven thirteenths the circumference of the boat rope
Painter	half the circumference of the boat rope
Yard rope	half the circumference of the halliards)

(Pinnace)	
(Main sheet	similar to the long boat's fore sheets
Fore sheets	seven eighths the circumference of the main sheets
Boat rope, cable laid	eight elevenths of the long boat's
Guest rope	seven eighths of the boat rope
Painter	five sevenths the circumference of the guest rope)

Table 62: Dimensions of standing and running rigging from Steel's *Elements of Mastmaking, Sailmaking & Rigging* (1794)

Tables 62,63 and 64 are reproduced as facsimiles from the 1932 edition of Steel's work. These tables are well enough known and approachable enough in style and spelling to warrant their inclusion in this form, and not now so readily available as to permit their omision altogether. Table 62 deals with ships of the Royal Navy, Table 63 with smaller naval vessels, and Table 64 with merchant vessels.

Names of the Standing and Running Rigging.	Ships from 110 to 74 Guns, or from 3000 to 1800 Tons.				Ships of 64 Guns, or from 1700 to 1300 Tons.				Ships from 50 to 36 Guns, or from 1200 to 800 Tons.				Ship from 32 to 28 Guns, or from 700 to 600 Tons.				Ships of 24 Guns, or from 550 to 450 Tons.				Ships from 22 to 20 Guns, or from 400 to 350 Tons.				Ships from 18 to 14 Guns, or from 300 to 250 Tons.			
		Blocks, &c.				Blocks, &c.				Blocks, &c.				Blocks, &c.				Blocks, &c.				Blocks, &c.				Blocks, &c.		
	Inches.	Species.	Inches.	Number.	Inches.	Species.	Inches.	Number.	Inches.	Species.	Inches.	Nuuber.	Inches.	Species.	Inches.	Number.	Inches.	Species.	Inches.	Number.	Inches.	Species.	Inches.	Number.	Inches.	Species.	Inches.	Number.
Bowsprit.																												
Woolding	3½	—	—	—	3	—	—	—	3	—	—	—	2½	—	—	—	2½	—	—	—	2½	—	—	—	2	—	—	—
Gammoning	8	—	—	—	7	—	—	—	6	—	—	—	5½	—	—	—	5½	—	—	—	5	—	—	—	4½	—	—	—
Shrouds	8	*H.	14	4	7½	*H.	13	4	6	*H.	9	2	6	*H.	9	2	5½	*H.	8	2	5	*H.	7	2	4½	*H.	6	2
Collars	6½	H.	14	4	6	H.	13	4	5	H.	9	2	5	H.	9	2	4½	H.	8	2	4	H.	7	2	3½	H.	6	2
Seizing	1	—	—	—	1	—	—	—	1	—	—	—	1	—	—	—	¾	—	—	—	¾	—	—	—	¾	—	—	—
Lashing	2	—	—	—	2	—	—	—	1½	—	—	—	1½	—	—	—	1½	—	—	—	1	—	—	—	1	—	—	—
Laniard	3½	—	—	—	3	—	—	—	2½	—	—	—	2½	—	—	—	2½	—	—	—	2½	—	—	—	2	—	—	—
Bobstays cabled	8	H.	14	3	8	H.	13	3	7	H.	9	2	7	H.	9	2	6½	H.	8	2	6	H.	7	2	5	H.	6	2
Collars	8½	H.	14	3	8	H.	13	3	7	H.	9	2	7	H.	9	2	6½	H.	8	2	6	H.	7	2	5	H.	6	2
Seizing	1½	—	—	—	1½	—	—	—	1	—	—	—	1	—	—	—	¾	—	—	—	¾	—	—	—	¾	—	—	—
Lashing	2	—	—	—	2	—	—	—	1½	—	—	—	1½	—	—	—	1½	—	—	—	1	—	—	—	1	—	—	—
Laniards	4	—	—	—	3½	—	—	—	3	—	—	—	3	—	—	—	3	—	—	—	3	—	—	—	2½	—	—	—
Horses	5	T.	—	6	4	T.	—	6	3½	T.	—	6	3	T.	—	6	3	T.	—	6	3	T.	—	6	3	T.	—	6
Straps	3½	—	—	4	3	—	—	4	3	—	—	4	2½	—	—	4	2½	—	—	4	2½	—	—	4	2	—	—	4
Laniards	2	—	—	—	2	—	—	—	1½	—	—	—	1½	—	—	—	1½	—	—	—	1½	—	—	—	1	—	—	—
Spritsail-yard.																												
Horses	4	—	—	—	3½	—	—	—	3	—	—	—	3	—	—	—	3	—	—	—	3	—	—	—	2½	—	—	—
Stirrups	3	T.	—	6	3	T.	—	6	2½	T.	—	6	2½	T.	—	6	2½	T.	—	6	2½	T.	—	6	2	T.	—	4
Braces	3½	D.	12	4	3	D.	11	4	2½	D.	9	4	2½	D.	9	4	2½	D.	9	4	2	D.	9	4	2½	D.	9	4
Pendents	4	S.	12	4	3½	S.	11	4	3	S.	9	4	3	S.	9	4	3	S.	9	4	3	S.	9	4	—	S.	9	4
Strapping	3½	—	—	—	3½	—	—	—	3	—	—	—	3	—	—	—	3	—	—	—	3	—	—	—	3	—	—	—
Lifts	3½	S.	12	4	3	S.	11	4	2½	S.	9	4	2½	S.	9	4	2½	S.	9	4	2½	S.	9	4	2	S.	7	4
Beckets	3½	—	—	—	3	—	—	—	2½	—	—	—	2½	—	—	—	2½	—	—	—	2½	—	—	—	2	—	—	—
Strapping	4	—	—	—	3½	—	—	—	2½	—	—	—	2½	—	—	—	2½	—	—	—	2½	—	—	—	2	—	—	—
Seizing	¾	—	—	—	¾	—	—	—	¾	—	—	—	—	—	—	—	—	—	—	—	—	—	—	—	—	—	—	—
Standing	4½	T.	—	4	4	T.	—	4	3	T.	—	4	3	T.	—	4	3	T.	—	4	3	T.	—	4	2½	T.	—	4
Straps	4½	—	—	—	4	—	—	—	3	—	—	—	3	—	—	—	3	—	—	—	2½	—	—	—	2	—	—	—
Laniards	2	—	—	—	1½	—	—	—	1½	—	—	—	1½	—	—	—	1½	—	—	—	1	—	—	—	¾	—	—	—
Haliard	3½	*L.t. S.	24 12	1 1	3	*L.t. S.	22 11	1 1	2½	*L.t. S.	18 9	1 1	2½	*L.t. S.	18 9	1 1	2½	*L.t. S.	18 9	1 1	2½	*L.t. S.	18 9	1 1	2	S.	7	2
Strapping	4½	—	—	—	4	—	—	—	3	—	—	—	3	—	—	—	3	—	—	—	3	—	—	—	2½	—	—	—
Seizing and Lashing	¾	—	—	—	¾	—	—	—	¾	—	—	—	—	—	—	—	—	—	—	—	—	—	—	—	—	—	—	—
Slings	6½	—	—	—	5	—	—	—	4½	—	—	—	4½	—	—	—	4	—	—	—	4	—	—	—	3½	—	—	—
Seizing and Racking	1½	—	—	—	1½	—	—	—	¾	—	—	—	¾	—	—	—	¾	—	—	—	¾	—	—	—	¾	—	—	—
Clue Lines	2½	St. bd. S.	10 10	4 2	2½	St. bd. S.	9 9	4 2	2	St. bd. S.	8 8	4 2	2	St. bd. S.	8 8	4 2	2	St. bd. S.	8 8	4 2	2	St. bd. S.	8 8	4 2	1½	St. bd.	6	2
Strapping	3	—	—	—	3	—	—	—	2½	—	—	—	2½	—	—	—	2½	—	—	—	2	—	—	—	1½	—	—	—
Bunt-lines	2	S.	8	2	2	S.	8	2	2	S.	8	2	2	S.	8	2	1½	S.	6	2	1½	S.	6	2	1	—	—	—
Strapping	2½	—	—	—	2½	—	—	—	2	—	—	—	2	—	—	—	1½	—	—	—	1½	—	—	—	1½	—	—	—
Ear-rings	1½	—	—	—	1½	—	—	—	1	—	—	—	1	—	—	—	1	—	—	—	1	—	—	—	1	—	—	—
Sheets cabled	5	—	—	—	4½	—	—	—	3½	—	—	—	3½	—	—	—	3½	—	—	—	3½	—	—	—	3	—	—	—
Jib.																												
Horses	4	—	—	—	3½	—	—	—	3½	—	—	—	3½	—	—	—	3	—	—	—	3	—	—	—	3	—	—	—
Seizings	—	—	—	—	—	—	—	—	—	—	—	—	—	—	—	—	—	—	—	—	—	—	—	—	—	—	—	—
Guy-pendents	4½	S.	13	2	4	S.	13	2	3½	S.	11	2	3½	S.	11	2	3½	S.	11	2	3	S.	11	2	2½	S.	10	2
Falls	3	S.	11	4	2½	S.	9	4	2½	S.	9	4	2½	S.	9	4	2½	S.	9	4	2½	S.	9	4	1½	S.	8	4
Strapping	4 3	—	—	—	4 2½	—	—	—	3½ 2½	—	—	—	3½ 2½	—	—	—	3½ 2½	—	—	—	3	—	—	—	2½	—	—	—
Lashers	1	—	—	—	1	—	—	—	1	—	—	—	1	—	—	—	¾	—	—	—	¾	—	—	—	¾	—	—	—
Out Hauler	4	—	—	—	3½	—	—	—	3	—	—	—	3	—	—	—	2½	—	—	—	2½	—	—	—	2	—	—	—
Tackle-Fall	3½	S.	9	2	2	S.	8	2	2	S.	7	2	2	S.	7	2	2	S.	7	2	2	S.	7	2	—	—	—	—
Strapping	2½	—	—	—	2	—	—	—	2	—	—	—	2	—	—	—	2	—	—	—	2	—	—	—	—	—	—	—
Stay	4½	S.	14	1	4	S.	13	1	3½	S.	11	1	3½	S.	11	1	3	S.	10	1	3	S.	10	1	3	S.	9	1
Strapping	4½	—	—	—	4	—	—	—	3½	—	—	—	3½	—	—	—	3	—	—	—	3	—	—	—	3	—	—	—
Tackle-Fall	2½	D. S.	9 9	1 1	2½	D. S.	9 9	1 1	2	D. S.	7 7	1 1	2	D. S.	7 7	1 1	2	S.	7	2	2	S.	7	2	1½	S.	6	2
Strapping	2½	—	—	—	2½	—	—	—	2	—	—	—	2	—	—	—	2	—	—	—	2	—	—	—	1½	—	—	—
Haliard	4	S.	13	1	3½	S.	12	1	3	S.	10	1	3	S.	10	1	2½	S.	9	1	2½	S.	9	1	2½	S.	9	1
Strapping	4	—	—	—	3½	—	—	—	3	—	—	—	3	—	—	—	2½	—	—	—	2½	—	—	—	2½	—	—	—
Down-Hauler	2½	S.	9	1	2½	S.	9	1	2	S.	7	1	2	S.	7	1	1½	S.	6	1	1½	S.	6	1	1½	S.	6	1
Sheets, single	3½	S.	11	2	3	S.	11	2	3½	S.	10	2	3½	S.	10	2	3	S.	10	2	2½	S.	9	2	2½	S.	9	2
Pendents	4½	S.	11	2	4	S.	11	2	4	S.	11	2	—	—	—	—	—	—	—	—	—	—	—	—	—	—	—	—
Flying-Jib	—	—	—	—	—	—	—	—	—	—	—	—	—	—	—	—	—	—	—	—	—	—	—	—	—	—	—	—
Haliard	2½	S.	8	1	2½	S.	7	1	2½	S.	7	1	2	S.	7	1	2	S.	7	1	2	S.	7	1	2	S.	7	1
Sheets	2½	—	—	—	2½	—	—	—	2½	—	—	—	2	—	—	—	2	—	—	—	2	—	—	—	2	—	—	—
Tack	2	—	—	—	2	—	—	—	2	—	—	—	1½	—	—	—	1	—	—	—	1	—	—	—	1	—	—	—
Down-Hauler	1½	S.	5	1	1½	S.	5	1	1	S.	5	1	1	S.	5	1	1	S.	5	1	1	S.	5	1	1	S.	5	1
Spritsail Topsail and Yard.																												
Horses	3	—	—	—	2½	—	—	—	2	—	—	—	2	—	—	—	2	—	—	—	2	—	—	—	2	—	—	—
Braces	2½	S.	9	2	2½	S.	9	2	2	S.	7	2	2	S.	7	2	2	S.	7	2	2	S.	7	2	1½	—	—	—
Strapping	2½	—	—	—	2½	—	—	—	2	—	—	—	2	—	—	—	2	—	—	—	2	—	—	—	—	—	—	—
Lifts, single	2½	S.	9	2	2	S.	8	2	2	S.	8	2	2	T.	—	2	1½	T.	—	2	1½	T.	—	2	1½	T.	—	2

Names of the Standing and Running Rigging.	110 to 74 Guns.				64 Guns.				50 to 36 Guns.				32 to 28 Guns.				24 Guns.				22 to 20 Guns.				18 to 14 Guns.			
		Blocks, &c.				Blocks, &c.				Blocks, &c.				Blocks, &c.				Blocks, &c.				Blocks, &c.				Blocks, &c.		
	In.	Species.	In	N	In.	Species.	In	N	In.	Species.	In	N	In.	Species.	In	N	In.	Species.	In	N	In.	Species.	In	N	In.	Species.	In	N
Strapping	2½				2				2				—				—				—				—			
Haliard	2½	S.	9	2	2½	S.	9	2	2	S.	7	2	2	S.	7	2	2	S.	7	2	2	S.	7	2	1½	S.	6	2
Strapping	2½				2½				2				2				2				2				1½			
Lashing	¾				¾				¾				¾				—				—				—			
Parral-Ropes	2½	Par.	12	1	2	Par.	12	1	2	Par.	10	1	2	Par.	10	1	1½	Par.	8	1	1½	Par.	8	1	1½	Par.	8	1
Clue-lines	2	S.	7	4	2	S.	7	4	1½	S.	6	4	1½	S.	6	4	1½	S.	6	4	1½	S.	6	2	1½	S.	6	2
Strapping	2				2				1½				1½				1½				1½				1½			
Lacing and Ear-rings	1				1				¾				¾				—				—				—			
Foremast.																												
Woolding	3½				3				2½				2½				2½				2½				2½			
Girtlines	5	S.	16	2	4	S.	15	2	3½	S.	13	2	3	S.	12	2	3	S.	11	2	3	S.	10	2	2½	S.	9	2
Strapping	5				4				3½				3½				3				3				2½			
Seizing	¾				¾				¾				¾				—				—				—			
Lashing	2½				2				1½				1½				1½				1				¾			
Pendts. of Tackles cabled	11	*S.c.	24	2	10½	*S.c.	22	2	8½	*S.c.	17	2	8	*S.c.	17	2	7½	*S.c.	15	2	7½	*S.c.	15	2	7	*S.c.	15	2
Strapping	7½	T.	—	8	6½	T.	—	8	5½	T.	—	8	5½	T.	—	4	5	T.	—	4	5	T.	—	4	4½	T.	—	4
Seizing	1½				1½				1				1				¾				¾				¾			
Runners of Tackles	7½	D.th.c.	21	4	6½	D.th.c.	21	4	5½	D.th.c.	17	4	5½	D.th.c.	16	2	5				4½				4			
Strapping	6				5½				5				5				4				4				3½			
Falls of Tackles	4	*S.th.c.	26	4	4	S.th.c.	24	4	3	*S.th.c.	18	4	3	*S.th.c.	17	2	3	*S.th.c.	15 11	2 2	3	*S.th.c.	15 11	2 2	2½	*S.th.c.	13 9	2 2
Strapping	5½	T.	—	2	5½	T.	—	2	5	T.	—	2	5	T.			4	T.			4	T.			3½	T.		
Seizing	1				1				¾				¾				¾				¾				¾			
Shrouds	11	D.E.	17	20	10½	D.E.	16	18	8½	D.E.	13	14	8	D.E.	13	14	7½	D.E.	11	14	7½	D.E.	11	12	7	D.E.	12	10
Seizings { Eye	1½				1½				1				1				1				1				1			
Seizings { Throat	1½				1½				1½				1½				1				1				1			
Seizings { End	1½				1½				1				1				¾				¾				—			
Laniard	5½				5				4				4				4				4				3½			
Ratling	1½				1½				1½				1½				1½				1½				1½			
Stay cabled 4 Strands	18	H.	26	1	16	H.	22	1	13	H.	17	1	12	H.	16	1	11	H.	14	1	11	H.	14	1	9½	H.	13	1
Seizings	2				1½				1½				1½				1½				1½				1			
Laniard	6				5				4½				4				4				4				3½			
Collar cabled 4 Strands double	9½	H.	26	1	8½	H.	22	1	6½	H.	17	1	6	H.	16	1	5½	H.	14	1	5½	H.	14	1	5	H.	13	1
Seizings	1½				1½				1½				1½				1½				1½				1			
Lashing	2½				2½				2				2				1½				1½				1½			
Preventer-Stay cabled 4 Strands	11½	H.	16	1	10½	H.	15	1	8½	H.	12	1	7½	H.	12	1	7	H.	11	1	7	H.	11	1	6	H.	10	1
Laniard	4½				4				3½				3½				3				3				2½			
Collar cabled 4 Strands double	6½	H.	16	1	5½	H.	15	1	4½	H.	12	1	4½	H.	12	1	4	H.	11	1	4	H.	11	1	4	H.	10	1
Lashing	2½				2				1½				1½				1				1				1			
Seizing	1½				1½				1½				1½				1				1				1			
Catharpin-Legs	7				6½				5				5				4½				4				3½			
Seizing	1½				1½				1				1				1				¾				¾			
Fore-yard.																												
Jears † { Tye Falls	7½	Tr.c. D.c.	26 26	2 2	6½	Tr.c. D.c.	24 24	2 2	5½	D.c	18	4	5	D.c.	18	4	7½ 3	S.do.sc. D.c. S.c.	20 12 9	3 4 2	7½ 3	S.do.sc. D.c. S.c.	20 12 9	3 4 2	7 2½	S.do.sc D.c. S.c.	20 12 9	3 4 2
Strapping	8½				8				6½				6				5½ 4 3				5½ 4 3				5 4 2½			
Seizings	2				2				1				1				1				1				1			
Lashing at the { Mast Head	4½				4				3½				3				3				3				3			
Lashing at the { Yard	3½				3				2				2				2				2				2			
Stoppers	6				5½				4½				—				—				—				—			
Horses	5½				5				4½				4				4				4				3½			
Stirrups	4	T.	—	12	4	T.	—	10	3	T.	—	10	3	T.	—	8	3	T.	—	8	3	T.	—	8	2½	T.	—	8
Seizings	¾				¾				—				—				—				—				—			
Laniard	2				2				2				2				1½				1½				1			
Yard-Tackle Pendts.	7	D.th.c.	17	2	6½	D.th.c.	16	2	5½	D.th.c.	14	2	5½	D.th.c.	14	2	5	D.th.c.	12	2	5	D.th.c.	12	2	4½	D.th.c.	11	2
Falls	3½	*D. S.	20 13	2 2	3½	*D. S.	18 13	2 2	3	*D. S.	16 11	2 2	3	*D. S.	16 11	2 2	3	*D. S.	13 11	2 2	3	*D. S.	14 10	2 2	2½	*D. S.	13 9	2 2
Strapping	5½				5				4				4				3½				3½				3			
Seizing	1				1				¾				¾				¾				¾				—			
Inner Tricing-Lines	2½	S.	8	2	2	S.	7	2	2	S.	7	2	2	S.	7	2	1½	S.	6	2	1½	S.	6	2	1½	S.	6	2
Outer Tricing-Lines	2	S.	7	4	2	S.	7	4	1½	S.	6	4	1½	S.	6	4	1½	S.	6	4	1½	S.	6	4	1½	S.	6	2
Strapping	2				2				2				2				1½				1½				1½			
Braces	4½	S.c.	16	4	4	S.c.	15	4	3½	S.c.	12	4	3½	S.c.	12	4	3	S.c.	10	4	3	S.c.	10	4	2½	S.c.	9	4
Pendents	5½	S.c.	16	2	5	S.c.	15	2	4½	S.c.	12	2	4½	S.c.	12	2	4	S.c.	10	2	4	S.c.	10	2	3½	S.c.	9	2
Preventer	4½				4				3½				3½				3				3				2½			
Strapping	5				4½				4				4				3				3				2½			
Seizing	¾				¾				¾				—				—				—				—			
Lashing	1½				1½				1½				1½				¾				¾				¾			
Preventers (in war only)	3½	S.	13	4	3½	S.	13	4	3	S.	11	4	3	S.	11	4	2½	S.	9	4	2½	S.	9	4	2	S.	8	4
Strapping	3½				3½				3				3				2½				2½				2			
Seizing	¾				¾				¾				—				—				—				—			
Lifts	4½	S.	16	6	4	S.	15	6	3½	S	12	6	3½	S.	12	6	3½	S.	11	6	3	S.	10	6	3	S.	9	6
Span for the Cap	6½	Sis.	24	2	5½	Sis.	22	2	4½	Sis.	20	2	4½	Sis.	20	2	4½	Sis.	18	2	4	Sis.	16	2	4	Sis.	14	2
Short Span	4½				4				3½				3½				3½				3				3			
Strapping	5				4½				4				4				3½				3				3			
Seizing	¾				¾				¾				¾				—				—				—			
Jigger-Tackle	2½	D. S.	10 10	2 2	2½	D. S.	9 9	2 2	2	D. S.	8 8	2 2	2	D. S.	8 8	2 2	2	D. S.	8 8	2 2	2	D. S.	8 8	2 2	—			
Strapping	3½				3½				3				3				3				3				—			
Truss-Pendents	8	T.	—	4	7½	T.	—	4	6	T.	—	4	5½	T.	—	4	5	T.	—	4	5	T.	—	4	4½	T.	—	4
Falls	3	*D.	11	4	3	*D.	11	4	2½	*D.	9	4	2½	*D.	9	4	2	*D.	8	4	2	*D.	8	4	2	*D.	8	4

† Ships of 28 Guns have their jears similar to 24 Gunships.

Names of the Standing and Running Rigging.	110 to 74 Guns.				64 Guns.				50 to 36 Guns.				32 to 28 Guns.				24 Guns.				22 to 20 Guns.				18 to 14 Guns.			
		Blocks, &c.				Blocks, &c.				Blocks, &c.				Blocks, &c.				Blocks, &c.				Blocks, &c.				Blocks, &c.		
	In.	Species.	In	N	In.	Species.	In	N	In.	Species.	In	N	In.	Species.	In	N	In.	Species.	In	N	In.	Species.	In	N	In.	Species.	In	N
Strapping	3½	—	—	—	3½	—	—	—	3	—	—	—	3	—	—	—	2½	—	—	—	2½	—	—	—	2½	—	—	—
Eye-Seizings	1½	—	—	—	1½	—	—	—	1	—	—	—	1	—	—	—	¾	—	—	—	¾	—	—	—	¾	—	—	—
Nave-Line	2	S.	7	1	2	S.	7	1	1½	S.	6	1	1½	S.	6	1	1½	S.	6	1	1½	S.	6	1	—	S.	—	—
Puddening the Yard	6½	—	—	—	6	—	—	—	6	—	—	—	6	—	—	—	5	—	—	—	5	—	—	—	5	—	—	—
Clue-Garnets	4	S.St.bd.	15	4	3½	S.St.bd.	13	4	3	S.St.bd.	11	4	3	S.St.bd.	11	4	2½	S.St.bd.	9	4	2½	S.St. bd.	9	4	2½	S.St. bd.	9	4
		S.	15	2		S.	13	2		S.	11	2		S.	11	2		S.	9	2		S.	9	2		S.	9	2
Straps ab. the Yard	4	—	—	—	3½	—	—	—	3	—	—	—	3	—	—	—	2½	—	—	—	2½	—	—	—	2½	—	—	—
Strapping	4	—	—	—	3½	—	—	—	3	—	—	—	3	—	—	—	2½	—	—	—	2½	—	—	—	2½	—	—	—
Seizing	¾	—	—	—	¾	—	—	—	¾	—	—	—		—	—	—		—	—	—		—	—	—	—	—	—	—
Lashing	1	—	—	—	1	—	—	—	1	—	—	—	1	—	—	—	¾	—	—	—	¾	—	—	—	—	—	—	—
Buntline-Legs	3	D.	11	4	3	D.	10	4	2½	D.	8	4	2½	D.	8	4	2	D.	8	4	2	D.	8	4	2	D.	8	4
Falls	3	S.	10	8	3	S.	10	8	2½	S.	8	8	2½	S.	8	8	2	S.	8	8	2	S.	8	8	2	S.	8	8
Strapping	3½	—	—	—	3½	—	—	—	2½	—	—	—	2½	—	—	—	2½	—	—	—	2½	—	—	—	2½	—	—	—
Leechline-Legs	2½	D.	11	4	2½	D.	10	4	2	D.	7	4	2	D.	7	4	2	D.	7	4	2	D.	7	4	2	D.	7	4
Falls	2½	S.	10	8	2½	S.	10	8	2	S.	7	8	2	S.	7	8	2	S.	7	8	2	S.	7	4	2	S.	7	4
Strapping	3½	—	—	—	3	—	—	—	2½	—	—	—	2½	—	—	—	2½	—	—	—	2½	—	—	—	2½	—	—	—
Slablines	2½	S.	9	2	2	S.	8	2	2	S.	7	2	2	S.	7	2	1½	S.	6	2	1½	S.	6	2	1½	S.	6	2
Strapping	2½	—	—	—	2	—	—	—	2	—	—	—	2	—	—	—	1½	—	—	—	1½	—	—	—	1½	—	—	—
Bowlines	4½	S.	16	4	4	S.	15	4	3½	S.	13	4	3½	S.	13	4	3	S.	12	2	3	S.	12	2	3	S.	11	2
Bridles	4½	T.	—	2	4	T.	—	2	3½	T.	—	2	3½	T.	—	2	3	T.	—	2	3	T.	—	2	3	T.	—	2
Strapping	4½	—	—	—	4	—	—	—	3½	—	—	—	3½	—	—	—	3	—	—	—	3	—	—	—	3	—	—	—
Seizing	¾	—	—	—	¾	—	—	—	¾	—	—	—	¾	—	—	—	¾	—	—	—	¾	—	—	—	—	—	—	—
Lashing	2½	—	—	—	2	—	—	—	2	—	—	—	2	—	—	—	1½	—	—	—	1½	—	—	—	1	—	—	—
Ear-rings	2	—	—	—	2	—	—	—	1½	—	—	—	1½	—	—	—	1½	—	—	—	1½	—	—	—	1½	—	—	—
Sheets cabled	7	S.c.	24	2	6	S.c.	22	2	5	S.c.	17	2	5	S.c.	17	2	5	S.c.	16	2	4½	S.c.	15	2	4	S.c.	14	2
Strapping	7½	T.	—	2	6	T.	—	2	5	T.	—	2	5	T.	—	2	5	T.	—	2	4½	T.	—	2	4	T.	—	2
Seizing	1	—	—	—	1	—	—	—	1	—	—	—	1	—	—	—	¾	—	—	—	¾	—	—	—	¾	—	—	—
Stoppers	5½	—	—	—	5	—	—	—	4	—	—	—	4	—	—	—	4	—	—	—	3½	—	—	—	3	—	—	—
Tacks taper and cabled	9½	Sho.	26	2	8	Sho.	20	2	6½	Sho.	16	2	6½	Sho.	16	2	5½	Sho.	14	2	5½	Sho.	14	2	5	Sho.	14	2
Strapping	6½	—	—	—	6	—	—	—	5	—	—	—	5	—	—	—	4½	—	—	—	4½	—	—	—	4	—	—	—
Seizing	1½	—	—	—	1½	—	—	—	1	—	—	—	1	—	—	—	¾	—	—	—	¾	—	—	—	¾	—	—	—
Stoppers	6	H. & T.	2	2	5½	H. & T.	2	2	5	H. & T.	2	2	5	H. & T.	2	2	4	H. & T.	2	2	4	H. & T.	2	2	—	—	—	—
Laniards	2	—	—	—	2	—	—	—	1½	—	—	—	1½	—	—	—	1½	—	—	—	1½	—	—	—	—	—	—	—
Gammoning Bumkin	4½	H. & T.	2	6	4	H. & T.	2	6	3½	H. & T.	2	6	3½	H. & T.	2	6	3	H. & T.	2	6	3	H. & T.	2	6	2½	H. & T.	2	6
Laniards	2	—	—	—	2	—	—	—	1½	—	—	—	1½	—	—	—	1½	—	—	—	1	—	—	—	1	—	—	—
Lnrds. for Pud. & Dol.	1½	T.	—	4	1½	T.	—	4	1	T.	—	4	1	T.	—	4	1	T.	—	4	1	T.	—	4	¾	T	—	4
Slings	12	T.	—	1	11	T.	—	—	9	T.	—	—	8½	T	—	—	—	T	—	—	—	T.	—	—	—	T	—	—
Strap	12	T.	—	1	11	T.	—	1	9	T.	—	1	8½	T.	—	1	—	T.	—	—	—	T	—	—	—	T.	—	—
Seizings	2	—	—	—	1½	—	—	—	1	—	—	—	1	—	—	—	—	—	—	—	—	—	—	—	—	—	—	—
Laniard	3½	—	—	—	3	—	—	—	2½	—	—	—	2½	—	—	—	—	—	—	—	—	—	—	—	—	—	—	—
Staysail Haliard	4	S.	13	2	3½	S.	12	2	3	S.	10	2	3	S.	10	2	2½	S	9	2	2½	S	9	2	2	S	8	2
Sheets	4	S.	13	2	4	S.	12	2	3	S	10	2	3	S.	10	2	2½	S	9	2	2½	S.	9	2	2½	S.	8	2
Tack	3½	—	—	—	3	—	—	—	2	—	—	—	2	—	—	—	2	—	—	—	2	—	—	—	1½	—	—	—
Downhauler	2½	S.	9	1	2	S.	8	1	2	S.	7	1	2	S.	7	1	1½	S.	6	1	1½	S.	6	1	1½	S.	6	1
Strapping	4	—	—	—	4	—	—	—	3	—	—	—	3	—	—	—	2½	—	—	—	2½	—	—	—		—	—	—
	2½	—	—	—	2½	—	—	—	2	—	—	—	2	—	—	—		—	—	—		—	—	—	2	—	—	—
Studdsl. Hal. Inner	3	S.	12	6	3	S.	11	6	2	S.	9	6	2	S.	9	6	2	S.	9	6	2	S.	9	6	2	S.	9	6
Studdsl. Hal. Outer	3½	S.	12	4	3½	S.	11	4	2½	S.	9	4	2½	S.	9	4	2½	S.	9	4	2½	S.	9	4	2½	S.	9	4
Sheets	3	—	—	—	3	—	—	—	2½	—	—	—	2	—	—	—	2	—	—	—	2	—	—	—	2	—	—	—
Tacks	3½	S.	12	2	3½	S.	11	2	3	S.	10	2	2½	S.	9	2	2½	S.	9	2	2½	S.	9	2	2½	S.	9	2
Strapping	3½	—	—	—	3½	—	—	—	3	—	—	—	2½	—	—	—	2½	—	—	—	2½	—	—	—	2½	—	—	—
Fore-Topmast.																												
Burton-Pendents	5½	T.	—	2	5	T.	—	2	4½	T.	—	2	4	T.	—	2	4	T.	—	2	3½	T.	—	2	3	T.	—	2
Falls	2½	*D.	11	2	2½	*D.	10	2	2½	*D.	9	2	2	*D.	9	2	2	*D.	8	2	2	*D.	8	2	1½	*D.	6	2
		*S.	11	2		*S.	10	2		*S.	9	2		*S	9	2		*S.	8	2		*S.	8	2		*S	6	2
Strapping	3½	—	—	—	3½	—	—	—	3	—	—	—	3	—	—	—	2½	—	—	—	2½	—	—	—	2	—	—	—
Shrouds	7	D.E.	11	12	6½	D.E.	10	12	5½	D.E.	8	8	5½	D.E.	8	8	5	D.E.	8	8	4½	—	7	8	4½	—	7	8
Seizings Eye	1	—	—	—	1	—	—	—	¾	—	—	—	¾	—	—	—	¾	—	—	—		Tarr'd Line			—	—	—	—
Seizings Throat	1	—	—	—	1	—	—	—	¾	—	—	—	¾	—	—	—	¾	—	—	—					—	—	—	—
Seizings End	¾	—	—	—	¾	—	—	—	¾	—	—	—	¾	—	—	—	Tarred Line			—	—	—	—	—	—	—	—	—
Laniard	3½	—	—	—	3½	—	—	—	2½	—	—	—	2½	—	—	—	2½	—	—	—	2½	—	—	—	2½	—	—	—
Ratline	1	—	—	—	1	—	—	—	1	—	—	—	1	—	—	—	1	—	—	—	1	—	—	—	1	—	—	—
Standing Backstays	7	D.E.	11	6	6½	D.E.	10	4	5½	D.E.	8	4	5½	D.E.	8	4	5	D.E.	8	4	4½	D.E.	7	4	4½	D.E.	7	2
Seizings Eye	1	—	—	—	1	—	—	—	¾	—	—	—	¾	—	—	—	½				½				½			
Seizings Throat	1	—	—	—	1	—	—	—	¾	—	—	—	¾	—	—	—	¾	Tarr'd Line			½	Tarr'd Line			½	Tarr'd Line		
Seizings End	¾	—	—	—	¾	—	—	—	¾	—	—	—	¾	—	—	—	¾				½				½			
Laniards	3½	—	—	—	3½	—	—	—	2½	—	—	—	2½	—	—	—	2½	—	—	—	2½	—	—	—	2½	—	—	—
Breast Bcksty. Runn.	5	S.	14	2	4	S.	12	2	3	S.	11	2	3½	S.	11	2	3½	S.	11	2	3	S.	10	2	3	S.	9	2
Falls	2½	D.	10	4	2½	D.	10	4	2½	D.	10	4	2½	D.	10	4	2	D.	8	4	2	D.	7	2	2	D.	7	2
																						S.	7	2		S.	7	2
Strapping	3	—	—	—	3	—	—	—	3	—	—	—	3	—	—	—	2½	—	—	—	2½	—	—	—	2½	—	—	—
Stay cabled 4 Strands	8½	—	—	—	7½	—	—	—	6	—	—	—	5½	—	—	—	5½	—	—	—	5	—	—	—	5	—	—	—
Collar	6½	S.	20	1	6	S.	18	1	5	S.	15	1	5	S.	15	1	5	S.	15	1	4½	S.	14	1	4	S.	13	1
Tackle	3½	L.t.	24	1	3	L.t.	20	1	2½	L.t.	18	1	2½	L.t.	18	1	2½	L.t.	18	1	2½	L.t.	16	1	2½	L.t.	15	1
		S.	14	1		S.	12	1		S.	10	1		S.	9	1		S.	9	1		S.	8	1		S.	8	1
Strapping	5	—	—	—	4	—	—	—	3½	—	—	—	3½	—	—	—	3½	—	—	—	3	—	—	—	3	—	—	—
Seizing	1½	—	—	—	1½	—	—	—	1	—	—	—	1	—	—	—		—	—	—		—	—	—		—	—	—
	¾	—	—	—	¾	—	—	—	¾	—	—	—	¾	—	—	—	¾	—	—	—	¾	—	—	—	¾	—	—	—
Preventer-Stay cabled 4 Strands	6½	—	—	—	5½	—	—	—	4½	—	—	—	4	—	—	—	4	—	—	—	4	—	—	—	4	—	—	—
Collar	5	S.	16	1	5	S.	15	1	4	S.	12	1	4	S.	12	1	4	S.	12	1	3½	S.	11	1	3	S.	10	1
Tackle	3	L.t.	18	1	2½	L.t.	18	1	2	L.t.	16	1	2	L.t.	16	1	2	L.t.	16	1	2	L.t.	15	1	2	L.t.	14	1
		S.	12	1		S.	10	1		S.	9	1		S.	8	1		S.	8	1		S.	7	1		S.	7	1
Strapping	4	—	—	—	3½	—	—	—	3	—	—	—	3	—	—	—	3	—	—	—	3	—	—	—	3	—	—	—
Seizing	1	—	—	—	1	—	—	—	¾	—	—	—	¾	—	—	—	¾	—	—	—	¾	—	—	—	¾	—	—	—
Lashing the Col.	2½	—	—	—	2½	—	—	—	2	—	—	—	2	—	—	—	1½	—	—	—	1	—	—	—	1	—	—	—
Shifting Backstays	7	T.	—	2	6½	T.	—	2	5½	T.	—	2	5½	T.	—	2	5	T.	—	2	4½	T.	—	2	4½	T.	—	1
Tackles	3	*D.	12	4	3	*D.	12	4	2½	*D.	10	4	2½	*D.	10	4	2	*D.	8	4	2	*D.	8	2	2	*D.	8	1
		*S.	12	4		*S.	12	4		*S.	10	4		*S.	10	4		*S.	8	4		*S.	8	2		*S	8	1

Names of the Standing and Running Rigging.	110 to 74 Guns.				64 Guns.				50 to 36 Guns.				32 to 28 Guns.				24 Guns.				22 to 20 Guns.				18 to 14 Guns.			
		Blocks, &c.				Blocks, &c.				Blocks, &c.				Blocks, &c.				Blocks, &c.				Blocks, &c.				Blocks, &c.		
	In.	Species.	In	N	In.	Species.	In	N	In.	Species.	In	N	In.	Species.	In	N	In.	Species.	In	N	In.	Species.	In	N	In.	Species.	In	N
Strapping	4	—	—	—	4	—	—	—	3½	—	—	—	3½	—	—	—	3	—	—	—	3	—	—	—	3	—	—	—
Futtock-Shrouds	7	*Pl.d.e	11	12	6½	*Pl.d.e	10	12	5½	*Pl.d.e.	8	8	5½	*Pl.d.e.	8	8	5	*Pl.d.e.	8	8	4½	*Pl.d.e.	7	8	4½	*Pl.d.e.	7	8
Seizing { Upper	1	—	—	—	1	—	—	—	¾	—	—	—	¾	—	—	—	¾	—	—	—	¾	—	—	—	¾	—	—	—
Seizing { Lower	¾	—	—	—	¾	—	—	—	¾	—	—	—	¾	—	—	—	¾	—	—	—	¾	—	—	—	¾	—	—	¾
Ratline	1½	—	—	—	1½	—	—	—	1	—	—	—	1	—	—	—	1	—	—	—	1	—	—	—	1	—	—	—
Top-Rope Pendents	9	S.Br.shi. I.bd.c.	26	2	8½	S.Br.shi. I.bd.c.	22	2	7	S.Br.shi. I.bd.c.	20	2	6½	S.Br.shi. I.bd.c.	18	2	6½	S.Br.shi. I.bd.c.	18	2	6	S.Br.shi. I.bd.c.	17	2	6	S.Br.shi. I.bd.c.	16	1
Falls	5	Tr.I.b.c.	24	2	4	Tr.I.b.c.	21	4	4	Tr.I.b.c.	15	4	3½	Tr.I.b.c.	15	4	3½	Tr.I.b.c.	15	4	3	Tr.I.b.c.	14	4	2	Tr.I.b.c.	14	2
Fore-topsail Yard.																												
Tie	6	Fl.si.c. D.c.	20 20	2 1	5½	Fl. si.c. D.c.	18 18	2 1	4½	Fl.si.c. D.c.	16 16	2 1	4½	Fl.si.c. D.c.	16 16	2 1	4½	Fl.si.c. D.c.	15 15	2 1	4½	Fl.si.c. D.c.	15 15	2 1	4	Fl.si.c. D.c.	14 —	1 —
Strapping	6½	—	—	—	5½	—	—	—	4½	—	—	—	4½	—	—	—	4½	—	—	—	4	—	—	—		—	—	—
Seizing	1	—	—	—	1	—	—	—	1	—	—	—	1	—	—	—	¾	—	—	—	¾	—	—	—	¾	—	—	—
Lashers { Mast-head	2½	—	—	—	2½	—	—	—	2	—	—	—	2	—	—	—	2	—	—	—	1½	—	—	—	1½	—	—	—
Lashers { Yard	2	—	—	—	2	—	—	—	1½	—	—	—	1½	—	—	—	1½	—	—	—	1	—	—	—	1	—	—	—
Haliards	3½	D.th.c. *S.th.c.	26 26	2 2	3½	D.th.c. *S.th.c.	24 24	2 2	3	D.th.c. *S.th.c.	19 19	2 2	3	D.th.c. *S.th.c.	18 18	2 2	2½	D.th.c. *S.th.c.	17 17	2 2	2½	D.th.c. *S.th.c.	16 16	2 2	2½	D.th.c. *S.th.c.	14 14	1 1
Strapping	5	—	—	—	4½	—	—	—	4	—	—	—	4	—	—	—	3½	—	—	—	3½	—	—	—	3½	—	—	—
Seizing	¾	—	—	—	¾	—	—	—	¾	—	—	—	¾	—	—	—	¾	—	—	—	¾	—	—	—	¾	—	—	—
Horses	4	—	—	—	4	—	—	—	3½	—	—	—	3½	—	—	—	3½	—	—	—	3	—	—	—	3	—	—	—
Stirrups	3	T.	—	6	3	T.	—	6	3	T.	—	6	3	T.	—	6	2½	T.	—	6	2	T.	—	4	2	T.	—	4
Braces	3½	S.	14	4	3½	S.	12	4	3	S.	10	4	3	S.	10	4	2½	S.	9	4	2½	S.	9	4	2	S.	8	4
Pendents	4½	S.	14	2	4½	S.	12	2	3½	S.	10	2	3½	S.	10	2	3½	S.	9	2	3½	S.	9	2	3	S.	8	2
Preventer	4	—	—	—	3½	—	—	—	3	—	—	—	3	—	—	—	2½	—	—	—	2½	—	—	—	2	—	—	—
Strapping	4	—	—	—	4	—	—	—	3	—	—	—	3	—	—	—	2½	—	—	—	2½	—	—	—	2	—	—	—
Lifts	3½	D. S.	12 12	2 4	3½	D. S.	12 12	2 4	3	D. S.	10 10	2 4	3	D. S.	10 10	2 4	3	Sis. S.	17 10	2 4	2½	Sis. S.	16 9	2 4	2½	Sis. S.	14 8	2 4
Beckets	4	—	—	—	3½	—	—	—	3	—	—	—	3	—	—	—	3	—	—	—	2½	—	—	—	2½	—	—	—
Strapping	4	—	—	—	4	—	—	—	3	—	—	—	3	—	—	—	3	—	—	—	2½	—	—	—	2½	—	—	—
Seizing	¾	—	—	—	¾	—	—	—	¾	—	—	—		—	—	—		—	—	—		—	—	—		—	—	—
Parral-Ropes	3½	Par.	24	1	3½	Par.	22	1	2½	Par.	16	1	2½	Par.	16	1	2½	Par	16	1	2½	Par.	15	1	2	Par.	14	1
Racking and Seizing	1	—	—	—	¾	—	—	—	¾	—	—	—	¾	—	—	—	¾	—	—	—	Tarred Line			—	—	—	—	—
Clue-lines	4	S.st.bd. S.	14 14	4 2	3½	S.st.bd. S.	12 12	4 2	3	S.st.bd. S.	10 10	4 2	3	S.st.bd. S.	10 10	4 2	2½	S.st.bd. S.	9 9	4 2	2½	S.st.bd. S.	9 9	4 2	2	S.st.bd. S.	8 8	4 2
Strapping	4	—	—	—	3½	—	—	—	3	—	—	—	3	—	—	—	2½	—	—	—	2½	—	—	—	2½	—	—	—
Bunt-lines	3	S.	11	4	3	S.	10	4	2½	S.	8	4	2½	S.	8	4	2	S.	7	4	2	S.	7	4	2	S.	7	4
Strapping	3	—	—	—	3	—	—	—	2½	—	—	—	2½	—	—	—	2	—	—	—	2	—	—	—	2	—	—	—
Leech lines	2½	S.	10	2	2½	S.	9	2	2	S.	7	2	2	S.	7	2	1½	S.	6	2	1½	S.	6	2	1½	S.	6	2
Strapping	2½	—	—	—	2½	—	—	—	2	—	—	—	2	—	—	—	1½	—	—	—	1½	—	—	—	1½	—	—	—
Bow-lines	3½	S.	12	2	3	S.	11	2	2½	S.	9	2	2½	S.	9	2	2½	S.	9	2	2½	S.	9	2	2	S.	8	2
Bridles	3½	T.	—	4	3	T.	—	4	2½	T.	—	4	2½	T.	—	4	2½	T.	—	4	2½	T.	—	4	2	T.	—	4
Strapping	3½	—	—	—	3	—	—	—	2½	—	—	—	2½	—	—	—	2½	—	—	—	2½	—	—	—	2	—	—	—
Lashing	1½	—	—	—	1	—	—	—	1	—	—	—	1	—	—	—	¾	—	—	—	¾	—	—	—	¾	—	—	—
Reef-tackle Pendents	4	*D.	9	4	3½	*D.	8	4	3	*D.	7	4	3	*D.	7	4	2½	*S.	7	4	2½	*S.	7	4	2½	*S.	7	4
Falls	2½	—	—	2	2	—	—	2	1½	—	—	2	1½	—	—	2	1½	—	—	2	1½	—	—	2	1½	—	—	2
Strapping	3	—	—	—	2½	—	—	—	2	—	—	—	2	—	—	—	2	—	—	—	2	—	—	—	2	—	—	—
Ear-rings	1½	—	—	—	1½	—	—	—	1½	—	—	—	1½	—	—	—	1½	—	—	—	1	—	—	—	1	—	—	—
Sheets	8	S. Sho. ¼ thk. & th. c.	26 26	2 2	7	S. Sho. ¼ thk. & th. c.	24 24	2 2	5½	S. Sho. ¼ thk. & th. c.	17 17	2 2	5½	S. Sho. ¼ thk. & th. c.	17 17	2 2	5	S. Sho. ¼ thk. & th. c.	16 16	2 2	4½	S. Sho. ¼ thk. & th. c.	15 15	2 2	4	S. Sho. ¼ thk. & th. c.	14 14	2 2
Straps for { Sheet Blocks	8	—	—	—	7½	—	—	—	6	—	—	—	6	—	—	—	5½	—	—	—	5	—	—	—	4½	—	—	—
Straps for { Quarter do.	6	—	—	—	5½	—	—	—	4½	—	—	—	4½	—	—	—	4	—	—	—	4	—	—	—	3½	—	—	—
Lashers for Quarter Blocks	2½	—	—	—	2	—	—	—	2	—	—	—	2	—	—	—	1½	—	—	—	1	—	—	—	1	—	—	—
Seizings	1½	—	—	—	1½	—	—	—	1	—	—	—	1	—	—	—	¾	—	—	—	¾	—	—	—	¾	—	—	—
Span	3½	—	—	—	3	—	—	—	3	—	—	—	3	—	—	—	3	—	—	—	3	—	—	—	3	—	—	—
Stoppers	6½	—	—	—	5½	—	—	—	4½	—	—	—	4½	—	—	—	3½	—	—	—	3	—	—	—	3	—	—	—
Slings	5	—	—	—	5	—	—	—	4	—	—	—	4	—	—	—	3½	—	—	—	3	—	—	—	3	—	—	—
Staysail-Stay	4	S.	13	1	4	S.	13	1	3½	S.	11	1	3½	S.	11	1	3	S.	10	1	3	S.	10	1	2½	S.	9	1
Tackle	2½	D. S.	10 9	1 1	2	D. S.	9 8	1 1	2	D. S.	8 7	1 1	2	D. S.	8 7	1 1	1½	D. S.	7 6	1 1	2	S. S.	8	2	2	S. S.	8	2
Haliard	3½	S.	12	1	3½	S.	12	1	3	S.	10	1	3	S.	10	1	2	S.	8	1	2	S.	8	1	2	S.	8	1
Strapping	3½	—	—	—	3½	—	—	—	3	—	—	—	3	—	—	—	2	—	—	—	2	—	—	—	2	—	—	—
Sheets	4	S.	13	2	3½	S.	12	2	3	S.	10	2	3	S.	10	2	2½	S.	9	2	2½	S.	9	2	2½	S.	9	2
Strapping	4	—	—	—	3½	—	—	—	3	—	—	—	3	—	—	—	2½	—	—	—	2½	—	—	—	2½	—	—	—
Out-Hauler	2½	S.	9	1	2½	S.	9	1	2	S.	7	1	2	S.	7	1	2	S.	7	1	2	S.	7	1	1½	S.	6	1
Down-Hauler	2½	S.	9	1	2	S.	9	1	2	S.	7	1	2	S.	7	1	1½	S.	6	1	1½	S.	6	1	1	S	5	1
Strapping	2½	—	—	—	2	—	—	—	2	—	—	—	2	—	—	—	1½	—	—	—	1½	—	—	—	1	—	—	—
Studdingsail-Haliards	3½	S.	12	6	3½	S.	12	6	3	S.	10	6	3	S.	10	6	2½	S.	9	6	2½	S.	9	6	2	S.	8	6
Sheets	5	S.	12	2	3	S.	12	2	2½	S.	10	2	2½	S.	10	2	2	S.	9	2	2	S.	9	2	1½	S.	8	2
Tacks	3½	S.	12	4	3½	S.	12	2	3	S.	10	2	3	S.	10	2	2½	S.	9	2	2½	S.	9	2	2	S.	8	2
†Down-Haulers	2	S.	9	2	2	S.	8	2	1½	S.	6	2	1½	S.	6	2	1½	S.	6	2	1½	S.	6	2	1	S.	—	—
Boom-Tackles	2	D. S.	8 8	2 4	2	D. S.	8 8	2 4	2	D. S.	7 7	2 4	2	D. S.	7 7	2 4	—	—	—	—	—	—	—	—	—	—	—	—
Tails and Straps	3½ 2½	—	—	—	3½ 2½	—	—	—	3 2	—	—	—	3 2	—	—	—	2	—	—	—	2	—	—	—	2	—	—	—
Fore Topgallant Mast.																												
Shrouds	4	T.	—	12	3½	T.	—	12	3	T.	—	12	3	T.	—	12	3	T.	—	12	2½	T.	—	12	2½	T.	—	12
Laniards	2	—	—	—	1½	—	—	—	1½	—	—	—	1½	—	—	—	1½	—	—	—	1	—	—	—	1	—	—	—
Standing-Backstays	4	D.E.	7	4	3½	D.E.	6	4	3	D.E.	6	4	3	D.E.	6	4	3	D.E.	6	4	2½	T.	—	4	2½	T.	—	4
Laniards	2	—	—	—	1½	—	—	—	1½	—	—	—	1½	—	—	—	1½	—	—	—	1	—	—	—	1	—	—	—
Stay cabled 4 Strands	4½	S.	12	1	4	S.	12	1	3½	S.	10	1	3½	S.	10	1	3½	S.	10	1	3	T.	—	1	3	T.	—	1
Strapping	3½	—	—	—	3	—	—	—	2½	—	—	—	2½	—	—	—	2½	—	—	—	2½	—	—	—	2½	—	—	—
Tackle	2	D. S.	7 7	1 1	2	D. S.	7 7	1 1	2	D. S.	7 7	1 1	—	—	—	—	—	—	—	—	—	—	—	—	—	—	—	—
Strapping	2½	—	—	—	2½	—	—	—	2½	—	—	—	—	—	—	—	—	—	—	—	—	—	—	—	—	—	—	—
Flagstaff Stay	2	T.	—	1	2	T.	—	1	2	T.	—	1	—	—	—	—	—	—	—	—	—	—	—	—	—	—	—	—
Haliards	1½	—	—	—	1½	—	—	—	1	—	—	—	1	—	—	—	¾	—	—	—	¾	—	—	—	¾	—	—	—
Royal Haliard	2½	S.	7	1	2	S	7	1	2	S	7	1	2	S.	7	1	1½	S.	6	1	1½	S.	6	1	1½	S.	6	1

† Studdingsail Down-haulers have six thimbles.

Names of the Standing and Running Rigging.	110 to 74 Guns.				64 Guns.				50 to 36 Guns.				32 to 28 Guns.				24 Guns.				22 to 20 Guns.				18 to 14 Guns.			
		Blocks, &c.				Blocks, &c.				Blocks, &c.				Blocks, &c.				Blocks, &c.				Blocks, &c.				Blocks, &c.		
	In.	Species.	In	N	In.	Species.	In	N	In.	Species.	In	N	In.	Species.	In	N	In.	Species.	In	N	In.	Species.	In	N	In.	Species.	In	N
Fore Topgallant Yard.																												
Tie	4				3½				3				3				3				2½				2½			
Haliard	2	D. S.	8 8	1 2	2	D. S.	8 8	1 2	1½	D. S.	6 6	1 2	1½	D. S.	6 6	1 2	1½	D. S.	6 6	1 2	1½	S.	6	2	1½	S.	6	2
Strapping	2½				2½				2				2				2				2				2			
Horses	3				3				2½				2½				2½				2½				2			
Braces	2	S.	8	6	2	S.	7	6	2	S.	7	4	2	S.	7	4	2	S.	7	4	2	S.	7	4	1½	S.	6	4
Pendents	3	S.	8	2	2½	S.	7	2	2	S.	6	2	2	S.	6	2		S.				S.				S.		
Strapping	2				2				2				2				2				2				1½			
Lifts single	2½	S.	8	2	2½	S.	7	2	2½	S.	7	2	2½	S.			2½	S.			2½	S.			2	S.		
Strapping	2½	T.		2	2½	T.		2	2½	T.		2		T.		2		T.		2		T.		2		T.		2
Parral-Ropes	2½	Par.	12	1	2	Par.	11	1	1½	Par.	8	1	1½	Par.	8	1	1½	Par.	8	1	1½	Par.	8	1	1½	Par.	6	1
Clue-lines	2½	S.	7	6	2	S.	7	6	1½	S	6	6	1½	S.	6	4	1½	S.	6	4	1½	S.	6	4	1	S.	5	4
Strapping	2½				2				2				2				2				1½				1			
Bow-lines	2½	S	7	2	2	S.	7	2	1½	S.	6	2	1½	T.		6	1½	T.		6	1	T.		6	1	T.		6
Bridles	2½	T.		6	2	T.		6	1½	T.		6	1½				1½				1				1			
Strapping	2½				2				2																			
Ear-rings	Tarr'd Line																											
Shifting Backstays	4	T.		2	3½	T.		2	3	T.		2	3	T.		2	3	T.		2	2½	T.		1	2½	T.		1
Tackles	2	*D. *S.	7 7	2 2	2	*D. *S.	7 7	2 2	1½	*D. *S.	6 6	2 2	1½	*D. *S.	6 6	2 2	1½	*D. *S.	6 6	2 2	1½	*S	6	2	1½	*S.	6	2
Strapping	2				2				2				2				2				1½				1½			
Studdingsail Haliards	2½	S.	7	6	2	S.	7	6	1½	S.	6	6	1½	S.	6	6	1½	S	6	6	1½	S.	6	6	1½	S.	6	6
Sheets	2				2				1				1				1				1				1			
Tacks	2	S.	7	4	2	S.	7	4	1½	S.	6	4	1½	S.	6	4	1½	S	6	4	1½	S.	6	4	1½	S.	6	4
Down-Haulers	1½	T.		1	1½	T.		1	1	T.		1	1	T.		1	1	T.		1	1	T.		1	1	T.		
Strapping	2				2				1½				1½				1½				1½				1½			
Main-Mast.																												
Wooldings	3½				3				2½				2½				2½				2½				2½			
Girtlines	5	S.	16	2	4	S.	15	2	3½	S.	13	2	3	S.	12	2	3	S.	11	2	3	S.	10	2	2½	S.	9	2
Strapping	5				4				3½				3½				3				3				2½			
Seizing	¾				¾				¾				¾															
Lashing	2½				2				1½				1½				1½				1				¾			
Pendents of Tackles cabled	11	*S.c.	24	2	10½	*S.c.	22	2	8½	*S.c.	17	2	8	*S.c.	17	2	7½	*S.c.	15	2	7½	*S.c.	15	2	7	*S.c.	15	2
Strapping	7½	T.		8	6½	T.		8	5½	T.		8	5½	T.		4	5	T.		4	5	T.		4	4½	T.		4
Seizing	1½				1½				1				1				¾				¾				¾			
Runners of Tackles	7½	D.th.c.	21	4	6½	D.th.c.	21	4	5½	D.th.c.	17	4	5½	D.th.c.	16	2	5	D.			4½	D.			4	D.		
Strapping	6				6				5				5				4				4				3½			
Falls of Tackles	4	*S.th.c.	26	4	4	*S.th.c.	24	4	3	*S.th.c.	18	4	3	*S.th.c.	17	4	3	*S.th.c.	15 11	2 2	3	*S.th.c.	15 11	2 2	2½	*S.th.c.	13 9	2 2
Strapping	5½				5½				5				5				4				4				3½			
Seizing	1				1				¾				¾				¾				¾				¾			
Shrouds cabled	11	D E.	17	20	10½	D.E.	16	18	8½	D.E.	13	14	8	D.E.	13	14	7½	D.E.	11	14	7½	D.E.	11	12	7 7	D.E.	10 10	12 10
Seizing – Eye	1½				1½				1				1				1				1				1			
Seizing – Throat	1½				1½				1½				1½				1				1				1			
Seizing – End	1½				1½				1				1				¾				¾				¾			
Laniard	5½				5				4				4				4				4				3½			
Ratline	1½				1½				1½				1½				1½				1½				1½			
Stay cabled 4 Strands	19	H.	26	1	17½	H.	22	1	14½	H.	19	1	12½	H.	17	1	11½	H.	16	1	11½	H.	16	1	10	H.	15	1
Seizing	2				2				1½				1½				1½				1½				1½			
Laniard	6				5				4½				4				4				4				3½			
Collar cabled 4 Strands	14	H.	26	1	12½	H.	22	1	10	H.	19	1	9	H.	17	1	9	H.	16	1	8½	H.	16	1	8	H.	15	1
Worming	2				2				1½				1				1				1				¾			
Seizing	2				2				1½				1½				1½				1½				1			
Lashing	4				3½				3				3				2½				2½				2			
Preventer-Stay cabled 4 Strands	13	H.	17	1	12	H.	16	1	10	H.	14	1	8½	H.	13	1	8	H.	12	1	8	H.	12	1	7	H.	11	1
Laniard	5				4				3				3				3				3				3			
Collar cabled 4 Strands	6	H.	17	1	6	H.	16	1	5	H.	14	1	5	H.	13	1	4½	H.	12	1	4½	H.	12	1	4½	H.	11	1
Lashing	2				2				2				2				2				2				2			
Seizing	1½				1½				1				1				1				1				1			
Catharpin-Legs	7				6½				5				5				4½				4				3½			
Seizings	1½				1½				1				1				1				¾				¾			
Stay Tackle Pendent	6	D.th.c.	18	1	6	D.th.c.	17	1	5	D.th.c.	14	1	5	D.th.c.	14	1	5	D.th.c.	12	1	5	D.th.c.	12	1	5	D.th.c.	12	1
Falls	3½	*S.th.c.	20 14	1 1	3½	*S.th.c.	18 14	1 1	3	*S.th.c.	16 12	1 1	3	*S.th.c.	16 12	1 1	3	*S.th.c.	13 11	1 1	3	*S.th.c.	14 11	1 1	3	*S.th.c.	13 11	1 1
Strapping	4½				4				4				4				4				3½				3½			
Seizing	¾				¾				¾				¾				¾				¾				¾			
Lashing	1½				1½				1½				1½				1½				1½				1½			
Fore-hatch Tackle Fall†	3½	D.th.c. *S.th.c. *S.th.c.	18 20 14	1 1 1	3½	D.th.c. *S.th.c. *S.th.c.	16 18 14	1 1 1	3	D.th.c. *S.th.c. *S.th.c.	14 16 12	1 1 1	3	D.th.c. *S.th.c. *S.th.c.	14 16 12	1 1 1	3	D.th.c. *S.th.c. *S.th.c.	12 13 11	1 1 1	3	D.th.c. *S.th.c. *S.th.c.	12 13 11	1 1 1	3	D.th.c. *S.th.c. *S.th.c.	12 13 11	1 1 1
Strapping	4½				4				4				4				3½				3½				3½			
Seizing	¾				¾				¾				¾				¾				¾				¾			
Main Yard.																												
Jears‡ – Tie, Falls	8	Tr.c. D.c.	28 28	2 2	7	Tr.c. D.c.	26 26	2 2	6	D.c.	19	4	5½	D.c.	19	4	7½ 3	S.d.sc.c. D. S.	20 12 9	3 4 2	7½ 3	S.d.sc.c. D. S.	20 12 9	3 4 2	7 2½	S.d.sc.c. D. S.	20 12 9	3 4 2
Strapping	9				8				6½				6				5½ 4½ 3				5 4 3				5 4 2½			
Seizing	2				2				1				1				1				1				1			
Lashers to the Mast-Head	4½				4				3½				3				3				3				3			
Lashers to the Yard	3½				3				2				2				2				2				2			
Stoppers	6				6				4½				4															

† The Stay and Fore-Hatch Tackle have 2 thimbles.

‡ Main Jears of a 28-gun ship have three 21-in. single blocks with double scores and coaked; two 13-in. treble blocks, coaked; and two 13-in. double blocks, coaked.

Names of the Standing and Running Rigging.	110 to 74 Guns.				64 Guns.			
		Blocks, &c.				Blocks, &c.		
	In.	Species.	In	N	In.	Species.	In	N
Horses	5½	T.	—	2	5	T.	—	2
Stirrups	4	T.	—	8	4	T.	—	8
Seizing	¾				¾			
Laniards	2				2			
Yard-Tackle Pendents	7	D.th.c.	17	2	6½	D.th.c.	16	2
Falls	3½	*S.th.c	20	2	3½	*S.th.c.	18	2
		S.	13	2		S.	13	2
Strapping	5½				5			
Seizing	1				1			
Inner tricing-Lines	2½	S.	8	2	2	S.	7	2
Outer tricing-Lines	2	S.	7	4	2	S.	7	4
Strapping	2				2			
Braces	4½	S.c.	16	2	4	S.c.	15	2
Pendents	5½	S.c.	16	2	5	S.c.	15	2
Preventer	4½				4			
Strapping	5				4½			
Seizing	¾				¾			
Preventers, in war only	3½	S.	13	4	3½	S.	13	4
Strapping	3½				3½			
Seizing	¾				¾			
Lifts	4½	S.	16	6	4	S.	15	6
Span for Cap	6½				6			
Short Span	4½				4			
Strapping	5				4½			
Seizing	¾				¾			
Jigger-Tackle	2½	D.	10	2	2½	D.	10	2
		S.	10	2		S.	10	2
Strapping	3½				3½			
Truss-pendents	8	T.	—	4	7½	T.	—	4
Falls	3	*D.	11	4	3	*D.	11	4
Strapping	3½				3½			
Eye-Seizings	1½				1½			
Nave-Line	2	S.	7	1	2	S.	7	1
Puddening the Yard	6½				6			
Clue-Garnets	4	St. bd.	15	4	3½	St. bd.	13	4
		S.	15	2		S.	13	2
Strap about the Yard	4				3½			
Strapping	4				3½			
Seizing	¾				¾			
Lashing	1				1			
Buntline-Legs	3	D.	11	2	3	D.	10	2
Falls	3	S.	11	10	3	S.	10	10
Strapping	3½				3½			
Leechline-Legs	2½	D.	11	4	2½	D.	10	4
Falls	2½	S.	10	10	2½	S.	10	10
Strapping	3½				3½			
Slablines	2½	S.	9	2	2	S.	8	2
Strapping	2½				2			
Bowlines	4½	D.	16	1	4½	D.	16	1
Bridles	4½	T.	—	6	4½	T.	—	6
Strapping	4½				4½			
Seizing	¾				¾			
Lashing	2				2			
Tackles	3	*L.t.	20	1	2½	*L.t.	16	1
		*S.	11	1		*S.	9	1
Strapping	3½				3			
Ear-rings	2				2			
Sheets cabled	7½	S.c.	24	4	6½	S.c.	22	4
Strapping	7½	T.	—	4	6½	T.	—	4
Seizing	1½				1½			
Lashers	2				2			
Stoppers	6				5½			
Tacks taper and cabled	10				8½			
Stoppers	6	H.& T.	2	2	5½	H. & T.	2	2
Laniards	2				2			
Lanids. for Pud. and Dol.	1½	T.	—	4	2	T.	—	4
Slings	12	—		1	11	—		1
Straps	12	—		1	11	—		1
Seizing	2				1½			
Laniard	3½				3			
Quarter Tack. Pendents	6	D.th.c.	18	2	6	D.th.c.	17	2
Falls	3½	*S.th.c.	20	2	3½	*S.th.c.	19	2
		*S.th.c.	14	2		*S.th.c.	14	2
Strapping	5				4½			
Seizing	1				1			
Luff-Tackles	3½	*D.	13	6	3½	*D.	13	6
		*S.	13	6		*S.	13	6
Strapping	4				4			
Seizing	¾				¾			
Staysail-stay	5	T.	—	2	4½	T.	—	2
Collar	4				3½			
Seizing	1				1			
Laniards	1½				1½			
†Haliards	3	S.	11	3	3	S.	10	3
†Sheets	4	S.	14	2	4	S.	13	2
Tacks	4				4			
Downhauler	2½	S.	9	1	2	S.	8	1

Names of the Standing and Running Rigging.	50 to 36 Guns.				32 to 28 Guns.			
		Blocks, &c.				Blocks, &c.		
	In.	Species.	In	N	In.	Species.	In	N
Horses	4½	T.	—	2	4	T.	—	2
Stirrups	3	T.	—	6	3	T.	—	6
Seizing		Tarred line						
Laniards	2				2			
Yard-Tackle Pendents	5½	D.th.c.	14	2	5¼	D.th.c.	14	2
Falls	3	*S.th.c.	16	2	3	*S.th.c.	16	2
		S.	12	2		S.	12	2
Strapping	4				4			
Seizing	¾				¾			
Inner tricing-Lines	2	S.	7	2	2	S.	7	2
Outer tricing-Lines	1½	S.	6	4	1½	S.	6	4
Strapping	2				2			
Braces	3½	S.c.	12	2	3½	S.c.	12	2
Pendents	4½	S.c.	12	2	4½	S.c.	12	2
Preventer	3½				3½			
Strapping	4				4			
Seizing	¾							
Preventers, in war only	3	S.	11	4	3	S.	11	4
Strapping	3				3			
Seizing	¾							
Lifts	3½	S.	12	6	3½	S.	12	6
Span for Cap	5				5			
Short Span	3½				3½			
Strapping	4				4			
Seizing	¾				¾			
Jigger-Tackle	2	D.	8	2	2	D.	8	2
		S.	8	2		S.	8	2
Strapping	3				3			
Truss-pendents	6	T.	—	4	5½	T.	—	4
Falls	2½	*D.	9	4	2½	*D.	9	4
Strapping	3				3			
Eye-Seizings	1				1			
Nave-Line	1½	S.	6	1	1½	S.	6	1
Puddening the Yard	6				6			
Clue-Garnets	3	St. bd.	12	4	3	St. bd.	12	4
		S.	12	2		S.	12	2
Strap about the Yard	3				3			
Strapping	3				3			
Seizing	¾							
Lashing	1				1			
Buntline-Legs	2½	D.	8	2	2½	D.	8	2
Falls	2½	S.	8	10	2½	S.	8	10
Strapping	3				3			
Leechline-Legs	2	D.	7	4	2	D.	7	4
Falls	2	S.	7	10	2	S.	7	10
Strapping	3				3			
Slablines	2	S.	7	2	2	S.	7	2
Strapping	2				2			
Bowlines	3½	D.	14	1	3½	D.	14	1
Bridles	3½	T.	—	6	3½	T.	—	6
Strapping	3				3			
Seizing	¾				¾			
Lashing	2				2			
Tackles	2	*L.t.	14	1	2	*D.	9	1
		*S.	8	1		*S.	9	1
Strapping	2½				2½			
Ear-rings	1½				1½			
Sheets cabled	5½	S.c.	17	4	5½	S.c.	17	4
Strapping	5½	T.	—	4	5½	T.	—	4
Seizing	1				1			
Lashers	2				2			
Stoppers	4½				4½			
Tacks taper and cabled	7				7			
Stoppers	4½	H. & T.	2	2	1½	H. & T.	2	2
Laniards	1½				1½			
Lanids. for Pud. and Dol.	1	T.	—	4	1	T.	—	4
Slings	9	—		1	8½			
Straps	9				8½			
Seizing	1				1			
Laniard	2½				2½			
Quarter Tack. Pendents	5	D.th.c.	14	2	5	D.th.c.	14	2
Falls	3	*S.th.c.	16	2	3	*S.th.c.	16	2
		*S.th.c.	12	2		*S.th.c.	12	2
Strapping	4				4			
Seizing	¾				¾			
Luff-Tackles	3	*D.	12	4	3	*D.	12	4
		*S.	12	4		*S.	12	4
Strapping	3½				3½			
Seizing	¾				¾			
Staysail-stay	3½	T.	—	2	3½	T.	—	2
Collar	3				3			
Seizing	¾				¾			
Laniards	1½				1½			
†Haliards	3	S.	10	3	3	S.	10	3
†Sheets	3	S.	10	2	3	S.	10	2
Tacks	3				3			
Downhauler	2	S.	8	1	2	S.	8	1

Names of the Standing and Running Rigging.	24 Guns.				22 to 20 Guns.				18 to 14 Guns.			
		Blocks, &c.				Blocks, &c.				Blocks, &c.		
	In.	Species.	In	N	In.	Species.	In	N	In.	Species.	In	N
Horses	4	T.	—	2	4	T.	—	2	3½	T.	—	2
Stirrups	3	T.	—	6	3	T.	—	6	2½	T.	—	6
Seizing												
Laniards	2				1				1			
Yard-Tackle Pendents	5	D.th.c.	12	2	5	D.th.c.	12	2	4½	D.th.c.	11	2
Falls	3	*S.th.c.	13	2	3	*S.th.c.	14	2	2½	*S.th.c.	13	2
		S.	11	2		S.	10	2		S.	9	2
Strapping	3½				3½				3			
Seizing	¾				¾							
Inner tricing-Lines	1½	S.	6	2	1½	S.	6	2	1½	S.	6	2
Outer tricing-Lines	1½	S.	6	4	1½	S.	6	4	1½	S.	6	4
Strapping	1½				1½				1½			
Braces	3	S.c.	10	2	3	S.c.	10	2	2½	S.c.	9	2
Pendents	4	S.c.	10	2	4	S.c.	10	2	3½	S.c.	9	2
Preventer	3				3				2½			
Strapping	3				3				2½			
Seizing												
Preventers, in war only	2½	S.	9	4	2½	S.	9	4	2	S.	8	4
Strapping	2½				2½				2			
Seizing												
Lifts	3½	S.	11	6	3	S.	11	6	3	S.	9	6
Span for Cap	4½				4½				4			
Short Span	3½				3½				3			
Strapping	3½				3½				3			
Seizing												
Jigger-Tackle	2	D.	8	2	—							
		S.	8	2								
Strapping	3											
Truss-pendents	5	T.	—	4	5	T.	—	4	4½	T.		4
Falls	2	*D.	8	4	2	*D.	8	4	2	*D.	8	4
Strapping	2½				2½				2½			
Eye-Seizings	¾				¾				¾			
Nave-Line	1½	S.	6	1	—							
Puddening the Yard	5½				5				5			
Clue-Garnets	2½	St bd.	9	4	2½	St. bd.	9	4	2½	St. bd.	9	4
		S.	9	2		S.	9	2		S.	9	2
Strap about the Yard	2½				2½				2½			
Strapping	2½				2½				2½			
Seizing												
Lashing	¾				¾							
Buntline-Legs	2	D.	7	2	2	D.	7	2	2	D.	7	2
Falls	2	S.	7	8	2	S.	7	8	2	S.	7	8
Strapping	2½				2½				2½			
Leechline-Legs	2	D.	7	4	2	D.	7	4	2	D.	7	4
Falls	2	S.	7	8	2	S.	7	4	2	S.	7	4
Strapping	2½				2½				2½			
Slablines	1½	S.	6	2	1½	S.	6	2	1½	S.		
Strapping	1½				1½				1½			
Bowlines	3	D.	12	1	3	D.	12	1	3	D.	12	1
Bridles	3	T.	—	6	3	T.	—	6	3	T.	—	6
Strapping	3				3				3			
Seizing	¾				¾				¾			
Lashing	1½				1½				1			
Tackles	2	*D.	8	1	2	*D.	8	1	2	*D.	8	1
		*S.	8	1		*S.	8	1		*S.	8	1
Strapping	2				2				2			
Ear-rings	1½				1½				1½			
Sheets cabled	5	S.c.	15	2	5	S.c.	15	2	4½	S.c.	14	2
Strapping	5	T.	—	2	5	T.	—	2	4½	T.	—	2
Seizing	¾				¾				¾			
Lashers												
Stoppers	4				3½				3			
Tacks taper and cabled	6				6				5½			
Stoppers	4	H. & T.	2	2	4	II. & T.	2	2	4	II. & T.	—	
Laniards	1½				1½				¾			
Lanids. for Pud. and Dol.	1	T.	—	4	1	T.	—	4				
Slings												
Straps												
Seizing												
Laniard												
Quarter Tack. Pendents	5	D.th.c.	12	2	5	D.th.c.	12	2	4½	D.th.c.	11	2
Falls	3	*S.th.c.	13	2	3	*S.th.c.	14	2	2½	*S.th.c.	13	2
		*S.th.c.	11	2		*S.th.c.	11	2		*S.th.c.	11	2
Strapping	4				4				3½			
Seizing												
Luff-Tackles	3	*D.	11	4	2½	*D.	10	3	2½	*D.	10	3
		*S.	11	4		*S.	10	3		*S.	10	3
Strapping	3½				3				3			
Seizing												
Staysail-stay	3½	T.	—	2	3	T.	—	2	3	T.	—	2
Collar	3				3				3			
Seizing	¾				¾				¾			
Laniards	1½				1				1			
†Haliards	2½	S.	9	3	2½	S.	9	3	2	S.	8	3
†Sheets	2½	S.	9	2	2½	S.	9	2	2	S.	8	2
Tacks	2				2				2			
Downhauler	2	S.	8	1	2	S.	8	1	1½	S.	6	1

† One hook and thimble to staysail haliard and sheet.

Names of the Standing and Running Rigging.	110 to 74 Guns.				64 Guns.				50 to 36 Guns.				32 to 28 Guns.				24 Guns.				22 to 20 Guns.				18 to 14 Guns.			
		Blocks, &c.				Blocks, &c.				Blocks, &c.				Blocks, &c.				Blocks, &c.				Blocks, &c.				Blocks, &c.		
	In.	Species.	In	N	In.	Species.	In	N	In.	Species.	In	N	In.	Species.	In	N	In.	Species.	In	N	In.	Species.	In	N	In.	Species.	In	N
Strapping	4				4				3				3				2½				2½				2			
	3				3				2½				2½				2				2				1½			
Studdingsail - Haliards																												
Inner	3	S.	12	6	3	S.	12	6	2	S.	9	6	2	S.	9	6	2	S.	9	6	2	S.	9	6	2	S.	9	6
Outer	3½	S.	12	4	3½	S.	12	4	2½	S.	9	4	2½	S.	9	4	2½	S.	9	4	2½	S.	9	4	2½	S.	9	4
Sheets	3				3				2½				2				2				2				2			
Tacks	3½	S.	12	2	3½	S.	12	2	2½	S.	9	2	2½	S.	9	2	2½	S.	9	2	2½	S.	9	2	2½	S.	9	2
Strapping	3½				3½				2½				2½				2½				2½				2½			
Main Topmast.																												
Burton-Pendents	5½	T.	—	2	5	T.	—	2	4½	T.	—	2	4	T.	—	2	4	T.	—	2	3½	T.	—	2	3	T.	—	2
Fall	2½	*D.	11	2	2½	*D.	11	2	2	*D.	9	2	2	*D.	9	2	2	*D.	8	2	2	*D.	8	2	1½	*D.	6	2
		*S.	11	2		*S.	11	2		*S.	9	2		*S.	9	2		*S.	8	2		*S.	8	2		*S.	6	2
Strapping	3½				3½				3				3				2½				2½				2			
Shrouds	7	D.E.	11	12	6½	D.E.	10	12	5½	D.E.	8	8	5½	D.E.	8	8	5	D.E.	8	8	4½	D.E.	7	8	4½	D.E.	7	8
Seizing — Eye	1				1				¾				¾	Tarr'd Line														
Seizing — Throat	1				1				¾				¾				¾	Tarred Line										
Seizing — End	¾				¾				¾				¾				¾											
Laniards	3½				3½				2½				2½				2½				2½				2½			
Ratline	1				1				1				1				1				1				1			
Standing Backstays	7	D.E.	11	6	6½	D.E.	10	4	5½	D.E.	8	4	5½	D.E.	8	4	5	D.E.	8	4	4½	D.E.	7	4	4½	D.E.	7	2
Seizings — Eye	1				1				¾				¾															
Seizings — Throat	1				1				¾				¾					Tarred Line										
Seizings — End	¾				¾				¾				¾															
Laniards	3½				3½				2½				2½				2½				2½				2½			
Breast Backstay Run	5	S.	14	2	4	S.	12	2	3½	S.	11	2	3½	S.	11	2	3½	S.	11	2	3	S.	10	2	3	S.	9	2
Falls	2½	D.	10	4	2½	D.	10	4	2½	D.	9	4	2½	D.	9	4	2	D.	8	4	2	D.	7	2	2	D.	7	2
																						S.	7	2		S.	7	2
Strapping	3				3				3				3				2½				2½				2½			
Stay cabled 4 Strands	8½				7½				6				6				6				5½				5½			
Collar	7	S.	20	1	6½	S.	18	1	5	S.	15	1	5	S.	15	1	5	S.	15	1	4½	S.	15	1	4	S.	14	1
Tackle	3½	L.t.	24	1	3	L.t.	20	1	2½	L.t.	18	1	2½	L.t.	18	1	2½	L.t.	18	1	2½	L.t.	16	1	2½	L.t.	15	1
		*S.	14	1		*S.	12	1		*S.	9	1		*S.	9	1		*S.	9	1		*S.	8	1		*S.	7	1
Strapping	4½				4				3½				3½				3				3				3			
Seizing	1½				1½				¾				¾				¾				¾				¾			
	¾				¾																							
Lashing	2				2				2				2				2				1½				1			
Preventer-Stay cabled 4 Strands	6½				6				4½				4½				4½				4				4			
Collar	5	S.	16	1	5	S.	15	1	3½	S.	11	1	3½	S.	1	1	3½	S.	11	1	3½	S.	11	1	3	S.	11	1
Tackle	3	L.t.	18	1	2½	L.t.	18	1	2	L.t.	16	1	2	L.t.	16	1	2	L.t.	16	1	2	*L.t.	15	1	1½	*L.t.		
		*S.	12	1		*S.	10	1		*S.	9	1		*S.	9	1		*S.	8	1		*S.	7	1		*S.		
Strapping	4				3½				3				3				3				3							
Seizing	1				1				¾				¾				¾				¾				¾			
Collar-Lashing	1½				1½				1½				1½				1½				1				¾			
Shifting Backstays	7	T.	—	2	6½	T.	—	2	5½	T.	—	2	5½	T.	—	2	5	T.	—	2	4½	T.	—	2	4½	T.	—	1
Tackles	3	*D.	12	4	3	*D.	12	4	2½	*D.	10	4	2½	*D.	10	4	2	*D.	8	4	2	*D.	8	2	2	*D.	8	1
		*S.	12	4		*S.	12	4		*S.	10	4		*S.	10	4		*S.	8	4		*S.	8	2		*S.	8	1
Strapping	4				4				3½				3½				3				3				3			
Futtock-Shrouds	7	*Pl.d.e.	11	12	6½	*Pl.d.e.	10	12	5½	*Pl.d.e.	8	8	5½	*Pl.d.e.	8	8	5	*Pl.d.e.	8	8	4½	*Pl.d.e.	7	8	4½	*Pl.d.e.	7	8
Seizings — Upper	1				1				¾				¾				¾				¾				¾			
Seizings — Lower	¾				¾				¾				¾				¾				¾				¾			
Ratline	1½				1½				1				1				1				1				1			
†Top-rope pendents	9	—	26	2	8½	—	22	2	7	—	20	2	6½	—	20	2	6½	—	18	2	6	—	17	2	6	—	16	1
Falls	5	Tr.I.b.c.	22	4	4	Tr.I.b.c.	21	4	4	Tr.I.b.c.	15	4	3½	D.I.b.c.	15	4	3½	D.I.b.c.	15	4	3	D.I.b.c.	14	4	3	D.I.b.c.	14	2
Main Topsail Yard.																												
Tie	6	Fl.si.	22	2	5½	Fl.si.	19	2	4½	Fl.si.	17	2	4½	Fl.si.	16	2	4½	Fl.si.	15	2	4½	Fl.si.	15	2	4	S.do.sc.	14	1
		D.	22	1		D.	19	1		D.	17	1		D.	16	1		D.	15	1		D.	15	1				
Strapping	6½				5½				4½				4½				4½				4				4			
Seizings	1				1				1				1				¾				¾				¾			
Lashing at the Mast-head	2½				2½				2½				2				2				1½				1½			
Lashing at the Yard	2				2				1½				1½				1½				1				1			
Haliards	3½	D.th.c.	26	2	3½	D.th.c.	24	2	3	D.th.c.	19	2	3	D.th.c.	18	2	2½	D.th.c.	17	2	2½	D.th.c.	16	2	2½	D.th.c.	14	1
		*S.th.c.	26	2		*St.h.c.	24	2		*S.th.c.	19	2		*S.th.c.	18	2		*S.th.c.	17	2		*S.th.c.	16	2		*S.th.c.	14	1
Strapping	5				4½				4½				4½				4				3½				3½			
Seizing	¾				¾				¾				¾				¾				¾				¾			
Horses	4½				4				3½				3½				3½				3				3			
Stirrups	3	T.	—	6	3	T.	—	6	3	T.	—	6	3	T.	—	6	2½	T.	—	4	2	T.	—	4	2	T.	—	4
Braces	3½	S.	14	2	3½	S.	12	2	3	S.	10	2	3	S.	10	2	2½	S.	9	2	2½	S.	9	2	2	S.	8	2
Pendents	4½	S.	14	2	4½	S.	12	2	3½	S.	10	2	3½	S.	10	2	3½	S.	9	2	3½	S.	9	2	3	S.	8	2
Preventers	4				3½				3				3				2½				2½				2			
Span about Mizen-mast	4½				4½				4				4				4				4				3½			
Strapping	4				4				3				3				2½				2½				2			
Lifts	3½	D.	12	2	3½	D.	12	2	3	D.	10	2	3	D.	10	2	3	D.	10	2	2½	S.	9	4	2½	S.	8	4
		S.	12	4		S.	12	4		S.	10	4		S.	10	4		S.	10	4								
Beckets	4				3½				3				3				3				2½				2½			
Strapping	4				4				3				3				3				2½				2½			
Seizing	¾				¾				¾																			
Parral-Ropes	4	Par.	25	1	3½	Par.	22	1	2½	Par.	18	1	2½	Par.	18	1	2½	Par.	18	1	2½	Par.	17	1	2	Par.	14	1
Racking and Seizing	1				¾				¾				¾				¾					Tarred Line						
Clue-Lines	4	S.st.bd.	14	4	3½	S.st.bd.	12	4	3	S.st.bd.	10	4	3	S.st.bd.	10	4	2½	S.st.bd.	9	4	2½	S.st.bd.	9	4	2	S.st.bd.	8	4
		S.	14	2		S.	12	2		S.	10	2		S.	10	2		S.	9	2		S.	9	2		S.	8	2
Strapping	4				3½				3				3				2½				2½				2			
Bunt-Lines	3	S.	11	6	3	S.	10	6	2½	S.	8	4	2½	S.	8	4	2	S.	7	4	2	S.	7	4	2	S.	7	4
Strapping	3				3				2½				2½				2				2				2			
Leech-Lines	2½	S.	10	2	2½	S.	9	2	2	S.	7	2	2	S.	7	2	1½	S.	7	2	1½	S.	6	2	1½	S.	6	2
Strapping	2½				2½				2				2				1½				1½				1½			

† Top-rope pendents have single iron-bound blocks, with brass sheaves and two thimbles.

Names of the Standing and Running Rigging.	110 to 74 Guns.	Blocks, &c.			64 Guns.	Blocks, &c.			50 to 36 Guns.	Blocks, &c.			32 to 28 Guns.	Blocks, &c.			24 Guns.	Blocks, &c.			22 to 20 Guns.	Blocks, &c.			18 to 14 Guns.	Blocks, &c.		
	In.	Species.	In	N	In.	Species.	In	N	In.	Species.	In	N	In.	Species.	In	N	In.	Species.	In	N	In.	Species.	In	N	In.	Species.	In	N
Bow-lines	4½	S.	15	2	4	S.	14	2	3	S.	11	2	3	S.	11	2	3	S.	11	2	3	S.	11	2	2½	S.	9	2
Bridles	4½	T.	—	6	4	T.	—	6	3	T.	—	6	3	T.	—	6	3	T.	—	6	3	T.	—	6	2½	T.	—	6
Strapping	4½				4				3½				3½				3½				3				2½			
Seizing	¾				¾				¾																			
Frapping and Lashing	2				2				1½				1½				1½				1½				1			
Reef-Tackle Pendents	4	*D.	9	4	3½	*D.	9	4	3	*D.	7	4	3	*D.	7	4	2½	*D.	7	2	2½	D.	7	2	2½	D.	7	2
Falls	2½				2				1½				1½				1½	*S.	7	2	1½	*S.	7	2	1½	*S.	7	2
Strapping	3				2½				2				2				2				2				2			
Ear-rings	1½				1½				1½				1½				1½				1				1			
Sheets	8½	S.Sho. ¼ D.thk. & th.c.	26 26	2 2	7½	S.Sho. ¼ D.thk. & th.c.	24 24	2 2	6	S.Sho. ¼ D.thk. & th.c.	17 17	2 2	6	S.Sho. ¼ D.thk. & th.c.	17 17	2 2	5½	S.Sho. ¼ D.thk. & th.c.	16 16	2 2	5	S.Sho. ¼ D.thk. & th.c.	15 15	2 2	4½	S.Sho. ¼ D.thk. & th.c.	14 14	2 2
Straps for Sheet Blocks	8½				8				6½				6½				6				5½				5			
Straps for Quarter do.	6				6				4½				4½				4				4				4			
Lashers for Quarter do.	2½				2				2				2				1				1				1			
Seizings	1½				1				1				1				¾				¾				¾			
Span	3½				3				3				3				3				3				3			
Stoppers	6½				6				4½				4½				4				3				3			
Slings	5½				5				4				4				4				3½				3½			
Staysail Haliards	3½	S.	13	1	3½	S.	13	1	3	S.	10	1	3	S.	10	1	3	S.	10	1	2½	S.	9	1	2½	S.	8	1
Strapping	4				4				3				3				3				2½				2½			
Sheets	3	S.	11	2	3	S.	11	2	3	S.	10	2	3	S.	10	2	3	S.	10	2	2½	S.	9	2	2½	S.	8	2
Strapping	3				3				3				3				3				2½				2½			
Pendent	4	S.	12	2	4	S.	12	2	4	S.	11	2																
Tack	3				3				3				3				2				2				2			
Down-Hauler	2½	S.	9	2	2	S.	8	2	2	S.	8	2	2	S.	8	2	1½	S.	6	2	1½	S.	6	2	1½	S.	6	2
Strapping	2½				2				1½				1½				1½				1½				1½			
†Brails	2	S.	7	2	2	S.	7	2	1½	S.	6	2	1½	S.	6	2	1½	S.	6	2	1½	S.	6	2				
Middle Staysail-Stay	4	S.	13	1	4	S.	12	1	3	S.	10	1	3	S.	10	1	3	S.	10	1	3	S.	10	1	2½	S.	9	1
Tackle	2	D. S.	8 8	1 1	2	D. S.	8 8	1 1	2	D. S.	8 8	1 1	2	D. S.	8 8	1 1	2	D. S.	8 8	1 1	2	S.	8	2	1½	S.	6	2
Haliard	3½	S.	12	1	3½	S.	12	1	2½	S.	9	1	2½	S.	9	1	2½	S.	9	1	2½	S.	9	1	2	S.	8	1
Sheets	3½	S.	12	2	3½	S.	12	2	2½	S.	9	2	2½	S.	9	2	2½	S.	9	2	2½	S.	9	2	2	S.	8	2
Tacks	3				3				2				2				2				2				2			
Mid. Staysl.-Downhauler	2½	S.	9	2	2	S.	8	2	2	S.	8	2	2	S.	8	2	1½	S.	7	1	1½	S.	6	1	1½	S.	6	1
Strapping	3 3				3 2				2½ 2½				2½				2½				2½				2			
Tricing-Line	2½	S.	9	1	2	S.	8	1	2	S.	8	1																
Studdingsail-Haliards	3½	S.	12	6	3½	S.	12	6	3	S.	10	6	3	S.	10	6	2½	S.	9	6	2½	S.	9	6	2	S.	8	6
Sheets	3	S.	12	2	3	S.	12	2	2½	S.	10	2	2½	S.	10	2	2	S.	9	2	2	S.	9	2	1½	S.	8	2
Tacks	3½	S.	12	4	3½	S.	12	4	3	S.	10	4	3	S.	10	4	2½	S.	9	2	2½	S.	9	2	2	S.	8	2
‡Down Haulers	2	S.	9	2	2	S.	8	2	1½	S.	6	2	1½	S.	6	2	1½	S.	6	2	1½	S.	6	2	1	T.	—	6
Boom-Tackles	2	D. S.	8 8	2 4	2	D. S.	8 8	2 4	2	D. S.	8 8	2 4	2	D. S.	8 8	2 4												
Lashing for Booms	2½				2½				2½				2½				2½				2				2			
Tailing and Strapping	3½ 2½				3½ 2½				3 2				3 2				2½ 2				2½ 2				2			
Main Topgallant Mast.																												
Shrouds	4	T.	—	12	3½	T.	—	12	3	T.	—	12	3	T.	—	12	3	T.	—	12	2½	T.	—	12	2½	T.	—	12
Laniard	2				1½				1½				1½				1½				1				1			
Standing-Backstays	4	D.E.	7	4	3½	D.E.	6	4	3	D.E.	6	4	3	D.E.	6	4	3	D.E.	6	4	2½	T.	—	4	2½	T.	—	4
Laniards	2				1½				1½				1½				1½				1				1			
Stay cabled 4 Strands	4½	S.	13	1	4	S.	12	1	3½	S.	11	1	3½	S.	10	1	3½	T.	—	1	3	T.	—	1	3	T.	—	1
Strapping	3½				3				3				3															
Flagstaff-Stay	2	T.	—	1	2	T.	—	1	2	T.	—	1																
Haliard	1½				1½				1				1				1				¾				¾			
Royal-Haliard	2				2				1½				1½				1				1				1			
Main Topgallant Yard.																												
Tie	4				3½				3½				3				3				2½				2½			
Haliards	2	D. S.	8 8	1 2	2	D. S	8 8	1 2	1½	D. S.	7 7	1 2	1½	D. S.	7 7	1 2	1½	D. S.	6 6	1 2	1½	S.	6	2	1½	S.	6	2
Strapping	3½				2½				2				2				2				2				2			
Horses	3				3				2½				2½				2½				2½				2			
Braces	2	S.	8	2	2	S.	7	2	1½	S.	6	2	1½	S.	6	2	1½	S.	6	2	1½	S.	6	2	1½	S.	6	2
Pendents	3	S.	8	2	2	S.	7	2	2	S.	7	2	2	S.	7	2												
Strapping	2				2				1				2				2				2							
Lifts, single	2½	S.	8	2	2½	S.	7	2	2½	T.	—	2	2½	T.	—	2	2½	T.	—	2	2½	T.	—	2	2	T.	—	2
Strapping	2½	T.	—	2	2½	T.	—	2	2½																			
Parral-Ropes	2	Par.	12	1	2	Par.	12	1	1½	Par.	9	1	1½	Par.	9	1	1½	Par.	8	1	1½	Par.	8	1	1½	Par.	7	1
Clue-lines	2	S.	7	6	2	S.	7	6	1½	S.	6	6	1½	S.	6	6	1½	S.	6	4	1½	S.	6	4	1	S.	5	4
Strapping	2				2				1½				1½				1½				1½				1			
Bow-Lines	2	S.	7	2	2	S.	7	2	1½	S.	6	2	1½	S.	6	2	1½	S.	6	2	1½	S.	6	2	1			
Bridles	2	T.	—	6	2	T.	—	6	1½	T.	—	6	1½	T.	—	6	1½	T.	—	6	1½	T.	—	4	1	T.	—	2
Strapping	2				2				1½				1½				1½				1½							
Ear-Rings	Tarred line																											
Shifting Backstays	4	T.	—	2	3½	T.	—	2	3	T.	—	2	3	T.	—	2	3	T.	—	2	2½	T.	—	1	2½	T.	—	1
Tackles	2	*D. *S.	7 7	2 2	2	*D. *S.	7 7	2 2	1½	*D. *S.	6 6	2 2	1½	*D. *S.	6 6	2 2	1½	*D. *S.	6 6	2 2	1½	*S.	6	2	1	*S.	5	2
Strapping	2½				2½				2				2				2				1½				1			
Staysail-Stay	3	S.	10	1	3	S.	10	1	2½	S.	8	1	2½	S.	8	1	2½	S.	8	1	2½	S.	8	1	2	S.	7	1
Haliards	2½	S.	9	1	2½	S.	9	1	2	S.	7	1	2	S.	7	1	2	S.	7	1	2	S.	7	1	1½	S.	6	1
Sheets	2½	S.	9	2	2½	S.	9	2	2	S.	7	2	2	S.	7	2	2	S.	7	2	1½	S.	7	2	1½	S.	6	2
Tack	2				2				1½				1½				1½				1				1			
Downhauler	2	S.	7	2	2	S.	7	2	1½	S.	6	2	1½	S.	6	2	1½	S.	6	2	1	S.	5	5	1	S.	5	2
Strapping	2½ 2				2½ 2				2				2				2				2				1½			
Studdingsail-Haliards	2	S.	7	6	2	S.	7	6	2	S.	7	6	2	S.	7	6	2	S.	7	6	1½	S.	6	6	1½	S.	6	6

† Staysail-Brails have likewise two thimbles. ‡ Studdingsail-Downhaulers, in ships down to 20 Guns, have 6 thimbles, besides the blocks above specified.

Names of the Standing and Running Rigging.	110 to 74 Guns.				64 Guns.				50 to 36 Guns.				32 to 28 Guns.				24 Guns.				22 to 20 Guns.				18 to 14 Guns.			
		Blocks, &c.				Blocks, &c.				Blocks, &c.				Blocks, &c.				Blocks, &c.				Blocks, &c.				Blocks, &c.		
	In.	Species.	In	N	In.	Species.	In	N	In.	Species.	In	N	In.	Species.	In	N	In.	Species.	In	N	In.	Species.	In	N	In.	Species.	In	N
Sheets	2	—	—	—	2	—	—	—	1½	—	—	—	1½	—	—	—	1½	—	—	—	1	—	—	—	1	—	—	—
Tacks	2	S.	7	4	2	S.	7	4	2	S.	7	4	2	S.	7	4	2	S.	7	4	1½	S.	6	4	1½	S.	6	4
Downhaulers ..	1½	T.	—	2	1½	T.	—	2	1	T.	—	2	1	T.	—	2	1	T.	—	2	1	T.	—	2	1	—	—	—
Strapping	2	—	—	—	2	—	—	—	2	—	—	—	2	—	—	—	2	—	—	—	1½	—	—	—	1½	—	—	—
Mizen-Mast.																												
Woolding	2½	—	—	—	2½	—	—	—	2½	—	—	—	2½	—	—	—	2	—	—	—	2	—	—	—	—	—	—	—
Girtlines	3½	S.	12	2	3½	S.	12	2	3	S.	11	2	2½	S.	10	2	2½	S.	10	2	2½	S.	10	2	2	S.	9	2
Strapping ..	3½	—	—	—	3½	—	—	—	3	—	—	—	2½	—	—	—	2½	—	—	—	2½	—	—	—	2	—	—	—
Seizing	¾	—	—	—	¾	—	—	—	¾	—	—	—	—	—	—	—	—	—	—	—	—	—	—	—	—	—	—	—
Lashing	1½	—	—	—	1½	—	—	—	1½	—	—	—	1½	—	—	—	1	—	—	—	1	—	—	—	¾	—	—	—
Burton-Pendents ..	5	T.	—	2	5	T.	—	2	4	T.	—	2	4	T.	—	2	4	T.	—	2	3½	T.	—	2	3½	T.	—	2
Falls	3	*D.	11	2	3	*D.	11	2	2½	*D.	9	2	2½	*D.	9	2	2½	*D.	9	2	2	*D.	8	2	2	*D.	8	2
		*S.	11	2		*S.	11	2		*S.	9	2		*S.	9	2		*S.	9	2		*S.	8	2		*S.	8	2
Strapping	3½				3½				3				3				3				2½				2½			
Shrouds	7	D.E.	11	12	6½	D.E.	10	12	5½	D.E.	8	10	5½	D.E.	8	10	5	D.E.	8	10	4½	D.E.	7	8	4½	D.E.	7	8
Seizing (Eye	1	—	—	—	1	—	—	—	1	—	—	—	1	—	—	—	¾	—	—	—		Tarr'd Line	—	—	—	—	—	—
Seizing (Throat	1	—	—	—	1	—	—	—	¾	—	—	—	¾	—	—	—	¾	—	—	—					—	—	—	—
Seizing (End	¾	—	—	—	¾	—	—	—	¾	—	—	—	¾	—	—	—	¾	—	—	—					—	—	—	—
Laniards	3½	—	—	—	3½	—	—	—	2½	—	—	—	2½	—	—	—	2½	—	—	—	2½	—	—	—	2½	—	—	—
Ratline	1	—	—	—	1	—	—	—	1	—	—	—	1	—	—	—	1	—	—	—	1	—	—	—	1	—	—	—
Stay cabled 4 Strands ..	8½	T.	—	2	7½	T.	—	2	6	T.	—	2	6	T.	—	2	6	T.	—	2	5½	T.	—	2	5½	T.	—	2
Seizing	1	—	—	—	1	—	—	—	1	—	—	—	1	—	—	—	¾	—	—	—	¾	—	—	—	¾	—	—	—
Laniards	3½	—	—	—	3½	—	—	—	3	—	—	—	3	—	—	—	3	—	—	—	2½	—	—	—	2½	—	—	—
Collar	7	T.	—	1	6½	T	—	1	5	T.	—	1	5	T.	—	1	5	T.	—	1	4½	T.	—	1	4	—	—	—
Seizing	1	—	—	—	1	—	—	—	1	—	—	—	1	—	—	—	1	—	—	—	¾	—	—	—	¾	—	—	—
Lashing	1½	—	—	—	1½	—	—	—	1½	—	—	—	1½	—	—	—	1½	—	—	—	1	—	—	—	¾	—	—	—
Yard or Gaff.																												
Jears	6	Tr.c.	22	1	5	Tr.c.	18	1	4½	Tr.c.	16	1	4	D.c.	16	1	3½	D.c.	13	1	3½	D.c.	13	1	3	D.c.	11	1
		D.c.	22	1		D.c.	18	1		D.c.	16	1		S.c.	16	1		S.c.	13	1		S.c.	13	1		S.c.	11	1
Strapping	6½	—	—	—	5½	—	—	—	4½	—	—	—	4	—	—	—	3½	—	—	—	3½	—	—	—	3	—	—	—
Seizing	1½	—	—	—	1½	—	—	—	1	—	—	—	1	—	—	—	¾	—	—	—	¾	—	—	—	¾	—	—	—
Lashing at the (Mast-Head ..	3	—	—	—	2½	—	—	—	2½	—	—	—	2½	—	—	—	2½	—	—	—	2½	—	—	—	2	—	—	—
Lashing at the (Yard ..	2½	—	—	—	2	—	—	—	2	—	—	—	1½	—	—	—	1½	—	—	—	1½	—	—	—	1	—	—	—
Derrick	4½	D.	15	1	4	D.	14	1	3½	D.	12	1	3½	D.	12	1	3	D.	11	1	3	D.	11	1	2½	D.	10	1
Span	4½	S.	15	1	4	S.	14	1	3½	S.	12	1	3½	S.	12	1	3½	S.	11	1	3	S.	11	1	2½	S.	9	1
Strapping	4½	T.	—	1	4	T.	—	1	3½	T.	—	1	3½	T.	—	1	3	T.	—	1	3	—	—	—	3	—	—	—
Seizing	1	—	—	—	1	—	—	—	¾	—	—	—	¾	—	—	—	—	—	—	—	—	—	—	—	—	—	—	—
Lashing	2	—	—	—	2	—	—	—	1½	—	—	—	1½	—	—	—	1	—	—	—	1	—	—	—	1	—	—	—
Vang-Pendents ..	4½	D.	9	2	4	D.	8	2	3½	—	—	—	3½	—	—	—	3	—	—	—	3	—	—	—	3	—	—	—
Falls	2½	*S.	9	2	2	*S.	8	2	2	*S.	8	4	2	*S.	8	4	2	*S.	8	4	2	*S.	8	4	1½	*S.	6	4
Strapping	3	—	—	—	3	—	—	—	2½	—	—	—	2½	—	—	—	2½	—	—	—	2	—	—	—	1½	—	—	—
Bow-lines	3	S.	10	4	3	S.	10	4	2½	S.	9	4	2½	S.	9	4	2½	S.	8	2	2½	S.	8	2	—	—	—	—
Strapping	3	T.	—	2	3	T.	—	2	2½	T.	—	2	2½	T.	—	2	2	T.	—	2	2	T.	—	2	—	—	—	—
Brail-Peek Legs ..	2	—	—	—	2	—	—	—	2	—	—	—	2	—	—	—	1½	—	—	—	1½	—	—	—	1½	—	—	—
Falls	2	S.	7	6	2	S.	7	6	2	S.	6	4	1½	S.	6	4	1½	S.	6	4	1½	S.	6	4	1½	S.	6	4
Throat.. ..	3	S.	10	2	3	S.	10	2	2½	S.	9	2	2½	S.	8	2	2	S.	7	2	2	S.	7	2	2	S.	7	2
Middle	2½	S.	9	2	2½	S.	9	2	2	S.	7	2	2	S.	7	2	2	S.	7	2	1½	S.	6	2	1½	S.	6	2
Foot	2½	S.	9	2	2½	S.	9	2	2	S.	7	2	2	S.	7	2	2	S.	7	2	1½	S.	6	2	1½	S.	6	2
Strapping ..	3½	T.	—	4	3	T.	—	4	2½	T.	—	4	2½	T.	—	4	2	T.	—	4	2	T.	—	4	2	T.	—	4
	2½				2½																							
Lacing Mizen to (Yards ..	1½	—	—	—	1½	—	—	—	1½	—	—	—	1½	—	—	—	1½	—	—	—	1	—	—	—	¾	—	—	—
Lacing Mizen to (Mast ..	3	—	—	—	3	—	—	—	2½	—	—	—	2½	—	—	—	2	—	—	—	2	—	—	—	2	—	—	—
Ear-rings	2	—	—	—	2	—	—	—	1½	—	—	—	1½	—	—	—	1	—	—	—	1	—	—	—	¾	—	—	—
Peek-Haliards ..	2	*S.	7	1	2	*S.	7	1	1½	*S.	6	1	1½	*S.	6	1	1½	*S.	6	1	1	*S.	5	1	1	*S.	5	1
Sheet	4½	*D.	15	1	4	*D.	14	1	3½	*D.	11	1	3	*D.	11	1	3	*D.	10	1	3	*D.	10	1	2½	S.	9	2
		S.	15	1		S.	14	1		S.	11	1		S.	11	1		S.	10	1		S.	10	1				
Strapping	4½	—	—	—	4	—	—	—	3	—	—	—	3	—	—	—	3	—	—	—	3	—	—	—	2½	—	—	—
Seizing	1	—	—	—	1	—	—	—	1	—	—	—	—	—	—	—	—	—	—	—	—	—	—	—	—	—	—	—
Tack	3	—	—	—	3	—	—	—	2	—	—	—	2	—	—	—	1½	—	—	—	1½	—	—	—	1½	—	—	—
Slings	6	—	—	—	5½	—	—	—	4	—	—	—	4	—	—	—	4	—	—	—	3½	—	—	—	3	—	—	—
Staysail-Stay ..	5	T.	—	2	4½	T.	—	2	4	T.	—	2	4	T.	—	2	4	T.	—	2	3½	T.	—	2	3	T.	—	2
Collar	4½	T.	—	1	4	T.	—	1	3½	T	—	1	3	T.	—	1	3	T.	—	1	2½	T.	—	1	2½	T.	—	1
Seizings	¾	—	—	—	¾	—	—	—	¾	—	—	—	¾	—	—	—	¾	—	—	—	¾	—	—	—	—	—	—	—
Lashing	1	—	—	—	1	—	—	—	¾	—	—	—	¾	—	—	—	¾	—	—	—	¾	—	—	—	—	—	—	—
Laniard	2	—	—	—	2	—	—	—	1½	—	—	—	1½	—	—	—	1	—	—	—	1	—	—	—	1	—	—	—
Haliard	3	S.	11	3	3	S.	11	3	2½	S.	8	3	2½	S.	8	3	2	S.	7	3	2	S.	7	3	2	S.	7	3
Sheets	3½	S.	11	2	3½	S.	11	2	2½	S.	8	2	2½	S.	8	2	2	S.	7	2	2	S.	7	2	2	S.	7	1
Tack	2½	—	—	—	2½	—	—	—	2½	—	—	—	2½	—	—	—	2	—	—	—	2	—	—	—	2	—	—	—
Down-Hauler ..	2½	S.	9	2	2	S.	7	2	1½	S.	6	2	1½	S.	6	2	1½	S.	6	2	1½	S.	6	2	1	S.	5	1
Strapping ..	3½	—	—	—	3½	—	—	—	2½	—	—	—	2½	—	—	—	2	—	—	—	2	—	—	—	1½	—	—	—
	2½	—	—	—	2½	—	—	—																				
Brails	2	*S.	7	2	2	*S.	7	2	2	*S	7	2	2	*S.	7	2	2	*S.	7	2	2	*S.	7	2	2	*S.	7	2
Boom-Driver.																												
Topping Lifts or Runner	5	—	—	—	5	—	—	—	4	—	—	—	4	—	—	—	4	—	—	—	3½	—	—	—	3	—	—	—
Span	5	D.	15	1	5	D.	15	1	4	D.	13	1	4	D.	13	1	3½	D.	13	1	3	D.	12	1	3	D.	10	1
Fall	3	D.	10	2	3	D.	10	2	2½	D.	9	2	2½	D.	9	2	2½	D.	9	2	2	D.	8	2	2	D.	8	2
		*S.	10	2		*S	10	2		*S	9	2		*S.	9	2		*S.	9	2		*S.	8	2		*S.	8	2
Strapping	3	—	—	—	3	—	—	—	2½	—	—	—	2½	—	—	—	2½	—	—	—	2	—	—	—	2	—	—	—
Guy-Pendents ..	3½	H. & T.	—	4	3½	H. & T.	—	4	3½	—	—	—	3	—	—	—	2½	—	—	—	2½	—	—	—	2½	—	—	—
Falls	2½	*D.	9	2	2½	*D.	9	2	2	*D.	8	2	2	*D.	8	2	2	*D.	8	2	2	*D.	8	2	2	*D.	7	2
		*S.	9	2		*S.	9	2		*S.	8	2		*S.	8	2		*S.	8	2		*S.	8	2		*S.	7	2
Boom-Sheets ..	3½	S.	12	2	3	S.	10	2	2½	S.	9	2	2½	S.	9	2	2½	S.	9	2	2½	S.	9	2	2	S.	8	2
Horses	3½	—	—	—	3½	—	—	—	3	—	—	—	3	—	—	—	3	—	—	—	2½	—	—	—	2½	—	—	—
Sheet Pendent ..	3½	—	—	—	3½	—	—	—	3½	—	—	—	3½	—	—	—	3½	—	—	—	3	—	—	—	3	—	—	—
Falls	2	*D.	8	1	2	*D.	8	1	2	*D.	8	1	2	*D.	8	1	2	*D.	8	1	2	*D.	7	1	2	*D.	7	1
		*S.	8	1		*S.	8	1		*S.	8	1		*S.	8	1		*S.	8	1		*S.	7	1		*S.	7	1
Haliards or Brails ..	2½	S.	9	6	2½	S.	9	6	2½	S.	9	6	2½	S.	9	6	2	S.	8	6	2	S.	8	6	2	S.	8	6
Lacing to the Yard ..	1½	—	—	—	1½	—	—	—	1½	—	—	—	1½	—	—	—	1½	—	—	—	1½	—	—	—	1½	—	—	—
Tack-Tackle ..	2	*D.	12	1	2	*D.	12	1	2	*D.	10	1	2	*D.	10	1	1½	*D.	10	1	1½	*D.	8	1	1½	*D.	7	1
		*S	12	1		*S.	12	1		*S.	10	1		*S.	10	1		*S.	10	1		*S.	8	1		*S.	7	1

Names of the Standing and Running Rigging.	110 to 74 Guns.				64 Guns.				50 to 36 Guns.				32 to 28 Guns.				24 Guns.				22 to 20 Guns.				18 to 14 Guns.			
		Blocks, &c.				Blocks, &c.				Blocks, &c.				Blocks, &c.				Blocks, &c.				Blocks, &c.				Blocks, &c.		
	In.	Species.	In	N	In.	Species.	In	N	In.	Species.	In	N	In.	Species.	In	N	In.	Species.	In	N	In.	Species.	In	N	In.	Species.	In	N
Down-Hauler	2½	—	—	—	2½	—	—	—	2	—	—	—	2	—	—	—	2	—	—	—	1½	—	—	—	1½	—	—	—
Lashing	¾	—	—	—	¾	—	—	—	¾	—	—	—	¾	—	—	—	¾	—	—	—	¾	—	—	—	¾	—	—	—
Cross-Jack yard.																												
Truss-Pendents	5	T.	—	2	4	T.	—	2	3½	T.	—	2	3	T.	—	2	3	T.	—	2	3	T.	—	2	—	—	—	—
Falls	2½	*L.t. *S.	18 9	1 1	2½	*L.t. *S.	18 9	1 1	2	*L.t. * S.	15 8	1 1	2	*L.t. *S.	15 8	1 1	1½	*L.t. *S.	14 6	1 1	2	*S.	7	2	—	—	—	—
Strapping	3	—	—	—	3	—	—	—	2½	—	—	—	2½	—	—	—	2½	—	—	—	2	—	—	—	—	—	—	—
Seizing	¾	—	—	—	¾	—	—	—	¾	—	—	—	¾	—	—	—	—	—	—	—	—	—	—	—	—	—	—	—
Span about the Cap	4	—	—	—	4	—	—	—	3	—	—	—	3	—	—	—	2½	—	—	—	2½	—	—	—	2	—	—	—
Braces	2½	S.	9	2	2½	S.	9	2	2	S.	8	2	2	S.	8	2	2	S.	8	2	1½	S.	6	2	1½	S.	6	2
Pendents	3½	S.	9	2	3½	S.	9	2	3	S.	8	2	2½	S.	8	2	2½	S.	8	2	2	S.	6	2	2	S.	6	2
Preventer	3	—	—	—	3	—	—	—	3	—	—	—	—	—	—	—	—	—	—	—	—	—	—	—	—	—	—	—
Strapping	2½	—	—	—	2½	—	—	—	2	—	—	—	2	—	—	—	2	—	—	—	2	—	—	—	2	—	—	—
Lifts, running	2½	S.	9	4	2½	S.	9	4	2	S.	8	4	2	S	8	4	2	S.	8	4	2	S.	7	4	1½	S.	6	4
Strapping	2½	—	—	—	2½	—	—	—	2	—	—	—	2	—	—	—	2	—	—	—	2	—	—	—	1½	—	—	—
Slings	5½	S.do.sc.	16	1	4½	S.do.sc.	14	1	3½	S.do.sc.	12	1	3½	S.do.sc.	12	1	3½	S.do.sc.	11	1	3½	S.do.sc.	11	1	3	T.	—	1
Strapping	4	—	—	—	4	—	—	—	3	—	—	—	3	—	—	—	3	—	—	—	3	—	—	—	2½	—	—	—
Seizing	1	—	—	—	1	—	—	—	¾	—	—	—	¾	—	—	—	—	—	—	—	—	—	—	—	—	—	—	—
Lashing	1½	—	—	—	1½	—	—	—	1	—	—	—	1	—	—	—	1	—	—	—	¾	—	—	—	—	—	—	—
Mizen Topmast.																												
Shrouds	4½	D.E.	8	8	4	D.E.	8	8	3½	D.E.	6	6	3½	D.E.	6	6	3½	D.E.	6	6	3	D.E.	5	6	2½	D.E.	5	6
Seizings	Tarred Line		—	—	—	—	—	—	—	—	—	—	—	—	—	—	—	—	—	—	—	—	—	—	—	—	—	—
Laniards	2½	—	—	—	2½	—	—	—	2	—	—	—	2	—	—	—	2	—	—	—	1½	—	—	—	1½	—	—	—
Ratline	1	—	—	—	1	—	—	—	1	—	—	—	1	—	—	—	1	—	—	—	¾	—	—	—	¾	—	—	—
Standing-Backstays	4½	D.E.	8	4	4	D.E.	8	4	3½	D.E.	6	2	3½	D.E.	6	2	3½	D.E.	6	2	3	D.E.	5	2	2½	D.E.	5	2
Seizing	Tarred Line		—	—	—	—	—	—	—	—	—	—	—	—	—	—	—	—	—	—	—	—	—	—	—	—	—	—
Laniards	2½	—	—	—	2½	—	—	—	2	—	—	—	2	—	—	—	2	—	—	—	1½	—	—	—	1½	—	—	—
Stay cabled 4 Strands	5	—	—	—	4½	—	—	—	4	—	—	—	4	—	—	—	4	—	—	—	3½	—	—	—	3	—	—	—
Laniard	2	T.	—	2	2	T.	—	2	2	T.	—	2	2	T.	—	2	1½	T.	—	2	1½	T.	—	2	1½	T.	—	2
Collar	4	S.	14	1	3½	S.	13	1	3	S.	12	1	3	S.	11	1	3	S.	10	1	2½	T.	—	1	2	T.	—	1
Seizing and Lashing	1	—	—	—	¾	—	—	—	¾	—	—	—	¾	—	—	—	¾	—	—	—	—	—	—	—	—	—	—	—
Flagstaff-stay	2	T.	—	1	1½	T.	—	1	1½	—	—	—	—	—	—	—	—	—	—	—	—	—	—	—	—	—	—	—
Haliards	1½	—	—	—	1½	—	—	—	1	—	—	—	1	—	—	—	¾	—	—	—	¾	—	—	—	¾	—	—	—
Shifting Backstays	4½	T.	—	1	4	T.	—	1	3½	T.	—	1	3½	T.	—	1	3½	T.	—	1	3	T.	—	1	2½	T.	—	1
Tackle	2½	*D. *S.	9 9	1 1	2	*D. *S.	8 8	1 1	2	*D. *S.	7 7	1 1	2	*D. *S.	7 7	1 1	2	*D. *S.	7 7	1 1	1½	*S.	6	2	1½	*S.	6	2
Strapping	3	—	—	—	2½	—	—	—	2½	—	—	—	2½	—	—	—	2½	—	—	—	2	—	—	—	2	—	—	—
Futtock-shrouds	4½	*Pl.d.e.	8	8	4	*Pl.d.e.	8	8	3½	*Pl.d.e.	6	6	3½	*P.ld.e.	6	6	3½	*Pl.d.e.	6	6	3	*Pl.d.e.	5	6	2½	*Pl.d.e.	5	6
Seizing	Tarred Line		—	—	—	—	—	—	—	—	—	—	—	—	—	—	—	—	—	—	—	—	—	—	—	—	—	—
Ratlines	1	—	—	—	1	—	—	—	1	—	—	—	1	—	—	—	1	—	—	—	1	—	—	—	¾	—	—	—
Top-Rope Pendents	5	*S.I.bd.	16	1	5	*S.I.bd.	14	1	4½	*S.I.bd.	12	1	4½	*S.I.bd.	12	1	4	*S.I.bd.	12	1	4	*S.I.bd.	12	1	4	*S.I.bd.	12	1
Falls	3	D.I.bd.	12	2	3	D.I.bd.	12	2	2½	D.I.bd.	10	2	2½	D.I.bd.	10	2	2½	D.I.bd.	10	2	2½	D.I.bd.	10	2	2½	D.I.bd.	10	2
Mizen Topsail Yard.																												
Tie	4	S.do.sc.	13	1	3½	S.do.sc.	12	1	3½	S.do.sc.	12	1	3½	S.do.sc.	12	1	3	S.do.sc.	10	1	3	S.do.sc.	10	1	3	S.do.sc.	10	1
Haliard	2½	D.th.c. *S.th.c.	12 12	1 1	2½	D.th.c. *S.th.c.	12 12	1 1	2	D.th.c. *S.th.c.	10 10	1 1	2	D.th.c. *S.th.c.	10 10	1 1	2	D.th.c. *S.th.c.	10 10	1 1	2	*S.th.c.	10	2 1	2	*S.th.c.	8	2 1
Strapping	3½	—	—	—	3½	—	—	—	3	—	—	—	3	—	—	—	3	—	—	—	3	—	—	—	2½	—	—	—
Lashing	1	—	—	—	1	—	—	—	¾	—	—	—	¾	—	—	—	¾	—	—	—	—	—	—	—	—	—	—	—
Horses	3	—	—	—	3	—	—	—	2½	—	—	—	2½	—	—	—	2½	—	—	—	2½	—	—	—	2	—	—	—
Stirrups	2½	T.	—	4	2½	T.	—	4	2	T.	—	4	2	T.	—	4	2	T.	—	2	—	—	—	—	—	—	—	—
Braces	2½	S.	9	2	2½	S.	9	2	2	S.	7	2	2	S.	7	2	1½	S.	6	2	1½	S.	6	2	1½	S.	5	2
Pendents	3	S.	9	2	3	S.	9	2	2½	S.	7	2	2½	S.	7	2	2	S.	6	2	2	S.	6	2	1½	—	—	—
Strapping	2½	—	—	—	2½	—	—	—	2	—	—	—	2	—	—	—	1½	—	—	—	1½	—	—	—	—	—	—	—
Lifts	2½	D. S.	9 9	2 2	2½	D. S.	9 9	2 2	2	D. S.	7 7	2 2	2	S.	7	4	2	S.	7	4	1½	S.	6	4	2	T.	—	2
Strapping	3	—	—	—	3	—	—	—	2½	—	—	—	2½	—	—	—	2	—	—	—	1½	—	—	—	—	—	—	—
Parral-Ropes	2½	Par.	16	1	2½	Par.	14	1	2	Par.	11	1	2	Par.	11	1	2	Par.	11	1	1½	Par.	11	1	1½	Par.	8	1
Clue-lines	2½	S.st.bd. S.	9 9	4 2	2½	S.st.bd. S.	9 9	4 2	2	S.st.bd. S.	7 7	4 2	2	S.st.bd. S.	7 7	4 2	2	S.st.bd. S.	7 7	4 2	1½	S.st.bd. S.	6 6	4 2	1½	S.st.bd. S.	5	4 2
Strapping	2½	—	—	—	2½	—	—	—	2	—	—	—	2	—	—	—	2	—	—	—	1½	—	—	—	1½	—	—	—
Bunt-Lines	2	S.	8	4	2	S.	8	4	1½	S.	6	2	1½	S.	6	2	1½	S.	6	2	1½	S.	6	1	1½	S.	6	1
Strapping	2	—	—	—	2	—	—	—	1½	—	—	—	1½	—	—	—	1½	—	—	—	1½	T.	—	1	1½	T.	—	1
Leech lines	2	S.	7	2	2	S.	7	2	2	S.	7	2	—	—	—	—	—	—	—	—	—	—	—	—	—	—	—	—
Strapping	2	—	—	—	2	—	—	—	2	—	—	—	—	—	—	—	—	—	—	—	—	—	—	—	—	—	—	—
Bow-lines	2½	S.	9	2	2	S.	8	2	1½	S.	6	2	1½	S.	6	2	1½	S.	6	2	1½	S.	6	2	1½	S.	5	2
Bridles	2½	T.	—	4	2	T.	—	4	1½	T	—	4	1½	T	—	4	1½	T.		4	1½	T.		4	1½	T.		4
Strapping	2½	—	—	—	2	—	—	—	1½	—	—	—	1½	—	—	—	1½	—	—	—	1½	—	—	—	1½	—	—	—
Reef-Tackle Pendents	3	—	—	—	3	—	—	—	3	—	—	—	—	—	—	—	—	—	—	—	—	—	—	—	—	—	—	—
Falls	1½	D. S.	7 7	2 2	1½	D. S.	7 7	2 2	1½	D. S.	7 7	2 2	—	—	—	—	—	—	—	—	—	—	—	—	—	—	—	—
Strapping	2	—	—	—	2	—	—	—	2	—	—	—	—	—	—	—	—	—	—	—	—	—	—	—	—	—	—	—
Ear-rings	1½	—	—	—	1½	—	—	—	1	—	—	—	1	—	—	—	¾	—	—	—	¾	—	—	—	¾	—	—	—
Sheets	5	S.Sho. S.	15 15	2 1	4½	S.Sho. S.	14 14	2 1	3½	S.Sho. S.	11 11	2 1	3½	S.Sho. S.	11 11	2 1	3½	S.Sho. S.	11 11	2 1	3	S.Sho. S.	10 10	2 1	3	S.Sho. S.	10 10	2 1
Strapping	5 4	—	—	—	5 4	—	—	—	4 3	—	—	—	4 3	—	—	—	4 3	—	—	—	3	—	—	—	3	—	—	—
Seizings	¾	—	—	—	¾	—	—	—	¾	—	—	—	—	—	—	—	—	—	—	—	—	—	—	—	—	—	—	—
Lashing	1½	—	—	—	1½	—	—	—	1	—	—	—	1	—	—	—	¾	—	—	—	¾	—	—	—	¾	—	—	—
Staysail Haliards	2½	S.	9	1	2	S.	8	1	2	S.	7	1	2	S.	7	1	1½	S.	6	1	1½	S.	6	1	1½	S.	6	1
Sheets	2½	S.	9	2	2	S.	8	2	2	S.	7	2	2	S.	7	2	1½	S.	6	2	1½	S.	6	2	1½	S.	6	2
Tacks	2	—	—	—	2	—	—	—	2	—	—	—	2	—	—	—	1½	—	—	—	1½	—	—	—	1½	—	—	—
Downhaulers	1½	S.	6	1	1½	S.	6	1	1	S.	5	1	1	S.	5	1	1	S.	5	1	1	S.	5	1	1	S.	5	1
Strapping	2½	—	—	—	2	—	—	—	2	—	—	—	2	—	—	—	1½	—	—	—	1½	—	—	—	—	—	—	—
Mizen Topgallant Mast.																												
Shrouds	2½	T.	—	8	2½	T.	—	8	2	T.	—	8	2	T.	—	8	2	T.	—	8	1½	T.	—	8	—	—	—	—
Laniards	1½	—	—	—	1½	—	—	—	1	—	—	—	1	—	—	—	1	—	—	—	¾	—	—	—	—	—	—	—
Backstays	2½	T.	—	4	2½	T.	—	4	2	T.	—	4	2	T.	—	4	2	T.	—	4	1½	T.	—	4	—	—	—	—
Laniards	1½	—	—	—	1½	—	—	—	1	—	—	—	1	—	—	—	1	—	—	—	¾	—	—	—	—	—	—	—

Names of the Standing and Running Rigging.	110 to 74 Guns.				64 Guns.				50 to 36 Guns.				32 to 28 Guns.				24 Guns.				22 to 20 Guns.				18 to 14 Guns.			
		Blocks, &c.				Blocks, &c.				Blocks, &c.				Blocks, &c.				Blocks, &c.				Blocks, &c.				Blocks, &c.		
	In.	Species.	In	N	In.	Species.	In	N	In.	Species.	In	N	In.	Species.	In	N	In.	Species.	In	N	In.	Species.	In	N	In.	Species.	In	N
Stay	3	T.	—	1	3	T.	—	1	2½	T.	—	1	2½	T.	—	1	2½	T.	—	1	2	T.	—	1	—	—	—	—
Laniard	1½	—	—	—	1½	—	—	—	1	—	—	—	1	—	—	—	1	—	—	—	1	—	—	—	—	—	—	—
Mizen Topgallant Yard.																												
Tie	3	S.	6	1	2½	—	—	—	2	S.	6	1	2	S.	6	1	2	S.	6	1	2	S.	6	1	—	—	—	—
Haliard	1½	S.	5	2	1½	S.	5	2	1½	S.	5	2	1½	S.	5	2	1½	S.	5	2	1	S.	5	2	—	—	—	—
Horses	2	T.	—	2	2	T.	—	2	2	—	—	—	2	—	—	—	2	—	—	—	2	—	—	—	—	—	—	—
Braces	1½	S.	5	2	1½	S.	5	2	1½	S.	5	2	1½	S.	5	2	1½	S.	5	2	1	T.	—	2	—	—	—	—
Lifts, single	2	T.	—	2	2	T.	—	2	2	T.	—	2	2	T.	—	2	2	T.	—	2	1½	T.	—	2	—	—	—	—
Parral-Ropes	1½	Par.	7	1	1½	Par.	6	1	1	Par.	6	1	1	Par.	6	1	1	Par.	6	1	1	Par.	6	1	—	—	—	—
Clue-Lines	1½	S.	5	2	1½	S.	5	2	1½	S.	5	2	1½	S.	5	2	1½	S.	4	2	1	S.	4	2	—	—	—	—
Bow-Lines	1½	T.	—	4	1½	T.	—	4	1	T.	—	4	1	T.	—	4	1	T.	—	4	1	T.	—	4	—	—	—	—
Bridles	1½	—	—	—	1½	—	—	—	1	—	—	—	1	—	—	—	1	—	—	—	1	—	—	—	—	—	—	—
Ear-Rings	Tarred Line		—	—	—	—	—	—	—	—	—	—	—	—	—	—		—	—	—		—	—	—	—	—	—	—
Strapping	2	—	—	—	1½	—	—	—	1½	—	—	—	1½	—	—	—	1½	—	—	—	1	—	—	—	—	—	—	—
Necessary Ropes.																												
Viol cabled	14	S.c.	56	1	13	S.c.	50	1	11	S.c.	42	1	—	—	—	—	—	—	—	—	—	—	—	—	—	—	—	—
Strapping	11½	—	—	—	11	—	—	—	8½	—	—	—	—	—	—	—	—	—	—	—	—	—	—	—	—	—	—	—
Seizing	2	—	—	—	2	—	—	—	1½	—	—	—	—	—	—	—	—	—	—	—	—	—	—	—	—	—	—	—
Lashing	4½	—	—	—	4	—	—	—	3½	—	—	—	—	—	—	—		—	—	—	—	—	—	—	—	—	—	—
Winding Tackle Pendents	13	Fourfold Coak	24	1	12	Fourfold Coak	22	1	9	Fourfold Coak	19	1	8½	Fourfold Coak	18	1	—	—	—	—	—	—	—	—	—	—	—	—
Fall	5	Tr.c. S.c.	24 24	1 1	4½	Tr.c. S.c.	22 22	1 1	3½	Tr.c. S.c.	19 19	1 1	3½	Tr.c. S.c.	18 18	1 1	—	—	—	—	—	—	—	—	—	—	—	—
Strapping	7	—	—	—	7	—	—	—	6	—	—	—	6	—	—	—	—	—	—	—	—	—	—	—	—	—	—	—
Seizing	1½	—	—	—	1½	—	—	—	1½	—	—	—	1½	—	—	—	—	—	—	—	—	—	—	—	—	—	—	—
Cat-Falls	6	Tr.I.bd. Br.sh.	26	2	5	Tr.I.bd. Br.sh.	22	2	4	Tr.I.bd. Br.sh.	18	2	4	Tr.I.bd. Br.sh.	16	2	3½	Tr.I.b.c.	15	2	3	Tr.I.b.c.	15	2	3	D.I.b.c.	14	2
Laniards	3	S. T. S Large Ragged Staples	10 — —	2 4 2	3	S. T. S.Large Ragged Staples	10 — —	2 4 2	2	S. T. S.Large Ragged Staples	8 — —	2 4 2	2	S. T. S. Large Ragged Staples	8 — —	2 4 2	2	S. T. S.Large Ragged Staples	8 — —	2 4 2	2	S. T. S.Large Ragged Staples	8 — —	2 4 2	2	—	—	—
Stoppers	5	—	—	—	5	—	—	—	4	—	—	—	3½	—	—	—	3½	—	—	—	2½	—	—	—	2	—	—	—
Guys. Mast-head	9½	—	—	—	8	—	—	—	7½	—	—	—	6½	—	—	—	5½	—	—	—	5½	—	—	—	5	—	—	—
Guys. Fore	9½	—	—	—	8	—	—	—	7½	—	—	—	6½	—	—	—	5½	—	—	—	5½	—	—	—	5	—	—	—
Guys. After	6	—	—	—	5	—	—	—	4½	—	—	—	3½	—	—	—	3½	—	—	—	3½	—	—	—	3	—	—	—
†Fish Tackle Pendents	10	*S.c.	28	1	9	*S.c.	26	1	7½	*S.c.	20	1	7	*S.c.	18	1	6	*S.c.	16	1	6	*S.c.	15	1	5½	*S.c.	15	1
Fall	4½	L.t.	38	2	4	L.t.	36	2	3½	L.t.	32	2	3½	L.t.	30	2	3½	L.t.	28	2	3	L.t.	26	2	3	L.t. S.	24 14	1 1
Strapping	7 6½	—	—	—	6 5½	—	—	—	5 4½	—	—	—	4½	—	—	—	4	—	—	—	3½	—	—	—	3½	—	—	—
Seizing	1½ 1	—	—	—	1½ 1	—	—	—	1½ 1	—	—	—	1	—	—	—	¾	—	—	—	¾	—	—	—	¾	—	—	—
Laniards	3	—	—	—	3	—	—	—	2½	—	—	—	2½	—	—	—	2½	—	—	—	2	—	—	—	1½	—	—	—
Anchor-Stock Tack	3	*D.	12	1	3	*D.	12	1	2½	*D.	9	1	2½	*D.	9	1	2½	*D.	9	1	2	*D.	8	1	2	*D.	8	1
Anchor-Stock Fall		*S.	12	1		*S.	12	1		*S.	9	1		*S.	9	1		*S.	9	1		*S.	8	1		*S.	8	1
Bill-Pendents	5½	T.	—	2	5	T.	—	2	4½	T.	—	2	4	T.	—	2	4	T.	—	2	3½	T.	—	2	3	—	—	—
Strapping	4	H.	—	1	4	H.	—	1	3	H.	—	1	3	H.	—	1	3	H.	—	1	2½	H.	—	1	2½	—	—	—
Seizing	¾	—	—	—	¾	—	—	—	¾	—	—	—	—	—	—	—	—	—	—	—	—	—	—	—	—	—	—	—
Stoppers, Sheet Anchor	9½	—	—	—	8	—	—	—	6½	—	—	—	6	—	—	—	5½	—	—	—	5	—	—	—	5	—	—	—
Best Bower	9½	—	—	—	8	—	—	—	6½	—	—	—	6	—	—	—	5½	—	—	—	5	—	—	—	5	—	—	—
Small Bower	9½	—	—	—	8	—	—	—	6½	—	—	—	6	—	—	—	5½	—	—	—	5	—	—	—	5	—	—	—
Spare	9½	—	—	—	8	—	—	—	6½	—	—	—	6	—	—	—	5½	—	—	—	5	—	—	—	5	—	—	—
Seizings	1	—	—	—	1	—	—	—	¾	—	—	—	¾	—	—	—	¾	—	—	—	¾	—	—	—	¾	—	—	—
Wing cabled	9	—	—	—	8	—	—	—	6½	—	—	—	6	—	—	—	5½	—	—	—	5	—	—	—	5	—	—	—
Dog	8	—	—	—	7½	—	—	—	6	—	—	—	6	—	—	—	5½	—	—	—	5	—	—	—	5	—	—	—
Seizing	1	—	—	—	1	—	—	—	¾	—	—	—	¾	—	—	—	¾	—	—	—	¾	—	—	—	¾	—	—	—
Stream Anchor	5½	—	—	—	4½	—	—	—	4	—	—	—	4	—	—	—	3½	—	—	—	3	—	—	—	3	—	—	—
Kedge	4½	—	—	—	3½	—	—	—	3	—	—	—	3	—	—	—	3	—	—	—	2½	—	—	—	2½	—	—	—
Deck & Bitt cabled	12	T.	—	13	10½	T.	—	12	8½	T.	—	10	8	T.	—	10	7½	T.	—	10	7½	T.	—	10	7	T.	—	10
Laniards	3½	—	—	—	3	—	—	—	2½	—	—	—	2½	—	—	—	2	—	—	—	2	—	—	—	2	—	—	—
Seizing	2 1	—	—	—	2 1	—	—	—	1½ 1	—	—	—	1½ 1	—	—	—	1	—	—	—	1	—	—	—	1	—	—	—
Stoppers Preventer	9½	—	—	—	9	—	—	—	7	—	—	—	7	—	—	—	6½	—	—	—	6	—	—	—	5	—	—	—
Ditto	9	—	—	—	8	—	—	—	6½	—	—	—	6½	—	—	—	6	—	—	—	5½	—	—	—	5	—	—	—
Ditto	8	—	—	—	7	—	—	—	6	—	—	—	6	—	—	—	5½	—	—	—	5	—	—	—	4½	—	—	—
Ditto	7½	—	—	—	6½	—	—	—	5½	—	—	—	5½	—	—	—	5	—	—	—	4½	—	—	—	4	—	—	—
Ditto	7	—	—	—	6	—	—	—	5	—	—	—	5	—	—	—	4½	—	—	—	4	—	—	—	3½	—	—	—
Ditto	6½	—	—	—	5½	—	—	—	4	—	—	—	4	—	—	—	4	—	—	—	3½	—	—	—	—	—	—	—
Ditto	5½	—	—	—	5	—	—	—	3½	—	—	—	3½	—	—	—	3½	—	—	—	3	—	—	—	—	—	—	—
Ditto	5	—	—	—	—	—	—	—	—	—	—	—	—	—	—	—	—	—	—	—	—	—	—	—	—	—	—	—
Ditto	4½	—	—	—	—	—	—	—	—	—	—	—	—	—	—	—	—	—	—	—	—	—	—	—	—	—	—	—
Shk. Painters Sheet Anchor, cabled	9	—	—	—	7½	—	—	—	6	—	—	—	5½	—	—	—	5	—	—	—	4½	—	—	—	4½	—	—	—
Best Bower	8	T.	—	1	7½	T.	—	1	6	T.	—	1	5½	T.	—	1	5	T.	—	1	4½	T.	—	1	4½	T.	—	1
Small Bower	8	T.	—	1	7½	T.	—	1	6	T.	—	1	5½	T.	—	1	5	T.	—	1	4½	T.	—	1	4½	T.	—	1
Spare, cabled	9	—	—	—	7½	—	—	—	6	—	—	—	5½	—	—	—	5	—	—	—	4½	—	—	—	4½	—	—	—
Seizing	1	—	—	—	1	—	—	—	¾	—	—	—	¾	—	—	—	¾	—	—	—	¾	—	—	—	¾	—	—	—
Buoy-Ropes Sheet-Anchor, cabled	8½	—	—	—	7½	—	—	—	6	—	—	—	5½	—	—	—	4½	—	—	—	4½	—	—	—	4	—	—	—
Best Bower, cabled	8½	—	—	—	7½	—	—	—	6	—	—	—	5½	—	—	—	4½	—	—	—	4½	—	—	—	4	—	—	—
Small Bower, cabled	8½	—	—	—	7½	—	—	—	6	—	—	—	5½	—	—	—	4½	—	—	—	4½	—	—	—	4	—	—	—
Seizing	1	—	—	—	1	—	—	—	¾	—	—	—	¾	—	—	—	¾	—	—	—	¾	—	—	—	—	—	—	—
Storm Anchor cabled	5	—	—	—	4½	—	—	—	3½	—	—	—	3½	—	—	—	3	—	—	—	3	—	—	—	3	—	—	—
Kedge cabled	4	—	—	—	4	—	—	—	3	—	—	—	3	—	—	—	2½	—	—	—	2½	—	—	—	2½	—	—	—
Ropes, Davit	3½	—	—	—	3	—	—	—	3	—	—	—	2½	—	—	—	2½	—	—	—	2	—	—	—	2	—	—	—
Bell	3½	T.	—	1	3	T.	—	1	2½	T.	—	1	2	T.	—	1	2	T.	—	1	2	T.	—	1	2	T.	—	1

† The Hook to the Fish Tackle-Pendent is to be large enough as with ease to hook the ring of the Anchor.

Names of the Standing and Running Rigging.	110 to 74 Guns.				64 Guns.				50 to 36 Guns.				32 to 28 Guns.				24 Guns.				22 to 20 Guns.				18 to 14 Guns.			
		Blocks, &c.				Blocks, &c.				Blocks, &c.				Blocks, &c.				Blocks, &c.				Blocks, &c.				Blocks, &c.		
	In.	Species.	In	N	In.	Species.	In	N	In.	Species.	In	N	In.	Species.	In	N	In.	Species.	In	N	In.	Species.	In	N	In.	Species.	In	N
Bucket	Lashing		—	—	—	—	—	—	—	—	—	—	—	—	—	—	—	—	—	—	—	—	—	—	—	—	—	—
Swab	3½	S.	12	2	3	S.	11	2	3	S.	10	2	3	S.	10	2	3	S.	10	2	3	S.	10	2	2½	S.	9	2
Entering	3½	—	—	—	3½	—	—	—	3½	—	—	—	3	—	—	—	3	—	—	—	3	—	—	—	3	—	—	—
Passing	5	T.	—	2	5	T.	—	2	4	T.	—	2	—	—	—	—	—	—	—	—	—	—	—	—	—	—	—	—
Laniard	2½	—	—	—	2½	—	—	—	2	—	—	—	—	—	—	—	—	—	—	—	—	—	—	—	—	—	—	—
Slip	3½	—	—	—	3	—	—	—	3½	—	—	—	—	—	—	—	—	—	—	—	—	—	—	—	—	—	—	—
Quarters, Poop and Stantions, in the Waste	3½	T.	—	16	3½	T.	—	16	3½	T.	—	16	3	T.	—	14	3	T.	—	14	3	T.	—	14	2½	T.	—	14
Fore, Main and Mizen Tops	Lashing		—	—	—	—	—	—	—	—	—	—	—	—	—	—	—	—	—	—	—	—	—	—	—	—	—	—
Wheel, or Tiller, White	4½	S.	14	2	4	S.	13	2	3½	S.	11	2	3½	S.	11	2	3	S.	10	2	3	S.	10	2	3	S.	10	2
Strapping	4½	T.	—	2	4	T.	—	2	3	T.	—	2	3	T.	—	2	3	T.	—	2	3	—	—	—	3	—	—	—
Seizing	1	—	—	—	1	—	—	—	1	—	—	—	—	—	—	—	—	—	—	—	—	—	—	—	—	—	—	—
Puddening of Anchors	4 3	Old Canvas	—	yd 20	4 3	Old Canvas	—	yd 20	3½ 2	Old Canvas	—	yd 20	3 2	Old Canvas	—	yd 15	3 2	Old Canvas	—	yd 14	3 2	Old Canvas	—	yd 14	2½ 2	Old Canvas	—	yd 12
Seizing	1½	—	—	—	1½	—	—	—	1	—	—	—	1	—	—	—	1	—	—	—	1	—	—	—	¾	—	—	—
Slings, Buoy	4	T.	—	6	4	T.	—	6	3½	T.	—	6	3½	T.	—	6	3	T.	—	6	3	T.	—	6	2½	—	—	—
Laniards	3	—	—	—	3	—	—	—	2½	—	—	—	2½	—	—	—	2	—	—	—	2	—	—	—	2	—	—	—
Seizings	1	—	—	—	1	—	—	—	Tarred Line		—	—	—	—	—	—	—	—	—	—	—	—	—	—	—	—	—	—
Gun	8	—	—	—	8	—	—	—	6½	—	—	—	6	—	—	—	5½	—	—	—	5½	—	—	—	4½	—	—	—
Nut	7	—	—	—	7	—	—	—	5	—	—	—	4½	—	—	—	4	—	—	—	4	—	—	—	3	—	—	—
Butt	5	T.	—	4	5	T.	—	4	5	T.	—	4	5	T.	—	4	5	T.	—	3	5	T.	—	3	5	T.	—	2
Hogshead	4	T.	—	4	4	T.	—	4	4	T.	—	4	4	T.	—	4	4	T.	—	3	4	T.	—	3	4	T.	—	3
Can-Hook	4½	—	—	—	4½	—	—	—	4½	—	—	—	4½	—	—	—	4½	—	—	—	4½	—	—	—	4	—	—	—
Straps for Wood Buoys..	4	—	—	—	4	—	—	—	3½	—	—	—	3½	—	—	—	3½	—	—	—	3½	—	—	—	3	—	—	—
Swabs	3½	—	—	—	3½	—	—	—	3½	—	—	—	3½	—	—	—	3½	—	—	—	3½	—	—	—	3	—	—	—
Cable-Bends	3	—	—	—	2½	—	—	—	2	—	—	—	2	—	—	—	2	—	—	—	2	—	—	—	1½	—	—	—
Rudder-Pendents, cabled	7½	T.	—	4	7	T.	—	4	5½	T.	—	4	5	T.	—	4	5	T.	—	4	5	T.	—	4	4	T.	—	4
Laniards	3	—	—	—	2½	—	—	—	—	—	—	—	—	—	—	—	—	—	—	—	—	—	—	—	—	—	—	—
Falls	3½	*L.t. *S.	28 15	2 2	3	*L.t. *S.	26 13	2 2	3	*L.t. *S.	24 12	2 2	—	—	—	—	—	—	—	—	—	—	—	—	—	—	—	—
Strapping	4½	—	—	—	4	—	—	—	4	—	—	—	—	—	—	—	—	—	—	—	—	—	—	—	—	—	—	—
Seizing	¾	—	—	—	¾	—	—	—	¾	—	—	—	—	—	—	—	—	—	—	—	—	—	—	—	—	—	—	—
Stern-Ladders 4 Strands	6	—	—	—	6	—	—	—	4½	—	—	—	4	—	—	—	4	—	—	—	4	—	—	—	4	—	—	—
Middle-Rope ..	2	—	—	—	2	—	—	—	1½	—	—	—	1½	—	—	—	1½	—	—	—	1½	—	—	—	1	—	—	—
Lashing	2	—	—	—	1	—	—	—	1	—	—	—	1	—	—	—	1	—	—	—	1	—	—	—	1	—	—	—
Futtock Staves ..	8 5	—	—	—	8 5	—	—	—	6 3½	—	—	—	5 3	—	—	—	5 3	—	—	—	4½ 3	—	—	—	4 2½	—	—	—
Swifters, for Capstan Bars	2	—	—	—	2	—	—	—	2	—	—	—	2	—	—	—	2	—	—	—	2	—	—	—	1½	—	—	—
Netting	1½	—	—	—	1½	—	—	—	1	—	—	—	1	—	—	—	1	—	—	—	¾	—	—	—	¾	—	—	—
Haliard for Top Lantern	1	S.	5	1	1	S.	5	1	1	—	—	—	—	—	—	—	—	—	—	—	—	—	—	—	—	—	—	—
Ensign..	2	—	—	—	1½	—	—	—	1	—	—	—	1	—	—	—	1	—	—	—	1	—	—	—	¾	—	—	—
Jack	1	—	—	—	1	—	—	—	¾	—	—	—	¾	—	—	—	¾	—	—	—	¾	—	—	—	¾	—	—	—
For Colours: Head-Line ..	White Line		—	—	—	—	—	—	—	—	—	—	—	—	—	—	—	—	—	—	—	—	—	—	—	—	—	—
For Colours: Rope-Bands ..	Marline		—	—	—	—	—	—	—	—	—	—	—	—	—	—	—	—	—	—	—	—	—	—	—	—	—	—
For Colours: Pendent-Slings	¾	—	—	—	¾	—	—	—	¾	—	—	—	¾	—	—	—	¾	—	—	—	¾	—	—	—	¾	—	—	—
Awnings Ridge and Side Ropes	3½	S.	12	3	3½	S.	12	3	3	S.	10	3	3	S.	10	3	3	S.	10	3	3	S.	10	3	2½	S.	9	3
Stops	1	—	—	—	1	—	—	—	1	—	—	—	1	—	—	—	1	—	—	—	1	—	—	—	¾	—	—	—
Crowfeet ..	1	Euphroe	34 24 22	1 1 1	1	Euphroe	32 22 20	1 1 1	1	Euphroe	26 16 14	1 1 1	1	Euphroe	26 16 14	1 1 1	1	Euphroe	24 14 12	1 1 1	1	Euphroe	24 14 12	1 1 1	¾	Euphroe	23 13 11	2 2 1
Haliard	1½	—	—	—	1½	—	—	—	1½	—	—	—	1½	—	—	—	1½	—	—	—	1½	—	—	—	1	—	—	—
Ridge-Tackle Fall	2½	D. S.	9 9	1 1	2½	D. S.	9 9	1 1	2½	D. S.	8 8	1 1	2½	D. S.	8 8	1 1	2½	D. S.	7 7	1 1	2	D. S.	7 7	1 1	2	D. S.	7 7	1 1
For different Uses of the Ship ..	— — —	*Sn.I.bd. Hks c.	24 16 14	4 4 4	— — —	*Sn.I.bd. Hks c.	22 15 14	4 4 4	— — —	*Sn.I.bd. Hks c.	17 13 12	4 4 4	— — —	*Sn.I.bd. Hks c.	16 13 12	4 4 4	— — —	*Sn.I.bd. Hks c.	14 12 11	4 4 4	— — —	*Sn.I.bd. Hks c.	14 11 10	4 4 4	— — —	*Sn.I.bd. Hks c.	12 10 9	4 3 3
Strapping	4½	—	—	—	4	—	—	—	4	—	—	—	4	—	—	—	3½	—	—	—	3	—	—	—	3	—	—	—
Seizing	¾	—	—	—	¾	—	—	—	¾	—	—	—	¾	—	—	—	—	—	—	—	—	—	—	—	—	—	—	—
Long Boat's Rigging.																												
Burton-Pendents ..	4	S.	11	2	4	S.	11	2	3½	S.	9	2	3½	S.	9	2	3	S.	8	2	3	S.	8	2	—	—	—	—
Runner	3	T.	—	2	3	T.	—	2	2½	T.	—	2	2½	T.	—	2	2½	T.	—	2	2½	T.	—	2	—	—	—	—
Falls	2½	*D. *S.	9 9	2 2	2½	*D. *S.	9 9	2 2	2	*D. *S.	8 8	2 2	2	*D. *S.	8 8	2 2	2	*D. *S.	8 8	2 2	2	*D. *S.	8 8	2 2	—	—	—	—
Strapping	3	—	—	—	3	—	—	—	2½	—	—	2	2½	—	—	—	2½	—	—	—	2½	—	—	—	—	—	—	—
Shrouds	4	D.E.	5	4	3½	D.E.	5	4	3	T.	—	4	3	T.	—	4	3	T.	—	4	2½	T.	—	4	—	—	—	—
Laniard	1½	—	—	—	1½	—	—	—	1	—	—	—	1	—	—	—	1	—	—	—	1	—	—	—	—	—	—	—
Stay	4½	D.E.	5	1	4	D.E.	5	1	3½	T.	—	1	3½	T.	—	1	3½	T.	—	1	3½	T.	—	1	2	—	—	—
Laniard	1½	—	—	—	1½	—	—	—	1	—	—	—	1	—	—	—	1	—	—	—	1	—	—	—	—	—	—	—
Tie	3	S.	9	1	3	S.	9	1	3	S.	9	1	—	—	—	—	—	—	—	—	—	—	—	—	—	—	—	—
Main Haliard ..	2	S.	7	2	2	S.	7	2	1½	S.	6	1	1½	S.	6	1	1½	S.	6	1	1	S.	5	1	1	—	—	—
Outer Haliard ..	2	D.I.bd. *S.	7 7	1 1	2	D.I.bd. *S.	7 7	1 1	1½	D.I.bd. *S.	6 6	1 1	1½	D.I.bd. *S.	6 6	1 1	1½	D.I.bd. *S.	6 6	1 1	1	D.I.bd. *S.	5 5	1 1	—	—	—	—
Sheet	2	D. *S.	8 8	1 1	2	D. *S.	8 8	1 1	1½	D. *S.	6 6	1 1	1½	S. T.	6 —	2 1	1½	S. T.	6 —	2 1	1	S. T.	5 —	2 1	1 —	—	—	—
Wooden Hoops	—	—	—	—	—	—	—	—	—	—	—	—	—	—	—	—	—	—	—	—	—	—	—	—	—	—	—	—
Downhauler ..	1½	S.	5	1	1½	S.	5	1	1	S.	5	1	1	S.	5	1	1	—	—	—	1	—	—	—	—	—	—	—
Strapping	2	—	—	—	2	—	—	—	1½	—	—	—	1½	—	—	—	1½	—	—	—	1	—	—	—	—	—	—	—
Topping-Lifts.. ..	2½	—	—	—	2	—	—	—	1½	—	—	—	1½	—	—	—	1½	—	—	—	1	—	—	—	—	—	—	—
Fore-Haliard	2	S.	7	1	2	S.	7	1	2	S.	7	1	2	S.	7	1	1½	—	—	—	1	—	—	—	1	—	—	—
Sheet	2	T.	—	2	2	T.	—	2	2	T.	—	2	2	T.	—	2	1½	T.	—	2	1	—	—	—	1	—	—	—
Tack	1	—	—	—	1	—	—	—	1	—	—	—	1	—	—	—	1	—	—	—	¾	—	—	—	—	—	—	—
Bowline	2½	T.	—	1	2½	T.	—	1	2	—	—	—	2	—	—	—	1½	—	—	—	1	—	—	—	—	—	—	—
Jib-Haliard ..	2	S. *	7 —	2 1	2	S. *	7 —	2 1	2	S.	7	1	2	S.	7	1	1½	—	—	—	1	—	—	—	—	—	—	—
Sheet	2½	T.	—	4	2½	—	—	—	2	—	—	—	2	—	—	—	1½	—	—	—	1	—	—	—	—	—	—	—
Out-Hauler	3	I.Tra.	—	1	2½	I.Tra.	—	1	2	I.Tra.	—	1	2	I.Tra.	—	1	1½	I.Tra.	—	1	1	I.Tra.	—	1	1	—	—	—

Names of the Standing and Running Rigging.	110 to 74 Guns.				64 Guns.				50 to 36 Guns.				32 to 28 Guns.				24 Guns.				22 to 20 Guns.				18 to 14 Guns.			
		Blocks, &c.				Blocks, &c.				Blocks, &c.				Blocks, &c.				Blocks, &c.				Blocks, &c.				Blocks, &c.		
	In.	Species.	In	N	In.	Species.	In	N	In.	Species.	In	N	In.	Species.	In	N	In.	Species.	In	N	In.	Species.	In	N	In.	Species.	In	N
In-Hauler	1	—	—	—	1	—	—	—	1	—	—	—	1	—	—	—	¾	—	—	—	¾	—	—	—	—	—	—	—
Boat-Rope cabled ..	9	—	—	—	7½	—	—	—	6½	—	—	—	6	—	—	—	4½	—	—	—	4½	—	—	—	—	—	—	—
Laniard	2½	—	—	—	2½	—	—	—	2	—	—	—	2	—	—	—	2	—	—	—	2	—	—	—	—	—	—	—
Guest-Rope cable ..	5	—	—	—	4	—	—	—	3	—	—	—	3	—	—	—	3	—	—	—	3	—	—	—	—	—	—	—
Grapnel-Rope cabled ..	4½	T.	—	1	4	T.	—	1	3½	T.	—	1	3½	T.	—	1	3½	T.	—	1	3½	T.	—	1	3	T.	—	1
Painter	4	T.	—	1	4	T.	—	1	3½	T.	—	1	3½	T.	—	1	3½	T.	—	1	3	T.	—	1	3	T.	—	1
Sternfast	3½	T.	—	1	3½	T.	—	1	3	T.	—	1	3	T.	—	1	2	T.	—	1	2	T.	—	1	—	—	—	—
Fenders cabled ..	6	—	—	—	6	—	—	—	5	—	—	—	4	—	—	—	4	—	—	—	3½	—	—	—	3	—	—	—
Laniards	2	—	—	—	2	—	—	—	1½	—	—	—	1½	—	—	—	1½	—	—	—	1	—	—	—	1	—	—	—
Rudder-Laniards ..	2	—	—	—	2	—	—	—	1½	—	—	—	1½	—	—	—	1	—	—	—	1	—	—	—	1	—	—	—
Pinnace.																												
Haliards { Main ..	1½	—	—	—	1½	—	—	—	1½	—	—	—	1½	—	—	—	1½	—	—	—	1½	—	—	—	—	—	—	—
Haliards { Fore ..	1½	—	—	—	1½	—	—	—	1½	—	—	—	1½	—	—	—	1½	—	—	—	1½	—	—	—	—	—	—	—
Sheets	1½	—	—	—	1½	—	—	—	1½	—	—	—	1½	—	—	—	1½	—	—	—	1½	—	—	—	—	—	—	—
Grapnel-Rope cabled ..	3½	T.	—	1	3½	T.	—	1	3½	T.	—	1	3½	T.	—	1	3½	T.	—	1	3½	T.	—	1	—	—	—	—
Painter	3½	T.	—	1	3½	T.	—	1	3½	T.	—	1	3½	T.	—	1	3½	T.	—	1	3½	T.	—	1	—	—	—	—
Sternfast	2½	T.	—	1	2½	T.	—	1	2	T.	—	1	2	T.	—	1	2	T.	—	1	2	T.	—	1	—	—	—	—
Slings	5½	H. & T.	8	12	5½	H. & T.	8	12	5	H. & T.	4	6	5	H. & T.	4	6	5	H. & T.	4	6	5	H. & T.	4	6	—	—	—	—
Seizings	1	—	—	—	1	—	—	—	¾	—	—	—	¾	—	—	—	¾	—	—	—	¾	—	—	—	—	—	—	—
Rudder-Laniards ..	1	—	—	—	1	—	—	—	1	—	—	—	1	—	—	—	1	—	—	—	1	—	—	—	—	—	—	—
Yawl, or Cutter, or Pinnace.																												
Main and Fore Haliards (if Cutters) ..	2	I.Tra.	—	2	2	I.Tra.	—	2	1	I.Tra.	—	2	1	I.Tra.	—	2	1	—	—	—	1	—	—	—	1½	—	—	—
Main } Sheets	2½	—	—	—	2	—	—	—	2	—	—	—	2	—	—	—	2	—	—	—	2	—	—	—				
Fore } Sheets	2½	—	—	—	1½	—	—	—	1½	—	—	—	1½	—	—	—	1½	—	—	—	1	—	—	—	1½	—	—	—
Grapnel-Rope cabled ..	3½	T.	—	1	3½	T.	—	1	3½	T.	—	1	3½	T.	—	1	3½	T.	—	1	3½	T.	—	1	3½	T.	—	1
Painter	3½	T.	—	1	3½	T.	—	1	3½	T.	—	1	3½	T.	—	1	3½	T.	—	1	3½	T.	—	1	3½	T.	—	1
Sternfast	—	—	—	—	—	—	—	—	—	—	—	—	—	—	—	—	—	—	—	—	—	—	—	—	2	T.	—	1
Slings	5½	H. & T.	8	12	5½	H. & T.	8	12	5½	H. & T.	8	12	5½	H. & T.	4	6	5½	H. & T.	4	6	5½	H. & T.	4	6	4½	H. & T.	4	6
Seizings	1	—	—	—	1	—	—	—	1	—	—	—	1	—	—	—	1	—	—	—	1	—	—	—	¾	—	—	—
Rudder-Laniards ..	1	—	—	—	1	—	—	—	1	—	—	—	1	—	—	—	1	—	—	—	1	—	—	—	1	—	—	—

Table 63: 'A Table of the Quantities and Dimensions of the Standing and Running Rigging of Brigs of 160 tons, Cutters of 200 tons, Sloops of 130 tons, and Ketches of 150 tons'.

A TABLE OF THE QUANTITIES AND DIMENSIONS OF THE STANDING AND RUNNING RIGGING OF BRIGS OF 160 TONS.

Names of the Standing and Running Rigging.	In.	Fath in Len.	Blocks &c. Species.	In	N
Bowsprit.					
Gammonings	4	26	—	—	—
Shrouds fine	4	8	—	—	—
Collars fine	3	2½	T.	—	2
Seizings	—	½	of a tarr'd Line		
Lashings	¾	5	—	—	—
Laniards fine	2	7	—	—	—
Bobstays cabled fine	4	11½	T.	—	2
Collars fine	4	3	—	—	—
Seizing	—	½	of a tarr'd Line		
Lashings	¾	6	—	—	—
Laniards fine	2	9	—	—	—
Horses	2½	9	T.	—	6
Straps	2	2	T.	—	4
Laniards	1	4	—	—	—
Spritsail.					
Horses	2	6	—	—	—
Braces	2	30	D.	7	4
Strapping	2½	4	—	—	—
Lifts	1½	20	S.	6	2
Beckets	1½	2	—	—	—
Strapping	1½	12	—	—	—
Standing	2	3	T.	—	4
Straps	2	1½	—	—	—
Laniards	¾	3	—	—	—
Haliard	1½	12	S.	6	2
Strapping	1½	1	—	—	—
Slings	3	3	—	—	—
Seizing and Racking	¾	5	—	—	—
Clue-lines	1½	16	St. bd.	6	2
Strapping	1½	1½	—	—	—
Bunt-Lines	1	14	—	—	—
Strapping	1½	1½	—	—	—
Ear-Rings	¾	10	—	—	—
Sheets cabled single	2½	22	—	—	—
Jib.					
Horses	2½	9	—	—	—
Guy-Pendents	2½	20	—	—	—
Out-Hauler	2	12	—	—	—
Stay	2½	14	—	—	—
Strapping	2½	1	—	—	—
Tackle-Fall	1½	12	S.	6	2
Strapping	1½	1	—	—	—
Haliard	2	22	S.	7	1
Strapping	2	1	—	—	—
Down-Hauler	1½	18	S.	6	1
Sheets single	2½	18	S.	8	2
Fore-mast.					
Girdlines	2	35	S.	8	2
Strapping	2	1	—	—	—
Lashings	¾	8	—	—	—
Pendents of Tackle cabled fine	6	5½	S.c. T.	13 —	2 2
Strapping	4	2	—	—	—
Seizing	¾	6	—	—	—
Runners of Tackles	4	14	*D.	10	2
Strapping	3	2	—	—	—
Falls of Tackles	2½	48	*S.th.c.	10 10	2 2
Strapping	3	5	—	—	—
Shrouds cabled, fine	5	47	D.E.	8	8
Seizing, Eye	¾	—	—	—	—
Throat	¾	60	—	—	—
Laniards, fine	3	40	—	—	—
Ratling	1	60	—	—	—
Stay cabled fine 4 Strands	7½	7½	H.	10	1
Seizings	¾	8	—	—	—
Laniard, fine	3	7	—	—	—
Collar cabled 4 Str. fine double	4	3½	H.	10	1
Seizings	¾	6	—	—	—
Lashing	1½	4	—	—	—
Preventer-Stay cabled 4 Strands, fine	5½	7½	H.	8	1
Laniard, fine	2½	7	—	—	—
Collar cabled 4 Strands fine double	3½	3½	H.d.sc.	8	1
Catharpin-Legs	3	6	—	—	—
Seizings	¾	25	—	—	—
Crowfoot for the Top	¾	40	Euphroe	12	1

Names of the Standing and Running Rigging.	In.	Fath in Len.	Blocks, &c. Species.	In	N
Tackle	1	4	—	—	—
Strapping	1	1	—	—	—
Jeers Tie	5	10	S.th.c.	14	2
Jeers Falls	2½	40	*D.co.	9	4
Strapping	2½	1	—	—	—
Lashing at the Mst.-Hd.	2	16	—	—	—
Horses	3	7½	T.	—	2
Stirrups	2	2	T.	—	2
Laniards	¾	2	—	—	—
Yard-Tackle-Pendents	4½	6	D.th.c.	10	2
Falls	2½	48	*S.th.c. S.	11 9	2 2
Strapping	4	2	—	—	—
Braces	2½	44	S.	8	4
Pendents	3	4	S.	8	2
Preventer	2½	5	—	—	—
Strapping	2½	3	—	—	—
Lifts	2	40	S.	7	4
Span for the Cap	3	3	—	—	—
Short Span	2½	1	—	—	—
Strapping	2½	3	—	—	—
Truss-Pendents	4	9	T.	—	4
Falls	1½	26	*D. *S.	6 6	2 2
Strapping	2	3	—	—	—
Eye-Seizings	¾	8	—	—	—
Puddening the Yard	4	4	—	—	—
Clue-Garnets	2	34	S. S.st.bd.	8 8	4 2
Straps about the Yard	2	4	—	—	—
Strapping	2	1	—	—	—
Buntline-Legs	1½	28	D.	6	4
Falls	1½	20	S.	6	6
Strapping	2	7	—	—	—
Leechline-Legs	1½	26	S.	6	6
Falls	1½	26	—	—	—
Strapping	1½	4	—	—	—
Slablines	1½	12	S.	5	1
Bowlines	2½	28	S.	9	2
Bridles	2½	3	—	—	—
Strapping	3	2	—	—	—
Lashing	1	4	—	—	—
Ear-Ring	1	20	—	—	—
Sheets cabled	3	36	S.c.	11	2
Strapping	3	2½	—	—	—
Seizing	¾	7	—	—	—
Tacks taper and cabled	4	22	—	—	—
Strapping	3	3	—	—	—
Seizing	¾	6	—	—	—
Studdingsail Hal. Inner	1½	25	S.	7	6
Outer	2	42	S.	7	4
Sheets	1½	7	—	—	—
Tacks	2	32	S.	7	2
Strapping	2	10	—	—	—
Fore Topmast.					
Burton-Pendents	2½	4	H.	—	2
Falls	1½	30	*S.	6 6	2 2
Strapping	2	2	—	—	—
Shrouds, fine	3½	28	D.E.	6	6
Seizings, Eye	—	½	of a tarr'd Line		
Throat / End	—	1	Tarr'd Line		—
Laniards, fine	2	24	—	—	—
Ratling	¾	35	—	—	—
Standing Backsts. fine	3½	38	D.E.	6	4
Seizings, Eye	—	½	of a tarr'd Line		
Throat / End	—	1	Tarr'd Line		—
Laniards, fine	2	16	—	—	—
Stay cabled 4 Strands fine	4	13	—	—	—
Collar, fine	3	2½	S.	12	1
Tackle	2	8	S.	7	2
Strapping	3	1½	—	—	—
Seizing	¾	7	—	—	—
Shifting-Backstays, fine	3½	9½	—	—	—
Tackles	1½	8	*D. *S.	6 6	1 1
Strapping	2	1	—	—	—
Futtock-Shrouds, fine	3½	11	*Fut.pla	6	6

Names of the Standing and Running Rigging.	In.	Fath in Len.	Blocks, &c. Species.	In	N
Seizing Upper / Lower	¾	20	—	—	—
Ratling	1	10	—	—	—
Top Rope-Pendents, fine	4½	11	S.I.bd.c.	13	1
Falls	2½	30	D.I.b.	11	2
Tie	3½	10	S.do.co.	11	1
Strapping	3½	1	T.	—	1
Lashers at the Mast-Head	1	7	—	—	—
Yard	¾	3	—	—	—
Haliards	2	32	D.th.c. *S.th.c.	10 10	1 1
Strapping	3	2½	—	—	—
Horses	2½	5½	—	—	—
Braces	2	44	S.	7	4
Pendents	2½	4½	S.	7	2
Strapping	2	3½	—	—	—
Lifts	2	34	Sis. S.	14 7	2 4
Beckets	2	2	—	—	—
Strapping	2	4	—	—	—
Parral-Ropes	2	7	Par.	11	1
Racking and Seizing	—	½	of a tarr'd Line		
Clue-Lines	2	44	S.st.bd. S.	7 7	4 2
Strapping	2	4	—	—	—
Bunt-Lines	1½	30	S.	6	4
Strapping	1½	1½	—	—	—
Leech-Lines	1	13	S.	5	2
Strapping	1	1	—	—	—
Bow-Lines	2	32	S.	7	2
Bridles	2	5	T.	—	6
Strapping	2	1½	—	—	—
Lashing	¾	5	—	—	—
Reef-Tackle Pendents	2	22	*S.	6	4
Falls	1	28	—	—	—
Strapping	1½	1	—	—	—
Ear-Rings	¾	30	—	—	—
Sheets	3½	26	S.Sho. ¼ do.thk. & th.	12 12	2 2
Straps for Sheet Blocks	4	3	—	—	—
Quarter Blocks	3	5	—	—	—
Lashers for Quarter Blocks	1	5	—	—	—
Seizings	¾	12	—	—	—
Span	2½	3	—	—	—
Slings	2½	4½	—	—	—
Staysail-Stay	2½	13	S.	9	1
Tackle	1	6	S.	5	2
Haliard	2	22	S.	7	1
Strapping	2	1	—	—	—
Sheets	2½	16	S.	8	2
Strapping	2½	1½	—	—	—
Out-Hauler	1½	8	S.	6	1
Down-Hauler	1	14	S.	5	1
Strapping	1	½	—	—	—
Studdingsail-Haliards	2	45	S.	7	6
Sheets	1½	9	S.	7	2
Tacks	2	40	S.	7	2
Down Haulers	¾	30	T	—	6
Tails and Straps	2	10	—	—	—
Fore-Topgallant-Mast.					
Shrouds	2	32	T.	—	8
Laniards	¾	8	—	—	—
Standing Backstays	2	24	T.	—	2
Laniards	¾	6	—	—	—
Stay	2½	19	T.	—	1
Strapping	2	½	—	—	—
Flagstaff-Haliards	¾	18	—	—	—
Tie	2	9	—	—	—
Haliard	1½	14	S.	5	2
Strapping	1½	1½	—	—	—
Horses	1½	4½	—	—	—
Braces	1	42	S.	5	4
Strapping	1	1	T.	—	2
Lifts single	1½	22	T.	—	2
Parral-Ropes	1	2½	Par.	6	1
Clue-Lines	1	34	S.	5	4
Strapping	1	1	—	—	—
Bow-Lines	1	42	T.	—	6

Names of the Standing and Running Rigging.	Inch	Fath in Len.	Blocks, &c. Species.	In	N
Bridles	1	2	—	—	—
Ear-Rings	—	½	of a tarr'd	Li	ne
Studdingsail-Haliards	1½	40	—	—	—
Sheets	1	12	—	—	—
Tacks	1½	34	—	—	—
Down-Haulers	¾	13	—	—	—
Royal Haliards	2	28	—	—	—
Main-Mast.					
Girt-Lines	2½	44	S.	8	2
Strapping	2	1	—	—	—
Lashings	¾	8	—	—	—
Pendents of Tackle cabled, fine	5	7	S.c.	3	2
Strapping	4	2	—	—	—
Seizing	¾	6	—	—	—
Runners of Tackles	4	16	*D.th.c.	10	2
Strapping	3	2	—	—	—
Falls of Tackles	2½	50	*S.th.c.	10	2
Strapping	3	5	—	—	—
Shrouds cabled, fine	5	70	D.E.	8	10
Throat	¾	40	—	—	—
End	¾	40	—	—	—
Laniards, fine	3	50	—	—	—
Ratling	1	85	—	—	—
Stay cabled 4 Strands, fine	8	11½	H.	12	1
Seizing	1	10	—	—	—
Laniards, fine	3	9	—	—	—
Collar cab. 4 Str., fine	6	4	H.	12	1
Worming	¾	25	—	—	—
Seizing	1	10	—	—	—
Lashing	2½	9	—	—	—
Catharpin-Legs	3	7	—	—	—
Seizings	¾	25	—	—	—
Stay-Tackle-Pendents	4	3	D.th.c.	11	1
Falls	2½	28	*S.th.c.	11	1
				9	1
Strapping	3	2½	T.	—	1
Seizing	¾	8	—	—	—
Lashing	1	7	—	—	—
Fore-Stay Tackle-Fall	2	28	*D.th.c.	11	1
			*S.th.c.	11	1
				9	1
Strapping	3	2½	—	—	—
Seizing	¾	8	—	—	—
Crowfoot for the Top	¾	50	Euphroe	14	1
Tackle	1	4	—	—	—
Strapping	1	1	T.	—	2
Jears { Tie	3	34	D.	11	1
Jears { Falls			S.	11	2
Strapping	3½	10	—	—	—
Lashers to the Mast-Head	2	18	—	—	—
Yard	1½	4	—	—	—
Horses	2½	8	T.	—	2
Yard-Tackle-Pendents	4½	6½	*D.th.c.	10	2
Falls	2½	54	*S.th.c.	11	2
			S.	9	2
Straps	4	2½	—	—	—
Strapping	3	3	—	—	—
Braces	2	38	S.c.	7	4
Pendents	2½	5	S.c.	7	2
Strapping	2½	2½	—	—	—
Lifts	2	46	S.	7	4
Span for the Cap	2½	3	—	—	—
Short Span	2½	1	—	—	—
Strapping	2	3	—	—	—
Truss-Pendents	4	11	T.	—	4
Falls	1½	30	*D.	6	2
			*S.	6	2
Strapping	4	7½	Horse down the Mast D.E.	7	1
Eye-Seizings	3	2	Strap D.E	7	1
Laniards	2	4	—	—	—
Seizings	¾	6	—	—	—
Sheets cabled	3	25	S.	11	2
Strapping	3½	3	—	—	—
Luff-Tackles	2½	40	*D.	10	2
			*S.	10	2
Strapping	2½	4	—	—	—
Staysail-Stay	4	11½	T.	—	2
Collar	3	2	—	—	—
Seizings	¾	6	—	—	—
Laniard	1	2	—	—	—
Haliard	2	21	*S.	7	3
Sheets	2½	7	*S.	7	2
Tacks	2½	2	—	—	—
Down-Hauler	1	10	S.	5	1
Strapping	2	3	—	—	—

Names of the Standing and Running Rigging.	Inch	Fath in Len.	Blocks, &c. Species.	In	N
Studdingsail-Haliards Inner	2	34	S.	9	6
Outer	2½	58	S.	9	4
Sheets	2	14	—	—	—
Tacks	2½	32	S.	9	2
Strapping	2½	12	—	—	—
Boom-Guy-Pendents	4½	7	T.	—	1
Tackle-Fall	2½	20	*D.	10	1
			*S.	10	1
Ditto	1½	5	*D.	6	2
Tricing-Line	1½	22	S.	6	2
Peek-Line	1	18	S.	5	1
Hanks	2½	20	—	—	—
Trucks			—	—	36
Lacing	1	30	—	—	—
Strapping	2½	3	—	—	—
	1½	3	—	—	—
Gaff-Topping-Lift	1	1			
	4	8½	S.I.bd.c.	13	1
Span	4	3	T.	—	1
Haliards	2½	24	D.	9	1
			*S.	9	1
Strapping	3½	1½	—	—	—
	2½	2	—	—	—
Tie	4	7½	S.I.bd.	13	1
Haliards	2½	28	D.th.	13	1
			*S.th.	13	2
Boom-Topping Lift	4½	16	S.I.bd.	14	1
			S.	11	1
Strapping	3	7	—	—	—
Fall	2	16	D.	8	1
			*S.	8	1
Strapping	3½	3	—	—	—
	3	3	—	—	—
	2	2	—	—	—
Main-Topmast.					
Burton-Pendents	2½	4	T.	—	2
Falls	1½	34	*D.	6	2
			*S.	6	2
Strapping	2	2	—	—	—
Shrouds, fine	3½	31	D.E.	6	6
Seizings, Eye	—	½	of a tarr'd	Li	ne
Throat / End	—	1	Tarr'd	Li	ne—
Laniards, fine	2	24	—	—	—
Ratling	¾	40	—	—	—
Standing-Backstays, fine	3½	46	D.E.	6	4
Seizings, Eye	—	½	of a tarr'd	Li	ne
Throat / End	—	1	Tarr'd	Li	ne—
Laniards, fine	2	16	—	—	—
Stay cabled 4 Str. fine	4	13	—	—	—
Collar, fine	3½	2	S.	12	1
Tackle	2	12	D.	7	1
			*S.	7	1
Strapping	3	1½	—	—	—
Seizing / Lashing	2	1½	—	—	—
Preventer Stay cabled 4 Strands, fine	3½	12	—	—	—
Lashing the Collar	1	6	—	—	—
Shifting Backstays, fine	3½	10½	T.	—	1
Tackles	1½	8	*D.	6	1
			*S.	6	1
Futtock-Shrouds, fine	3½	12	*Pl.d.e.	6	6
Seizings, Upper / Lower	¾	20	—	—	—
Ratling	1	12	—	—	—
Top-Rope-Pendents, fine	4½	12	S.I.bd.	13	1
Falls	2½	32	D.I.bd.c.	11	2
Tie	3½	11	S.do.sc.	11	1
Strapping	3½	1	T.	—	1
Lashers at the Mast-Head	1½	7	—	—	—
Yard	¾	3	—	—	—
Haliards	2	34	D.th.c.	10	1
			*S.th.c.	10	1
Strapping	2½	2½	—	—	—
Horses	2½	5½	—	—	—
Braces	2	45	S.	7	4
Pendents	2½	5	S.	7	2
Strapping	2	2½	—	—	—
Lifts	2	36	Sis.	14	2
			S.	7	4
Beckets	2	2	—	—	—
Strapping	2	4	—	—	—
Parral-Rope	2	7	Par.	11	1
Racking and Seizing	—	½	of a tarr'd	Li	ne

Names of the Standing and Running Rigging.	Inch	Fath in Len.	Blocks, &c. Species.	In	N
Clue-Lines	2	48	S.st.bd.	7	4
			S.	7	2
Strapping	2	4	—	—	—
Bunt-Lines	1½	34	S.	6	4
Strapping	1½	1½	—	—	—
Leech-lines	1	14	S.	5	2
Strapping	1	1	—	—	—
Bow-lines	2	34	S.	7	2
Bridles	2	8	T.	—	6
Strapping	2	1	—	—	—
Frapping and Lashing	1	7	—	—	—
Reef-Tackle Pendents	2	24	*S.	6	4
Falls	1½	32	—	—	—
Strapping	1½	1	—	—	—
Ear-rings	¾	30	—	—	—
Sheets	3½	30	Sho.c.	12	2
			¼ D.thk. & th.	12	2
Straps for Sht.-Block	4	3	—	—	—
Quarter-Blocks	3	5	—	—	—
Lashers for Quarter-Bl.	1	5	—	—	—
Seizings	¾	12	—	—	—
Span	2½	3	—	—	—
Slings	2½	4½	—	—	—
Staysail Haliards	2	22	S.	7	1
Strapping	2	1	—	—	—
Sheets	2	19	S.	7	2
Strapping	2	1½	—	—	—
Tack	2	2	—	—	—
Down-Hauler	1	12	S.	5	1
Strapping	1	1	—	—	—
Middle Staysail-Stay	2½	15	S.	9	1
Tackle	1½	27	S.	6	2
Haliard	2	21	S.	7	1
Sheets	2½	24	S.	7	2
Tack	1½	6	—	—	—
Down-Hauler	1	15	S.	5	1
Strapping	2	4	—	—	—
Studdingsail Haliards	2	50	S.	7	6
Sheets	1½	10	S.	7	2
Tacks	2	42	S.	7	2
Down-Haulers	¾	32	T.	—	6
Lashers for Booms	2	16	—	—	—
Tailing and Strapping	2	10	—	—	—
Main Topgallant Mast.					
Shrouds	2	32	T.	—	8
Laniard	¾	8	—	—	—
Standing-Backstays	2	27	T.	—	12
Laniards	¾	6	—	—	—
Stay	2½	14	T.	—	1
Strapping	2	½	—	—	—
Flagstaff-Haliards	¾	18	—	—	—
Tie	2	9	—	—	—
Haliard	1½	16	S.	5	2
Strapping	1½	1½	—	—	—
Horses	1½	4½	—	—	—
Braces	1	44	S.	5	4
Strapping	1	1	—	—	—
Lifts, single	1½	23	T.	—	2
Parral-Ropes	1	2½	Par.	6	1
Clue-lines	1	36	S.	5	4
Strapping	1	1½	—	—	—
Bow-lines	1	42	T.	—	4
Bridles	1	4	T.	—	2
Ear-rings	—	½	of a tarr'd	Li	ne
Staysail-stay	2	17	S.	7	1
Haliard	1	24	S.	5	1
Sheets	1	26	S.	5	2
Tacks	1	4	—	—	—
Down-Hauler	¾	18	S.	4	2
Strapping	1	2	—	—	—
Studdingsail-Haliards	1½	48	S.	6	6
Sheets	1	30	—	—	—
Tacks	1½	38	S.	6	4
Down-Haulers	¾	28	—	—	—
Royal-Haliards	2	3½	—	—	—
Necessary Ropes.					
Cat-Falls	2½	36	D.I.bd.c.	11	2
Laniards	1	4	—	—	—
Stoppers	1½	2	—	—	—
Guys, Mast-Head	4	20	—	—	—
Fore	4	12	—	—	—
After	2½	14	—	—	—
Fish-Tackle-Pendent	4	3	H. 1 & T.	—	2
Stoppers, Sheet Anchor	4	4	—	—	—
Best Bower	4	4	—	—	—
Small Bower	4	4	—	—	—
Seizings	¾	10	—	—	—

Names of the Standing and Running Rigging.	In.	Fath in Len.	Blocks, &c. Species.	In	N
Stream-Anchor ..	2½	3	—	—	—
Kedge	2	3	—	—	—
Deck and Bit cabled ..	6	12	T.	—	6
Laniards	2	16	—	—	—
Seizing	¾	36	—	—	—
Stoppers Preventer ..	4½	9	—	—	—
Ditto	4	9	—	—	—
Ditto	3½	9	—	—	—
Ditto	3	9	—	—	—
Ditto	2½	11	—	—	—
Shank - Painters Sheet - Anchor cabled ..	4	4	—	—	—
Best Bower	4	4	T.	—	1
Small Bower	4	4	T.	—	1
Seizing	¾	10	—	—	—
Buoy - Rope's Sheet Anchor cabled ..	4	18	—	—	—
Best Bower cabled ..	4	18	—	—	—
Small Bower cabled ..	4	18	—	—	—
Storm Anchor cabled ..	3	18	—	—	—
Kedge cabled	2½	16	—	—	—
Ropes, Entering ..	2½	9	—	—	—

Names of the Standing and Running Rigging.	In.	Fath in Len.	Blocks, &c. Species.	In	N
Quarts. and Stantns. in the Waste ..	2½	130	—	—	—
Wheel, or Tiller, White ..	2	10	S.	7	4
Strapping	2	2	—	—	—
Puddening of Anchors {	2½	40	—	—	—
	2	14	—	—	—
Seizings	¾	36	—	—	—
Slings, Buoy	2½	30	—	—	—
Laniards	1	6	—	—	—
Seizing	¾	16	—	—	—
Butt	4½	5	T.	—	2
Hogshead	3½	8	T.	—	3
Can-Hook	3½	4	—	—	—
Straps for wood Buoys ..	3	4	—	—	—
Swabs	3	9	—	—	—
Cable-Bends	1½	32	—	—	—
Futtock-Staves .. {	4	4	—	—	—
	2½	2	—	—	—
Lashing Cables between Decks, Sheep Pens, Steep-Tubs, Tubs in the Tops, & Booms ..	¾	220	—	—	—

Names of the Standing and Running Rigging.	In.	Fath in Len.	Blocks, &c. Species.	In	N
Haliard for Ensigns ..	¾	12	—	—	—
Jacks	¾	8	—	—	—
For Colours { Head-Line ..	—	1	White Line	—	—
Rope-Bands ..	—	2lb.	Marline	—	—
Pendent-Sligs.	¾	18	—	—	—
For different Uses of the Ship .. {	—	—	Sn.I.bd. Hk. c.	12	2
	—	—		10	2
	—	—		9	2
Pinnace, Yawl or Cutter.					
Main and Fore Haliards..	1½	10	—	—	—
Sheets	1½	10	—	—	—
Grapnel Rope cabled ..	3	35	—	—	—
Painter	3	5	T.	—	2
Sternfast	2	5	—	—	—
Slings	4½	7	H. & T.	4	6
Seizings	¾	10	—	—	—
Rudder-Laniards ..	¾	1	—	—	—

A TABLE OF THE QUANTITIES AND DIMENSIONS OF THE STANDING AND RUNNING RIGGING OF CUTTERS OF 200 TONS.

Names of the Standing and Running Rigging.	Inch	Fath in Len.	Blocks, &c. Species.	In	N
Bowsprit.					
Shrouds	5½	14	—	—	—
Tackle-Falls	2½	38	*D.	10	4
Jib.					
Haliard	5	35	S.c.	14	2
Tackle-Fall	3½	40	Tr.c.	12	2
Jack	6½	24	Sn.I.bd.c.	15	1
Sheets cable-laid ..	5	18	*S.	14	2
Downhauler	2	24	S.	8	1
Inhauler .. .. {	2½	15	D.c.	8	1
			*Sn.c.c.	8	1
Heelrope	3	21	Sn.I.bd.	13	1
Flying-Jib-Haliard ..	2½	35	—	—	—
Sheets	2½	24	—	—	—
Jack	2½	30	—	—	—
Down Hauler	2	24	—	—	—
Foremast.					
Sail-Haliards .. {	2½	50	D.c.	10	1
			*S.	10	1
Down-Hauler	2	30	*S.	8	1
Jack-Tackle	2½	10	*S.	8	2
Bow-Line	3	14	*S.	8	2
Sheets	3	7	*S.	10	2
Mainmast.					
Girtlines	2½	50	S.	9	2
Lashings	1	10	—	—	—
Pendents of Tackles ..	6	26	S.c.	15	2
Runners of Tackles ..	5½	24	*L.t.c.	20	2
Falls	3½	45	*S.c.	11	2
Shrouds cabled, fine ..	8	84	D.E.	12	8
Seizings, Eye / Throat / End .. }	1	68	—	—	—
Laniards, fine	4	40	—	—	—
Ratline	1½	120	—	—	—
Stay cabled 4 Strds., fine	13	16	D.E.	18	1
Seizings, fine	1½	10	—	—	—
Laniard, fine	4	10	—	—	—
Worming	1½	60	—	—	—
Lashing	2	25	—	—	—
Preventer-Stay cabled 4 Strands .. }	6½	18	D.E.	9	1
			I.b.	9	1
Laniards, fine	3	4	—	—	—
Seizing	1	9	—	—	—
Boom-Topping Lifts ..	4½	26	S.I.bd.c.	12	1
Runner .. .. {	4	12	S.c.	12	1
			S.c.	13	1
Fall .. .. {	2½	32	D.c.	12	1
			*S.c.	9	1
Guy-Pendent ..	5	9	T.	—	1
Tackle-Fall .. {	3	30	*D.	12	1
			*S.	12	1
Gaff-Span	4½	5	—	—	—

Names of the Standing and Running Rigging.	Inch	Fath in Len.	Blocks, &c. Species.	In	N
Down-Hauler-Peek ..	2	28	S.	8	1
Throat	2	38	S.	8	1
Inner-Tie	6	13	S.I.bd.c.	14	2
Haliard	3	56	D.c.	14	2
Outer Tie	6	19½	—	—	—
Haliards .. {	3½	90	S.I.bd.c.	12	3
			D.c.	12	2
Ear-rings, Inner / Outer .. {	1½	18	—	—	—
Sheet	3	60	Tr.c.	15	2
Tack-Tackle .. {	2½	20	*D.	8	1
			*S.	8	1
Luff-Tackles .. {	2½	40	*D.	11	2
			*S.	11	2
Main-Reef-Pendents, fine .. {	5	7	—	—	—
	5	7	—	—	—
	5	9	—	—	—
	5	10	—	—	—
Topmast.					
Tackle-Fall .. {	3	30	*L.t.	20	1
			*S.	11	1
Tie	3	26	—	—	—
Haliard .. {	2	32	D.	9	1
			*S.	9	1
Horses	2½	8	—	—	—
Braces	2	60	S.	8	2
Lifts	2	45	S.	8	4
Parral-Ropes	2	8	Par.	12	1
Racking and Seizing ..	¾	10	—	—	—
Clue-lines	1½	44	S.	6	4
Bunt lines..	1½	44	S.	6	2
Bow-lines	2	66	Tr.	9	1
Bridles	2	7	T.	—	6
Sheets	3½	42	S.Sho.	10	2
Quarter-Blocks ..	—	—	D&Dsc.c.	12	1
Trysail-Sheet ..	2	33	*Tr.	11	2
Down-Hauler	2	24	—	—	—
Lacing	2	24	—	—	—
Studdingsail-Haliards ..	2	96	—	—	—
Sheets	2	38	—	—	—
Tacks	2	90	—	—	—
Down-Haulers ..	2	30	—	—	—
Topgallant-mast.					
Standing Backstays ..	2½	60	—	—	—
Tackles	2	28	*S.	8	8
Stay	3	30	—	—	—
Haliards	2	40	S.	9	2
Top-Rope	4	28	D.	9	1
Fall	2	45	*S.	9	1
Tricing-Line	2	30	S.	8	2
Cross-jack-yard.					
Clue-Lines	2	40	—	—	—
Braces	2½	60	S.	8	2
Sheets	2½	14	—	—	—

Names of the Standing and Running Rigging.	Inch	Fath in Len.	Blocks, &c. Species.	In	N
Haliard .. {	2½	50	*D.c.	9	1
			*S.c.	9	1
Lifts, running ..	3	35	S.	8	2
Bunt-Lines	2	40	—	—	—
Tacks	2½	14	—	—	—
Horses	3	12	—	—	—
Stirrups	2	8	—	—	—
Horse down the Mast ..	5	25	D.E.	7	2
Strap	3½	4	—	—	—
Laniard	2½	5	—	—	—
Necessary Ropes.					
Cat-Falls	3	44	D.I.bd.c.	12	2
Fish-Tackle-Pendent ..	4	3	H. & T.	1	1
Stoppers, Sheet-Anchor ..	5	5	—	—	—
Best Bower	5	5	—	—	—
Small Bower	5	5	—	—	—
Stream Anchor ..	2½	3	—	—	—
Kedge	2	3	—	—	—
Deck and Bit cabled ..	6½	7½	T.	—	4
Laniards	2	10	—	—	—
Seizings	¾	20	—	—	—
Shank Painters Sheet-Anchor cabled ..	4	6	—	—	—
Best Bower	4	4	—	—	—
Small Bower	4	4	—	—	—
Buoy-Rope Sheet-Anchor cabled	4	18	—	—	—
Best Bower cabled ..	4	18	—	—	—
Small Bower cabled ..	4	18	—	—	—
Stream-Anchor cabled	3	20	—	—	—
Kedge cabled	2½	20	—	—	—
Entering	3	10	—	—	—
Wheel, White	2	12	S.	7	2
Puddening of Anchors {	2½	50	—	—	—
	2	18	—	—	—
	¾	42	—	—	—
	2½	30	—	—	—
	1½	6	—	—	—
Slings, Buoy	4	12	—	—	—
Hogshead ..	3	3	—	—	—
Can-Hooks ..	4	6	—	—	—
Cable-Bends	1½	34	—	—	—
Nettings for the Tops, Quarters, Waste, and Barricadoes	¾	540	—	—	—
Haliard for Ensigns ..	¾	20	—	—	—
Slings, Pendent, &c. ..	¾	24	—	—	—
For different Uses of the Ship .. {	—	—	*Sn.I. bd.c. Bulkhd.	12	2
	—	—		10	2
	—	—		9	2
Ridge Ropes for the Quarters ..	2½	108	—	—	—

Names of the Standing and Running Rigging.	In.	Fath in Len.	Blocks, &c. Species.	In	N
Boat.					
Main and Fore Haliards . .	1½	10	—	—	—
Main and Fore Sheets . .	1½	10	—	—	—
Grapnel-Ropes cabled . .	3	35	—	—	—
Painter	3	5	—	—	—
Slings	4½	5	—	—	—
Seizings	¾	10	—	—	—
Rudder-Laniards . .	¾	1	—	—	—
Sternfast	2	5	—	—	—

Fitting the Rigging in the House.

Spunyarn	20 Cwt.
Lines, tarred	8 No.
Marline	15 lb.
Old Canvas	140 Yards.
Tar	1½ Barrels.
Tallow	28 lb.
Twine Ordinary	2 lb.

Length of the Warp of the Main Shrouds, 18 Yards

A TABLE OF THE QUANTITIES AND DIMENSIONS OF THE STANDING AND RUNNING RIGGING OF SLOOPS OF 130 TONS.

Names of the Standing and Running Rigging.	Inch	Fath in Len.	Blocks, &c. Species.	In	N
Bowsprit.					
Gammoning	3½	10	H.	8	2
Laniards	2½	2½	—	—	—
Shrouds	3½	12	—	—	—
Tackle-Falls	2	12	*S.	8	2
Horses	3	13	T.	—	2
Laniards	1	3½	—	—	—
Strap	2½	3	T.	—	2
Jib.					
Stay cabled 4 Strds., fine	4½	21	—	—	—
Haliard	2½	33	*S.	9	2
Sheets cable-laid . .	3½	20	*S.	10	2
Downhauler	1½	22	S.	6	1
Fore.					
Sail-Haliards	2½	32	*S.	9	2
Down-Hauler	1½	15	—	—	—
Tack-Tackle	1	4	—	—	—
Bow-Line	3	12	—	—	—
Bridles	3	1½	—	—	—
Sheets	3	5	S.	9	1
Fore Burton-Pendents	3	5½	S.	10	1
Falls . . . . }	2½	24	D.	9	1
			*S.	9	2
Mainmast.					
Girt Lines	2½	50	S.	9	2
Lashings	1	8	—	—	—
Pendents of Tackle . .	4½	12	S.c.	14	2
Runners of Tackles . .	4½	22	*L.t.c.	24	2
Falls	2½	46	*S.th.c.	14	2
Shrouds cabled, fine . .	6	78	D.E.	10	8
Seizings, Eye . . / Throat / End . . }	1	60	—	—	—
Laniards, fine	2½	32	—	—	—
Ratline	1	146	—	—	—
Stay cabled 4 Str. fine . .	9	12	D.E.	17	1
Seizings, fine	1½	9	—	—	—
Laniard, fine	3	10	—	—	—
Worming	1	50	—	—	—
Boom-Topping-Lift . .	3½	27	*S.I.bd.	13	1
Runner . . }	3	10	S.	13	1
			S.	11	1

Names of the Standing and Running Rigging.	Inch	Fath in Len.	Blocks, &c. Species.	In	N
Fall . . . . }	2	20	D.	9	1
			*S.	9	1
Guy-Pendent . .	4½	7	—	—	—
Tackle-Fall . . }	2½	20	*D.	9	1
			*S.	9	1
Gaff-Span	4½	5	—	—	—
Down-Hauler Peek . .	2	22	S.	7	2
Throat	2	32	S.	7	1
Inner Tie	4½	14	S.I.bd.c.	15	1
Haliard	2½	36	*D.th.c.	19	2
Outer Tie	4½	15	S.I.bd.c.	15	1
Haliard	2½	40	*D.th.c.	19	2
Ear-Rings, Inner }	1½	4	—	—	—
Outer }	1	8	—		—
Reef . .	3½	9	—	—	—
Lacing	1	14	—	—	—
Sheet	3	36	D.	14	2
Rope	2	5	—	—	—
Tack-Tackle . . }	1½	7	*D.	7	1
			*S.	7	1
Luff-Tackles . . }	2	16	*D.	8	1
			*S.	8	1
Topmast.					
Haliard	2	30	—	—	—
Braces	1½	36	—	—	—
Lifts	1½	34	—	—	—
Bow-Lines	1½	34	—	—	—
Bridles	1½	4	T.	—	4
Sheets	1½	34	—	—	—
Cross-Jack-Yard.					
Braces	1½	40	S.	6	2
Sheets	2½	15	—	—	—
Haliard	2	48	*D.	8	2
Tacks	2½	15	H. & T.	2	2
Horse down the Mast . .	4	15	T.	—	4
Yard-Ropes	2	40	S.	8	2
Necessary Ropes.					
Cat Falls	3½	8	D.I.b.	12	2
Fish-Tackle-Pendents . .	3½	5	H. & T.	1	1
Stoppers, Sheet-Anchor	3	8	—	—	—
Best Bower	3	8	—	—	—
Small Bower	3	8	—	—	—

Names of the Standing and Running Rigging.	Inch	Fath in Len.	Blocks, &c. Species.	In	N
Stream-Anchor . .	3	8	—	—	—
Kedge	3	8	—	—	—
Shank - Painter Sheet-Anchor cable . .	3	8	—	—	—
Best Bower	3	8	—	—	—
Small Bower	3	8	—	—	—
Buoy - Rope Sheet-Anchor cabled . .	2½	18	—	—	—
Best Bower	2½	18	—	—	—
Small Bower	2½	16	—	—	—
Puddening of Anchors . .	2	25	worn	—	—
	2	25	worn	—	—
	2	25	worn	—	—
	2	25	worn	—	—
	2	25	worn	—	—
Buoy-Slings	2½	10	—	—	—
Hogshead-Slings	3	3	T.	—	1
Cable-Bends	1	12	—	—	—
Boat.					
Painter	3	7	—	—	—

Fitting the Rigging in the House.

Spunyarn	8 Cwt.
Lines, tarred	4 No.
Marline	10 lb.
Old Canvas	50 Yards.
Tar	½ Barrel.
Tallow	28 lb.
Twine Ordinary	1 lb.

Length of the Warp of the Main Shrouds, 17 Yards

A TABLE OF THE QUANTITIES AND DIMENSIONS OF THE STANDING AND RUNNING RIGGING OF KETCHES OF 150 TONS.

Names of the Standing and Running Rigging.	Inch	Fath in Len.	Blocks, &c. Species.	In	N
Bowsprit.					
Strap	2½	3	—	—	—
Gammoning	3½	20	—	—	—
Shrouds, fine	3½	14	*D.E.	7	4
Laniards, fine	1	6	H. & T.	—	2
Bobstays cabled, fine . .	3	6	D.E.	6	2

Names of the Standing and Running Rigging.	Inch	Fath in Len.	Blocks, &c. Species.	In	N
Collars, fine	2½	3	—	—	—
Laniards, fine	1½	3½	—	—	—
Horses	3	14	worn	—	—
Laniards	1½	4	—	—	—
Sheets, cabled	2	16	—	—	—
Ropebands & Ear-Rings . .	¾	25	—	—	—

Names of the Standing and Running Rigging.	Inch	Fath in Len.	Blocks, &c. Species.	In	N
Haliards . . }	2½	16	S.	8	1
			I.Cring.	—	30
Jib.					
Stay cabled 4 Strands . .	4½	24	S.	10	2
Sheets	4½	16	H. & T.	1	1
Tack	2	28	—	—	—

Names of the Standing and Running Rigging.	In.	Fath in Len.	Blocks, &c. Species.	In	N
Haliard	2	36	*S.	10	1
Haliard	1½	55	S.	7	2
Down-Hauler	2	20	*S.	7	1
Brails	1½	20	—	—	—
Laniards	1½	4	—	—	—
Boom-Haliards	2	17	—	—	—
Inhauler	1½	18	—	—	—
Fore-sail.					
Haliard	2½	30	I. Cring.	—	30
			*S.	8	2
Down-Hauler	1½	16	S.	7	1
Bow-Lines	2	20	—	—	—
Sheets tapered and cabled	2½	40	*D.	10	4
Main-mast.					
Girtlines	2½	40	S.	9	2
Lashings	1	8	—	—	—
Pendents of Tackles cabled, fine	4	12	*L.t.	20	1
			*S.	12	1
Seizings	¾	10	—	—	—
Runners of Tackles	3½	12	*L.t.	20	1
			*S.	11	2
Falls	2	64	—	—	—
Shrouds cabled, fine	6	120	D.E.	9	12
Seizings, Eye	1	100	—	—	—
Throat	¾	50	—	—	—
End	¾	50	—	—	—
Worming	¾	540	—	—	—
Laniards, fine	3	48	—	—	—
Ratline	1	160	—	—	—
Stay cabled 4 Strds., fine	10	12	D.E.	16	1
Seizings	1½	8	—	—	—
Worming	1	64	—	—	—
Laniards, fine	3½	18	—	—	—
Mousing	¾	45	—	—	—
Catharpin-Legs	1½	16	—	—	—
Burton-Tackle-Runners	3½	11	S.	11	2
Falls	2	24	*S.	13	1
Crowfoot for the Top	1	24	Euphroe	14	1
Tackle	1½	3	S.	7	2
Jeers { Tie	4½	44	D.	15	1
Jeers { Falls			S.	15	3
Lashers to the Mast-Head	2	10	—	—	—
Yard	2	10	—	—	—
Horses	3	10	T.	—	2
Laniards	1½	4	—	—	—
Braces	2½	50	S.	10	2
Pendents	3	8	S.	10	2
Lifts	2½	60	S.	10	2
Span for the Cap	3½	3	S.	10	2
Parral-Ropes	3	14	Par.	16	1
Nave-Lines	1½	18	—	—	—
Racking and Seizings	1	18	—	—	—
Main	1½	26	S.	7	2
Brails, Middle	1½	26	T.	—	4
Lower	1½	24	T.	—	4
Hanks	2	15	—	—	—
Truss { Pendents	4	9	—	—	—
Truss { Falls	2	12	D.	7	1
Clue-Garnets	2½	60	S.str.	10	4
Buntline-Leg-Falls	2	80	L.Buntl.	14	2
			S.	8	8
Leechline-Leg-Falls	2	40	S.	8	4
Bow-Lines	4	36	D.	14	1
Bridles	4	12	T.	—	4
Ropebands & Ear-Rings	1	60	—	—	—
Sheets taper'd & cabled	4½	18	S.Sho.	14	2
Tacks tapered & cabled	5½	26	—	—	—
Wingsail Guys and Lee Fangs	1½	30	—	—	—
Haliard	2	40	D.	10	2
Sheets	2	44	D.	9	4
Topping-Lift	2	36	—	—	—
Span	2	3	—	—	—
Luff-Tackle	2	20	*L.t.	16	1
			*S.	10	1
Studdingsail-Haliards, Inner and Outer	2	40	S.	8	4
Sheets	1½	10	S.	8	2
Tacks	1½	20	—	—	—
Main-topmast.					
Shrouds, fine	3½	33	*D.E.	6	6
Seizings, Eye	¾	6	—	—	—
Throat	¾	9	—	—	—
End	¾	9	—	—	—
Laniards fine	2	18	—	—	—
Ratline	¾	50	—	—	—

Names of the Standing and Running Rigging.	In.	Fath in Len.	Blocks, &c. Species.	In	N
Standing-Backstays, fine	3½	30	D.E.I.bd. with H.ks	6	2
Seizings, Eye, Throat, End	¾	10	—	—	—
Laniards, fine	2	10	—	—	—
Stay cabled 4 Strands, fine	3½	26	S.	10	1
				9	1
Runners, fine	3½	14	S.	8	2
				7	1
Laniards	1½	4	—	—	—
Shifting-Backstays, fine	3½	16	T.	—	1
Tackle	1½	10	*S.	7	2
Seizings	¾	5	—	—	—
Futtock-Shrouds, fine	3½	10	*Pl. de	6	6
Seizings, Upper, Lower	¾	12	—	—	—
Ratline	¾	12	—	—	—
Top-Rope-Pendent, fine	4½	13	T.	—	1
Fall	2	40	*D.	9	1
			*S.	8	2
Tie	3½	14	S.	12	2
Lashers at the Mast Head	1	4	—	—	—
Yard	1	2	—	—	—
Haliard	2	34	D.	11	1
			*S.	11	1
Horses	2½	7	—	—	—
Braces	2	50	S.	8	2
Pendents	2½	6	S.	8	2
Lifts	2	52	S.	8	4
Beckets	2	2	—	—	—
Parral-ropes	2	8	Par.	11	1
Racking and Seizing	¾	8	—	—	—
Clue-Lines	2	60	S.st.bd.	7	4
				7	4
Bunt-Lines	½	30	S.	7	2
Leech-Lines	1½	18	S.	7	2
Bow-Lines	2	50	S.	8	2
Bridles	2	12	T.	—	4
Ropebands & Ear-Rings	¾	35	—	—	—
Sheets tapered	3½	44	S.Sho.	14	2
			D. quart	14	2
Lashers for Quarter Block	1	5	—	—	—
Span	3	3	—	—	—
Studdingsail-Haliards	1½	70	S.	7	8
Sheets	1½	6	—	—	—
Tacks	1½	46	S.	7	4
Bow-Lines	1	28	—	—	—
Topgallant Mast.					
Haliards	1½	40	S.	7	2
Braces	1	40	T.	—	4
Clue-Lines	1	40	S.	6	4
Mizen Mast.					
Girt-Lines	2	30	S.	7	2
Lashings	1	4	—	—	—
Shrouds, fine	3	36	D.E.	6	6
Seizings, Eye	¾	6	—	—	—
Throat	¾	9	—	—	—
End	¾	9	—	—	—
Laniards, fine	1½	10	—	—	—
Ratline	¾	50	—	—	—
Stay cabled 4 Str. fine	3½	12	D.E.	6	2
Laniards, fine	1½	6	—	—	—
Haliard and Strap	1½	35	—	—	—
Jears	2	16	D.	7	1
			S.	7	1
Lashers at the Mast-Head	1	5	—	—	—
Yard	1	2	—	—	—
Derrick	2	20	S.	8	3
Span	2	3	S.sn.	12	1
Yard-Rope-Pendents	2	6	S.	6	2
Fall	2	20	S.	6	2
Backstays	2½	12	D.E.	6	2
Laniards	1½	7	—	—	—
Brail-Peek-Legs	¾	28	S.	6	4
Main	1½	12	S.	6	4
Lower	1	30	T.	—	10
Lacing Mizen to Yard	¾	10	—	—	—
Ear-Rings and Ropebands	1	4	—	—	—
Sheet	2	7	*S.	8	2
Staysail-Haliard	1½	24	S.	7	2
Sheet	2	4	—	—	—
Tack	2	4	—	—	—
Cross-Jack-Yard.					
Braces	1½	24	S.	6	4
Laniards	¾	3	—	—	—
Lifts standing	1½	4	—	—	—
Strap	1½	2	T.	—	2
Slings	2	2	T.	—	1

Names of the Standing and Running Rigging.	In.	Fath in Len.	Blocks, &c. Species.	In	N
Worming for the Bowsprit, Main - Top & Mizen - Shrouds, & Backstays	No. 36, whitelines of 12 threads.	—	—	—	—
Mizen-Topmast.					
Shrouds, fine	2	12	T.	—	8
Laniards, fine	¾	5	—	—	—
Standing Backstays, fine	1½	20	—	—	—
Laniards, fine	1	4	—	—	—
Stay cabled 4 Str. fine	1	14	—	—	—
Futtock-Shrouds, fine	1½	4	T.	—	4
Tie	2	4	S.	7	1
Haliard	1½	16	S.	7	2
Horses	2	5	—	—	—
Braces	1½	22	S.	6	2
Lifts	2	18	S.	6	2
Parral-Rope	1½	4	—	—	—
Clue-Lines	1½	22	S.	6	2
Bow-Line	1½	24	S.	6	2
Bridles	1	6	T.	—	6
Ear-Rings & Ropebands	¾	25	—	—	—
Sheets	2	20	S.	8	4
Necessary Ropes.					
Cat-Falls	2½	14	D.I.bd.	12	2
Fish-Tackle-Pendent	4	3	*S.	12	1
			*D.sc.		
Fall	2	22	*D.	10	1
			*S.	10	1
Stoppers { Best Bower / Small ditto	4	8	—	—	—
Deck & Bit. cab. No. 10	5	3	T.	—	8
Laniards	1½	4	—	—	—
Seizings	¾	10	—	—	—
Shank-Painter { Best Bower / Small ditto	3	6	—	—	—
Buoy Rope Sheet Anchor cabled	4	18	—	—	—
Best Bower cabled	4	18	—	—	—
Small ditto ditto	4	18	—	—	—
Puddening of Anchors	2	30	worn	—	—
Seizings	¾	30	—	—	—
Slings, Buoy	2½	16	T.	—	1
Straps	2	4	—	—	—
Butt	4	4½	T.	—	1
Hogshead	3½	3½	T.	—	1
Cable-Bends	1½	30	—	—	—
Futttock-Staves	5	3	—	—	—
	3	1½	—	—	—

Length of the first Warp of

Main-Shrouds	18 Yards	1 Foot.
Main-Topmast Shrouds	10 ,,	2 ,,
Main-Backstays	28 ,,	2 ,,
Mizen-Shrouds	11 ,,	2 ,,

For Fitting the Rigging in the House.

Species.	Quantity.
Spunyarn	16 Cwt.
Lines, tarred	12 Number.
Marline, ditto	12 lb.
Old Canvas	60 Yards.
Tar	1 Barrel.
Tallow	28 lb.
Twine Ordinary	1 lb.

Table 64: 'Dimensions of the Standing and Running Rigging of Merchant-Shipping; with the Species, Size, and Number, of Blocks, Hearts, Dead-eyes, etc'.

DIMENSIONS OF THE STANDING AND RUNNING RIGGING OF MERCHANT - SHIPPING ;

WITH THE SPECIES, SIZE, AND NUMBER, OF BLOCKS, HEARTS, DEAD-EYES, ETC.

Bowsprit and Spritsail-yard.	1257 Tons.					818 Tons.					544 Tons.					330 Tons.					Sloop of 60 Tons.				
	Size in Inch	Fath in Len.	Blocks, &c. Species.	Sz in In	Numb.	Size in Inch	Fath in Len.	Blocks, &c. Species.	Sz in In	Numb.	Size in Inch	Fath in Len.	Blocks, &c. Species.	Sz in In	Numb.	Size in Inch	Fath in Len.	Blocks, &c. Species.	Sz in In	Numb.	Size in Inch	Fath in Len.	Blocks, &c. Species.	Sz in In	Numb.
Bowsprit.																									
Gammoning	7	122	—	—	—	6½	119	—	—	—	5½	75	—	—	—	5	63	—	—	—	—	—	—	—	—
Shrouds	7	30	*Hearts	13	4	6½	13½	*Hearts	11	2	5½	12	*Hearts	9	2	5	10½	*Hearts	7	2	3½	9	—	—	—
Collars	6	10	Hearts	13	4	5½	4½	Hearts	11	2	4½	4	Hearts	9	2	4	3½	Hearts	7	2	—	—	—	—	—
Seizings	1	15	—	—	—	1	7	—	—	—	1	6	—	—	—	¾	5	—	—	—	—	—	—	—	—
Lashings	2	12	—	—	—	1½	6	—	—	—	1½	5	—	—	—	1	4½	—	—	—	—	—	—	—	—
Laniards	3	20	—	—	—	2½	9	—	—	—	2½	8	—	—	—	2½	7	—	—	—	2	9	*Single	6	2
Bobstays 4 Strands cabld.	9	30	Hearts	14	3	8½	26½	Hearts	11	3	7	16	Hearts	9	2	6	14	Hearts	7	2	—	—	—	—	—
Collars	9	8	Hearts	14	3	8½	6½	Hearts	11	3	7	4	Hearts	9	2	6	3½	Hearts	7	2	—	—	—	—	—
Seizings	1½	30	—	—	—	1	26½	—	—	—	1	16	—	—	—	¾	14	—	—	—	—	—	—	—	—
Lashings	2	9	—	—	—	1½	7½	—	—	—	1½	5	—	—	—	1	4½	—	—	—	—	—	—	—	—
Laniards	4½	16	—	—	—	4	13	—	—	—	3	8	—	—	—	3	7	—	—	—	—	—	—	—	—
Horses	4½	17	Thimbls.	—	6	4	15	Thimbls.	—	6	3	14	Thimbls.	—	6	3	12	Thimbls.	—	6	3	11	Thimbls.	—	2
Straps	3	3½	Thimbls.	—	4	3	3	Thimbls.	—	4	2½	2½	Thimbls.	—	4	2½	2	Thimbls.	—	4	2½	2	Thimbls.	—	2
Laniards	2	6½	—	—	—	1½	5½	—	—	—	1½	5	—	—	—	1½	4½	—	—	—	¾	3½	—	—	—
Spritsail-yard.																									
Horses	3½	11	—	—	—	3	10	—	—	—	3	9½	—	—	—	3	7	—	—	—	—	—	—	—	—
Stirrups	3	5½	Thimbls.	—	6	2½	5	Thimbls.	—	6	2½	4½	Thimbls.	—	6	2½	3½	Thimbls.	1	6	—	—	—	—	—
Braces	3	70	Double	11	4	2½	62½	Double	9	4	2½	60	Double	9	4	2	42½	Double	9	4	—	—	—	—	—
Pendents..	3½	5½	Single	11	4	3	5	Single	9	4	3	5	Single	9	4	3	3½	Single	9	4	—	—	—	—	—
Strapping	3½	10	—	—	—	3	9	—	—	—	3	8½	—	—	—	3	6	—	—	—	—	—	—	—	—
Lifts	3	49	Single	11	4	2½	44	Single	9	4	2½	42	Single	9	4	2½	30	Single	9	4	—	—	—	—	—
Beckets	3	3	—	—	—	2½	2½	—	—	—	2½	2	—	—	—	2½	2	—	—	—	—	—	—	—	—
Strapping	3½	4	—	—	—	2½	3	—	—	—	2½	2	—	—	—	2½	2	—	—	—	—	—	—	—	—
Seizing	¾	8	—	—	—	¾	6	—	—	—	¾	4	—	—	—	¾	4	—	—	—	—	—	—	—	—
Standing	4	8	Thimbls.	—	4	3	7	Thimbls.	—	4	3	7	Thimbls.	—	4	3	5	Thimbls.	—	4	—	—	—	—	—
Straps	4	4	—	—	—	3	3½	—	—	—	3	3½	—	—	—	2½	2½	—	—	—	—	—	—	—	—
Laniards	1½	6	—	—	—	1½	5	—	—	—	1½	5	—	—	—	1	4	—	—	—	—	—	—	—	—
Haliards	3	33½	*Long.t. Single	24 12	1 1	2½	30	*Long.t. Single	18 9	1 1	2½	28	*Long.t. Single	18 9	1 1	2½	20	*Long.t. Single	18 9	1 1	—	—	—	—	—
Strapping	3½	3	—	—	—	3	2½	—	—	—	3	2	—	—	—	3	2	—	—	—	—	—	—	—	—
Seizing and Lashing	¾	6½	—	—	—	¾	6	—	—	—	¾	5½	—	—	—	—	—	—	—	—	—	—	—	—	—
Slings	5	5½	—	—	—	4½	5	—	—	—	4	4½	—	—	—	4	3½	—	—	—	—	—	—	—	—
Seizing and Racking ..	1½	11	—	—	—	1	10	—	—	—	¾	9	—	—	—	¾	7	—	—	—	—	—	—	—	—
Clue-Lines	2½	37	Strap bd. Single	10 10	4 2	2	33	Strap bd. Single	8 8	4 2	2	32	Strap bd. Single	8 8	4 2	2	23	Strap bd. Single	8 8	4 2	—	—	—	—	—
Strapping	3	4	—	—	—	2½	3½	—	—	—	2½	3	—	—	—	2	2½	—	—	—	—	—	—	—	—
Bunt-Lines	2	28	Single	8	2	2	25	Single	8	2	1½	24	Single	6	2	1½	17	Single	6	2	—	—	—	—	—

Spritsail-Yard, Jib, Flying-Jib, Spritsail, Topsail and Yard.	1257 Tons.					818 Tons.					544 Tons.					330 Tons.					Sloop of 60 Tons.				
	Size in Inch	Fath in Len.	Blocks, &c. Species.	In	N	Size in Inch	Fath in Len.	Blocks, &c. Species.	In	N	Size in Inch	Fath in Len.	Blocks, &c. Species.	In	N	Size in Inch	Fath in Len.	Blocks, &c. Species.	In	N	Size in Inch	Fath in Len.	Blocks, &c. Species.	In	N
Strapping	2½	1	—	—	—	2	1	—	—	—	1½	1	—	—	—	1½	1	—	—	—	—	—	—	—	—
Ear-Rings	1½	14	—	—	—	1	12	—	—	—	1	12	—	—	—	1	8½	—	—	—	—	—	—	—	—
Sheets cabled	4½	35	—	—	—	3½	31	—	—	—	3½	30	—	—	—	3½	22	—	—	—	—	—	—	—	—
Jib.																									
Horses	3½	18	—	—	—	3½	15	—	—	—	3	14	—	—	—	3	13	—	—	—	—	—	—	—	—
Seizings	1	9	—	—	—	1	7	—	—	—	¾	7	—	—	—	—	—	—	—	—	—	—	—	—	—
Guy-Pendents	4	27½	Single	13	2	3½	23	Single	11	2	3½	22	Single	11	2	3	21	Single	9	2	—	—	—	—	—
Falls	2½	24	Single	9	4	2½	20	Single	9	4	2½	19	Single	9	4	2½	18	Single	7	4	—	—	—	—	—
Strapping ..	4 2½	4 3	—	—	—	3½ 2½	3 2	—	—	—	3½ 2½	3 2	—	—	—	3	3	—	—	—	—	—	—	—	—
Lashers	1	13½	—	—	—	1	11½	—	—	—	¾	9½	—	—	—	¾	9	—	—	—	—	—	—	—	—
Out-Hauler..	3½	11	—	—	—	3	9½	—	—	—	2½	9	—	—	—	2½	8½	—	—	—	—	—	—	—	—
Tackle-Fall	2½	27	Single	8	2	2	23	Single	7	2	2	22	Single	7	2	2	21	Single	7	2	—	—	—	—	—
Strapping	2	1	—	—	—	2	1	—	—	—	2	1	—	—	—	2	1	—	—	—	—	—	—	—	—
Stay 4 Strands cabled ..	5	27½	Single	14	1	4½	23	Single	12	1	3½	22	Single	10	1	3	21	Single	10	1	3½	13	—	—	—
Strapping	4½	1	—	—	—	4½	1	—	—	—	3½	1	—	—	—	3	1	—	—	—	—	—	—	—	—
Tackle-Fall ..	2½	20	Double Single	10 10	1 1	2	19	Double Single	9 9	1 1	2	18	Single	7	2	2	17	Single	7	2	—	—	—	—	—
Strapping	2½	2	—	—	—	2	1½	—	—	—	2	1	—	—	—	2	1	—	—	—	—	—	—	—	—
Haliards	3½	40	Single	12	1	3	35	Single	11	1	2½	33	Single	9	1	2½	31	Single	8	1	2	22	*Single	7	2
Strapping	3½	1	—	—	—	3	1	—	—	—	2½	1	—	—	—	2½	1½	—	—	—	—	—	—	—	—
Downhauler	2½	35	Single	9	1	2	31	Single	9	1	1½	28	Single	7	1	1½	27	Single	6	1	1½	14½	Single	5	1
Sheets, single	3	38½	Single	11	2	3	33	Single	11	2	3	31½	Single	10	2	2½	—	Single	9	2	2	13	*Single	9	2
Pendents..	4	9½	Single	11	2	4	8	Single	11	2	—	—	—	—	—	—	—	—	—	—	—	—	—	—	—

Spritsail-Yard, Jib, Flying-Jib, Spritsail, Topsail and Yard.—*continued*.	1257 Tons.					818 Tons.					544 Tons.					330 Tons.					Sloop of 60 Tons.				
	Size in Inch	Fath in Len.	Blocks, &c. Species.	In	N	Size in Inch	Fath in Len.	Blocks, &c. Species.	In	N	Size in Inch	Fath in Len.	Blocks, &c. Species.	In	N	Size in Inch	Fath in Len.	Blocks, &c. Species.	In	N	Size in Inch	Fath in Len.	Blocks, &c. Species.	In	N
Flying-Jib.																									
Haliards	2½	48	Single	7	1	2½	43	Single	7	1	2	40	Single	7	1	2	37	Single	7	1	—	—	—	—	—
Sheets	2½	44	—	—	—	2½	38	—	—	—	2	36	—	—	—	2	30	—	—	—	—	—	—	—	—
Tack	2	2	—	—	—	2	2	—	—	—	1½	2	—	—	—	1	2	—	—	—	—	—	—	—	—
Downhauler	1½	42	Single	5	1	1½	38	Single	5	1	1	35	Single	5	1	1	32	Single	5	1	—	—	—	—	—
Spritsail, Topsail and Yard.																									
Horses	2½	7½	—	—	—	2	7	—	—	—	—	—	—	—	—	—	—	—	—	—	—	—	—	—	—
Braces	2½	53	Single	9	2	2	52	Single	8	2	—	—	—	—	—	—	—	—	—	—	—	—	—	—	—
Strapping	2½	2	—	—	—	2	2	—	—	—	—	—	—	—	—	—	—	—	—	—	—	—	—	—	—
Lifts, single	2	38	Single	8	2	2	37	Single	7	2	—	—	—	—	—	—	—	—	—	—	—	—	—	—	—
Strapping	2	1½	—	—	—	2	1½	—	—	—	—	—	—	—	—	—	—	—	—	—	—	—	—	—	—
Haliard	2½	28½	Single	9	2	2	27½	Single	8	2	—	—	—	—	—	—	—	—	—	—	—	—	—	—	—
Strapping	2½	2	—	—	—	2	2	—	—	—	—	—	—	—	—	—	—	—	—	—	—	—	—	—	—
Lashing	¾	7	—	—	—	¾	6½	—	—	—	—	—	—	—	—	—	—	—	—	—	—	—	—	—	—
Parral-Ropes	2½	4	Parral	12	1	2	3½	Parral	10	1	—	—	—	—	—	—	—	—	—	—	—	—	—	—	—
Clue-Lines	2	47½	Single	7	4	1½	46	Single	6	4	—	—	—	—	—	—	—	—	—	—	—	—	—	—	—
Strapping	2	3	—	—	—	1½	2½	—	—	—	—	—	—	—	—	—	—	—	—	—	—	—	—	—	—
Lacing and Ear-Rings	1	38	—	—	—	¾	37	—	—	—	—	—	—	—	—	—	—	—	—	—	—	—	—	—	—

Fore-mast and Fore-yard.	1257 Tons.					818 Tons.					544 Tons.					330 Tons.					Sloop of 60 Tons.				
	Size in Inch	Fath in Len.	Blocks, &c. Species.	In	N	Size in Inch	Fath in Len.	Blocks, &c. Species.	In	N	Size in Inch	Fath in Len.	Blocks, &c. Species.	In	N	Size in Inch	Fath in Len.	Blocks, &c. Species.	In	N	Size in Inch	Fath in Len.	Blocks, &c. Species.	In	N
Foremast.*																									
Pendents of Tackles cabld.	10½	17	*Sing.co.	22	2	9½	15	*Sing.co.	17	2	7½	7	*Sing.co.	15	2	Fore Burton Pendents					2½	3½	Single Double	9 7	1 1
Strapping	6½	3½	Thimbls.	—	8	5½	3	Thimbls.	—	8	5	2	Thimbls.	—	4	—	—	Falls	—	—	2	16	*Single	7	2
Seizing	1½	8½	—	—	—	1	7½	—	—	—	¾	3	—	—	—	—	—	—	—	—	—	—	—	—	—
Runners of Tackles cabled	6½	35	D.th.co.	21	4	5½	30	D.th.co.	17	4	5	28	—	—	—	—	—	—	—	—	—	—	—	—	—
Strapping	5½	7	—	—	—	5	6	—	—	—	—	—	—	—	—	—	—	—	—	—	—	—	—	—	—
Falls of Tackles	4	176	*S.th.co.	24	4	3½	154	*S.th.co.	20	4	3	73	*S.th.co.	15 11	2 2	—	—	—	—	—	—	—	—	—	—
Strapping	5½	8	Thimbls.	—	2	5	7	Thimbls.	—	2	4	6	—	—	—	—	—	—	—	—	—	—	—	—	—
Seizing	1	32	—	—	—	¾	30	—	—	—	¾	13	—	—	—	—	—	—	—	—	—	—	—	—	—
†Shrouds 4 Strands cabled	10½	140	D. Eyes	16	14	9½	124	D. Eyes	14	14	7½	84	D. Eyes	11	12	7	62	D. Eyes	11	10	—	—	—	—	—
Seizings — Eye	1½	35	—	—	—	1½	31	—	—	—	1	21	—	—	—	1	15	—	—	—	—	—	—	—	—
Seizings — Throat	1½	70	—	—	—	1½	62	—	—	—	1	42	—	—	—	1	31	—	—	—	—	—	—	—	—
Seizings — End	1½	70	—	—	—	1½	62	—	—	—	¾	42	—	—	—	¾	31	—	—	—	—	—	—	—	—
Laniards	5	87	—	—	—	4½	77	—	—	—	4	52	—	—	—	4	39	—	—	—	—	—	—	—	—
Ratline	1½	227	—	—	—	1½	200	—	—	—	1½	136	—	—	—	1½	100	—	—	—	—	—	—	—	—
Catharpin Legs	6	15	—	—	—	6	13	—	—	—	5	12	—	—	—	4	10½	—	—	—	—	—	—	—	—
Seizings	1½	60	—	—	—	1½	52	—	—	—	1	48	—	—	—	1	41	—	—	—	—	—	—	—	—
Fore-yard.																									
Stay cabled 4 Strands	16	14	Heart	22	1	14½	12	Heart	20	1	12	10	Heart	16	1	11	9	Heart	14	1	—	—	—	—	—
Seizing	1½	21	—	—	—	1½	17	—	—	—	1½	15	—	—	—	1½	13	—	—	—	—	—	—	—	—
Laniard	5	14	—	—	—	4½	11	—	—	—	4	9	—	—	—	4	8	—	—	—	—	—	—	—	—
Collar cabled 4 Strs. double	9	7	Heart	22	1	7½	6	Heart	20	1	6½	5	Heart	16	1	5½	4	Heart	14	1	—	—	—	—	—
Seizings	1½	11	—	—	—	1½	9	—	—	—	1½	8	—	—	—	1½	6	—	—	—	—	—	—	—	—
Lashing	2½	7	—	—	—	2	6	—	—	—	2	5	—	—	—	1½	4	—	—	—	—	—	—	—	—
Jears Falls	6½	112	Tr. coak Do.coak	24 24	2 2	5½	77	Do.coak	18	4	—	—	—	—	—	—	—	—	—	—	—	—	—	—	—
Tie	—	—	—	—	—	—	—	—	—	—	7½	14½	S.do.sc.	20	3	7	12½	S.do.sc.	20	3	—	—	—	—	—
Falls	—	—	—	—	—	—	—	—	—	—	3½ 5½	73 3	Do. coak S. coak	12 9	4 2	3 5½	63 3	Do. coak S. coak	12 9	4 2	—	—	—	—	—
Strapping	8	11	—	—	—	6½	9	—	—	—	4 3	2 1½	—	—	—	4 3	2 1½	—	—	—	—	—	—	—	—
Seizings	2	22				1½	18	—	—	—	1	13	—	—	—	1	13	—	—	—	—	—	—	—	—
Lashing at the Mast Head	4	30	—	—	—	3½	28	—	—	—	3	23	—	—	—	2½	20	—	—	—	—	—	—	—	—
Lashing at the Yard	3	10	—	—	—	2½	9	—	—	—	2	7	—	—	—	2	6	—	—	—	—	—	—	—	—
Stoppers	5½	5	—	—	—	4½	4½	—	—	—	—	—	—	—	—	—	—	—	—	—	—	—	—	—	—
Horses	5	14	—	—	—	4½	13	—	—	—	4	9½	—	—	—	4	8	—	—	—	—	—	—	—	—
Stirrups	4	14	Th.	—	10	3	13	Th.	—	10	3	9½	Th.	—	8	3	8	Th.	—	8	—	—	—	—	—
Seizings	¾	14	—	—	—	¾	13	—	—	—	Tarred Line			—	—	—	—	—	—	—	—	—	—	—	—
Laniard	2	4½	—	—	—	2	4	—	—	—	1½	3	—	—	—	1½	2½	—	—	—	—	—	—	—	—
Yard-Tackle Pendents	6½	8	D.th.co.	16	2	5½	7	D.th.co.	14	2	5	5	D.th.co.	12	2	5	4½	D.th.co.	12	2	—	—	—	—	—
Falls	3½	90	*D. S.	18 13	2 2	3	75	*D. S.	16 11	2 2	3	55	*D. S.	14 10	2 2	3	50	*D. S.	13 9	2 2	—	—	—	—	—
Strapping	5	8	—	—	—	4	7	—	—	—	3½	5	—	—	—	3	4½	—	—	—	—	—	—	—	—
Seizing	1	16	—	—	—	1	14	—	—	—	¾	10	—	—	—	¾	9	—	—	—	—	—	—	—	—
Inner Tricing-Lines	2	35	Single	7	2	2	29	Single	7	2	1½	23½	Single	6	2	1½	20	Single	6	2	—	—	—	—	—
Outer Tricing-Lines	2	35	Single	7	4	2	29	Single	7	4	1½	23½	Single	6	4	1½	20	Single	6	4	—	—	—	—	—
Strapping	2	6	—	—	—	2	5	—	—	—	2	4	—	—	—	1½	3½	—	—	—	—	—	—	—	—
Braces	4	95	Sing.co.	15	4	3¾	80	Sing.co.	13	4	3½	64	Sing.co.	12	4	3	55	Sing.co.	10	4	—	—	—	—	—
Pendents	5	12	Sing.co.	15	2	4¾	10	Sing.co.	13	2	4½	8	Sing.co.	12	2	4	7	Sing.co.	10	2	—	—	—	—	—
Strapping	4½	6	—	—	—	4	5	—	—	—	4	4	—	—	—	3	3½	—	—	—	—	—	—	—	—
Seizing	¾	16	—	—	—	¾	13	—	—	—	—	Tarred Line		—		—	—	—	—	—	—	—	—	—	—
Lashing	1½	20	—	—	—	1½	17	—	—	—	1½	13	—	—	—	1	12	—	—	—	—	—	—	—	—

* As no Rope is particularly allowed for Girt-Lines, take any of the 5 inch cordage in large, and 3 inch for small Ships.

† The length of the first warp is taken from the upper side of the bolsters, on the trestle-trees, to the foremost dead-eye in the channel; or, from the middle of the opposite side of the mast-head down to the deck.

Fore-yard—*continued.*	1257 Tons.					818 Tons.					544 Tons.					330 Tons.					Sloop of 60 Tons.				
	Size in Inch	Fath in Len.	Blocks, &c. Species.	In	N	Size in Inch	Fath in Len.	Blocks, &c. Species.	In	N	Size in Inch	Fath in Len.	Blocks, &c. Species.	In	N	Size in Inch	Fath in Len.	Blocks, &c. Species.	In	N	Size in Inch	Fath in Len.	Blocks, &c. Species.	In	N
Lifts	4	100	Single	15	6	3½	85	Single	13	6	3½	69	Single	12	6	3	58	Single	10	6	—	—	—	—	—
Span for the Cap	5½	8	Sister	22	2	5	7	Sister	20	2	4¼	5½	Sister	18	2	4½	4	Sister	16	2	—	—	—	—	—
Short Span	4	3	—	—	—	3½	2½	—	—	—	3½	2	—	—	—	3	2	—	—	—	—	—	—	—	—
Strapping	4½	5	—	—	—	4	5	—	—	—	3½	3	—	—	—	3	3	—	—	—	—	—	—	—	—
Seizing	¾	10	—	—	—	¾	8	—	—	—	¾	7	—	—	—	Tarred Line					—	—	—	—	—
Truss-Pendents	7½	16	Thimbls.	—	4	6	13	Thimbls.	—	4	5½	11	Thimbls.	—	4	4	8	Thimbls.	—	4	—	—	—	—	—
Falls	3	51	*Double	11	4	2½	42	*Double	10	4	2½	34	*Double	9	4	2	29	*Double	8	4	—	—	—	—	—
Strapping	3½	5	—	—	—	3	4	—	—	—	3	3½	—	—	—	2½	2½	—	—	—	—	—	—	—	—
Eye-Seizings	1½	15	—	—	—	1	12	—	—	—	1	10½	—	—	—	¾	7½	—	—	—	—	—	—	—	—
Nave-Line	2	16	Single	7	1	1½	13	Single	7	1	1½	11	Single	6	1	1½	9	Single	6	1	—	—	—	—	—
Clue-Garnets	3½	61	S.St.bd.	13	4	3	55	S.St.bd.	12	4	3	44	S.St.bd.	11	4	2½	34	S.St.bd.	9	4	—	—	—	—	—
			Single	13	2			Single	12	2			Single	11	2			Single	9	2					
Straps about the Yard	3½	7½	—	—	—	3	7	—	—	—	3	5½	—	—	—	4	2½	—	—	—	—	—	—	—	—
Strapping	3½	2	—	—	—	3	2	—	—	—	3	2	—	—	—	1½	2½	—	—	—	—	—	—	—	—
Seizing	¾	7½	—	—	—	¾	7	—	—	—	Tarred Line					—	—	—	—	—	—	—	—	—	—
Lashing	1	10	—	—	—	1	9	—	—	—	1	7	—	—	—	¾	6	—	—	—	—	—	—	—	—
Bunt-line Legs	3	47	Double	10	4	2½	39	Double	8	4	2½	30	Double	8	4	2	26	Double	8	4	—	—	—	—	—
Falls	3	47	Single	10	8	2½	39	Single	8	8	2½	30	Single	8	8	2	26	Single	8	8	—	—	—	—	—
Strapping	3½	11	—	—	—	3	9	—	—	—	2½	7½	—	—	—	2½	6½	—	—	—	—	—	—	—	—
Leech-line Legs	2½	47	Double	10	4	2	39	Double	9	4	2	30	Double	8	4	2	26	Double	7	4	—	—	—	—	—
Falls	2½	47	Single	10	8	2	39	Single	9	8	2	30	Single	8	8	2	26	Single	7	8	—	—	—	—	—
Strapping	3	11	—	—	—	2½	9	—	—	—	2½	7½	—	—	—	2½	6½	—	—	—	—	—	—	—	—
Slablines and	2	35	Single	8	2	2	29	Single	8	2	2	23½	Single	7	2	1½	20	Single	7	2	—	—	—	—	—
Strapping	2		—			2		—			2		—			1½		—			—	—	—	—	—
Bow-Lines	4½	54	Single	16	4	4	45	Single	15	4	3½	36	Single	13	4	3	31	Single	12	2	—	—	—	—	—
Bridles	4½	5½	Thimble	—	2	4	4½	Thimble	—	2	3½	3½	Thimble	—	2	3	3	Thimble	—	2	—	—	—	—	—
Strapping	4½	4	—	—	—	4	3½	—	—	—	3½	3	—	—	—	3	2½	—	—	—	—	—	—	—	—
Seizing	1	11	—	—	—	¾	9	—	—	—	¾	7	—	—	—	¾	6	—	—	—	—	—	—	—	—
Lashing	2½	11	—	—	—	2	9	—	—	—	2	7	—	—	—	1½	6	—	—	—	—	—	—	—	—
Ear-Rings	2	27	—	—	—	2	22	—	—	—	1½	18	—	—	—	1½	15	—	—	—	—	—	—	—	—

Fore-yard and Fore-topmast.	1257 Tons.					818 Tons.					544 Tons.					330 Tons.					Sloop of 60 Tons.				
	Size in Inch	Fath in Len.	Blocks, &c. Species.	In	N	Size in Inch	Fath in Len.	Blocks, &c. Species.	In	N	Size in Inch	Fath in Len.	Blocks, &c. Species.	In	N	Size in Inch	Fath in Len.	Blocks, &c. Species.	In	N	Size in Inch	Fath in Len.	Blocks, &c. Species.	In	N
Sheets cabled and	5	60	Sing.co.	18	2	4½	55	Sing.co.	17	2	4	44	Sing.co.	16	2	3½	34	Sing.co.	15	2	—	—	—	—	—
Strapping	5		Thimble	—	2	4½		Thimble	—	2	4		Thimble	—	2	3½		Thimble	—	2	—	—	—	—	—
Seizing	1	13	—	—	—	1	11	—	—	—	1	9	—	—	—	¾	8	—	—	—	—	—	—	—	—
Stoppers	4	4	—	—	—	3½	3½	—	—	—	3	3	—	—	—	2½	2½	—	—	—	—	—	—	—	—
Tacks cabled and	5	50	Sho.	18	2	4½	45	Sho.	17	2	4	37	Sho.	16	2	3½	29	Sho.	15	2	—	—	—	—	—
Strapping	5		Sing.co.	18	2	4½		Sing.co.	17	2	4		Sing.co.	16	2	3½		Sing.co.	15	2	—	—	—	—	—
Seizing	1½	—	—	—	—	1½	—	—	—	—	1½		—	—	—	1		—	—	—	—	—	—	—	—
Stoppers	4½	—	H. & T.	2	2	4	—	H. & T.	2	2	3½	—	H. & T.	2	2	3	3	H. & T.	2	2	—	—	—	—	—
Slings, 4 Strand, cable	13½	14	Thimble	—	1	13	13	Thimble	—	1	9	10	Thimble	—	1	—	—	—	—	—	—	—	—	—	—
laid, and Strap.	13½		Thimble	—	1	13		Thimble	—	1	9		Thimble	—	1	—	—	—	—	—	—	—	—	—	—
Seizing	2	7	—	—	—	2	6	—	—	—	1½	4	—	—	—	—	—	—	—	—	—	—	—	—	—
Laniard	3½	6	—	—	—	3	5	—	—	—	2½	3	—	—	—	—	—	—	—	—	—	—	—	—	—
Staysail-Stay	6	14	—	—	—	6	12	—	—	—	5	10	—	—	—	4½	9	—	—	—	—	**Fore-sail.**			
Haliard	3½	30	Single	12	2	3	26	Single	11	2	3	22	Single	10	2	2½	20	Single	9	2	2	21	*Single	9	2
Sheets	3½	28	Single	12	2	3	24	Single	11	2	3	20	Single	10	2	2½	18	Single	9	2	2½	3½	—	—	—
Tack	2	2	—	—	—	2½	2	—	—	—	2½	2	—	—	—	2	2	—	—	—	1	2½	—	—	—
Down-Hauler	2½	28	Single	8	1	2	24	Single	7	1	2	20	Single	7	1	1½	18	Single	6	1	1½	10	—	—	—
Strapping	3½	5	—	—	—	3	4	—	—	—	3	3½	—	—	—	2½	4	—	—	—	2½	8	Bow-Line	—	—
	2½	1½				2	1				2	1													
Studdingsail Inner Haliards	3	49	Single	11	6	2½	40	Single	10	6	2	33	Single	9	6	2	27	Single	9	6	2½	1	Bridles	—	—
Outer	3½	80	Single	11	4	3	68	Single	10	4	2½	55	Single	9	4	2½	46	Single	9	4	—	—	—	—	—
Sheets	3	13½	—	—	—	2½	11	—	—	—	2	9	—	—	—	2	8	—	—	—	—	—	—	—	—
Tacks and	3½	60	Single	11	2	3	50	Single	10	2	2½	40	Single	9	2	2½	33	Single	9	2	—	—	—	—	—
Strapping	3½		—			3		—			2½		—			2½		—			—	—	—	—	—
Fore Topmast.																									
Burton-Pendents	5	6	Thimbls.	—	2	4½	5	Thimble	—	2	—	—	—	—	—	—	—	—	—	—	—	—	—	—	—
Falls	2½	55	*Double	10	2	2½	47	*Double	9	2	—	—	—	—	—	—	—	—	—	—	—	—	—	—	—
			*Single	10	2			*Single	9	2															
Strapping	3½	4½	—	—	—	3	4	—	—	—	—	—	—	—	—	—	—	—	—	—	—	—	—	—	—
Shrouds cabled 4 Strands	6½	64	D. Eyes	10	8	6	55	D. Eyes	8	8	5½	50	D. Eyes	8	8	4½	32	D. Eyes	7	6	—	—	—	—	—
Seizings Eye	1	16	—	—	—	1	14	—	—	—	¾	12	—	—	—	—	—	—	—	—	—	—	—	—	—
Throat	1	28	—	—	—	1	24	—	—	—	¾	21	—	—	—	Tarred Line					—	—	—	—	—
End	¾	24	—	—	—	¾	21	—	—	—	¾	18	—	—	—	—	—	—	—	—	—	—	—	—	—
Laniards	3½	43	—	—	—	3	36	—	—	—	2½	33	—	—	—	2½	21	—	—	—	—	—	—	—	—
Ratline	1	85	—	—	—	1	75	—	—	—	1	69	—	—	—	1	44	—	—	—	—	—	—	—	—
Futt.-Shrouds cabled 4 Sts.	6½	24	Pl.d.e.	10	8	6	21	Pl.d.e.	8	8	5½	19	Pl.d.e.	8	8	4½	12	Pl.d.e.	7	6	—	—	—	—	—
Seizing Upper	1	30	—	—	—	1	26	—	—	—	¾	23	—	—	—	¾	15	—	—	—	—	—	—	—	—
Lower	¾	26	—	—	—	¾	23	—	—	—	¾	20	—	—	—	¾	13	—	—	—	—	—	—	—	—
Ratline	1	30	—	—	—	1	26	—	—	—	1	23	—	—	—	1	15	—	—	—	—	—	—	—	—
Back-Stays cabled 4 Stds.	6½	110	D. Eyes	10	6	6	94	D. Eyes	9	6	5½	86	D. Eyes	8	6	4½	50	D. Eyes	7	4	—	—	—	—	—
Seizings Eye	1	13	—	—	—	1	12	—	—	—	¾	10	—	—	—	Tarred Line					—	—	—	—	—
Throat	1	22	—	—	—	1	21	—	—	—	¾	17	—	—	—						—	—	—	—	—
End	¾	19	—	—	—	¾	18	—	—	—	¾	15	—	—	—						—	—	—	—	—

Fore-topmast and Fore-topsail Yard.	1257 Tons.					818 Tons.					544 Tons.					330 Tons.					Sloop of 60 Tons.				
	Size in Inch	Fath in Len.	Blocks, &c. Species.	In	N	Size in Inch	Fath in Len.	Blocks, &c. Species.	In	N	Size in Inch	Fath in Len.	Blocks, &c. Species.	In	N	Size in Inch	Fath in Len.	Blocks, &c. Species.	In	N	Size in Inch	Fath in Len.	Blocks, &c. Species.	In	N
Laniards	3½	22	—	—	—	3	19	—	—	—	2½	17	—	—	—	2½	10	—	—	—	—	—	—	—	—
Stay cabled 4 Strands	7½	22	—	—	—	7	19	—	—	—	6	17	—	—	—	5	15	—	—	—	—	—	—	—	—
Collar	6	3½	Single	18	1	5½	3	Single	16	1	5	2½	Single	15	1	4	2½	Single	13	1	—	—	—	—	—
Tackle	3½	22	Long.t. *Single	20 12	1 1	3	19	Long.t. *Single	18 10	1 1	2½	17	Long.t. *Single	18 9	1 1	2½	15	Long.t. *Single	16 8	1 1	—	—	—	—	—
Strapping	4½	3	—	—	—	4	2½	—	—	—	3	2½	—	—	—	3	2	—	—	—	—	—	—	—	—
Seizing	1½ ¾	6	—	—	—	1 ¾	5	—	—	—	1 ¾	5	—	—	—	¾	4	—	—	—	—	—	—	—	—
Top-Rope Pendents cabled 4 Strands	8½	33	S.Br.shi. I.bd.co.	22	2	7½	31	S.Br.shi. I.bd.co.	20	2	6½	29	S.Br.shi. I.bd.co.	18	2	6	25	S.Br.shi. I.bd.co.	16	2	—	—	—	—	—
Falls	4½	107	Tr.I.bd.c.	20	4	4	100	Tr.I.bd.c.	18	4	3½	94	Tr.I.bd.co	16	4	3	81	Tr.I.bd.co	14	4	—	—	—	—	—
Fore Topsail Yard.																									
Tie cabled 4 Strands	5½	40	Fl.si.co. Dou.co.	18 18	2 1	5	35	Fl.si.co. Dou.co.	16 16	2 1	4½	32	Fl.si.co. Dou.co.	15 15	2 1	4	28	Fl.si.co. Dou.co.	14 14	2 1	—	—	—	—	—
Strapping	5½	8	—	—	—	5	7	—	—	—	4½	6	—	—	—	4	5½	—	—	—	—	—	—	—	—
Seizing	1	12	—	—	—	1	10	—	—	—	1	9	—	—	—	¾	8	—	—	—	—	—	—	—	—
Lashers Mast-Head	2½	11	—	—	—	2½	9	—	—	—	2	8	—	—	—	1½	7	—	—	—	—	—	—	—	—
Lashers Yard	2	5½	—	—	—	2	4½	—	—	—	1½	4	—	—	—	1	3½	—	—	—	—	—	—	—	—
Haliards	3½	110	Do.th.co. *S.th.co.	22 22	2 2	3½	94	Do.th.co. *S.th.co	20 20	2 2	3	86	Do.th.co. *S.th.co.	18 18	2 2	2½	74	Do.th.co. *S.th.co.	16 16	2 2	—	—	—	—	—
Strapping	4½	8	—	—	—	4½	7	—	—	—	4	6	—	—	—	3½	5	—	—	—	—	—	—	—	—
Seizing	¾	16	—	—	—	¾	14	—	—	—	¾	12	—	—	—	¾	10	—	—	—	—	—	—	—	—
Horses	4	11	—	—	—	3½	10	—	—	—	3	9	—	—	—	3	7	—	—	—	—	—	—	—	—
Stirrups	3	8	Thimble	—	6	3	7	Thimble	—	6	2½	6½	Thimble	—	6	2	5	Thimble	—	4	—	—	—	—	—
Braces	3½	91	Single	12	4	3½	80	Single	10	4	3	71	Single	9	4	3	55	Single	8	4	—	—	—	—	—
*Pendents	4½	9	Single	12	2	4½	8	Single	10	2	4	7	Single	10	2	3½	5½	Single	9	2	—	—	—	—	—
Strapping	14	4½	—	—	—	3½	4	—	—	—	3	3½	—	—	—	2½	2½	—	—	—	—	—	—	—	—
Lifts	3½	62	Double Single	12 12	2 4	3½	54	Double Single	10 10	2 4	3	48	Sister Single	17 10	2 4	2½	37	Sister Single	14 8	2 4	—	—	—	—	—
Beckets	3½	3	—	—	—	3½	3	—	—	—	3	2½	—	—	—	2½	2	—	—	—	—	—	—	—	—
Strapping	3½	9	—	—	—	3½	9	—	—	—	3	7	—	—	—	2½	6	—	—	—	—	—	—	—	—
Seizing	¾	27	—	—	—	¾	27	—	—	—	Tarred Line		—	—	—	—	—	—	—	—	—	—	—	—	—
Parral-Ropes	3	11	Parral	22	1	2½	10	Parral	20	1	2½	9	Parral	16	1	2	7	Parral	14	1	—	—	—	—	—
Racking and Seizing	1	15	—	—	—	¾	13	—	—	—	¾	12	—	—	—	Tarred Line			—	—	—	—	—	—	—
Clue-lines	3½	82	S.st.bd. Single	12 12	4 2	3½	72	S.st.bd. Single	10 10	4 2	3	64	S.st.bd. Single	10 10	4 2	2½	49	S.st.bd. Single	8 8	4 2	—	—	—	—	—
Strapping	3½	8	—	—	—	3½	7	—	—	—	3	6	—	—	—	2½	5	—	—	—	—	—	—	—	—
Bunt-Lines and Strapping	3 3	66	Single —	10 —	4 —	3 3	58	Single —	10 —	4 —	2½ 3½	52	Single —	8 —	4 —	2 2	40	Single —	7 —	4 —	—	—	—	—	—
Leech-Lines and Strapping	2½ 2½	27	Single —	9 —	2 —	2½ 2½	24	Single —	9 —	2 —	2 2	21	Single —	7 —	2 —	2 2	16	Single —	6 —	2 —	—	—	—	—	—
Bow-lines	3	60	Single	11	2	3	52	Single	11	2	2½	47	Single	9	2	2	36	Single	8	2	—	—	—	—	—
Bridles	3	12	Thimble	—	4	3	10	Thimble	—	4	2½	9	Thimble	—	4	2	7	Thimble	—	4	—	—	—	—	—
Strapping	3	5	—	—	—	3	4	—	—	—	2½	3	—	—	—	2	2	—	—	—	—	—	—	—	—
Lashing	1	12	—	—	—	1	10	—	—	—	1	9	—	—	—	¾	7	—	—	—	—	—	—	—	—
Reef-tackle Pendents	3½	41	*Double	8	4	3½	38	*Double	8	4	3	32	*Single	7	4	2½	25	*Single	7	4	—	—	—	—	—
Falls	2	55	—	—	—	2	48	—	—	—	1½	43	—	—	—	1½	33	—	—	—	—	—	—	—	—
Strapping	2½	3	—	—	—	2½	3	—	—	—	2	2	—	—	—	2	1½	—	—	—	—	—	—	—	—
Ear-Rings	1½	41	—	—	—	1½	38	—	—	—	1½	32	—	—	—	1	25	—	—	—	—	—	—	—	—
Sheets	5½	49	Sin.Sho. ¼ thk. & thin. co.	20 20	2 2	5	43	Sin.Sho. ¼ thk. & thin. co.	18 18	2 2	4½	38	Sin.Sho. ¼ thk. & thin.co.	16 16	2 2	4	30	Sin.Sho. ¼ thk. & thin.co.	14 14	2 2	—	—	—	—	—
Strapping for Sheet Blks.	6	5	—	—	—	5½	4	—	—	—	5	3½	—	—	—	4½	3	—	—	—	—	—	—	—	—
Strapping for Quarter do.	4½	10	—	—	—	4½	8	—	—	—	4	7	—	—	—	3½	6	—	—	—	—	—	—	—	—
Lashers for Quarter do.	2	10	—	—	—	2	8	—	—	—	2	7	—	—	—	1½	6	—	—	—	—	—	—	—	—
Seizings	1½	18	—	—	—	1	16	—	—	—	1	14	—	—	—	¾	11	—	—	—	—	—	—	—	—
Span	3	8	—	—	—	3	7	—	—	—	3	6	—	—	—	2½	5	—	—	—	—	—	—	—	—
Stoppers	4½	4	—	—	—	4½	3½	—	—	—	3½	3	—	—	—	3	2½	—	—	—	—	—	—	—	—
Slings	5	11	—	—	—	4½	10	—	—	—	3½	9	—	—	—	3	7	—	—	—	—	—	—	—	—
Staysail-Stay cabled 4 Strands	7	22	Single	14	1	7	19	Single	14	1	5½	17	Single	12	1	4½	15	Single	10	1	—	—	—	—	—
Tackle	3½	14	Double Single	10 9	1 1	3½	12	Double Single	10 9	1 1	2½	11	Double Single	8 7	1 1	2	10	Single Single	8	2	—				
Haliard and Strapping	3½ 3½	38	Single	12	1	3½ 3½	33	Single	12	1	2½ 2½	29	Single	10	1	2 2	26	Single	9	1	—	—	—	—	—
Sheets and Strapping	3½ 3½	33	Single	12	2	3½ 3½	28	Single	12	2	2½ 2½	25	Single	10	2	2 2	22	Single	9	2	—	—	—	—	—
Out-Hauler	2½	19	Single	9	1	2½	17	Single	9	1	2	15	Single	8	1	1½	13	Single	7	1	—	—	—	—	—
Down-Hauler and Strapping	2½ 2½	29	Single	9	1	2½ 2½	25	Single	9	1	2 2	23	Single	8	1	1½ 1½	20	Single	7	1	—	—	—	—	—
Studding-Sail-Haliards	3½	90	Single	12	6	3	80	Single	10	6	2½	70	Single	9	6	2	55	Single	8	6	—	—	—	—	—
Sheets	3	46	Single	12	2	2½	40	Single	10	2	2	36	Single	9	2	1½	28	Single	8	2	—	—	—	—	—
Tacks	3½	101	Single	12	2	3	88	Single	10	2	2½	79	Single	9	2	2	61	Single	8	2	—	—	—	—	—
Down-Haulers	2	55	Thimble Single	— 9	6 2	2	48	Thimble Single	— 9	6 2	1½	43	Thimble Single	— 8	6 2	1½	33	Thimble Single	— 6	6 2	—	—	—	—	—
Boom-Tackles	2	77	Double Single	8 8	2 4	2	67	Double Single	8 8	2 4	—	—	—	—	—	—	—	—	—	—	—	—	—	—	—
Tails and Straps	3½ 2½	18 6	—			3½ 2½	16 5	—	—	—	3 2	14 4	—	—	—	2	14	—	—	—	—	—	—	—	—

* Braces may rig with or without Pendents.

Fore-topgallant Mast, Fore-topgallant Yard and Main Mast.	1257 Tons.					818 Tons.					544 Tons.					330 Tons.					Sloop of 60 Tons.				
	Size in Inch	Fath in Len.	Blocks, &c. Species.	In	N	Size in Inch	Fath in Len.	Blocks, &c. Species.	In	N	Size in Inch	Fath in Len.	Blocks, &c. Species.	In	N	Size in Inch	Fath in Len.	Blocks, &c. Species.	In	N	Size in Inch	Fath in Len.	Blocks, &c. Species.	In	N
Fore Topgallant Mast.																									
Shrouds cabled, 4 Strands	4½	73	Thimble	—	12	3¾	62	Thimble	—	12	3	57	Thimble	—	12	2½	49	Thimble	—	8	—	—	—	—	—
Laniards	2	12	—	—	—	1½	10	—	—	—	1½	9	—	—	—	1	8	—	—	—	—	—	—	—	—
Backstays cabled 4 Strands	4½	42	D. Eyes	7	4	3¾	38	D. Eyes	6	4	3	34	D. Eyes	6	4	2½	28	Thimble	—	4	—	—	—	—	—
Laniards	2	10	—	—	—	1½	9	—	—	—	1½	8	—	—	—	1	7	—	—	—	—	—	—	—	—
Stay cabled, 4 Strands ..	4½	28	Single	12	1	3¾	25	Single	12	1	3	23	Single	10	1	2½	21	Thimble	—	1	—	—	—	—	—
Strapping	3½	1½	—	—	—	3	1½	—	—	—	2½	1½	—	—	—	—	—	—	—	—	—	—	—	—	—
Royal or Flagstaff-Stay ..	2½	28	Thimble	—	1	2	25	Thimble	—	1	—	—	—	—	—	—	—	—	—	—	—	—	—	—	—
Haliards	1½	62	—	—	—	1½	53	—	—	—	1	48	—	—	—	¾	41	—	—	—	—	—	—	—	—
Royal-Haliard ..	2	62	Single	7	1	2	53	Single	7	1	1½	48	Single	6	1	1½	41	Single	6	1	—	—	—	—	—
Fore Topgallant Yard.																									
Tie	3½	16	—	—	—	3½	14	—	—	—	3	12	—	—	—	2½	11	—	—	—	—	—	—	—	—
Haliards	2	28	Double Single	8 8	1 2	2	24	Double Single	8 8	1 2	1½	21	Double Single	7 7	1 2	1½	19	Single	6	2	—	—	—	—	—
Strapping	2½	1½	—	—	—	2½	1½	—	—	—	2	1½	—	—	—	2	1	—	—	—	—	—	—	—	—
Horses	3	7	—	—	—	3	6	—	—	—	2½	5½	—	—	—	2½	5	—	—	—	—	—	—	—	—
Braces and Strapping ..	2 2	108	Single	7	6	2 2	93	Single	7	6	2 2	87	Single	6	4	1½ 1½	75	Single	6	4	—	—	—	—	—
Lifts, single and Strapping ..	2½ 2½	54	Single Thimble	7 —	2 2	2½ 2½	46	Single Thimble	7 —	2 2	2½	43	Thimble	—	2	2	37	Thimble	—	2	—	—	—	—	—
Parral-Ropes	2	4½	Parral	11	1	2	4	Parral	9	1	1½	3½	Parral	8	1	1	3	Parral	6	1	—	—	—	—	—
Clue-Lines and Strapping ..	2 2	72	Single	7	6	2 2	62	Single	7	6	1½ 2	58	Single	6	4	1 1½	30	Single	5	4	—	—	—	—	—
Bow-Lines	2	72	Single	7	2	2	62	Single	7	2	2½	58	Thimble	—	6	1	30	Thimble	—	4	—	—	—	—	—
Bridles	2	4½	Thimble	—	6	2	4	Thimble	—	6	1½	3½	Thimble	—	6	1	3	Thimble	—	4	—	—	—	—	—
Strapping	2	1	—	—	—	2	1	—	—	—	—	—	—	—	—	—	—	—	—	—	—	—	—	—	—
Ear-Rings, Tarred-Line ..	—	27	—	—	—	—	24	—	—	—	—	21	—	—	—	—	18	—	—	—	—	—	—	—	—
Studdingsail-Haliards ..	2½	73	Single	7	6	2	62	Single	7	6	1½	57	Single	6	6	1½	49	Single	6	6	—	—	—	—	—
Sheets	2	36	—	—	—	2	31	—	—	—	1	27	—	—	—	1	24	—	—	—	—	—	—	—	—
Tacks	2	46	Single	7	4	2	41	Single	7	4	1½	37	Single	7	4	1½	34	Single	7	4	—	—	—	—	—
Down-Haulers ..	1½	36	Thimble	—	1	1½	31	Thimble	—	1	1	27	Thimble	—	1	1	24	Thimble	—	1	—	—	—	—	—
Strapping	2	9	—	—	—	2	8	—	—	—	1½	7.	—	—	—	1½	6	—	—	—	—	—	—	—	—
Mainmast.	For Girt Lines, see Fore-Mast.																								
Pendts. of Tackles cabled ..	10½	17	*Sin.co.	22	2	9½	15	*Sin.co.	17	2	7½	7	*Sin.co.	15	2	—	—	—	—	—	4	8	Sin.co.	10	2
Strapping	6½	3½	Thimble	—	8	5½	3	Thimble	—	8	5	2	Thimble	—	4	—	—	—	—	—	—	—	—	—	—
Seizing	1½	8½	—	—	—	1	7½	—	—	—	¾	3	—	—	—	—	—	—	—	—	—	—	—	—	—
Runners of Tackles cabled	6½	35	D.th.co.	21	4	5½	30	D.th.co.	17	4	5	28	—	—	—	—	—	—	—	—	4	14	*L.t.c.	20	2
Strapping	5½	7	—	—	—	5	6	—	—	—	—	—	—	—	—	—	—	—	—	—	—	—	—	—	—
Falls of Tackles ..	4	190	*S.th.co.	24	4	3½	170	*S.th.co.	20	4	3	150	*S.th.co.	15 11	2 2	—	—	—	—	—	2	30	*S.th.c.	10	2
Strapping	5½	8	Thimble	—	2	5	7	—	—	—	4	6	—	—	—	—	—	—	—	—	—	—	—	—	—
Seizing	1	32	—	—	—	¾	30	—	—	—	¾	13	—	—	—	—	—	—	—	—	—	—	—	—	—
*Shrouds, 4 Strand cabled	10½	155	D. Eyes	16	14	9½	136	D. Eyes	14	14	7½	96	D. Eyes	11	12	7	72	D. Eyes	11	10	6½	70	D. Eyes	9	8
Seizings, Eye ..	1½	36	—	—	—	1½	32	—	—	—	1	22	—	—	—	1	16	—	—	—	¾	—	—	—	—
Throat ..	1½	70	—	—	—	1½	62	—	—	—	1	42	—	—	—	1	31	—	—	—	¾	45	—	—	—
End ..	1½	70	—	—	—	1½	62	—	—	—	¾	42	—	—	—	¾	31	—	—	—	¾	—	—	—	—
Laniards	5	87	—	—	—	4½	77	—	—	—	4	52	—	—	—	4	39	—	—	—	2½	25	—	—	—
Ratline	1½	230	—	—	—	1½	203	—	—	—	1½	138	—	—	—	1½	102	—	—	—	1	96	—	—	—
Catharpin-Legs	6	16	—	—	—	6	14	—	—	—	5	13	—	—	—	4	12	—	—	—	3	9	—	—	—
Seizings	1½	60	—	—	—	1½	52	—	—	—	1	48	—	—	—	1	41	—	—	—	¾	25	—	—	—
Stay cabled, 4 Strands ..	16	21	Heart	22	1	14½	18½	Heart	20	1	12	—	Heart	16	1	11	—	Heart	14	1	9	9½	D. Eyes	15	1
Seizing	1½	21	—	—	—	1½	17	—	—	—	1½	15	—	—	—	1½	13	—	—	—	1½	7	—	—	—
Laniard	5	14	—	—	—	4½	11	—	—	—	4	9	—	—	—	4	8	—	—	—	3	8	—	—	—
Collar cabled, 4 Strands ..	12	10	Heart	22	1	10	9	Heart	22	1	9	7	Heart	16	1	8	6	Heart	14	1	—	—	—	—	—
Lashing	3	17	—	—	—	2½	16	—	—	—	2	12	—	—	—	2	10	—	—	—	2	—	—	—	—
Seizing	1½	15	—	—	—	1½	14	—	—	—	1½	10	—	—	—	1½	9	—	—	—	1	—	—	—	—
Jeers .. .. Falls	7	119	Tr.co. Do.co.	24 24	2 2	6	85	Do.co.	18	4	—	—	—	—	—	—	—	—	—	—	—	—	—	—	—

* See Fore Mast Shrouds.

Main-yard.	1257 Tons.					818 Tons.					544 Tons.					330 Tons.					Sloop of 60 Tons.				
	Size in Inch	Fath in Len.	Blocks, &c. Species.	In	N	Size in Inch	Fath in Len.	Blocks, &c. Species.	In	N	Size in Inch	Fath in Len.	Blocks, &c. Species.	In	N	Size in Inch	Fath in Len.	Blocks, &c. Species.	In	N	Size in Inch	Fath in Len.	Blocks, &c. Species.	In	N
Main-yard.																									
Tie	—	—	—	—	—	—	—	—	—	—	8	15	S.do.sc.	20	3	7½	10	S.do.sc.	20	3	—	—	—	—	—
Falls	—	—	—	—	—	—	—	—	—	—	4 5½	75 3	Do.co. Sin.co.	12 9	4 2	3½ 5½	65 3	Do. coak S. coak	12 9	4 2	—	—	—	—	—
Strapping ..	8	11	—	—	—	6½	9	—	—	—	4 3	2 1½	—	—	—	4 3	2 1½	—	—	—	—	—	—	—	—
Seizing	2	22	—	—	—	1½	18	—	—	—	1	13	—	—	—	1	13	—	—	—	—	—	—	—	—
Lashing at the Mast-Head ..	4	30	—	—	—	3½	28	—	—	—	3	25	—	—	—	2½	20	—	—	—	—	—	—	—	—
Lashing at the Yard ..	3	10	—	—	—	2½	9	—	—	—	2	7	—	—	—	2	6	—	—	—	—	—	—	—	—
Stoppers	5½	5	—	—	—	4½	5	—	—	—	—	—	—	—	—	—	—	—	—	—	—	—	—	—	—
Horses	5	14	—	—	—	4½	13	—	—	—	4	9½	—	—	—	4	8	—	—	—	—	—	—	—	—
Stirrups	4	14	Thimble	—	10	3	13	Thimble	—	10	3	9½	Thimble	—	8	3	8	Thimble	—	8	—	—	—	—	—
Seizings	¾	14	—	—	—	¾	13	—	—	—	Tarred Line.			—	—	—	—	—	—	—	—	—	—	—	—
Laniard	2	5	—	—	—	2	4	—	—	—	1½	3	—	—	—	1½	2½	—	—	—	—	—	—	—	—
Yard-Tackle-Pendents ..	6½	8½	D.th.co.	16	2	5½	7½	D.th.co.	14	2	5	5½	D.th.c.	12	2	5	4½	D.th.co.	12	2	—	—	—	—	—
Falls	3½	99	*Double Single	18 13	2 2	3	87	*Double Single	16 11	2 2	3	64	*Double Single	14 10	2 2	3	52	*Double Single	13 9	2 2	—	—	—	—	—

Main-yard.—*continued.*	1257 Tons.					818 Tons.					544 Tons.					330 Tons.					Sloop of 60 Tons.				
	Size in Inch	Fath in Len.	Blocks, &c. Species.	In	N	Size in Inch	Fath in Len.	Blocks, &c. Species.	In	N	Size in Inch	Fath in Len.	Blocks, &c. Species.	In	N	Size in Inch	Fath in Len.	Blocks, &c. Species.	In	N	Size in Inch	Fath in Len.	Blocks, &c. Species.	In	N
Strapping	5	8	—	—	—	4	7	—	—	—	3½	5	—	—	—	3	4½	—	—	—	—	—	—	—	—
Seizing	1	16	—	—	—	1	14	—	—	—	¾	10	—	—	—	¾	9	—	—	—	—	—	—	—	—
Inner Tricing-Lines ..	2	36	Single	7	2	2	30	Single	7	2	1½	24	Single	6	2	1½	20	Single	6	2	—	—	—	—	—
Outer Tricing-Lines ..	2	36	Single	7	4	2	30	Single	7	4	1½	24	Single	6	4	1½	20	Single	6	4	—	—	—	—	—
Strapping	2	6	—	—	—	2	5	—	—	—	2	4	—	—	—	3	3½	—	—	—	—	—	—	—	—
Braces	4	85	Single	15	4	3¾	75	Single	13	4	3½	55	Single	12	4	3	45	Single	10	4	—	—	—	—	—
Pendents..	5	8½	Single	15	2	4¾	7½	Single	13	2	4½	5½	Single	12	2	4	4½	Single	10	2	—	—	—	—	—
Strapping	4½	6	—	—	—	4	5	—	—	—	4	4	—	—	—	3	3½	—	—	—	—	—	—	—	—
Seizing	¾	16	—	—	—	¾	13	—	—	—	Tarred Line			—	—	—	—	—	—	—	—	—	—	—	—
Lashing	1½	6	—	—	—	1½	5	—	—	—	1½	4	—	—	—	1	3	—	—	—	—	—	—	—	—
Lifts	4	102	Single	15	6	3½	90	Single	13	6	3½	66	Single	12	6	3	54	Single	10	6	—	—	—	—	—
Span for cap	6	8	Sister	22	2	5½	7	Sister	20	2	5	6	Sister	18	2	4½	5	Sister	16	2	—	—	—	—	—
Short span	4	3	—	—	—	3½	2½	—	—	—	3½	2	—	—	—	3	2	—	—	—	—	—	—	—	—
Strapping	4½	5	—	—	—	4	5	—	—	—	3½	3	—	—	—	3	3	—	—	—	—	—	—	—	—
Seizing	¾	10	—	—	—	¾	8	—	—	—	¾	7	—	—	—	Tarred Line			—	—	—	—	—	—	—
Truss-Pendents	7½	16	Thimble	—	4	6	14	Thimble	—	4	5½	12	Thimble	—	4	4½	9	Thimble	—	4	—	—	—	—	—
Falls	3	52	*Double	11	4	2½	43	*Double	10	4	2½	35	*Double	9	4	2	30	*Double	8	4	—	—	—	—	—
Strapping	3½	5	—	—	—	3	4	—	—	—	3	3½	—	—	—	2½	2½	—	—	—	—	—	—	—	—
Eye-Seizings	1½	15	—	—	—	1	12	—	—	—	1	10½	—	—	—	¾	7½	—	—	—	—	—	—	—	—
Nave Lines	2	17	Single	7	1	1½	14	Single	7	1	1½	12	Single	6	1	1½	10	Single	6	1	—	—	—	—	—
Clue-Garnets ..	3½	70	S.st.bo. Single	13 13	4 2	3	62	S.st.bo. Single	12 12	4 2	3	46	S.st.bo. Single	11 11	4 2	2½	37	S.st.bo. Single	9 9	4 2	—	—	—	—	—
Strap about the Yard ..	3½	8	—	—	—	3	7	—	—	—	3	6	—	—	—	2½	4	—	—	—	—	—	—	—	—
Strapping	3½	2	—	—	—	3	2	—	—	—	3	2	—	—	—	2½	1½	—	—	—	—	—	—	—	—
Seizing	¾	7	—	—	—	¾	7	—	—	—	Tarred Line			—	—	—	—	—	—	—	—	—	—	—	—
Lashing	1	10	—	—	—	1	9	—	—	—	1	7	—	—	—	¾	6	—	—	—	—	—	—	—	—
Buntline-Legs	3	47	Double	10	2	2½	39	Double	9	2	2½	30	Double	8	2	2	26	Double	8	2	—	—	—	—	—
Falls	3	47	Single	10	10	2½	39	Single	9	10	2½	30	Single	8	8	2	26	Single	8	8	—	—	—	—	—
Strapping	3½	11	—	—	—	3	9	—	—	—	2½	8	—	—	—	2½	7	—	—	—	—	—	—	—	—
Leechline-Legs	2½	47	Double	10	4	2½	39	Double	9	4	2	30	Double	8	2	2	26	Double	8	2	—	—	—	—	—
Falls	2½	47	Single	10	10	2½	39	Single	9	10	2	30	Single	8	8	2	26	Single	8	8	—	—	—	—	—
Strapping	3	12	—	—	—	2½	10	—	—	—	2½	9	—	—	—	2½	8	—	—	—	—	—	—	—	—
Slablines and ..	2	36	Single	8	2	2	32	Single	8	2	2	24	Single	7	2	1½	21	Single	7	2	—	—	—	—	—
Strapping ..	2					2					2					1½					—	—	—	—	—
Bowlines	4½	54	Double	16	1	4	45	Double	15	1	3½	36	Double	14	1	3	31	Double	12	1	—	—	—	—	—
Bridles	4½	13	Thimbls.	—	6	4	11	Thimbls.	—	6	3½	9	Thimbls.	—	6	3	8	Thimbls.	—	6	—	—	—	—	—
Strapping	4½	4	—	—	—	4	3	—	—	—	3½	2	—	—	—	3	2	—	—	—	—	—	—	—	—
Seizing	1	4	—	—	—	¾	3	—	—	—	¾	2	—	—	—	¾	2	—	—	—	—	—	—	—	—
Lashing	2½	4	—	—	—	2	3	—	—	—	2	2	—	—	—	1½	2	—	—	—	—	—	—	—	—
Ear-Rings	2	24	—	—	—	2	21	—	—	—	1½	16	—	—	—	1½	13	—	—	—	—	—	—	—	—
Sheets cabled and	5	85	Sing.co.	18	2	4½	75	Sing.co.	17	2	4	55	Sing.co.	16	2	3½	45	Sing.co.	15	2	—	—	—	—	—
Strapping ..	5		Thimbls.	—	2	4½		Thimbls.	—	2	4		Thimble	—	2	3½		Thimble	—	2	—	—	—	—	—
Seizing	1	14	—	—	—	1	12	—	—	—	1	9	—	—	—	¾	7	—	—	—	—	—	—	—	—
Stoppers	4	4	—	—	—	3½	4	—	—	—	3	3½	—	—	—	2½	3	—	—	—	—	—	—	—	—
Tacks, cabled and	5	85	Sing.co.	18	2	4½	75	Sing.co.	17	2	4	55	Sing.co.	16	2	3½	45	Sing.co.	15	2	—	—	—	—	—
Strapping ..	5					4½					4					3½					—	—	—	—	—
Seizing	1½	8	—	—	—	1½	7	—	—	—	1½	6	—	—	—	1	5	—	—	—	—	—	—	—	—
Stoppers	4½	4	H. & T.	2	2	4	4	H. & T.	2	2	3½	3½	H. & T.	2	2	3	3	H. &. T.	2	2	—	—	—	—	—
Slings, 4 Strands, cable	13½	13	Thimble	—	1	13	11	Thimble	—	1	9	8	Thimble	—	1	—	—	—	—	—	—	—	—	—	—
laid and Strap	13½		Thimble	—	1	13		Thimble	—	1	9		Thimble	—	1	—	—	—	—	—	—	—	—	—	—
Seizing	2	9	—	—	—	2	8	—	—	—	1½	6	—	—	—	—	—	—	—	—	—	—	—	—	—
Laniard	3½	9	—	—	—	3	8	—	—	—	2½	6	—	—	—	—	—	—	—	—	—	—	—	—	—
Staysail-Stay	6	13	—	—	—	6	12	—	—	—	5	10½	—	—	—	4½	9	—	—	—	—	—	—	—	—
Haliard	3½	40	*Single	12	2	3	34	*Single	11	2	3	30	*Single	10	2	2½	26	*Single	9	2	—	—	—	—	—
Sheets	3½	13	*Single	12	2	3	12	*Single	11	2	3	10	*Single	10	2	2½	9	*Single	9	2	—	—	—	—	—
Tack	3	4½	—	—	—	2½	4	—	—	—	2½	3½	—	—	—	2	3	—	—	—	—	—	—	—	—
Downhauler ..	2½	20	Single	8	1	2	18	Single	7	1	2	15	Single	7	1	1½	13	Single	6	1	—	—	—	—	—
Strapping ..	3½ 2½	4 1	—	—	—	3 2	4 1	—	—	—	3 2	3 1	—	—	—	2½	3	—	—	—	—	—	—	—	—
Studdingsail-Haliards Inner ..	3	51	Single	11	6	2½	45	Single	10	6	2	33	Single	9	6	2	27	Single	9	6	—	—	—	—	—
Studdingsail-Haliards Outer ..	3½	87	Single	11	4	3	77	Single	10	4	2½	57	Single	9	4	2½	47	Single	9	4	—	—	—	—	—
Sheets	3	25½	—	—	—	2½	22½	—	—	—	2	16½	—	—	—	2	13	—	—	—	—	—	—	—	—
Tacks and ..	3½	47	Single	11	2	3	41½	Single	10	2	2½	30½	Single	9	2	2½	25	Single	9	2	—	—	—	—	—
Strapping ..	3½					3					2½					2½					—	—	—	—	—

Main-top Mast	1257 Tons.					818 Tons.					544 Tons.					330 Tons.					Sloop of 60 Tons.				
	Size in Inch	Fath in Len.	Blocks, &c. Species.	In	N	Size in Inch	Fath in Len.	Blocks, &c. Species.	In	N	Size in Inch	Fath in Len.	Blocks, &c. Species.	In	N	Size in Inch	Fath in Len.	Blocks, &c. Species.	In	N	Size in Inch	Fath in Len.	Blocks, &c. Species.	In	N
Main Topmast.																									
Burton-Pendents	5	6	Thimble	—	2	4½	5	Thimble	—	2	—	—	—	—	—	—	—	—	—	—	—	—	—	—	—
Falls	2½	55	*Double *Single	10 10	2 2	2½	50	*Double *Single	9 9	2 2	—	—	—	—	—	—	—	—	—	—	—	—	—	—	—
Strapping	3½	5	—	—	—	3	4	—	—	—	—	—	—	—	—	—	—	—	—	—	—	—	—	—	—
Shrouds cabled 4 Strands..	6½	65	D. Eyes	10	8	6	56	D. Eyes	8	8	5½	51	D. Eyes	8	8	4½	33	D. Eyes	7	6	—	—	—	—	—
Seizings Eye	1	16	—	—	—	1	14	—	—	—	¾	12	—	—	—	—	—	—	—	—	—	—	—	—	—
Seizings Throat ..	1	28	—	—	—	1	24	—	—	—	¾	21	—	—	—	Tarred Line			—	—	—	—	—	—	—
Seizings End ..	¾	24	—	—	—	¾	21	—	—	—	¾	18	—	—	—	—	—	—	—	—	—	—	—	—	—
Laniards	3½	44	—	—	—	3	36	—	—	—	2½	33	—	—	—	2½	21	—	—	—	—	—	—	—	—
Ratline	1	86	—	—	—	1	76	—	—	—	1	70	—	—	—	1	45	—	—	—	—	—	—	—	—
Futtock Shrouds cabled ..	6½	24	Pl.d.e.	10	8	6	21	Pl.d.e.	8	8	5½	19	Pl.d.e.	8	8	4½	12	Pl.d.e.	7	6	—	—	—	—	—
Seizing Upper ..	1	30	—	—	—	1	26	—	—	—	¾	23	—	—	—	¾	15	—	—	—	—	—	—	—	—
Seizing Lower ..	¾	26	—	—	—	¾	23	—	—	—	¾	20	—	—	—	¾	13	—	—	—	—	—	—	—	—

Main-topsail Yard, Main-top-gallant Mast and Main-top-gallant Yard.	1257 Tons.					818 Tons.					544 Tons.					330 Tons.					Sloop of 60 Tons.				
	Size in Inch	Fath in Len.	Blocks, &c. Species.	In	N	Size in Inch	Fath in Len.	Blocks, &c. Species.	In	N	Size in Inch	Fath in Len.	Blocks, &c. Species.	In	N	Size in Inch	Fath in Len.	Blocks, &c. Species.	In	N	Size in Inch	Fath in Len.	Blocks, &c. Species.	In	N
Tacks	3½	60	Single	12	4	3	53	Single	10	4	2½	47	Single	9	4	2	37	Single	8	4	—	—	—	—	—
Downhaulers	2	54	Single Thimble	9	2 6	2	48	Single Thimble	9	2 6	1½	43	Single Thimble	8	2 6	1½	33	Single Thimble	7	2 6	—	—	—	—	—
Boom Tackles	2	72	Double Single	8 8	2 4	2	64	Double Single	8 8	2 4	—	—	—	—	—	—	—	—	—	—	—	—	—	—	—
Lashing for Booms	2½	20	—	—	—	2½	20	—	—	—	2	15	—	—	—	1½	12	—	—	—	—	—	—	—	—
Tailing and Strapping	3½ 2½	18 6	—	—	—	3½ 2½	16 5	—	—	—	3 2	14 4½	—	—	—	2	11 3½	—	—	—	—	—	—	—	—
Main Topgallant Mast.																									
Shrouds cabled 4 Strands	4½	76	Thimbls.	—	12	3¾	65	Thimbls.	—	12	3	60	Thimbls.	—	12	2½	50	Thimbls.	—	8	—	—	—	—	—
Laniards	2	12	—	—	—	1½	10	—	—	—	1½	9	—	—	—	1	8	—	—	—	—	—	—	—	—
Backstays cabled 4 Strands	4½	42	D. Eyes	7	4	3¾	39	D. Eyes	6	4	3	35	D. Eyes	6	4	2½	30	Thimbls.	—	4	—	—	—	—	—
Laniards	2	10	—	—	—	1½	9	—	—	—	1½	8	—	—	—	1	7	—	—	—	—	—	—	—	—
Stay cabled 4 Strands	4½	30	Single	12	1	3¾	27	Single	12	1	3	25	Single	10	1	2½	22	Thimble	—	1	—	—	—	—	—
Strapping	3½	1½	—	—	—	3	1½	—	—	—	2½	1½	—	—	—	—	—	—	—	—	—	—	—	—	—
Royal or Flagstaff Stay	2½	30	Thimble	—	1	2	27	Thimble	—	1	—	—	—	—	—	—	—	—	—	—	—	—	—	—	—
Haliards	1½	64	—	—	—	1½	55	—	—	—	1	50	—	—	—	¾	42	—	—	—	—	—	—	—	—
Royal Haliards	2	64	Single	7	1	2	55	Single	7	1	1½	50	Single	6	1	1½	42	Single	6	1	—	—	—	—	—
Main Topgallant Yard.																									
Tie	3½	17	—	—	—	3½	15	—	—	—	3	13	—	—	—	2½	12	—	—	—	—	—	—	—	—
Haliard	2	30	Double Single	8 8	1 2	2	26	Double Single	8 8	1 2	1½	22	Double Single	7 7	1 2	1½	20	Single	6	2	—	—	—	—	—
Strapping	2½	1½	—	—	—	2½	1½	—	—	—	2	1½	—	—	—	2	1	—	—	—	—	—	—	—	—
Horses	3	7½	—	—	—	3	7	—	—	—	2½	6	—	—	—	2½	5	—	—	—	—	—	—	—	—
Braces and Strapping	2 2	66	Single	7	2	2 2	59	Single	7	2	2	54	Single	6	2	1½	45	Single	6	2	—	—	—	—	—
Lifts, single and Strapping	2½ 2½	57	Single Thimbls.	7	2 2	2½ 2½	51	Single Thimbls.	7	2 2	2½	46	Thimbls.	—	2	2	39	Thimbls.	—	2	—	—	—	—	—
Parral-Ropes	2	5	Parral	12	1	2	4	Parral	10	1	1½	3½	Parral	9	1	1	3	Parral	7	1	—	—	—	—	—
Clue-Lines and Strapping	2 2	76	Single	7	6	2 2	68	Single	7	6	1½ 2	62	Single	6	4	1 1½	52	Single	5	4	—	—	—	—	—
Bow-Lines	2	75	Single	7	2	2	67	Single	7	2	1½	62	—	—	—	1	52	—	—	—	—	—	—	—	—
Bridles	2	9	Thimbls.	—	6	2	8	Thimbls.	—	6	1½	7½	Thimbls.	—	6	1	6½	Thimbls.	—	4	—	—	—	—	—
Strapping	2	1	—	—	—	2	1	—	—	—	—	—	—	—	—	—	—	—	—	—	—	—	—	—	—
Ear-Rings, Tarred-Line	—	28	—	—	—	—	25	—	—	—	—	23	—	—	—	—	19	—	—	—	—	—	—	—	—
Staysail-Stay	3	38	Single	10	1	3	32	Single	10	1	2½	30	Single	8	1	2	27	Single	7	1	—	—	—	—	—
Haliards	2½	48	Single	9	1	2½	42	Single	9	1	2	40	Single	7	1	1½	37	Single	6	1	—	—	—	—	—
Sheets	2½	48	Single	9	2	2½	42	Single	9	2	2	40	Single	7	2	1½	37	Single	6	2	—	—	—	—	—
Tack	2	7	—	—	—	2	7	—	—	—	1½	6	—	—	—	1	5	—	—	—	—	—	—	—	—
Down-Hauler	2	35	Single	7	2	2	29	Single	7	2	1½	27	Single	6	2	1	24	Single	5	2	—	—	—	—	—
Strapping	2½ 2	3 1	—	—	—	2½ 2	3 1	—	—	—	2	4	—	—	—	1½	4	—	—	—	—	—	—	—	—
Studding-Sail-Haliards	2	74	Single	7	6	2	63	Single	7	6	2	58	Single	7	6	1½	50	Single	6	6	—	—	—	—	—
Sheets	2	37	—	—	—	2	31	—	—	—	1½	29	—	—	—	1	25	—	—	—	—	—	—	—	—
Tacks	2	47	Single	7	4	2	41	Single	7	4	2	39	Single	7	4	1½	35	Single	6	4	—	—	—	—	—
Down-Hauler	1½	37	Thimbls.	—	2	1½	31	Thimbls.	—	2	1	29	Thimbls.	—	2	1	25	Thimbls.	—	2	—	—	—	—	—
Strapping	2	9	—	—	—	2	8	—	—	—	2	7	—	—	—	1½	6	—	—	—	—	—	—	—	—
Mizen-mast.																									
Burton-Pendents	5	8	Thimbls.	—	2	4½	8	Thimbls.	—	2	4	7	Thimbls.	—	2	3½	6	Thimbls.	—	2	—	—	—	—	—
Falls	3	58	*Double *Single	11 11	2 2	2¾	56	*Double *Single	10 10	2 2	2½	49	*Double *Single	9 9	2 2	2	43	*Double *Single	8 8	2 2	—	—	—	—	—
Strapping	3½	4	—	—	—	3	4	—	—	—	3	3	—	—	—	2½	3	—	—	—	—	—	—	—	—
Shrouds cabled 4 Strands	7	80	D. Eyes	10	12	6½	75	D. Eyes	8	12	5½	65	D. Eyes	8	10	4½	44	D. Eyes	7	8	—	—	—	—	—
Seizings: Eye	1	20	—	—	—	1	19	—	—	—	1	16	—	—	—				—	—	—	—	—	—	—
Seizings: Throat	1	40	—	—	—	1	38	—	—	—	¾	32	—	—	—	Tar	red	Line	—	—	—	—	—	—	—
Seizings: End	¾	30	—	—	—	¾	28	—	—	—	¾	24	—	—	—				—	—	—	—	—	—	—
Laniards	3½	40	—	—	—	3	38	—	—	—	2½	32	—	—	—	2½	22	—	—	—	—	—	—	—	—
Ratline	1	120	—	—	—	1	113	—	—	—	1	97	—	—	—	1	66	—	—	—	—	—	—	—	—
Stay 4 Strands cabled	8	14	Thimbls.	—	2	7¼	13	Thimbls.	—	2	6	11	Thimbls.	—	2	5	9	Thimbls.	—	2	—	—	—	—	—
Seizing	1	7	—	—	—	1	6	—	—	—	1	5	—	—	—	¾	4	—	—	—	—	—	—	—	—
Laniard	3½	6	—	—	—	3	6	—	—	—	2½	5	—	—	—	2½	4	—	—	—	—	—	—	—	—
Collar	7	3½	Thimble	—	1	6	3	Thimble	—	1	5	2½	Thimble	—	1	4	2	Thimble	—	1	—	—	—	—	—
Seizing	1	3	—	—	—	1	3	—	—	—	1	2½	—	—	—	¾	2	—	—	—	—	—	—	—	—
Lashing	1½	3	—	—	—	1½	3	—	—	—	1½	2½	—	—	—	1	2	—	—	—	—	—	—	—	—

Yard or Gaff	1257 Tons.					818 Tons.					544 Tons.					330 Tons.					Sloop of 60 Tons.				
	Size in Inch	Fath in Len.	Blocks, &c. Species	In	N	Size in Inch	Fath in Len.	Blocks, &c. Species.	In	N	Size in Inch	Fath in Len.	Blocks, &c. Species.	In	N	Size in Inch	Fath in Len.	Blocks, &c. Species.	In	N	Size in Inch	Fath in Len.	Blocks, &c. Species.	In	N
Yard or Gaff.																									
Jeers	5	58	Treb.co. Dou.co.	18 18	1 1	4½	56	Treb.co. Dou.co.	16 16	1 1	4	32	Dou.co. Sing.co.	16 16	1 1	3	29	Dou.co. Sing.co.	12 12	1 1	—	—	—	—	—
Strapping	5½	8	—	—	—	5	7	—	—	—	4	6	—	—	—	3	5	—	—	—	—	—	—	—	—
Seizing	1½	10	—	—	—	1½	9	—	—	—	1	8	—	—	—	¾	7	—	—	—	—	—	—	—	—
Lashing at the Mast Hd.	2½	10	—	—	—	2½	9	—	—	—	2½	8	—	—	—	2	7	—	—	—	—	—	—	—	—
Lashing at the Yard	2	5	—	—	—	2	4½	—	—	—	1½	4	—	—	—	1	3½	—	—	—	—	—	—	—	—
Derrick	4	39	Double	14	1	3½	37	Double	12	1	3	32	Double	11	1	2½	29	Double	10	1	—	—	—	—	—
Span	4	4	Single	14	1	3½	4	Single	12	1	3	3	Single	11	1	2½	3	Single	9	1	—	—	—	—	—
Strapping	4	2½	Thimble	—	1	3½	2½	Thimble	—	1	3	2	Thimble	—	1	2½	2	Thimble	—	1	—	—	—	—	—
Seizing	1	9	—	—	—	1	9	—	—	—	¾	8	—	—	—	Tarr	ed	Line	—	—	—	—	—	—	—
Lashing	2	8	—	—	—	2	8	—	—	—	1½	7	—	—	—	1	6	—	—	—	—	—	—	—	—

Main-top mast and Main-topsail Yard—	1257 Tons.					818 Tons.					544 Tons.					330 Tons.					Sloop of 60 Tons.				
	Size in Inch	Fath in Len.	Blocks, &c. Species.	In	N	Size in Inch	Fath in Len.	Blocks, &c. Species.	In	N	Size in Inch	Fath in Len.	Blocks, &c. Species.	In	N	Size in Inch	Fath in Len.	Blocks, &c. Species.	In	N	Size in Inch	Fath in Len.	Blocks, &c. Species.	In	N
Ratline	1	30				1	26				1	23				1	15				—	—	—		
Backstays cabled 4 Strands	6½	120	D. Eyes	10	6	6	100	D. Eyes	9	6	5½	90	D. Eyes	8	6	4½	56	D. Eyes	7	4	—	—	—		
Seizings Eye	1	13				1	12				¾	10									—	—	—		
Seizings Throat	1	22				1	21				¾	17				Tarred Line					—	—	—		
Seizings End	¾	19				¾	18				¾	15									—	—	—		
Laniards	3½	22				3	19				2½	17				2½	10				—	—	—		
Stay cabled 4 Strands	7½	24				7	21				6	20				5	16				—	—	—		
Collar	6	4	Single	18	1	5½	3½	Single	16	1	5	3	Single	15	1	4	2½	Single	13	1	—	—	—		
Tackle	3½	22	Long.t. *Single	20 12	1 1	3	19	Long.t. *Single	18 10	1 1	2½	17	Long.t. *Single	18 9	1 1	2½	15	Long.t. *Single	16 8	1 1	—	—	—		
Strapping	4½	3				4	2½				3	2½				3	2				—	—	—		
Seizing	1½ ¾	6				1 ¾	5				1 ¾	5				¾	4				—	—	—		
Top Rope Pendents cab. 4 Strands	8½	34	S.Br.Sh. I.bo.co.	22	2	7½	32	S.Br.Sh. I.bo.co.	20	2	6½	30	S.Br.Sh. I.bo.co.	18	2	6	26	S.Br.Sh. I.bo.co.	16	2	—	—	—		
Falls	4½	108	Tr.I.b.co.	20	4	4	100	Tr.I.bo.co	18	4	3½	95	Tr.I.bo. co	16	4	3	82	Tr.I.bo. co	14	4	—	—	—		
Main Topsail-yard.																									
Tie cabled 4 Strands	5½	40	Fl.si.co. Dou.co	18 18	2 1	5	35	Fl.si.co. Dou.co.	16 16	2 1	4½	32	Fl.si.co. Dou.co.	15 15	2 1	4	28	Fl.si.co. Dou.co.	14 14	2 1	—	—	—		
Strapping	5½	8				5	7				4½	6				4	5½				—	—	—		
Seizing	1	12				1	10				1	9				¾	8				—	—	—		
Lashers at the Mast Hd.	2½	11				2½	9				2	8				1½	7				—	—	—		
Lashers at the Yard	2	5½				2	4½				1½	4				1	3½				—	—	—		
Haliards	3½	110	Do.th.co. *Sth.co.	22 22	2 2	3½	100	Do.th.co. *S.th.co.	20 20	2 2	3	90	Do.th.co. *S.th.co.	18 18	2 2	2½	76	Do.th.co. *S.th.co.	16 16	2 2	2	20	—		
Strapping	4½	8				4½	7				4	6				3½	5				—	—	—		
Seizing	¾	16				¾	14				¾	12				¾	10				—	—	—		
Horses	4	11				3½	10				3	9				3	7				—	—	—		
Stirrups	3	8	Thimbls.	—	6	3	7	Thimbls.	—	6	2½	6½	Thimbls.	—	6	2	5	Thimbls.	—	4	—	—	—		
Braces	3½	70	Single	12	4	3½	60	Single	10	4	3	50	Single	9	4	3	42	Single	8	2	1½	23	—		
*Pendents	4½	9	Single	12	2	4½	8	Single	10	2	4	7	Single	10	2	3½	5½	Single	9	2	—	—	—		
Strapping	4	4½				3½	4				3	3½				2½	2½				—	—	—		
Lifts	3½	64	Double Single	12 12	2 4	3½	55	Double Single	10 10	2 4	3	48	Sister Single	17 10	2 4	2½	38	Sister Single	14 8	2 4	1½	21	—		
Beckets	3½	3				3½	3				3	2½				2½	2				—	—	—		
Strapping	3½	9				3½	9				3	7				2½	6				—	—	—		
Seizings	¾	27				¾	27				Tarred Line					—	—				—	—	—		
Parral Ropes	3	11	Parral	22	1	2½	10	Parral	20	1	2½	9	Parral	16	1	2	7	Parral	14	1	—	—	—		
Racking and Seizing	1	15				¾	13				¾	12				Tarred Line					—	—	—		
Clue Lines	3½	83	S.st.bo. Single	12 12	4 2	3½	73	S.st.bo. Single	10 10	4 2	3	65	S.st.bo. Single	10 10	4 2	2½	50	S.st.bo. Single	8 8	4 2	—	—	—		
Strapping	3½	8				3½	7				3	6				2½	5				—	—	—		
Bunt Lines and Strapping	3 3	66	Single	10	4	3 3	58	Single	10	4	2½ 2½	52	Single	8	4	2 2	40	Single	7	4	—	—	—		
Leech lines and Strapping	2½ 2½	29	Single	9	2	2½ 2½	25	Single	9	2	2 2	22	Single	7	2	2 2	17	Single	6	2	—	—	—		
Bow lines	3	60	Single	11	2	3	52	Single	11	2	2½	47	Single	9	2	2	36	Single	8	2	1½	21	—		
Bridles	3	12	Thimbls.	—	4	3	10	Thimbls.	—	4	2½	9	Thimbls.	—	4	2	7	Thimbls.	—	4	1½	3	Thimbls.	—	4
Strapping	3	5				3	4				2½	3				2	2				—	—	—		
Lashing	1	12				1	10				1	9				¾	7				—	—	—		
Reef Tackle Pendents	3½	42	*Double	8	4	3½	38	*Double	8	4	3	32	*Single	7	4	2½	25	*Single	7	4	—	—	—		
Falls	2	64				2	56				1½	48				1½	37				—	—	—		
Strapping	2½	3				2½	3				2	2				2	1½				—	—	—		
Ear Rings	1½	42				1½	38				1½	32				1	25				—	—	—		
Sheets	5½	52	S.Sho. ¼ thk. & thin.co.	20 20	2 2	5	45	S.Sho. ¼ thk. & thin.co.	18 18	2 2	4½	40	S.Sho. ¼ thk. & thin.co.	16 16	2 2	4	32	S.Sho. ¼ thk. & thin.co.	14 14	2 2	1½	21	—		
Strapping for Sheet Blk.	6	5				5½	4				5	3½				4½	3				—	—	—		
Strapping for Quart. do.	4½	10				4½	8				4	7				3½	6				—	—	—		
Lashers for Quarter do.	2	10				2	8				2	7				1½	6				—	—	—		
Seizing	1½	18				1	16				1	14				¾	11				—	—	—		
Span	3	8				3	7				3	6				2½	5				—	—	—		
Stoppers	4½	4				4½	3½				3½	3				3	2½				—	—	—		
Slings	5	11				4½	10				3½	9				3	7				—	—	—		
Staysail Stay cabled 4 St.	7	24	Single	14	1	7	21	Single	14	1	5½	20	Single	13	1	4½	16	Single	10	1	—	—	—		
Tackle	3¼	14	Double Single	10 9	1 1	3½	12	Double Single	10 9	1 1	2½	11	Double Single	8 7	1 1	2	10	Single	8	2	—	—	—		
Haliard and Strapping	3½ 3½	48	Single	12	1	3½ 3½	42	Single	12	1	2½ 2½	40	Single	10	1	2 2	32	Single	9	1	—	—	—		
Sheets and Strapping	3½ 3½	64	Single	12	2	3½ 3½	56	Single	12	2	2½ 2½	53	Single	10	2	2 2	41	Single	9	2	—	—	—		
Tack	2½	5				2½	5				2	4				1½	4				—	—	—		
Downhauler and Strapping	2½ 2½	29	Single	9	1	2½ 2½	25	Single	9	1	2 2	24	Single	8	1	1½ 1½	18	Single	7	1	—	—	—		
Brails	2	47	Thimble Single	7	2 2	1½	41	Thimble Single	6	2 2	1½	39	Thimble Single	6	2 2	—	—	—			—	—	—		
Middle Staysail Stay	5	24	Single	12	1	4½	21	Single	12	1	3½	20	Single	10	1	3	16	Single	9	1	—	—	—		
Tackle	2½	22	Double Single	8 8	1 1	2½	19	Double Single	8 8	1 1	2	18	Double Single	8 8	1 1	2	14	Single	7	2	—	—	—		
Haliard	3½	46	Single	12	1	3½	40	Single	10	1	3	38	Single	9	1	2	30	Single	8	1	—	—	—		
Sheets	3½	46	Single	12	1	3	40	Single	10	1	2½	38	Single	9	1	2	30	Single	8	1	—	—	—		
Tack	3	5				2½	5				2	4				2	4				—	—	—		
Down-Hauler	2½	30	Single	8	2	2	26	Single	8	2	2	25	Single	7	1	1½	20	Single	6	1	—	—	—		
Strapping	3	8				2½	8				2½	6				2	5				—	—	—		
Tricing Lines	2	18	Single	8	1	2	15	Single	8	1	—	—	—			—	—	—			—	—	—		
Studdingsail-Haliards	3½	90	Single	12	6	3	80	Single	10	6	2½	70	Single	9	6	2	55	Single	8	6	—	—	—		
Sheets	3	46	Single	12	2	2½	40	Single	10	2	2	36	Single	9	2	1½	28	Single	8	2	—	—	—		

Yard or Gaff—*continued.*	1257 Tons.					818 Tons.					544 Tons.					330 Tons.					Sloop of 60 Tons.				
	Size in Inch	Fath in Len.	Blocks, &c. Species	In	N	Size in Inch	Fath in Len.	Blocks, &c. Species.	In	N	Size in Inch	Fath in Len.	Blocks, &c. Species.	In	N	Size in Inch	Fath in Len.	Blocks, &c. Species.	In	N	Size in Inch	Fath in Len.	Blocks, &c. Species.	In	N
Vang-Pendents	4	11	Double	8	2	3½	10	Double	8	2	3	9	—	—	—	3	8	—	—	—	—	—	—	—	—
Falls	2	33	*Single	8	2	2	30	*Single	8	2	2	27	*Single	8	4	1½	24	*Single	7	4	—	—	—	—	—
Strapping	3	1	—	—	—	2½	1	—	—	—	2	1	—	—	—	1½	1	—	—	—	—	—	—	—	—
Bowlines	3	18	Single	10	4	2½	16	Single	9	4	2½	14	Single	7	4	—	—	—	—	—	—	—	—	—	—
Strapping	3	3	Thimbls.	—	2	2½	3	Thimbls.	—	2	2½	2	Thimbls.	—	2	—	—	—	—	—	—	—	—	—	—
Brail-Peek-Legs	2	12	—	—	—	2	11	—	—	—	1½	10	—	—	—	1½	8	—	—	—	—	—	—	—	—
Falls	2	18	Single	7	6	2	17	Single	7	6	1½	16	Single	6	4	1½	12	Single	6	4	—	—	—	—	—
Throat	3	24	Single	10	2	3	22	Single	9	2	2½	20	Single	8	2	2	17	Single	7	2	—	—	—	—	—
Middle	2½	19	Single	9	2	2½	18	Single	8	2	2	16	Single	7	2	1½	14	Single	6	2	—	—	—	—	—
Foot	2½	32	Single	9	2	2½	30	Single	8	2	2	28	Single	7	2	1½	23	Single	6	2	—	—	—	—	—
Strapping ..	3 2½	4 3	Thimble	—	4	3 2½	4 3	Thimbls.	—	4	2½	6	Thimbls.	—	4	2	5	Thimbls.	—	4	—	—	—	—	—
Lacing Mizen to Yard ..	1½	19	—	—	—	1½	18	—	—	—	1½	16	—	—	—	¾	14	—	—	—	—	—	—	—	—
Lacing Mizen to Mast ..	3	26	—	—	—	2	25	—	—	—	2	23	—	—	—	2	21	—	—	—	—	—	—	—	—
Ear-Rings	2	10	—	—	—	1½	10	—	—	—	1½	9	—	—	—	1	8	—	—	—	—	—	—	—	—
Peek-Haliards	2	19	*Single	7	1	1½	18	*Single	6	1	1½	16	*Single	6	1	1	14	*Single	5	1	—	—	—	—	—
Sheet	4	26	*Double Single	14 14	1 1	3½	25	*Double Single	12 12	1 1	3	21	*Double *Single	10 10	1 1	2½	19	*Single	9	2	—	—	—	—	—
Strapping	4	3½	—	—	—	3½	3	—	—	—	3	2½	—	—	—	2½	2	—	—	—	—	—	—	—	—
Seizing	1	7	—	—	—	1	6	—	—	—	1	5	—	—	—	Tarred Line			—	—	—	—	—	—	—
Tack	3	5	—	—	—	2½	5	—	—	—	2	4	—	—	—	1½	3	—	—	—	—	—	—	—	—
Slings	5½	6	—	—	—	5	6	—	—	—	4	5	—	—	—	3	4	—	—	—	—	—	—	—	—
Staysail-Stay	5	13	Thimbls.	—	2	4½	12	Thimbls.	—	2	4	11	Thimbls.	—	2	3	9	Thimbls.	—	2	—	—	—	—	—
Collar	4½	4	Thimble	—	1	4	3½	Thimble	—	1	3½	2½	Thimble	—	1	2½	2	Thimble	—	1	—	—	—	—	—
Seizing	1	4	—	—	—	¾	3½	—	—	—	¾	2½	—	—	—	Tarred Line			—	—	—	—	—	—	—
Lashing	1	4	—	—	—	1	3½	—	—	—	¾	2½	—	—	—	¾	2	—	—	—	—	—	—	—	—
Laniard	2	8	—	—	—	2	7	—	—	—	1½	5	—	—	—	1	4	—	—	—	—	—	—	—	—
Haliard	3	26	Single	11	3	2½	24	Single	9	3	2	22	Single	8	3	2	18	Single	7	3	—	—	—	—	—
Sheets	3½	13	Single	11	2	3	12	Single	9	2	2½	11	Single	8	2	2	9	Single	7	2	—	—	—	—	—
Tack	2½	4	—	—	—	2½	4	—	—	—	2½	3	—	—	—	2	2	—	—	—	—	—	—	—	—
Down-Hauler	2	13	Single	7	2	2	12	Single	7	2	1½	11	Single	6	2	1	9	Single	5	1	—	—	—	—	—
Strapping ..	3½ 2½	3 2	—	—	—	2½	5	—	—	—	2	4	—	—	—	2	4	—	—	—	—	—	—	—	—
Brails	2	11	Single	7	2	2	10	Single	7	2	2	9	Single	7	2	1½	7	Single	6	2	—	—	—	—	—

Boom-driver, Cross-Jack Yard and Mizen Top-mast.	1257 Tons.					818 Tons.					544 Tons.					330 Tons.					Sloop of 60 Tons.				
	Size in Inch	Fath in Len.	Blocks, &c. Species.	In	N	Size in Inch	Fath in Len.	Blocks, &c. Species.	In	N	Size in Inch	Fath in Len.	Blocks, &c. Species.	In	N	Size in Inch	Fath in Len.	Blocks, &c. Species.	In	N	Size in Inch	Fath in Len.	Blocks, &c. Species.	In	N
Boom-Driver.																					or, **Main-Sail.**				
Topping-Lifts or Runner..	5	34	—	—	—	4½	29	—	—	—	4	27	—	—	—	3	21	—	—	—	4	30	*S.I.b.d.	10	1
Span	5	6	Double	15	1	4½	6	Double	15	1	4	5	Double	13	1	3	4	Double	11	1	—	—	Single Single	10 8	1 1
Fall	3	46	Double *Single	10 10	2 2	3	40	Double *Single	10 10	2 2	2½	37	Double *Single	9 9	2 2	2	28	Double *Single	8 8	2 2	2	20	Double *Single	8 8	1 1
Strapping	3	5	—	—	—	3	5	—	—	—	2½	4½	—	—	—	2	4	—	—	—	4	4	Gaff-Span		
Guy-Pendents	3½	12	H. & T.	—	4	3½	11	H. & T.	—	4	3	10	—	—	—	2½	9	—	—	—	4	6	—	—	—
Falls	2½	60	*Double *Single	9 9	2 2	2½	58	*Double *Single	9 9	2 2	2	54	*Double *Single	8 8	2 2	2	50	*Double *Single	7 7	2 2	2	20	*Double *Single	8 8	1 1
Boom-Horses	3½	14	—	—	—	3½	12	—	—	—	3	11	—	—	—	2½	10	—	—	—	2	8	—	—	—
Sheet	3	26	Single	10	2	3	25	Single	10	2	2½	26	Single	9	2	2	22	Single	8	2	2	16	Single	8	2
Sheet-Pendent	3½	14	—	—	—	3½	12	—	—	—	3½	11	—	—	—	3	10	—	—	—	2½	7	—	—	—
Falls	2	20	*Double *Single	8 8	1 1	2	20	*Double *Single	8 8	1 1	2	18	*Double *Single	8 8	1 1	2	16	*Double *Single	7 7	1 1	2	14	*Double *Single	6 6	1 1
Hald. or Brails Inner Tie Haliard	2½	108	Single	9	6	2½	100	Single	9	6	2	95	Single	8	6	2	90	Single	7	6	3½ 2	9 20	S.I.bd.co *D.th.co.	13 17	1 2
Lac. to Yard Outer Tie Haliard	1½	30	—	—	—	1½	28	—	—	—	1½	26	—	—	—	1	24	—	—	—	3½ 2	10 26	S.I.bd.co. *D.th.co.	13 17	1 2
Tack-Tackle ..	2	12	*Double *Single	12 12	1 1	2	12	*Double *Single	12 12	1 1	2	10	*Double *Single	10 10	1 1	1½	9	*Double *Single	8 8	1 1	—	—	—	—	—
Down-Haulers Peek / Throat ..	2½	19	—	—	—	2½	18	—	—	—	2	16	—	—	—	1½	14	—	—	—	1½ 1½	14 20	Single Single	6 6	2 1
Lashing	¾	8	—	—	—	¾	8	—	—	—	¾	7	—	—	—	¾	6	—	—	—	—	—	—	—	—
Ear-Rings Inner / Outer ..	—	—	—	—	—	—	—	—	—	—	—	—	—	—	—	—	—	—	—	—	1 ¾	3 5	—	—	—
Reef	—	—	—	—	—	—	—	—	—	—	—	—	—	—	—	—	—	—	—	—	3	6	—	—	—
Cross-jack Yard.																									
Truss-Pendents	4	6	Thimbls.	—	2	3½	6	Thimbls.	—	2	3	5	Thimbls.	—	2	2¾	4	Thimbls.	—	2	—	—	—	—	—
Falls	2½	19	*L.t. *Single	18 9	1 1	2½	18	*L.t. *Single	18 9	1 1	2	15	*L.t. *Single	16 8	1 1	2	12	*L.t. *Single	14 7	1 1	—	—	—	—	—
Strapping	3	3	—	—	—	2½	2	—	—	—	2	2	—	—	—	2	2	—	—	—	—	—	—	—	—
Seizing	¾	8	—	—	—	¾	8	—	—	—	¾	7	—	—	—	Tarred Line			—	—	—	—	—	—	—
Span about the Cap ..	4	2	—	—	—	3½	2	—	—	—	3	2	—	—	—	2	1½	—	—	—	—	—	—	—	—
Braces and Strapping ..	2½	60	Single	9	2	2½	54	Single	9	2	2	49	Single	8	2	1½	38	Single	6	2	1½	20	—	—	—
Lifts running, and Strap...	2½	50	Single	9	4	2½	45	Single	9	4	2	42	Single	8	4	1½	32	Single	6	4	1½	18	Single	6	4
Slings	4½	5½	S.do.sc.	14	1	4½	5	S.do.sc.	14	1	3½	4½	S.do.sc.	12	1	3	3½	Thimble	—	1	—	—	—	—	—
Strapping	4	2	—	—	—	3½	2	—	—	—	3	2	—	—	—	2½	1½	—	—	—	—	—	—	—	—
Seizing	1	3	—	—	—	¾	3	—	—	—	¾	3	—	—	—	Tarred Line			—	—	—	—	—	—	—
Lashing	1½	5	—	—	—	1	5	—	—	—	1	4	—	—	—				—	—	—	—	—	—	—
Mizen Topmast.																									
Shrouds cabled 4 Strands..	5	47	D. Eyes	8	8	4¾	42	D. Eyes	8	8	4	30	D. Eyes	7	6	3	28	D. Eyes	6	6	—	—	—	—	—
Seizings	¾	70	—	—	—	¾	63	—	—	—	Tarred Line			—	—	—	—	—	—	—	—	—	—	—	—
Laniards	2½	31	—	—	—	2½	28	—	—	—	2	20	—	—	—	1½	19	—	—	—	—	—	—	—	—

Mizen Top-mast and Mizen-topsail Yard.	1257 Tons.					818 Tons.					544 Tons.					330 Tons.					Sloop of 60 Tons.				
	Size in Inch	Fath in Len.	Blocks, &c. Species.	In	N	Size in Inch	Fath in Len.	Blocks, &c. Species.	In	N	Size in Inch	Fath in Len.	Blocks, &c. Species.	In	N	Size in Inch	Fath in Len.	Blocks, &c. Species.	In	N	Size in Inch	Fath in Len.	Blocks, &c. Species.	In	N
Ratline	1	59	—	—	—	1	52	—	—	—	1	37	—	—	—	¾	35	—	—	—	—	—	—	—	—
Backstays	5	80	D. Eyes	8	6	4¾	70	D. Eyes	8	6	4	40	D. Eyes	7	4	3	37	D. Eyes	5	4	—	—	—	—	—
Seizings	Tarred Line			—	—	—	—	—	—	—	—	—	—	—	—	—	—	—	—	—	—	—	—	—	—
Laniard	2½	20	—	—	—	2½	20	—	—	—	2	12	—	—	—	1½	12	—	—	—	—	—	—	—	—
Stay cabled 4 Strands	5	12	—	—	—	4¾	10	—	—	—	4	9	—	—	—	3½	8	—	—	—	—	—	—	—	—
Laniard	2½	6	Thimbls.	—	2	2½	5	Thimbls.	—	2	2	4½	Thimbls.	—	2	1½	4	Thimbls.	—	2	—	—	—	—	—
Collar	4	3	Single	14	1	3½	3	Single	13	1	3	2½	Single	11	1	2½	2	Thimbls.	—	1	—	—	—	—	—
Seizing and Lashing	1	6	—	—	—	1	6	—	—	—	¾	5	—	—	—	Tarred Line			—	—	—	—	—	—	—
Flagstaff-Stay	2	12	Thimble	—	1	2	10	Thimble	—	1	1½	9	Thimble	—	1	—	—	—	—	—	—	—	—	—	—
Haliards	1½	20	—	—	—	1½	18	—	—	—	1	15	—	—	—	¾	14	—	—	—	—	—	—	—	—
Futtock-Shrouds	4½	14	*Pl.d.e.	8	8	4	12	Pl.d.e.	8	8	3½	9	Pl.d.e.	7	6	2½	8	Pl.d.e.	6	6	—	—	—	—	—
Seizings	Tarred Line			—	—	—	—	—	—	—	—	—	—	—	—	—	—	—	—	—	—	—	—	—	—
Ratlines	1	15	—	—	—	1	13	—	—	—	1	10	—	—	—	¾	9	—	—	—	—	—	—	—	—
Top-Rope Pendents	5½	15	*S.I.bd.	14	1	5	13½	*S.I.bd.	14	1	4½	11	*S.I.bd.	12	1	4	10	S.I.bd.	12	1	—	—	—	—	—
Falls	3	37	D.I.bd.	12	2	3	32	D.I.bd.	12	2	2½	27	D.I.bd.	10	2	2½	25	D.I.bd.	10	2	—	—	—	—	—
Mizen Topsail Yard.																									
Tie	4	13	S.do.sc.	13	1	4	12	S.do.sc.	13	1	3½	10	S.do.sc.	12	1	3	9	S.do.sc.	10	1	—	—	—	—	—
Haliard	2½	40	D.th.co. *S.th.co.	12 12	1 1	2½	36	D.th.co. *S.th.co.	12 12	1 1	2	30	D.th.co. *S.th.co.	10 10	1 1	2	28	*S.th.co.	8	2	—	—	—	—	—
Strapping	3½	4	—	—	—	3½	4	—	—	—	3	4	—	—	—	2½	2	—	—	—	—	—	—	—	—
Lashing	1	3	—	—	—	1	3	—	—	—	¾	3	—	—	—	Tarred Line			—	—	—	—	—	—	—
Horses	3	8	—	—	—	3	7	—	—	—	2½	6	—	—	—	2	5½	—	—	—	—	—	—	—	—
Stirrups	2½	4	Thimbls.	—	2	2½	3½	Thimbls.	—	2	2	3	Thimbls.	—	2	—	—	—	—	—	—	—	—	—	—
Braces and Strapping	2½	48	Single	9	2	2½	44	Single	9	2	2	39	Single	7	2	1½	37	Single	5	2	—	—	—	—	—
Lifts	2½	50	Double Single	9 9	2 2	2½	45	Double Single	9 9	2 2	2	40	Single	7	4	2	38	Thimbls.	—	2	—	—	—	—	—
Strapping	3	3	—	—	—	3	3	—	—	—	2	2	—	—	—	—	—	—	—	—	—	—	—	—	—
Parral-Ropes	2½	5	Parral	14	1	2½	4½	Parral	14	1	2	3½	Parral	11	1	1½	3	Parral	9	1	—	—	—	—	—
Clue-lines and Strapping	2½	58	S.st.bo. Single	9 9	4 2	2½	52	S.st.bo. Single	9 9	4 2	2	45	S.st.bo. Single	7 7	4 2	1½	42	S.st.bo. Single	5 5	4 2	—	—	—	—	—
Bunt-lines and Strapping	2	41	Single	8	4	2	37	Single	8	4	1½	32	Single	7	2 2	1½	30	Single Thimble	6 —	1 1	—	—	—	—	—
Leech-lines and Strapping	2	16	Single	7	2	2	15	Single	7	2	—	—	—	—	—	—	—	—	—	—	—	—	—	—	—
Bowlines and Strapping	2	39	Single	8	2	2	35	Single	8	2	1½	30	Single	6	2	1½	28	Single	5	2	—	—	—	—	—
Bridles	2	8	Thimbls.	—	4	2	7	Thimbls.	—	4	1½	6	Thimbls.	—	4	1½	5½	Thimbls.	—	4	—	—	—	—	—
Reef-Tackle Pendents	3	29	—	—	—	3	26	—	—	—	—	—	—	—	—	—	—	—	—	—	—	—	—	—	—
Falls	1½	35	Double Single	7 7	2 2	1½	31	Double Single	7 7	2 2	—	—	—	—	—	—	—	—	—	—	—	—	—	—	—
Strapping	2	1	—	—	—	2	1	—	—	—	—	—	—	—	—	—	—	—	—	—	—	—	—	—	—
Ear-Rings	1½	34	—	—	—	1½	30	—	—	—	1	26	—	—	—	¾	24	—	—	—	—	—	—	—	—
Sheets	4½	34	S.Sho. Single	14 14	2 1	4½	30	S.Sho. Single	13 13	2 1	3½	26	S.Sho. Single	12 12	2 1	3	24	S.Sho. Single	10 10	2 1	—	—	—	—	—
Strapping	5 4	2 3	—	—	—	5 4	2 3	—	—	—	4 3	2 3	—	—	—	3	4	—	—	—	—	—	—	—	—
Seizings	¾	14	—	—	—	¾	14	—	—	—	Tarred Line			—	—	—	—	—	—	—	—	—	—	—	—
Lashing	1½	5	—	—	—	1½	5	—	—	—	1	4	—	—	—	¾	3	—	—	—	—	—	—	—	—
Staysail-Haliards	2½	24	Single	8	1	2	21	Single	8	1	1½	18	Single	7	1	1½	16	Single	6	1	—	—	—	—	—
Sheets	2½	26	Single	8	2	2	23	Single	8	2	1½	20	Single	7	2	1½	18	Single	6	2	—	—	—	—	—
Tacks	2	2	—	—	—	2	2	—	—	—	2	2	—	—	—	1½	2	—	—	—	—	—	—	—	—
Downhaulers	1½	17	Single	6	1	1½	15	Single	6	1	1	12	Single	6	1	1	10	Single	5	1	—	—	—	—	—
Strapping	2	2	—	—	—	2	2	—	—	—	1½	1½	—	—	—	1½	1	—	—	—	—	—	—	—	—

Mizen-topgallant Mast, and Mizen-topgallant Yard	1257 Tons.					818 Tons.					544 Tons.					330 Tons.					Sloop of 60 Tons.				
	Size in Inch	Fath in Len.	Blocks, &c. Species.	In	N	Size in Inch	Fath in Len.	Blocks, &c. Species.	In	N	Size in Inch	Fath in Len.	Blocks, &c. Species.	In	N	Size in Inch	Fath in Len.	Blocks, &c. Species.	In	N	Size in Inch	Fath in Len.	Blocks, &c. Species.	In	N
Mizen Topgallant Mast.																									
Shrouds 4 Strands cabled	3¼	40	Thimbls.	—	6	3	36	Thimble	—	6	2½	30	Thimbls	—	6	1½	20	Thimbls.	—	4	—	—	—	—	—
Laniards	2	13	—	—	—	2	12	—	—	—	1½	10	—	—	—	¾	7	—	—	—	—	—	—	—	—
Backstays 4 Strands cabled	3¼	40	Thimbls.	—	4	3	36	Thimbls.	—	4	2½	30	Thimbls.	—	4	1½	30	Thimbls.		2	—	—	—	—	—
Laniards	2	6	—	—	—	2	6	—	—	—	1½	5	—	—	—	¾	4	—	—	—	—	—	—	—	—
Stay	3½	15	Thimbls.	—	1	3½	14	Thimble	—	1	3	13	Thimble	—	1	2	11	Thimble	—	1	—	—	—	—	—
Laniard	2	4	—	—	—	2	4	—	—	—	1½	3	—	—	—	1	2	—	—	—	—	—	—	—	—
Mizen Topgallant Yard.																									
Tie	2½	7	Single	7	1	2½	6½	Single	7	1	2	6	Single	6	1	1¾	5	Single	6	1	—	—	—	—	—
Haliard	2	28	Single	6	2	2	26	Single	6	2	1½	24	Single	5	2	1	20	Single	5	2	—	—	—	—	—
Horses	2	5	Thimbls.	—	2	2	4½	Thimbls.	—	2	2	4	Thimbls.	—	2	—	—	—	—	—	—	—	—	—	—
Braces	1½	52	Single	5	2	1½	46	Single	5	2	1½	44	Single	5	2	1	32	Single	5	2	—	—	—	—	—
Lifts, single	2	26	Thimbls.	—	2	2	26	Thimbls.	—	2	2	22	Thimbls.	—	2	1	16	Thimbls.	—	2	—	—	—	—	—
Parral-Ropes	1½	4	Parral	6	1	1½	3½	Parral	6	1	1	3	Parral	6	1	1	2½	Parral	5	1	—	—	—	—	—
Clue-Lines	1½	57	Single	5	2	1½	50	Single	5	2	1	48	Single	5	2	1	36	Single	4	2	—	—	—	—	—
Bow-Lines	1½	55	Thimble	—	4	1½	48	Thimbls.	—	4	1	46	Thimbls.	—	4	1	34	Thimbls.	—	4	—	—	—	—	—
Bridles	1½	3	—	—	—	1½	3	—	—	—	1	3	—	—	—	1	2	—	—	—	—	—	—	—	—
Ear-Rings	Tarred Line			—	—	—	—	—	—	—	—	—	—	—	—	—	—	—	—	—	—	—	—	—	—
Strapping	1½	3	—	—	—	1½	3	—	—	—	1	2½	—	—	—	1	2	—	—	—	—	—	—	—	—

Necessary Ropes.	1257 Tons.					818 Tons.					544 Tons.					330 Tons.					Sloop of 60 Tons.				
	Size in Inch	Fath in Len.	Blocks, &c. Species.	In	N	Size in Inch	Fath in Len.	Blocks, &c. Species.	In	N	Size in Inch	Fath in Len.	Blocks, &c. Species.	In	N	Size in Inch	Fath in Len.	Blocks, &c. Species.	In	N	Size in Inch	Fath in Len.	Blocks, &c. Species.	In	N
Necessary Ropes.																									
Cat-Falls	5	96	Tr.I.bo. Br.sh.	22	2	5	80	Tr.I.bo. Br.sh.	20	2	4	66	Tr.I.bo. Br.sh.	18	2	3	58	D.I.bo.c.	15	2	— —	— —	— —	— —	— —
Fish-Tackle Pendents ..	18	12	*S.co.	20	1	7	10	*S.co.	20	1	6	9	*S.co.	18	1	5	8	*S.co.	15	1	—	—	—	—	—
Falls	4	40	L.t.	30	2	4	40	L.t.	30	2	3½	35	L.t.	28	2	3	30	L.t. Single	24 24	1 1	— —	— —	— —	— —	— —
Strapping ..	6 5	6 5	—	—	—	6 5	6 5	—	—	—	4½	9	—	—	—	3½	7	—	—	—	—	—	—	—	—
Seizing	1½ 1	10	—	—	—	1½ 1	10	—	—	—	1	10	—	—	—	¾	12	—	—	—	—	—	—	—	—
Deck-Stoppers cabled ..	8½	36	Thimbls.	—	10	8	36	Thimbls.	—	10	8	30	Thimbls.	—	10	7	24	Thimbls.	—	8	—	—	—	—	—
Buoy-Ropes cabled ..	9	50	—	—	—	8¼	50	—	—	—	6	50	—	—	—	5	50	—	—	—	4	30	—	—	—
Masting Fall	6½	130	—	—	—	6	130	—	—	—	5	120	—	—	—	4½	108	—	—	—	—	—	—	—	—
Messengers, two each ..	11½	45	—	—	—	10	40	—	—	—	9	36	—	—	—	7	34	—	—	—	—	—	—	—	—
Hawsers, one of each	6 7 10	120 120 120	—	—	—	6 7 9	120 120 120	—	—	—	5 7 9	120 120 120	—	—	—	5 7 9	120 120 120	—	—	—	4½ 5½ —	80 70 —	—	—	—
Warp	—	—	—	—	—	—	—	—	—	—	—	—	—	—	—	4	120	—	—	—	2½	120	—	—	—
Stream Cable, one ..	16½	120	—	—	—	15	120	—	—	—	14	120	—	—	—	—	—	—	—	—	—	—	—	—	—
Cables	20	120	Six each	—	—	18	120	Six each	—	—	15	120	Four each	—	—	13½	120	Two each	—	—	9	90	Two each	—	—

FINIS.

Table 65: 'The Proportion of the Rigging of a Ship of near Six Hundred Tuns, according to the Customary Allowance; with the Sizes of the Blocks, and also the Rigging for the Long-boat and Pinnace; with the other Utensils proper to compleat the Rigging; from whence the Rigging of all other Three-Mast Ships may be known [Sutherland, 1711]'.

Column heading abbreviations

A	*Circumference of rope*
B	*Length of rope*
C	*Type of block*
D	*Size of block*
E	*Number of blocks*
F	*Type of sheave: Ash, Elm or Lignum vitae*

Abbreviations for types of block

S.	single sheaved block
D.	doubled sheaved block
Tr.	Treble sheaved block
R.	Round or sheet block
L.t.	Long tackle block
Sh.	Shoulder block
Sn.	Snatch block
D.E.	Deadeye
Par.	Parrel
I.bd.	Iron bound

	Ropes		*Blocks*			
	A	*B*	*C*	*D*	*E*	*F*
	Inch	*Fathom*		*Inch*	*No*	*A/E/L*
Bowsprit						
Horses	3	8	D.E.	8	14	
Lanyard	1½	3				
Straps						
Gammoning	5½	60	D.E.	46	2	
Woolding	2½	65				
Bobstay, worn		6				
Lanyard						
Sheets, cabled	3	60	R.	11	2	
Pendents, cabled						
Tye						
Halliards	3	18	L.t.	23	1	L.
			S.	13	1	A.
Lifts	3	38	S.	9	6	A.
Standing lifts	3	6	D.E.	7	4	
Lanyards	1½	5				
Braces	2	58	S.	9	10	A.
Pendents	2½	3				
Slings	4	4				
Seizing & racking	1	4				
Horses for yard, worn	3	6	D.E.	8	14	
Lanyards	1	5				
Clewlines	2½	32	S.	9	6	A.
Buntlines	2	22	L.t.	10	1	A
Reef lines	1	36				
Flying jib halliards	3	40				
—Sheets	2½	26				
—Jack	3	16				
Spritsail topmast						
Shrouds	2	14	D.E.	5	8	
Lanyards	1	8				
Backstay pendents	2	3	S.	6	10	A.
—Falls	1	18	S.	6	10	A.
Tye	2	2½				
Halliards	1½	7	S.	7	2	A.
Lifts	1½	14	S.	6	4	A.
Braces	1	30	S.	6	4	A.
Pendents	1½	3				
Parrel ropes	1½	1½	Par.	10	1	
Clewlines	1½	36	S.	6	6	A.
Robands & earings						
Gaskets						
Foremast						
Pendents of tackles, fine	6	6	S.	16	2	L.
Runners of tackles	5½	24	L.t.	28	2	L.
Falls of tackles	3	56	S.	16	2	L.
Shrouds, fine	6	110	D.E.	12	12	E.
Lanyards	3	42				
Ratline	1	200				
Worming	¾	450				
Catharpin legs & falls	1½	20	S.	8	12	A.
Stay, cabled, fine	10	11	D.E.	16	1	E.
Lanyard	3½	8				
Worming	1½	54				
Collar, cabled	9	2½	D.E.	16	1	E.
Mast woolding						
& puddening	2½	95				
Crowsfeet for the top	1	20	S.	6	2	A
Tackle for ditto	1	3	S.	6	2A	
Jeers	5	70	D.	18	3	L.
Lashings to masthead,						
yard & blocks	3	21				
Tackles for boats						
Pendents & falls	2½	28	S.	10	2	A.
Lifts	3	50	S.	11	4	L.
Straps for cap	3	2½				
Straps for blocks						
Braces	3	48	S.	9	4	A.
Pendents	3	5				
Parrel ropes	4	12	Par.	24	1	
Naveline	1½	20	S.	6	2	A.
Racking & seizing	1	16				
Horses for yard, worn	4½	10	D.E.	8	14	E.
Lanyards	2	5				
Puddening of yard, worn	5	8				
Sheet, cabled	5	60	S.	17	2	L.
Stoppers, worn	5	3				
Lanyards	¾	2				
Tacks, tapered & cabled	6	28				
Luff hook ropes	5	7				
Bowlines	3	42	S.	12	2	A.
Lashings for blocks	2½	6				
Bridles	3	4				
Clewgarnets	3	48	Sh.	10	8	L.
Buntlines	2	70	L.t.	16	2	A.
			S.	8	8	A.
Leechline legs & falls	2	36	S.	8	6	A.
Reeflines	1	46				
Earings						
Robands & earings	1½	80				
Gaskets						
Fore topmast						
Shrouds, fine	4	58	D.E.	8	8	E.
Lanyard	2½	26				
Ratline	¾	40				
Top rope pendent	6	11	S.	16	12	L.I.bd.
—Fall	3½	38	D.	15	2	L.I.bd.
Burton tackle pendents	3	5	S.	9	4	A.
—Falls	2	34	S.	9	4	A.
Futtock shrouds	3½	24	D.E.	8	8	
Standing backstays, fine	4	58	D.E.	8	4	
Lanyard	2	12				

	Ropes		*Blocks*			
	A	*B*	*C*	*D*	*E*	*F*
	Inch	*Fathom*		*Inch*	*No*	*A/E/L*
Stay	4½	14	L.t.	18	1	A.
			S.	12	1	A.
Lanyards	2½	8				
Runner, fine	4½	16	S.	14	2	L.
Halliards	2½	36	S.	20	1	L.
			S.	18	1	L.
Lifts	2½	56	S.	9	4	L.
Beckets at cap	2½	1½				
Slings, worn	3½	9				
Parrel ropes	3	6	Par.	18	1	
Racking	¾	7				
Horses for yard, worn	3	6	D.E.	7	4	
Sheet, fine	5	42	Sh.	17	4	L.
Span, worn	2	3				
Lashings for quarter blocks	¾	4				
Lashings for sheet blocks						
Bowlines	2	36	S.	8	2	A.
			Tr.	9	1	
Bridles	2	10				
Clewlines	2½	66	S.	9	4	
Buntlines	2	32	L.t.	14	1	
			S.	8	2	A.
Reef tackle tie	2½	5	S.	8	4	A.
—Fall	1½	12	S.	8	4	A.
Leechlines	2	12	S.	7	4	
Braces	2½	50	S.	8	6	A.
Pendents	2½	3	S.	8	6	A.
Staysail stay, worn	3½	6	D.E.	6	2	E.
—Cringles, worn	2½	3				
—Lanyards	1	3				
—Halliards	1½	16	S.	9	2	
—Sheet	1½	10				
—Tack	2	2				
Reeflines	¾	36				
Earings						
Robands & earings	1	40				
Fore topgallant mast						
Shrouds	2	12	D.E.	5	4	E.
Lanyards	1	10				
Futtock shrouds	2	4	D.E.	5	4	
Stay	1½	15	S.	7	1	A.
Tye	2	2½	S.	6	3	A.
Halliards	1½	24	S.	6	3	A.
Lifts	1	16	S.	6	4	A.
Braces	1	46	S.	6	8	A.
Pendents	1	4				
Parrel ropes	1	2	Par.	9	1	
Bowlines	1	34	S.	6	6	A.
Bridles	1	8				
Clewlines	1½	46	S.	6	4	A.
Mainmast						
Pendents of tackles	6½	7	S.	17	2	L.
Runners of tackles	6	24	L.t.	28	2	L.
Falls of tackles	3½	56	S.	18	2	L.
Garnet pendent	5	11	L.t.	30	1	L.
—Guy	4	8	S.	18	1	L.
—Fall	3½	28	Sn.	26	1	L.
Shrouds, fine	6½	142	D.E.	13	14	E.
Lanyards	3½	50				
Ratline	1½	210				
Worming	1	700				
Catharpin legs & falls	2	24	S.	8	14	A.

	Ropes		*Blocks*			
	A	*B*	*C*	*D*	*E*	*F*
	Inch	*Fathom*		*Inch*	*No*	*A/E/L*
Stay, cabled (4 strands), fine	12	16	D.E.	18	1	E.
Lanyard	4½	10				
Lashing to foremast	2½	6				
Worming	2	80				
Collar, cabled	9½	7	D.E.	18	1	E.
Mast woolding	2½	130				
Crowsfeet for top	1	24	S.	10	1	A.
Tackle for ditto	1	3	S.	6	1	A.
Jeers, fine	5½	80	D.	20	3	L.
Lashings for masthead, yard & blocks	3	23				
Tackles for boats						
Pendents & falls	2½	28	S.	10	2	A.
Lifts	3½	58	S.	11	4	L.
Straps for cap	3	4				
Braces	3	60	S.	9	4	A.
Pendents	3	5				
Parrel ropes	4	12	Par.	27	1	
Naveline	1½	24	S.	8	2	A.
Racking & seizing	1	18				
Horses for yard, worn	4½	12	D.E.	9	4	E.
Lanyards	2½	4				
Puddening for yard, worn	5½	10				
Sheet, cabled	6	64	S.	18	2	L.
Stoppers	5	3				
Lanyards	¾	2				
Tacks, tapered & cabled	6½	30				
Luff tackles	3	26	L.t.	18	2	A.
			S.	11	2	A.
Bowlines	3½	36	Sn.	24	1	L.
Bridles	3	10				
Tackle	2½	10	S.	10	2	A.
Clewgarnets	2½	50	Sh.	10	6	L.
Buntlines	2	75	L.t.	18	2	A.
			S.	8	8	A.
Leechline pendents, legs & falls	2	38	S.	9	4	A.
Staysail stay, worn	4	11	D.E.	8	2	E.
Lanyards	2	3				
Cringles, worn	2½	3				
Halliards	2	20	S.	8	1	A.
Sheet	2½	4				
Tack	3	2				
Studdingsail halliards	3	46	S.	10	4	A.
—Sheet	2½	12				
—Tack	3	24				
Reeflines	1½	60				
Gaskets	1½	90				
Main topmast						
Shrouds	4½	64	D.E.	9	8	
Lanyards	2½	26				
Ratline	1	60				
Top rope pendent	6	13	S.	18	1	I.bd.
—Fall	4	40	D.	16	2	L.
Burton tackle pendents	3½	6				
—Fall	2	36	S.	10	6	A.
Futtock shrouds	4	26	D.E.	9	8	E.
Standing backstays, fine	4½	66	D.E.	8	6	E.
Lanyards	2½	12				
Stay, cabled (4 strands), fine	4½	19	S.	14	1	A.

	Ropes		Blocks			
	A	B	C	D	E	F
	Inch	Fathom		Inch	No	A/E/L
			L.t.	20	1	L.
			D.	12	1	A.
Tye runner	5	18	S.	15	2	L.
Halliards	3	45	S.	21	1	L.
			S.	19	1	L.
Lifts	2½	58	S.	9	4	A.
Beckets at the cap	2½	2				
Braces	2½	54	S.	9	4	A.
Pendents	2½	5				
Beckets at mizzen mast	3	2				
Slings, worn	3	9				
Parrel ropes	3	7	Par.	20	1	
Racking	¾	9				
Yard horses, worn	2½	7	D.E.	8	4	
Sheet	5½	46	Sh.	18	4	L.
Span, worn	2½	3				
Lashings for quarter & sheet blocks	¾	4				
Bowlines	3	42	S.	10	2	A.
Bridles	3	16				
Clewlines	3	70	S.	10	6	A.
Buntlines	2	34	L.t.	18	2	A.
			S.	9	2	A.
Leechlines	2	14	S.	8	2	A.
Reef tackle tie	2½	5	S.	10	A.	
—Fall	1½	16	S.	10	4	A.
Staysail stay, worn	2½	8	D.E.	6	2	
—Lanyards	1½	3				
—Cringles, worn	2	3				
—Halliards	1½	18	S.	8	2	A.
—Sheet	1½	15	S.	6	1	A.
—Tack	2	1				
Studdingsail halliards	2½	40	S.	10	4	A.
—Sheet	2	6				
—Tack	2	10				
Reeflines	1	36				
Earings						
Robands & earings	1	56				
Main topgallant mast						
Shrouds	2½	14	D.E.	6	4	E.
Lanyards	1	10				
Futtock shrouds	2½	5	D.E.	6	4	
Stay	1½	18	S.	6	4	A.
Tye	2½	3				
Halliards	1½	28	S.	7	2	A.
Lifts	1½	16	S.	6	6	A.
Braces	1	48	S.	6	6	A.
Pendents	1	4½				
Parrel ropes	1	2	Par	10	1	
Bowlines	1½	48	S.	6	6	A.
Bridles	1	4				
Clewlines	1½	54	S.	7	4	A.
Mizzen mast						
Shrouds, fine	4½	64	D.E.	9	8	E.
Lanyards	2½	24				
Ratline	1	70				
Burton tackle pendents	3½	5	S.	10	6	A.
—Falls	2½	36	S.	10	6	A.
—Straps for blocks						
Stay, cabled, (4 strands), fine	4½	12	D.E.	9	2	E.
Lanyards	2½	3½				
Collar	4	2				
Crowsfeet for top	¾	12	S.	4	2	A.
Tackle for ditto	¾	3				
Jeer, fine	4	30	D.	15	1	L.
			S.	15	1	L.
Lashings for masthead, yard & blocks	2½	6				
Parrel ropes	2½	4	Par.	21	1	
Truss	2½	18	L.t.	16	1	A.
			S.	9	1	A.
Slings	4½	8	S.	13	2	A.
Sheet	3½	18				
Tack	2½	3				
Bowlines	2½	8	S.	9	2	A.
Peak, middle & main brails	2	90	S.	8	12	A.
Lacing to mizzen	1	34				
Staysail halliards	2	18	S.	8	1	A.
—Sheet	2½	4				
—Tack	2	3				
Crossjack yard						
Standing lifts	2½	6	D.E.	5	4	E.
Lanyards	1½	4				
Braces	1½	36	S.	8	6	A.
Pendents	2	2½				
Slings	3½	4	S.	12	1	A.
Mizzen topmast						
Shrouds	2½	24	D.E.	5	6	E.
Lanyards	1	10				
Futtock shrouds	1½	10	D.E.	5	6	E.
Stay	2½	8	S.	8	9	A.
Lanyards	1	6				
Tye	2½	3½	L.t.	15	1	A.
Halliards	2	18	S.	9	1	A.
Straps for blocks						
Lifts	1½	20	S.	7	4	A.
Braces	1½	34	S.	7	4	A.
Pendents	1½	2½				
Parrel ropes	1½	2	Par.	11	1	
Sheets	2½	32	S.	9	4	A.
Bowlines	1	32	S.	9	4	A.
Bridles	1	5				
Clewlines	1½	32	S.	7	4	A.
Additional ropes						
Viol, cabled	10	30	S.	43	1	L.
Straps	7	6				
Lashings	3	12				
Lashings & seizings for blocks	1½	7				
Winding tackle pendent	8	11	Tr.	22	1	L.
—Fall	5	35	D.	22	1	L.
			Sn.	20	1	L.
Fish tackle pendent	6½	7	L.t.	28	1	L.
—Fall	3½	8	S.	16	1	L.
—Straps at mainmast						
—Lanyards	2	5				
Cat ropes	4½	40	D.	16	2	L.I.bd.
Lanyards	2	7				
Stoppers for sheet, best bower & small anchors	5½	21				
Shank painters for sheet, best bower & small anchors	5½	15				
Stoppers at bitts	8	12				
	7½	4				

	Ropes		*Blocks*			
	A	*B*	*C*	*D*	*E*	*F*
	Inch	*Fathom*		*Inch*	*No*	*A/E/L*
Seizings	¾	40				
Lanyards	3	23				
Buoy ropes, cabled for sheet, best bower & small anchors	6½	60				
Preventers	7	6				
	5½	18				
	4	4				
	3½	12				
Lanyards	1	8				
	¾	64				
Slings, buoy	3½	30				
Gun	7	6				
Butt	6	9				
Hogshead	4	8				
Nutt	3	2				
Horses at head, worn	3½	8	D.E.	9	4	E.
Lanyards	1½	5				
Ladder at poop, worn	4	18				
Middle rope	1½	14				
Lashing	1	8				
Futtock staves, worn	4½	7				
Cable bends	2½	30				
Ropes, bell, worn	2	2				
Can hook	3	2½				
Davit	-	-	S.	19	1	L.
Entering	3	12	Seiz.			
Port	2½	52	S.	11	12	L.
	2	50		10	11	L.
Slip	-	-		9	7	L.
Waist stanchion, worn	2	56		6	10	L.
Tiller	3	3				
Waist, worn	4½	14				
Lanyards	2	6				
Selvagees for shrouds, worn	3½	3				
Puddening of anchors	2½	60				
Seizings	¾	50				

Long boat

Burton pendents	3	2				
—Falls	2	16				
Fore sheets	1	2				
Halliards	1	7				
Main stay	2½	4				
—Tye	2½	3	S.	7	2	A.
—Halliards	1½	7	S.	4	2	A.
—Sheet	2	5				
—Tack	2	1				
Guest rope, cabled	6½	25				
Grapnel rope, cabled	3½	30				
Painter	3	5				
Yard rope	¾	9				

Pinnace

Fore sheet	1	2
Main sheet	1	
Boat rope, cabled	4	30
Guest rope, cabled	3½	26
Painter	2½	4

Skiff

[Fore sheet, main sheet, hanks, swifter, grapnel rope, painter, sternfast, fenders, lanyards and 'rother-rope' mentioned, but no data supplied].

'An Abstract of Iron work'

Tackle hooks	30	Staples	24
Futtock hooks	24	Thimbles, large	34
Fish	1	Thimbles, ordinary	118
Futtock plates	24		

'The whole Allowance for Seizings and Strapping of Blocks'

	A	*B*		*A*	*B*
Strapping for all blocks	6	40	All seizings	2	30
	5	20		1	20
	5	30		1	216
	4	40		¾	216
	4	40			
	3	40			
	3	40			
	2	50			
	2	50			
	1	40			

'The Number and Nature of the Blocks taken by themselves for the more convenient Receiving and Providing them'

C	*D*	*E*	*F*	*C*	*D*	*E*	*F*
Deadeyes	5	32	A.	Snatch blocks	20	1	
	6	12	A.		24	1	
	7	8	A.		26	1	
	8	44	A.	Long-headed	46	2	A.
	9	30	A.	Round	11	2	L.
	12	12	A.	Strapped board	8	6	A.
	13	14	A.	blocks	10	22	A.
	16	2	A.	Shoulder blocks	17	2	L.
	18	2	A.		18	2	L.
Single blocks	6	100	A.	Long tackle	15	1	A.
	7	35	A.	blocks	16	1	A.
	8	100	A.		18	1	A.
	9	75	A.		20	1	A.
	10	44	A.		23	1	L.
	12	4	A.		24	2	L.
	13	3	A.		28	6	L.
	14	3	A.		30	1	L.
	15	3	A.	Treble blocks	9	1	A.
	16	5	A.		22	1	L.
	17	6	A.	Double blocks	15	1	L.
	18	7	L.		18	3	L.
	19	2	L.		20	3	L.
	20	1	L.	Large buntline	14	1	A.
	21	1	L.	blocks	16	3	A.
Parrels	9	1			18	2	A.
	10	1		Cat blocks,	16	2	L.
	11	1		double iron-bound			
	18	1		Top block, iron-	16	1	
	20	1		bound, single	18	1	
	21	1		Top tackle	15	2	L.
	24	1		blocks, iron-	16	2	L.
	27	1		bound single			
Long crows-	12	1		Viol blocks	43	1	L.
foot blocks	13	1					
	14	1					

John Davis' rigging tables (published in 1711) consist of circumferences and lengths of ropes for the following ships:

East Indiaman *Degrave* of 760 tons (1676);

HMS *Swan*, a Sixth Rate of 24 guns, 249 tons, built in Deptford (1694);

'One of Her Majesty's New Fifth Rates';

HMS *Bonadventure*, a Fourth Rate of 48 guns, 569 tons, built in Portsmouth (1683, rebuilt 1699);

HMS *Lenox*, a Third Rate of 70 guns, 1089 tons, built in Deptford (1701);

HMS *Vanguard*, a Second Rate of 90 guns, 1357 tons, built in Portsmouth (1678, rebuilt 1710);

HMS *Royal Sovereign*, a First Rate of 100 guns, 1883 tons, built in Woolwich (1701).

A comparison of these rigging data for Sixth to First Rates with those provided in Sir Anthony Deane's *Doctrine of Naval Architecture* of 1670 provides clear evidence of the development in rigging over this period. John Davis described himself thus:

> I have used the Sea this 30 Years; and some part of that time I was Boatswain in several of the Ships belonging to the Royal Navy, as well as some East and West India Men, where I made it part of my Study not only to facilitate the work of a Seaman in Rigging a Ship, but likewise to make it more plain and familiar to the very meanest Capacity, which I hope without Ostentation, I have contrived to the general satisfaction of the Navy.

He recorded meticulously the rigging details of the ships he encountered, and his small book has been saved for posterity thanks to a limited reprint published by the Nautical Research Guild.

Obvious errors in the originals of the following tables, chiefly in comparisons of the length of ropes in fathoms with the length as a multiple of the length of a main beam, have been corrected, and the spellings have been modernised.

Table 66: 'This is the ... Ship's Rigging' [East Indiaman *Degrave*, or an Indiaman of 84 feet keel, 28 feet beam and 12 feet in the hold]. (Davis, 1711)

	Circumference (inches)	*Length (fathoms/feet)*
Bowsprit & spritsail yard		
Bowsprit horse (one piece)	2¼	5
2 standing lifts	2¼	6
2 running lifts	2¼	33
A pair of halliards with single blocks	2¼	13
Ditto with a long tackle block	2	29
2 braces, double	2	40
—Triple	1¾	45 /3
2 sheets	2½	40
2 clewlines with blocks in the clews	1¼	23
—Ditto with none	1½	18
2 buntlines	1¼	16
2 reeflines	¾	18
Sprit topmast		
First pair of shrouds	1½	4
A single tye	2	2 /½
Halliards for ditto	1½	5
A pair with 2 single blocks	1½	7
2 single lifts	1½	8
Ditto with blocks at the yardarms	1¼	12
2 single braces	1½	24
Ditto with pendents	1¼	26
2 single clewlines	1½	28
Ditto with blocks in the clews	1¼	36
2 reeflines	¾	10 /3
Foremast		
Pendents for tackles with a hitch	5¼	3
2 runners for tackles	4½	20
2 falls	3	60
Half the first pair of shrouds (15 yards 1 foot)	5¼	5½
Fore stay, cable laid	9	9
Tye	5¾	14
1 of the 2 pairs of halliards for ditto	2½	35
1 jeer with 2 double breast blocks	4¼	40
2 lifts to the cap	2¼	52
1 pendent for yard tackle (8 feet)	2¾	-
1 fall for ditto with a long tackle block aloft	2¼	23
2 pendents for braces	2¼	2½
2 braces	2	31
1 single naveline	1½	12
Ditto with a block at the parrel	1¼	18
2 sheets	3¾	36
2 tacks	5¾	27
2 bowlines	2¾	28
2 horses for yard	2½	8
2 parrel ropes	3	9
2 clewgarnets	2	43
2 buntline legs	1½	27
2 falls for ditto	1¼	28
1 fall in the new fashion	1¼	23
1 topping lift for ditto	1	13
2 studdingsail halliards	2	32
2 legs for leechlines, double	1¼	24
2 falls for ditto	1¼	36
2 pendents for ditto	2½	3
2 single leechline legs	1½	12
2 single leechlines with the ends to the cringles	1½	43
2 pendents for ditto	2½	5
Fore topmast		
1 pair of pendents for Burtons	2¾	3½
2 falls for ditto	1½	24
2 with long tackle blocks aloft	1¾	30
1 top rope	4½	10
1 ditto with a double and a triple block	2¾	38
1 ditto with 2 double breast blocks	-	31
1 ditto with a double and a single block	3	20
Half the first pair of shrouds	3¼	5
1 pair of breast stays	3	25
Stay	3½	11
1 runner for yard	3	12
1 pair of halliards with 2 single blocks	2	25
1 ditto with a double and a single block	1¾	31
2 lifts	2	44
2 braces	2	38
2 bowlines	2¼	36
2 clewlines	2¼	55
2 parrel ropes	2¼	6
2 horses for yard	2½	5
1 pair of buntline legs	1¼	12½

	Circumference (inches)	Length (fathoms/feet)
1 fall for ditto	-	13
2 leechlines	1¼	9
2 reef tackle ties	2	5
2 falls to haul out in the top	1½	12
2 to haul upon deck	1½	30
2 reeflines	¾	15
Fore topsail staysail stay	2½	5½
1 pair of single halliards	1½	13
Ditto to haul upon the forecastle	1½	35
Sheet	1	9
1 pair of topsail studdingsail halliards	2	28
1 pair to hoist upon the forecastle	2	55
2 tacks	1¼	16
2 sheets	1	4
Fore topgallant mast		
Half the first pair of shrouds	2	2½
Stay	2½	14
1 pair of backstays	2	30
1 single tye	2	2
1 pair of halliards for ditto	1½	16
1 tye with a block at the yard	2	4
1 pair of halliards for ditto	1¾	22
2 single lifts	1½	9
2 double ditto	1¼	15
2 single braces	1½	34
2 double ditto	1¼	37
2 pendents for ditto	1¾	1½
2 bowlines	¾	29
2 lines to the forecastle	-	40
2 single clewlines	1½	38
2 double ditto	1¼	44
2 reeflines	¾	9
2 horses for yard	2	3
Mainmast		
1 pair of pendents for tackles	6	3½
2 runners for ditto	4½	21
2 falls for ditto	3	60
Half the first pair of shrouds (17 yards)	6	8½
Stay	11½	12
1 runner to hoist the yard	6½	16
2 falls	2½	78
1 jeer	4¾	40
1 single Burton	2½	25
1 runner for a Flemish or double Burton	3	9
1 fall for ditto	2½	26
1 Spanish Burton fall	3½	43
1 pendent for yard tackle (8 feet)	3½	-
1 fall for ditto with 1 long tackle block	2½	29
2 lifts to the cap	3	59
2 pendents for braces	2¾	6
2 braces for ditto	2½	40
2 sheets	4½	38
2 tacks	6½	27
2 bowlines	2½	23
2 clewgarnets	2¼	44
2 parrel ropes	3	9
2 horses for yard	3	9
2 buntline legs	1¼	29
2 falls for ditto	1¼	30

	Circumference (inches)	Length (fathoms/feet)
1 fall for ditto, new fashion	1	28
1 topping lift for ditto	1	17
2 leechline legs	1½	28
2 falls for ditto	1¼	54
2 single legs for ditto	1½	14
2 reeflines (3 times the yard)	-	-
1 pair of main studdingsail halliards	2½	40
2 sheets	2¼	7
2 tackles	2½	14
2 guys for the booms	2½	12
1 pair of main staysail halliards, single	2¼	17
1 pair double	2	24
1 sheet	2½	3
1 tack	2	1½
Main topmast		
Pendents for Burtons	3	3½
2 single falls	1½	28
2 ditto with long tackle blocks aloft	-	34
Half the first pair of shrouds (11 yards 2 feet)	3½	5½
Half a pair of breast stays	3¼	28
Half a pair of backstays	-	28½
Stay	3¾	16
1 top rope	4¾	11
1 fall with 2 double breast blocks	3	34
1 ditto with 1 triple block	-	43
1 runner	3½	13
1 pair of halliards	2¼	30
1 pair with 1 long tackle block	-	42
2 lifts	2½	48
2 braces	-	41
2 bowlines	-	33
2 clewlines	-	53
2 sheets	3¼	30
1 buntline leg	1½	13
1 fall for ditto	-	14
Main topsail staysail halliards	2	16
1 pair with the end fast to the sail	2	10
1 sheet	1½	11
1 tack	1¾	1½
2 studdingsail halliards to hoist in the top	2¼	29
2 to hoist upon the deck	-	46
2 tacks	2½	6
2 to haul out upon the poop	1½	33
2 sheets	2	9
Half the first pair of main topgallant shrouds	1¾	2½
Half a pair of backstays	1½	15
Stay	2½	13
2 single lifts	1¾	11
2 double ditto	1½	16
1 single tye	2¼	2½
Halliards for ditto	1½	19
1 pair with one end fast in the main top	-	25
2 pendents for braces	1¾	5
2 braces for ditto	1¼	43
2 single braces	1½	35
2 bowlines	1	40
2 single clewlines	-	40
2 ditto with blocks in the clews	-	46
2 reeflines	-	10½
Half the first pair of mizzen shrouds	4	6
1 pair of pendents for Burtons	3½	4
2 falls for ditto	1¼	27
1 jeer	3	19
or	3	24

	Circumference (inches)	Length (fathoms/feet)
or	2¾	36
1 sheet with 1 double breast block	1¾	15
1 ditto with 2 single blocks	1¾	11
1 truss tackle fall	1½	13
1 ditto with a long tackle block	1¼	18
2 bowlines, double	1½	22
1 single with a hook	1¾	6
2 legs for peak brails	1	7
2 brails for ditto	1	11
1 pair of brails, or the uppermost pair below the mast	1¼	16
Second pair, or clew brails	-	16
Third pair, or lower brails to the quarterdeck	-	13
The first pair below the mast to the main shrouds	1¼	24
Second pair, or clew brails	1¼	24
Third pair	1	19
2 lacings for the mizzen	¾	17
2 reeflines	-	20
Stay	3¾	8
1 pair of pendent halliards to the mizzen peak	1	17
1 pair of mizzen staysail halliards to the top, double	1	25
Ditto, single	1¼	17
1 topping lift for the staysail	1	16
2 crossjack braces, double	1	36
2 single	1¼	27
1 pair of pendents	1¾	3
1 pair of slings	3	3½
1 pair of crossjack studdingsail halliards	1½	24
2 standing lifts	2½	6
Half the first pair of mizzen topmast shrouds	1¾	3
Half a pair of backstays	1½	9
Stay	2¼	6
2 single lifts	1½	13
2 ditto with blocks at yardarms	2	20
2 sheets	2	20
1 single tye	2	3½
Halliards for ditto with one end in the top	1½	14
1 pair with 2 single blocks	1¼	20
2 single braces forward	1½	16
2 ditto to the mizzen peak	1½	19
2 pendents for braces aft	-	27
2 braces ditto	-	28
2 single clewlines	-	34
2 ditto with blocks in the clews	-	24
2 bowlines	-	-
2 reeflines	-	-
Additional ropes		
Viol	6½	25
Pendent for the fish tackle	4½	5
Fall for ditto, with a long tackle and a single block	2½	27
Ditto with 2 long tackle blocks	2¼	36
Winding tackle pendent	7	8
Fall for ditto	2¾	31
Ditto with a double and a triple block	2½	36
2 cat ropes	2½	36
2 cat ropes	2¼	43
2 cat ropes	-	51

Table 67: 'Her Majesty's Ship the *Swan*'s Rigging, a Sixth Rate, is all exactly cut out by the length of the Main Beam, which is 24 Feet, 1 Inch, and not by the extream breadth from outside to outside; and the same Rule following will Rig the Navy for ever'.

Column heading abbreviations
A Circumference of cordage in inches
B Length in fathoms
C Length as a multiple of main beam length
D Feet added to or subtracted from C

	A	B	C	D
Bowsprit				
1 horse, all in a piece	2	5	1¼	
2 running lifts	2	31	7½	+6
2 standing lifts	2	5	1¼	
1 pair of halliards with 2 single blocks	2	14	3¼	
2 braces	1¾	39	10	-6
2 pendents	2	14 feet	½	+2
2 horses for yard	2	5½	1½	-3
2 sheets, cable laid	2 full	40	10	
2 single clewlines	1¼	34	8½	
2 buntlines	1¼	32	8	
2 reeflines	¾	15½	4	-3
Sprit topmast				
Half the first pair of shrouds	1¼	9 feet	⅓	+1
1 single tye	2	9 feet	⅓	+1
1 pair of single halliards	1	3	⅔	+2
2 single lifts	1 full	7	1⅔	+2
2 double ditto	1	10	2½	
2 single braces	1 full	17	4½	-6
2 double ditto	1 bare	21	5	
2 single clewlines	1 ditto	26	6½	
2 reeflines	Line	8	2	
Foremast				
1 pair of pendents for tackles	4¾	3	⅔	+2
2 runners for ditto	3½ full	16	4	
2 falls for ditto	2½	54	13	
Half the first pair of shrouds	4¾	7	2	-6
Stay	7¼	7½	2	-3
1 jeer, with a double block aloft and a double block on the yard	3¾	20	5	
1 pendent for yard tackle	3	3	⅔	+2
1 fall for ditto	2½	20	5	
2 lifts	2½	42	10½	
2 braces	2 bare	33	8¼	
2 pendents	2¼	15 feet	¼	+3
2 parrel ropes	2½	8½	2	+3
1 naveline with 1 block	1	11½	3	-3
2 horses for yard	2½	7½	2	-3
2 sheets	3½	43	10½	+6
2 tacks at the head	5½	24	6	
2 bowlines	2½	27	6⅔	+2
2 bridles	2½	3	⅔	+2
2 buntline legs	1¼	26	6½	
2 falls for ditto	1¼	28½	7	
2 single legs for leechlines	1¼	11	3	-6
2 falls with one end of each to the topmmast head	1¼	41	10	+6
2 reeflines	1¼	23	6	+6
1 luff hook rope	3	7½	2	-3

	A	B	C	D
Fore topmast				
1 pair of pendents for Burtons	3½	3	⅔	+2
2 falls for ditto	1¼	24	6	
Half the first pair of shrouds	1	4ft 4 in	1	+4
1 top rope	4 9	2	+6	
1 fall for ditto	2½	29	7	+6
Half the breast backstays, with				
deadeyes in the chains	3	20	5	
Half the after backstays	3¾	20ft 4in	5	+4
Stay	3¾	10	2⅔	-1
1 single runner with 2 single blocks	3	8	2	
1 pair of halliards with				
2 single blocks	2	24	6	
2 lifts	2	39	10	-6
2 braces	2	34	8½	+2
2 pendents	2¼	14 feet	½	+2
1 pair of slings for the yard	3	8½	2	+3
2 parrel ropes	2¼	5½	1½	-3
2 horses for the yard	2¼	4	1	
2 sheets	4½	49	12	+6
2 bowlines	2	32	8	
4 bridles	2	6½	1½	+3
2 clewlines	2	48	12	
1 pair of buntline legs	1¼	10½	24	
1 fall for ditto	1¼	12	3	
2 reef tackle falls at				
the top	1¼	12	3	
2 pendents for 1 reef	1¾	3	½	+2
2 reeflines	¾	13	3	+6
2 leechlines	1	9	2	+6
Fore topsail staysail stay	2½	6	1½	
1 pair of single halliards	1¼	13	3	+6
1 sheet	1	7	2	-6

	A	B	C	D
Fore topgallant				
First pair of shrouds	1¾	10 feet	½	-2
1 single tye	1¼	10 feet	½	-2
1 pair of halliards for ditto	1	15	4	-6
2 lifts with 4 blocks	1	12	3	
2 single braces	1 full	29	7½	-6
2 double ditto	1	34	8	
2 pendents	1¼	14 feet	½	+2
2 single clewlines	1	33	8	+6
2 bowlines in the spritsail top	¾ bare	26	6½	
4 bridles	¾	5	1⅙	
Stay	2 bare			

	A	B	C	D
Mainmast				
1 pair of pendents for tackles	4½	4	1	
2 runners for ditto	3½ full	17½	4½	-3
2 falls for ditto	2½	56	14	
First pair of shrouds	4½ full	72	2	-3
Pendent	5½ full	7½	2	-3
Garnet guy	3½ full	7	2	(-6)
Garnet fall	3	24	6	
1 single Burton with				
2 single blocks	3	22	5½	
1 Spanish Burton fall	3½	30	7½	
Stay	8¼	12	3	
The collar at the Stem	8½	4	1	
1 jeer with 2 double				
breast blocks	3¼	30	7½	
1 yard sail pendent	3½	2½	½	+3
1 ditto fall	2½	24	6	
2 lifts	2½	48	12	
2 braces	2¼	40	10	
2 pendents	2½	5	1	+6
2 parrel ropes	3	9	2	+6
1 naveline with 1 single block	1	12	3	
2 horses for yard	2½	9	2	+6
2 sheets, cable laid	3½	40	10	
2 tacks	5½	24	6	
2 bowlines	2¼	24	6	
4 bridles	2¼	6	1	
2 clewgarnets	2½ bare	44	11	
2 buntline legs	1¾	26	6½	
2 falls for ditto	1¼	36	9	
2 leechline legs, double	1½	24	6	
2 falls with one end of				
each fast at the topmmast head	1¼	48	12	
1 pair of pendents for				
ditto at the masthead	2¼	1½	⅓	-1
Staysail stay, up to the mouse	3	8	2	
Halliards with one end fast				
to the head of the sail	1½	18	4½	
Sheet	2	3	⅔	+2
2 studdingsail halliards	2	30	7½	
2 sheets	2	14	3½	
2 tacks, twice the boom if with board	2			
2 guys for the booms, with one end				
at the forecastle or fore chains,				
the other aft	2	26	6½	

	A	B	C	D
Main topmast				
Pendents for Burtons	3¼	3½	⅔	+3
2 falls with single blocks	1¼	24	6	
First pair of shrouds	3¼	10ft 4in	2⅔	
Top rope pendent	3½	10	2½	
1 fall	2¾	29	7	+6
Half a pair of breast stays, with				
deadeyes in the chains	3¼	25ft 4in	6½	-2
Half a pair of backstays, with deadeyes				
commonly higher up than the chains	3¼	26½	6	-4
Stay	3½ full	16	4	
1 runner for the yard				
with 2 single blocks	3¼	12	3	
1 pair of halliards with				
2 single blocks	2	27	7	-6
2 lifts	2	56	14	
2 braces	2	41	10	+6
2 pendents	2¼	5	1½	-6
2 pair of slings	3	9½	2½	-3
2 parrel ropes	2½	6	1½	
2 horses for the yard	2¼	5	1½	-3
2 sheets	5	30	7½	
2 bowlines	2¼	39	9	
6 bridles	2¼	8	2	
2 clewlines	2¼	52	13	
1 pair of buntline legs,				
with fall	2¼	26	6½	
2 leechlines	1	10	2½	
2 reef tackle ties	2¼	3½	1	-3
2 falls in the top	1½	13	3	+6
Staysail stay	2¾	6½	1½	+3
Halliards, with one end				
to the head of the sail	1½	15	4	-6
Sheet	1¼	7½	2	-3
2 studdingsail halliards	1½	40	10	
2 sheets	1	5	1½	-6
2 tacks at the poop	1	24	6	
2 ditto at the yard	1	14	3½	

	A	B	C	D
Main topgallant mast				
Half the first pair of shrouds	1½	13 feet	½	+1
Half the pair of breast stays,				
with deadeyes in the chains	1½	14ft 4in	3½	+4

	A	B	C	D
Half a pair of backstays	1½	15	3½	+6
Stay	2	14½	3½	+3
A single tye	2	14	3½	
1 pair of single halliards,				
with one end to the cross trees	1¼	18	4½	
2 lifts, double	1¼	13½	3½	-3
2 braces	1¼	40	10	
2 pendents	1½	5	1½	-6
1 parrel rope	1½	1½	½	-3
2 bowlines	¾	40	10	
4 bridles	1	2¼	½	+3
2 single clewlines	1¼	37	9½	-6
2 double clewlines	1	42	10½	
2 reeflines	line	6½	1½	+3

Mizzen mast				
Pendents for Burtons	3½	4	1	
2 falls for ditto	1½	23	5½	+6
Half the first pair of shrouds	3½	5½	1½	-3
Stay	4 bare	8½	2	+3
1 jeer with 1 double block				
& 1 single	3½ bare	21	5	+6
1 parrel rope	2½	2½	½	+3
1 truss tackle fall with a long				
tackle block & a single block	1½	16	4	
Ditto with 2 single blocks	1½ full	14	3½	
1 pair of slings for the yard	3¾	4	1	
1 sheet with 2 single blocks	2¼	10	2½	
Ditto with a double & a single block	2	13	3	+6
2 bowlines with 4 single blocks	1¼	9½	2½	-3
2 pairs of legs for peak brails	1½	6½	1½	+3
2 brails for ditto	1	10	2½	
The upper pair of brails below				
the Mast, down by the mast	1¼	14½	3½	+3
The second pair, or clew braces	1¼	14½	3½	+3
The lower pair of braces	1¼	12½	3	+3
Lacing for the mizzen	¾	16½	4	+3
Reefline for the mizzen	¾	10½	2½	+3
Staysail halliards, with one end				
fast to the head of the sail, and				
a block under the top	1	16	4	
1 sheet	2	3	¾	
2 crossjack braces, single	1¼	25	6	+6
2 ditto, double	1	36	9	
2 standing lifts to the yardarm	2	5	1¼	
Slings for yard	3	3	¾	

Mizzen topmast				
Half the first pair of shrouds	1¾	17 feet	¾	-1
Half the pair of backstays,				
with deadeyes in the chains	1	15ft 2in	4	-4
A stay, in the main top	2	6	1¼	
1 single tye	2	17 feet	¾	-1
Halliards for ditto with				
one end fast in the top	1	11	2½	+6
2 single lifts	1 full	11½	3	-3
2 double lifts	1	16½	4	+3
2 single braces, forward	1 full	25	6	+6
2 double braces, forward	1	35	9	-6
2 single braces, aft	1 full	20	5	
2 double	braces, aft	1	24	6
2 pendents for ditto	1¼	15 feet	½	+3
1 parrel rope	1½	1¾	½	-1
2 sheets	2	23	5¾	
2 bowlines	¾	25	6	+6
4 bridles	1	5	1¼	
2 single clewlines	1 full	24	6	
2 double ditto	1	30	7½	

	A	B	C	D
2 reeflines	Line	8½	2	

Additional ropes				
Winding tackle pendent	5	8	2	+3
Fall for ditto, with 2 double blocks	3	26	6	
Fish tackle pendent	4	4¼	4	+1
Fall for ditto, with a long tackle				
block & a single block	2¼	24	6	
Pinnace rope	3½	24	6	
Guy rope	3	24	6	

Table 68: 'This Rigging following is for one of her Majesty's New Fifth Rates, cut out by the length of the Main Beam, which is 26 Foot 6 Inches, but not from Outside to Outside.'

	A	B	C	D
Bowsprit				
1 horse, all in a piece	2¼	5ft 4in	1⅓	-1
2 running lifts	2	36	8	+4
2 standing ditto to the yardarm	2¾	6	1⅓	-1
1 pair of halliards with a long tackle block at the bowsprit end, and a single block at the yard	2¼	20	4½	-1
2 braces, double	2¼	41	9¼	-6
2 pendents for ditto	2½ full	14	3	+4
Slings for yard	4	2	½	-1
2 horses for yard	2½	6	1⅓	
2 sheets	2½	48	11	-3
2 single clewlines	1½	20	4½	-1
2 buntlines	1½	18	4	+2
2 reeflines	¾	18	4	+2
Sprit topmast				
Half the first pair of shrouds	1¼	10 feet	½	-3
1 single tye	2¼	10	2¼	
Halliards for ditto with one end to the bowsprit	1¼	9	2	+1
2 single lifts	1¼	9	2	+1
2 double ditto	1¼ bare	12	3	-4
2 single braces	1¼	20	4½	-1
2 double ditto	1 bare	30	7	-5
1 parrel rope	1½	2	½	-1
2 single clewlines	1¼	30	7	-5
2 double ditto	1¼ bare	39	9	-4
2 reeflines	Line	9	2	+1
Foremast				
Pendents for tackles	4¼	5	1	+4
Half the first pair of shrouds	4¼	7	1½	+2
2 runners for tackles	4	17	4	-4
2 falls for ditto	2½	60	13½	+2
Stay	8½	8	1¾	+2
Bowsprit collar	6¾	11	2½	
1 yard tackle pendent	3½	2½	½	+2
1 fall for ditto with a long tackle block and a single block	2½	20	4½	-1
1 jeer with 2 double breast blocks	4	29	6½	+2
2 lifts to the cap	2½	48	11	-3
2 braces	2¼	35	8	-2
2 pendents for ditto	2½	14	3	+4
2 parrel ropes	3	9	2	+1
1 naveline with 1 block	1	12	2¾	-1
2 horses for yard	3	8½	2	-2
2 sheets, cable laid	2½ full	41	9½	-6
2 tacks	6	26	6	-3
2 bowlines	2½	32	7⅓	-2
2 bridles	2½	3½	¾	+2
2 clewgarnets	2¼	44	10	-1
2 buntline legs	1¼	25	5⅔	
1 fall for ditto with a long tackle block and a single block	1¼ full	23	5	+6
2 double legs for leech lines	1½ full	22	5	
2 falls for ditto with 1 end of each fast at the topmast head	1¼	50	11⅓	-2
2 studdingsail halliards	2½ bare	29	6½	+2
2 sheets	1½	8	1¾	+2
2 tacks, twice the boom	2½			
2 guys for the boom	2¼ bare	14	3	+4

	A	B	C	D
Fore topmast				
Pendents for Burtons	3¼	4	1	-2
2 falls for ditto	1½	26	6	-3
Half the first pair of shrouds	3½	5½	1½	-7
Half the pair of breast stays, with deadeyes in the chains	3½	25½	5⅔	+3
Half the pair of backstays, with deadeyes normally higher than the chains	3½	26	6	-3
Stay	4 full	12	2¾	-1
1 runner with 2 single blocks	3½	13½	3	+3
1 pair of halliards with 2 single blocks	2¼	25	5⅔	
2 lifts	2¼	44	10	-1
2 braces	2¼	42	9⅔	-4
2 pendents for ditto	2¼ full	14	3	+6
1 pair of yard slings	3½	6½	1½	
2 parrel ropes	2½	7	1½	+2
2 horses for yard	2½	5	1	+4
2 sheets	3¾	30	7	-5
2 bowlines	2¼	38	8½	+3
4 bridles	2¼	5	1	+4
2 clewlines	2¼	54	12⅓	-3
Buntline legs	1½ full	43	9½	+6
Falls for ditto	1¼ full	14	3	+6
2 leechlines	1¼	9	2	+1
Reef tackle fall in the top	1¼	13	3	-1
2 reeflines	¾	14½	3⅓	-1
Staysail stay	2¾	7	1½	+2
Staysail halliards, with 1 block	1¼ full	17	4	-4
Sheet	1¼ bare	7	1½	+2
2 studdingsail halliards	2	40	9	+2
2 sheets	1¼	3½	¾	+2
4 tacks at the mast	1¼	30	7	-5
2 ditto at the yard	1	20	4½	-1
Fore topgallant mast				
Half the first pair of shrouds	1½ full	10ft 3in	½	-3
Stay	2 full	14½	3⅓	-1
1 single tye	2	13 feet	½	
Halliards for ditto with 1 block	1¼	17	4	-4
2 single lifts	1¼	9	2	+1
2 double ditto	1¼	14	3	+6
2 single braces	1¼	36	8	+4
2 double ditto	1¼ bare	43	9⅔	-2
2 pendents for ditto	1¼ full	13 feet	½	
1 parrel rope	1½ full	1¾	½	-3
2 bowlines	¾	44	10	-1
2 bridles	¾	5½	1½	-7
2 single clewlines	1¼	39	8¾	+3
2 reeflines	Line	7½	1¾	-1
Mainmast				
Pendents for tackles	5½	5⅓	1½	-1
2 falls for ditto	2¾	62	14	+1
Half the first pair of shrouds	5¼	7ft 5in	1¾	+1
2 runners for tackles	4	20	4½	-1
Garnet pendent	6	8	1¾	+2
Garnet fall	2¾	26	6	-3
Guy	3¾	6½	1½	
Stay, cable laid	9½	13½	3	+2
1 yard tackle pendent	3¾	3	⅔	+1
1 fall for ditto	2¾	25½	6	-6
1 jeer with 2 double breast blocks	4¾	34	8	-4
2 lifts	2¾	52	12	-6
2 braces	2¾ bare	44	10	-1
2 pendents for ditto	3	5½	1½	-7
2 parrel ropes	2¾	9½	2	+4
1 naveline with 2 blocks	1	21	4¾	
Ditto with 1 block	1¼	14	3	+5

	A	B	C	D
2 yard horses	3	10	2¼	
2 sheets, cable laid	4	47	10½	+4
2 tacks, cable laid	6½	29	6½	+2
2 bowlines	2½	29	6½	+2
4 bridles	2½ bare	7	1⅔	-2
2 clewgarnets	2½ bare	46	10½	-2
2 buntline legs	1¼	28	6⅓	
1 fall with 1 long tackle block & 1 single, for ditto	1¼ bare	27	6	+3
1 uphauler	1	17	4	-4
2 double legs for leechlines	1¼	26	6	-6
2 falls for ditto with one end of each at the masthead	1¼	56	12½	+5
1 pair of pendents for ditto	2½	1½	⅓	
Staysail stay	3½	8½	2	-2
1 pair of halliards with 1 block	1¾	19	4⅓	
1 sheet	2½	3	⅔	
2 studdingsail halliards	2¼	30	7	-5
2 sheets	2¼	4	1	-2½
2 tacks, 5 times the length of the boom if they go without board	2			
4 guys	2	28	6⅓	

Main topmast				
2 pendents for Burtons	3½	4	1	-2½
2 falls for ditto	1¾	29	6½	+2
Half the first pair of shrouds	3½	6½	1½	
Half the pair of breast stays, with deadeyes in the chains	3½	28ft 4in	6⅓	+5
Half the pair of back stays	3¾	29	6½	+2
1 top rope	5	11½	2½	+3
1 fall for ditto	3	34	8	-8
Stay	4½	17	4	-4
1 runner for the topsail yard, with 2 single blocks	3¼	15½	3½	
1 pair of halliards with 2 single blocks	2¼	30	7	-5
2 lifts	2¼	52	12	-6
2 braces	2¼	49	11	+4
2 pendents for ditto	2(½)	5½	1⅓	-1½
1 pair of yard slings	3	11½	2½	+3
2 parrel ropes	2	6½	1½	
2 horses for yard	3	5½	1½	-7
2 sheets	5	35	8	-2
2 bowlines	2	42	9⅔	-4
6 bridles	2	10	2	+7
2 clewlines	2	63	14	+7
2 buntline legs	1	14	3	+5
1 fall for ditto	1	15	3½	-2
2 leechlines	1	10	2	+7
2 reeflines	¾	16½	3¾	
2 reef tackle ties with 1 reef	2½	40 feet	1½	-1
2 falls for ditto, upon deck	1¾	20	4½	
2 falls in the top	1¼	15	3½	-2

Main topgallant mast				
Half the first pair of shrouds	1¾	16½	3¾	
Half the pair of backstays	1¾	17	4	-4
Stay	2¼	17	4	-4
1 single tye	2¼	17 feet	⅔	-1
1 pair of halliards for ditto	1¼ full	20	4½	
2 lifts, double	1¼ full	18	4	+2
2 braces	1½ full	50	11⅓	
2 pendents for ditto	1½ full	5	1	+4
2 bowlines	1 bare	47	10½	+4
4 bridles	1 bare	5	1	+4
2 single clewlines	1¼ full	43	9⅔	+2

	A	B	C	D
Mizzen mast				
Pendents for Burtons	3¾	4½	1	
2 falls with single blocks	1¾	33	7½	
Half the first pair of shrouds	4	6¼	1½	-2
Stay	4	8½	2	-2
1 jeer with a double & a single block	4	24	5½	-1
1 parrel rope	2¾	3	⅔	
1 truss tackle fall with a long tackle block & a single block	1¾	18½	4	+5
1 pair of yard slings	4	14	3	+5
1 sheet with a breast block & a single block	2½	16	3⅔	-1
2 bowlines with 4 single blocks	1½	16½	3¾	
2 pairs of legs for peak brails	1	7½	1⅔	+1
2 brails for ditto	1¼	11½	2½	+3
The upper pair below the mast down by the mast	1½	18½	4	+5
The clew brails	1½	18½	4	+5
The lower pair of brails	1½	16½	3¾	
Lacing for the mizzen	1	18	4	+2
2 reeflines	1 bare	21	4¾	
Staysail halliards with a block under the mizzen top, one end bent to the head of the sail	1½	17½	4	-1
1 sheet	2¼	3	⅔	
1 topping lift for the crowsfoot of the staysail, with a block under the main top to top the sail upon a wind	1	16	3⅔	-1
1 pair of crossjack braces, single	1½	29	6½	+2
1 pair double	1¼	41	9⅓	-1
2 pendents for ditto	1¾	3	⅔	
2 standing lifts to the yardarms	2¼	5½	1⅓	-1½
1 pair of yard slings	3½	3½	¾	+1

Mizzen topmast				
Half the first pair of shrouds	2 bare	19 feet	⅔	+1
Half the pair of backstays	2 bare	8½	2	-2
1 stay in the main top	2½	6ft 6in	1	-3
1 single tye with 1 end about the yard	2	3ft 2in	¾	
1 pair of halliards for ditto, with 1 end in the top	1¼	13	3	-1
2 blocks	1¼ bare	20	4½	
2 lifts, double	1½	21	4¾	
2 braces, double, forward	1¼	34	7⅔	+1
2 pendents for ditto	1½	3	⅔	
2 mizzen peak braces	1¼	29	6⅔	-2
2 pendents for ditto	1½	14 feet	½	+1
1 parrel rope	1¾	2½	½	+2
2 bowlines	1 full	27	6	+3
4 Bridles	1 full	5½	1⅓	-1½
2 single clewlines	1¼	30	7	-5
2 reeflines	Line	9	2	+1

Additional ropes				
Viol, cable laid	7	27	6	+3
Winding tackle pendent	6	6	1⅓	
1 fall for ditto with 2 double breast blocks	2¼	32	7	+7
1 fish tackle pendent	5¼	5½	1⅓	-1½
1 fall for ditto with a long tackle block & a single block	2½	28	6⅓	
2 cat ropes with 8 sheaves in the catheads and blocks	2½	37	8⅓	+2
Sheet anchor buoy rope	6	26	6	-3
Best bower ditto	5½	20	4½	
Small bower ditto	5½	20	4½	

	A	B	C	D
Longboat rope	5½	28	6⅓	
Guy rope	4	28	6⅓	
Pinnace rope	4	24	5½	-1
Guy rope	3	24	5½	-1

Table 69: 'This Rigging following is for Her Majesty's Ship the *Bonadventure*, and the length of every Rope both Standing and Running, is cut out by the length of the Main Beam from Outside to Outside, and the same Rules will serve to Rig any Ship of the same Rate. This is a 33 feet Beam.'

	A	B	C	D
Bowsprit				
1 horse, all in a piece	3¼	8	1½	-1
2 sheets, cable laid	2½	50	9	+3
1 pair of halliards with a long tackle block & a single block	2½	30	5½	-1
2 lifts	2½	40	7⅓	-2
2 standing lifts	2½	6½	1¼	-2
Gammoning	5½	60	11½	-3
2 braces	2½	59	10⅔	+2
2 pendents for ditto	2½	3	½	+1½
2 yard horses	2½	7	1⅓	-2
Yard slings	5	4	⅔	+2
2 single clewlines	2	28	5	+3
2 buntlines	1½	22	4	
2 reeflines	1	22	4	
Sprit topmast				
Half the first pair of shrouds	2	11ft 6in	⅓	
1 single tye	2	2½	½	+1½
1 pair of single halliards for ditto	1½	8	1½	+2
2 lifts	1½	14	2⅔	-4
2 braces	1½	33	6	
2 pendents for ditto	2	⅓	+1	
1 parrel rope	1½	2	⅓	+1
2 reeflines	¾ or line	10½	2	-3
2 single clewlines	1½	33	6	
Foremast				
Pendents for tackles	6	6	1	
2 runners for ditto	5¼	24	4⅔	+1
2 falls for ditto	3	62	11⅓	-2
Half the first pair of shrouds	6	8	1½	-1
Stay, cable laid	10	11	2	
2 jeers with 2 double breast blocks	5½	60	11	-3
Fore yard tackle pendent	3½	3	½	+1½
1 fall for ditto	2½	35	6½	-1
2 lifts	3	60	11	-3
2 braces	2½	46	8⅔	+1
2 pendents for ditto	2¾	5	1	-3
1 breast rope	4	5	1	-3
2 parrel ropes	3½	10	2	-6
1 naveline with 2 blocks	1½	20	2⅔	-1
2 yard horses	3½	12	2	+6
2 sheets, cable laid	5	60	11	-3
2 tacks	6	37	6¼	+2
2 bowlines	3½	41	7⅓	+4
2 bridles	3½	4	⅔	+2
2 clewgarnets	2½	51	9⅓	-2
2 pairs of buntline legs	2	27	5½	-3
2 falls for ditto		30	5½	-1
1 fall with a long tackle block aloft and a single by the belfry	1½	25	4⅔	-4
1 uphauler for the buntlines, with one end fast to the strap of the long tackle block, & the other rove through a single block under the foretop & down upon the forecastle, so when you set your sail, let go your buntlines and leechlines, & overhaul them, & let go the clewgarnets, & the tack will come aboard, without going upon the yard to overhaul them; & more purchase & less rope	1	14	3⅔	-4
2 double legs for leechlines	2	26	4⅔	+2

	A	B	C	D
2 falls for ditto with 1 end of each fast to the topmmast head	1¼	55	10	
2 reeflines	1½	31	5⅔	-1
2 fore studdingsail halliards	2¼	46	8⅓	+1
2 sheets	2¼	16	3	-3
2 tacks 4 times the length of one boom				

Fore topmast

	A	B	C	D
1 pair of Burton pendents	3½	4½	¾	
2 falls for ditto with single blocks	2	29	5⅓	-2
Half the first pair of shrouds	4	6ft 4in	1	+7
Half the pair of breast stays, with deadeyes in the chains	3¾	14ft 4in	2¾	-3
Half the pair of backstays, with deadeyes normally higher than the chains		15		
1 top rope	6	12	2	+6
1 ditto fall	4	37	6⅔	+2
Stay	4½	15	2¾	-3
1 single runner with 1 single block upon the yard	4	14	2⅔	-4
1 pair of halliards for ditto, with 2 single blocks	2	28½	5	+6
Ditto with a long tackle block aloft & a single below		36	6⅔	-4
2 lifts	2	53	9⅔	-1
2 braces	2	26	4⅔	+2
2 pendents for ditto	2½	2¼ or ½		
1 pair of yard slings	4	9	1⅔	+1
2 parrel ropes	2½	9	1⅔	+1
2 yard horses	3	8	1½	-1
2 sheets	5½	40	7	
2 bowlines	2½	48	8	+2
4 bridles	2½	7	1⅓	-2
2 clewlines	2½	64	11⅔	-1
1 pair of buntline legs	2	14½	2⅔	-1
1 fall for ditto	2	15½	2¾	
2 leechlines	2½	12	2	-6
2 reef tackle ties for 1 reef	2	4½	¾	
2 falls to haul out in the top	1¾	16	3	-3
2 falls to haul out upon the forecastle	1	44	8	
2 reeflines	3½	19	3½	-1
Staysail stay	2	7	1⅓	-2
1 pair of single halliards with 1 end fast to the head of the sail	2	16	3	-3
1 sheet	2½	9	1⅔	-1
2 studdingsail halliards to hoist in the top	2½	30	5½	-1
2 to hoist upon the forecastle	2½	48	9	-9
2 tacks in the waist to secure the booms	2½	30	5½	-1
2 sheets	1½	5	1	-3

Fore topgallant mast

	A	B	C	D
Half the first pair of shrouds	2	16 feet or	½	
Half the pair of backstays	1¾	17ft 2in	3	+5
Stay	2	17	3	+3
1 single tye	2½	2½	½	-1½
Halliards for ditto with 1 end fast to a backstay or cross tack	1½	24	4⅓	+1
2 lifts, double	1½	16	3	-3
2 braces	1	52	9	-7
2 pendents for ditto	1½	3	½	+1½
2 parrel ropes	1½	3	½	+1
2 bowlines to the bowsprit end	1	33	6	
Ditto upon the forecastle		44	8	
4 bridles		5	1	-3
2 single clewlines	1½	49	9	-3
2 reeflines	¾	9	1⅔	-1

Mainmast

	A	B	C	D
Pendents for tackles with a hitch	6½	7	1⅓	+1
2 runners for ditto	5½	26½	4⅔	+2
2 falls for ditto	3	64	11⅔	-1
Half the first pair of shrouds	6½	8ft 4in	1⅔	-3
1 garnet pendent	7	9	1⅔	-1
1 ditto guy	4½	8½	13⅔	-4
1 ditto fall	3½	29	5⅓	-2
1 single Burton fall with 2 single blocks to the stay, and 2 single blocks below	2½	27	5	-3
1 runner for a double Burton	3	9	1⅔	-1
1 fall for ditto with 1 end to the stay or guy	2½	28	5	+3
1 Spanish Burton fall with 2 blocks upon the guys, 1 to one end of the fall, and the bight of the fall to the tail of the single block below	3½	41	7⅓	+4
Stay, cable laid	12	16	3	-3
Worming for the stay	1½	70	13	-9
Collar at the stem	10	7	1⅓	-2
Jeers with 2 double breast blocks	5¾	38	6¾	+3
1 yard tackle pendent	3½	3½	⅔	-1
1 fall for ditto with a long tackle block & a single block	2½	36	6⅔	-4
2 lifts	3½	64	11⅔	-1
2 braces	2¾	66	12	
2 pendents for ditto	3½	7	1⅓	-2
2 parrel ropes	4	12	2	+6
1 naveline with 2 single blocks	1½	24	4⅓	+1
2 yard horses	4	14	2⅔	-4
2 sheets, cable laid	5¼	59	11	-9
2 tacks ditto	6¾	33	6	
2 bowlines	3¼	38	6¾	+5
4 bridles	3½	10	2	-6
2 clewgarnets	2¾	56	10	+6
2 buntline legs, double	2¾	38	6¾	+3
1 fall for the fore buntline	2	34	6	+6
1 uphauler	1½	22	4	
2 leechline legs		29	5⅓	-2
2 falls for ditto with 1 end of each fast to the topmast head	1¾	60	11	-3
1 pair of leechline pendents	2½	1½	⅓	-2
2 reeflines	1	36½	6⅔	-1
1 quarter tackle fall with a long tackle block aloft & a single below	2½	30	5¼	-1
1 pendent for the tackle	3	2½	½	-1½

Main topmast

	A	B	C	D
Pendents for Burtons	3½	5½	1	
2 falls for ditto with 2 single blocks	2	33	6	
1 top rope	6	14⅓	2⅔	-1
1 ditto fall	4	44	8	
Half the first pair of shrouds	4¼	7ft 4in	1⅔	+2
Half the pair of breast stays with deadeyes in the chains	4	16	3	-3
Half the pair of after backstays	4	16½	3	
Stay	5	19	3¼	-1
1 single jeer with a single block at the yard	4½			
1 pair of halliards for ditto, with 2 single blocks	2½	33½	6	+3
Ditto with a long tackle block aloft, and a single below		46	8⅓	+1
2 lifts	2½	56	10	+6

	A	B	C	D
2 braces	2½	53	9⅔	-1
2 pendents for ditto	3	6	1	-3
1 pair of yard slings	4	10	2	-6
2 parrel ropes	2½	9	1⅔	-1
2 yard horses	2½	8	1½	-1
2 sheets	6	44	8	
2 bowlines	2¾	43	8	-6
6 bridles	3	10	2	-6
2 clewlines	2¾	70	13	-9
1 pair of buntline legs	2	16	3	-3
1 fall for ditto		18	3⅓	-3
2 leechlines	1¾	13	2⅓	+1
2 reef tackle ties	2½	5	1	-3
2 falls to hoist in the top	1¾	17	3	+3
2 to hoist upon deck		47	8½	+2
2 reeflines	1	23½	4⅓	-2
1 staysail stay	3½	9	1⅔	-1
1 pair of halliards with 1 end fast to the head of the sail	1¾	24	4⅓	
1 sheet	1¾	16	3	-3
2 studdingsail halliards to hoist in the top	2¼	39	7	+3
2 to hoist upon deck		55	10	
2 sheets	2	6	1	+3
2 tacks to be laid upon the yard	2	20	3⅓	-1
2 to come aft upon the poop	2	39	7	+3

Main topgallant mast

	A	B	C	D
Half the first pair of shrouds	2	18ft 6 in	½	+2
Half the pair of backstays	1¾	20½	3⅔	+2
Stay	2	17½	3	+6
1 single tye with 1 end above the yard	2¾	17 feet	½	+½
1 pair of halliards with 1 end fast to the cross trees or backstay	1½	24	4⅓	-1
2 lifts	1½	23	4	+6
2 braces	1½	60	11	-3
2 pendents for ditto	2	5	1	-3
2 parrel ropes	2	3½	⅔	-1½
2 bowlines	1	48	9	-9
4 bridles	1	6	1	+3
2 single clewlines withthe eyes about the clews of the sail	1½	50	9	+3
2 reeflines	¾	10	2	-6

Mizzen mast

	A	B	C	D
Pendents for Burtons	3¾	4	⅔	+2
2 falls for ditto with single blocks	2¼	34	6	+6
Half the first pair of shrouds	4¼	7ft 2in	1⅓	
Stay	4½	11	2	
1 jeer with a double and a single block	4	28	5	+3
1 parrel rope	3	3¾ or ⅔		
1 truss tackle fall with a long tackle block & a single	2¼	17	3	+3
1 pair of yard slings	4½	4	⅔	+2
1 sheet with a double breast & a single block	2¼	17	3	+3
2 bowlines with 4 blocks	2	19	3	-1½
2 legs for peak brails	1¼	8	1½	-1
2 brails for ditto	1¼	16	3	-3
The upper pair below the mast & the second pair downward	1¼	19	3½	-1
The third pair or foot brails	1¾	14	2⅔	-4
(These brails come down upon the quarterdeck)				
The following brails to the main shrouds:				
The upper pair below the mast & the second pair downward	2	26½	4¾	+2
The lower pair	1½	19	3½	-6
Halliards for the mizzen peak	1½	22	4	

	A	B	C	D
Lacing for the mizzen course	1	17½	3	+6
2 reeflines for the mizzen	1	24	4⅔	+1
Mizzen staysail halliards with a block under the mizzen top & 1 end to the head of the sail	1½	19	3½	-1½
1 topping lift with a block under the main top		18	3½	-2
1 sheet	3½	3	⅔	-1
The crossjack braces, double	1½	44	8	
Pendents	2	3½	⅔	-1
Standing lifts to the yardarms, without cleats	2	8	1½	-1

Mizzen topmast

	A	B	C	D
Half the first pair of shrouds	2¼	20 feet or ⅔		-2
Half the pair of backstays	1¾	10ft 5in	2	-1
The stay, in the main top, with deadeyes	2	7½	1⅓	+1
1 single tye with 1 end about the yard	2½	3½	⅔	-1
1 pair of halliards with 1 end fast in the top	1¾	15	2¼	-3
2 sheets				
2 lifts	1½	24	4⅓	+1
2 braces forward, double	1¼	43	8	-6
2 braces aft	1½	29	5⅓	-2
2 pendents for ditto	1½	14 feet	½	-2
2 parrel ropes	2¼	4	⅔	+2
2 bowlines	1½	32	6	-6
4 bridles		5	1	-3
2 clewlines	1½	32	6	-6
2 reeflines	¾	11	2	

Additional ropes

	A	B	C	D
Viol, cable laid	8½	30	5⅓	+4
Winding tackle pendent	8	10	2	-6
1 fall with 2 double breast blocks	4	30	5⅓	+4
Ditto with 1 treble block aloft		35	6½	-4
Fish tackle pendent	6½	5¼	1	+1
Fall for ditto with a long tackle block and a single block	2¼	29	5⅓	-2
2 cat ropes with 8 sheaves in the catheads & blocks	3½	46	8⅓	+1
Buoy rope for sheet anchor	6	28	5	+3
For best bower		20	3⅔	-1
For small bower		20	3⅔	-1

Table 70: 'This Rigging following is for Her Majesty's Ship the *Lenox*, and the exact Length of every Rope is found out by the extreme breadth at the Main Beam, which is 40 Feet, and the same Rules will Rig any of the third Rates.'

	A	*B*	*C*	*D*
Bowsprit				
One horse all in a piece	3½	8	1⅕	
1 bobstay	5	8½	1¼	+1
1 bowsprit collar	4	14 feet or ⅓		
2 sheets, cable laid	3½	60	9	
1 pair of halliards with 1 long tackle block & 1 single block	3	32	5	-8
2 lifts	3	54	8	+4
2 double braces	2½	64	9	-2
2 treble braces	2½	76	11⅓	+3
2 pendents	2¾	9	1⅓	+2
Slings for yard	5	3¾	½	+2
2 horses for yard	2½	9	1⅓	+2
2 clewlines	2	36	5⅔	+3
2 buntlines	2	34	5	+4
2 reeflines	1	27	4	+2
Sprit topmast				
Half the first pair of shrouds	3	16 feet	⅓	+2½
1 tye	3	2½	⅓	
Halliards for ditto with 2 blocks	2	10	½	
2 lifts	1¾	20	3	
2 braces	1½ full	42	6⅓	-1
2 pendents	2	4	½	+4
2 parrel ropes	2½	5	¾	
2 clewlines	1½	60	9	
2 reeflines	1	14	2	+4
Foremast				
2 pairs of pendents for tackles	7½	17½	2½	+5
2 runners for tackles	6½	28	4	+8
4 falls	3½	163	24	
Half the first pair of shrouds	7½	9ft 5in	1½	-1
Stay	13¼	13½	2	+1
The collar about the bowsprit	8½	2½	⅓	-1½
1 jeer with 2 double blocks	6	48	7	+8
1 pendent for a yard tackle	5	3	½	-2
1 fall for ditto	2¾	36	5	+3
2 lifts	3½	70	10⅔	-7
2 braces	3	54	8	+4
2 pendents	3½	4	½	+4
2 parrel ropes	4	18	2⅓	+2
1 naveline with 2 blocks	1½	27½	4	+5
2 yard horses	4	17	2½	+2
2 sheets, cable laid	5½	68	10	+8
2 tacks	7¾	46	6¾	+6
2 bowlines	4	52	8	-8
2 bridles	4	6	¾	+6
2 clewgarnets	2¾	72	10¾	+2
2 buntline legs	2½	44	6⅔	-2
(I used to have but 1 fall in her, with 1 long tackle block & 1 single)	2	30	4½	
2 leechline legs	2	17	2½	+2
2 falls with 1 end of each to the masthead 2	50	9		
2 reeflines	1½	40	6	
Luff hook rope	4	9	1⅓	+1
Fore topmast				
1 pair of pendents for Burtons	4	5½	¾	+3
2 falls for ditto with 2 long tackle blocks	2	46	7	-4
Half the first pair of shrouds	4½	7ft 4in	1	+6½
2 top ropes	6	32	5	-8
2 falls	4	90	13½	
Half the pair of breast stays & half the pair of backstays, both with deadeyes in the chains	4½	18½	2⅔	
Stay	5½	18	2⅔	+1½
2 runners with a double block at the yard	3½	35	5¼	
2 pairs of halliards with 4 single blocks	2½	84	12⅔	-2½
2 lifts	2½	64	9⅔	-2½
2 braces	2½	56	8⅓	+3
2 pendents	3	4	½	+4
1 pair of yard slings	4½	10	1½	
2 parrel ropes	3½	12	2	-8
2 horses for yard	3½	14	2	+4
2 sheets	6¼	48	7	+8
2 bowlines	2½	54	8	+4
4 bridles	2½	9	1⅓	+1
2 clewlines	3	76	11⅓	+3
1 pair of buntline legs	2	18	2⅔	+1½
1 fall for ditto	2	20	3	
2 leechlines	2	50	7⅔	-7
2 reeflines	1	23	3¼	+5
2 reef tackle ties	2½	3	½	-2
2 falls to haul out in the top	2	18	2⅔	+1½
Ditto to haul upon the deck	2	58	8⅔	+1
Staysail stay	3½	9	1⅓	-½
1 pair of halliards with 1 block	2	20	3	
Ditto with 2 blocks	2	27	4	+2
1 sheet	2	18	2⅔	+1
Fore topgallant mast				
Half the first pair of shrouds	2½	3ft 1in	½	-1
Half the pair of breast stays	2	21	3	+6
Half the pair of back stays	2	21ft 2in	3	+8
Stay	2½	22	3⅓	-1
1 single tye	2½	6½	1	+1
1 pair of halliards with 1 end fast to a backstay	1½	25	3¾	
2 lifts	2	22	3½	-1
2 braces	1½	54	8	+4
2 pendents	2	3	½	-2
2 parrel ropes	2	4	⅔	-2
2 bowlines	1	42	6⅓	-1
4 bridles	1½	6	¾	+6
2 clewlines	1	59	8¼	+4
2 reeflines	¾	10½	1½	+3
Mainmast				
2 pairs of pendents for tackles	8	18½	2¾	+1
2 runners for ditto	6½	30	4½	
4 falls	3½	148	22	+8
Half the first pair of shrouds	8	11	1½	+6
For eight pairs	8	178	26½	+8
Garnet pendent	8	12	2	-8
Ditto guy	4½	11	1½	+6
Ditto fall	3½ full	36	5¼	+6
1 quarter tackle fall	2	33	5	-2
1 single Burton fall with 2 blocks	2½	32	5	-8
1 Spanish Burton fall	3½	48	7	+8
1 double Burton fall	2½	33	5	-2
1 runner for ditto	3	11	1½	+6
Allowance for ratlines	1½	160	24	
Catharpin legs & falls	2	48	7	+8
Stay	15¼	19	3	-6
The collar at the stem 12	8	1⅕		

	A	B	C	D
1 crowsfoot for the main top	1½	92	14	-8
2 jeers with triple blocks aloft & with double blocks upon the yard	6	124	18⅔	-3
1 yard tackle pendent	5	3½	½	+1
1 fall for ditto	3	36	5⅓	+3
2 lifts	3½	76	11⅓	+3
2 braces	3½	76	11⅓	+3
2 pendents	4	8	1⅕	
2 parrel ropes	4½	20	3	
1 breast rope	6	5¾	¾	+4½
1 naveline with 2 blocks	2	30	4½	
2 horses for yard	4½	19	3	-6
2 sheets, cable laid	6	70	10½	
2 tacks	8¼	45	6¾	
2 bowlines	4	42	6½	
4 bridles	3½	10	2	-6
2 clewgarnets	3	72	10¾	+2
2 buntline legs	2	44	6⅔	-3
(I used to have 1 fall in her with 1 long tackle block and 1 single block)	2	36	5⅓	+3
The uphauler, to come down by the main stay	1½	23	3⅓	+5
2 leechline legs	2	36	5⅓	+3
2 falls for ditto with 1 end of each to the masthead	2	70	10½	
2 reeflines	1½	45	6¾	
Staysail stay, to the mouse	4½	12	1¾	+2
1 pair of halliards with 2 blocks	2½	30	4½	
1 sheet	3½	4	½	+4
2 studdingsail halliards	2¾	60	9	
2 sheets	3	26	4	-4
2 guys	3	29	4⅓	+1
2 tacks	2¼	38	5¾	-2

Main topmast

	A	B	C	D
Pendents for Burtons	4	6	1	-4
2 falls with single blocks	2½	36	5⅓	+3
2 ditto with long tackle blocks aloft & single blocks below	2¼	43	6⅓	+5
Half the first pair of shrouds	4½ full	35	5¼	
2 top ropes	7	96	14⅓	+3
2 falls	4	20	3	
Half the pair of breast stays, with deadeyes	4½	20½	3	+3
Half the pair of backstays, with deadeyes in the chains	4½	20½	3	+3
Stay	5½	25	3¾	
2 runners with a double block on the yard	3½	39	6	-6
2 pairs of halliards with 4 single blocks	2½	98	14½	+8
2 lifts	3	70	10½	
2 braces	2½	66	10	-4
2 pendents	3	7	1	+2
1 pair of slings	4½	12	1¾	+2
2 parrel ropes	3¾	12	1¾	+2
2 yard horses	3½	11	1½	+6
2 sheets	6½	52	8	-8
2 bowlines	3½	56	8⅓	+3
6 bridles	3⅔	16	2⅓	+3
2 clewlines	3¼	82	12¼	+2
1 pair of buntline legs	2	22	3¼	+2
1 fall for ditto	2	23	3⅓	+5
2 leechlines to the masthead	2	59	9	-6
2 reef tackle ties	3		¾	+6
2 falls to haul out in the top	2	16	2⅓	+3
Ditto to haul out upon deck	2	60	9	
2 reeflines	1	25	3¾	
Main top staysail stay	3½	9	1⅓	+1

	A	B	C	D
1 pair of halliards for ditto with 2 blocks	2	26	4	-4
1 sheet	2	18	2⅔	+1
2 studdingsail halliards	2½	70	10½	
2 sheets	2	10	1½	
2 tacks to haul out upon the poop	2	56	8⅓	+3
2 bowlines	¾	26	4	-4

Main topgallant mast

	A	B	C	D
Half the first pair of shrouds	2½	3ft 4in	½	+2
Half the pair of backstays	2¼	22	3⅓	-1
Stay	2½ full	22	3⅓	-1
1 double tye	2½	9	1⅓	+1
(I had the halliards with a block fast in the main top)	1¾	37	5½	+2
2 lifts	2	22	3⅓	-1
2 braces	1¾	66	10	-4
2 pendents	2	6	¾	+6
2 parrel ropes	2¼	8	1⅕	
2 bowlines	1½	56	8⅓	+3
4 bridles	1½	7	1	+2
2 clewlines, double	1½	66	10	-4
2 reeflines	¾	18	2⅔	+1

Mizzen mast

	A	B	C	D
Pendents for tackles or Burtons	5	6	¾	+6
2 tackle falls	2½	60	9	
2 Burton falls	2½	36	5⅓	+3
Half the first pair of shrouds	5½	9	1⅓	+1
Stay	6	13½	2	+1
1 pair of jeers with 2 double blocks	4	44	6⅔	-3
Pendent halliards for the mizzen peak	2	26	4	-4
1 parrel rope	4½	6	¾	+6
1 truss tackle fall with 1 long tackle block & 1 single block	2½	28	4	+8
1 sheet with 1 double breast block & 1 single	3½	30	4½	
2 bowlines	2	19	3	-6
2 legs for peak brails	1½	11	1½	+6
2 brails for ditto	2	16	2⅓	+3
The lower pair above the mast	2	27	4	+2
The upper pair below the mast	2	34	5	+4
The second pair or clew brails	2	34	5	+4
The third pair	2	32	4⅔	+2
(All these brails come to the main shrouds)				
Lacing for the mizzen	1½	42	6⅓	-1
The reeflines for ditto	1½	39	6	-6
The topping lift for the mizzen staysail with a block under the main top	2	20	3	
Halliards, with a block under the mizzen top, and 1 in the sail	2	36	5⅓	+3
Sheet	3½	4	½	+4
2 crossjack braces, full	1¾	54	8	+4
2 pendents	2	4	½	+4
2 standing lifts to the yardarms	3½	10	1½	
1 pair of slings	4	5	¾	

Mizzen topmast

	A	B	C	D
Half the first pair of shrouds	3	5	¾	
Half the pair of backstays	2	13	2	-2
Stay, in the main top	2½	10½	1½	+3
1 tye with 3 blocks	2	10½	1½	+3
1 pair of halliards with 2 blocks	1¾	30	4½	
2 lifts	2½	28	4	+8
2 braces	1½	44	6½	+4
2 pendents to the mizzen peak	2	3	½	-2
2 parrel ropes	2	5	¾	
2 bowlines	1½	38	5⅔	+1

	A	B	C	D
4 bridles	1½	6	¾	+6
2 clewlines	1½	50	7½	
2 reeflines	¾	13	2	-2
2 sheets	2½	40	6	
Additional ropes				
Viol, cable laid	4½ (11?)	37	5½	+2
Winding tackle pendents	4 (10?)	13	2	-2
Fall for ditto	5	48	7	+8
Fish tackle pendent	8	8	1	+8
Fall for ditto	4	46	7	-4
2 cat Ropes	5½	56	8⅓	+3

Table 71: 'This Rigging following is for Her Majesties Ship the *Vanguard*, and the Exact Length of every Rope, is found out by the extream Breadth at the Main Beam, which is 45 Feet, and the same Rules will Rig all the Second Rates.'

	A	B	C	D
Bowsprit				
1 horse, all in a Piece	3½	8	1	+3
1 bobstay	6	6⅓	1	-7
1 collar	4	5	⅔	
2 sheets	3¾	64	8½	+2
1 pair of halliards	3½	32	4¼	+1
2 running lifts	3½	55	7⅓	
2 standing lifts	3	8	1	+3
2 braces, treble	2¾	78	10⅓	+3
2 pendents	3	3	⅓	+3
Slings for the yard	6	4	½	+1½
2 horses for ditto	3	8	1	+3
2 clewlines	2½	32	4¼	+1
2 buntlines	2¼	25	3⅓	
2 reeflines	1¼	29	4	-6
Sprit topmast				
Half the first pair of shrouds	2¾	3	⅓	+3
1 single tye	3	3	⅓	+3
1 pair of halliards for ditto with 2 blocks	1¾	8	1	+3
2 lifts	1¾	22	3	-3
2 braces	1¾	44	6	-6
2 pendents	2¼	3	⅓	+3
2 clewlines	2	44	6	-6
2 parrel ropes	2	2½	⅓	
2 reeflines	¾	15	2	
2 horses for the yard	2	5	⅔	
Foremast				
2 pairs of pendents for tackles	8	9	1	+9
2 runners for ditto	6½	27	3½	+5
There are 4 falls for 4 tackles, & half of them are:	3½	92	12⅓	-3
Half the first pair of shrouds	8	10ft 4in	1⅓	+4
Stay, cable laid	14	15	2	
Collar on the bowsprit	9	3½	½	-1½
1 jeer with a double & a treble block	6½	54	7¼	-2
1 yard tackle pendent	4½	3½	½	-1½
1 fall for ditto	3½	36	4¾	+2¼
2 lifts	3½	64	8½	+2
2 braces	3½	57	7⅔	-3
2 pendents	3¾	5½	⅔	+3
2 parrel ropes	4½	18	2⅓	+3
1 breast rope	6	4½	½	+4
1 naveline	2	30	4	
2 horses for the yard	4½	16	2	+6
2 sheets, cable laid	5½	72	9⅔	-3
2 tacks, tapered & ditto	8¾	42	5⅔	-3
2 bowlines	4	50	6⅔	
2 bridles	4	5	⅔	
2 clewgarnets	3	54	7¼	-2
2 buntline legs	2½	44	6	-6
1 fall with 1 long tackle block & 1 single	2¼	36	4¾	+2
1 topping lift with a single block under the fore top	1¾	30	4	
1 pair of pendents for leechlines	3	3	⅓	+3
2 double legs for leechlines	2¼	30	4	
2 falls for ditto	2¼	60	8	
2 reeflines	1½	45	6	
Fore topmast				
1 pair of pendents for Burtons	4	4½	⅔	-3

	A	B	C	D
2 falls for ditto 1 long tackle block	2½	46	6	+6
Half the first pair of shrouds	4¾	8	1	+3
2 top ropes	7½	32	4¼	+1
2 falls with a treble & a double block	4½	84	11	+9
Half the pair of breast stays	4½	18	2½	+3
& half the pair of backstays, both with deadeyes in the chains	4½	18½	2½	+6
Stay	5½	17	2¼	+1
2 runners or jeers with a double block upon the yard, & 2 single blocks at the masthead	4	36	4¾	+2
2 pairs of halliards with 4 single blocks	2¼	70	9⅓	
2 lifts	2¼	63	8¼	+7
2 braces	2¼	56	7½	-1
2 pendents	3	3	⅓	+3
1 pair of slings for the yard	4	14	2	-6
2 parrel ropes	3	11	1½	-1½
2 horses for the yard	3	7	1	-3
2 sheets	7½	44	6	+6
2 bowlines	2¼	53	7	+3
4 bridles	2¼	8	1	+3
2 clewlines	3½	78	10⅓	+3
1 pair of buntline legs	2½	19	2½	+1½
1 fall for ditto	2½	19	2½	+1½
2 leechlines in the top	2	16	2	+6
2 ditto to the foot of the sail, & so to the topmast head, & down upon the deck	2½	50	6⅔	
2 reef tackle tyes	3	5	⅔	
2 falls to haul out upon the deck	2½	60	8	
2 reeflines	1	25	3⅓	

Fore top staysail

	A	B	C	D
Stay	3	10	1⅓	
1 pair of halliards with 2 blocks	2¼	24	3¼	-2
1 sheet	2	15	2	

Fore top studdingsail

	A	B	C	D
2 pairs of halliards with blocks at the masthead, & then down upon deck in the forecastle	2¾	55	7⅓	
2 tacks in the waist	2	40	5⅓	
2 sheets	2	8	1	+3

Fore topgallant mast

	A	B	C	D
Half the first pair of Shrouds	2½	3½	½	-1½
Half the pair of breast stays	2½	19½	2⅔	-3
Stay	2½	19½	2⅔	-3
1 tye with a block upon the yard	3	4½	⅔	-3
1 pair of halliards for ditto	1¾	33	4⅓	+3
2 lifts	2¼	22	3	-3
2 braces	1¾	58	7¾	+3
2 pendents	2	4	½	+1½
2 parrel ropes	2	4	½	+1½
2 bowlines in the top, 38 fathoms 5 times add 3 upon the forecastle	1½	63	8⅓	+3
4 bridles	1½	6	¾	+2½
2 clewlines	1¾	62	8¼	+1
2 reeflines	1	12	1⅔	-3
1 flagstaff stay	2¼	22	3	-3
1 pair of halliards for ditto	2	11	1½	-1

Mainmast

	A	B	C	D
2 pairs of pendents for tackles	8½	21	2⅔	+2
2 runners for ditto	6½	28	3¾	
There are 4 falls for these tackles, & half of them are	3½	90	12	
Half the first pair of shrouds	8½	11½	1½	+1½
Pendent of the garnet	7½	11	1½	-1½
1 guy ditto	5	10	1½	-7½
1 fall ditto	4	36	4¾	+2
1 pendent for a quarter tackle	3½	3	⅓	+3
1 fall for ditto with 1 long tackle block & 1 single	3	38	5	+3
1 single Burton with 2 blocks	2¼	36	4¾	+2
1 pendent for a yard tackle	4	3½	½	-1½
1 fall for ditto	3¼	36	4¾	+2
Stay, cable laid	17	20	2⅔	
Collar at the stem for ditto	11½	9½	1¾	+1
1 jeer with a double & a treble block	7½	55	7⅓	
2 lifts	4	78	10⅓	+3
2 braces	3¾	80	10⅔	
2 pendents	4½	9	1¼	-2
2 parrel ropes	5	19	2¾	+1½
1 breast rope	6	5½	⅔	+3
1 naveline	2	30	4	
2 horses for the yard	5	17	2¼	+1
2 sheets, cable laid	6¼	78	10⅓	+3
2 tacks, tapered ditto	9	40	5⅓	
2 main bowlines	3½	52	7	-3
4 bridles for ditto	3½	9½	1¼	+1
2 clewgarnets	3½	58	7¾	
2 buntline legs	2½	45	6	
2 falls	2½	49	6½	+1
1 fall with a long tackle block & a single block	2½	38	5	+3
1 uphauler with a block under the top, & down by the main stay (as I used to have when I was in her)	2	24	3¼	-2
2 leechline legs	2½	37	5	-3
2 falls to the masthead	2½	73	9¾	
2 reeflines	1¾	51	7	-9
1 pair of staysail halliards with 2 blocks	2½	31	4	+6
False stay	4	12	1⅓	-3
1 sheet	3½	6	¾	+3
2 studdingsail halliards	3½	58	7¾	
2 tacks (if the booms are without board, they must be 5 times the length of the board)	2½			
2 sheets	2¾	16	2	+6
2 guys	3½	28	3¼	

Main topmast

	A	B	C	D
1 pair of pendents for Burtons	4	6	¾	+2½
2 falls with 2 long tackle blocks, & with 2 single blocks	2½	53	7	+3
& with 4 single blocks	2½	37	5	+3
Half the first pair of shrouds	5	9ft 2in	1¼	
Half the pair of breast stays	5	20½	2¾	
& half the pair of backstays, both with deadeyes in the chains	5	21	2¾	+3
Stay	7½	25½	3½	-4½
2 single runners or jeers with a double block on the yard	4	43	5¼	
2 pairs of halliards with 4 single blocks	3	70	9⅓	
2 lifts	2¾	71	9⅓	+6
2 braces	2¾	66	9	-9
2 pendents	3	6	¾	+2½
1 pair of slings for the yard	4	14	2	-6
2 parrel ropes	3½	10	1⅓	
2 horses for the yard	2¼	9	1¼	-2

	A	B	C	D
2 sheets	8	54	7¼	-2
2 bowlines	3½	56	7¼	-1½
6 bridles	3⅔	12	1⅔	-3
2 clewlines	3½	88	11¼	
2 buntline legs	2½	21	2¼	+3
1 fall for ditto	2½	22	3	-3
2 leechlines in the top	2½	19	2½	+1½
2 to the foot of the sail & to the topmast head & so down upon deck	2½	60	8	
2 reef tackle ties	3½	6	¾	+3
2 falls in the top	2½	28	3¼	
2 falls to haul out upon deck with 2 long tackle blocks at the pendent	2½	98	13	+3
Staysail stay	3¾	12	1⅔	-3
1 pair of halliards with 2 blocks	2¼	28	3¼	
1 sheet, double	2¼	29	4	-6
2 studdingsail halliards with 2 blocks at the topmast head, & so down & to hoist upon the deck	3	64	8½	+1½
2 tacks upon the poop	2½	50	6⅔	
2 sheets	2	8	1	+3

Main topgallant mast				
Half the first pair of shrouds	3	3½	½	-1½
Half the pair of backstays	2½	22	3	-3
Stay	3	24	3¼	-2
1 tye with 1 block upon the yard	3	9	1¼	-2
1 pair of halliards with 2 blocks; that is 1 of them in the top	2	41	5½	-1½
2 lifts	2¼	25	3½	
2 braces	1¼	70	9⅓	
2 pendents	2	6	¾	+3
2 bowlines	1½	53	7	+3
4 bridles	1½	7	1	-3
2 clewlines	1¼	70	9⅓	
2 reeflines	¾	14½	2	-3

Mizzen mast				
1 pair of pendents for tackles	5	5	⅔	
2 falls with 2 long tackle blocks aloft	2¼	48	6⅓	+3
Half the first pair of shrouds	5	9	1¼	-2
Stay	6	14	1¾	+6
1 jeer with 1 double block & 1 triple	5	39	5¼	-2
1 parrel rope	4	5	⅔	
1 truss tackle fall with 1 long tackle block	2¼	24	3¼	-2
1 pair of slings for the yard	5½	5½	¾	
2 bowlines with 4 blocks	2½	36	4¼	+3
1 topping lift for the yard	2½	27	3⅔	-3
2 legs for peak brails	2	10	1⅓	
2 brails for ditto	2	18	2½	-4½
The second pair above the mast	2½	25½	3⅓	+3
These brails following come to the main shrouds:				
The upper pair below the mast	2½	30	4	
The second pair below the mast	2½	30	4	
The clew brails	2½	32	4¾	+1
The fourth pair	2	24	3¼	-2
Pendent halliards	2	24	3¼	-2
The lacing for the mizzen	1½	42	5⅔	-3
2 reeflines	1½	46	6	+6
Sheet with a double breast block & a single block	3¾	34	4½	+1½
The mizzen staysail hHalliards to the mizzen top with 2 single blocks	2	32	4¼	+1
1 topping lift for the sail to the main top	1¾	19	2½	+1½

	A	B	C	D
1 sheet	3½	5	⅔	
2 crossjack braces	2	52	7	-3
2 pendents	2¼	4	½	+1½
2 standing lifts	3½	9½	1¼	+1
1 pair of slings for the yard	4¾	5½	¾	

Mizzen topmast				
Half the first pair of shrouds	2½	5ft 2in	¾	-1
Half the pair of backstays	2½	13¼	3	
The stay in the main top	3	9	1¼	-2
1 tye or jeer with a block upon the yard	2½	11	1½	-1
1 pair of halliards with 2 blocks	2	30	4	
2 lifts to be made fast below	2	30	4	
2 braces to the mizzen peak	1¼	46	6	+6
2 pendents for ditto	2	3	⅓	+3
2 clewlines	2	54	7¼	-2
2 parrel ropes	2¼	5	⅔	
2 sheets	3½	32	4¼	+1
2 bowlines	1½	38	5	+3
4 bridles	1½	7	1	-3
2 reeflines	¾	15	2	+2
1 flagstaff stay	2	12	1⅔	-3
1 pair of halliards	2	25	3⅓	

Additional ropes				
1 viol, cable laid	12	40	5⅓	
Winding tackle pendents	12	14	1¾	+6
Fall for ditto	6¼	54	7¼	-2
Fish tackle pendent	8	10	1⅓	
Fall for ditto	4	40	5⅓	
2 cat ropes with 6 sheaves in the catheads & 4 in the blocks	4½	86	11½	-1½

Table 72: 'Her Majesty's Ship the *Royal Sovereign* was Built at Woolwich, and Lanched the 25th of July 1701, and here is the exact Length of all her Rigging, both Standing and Running, Cut out by her extream Breadth at the Main Beam, which is 50 Feet 3½ Inches, her Main Beam being 46 Feet 8 Inches; and the same Rules will Rig any of the first Rates for ever, as long as Ships go to Sea.'

	A	*B*	*C*	*D*
Bowsprit				
Horse, all in a piece, from the knee to the cross piece at the beakhead	4½	9½	1	+7
2 sheets	4¼	77	9¼	-2
1 pair of halliards with a treble block at the bowsprit-end, and a double block on the yard	3¾	58	7	-2
2 lifts	4½	60	7¾	-2½
2 standing lifts	4¼	10	1¼	-2½
2 horses for the yard	3¾	10½	1¼	+2
2 braces, treble	3¼	90	10¾	+2½
1 pair of slings for the yard	6½	6	¾	-1½
2 clewlines	2¾	36	4¼	-½
2 pairs of buntline legs	2½	17½	2	+5
2 falls for ditto	2½	36	4¼	-½
2 reeflines	1½	31½	3¾	+1½
Bobstay	6½	9	1	+4
Sprit topmast				
Half the first pair of shrouds	3¾	4½	½	+2
1 tye with a block upon the yard	3½	3½	½	-3
1 pair of halliards with a double breast block aloft, & a single below	2½	19	2¼	+1½
2 lifts	2½	27½	3	-6
2 braces	2¼	52	6¼	-6
2 pendents	2½	4½	½	+2
2 parrel ropes	2¼	5	⅔	-1½
2 clewlines	2¼	50	6¾	
2 reeflines	1	18	2	+8
2 horses for the yard	2½	6	¾	-1½
Foremast				
2 pairs of pendents for fore tackles	8½	21	2½	+1
2 runners for ditto	7	34	4	+4
Half the 4 falls is	3¾	85	10¼	-2½
Half the first pair of shrouds	8½	11½	1½	-6
Stay, cable laid	15½	14½	1¾	
Collar about the bowsprit	11½	3½	½	-4
1 of the jeers with 2 treble blocks	7	58	7	-2
1 yard tackle pendent	6	3½	½	-4
1 fall for ditto	3½	40	4¾	+2½
2 lifts	4	76	9	+6
2 braces	3¾	66	8	-4
2 pendents	4½	6	¾	-1½
2 parrel ropes	5½	28	3¼	+5½
1 naveline with a long tackle block aloft, & a single block to the parrel	2¼	34	4	+4
2 horses for the yard	5½	15½	2	-7
2 sheets, cable laid	6½	78	9⅓	-1
2 tacks, tapered and ditto	9	48	5¾	+6
2 bowlines	5	58	7	-2
2 bridles	5	5½	⅔	-4
2 clewgarnets	3½	66	8	-4
2 buntline legs	2¾	42	5	+2
1 fall with a long tackle block aloft & a single below	2¾	39	4⅔	+3/4
1 topping lift for the buntlines with a block under the fore top	2¼	19	2¼	+1½
2 leechline legs	2¾	36	4⅓	+6
1 fall for ditto	2¾	80	9⅔	-3
2 reeflines	1¾	46	5½	+4
Fore topmast				
Half the first pair of shrouds	5¼	9½	1	+7
1 pair of pendents for tackles or Burtons	4½	6	¾	-1½
2 falls with long tackle blocks aloft & single below	2¾	54	6½	
Half a pair of breast stays	5	20ft 2in	2½	-3
And half a pair of after backstays, all with deadeyes in the chains	5	20ft 4in	2½	-1
Stay	6½	18	2	+8
1 of the runners or jeers for the yard, with a double block upon the yard	5	22	2⅔	-1
1 of the 2 pairs of halliards with a long tackle block aloft & 2 single blocks below	3	49	6	-6
1 top rope	8½	16	2	-4
1 of the 2 falls	4¼	56	6½	
2 lifts	3½	74	9	-6
2 braces	3½	69	8¼	+1½
2 pendents, full	3½	5	⅔	-3
1 pair of slings for the yard	5½	19	2¼	+1½
2 parrel ropes	3½	14	1¾	-3½
2 horses for the yard	3½	9½	1	+7
2 sheets	7¼	60	7¼	-2½
2 bowlines	3¼	60	7¼	-2½
4 bridles	3¼	11	1⅓	-8
2 clewlines	3¾	90	10¾	+2½
1 pair of buntline legs	2¾	22	2⅔	-1
1 fall for ditto	2¾	26	3	+6
2 reeflines	1½	30	3⅔	-3
2 reef tackle ties with a double breast & a treble block for each	3¼	73½	1	-5
2 falls upon the forecastle	2¾	70	8½	-5
Fore top staysail stay	4	13	1½	+3
1 pair of halliards with 2 blocks	2¾	39	4⅔	-3
Sheet	1½	17	2	+2
Tack	2¼	3	⅓	+1½
2 fore studdingsail halliards	3¼	56	6¾	-1½
2 sheets	2½	20	2½	-5
2 tacks, 4 times the length of the boom	2¾			
2 guys for the boom	3	30	3⅔	-3½
2 fore top studdingsail halliards down upon the forecastle & 2 blocks at the fore topmast head	3	67	8	+2
2 tacks in the waist	2½	46	5½	+1
2 sheets	2½	15	1¾	+3
Fore topgallant mast				
Half the first pair of shrouds	3	4½	½	+2
Half a pair of breast stays	3	24	3	-6
And half a pair of backstays, both with deadeyes in the chains	3	24ft 5in	3	-1
Stay	3½	24	3	-6
1 tye with a single block upon the yard & 2 at the masthead	3	10	1¼	-2½
1 pair of halliards with 1 of the 2 blocks in the fore top	2½	39	4⅔	-8
2 lifts	2¼	32	4	-8
2 braces	2	60	7¼	-2½
2 pendents	2½	3½	½	-4
2 parrel ropes	2¼	5	⅔	-3½
2 bowlines	1¾	68	8	+8
4 bridles	1¾	7	¾	+4½
2 clewlines	2	66	8	-4
2 reeflines	1	14½	1¾	-6
Flagstaff stay	2½	26	3	+6
1 pair of halliards to hoist in the fore top	2	32	4	-8

	A	B	C	D
Ditto at the topmast head	2	16	2	-4
Mainmast				
2 pairs of pendents for the tackles	9	22	2⅔	-1
Half of the 4 falls	4	88	10½	+3
2 runners	6¼	36	4⅓	-8
Half the first pair of shrouds	9¼	13	1½	+3
Pendent of the garnet	8½	15	1¾	+2½
Fall of ditto	4½	45	5½	-5
Guy of ditto	5¼	13	1⅓	+3
1 Burton fall with a long tackle block aloft, & a single below (but if you have a pendent it must be shorter)	3	56		-1½
1 pendent for 1 of the quarter tackles in long tackle blocks & a single	4½	3½	½	-4
1 fall for ditto	3¾	40	4¾	+2½
1 pendent for 1 of the yard tackles	6	3½	½	-4
1 fall for ditto	3¾	44	5¼	+1½
1 of the jeers with 3 sheaves aloft, & 3 in the block on the yard	7¼	83	10	-2
Stay, cable laid	19	23	2⅔	+2
Collar at the stem	14½	12½	1½	
2 lifts	4¾	92	11	+2
2 false braces forward	3¾	86	10⅔	-8
2 braces aft	3¾	90	10¾	+2½
2 pendents aft	4½	9	1	+4
2 parrel ropes	5½	27	3¼	-6½
1 naveline with a long tackle block under the main top & a single block to the parrel	2¼	43	5	+8
2 horses for the yard	5¾	18½	2¾	-1½
2 sheets, cable laid	7	83	10	-2
2 tacks, tapered & ditto	10	48	5¾	+6
2 bowlines	4½	59	7	+4
4 bridles	4½	12	1½	-6
2 clewgarnets	3¾	79	9½	-1
2 buntline legs	2¼	46½	5½	+4
2 falls with long tackle blocks aloft & single below	2½	98	11¾	+6
1 fall with a long tackle block aloft, & 2 single in the straps, for the legs to run in, & a single block below fast by the belfry	2¼	50	6	
1 topping lift for those buntlines with a block under the main top & so down by the main stay, then let go your leechlines & clewgarnets, when you have topped your buntlines close up, & the tack will come aboard with ease	2	30	3⅔	-3½
1 pair of pendents for leechlines at the masthead	3¾	2½	⅓	-1½
2 leechline legs	2¾	11	5¼	+1½
2 falls to the topmast head	2¾	92	11	+2
2 reeflines	2	54½	6½	+2
False stay for the staysail	4¾	14½	1¾	-6
1 pair of halliards with 2 long tackle blocks aloft, & a single in the head of the sail	3	55	6⅔	-3½
1 sheet, double	3½	14	1⅔	-½
2 main studdingsail halliards	3½	60	8	+2
2 tacks 5 times the length of the boom, with a block at the quarter of the yard	3½			
2 guy for the booms	3½	43	5	+8
2 sheets	3	26	3	+6
Main topmast				
1 pair of pendents for the tackles in place of Burtons	5½	8	1	-2

	A	B	C	D
2 falls with long tackle blocks aloft & single below	2¾	56	6¾	-1½
Half the first pair of shrouds	5½	10½	1¼	+6
Half a pair of breast stays	5½	23ft 5in	2¾	-½
Half of an after backstay, all with deadeyes in the chains	5½	23ft 5in	3	-7
1 of the top ropes	8½	20	2½	-5
1 of the falls with 3 sheaves below, & 3 in the block aloft	4¾	60	7¼	-2½
Stay	7½	25½	3	+3
1 of the jeers or runners with a double block upon the yard, and 2 single aloft	5½	24	3	-6
1 of the 2 pairs of halliards with 2 long tackle blocks	3	80	9⅔	-3½
2 lifts	3½	80	9⅔	-3½
2 braces	3½	78	9⅓	+1½
2 pendents	3¾	6	¾	-1½
1 pair of slings for the yard	5½	18	2	+8
2 parrel ropes	3½	15½	2	-7
2 horses for the yard	3½	10	1¼	-2½
2 sheets	8	70	8⅓	+3¼
2 bowlines	3½	69	8¼	
6 bridles	3½	22	2⅔	-1½
2 clewlines	3¾	98	11¾	+6
1 pair of buntline legs	2¾	23	2¾	+6
1 fall for ditto	2¾	25	3	
2 leechlines to the leech & foot of the sail, & so to the other ends to the topmast head & down upon deck	2¾	76	9	+6
2 reef tackle ties	3¾	7½	1	-5
2 falls with double breast blocks to the end of the tye, and a treble breast block fast to the parrel	2¾	112	13½	-3
2 reeflines	1½	16	2	-4
Staysail stay	4	15	1¾	+2½
1 pair of halliards with a long tackle block to the Stay & a single to the sail	2¾	65	8	-10
1 sheet, double	2¾	36	4⅓	-½
1 tack	3½	2½	⅓	-1½
2 studdingsail halliards with 2 blocks at the masthead	2¾	59	7	+4
2 tacks upon the poop	2¾	64	7⅔	+1
2 sheets	2½	12	1½	-3
Main Topgallant mast				
Half the first pair of shrouds	3½	5	⅔	-3
Half the pair of breast stays	3½	27ft 4in	3⅓	-8
Half a pair of backstays, all with deadeyes in the chains	3½	28ft 4in	3½	-3
Stay, bare	4	25	3	
1 tye with a block upon the yard	3½	12½	1½	
1 pair of halliards for the tye, with a long tackle block at the end of the tye & a single block fast in the main top	2½	58	7	-2
2 lifts	2½	26	3	+6
2 braces, full	2	80	9⅔	-3½
2 pendents	2½	6	⅔	+3
2 parrel ropes	2½	5½	⅔	-4
2 bowlines	1¾	66	8	-4
4 bridles	1¾	7	1	-8
2 clewlines	2	80	9½	+5
2 reeflines	1	6½	2	-1
1 flagstaff stay	3¼	26	3	+6
1 pair of halliards to hoist in the top	2½	36	4⅓	-8
Ditto to hoist at the topmast head	2½	19	2¼	+1½

	A	B	C	D
Mizzen mast				
1 pair of pendents for tackles	5½	5½	¾	-4
2 falls with long tackle blocks aloft & single below	3	55	6⅔	-3½
Half the first pair of shrouds	5½	8ft 2in	1	
Stay	6½	15½	1¾	+5½
1 jeer with 4 sheaves aloft & 3 in the block upon the yard	5½	62	7½	-3
1 parrel rope	4¾	8½	1	+1
1 truss tackle with a long tackle block & a single block	2¾	33	4	-2
1 pair of slings	6	7	¾	+4½
1 sheet with a treble & a double breast block	4¾	44	5¼	+1½
2 bowlines with 2 double breast blocks at the mizzen yard, & 2 single in the main shrouds	2½	49	6	-6
2 legs for peak brails	2½	11½	1⅓	+2½
2 brails for ditto	2½	22	2⅔	-1½
Second pair downward above the mast	2½	28	3⅓	+1½
Third pair	2½	27	3¼	-6
These brails following come to the main shrouds:				
The upper pair below the mast downwards	2¾	37	4½	-3
The second pair	2¾	39	4⅔	+½
The third pair or clew brails	2¾	38	4½	+3
The fourth pair	2¼	33	4	-2
The fifth pair, or foot brails	2¼	24	3	-6
Lacing for the mizzen	2¼	33	4	-2
Reeflines	1¼	35	4¼	-2½
Topping lift for the crowsfoot for the mizzen yard	2¾	33	4	-2
Pendent halliards for the peak	2¼	28	3⅓	+1½
Lantern halliards for main top	1½	24	3	-6
Mizzen staysail halliards with a long tackle block under the mizzen top, & a single block in the head of the sail	2½	54	6½	-1
Sheet	3¾	14	1⅔	+½
1 topping lift for the crowsfoot with a block under the main top for the staysail	2½	33	4	-2
2 crossjack braces	2½	62	7½	-3
2 lifts, treble, with 1 end of each fast to the quarter of the yard, & the other made fast in the mizzen shrouds	2¾	63	7½	+3
1 pair of yard slings	5½	6½	¾	+1½
Mizzen topmast				
Half the first pair of shrouds	3½	6	¾	-1½
Half the pair of breast stays	3½	14½	1¾	-4
And half a pair of backstays, both with deadeyes in the chains	3¾	15	1¾	+3
Stay, in the main top	3¾	10½	1¼	
1 tye with 2 single blocks at the masthead & upon the yard	3	14	1⅔	+½
1 pair of halliards for ditto with a long tackle block at the end of the tye, & a single below	2	53	6⅓	+1½
2 lifts made fast in the mizzen shrouds	2¼	48	5¾	+6
2 braces to the mizzen peak	2	49	6	-6
2 pendents	2¼	3	⅓	+1¼
2 parrel ropes	2¾	6	¾	-1½
2 sheets	4	40	4¼	+2½
2 bowlines	2	42	5	+2
4 bridles	2	7½	1	-5
2 clewlines	2	57	7	-8
2 reeflines	1	17	2	+2
The flagstaff stay in the main top	2½	12	1½	-3
1 pair of halliards to hoist in the mizzen top	1¾	17	2	+2
Ditto to hoist upon the quarterdeck	1¾	36	4⅓	-8

	A	B	C	D
Additional ropes				
Viol, cable laid	13½	44	5¼	+1½
Winding tackle pendents, with 5 sheaves	12½	16	2	-4
1 fall	6½	58	7	-2
1 fish tackle pendent, with 4 sheaves	9½	9½	1	+7
1 fall for ditto	4¾	56	6¾	-1½
Cat ropes, with 7 sheaves	6½	85	10¼	-2

Bibliography

Allen O E *The Pacific Navigators* Amsterdam 1981

Anderson R & R C *The Sailing Ship* London 1926, reprinted London 1980

Anderson R C *Seventeenth Century Rigging* London 1955

Anderson R C *Oared Fighting Ships* London 1962

Anderson R C *The Rigging of Ships in the Days of the Spritsail Topmast 1600-1720* London 1927, reprinted London 1982

Anderson R C & Salisbury W (ed) *A Treatise on Shipbuilding and a Treatise on Rigging, written about 1620-1625* SNR London 1958

Archibald E H H *Dictionary of Sea Painters* London 1982

Archibald E H H *The Fighting Ship in the Royal Navy 1897-1984* Poole, Dorset 1984

Armstrong R *The Merchantmen* London 1969

Baker W A *The Mayflower and other Colonial Vessels* London 1983

Baron Bowen, jnr R le 'Arab Dhows of Eastern Arabia' AN Vol 9/2

Baron Bowen, jnr R le 'The Dhow Sailor' AN Vol 11/3

Baron Bowen, jnr R le 'Primitive Watercraft of Arabia' AN Vol 12/3

Bathe B W *Ship Models* London 1966

Blanckley T R *A Naval Expositor* London 1750, reprinted Rotherfield 1988

Bobrik E *Handbuch der Praktischen Seefahrtskunde* 2 Volumes 1848, reprinted Kassel 1978

Boudriot J *The Seventy-four Gun Ship* 4 volumes, Paris and Rotherfield 1986-88

Boudriot J (trans Wells H B) 'Identification of lateen-rigged craft: Barque, Polacre and Pinque' NRJ Vol 25/4

Boudriot J (trans Wells H B) 'The Chasse-Marée and the Lugger' NRJ Vol 26/2

Boudriot J (trans Wells H B) 'The Barques Longues' NRJ Vol 27/2

Boudriot J (trans Wells H B) 'The Frigate La Flore' NRJ Vol 27/4

Boudriot J (trans Wells H B) 'Vessels carrying Mortars' NRJ Vol 29/4

Boven F C *From Carrack to Clipper* London 1948

Bracker J, North M & Tamm P *Maler der See, Marinemalerei in Dreihundert Jahren* Herford 1980

Brady W N *The Kedge Anchor or Young Sailor's Assistant* New York 1876, reprinted London 1974

Buys C J & Smith S O 'Chinese Batten Lug Sails' MM Vol 66/3

Cairo R F 'A note on the Basket Boats of Indochina' NRJ Vol 27/1

Cannenburg W V (ed) *Beschrijvende Catalogus der Scheepsmodellen en Scheepsbouwkundige Teekeningen 1600-1900* Amsterdam 1943

Cannenburg W V (ed) *Nederlandsch Historisch Scheepvaart Museum, Platen-Album* Amsterdam 1941

Chapelle H I *The History of the American Sailing Navy* New York 1949

Chapelle H I *The History of American Sailing Ships* New York 1935

Chapelle H I *The Search for Speed under Sail* London 1983

Chapman F H af *Architectura Navalis Mercatoria* Stockholm 1768, reprinted Rostock 1962-84

Crone G C E *De Jachten der Oranjes* Amsterdam 1937

Curti O *Schiffsmodellbau* Bielefeld 1975

Curti O *Masten, Rahen, Takelwerk* Bielefeld 1980

Davis C G *Ships of the Past* New York 1929

Davis J *The Seaman's Speculum or Compleat School-Master* London 1711, reprinted NRG Washington 1985

Duhamel du Monceau H L *Anfangsgründe der Schiffbaukunst, oder praktische Abhandlung über den Schiffbau* German translation by Müller C G D, Berlin 1791, reprinted Kassel 1973

Eichler C *Vom Bug zum Heck* Berlin 1943

Edson, jnr M 'The Schooner Rig, a Hypothesis' AN 25/2

Færøvik B & O *Inshore Craft of Norway* London 1979

Falconer W *An Universal Dictionary of the Marine* London 1769, reprinted Newton Abbot 1970

Falconer W *A New Universal Dictionary of the Marine* enlarged by Burney W, London 1815, reprinted New York 1974

Fincham J *Masting Ships and Mastmaking* London 1829-54, reprinted London 1982

Fincham J *A History of Naval Architecture* London 1851, reprinted London 1979

Foster Smith P C *The Artful Roux* Salem 1978

Fox F *Great Ships, the Battlefleet of King Charles II* London 1980

Fox Smith C *Ship Models* London 1972

Franklin J *Navy Board Ship Models 1650-1750* London 1989

Freeston E C *Prisoner of War Ship Models 1775-1825* Annapolis 1973

Furttenbach J *Architectura Navalis* Ulm 1629, reprinted Hamburg 1968

Gardner J *Warships of the Royal Navy* London 1968

Gaunt W *Marine Painting* New York 1976

Goldsmith-Carter G *Sailing Ships and Sailing Craft* London 1969

Goodenough S *Sailing Ships* London 1981

Groenewegen G *Versameling van vier ent achtig Stuks Hollandsche Schepen* Rotterdam 1789, (reprint)

Groot I de & Vorstman R *Sailing Ships* New York 1980

Gruppe H E *The Frigates* Amsterdam 1982

Harland J *Seamanship in the Age of Sail* London 1984

Harris D G *F H Chapman, The First Naval Architect and his Work* London 1989

Hawkins C W *The Dhow* Lymington 1977

Höckel R *Modellbau von Schiffen des 16. und 17. Jahrhunderts* Rostock and Bielefeld 1971

Hornell J 'A Tentative Classification of Arab Sea-craft' MM Vol 28/1

Höver O *Von der Galiot zum Fünfmaster* Bremen 1934

Howard Dr F *Sailing Ships of War 1400-1860* London 1979

Howarth D *The Men-of-War* Amsterdam 1979

Hutchinson W *A Treatise on Naval Architecture* Liverpool 1794, reprinted London 1969

Jobé J (ed) *The Great Age of Sail* London 1979

Kaufmann Dr G *Zeichner der Admiralität* Herford 1981

Kemp P (ed) *The Oxford Companion to Ships and the Sea* London 1976

Kerchove R de *International Maritime Dictionary* New York 1961

Ketting H *Prins Willem* Rostock and Bielefeld 1981

Klawitter G D *Vorlege-Blätter für Schiff-Bauer* Berlin 1835 reprinted Kassel 1975

Korth J W D *Die Schiffbaukunst* Berlin 1826, reprinted Kassel 1980

Landström B *Sailing Ships* London 1978

Lavery B (ed) *Deane's Doctrine of Naval Architecture 1670* London 1981

Lavery B *The Ship of the Line* 2 volumes, London 1983

Lavery B *The 74-Gun Ship Bellona* London 1985

Leather J *Gaff Rig* London 1983

Leather J *Spritsails and Lugsails* London 1979

Lees J *The Masting and Rigging of English Ships of War 1625-1860* London 1979

Lever D *The Young Officer's Sheet Anchor* London 1811-19, reprinted New York 1963

Longridge C N *The Anatomy of Nelson's Ships* London 1972

MacGregor D R *Fast Sailing Ships 1775-1885* Lymington 1973

MacGregor D R *Merchant Sailing Ships* Watford 1980

MacGregor D R *Schooners in four Centuries* Hemel Hempstead 1982

Maddocks M *The Atlantic Crossing* Amsterdam 1982

Magoun F A *The Frigate Constitution and other Historic Ships* New York 1928

Marquardt K H *Arab Dhow (Baghla) Model Plan* Westernhausen 1958

Marquardt K H *Chinese junk (Footchow pole) Model Plan* Westernhausen 1958

Marquardt K H *Malayan Prow (Prao mayang) Model Plan* Westernhausen 1957

Marquardt K H *Bemastung und Takelung von Schiffen des 18. Jahrhunderts* Rostock and Bielefeld 1986

Marquardt K H 'The Origin of Schooners' GC Volume 10/1

Marquardt K H *Schoner in Nord und Süd* Rostock and Bielefeld 1989

Masefield J *Sea Life in Nelson's Time* London 1905, reprinted London 1972

Middendorf F L *Bemastung und Takelung der Schiffe* Berlin 1903, reprinted Kassel 1977

Miller R *The East Indiamen* Amsterdam 1981

Mondfeld W zu *Historische Schiffsmodelle* Munich 1978

Mondfeld W zu *Die Schebecke und andere Schiffstypen des Mittelmeerraumes* Rostock and Bielefeld 1974

Morton H *The Wind Commands* Middletown and Vancouver 1975

Mountaine W *The Seaman's Vade-Mecum and Defensive War by Sea* London 1756, reprinted London 1971

Napier R 'The Bermuda Sloop circa 1740' NRJ Vol 30/4

Nares G S *Seamanship* 1862, reprinted Old Woking 1979

Paasch H *From Keel to Truck* Hamburg 1901, reprinted Hamburg 1978

Paasch H *Illustrated Marine Encyclopedia* 1890, reprinted Watford 1977

Paris E *Souvenirs de Marine* Paris 1882, 6 volumes (Vols 1-3 reprinted)

Paris E *Segelkriegsschiffe des 17. Jahrhunderts* Rostock and Bielefeld 1975

Paris E *Die grosse Zeit der Galeeren und Galeassen* Rostock and Bielefeld 1973

Paris E *Linienschiffe des 18. Jahrhunderts* Rostock and Bielefeld 1983

Petrejus E W *Model van de Oorlogsbrik Irene* Hengelo 1970

Petrejus E W *Das Modell der Brigg Irene* German revised translation, Bielefeld 1988

Rees A *Naval Architecture 1819-20* reprinted Newton Abbot 1970

Röding J H *Allgemeines Wörterbuch der Marine* 4 volumes, Hamburg 1793-98, reprinted Amsterdam 1969

Rousmanière J *The Luxury Yachts* Amsterdam 1982

Segditsas P E *Elsevier's Nautical Dictionary* 3 volumes New York 1966

Serres D & J T *Liber Nauticus* London 1805, reprinted London 1979

Smith J & Goell K (eds) *A Sea Grammar, with the plaine Exposition of Smiths Accidence for young Sea-Men, enlarged* London 1627, reprinted London 1970

Sopers P J V M *Schepen die verdwijnen* Amsterdam 1941

Steel D *Elements of Mastmaking, Sailmaking and Rigging (from the 1794 edition)* New York 1932

Steel D *The Art of Rigging* London 1818, reprinted Brighton 1974

Steinhaus C F *Die Schiffbaukunst in ihrem ganzen Umfange* 2 volumes, Hamburg 1858

Steinhaus C F *Die Construction und Bemastung der Segelschiffe* Hamburg 1869; the original two titles were reprinted in one volume, Kassel 1977

Sutherland W *The Ship-builder's Assistant* London 1711, reprinted Rotherfield 1989

Sutton J *Lords of the East* London 1981

Szymanski H *Deutsche Segelschiffe* Berlin 1934

Szymanski H *Die Segelschiffe der Deutschen Kleinschiffahrt* Lübeck 1929

Thubron C *The Venetians* Amsterdam 1982

Toelzel B *Pierre Ozanne, Auf See und vor Anker* Hamburg 1986

Tryckare T *The Lore of Ships* New York 1973

Underhill H A *Masting and Rigging the Clipper ship & Ocean Carrier* Glasgow 1958

Underhill H A *Sailing Ships' Rigs and Rigging* Glasgow 1969

Unknown *Der geöffnete See-Hafen* Hamburg 1702 and 1706, reprinted Hamburg 1989

US Naval Academy Museum *The Henry Huddleton Rogers Collection of Ship Models* Annapolis 1971

Vocino M *La Nave nel Tempo* Rome 1942

Wagner E *Decksarbeit* Hamburg 1944

Walker B *The Armada* Amsterdam 1982

Williams G R *The World of Model Ships and Boats* London 1971

Winter H *Der Holländische Zweidecker von 1660-70* Rostock and Bielefeld 1967

Worcester G R *The Junks and Sampans of the Yangtze* Annapolis 1971